CAMBRIDGE AERONAUTICAL SERIES

General Editors

ERNEST F. RELF, C.B.E., F.R.S.
PROFESSOR W. A. MAIR

III
THE THEORY OF
SUBSONIC PLANE FLOW

THE THEORY OF SUBSONIC PLANE FLOW

BY

L. C. WOODS, M.A., D.Sc.

Fellow of Balliol College, Oxford.
Formerly Nuffield Research Professor of
Mechanical Engineering in the University of
New South Wales, Australia

CAMBRIDGE
AT THE UNIVERSITY PRESS
1961

CAMBRIDGE UNIVERSITY PRESS
Cambridge, New York, Melbourne, Madrid, Cape Town,
Singapore, São Paulo, Delhi, Tokyo, Mexico City

Cambridge University Press
The Edinburgh Building, Cambridge CB2 8RU, UK

Published in the United States of America by Cambridge University Press, New York

www.cambridge.org
Information on this title: www.cambridge.org/9780521283199

First published 1961
First paperback edition 2011

A catalogue record for this publication is available from the British Library

ISBN 978-0-521-06858-1 Hardback
ISBN 978-0-521-28319-9 Paperback

CONTENTS

CONTENTS

CONTENTS

vii

PART II. COMPLEX VARIABLE THEORY

Chapter 3. CAUCHY INTEGRALS AND THEIR APPLICATION TO HARMONIC FUNCTION BOUNDARY VALUE PROBLEMS

CONTENTS

Chapter 4. MIXED AND PERIODIC BOUNDARY CONDITIONS

PART III. APPLICATIONS

Chapter 6. GENERAL ACCOUNT OF METHODS:
SOURCES, DOUBLETS AND VORTICES

Chapter 7. FLOW IN CHANNELS

Chapter 11. WAKES AND CAVITIES

GENERAL DESCRIPTION

THEORY OF STEADY WAKE FLOW

THEORY OF UNSTEADY WAKE FLOW

Chapter 12. CASCADES OF AEROFOILS

Chapter 13. LIFTING AEROFOILS IN CHANNELS AND JETS

THE CONFORMAL TRANSFORMATIONS

AEROFOILS IN CHANNELS

AEROFOILS IN JETS

AEROFOILS BETWEEN POROUS WALLS

Chapter 14. UNSTEADY MOTION OF AEROFOILS
IN CHANNELS AND JETS

UNSTEADY MOTION OF AN AEROFOIL
IN A CLOSED WIND TUNNEL

PREFACE

The main objective of this account of two-dimensional, subsonic, inviscid fluid motion is to provide a concise and systematic treatment of the subject by employing the same mathematical methods and variables throughout. Another aim is to present some recent developments—such as the theory of ventilated wind tunnels, jet-flaps and unsteady Helmholtz motions—in a text for the first time. At a late stage in planning the book it was realized that it would be useful to a wider group of readers if the purely mathematical techniques were separated from their aerodynamic applications. To achieve these ends the book has been divided into three parts, namely:

>   I. Fluid motion theory
> II. Complex variable theory, and
> III. Applications.

In part I the differential equations and boundary conditions for various types of fluid motion are derived briefly, with special emphasis placed on those aspects that are required in the applications of part III. The reader is assumed to have had an introductory course on fluid dynamics. Fluid speeds are assumed to be sufficiently small compared with sonic speed to permit the use of a linear relation between pressure and specific volume—the 'tangent' gas approximation of Kàrmàn, Tsien and Chaplygin. This approximation leads naturally to the choice of the complex number $\tau = \Omega + i\theta$ (see List of Symbols on p. xxi) as the dependent variable, and to the use of the complex stream-function $w = \phi + i\psi$ as the independent variable. Almost all of the book is concerned with the problem of calculating $\tau = \tau(w)$ subject to a variety of boundary conditions on the real and imaginary parts of τ.

Part II deals with the problem just stated without any reference to the aerodynamic applications. The methods are those of Cauchy integrals and conformal mapping, although the reader is expected to have no more than an elementary understanding of complex variable theory. These methods are used to solve a large number of boundary value problems, especially those involving mixed boundary conditions. There seems to be a lack of accounts of this subject suitable for engineers and scientists, and so it is hoped that part II will have some value other than providing the foundation to part III.

The range of the applications made in part III can be appreciated by glancing at the list of contents on p. x. Some of this work is new, or very recent, while the rest is now classical material presented in a new way. The choice of applications has been considerably influenced by my own researches, and it is hoped that this has not distorted the account too much. However, the great volume of research papers which have employed complex variable methods in fluid dynamics make it very difficult to do justice to all but a small fraction of the work and its authors.

I wish to express my gratitude to those people who have assisted me directly or indirectly to write this work. There is space to mention only a few of these. First, thanks are due to Professors A. Thom and G. Temple of Oxford University; to the former for showing me the importance of the variables τ and w used in this book, and to the latter for introducing me to the essential principle of part II, i.e. the use of Cauchy integrals to calculate $\tau(w)$. Dr W. P. Jones (Superintendent of the Aerodynamics Division of the National Physical Laboratory) also deserves thanks for giving me early guidance in unsteady aerofoil theory. I am grateful to Mr W. B. Smith-White, Dr S. Rosenblat and Dr G. Csanady for a number of suggestions which have helped me to improve the presentation of parts I and II. And finally, I thank the editors of the Cambridge Aeronautical Series for inviting me to write this book and the Syndics of the Cambridge University Press for undertaking to publish it.

L. C. W.

BALLIOL COLLEGE
OXFORD

LIST OF PRINCIPAL SYMBOLS

The following symbols are used consistently throughout the book:

c	aerofoil chord
$C_L = L/(\tfrac{1}{2}\rho_a c U^2)$	lift coefficient
$C_D = D/(\tfrac{1}{2}\rho_a c U^2)$	drag coefficient
$C_M = M/(\tfrac{1}{2}\rho_a c^2 U^2)$	moment coefficient
$C_p = (p - p_a)/(\tfrac{1}{2}\rho_a c U^2)$	pressure coefficient
D	drag force
L	lift force
$\mathscr{L}$	Laplace transform (see §4.23)
M	Mach number; nose-up moment
p	pressure; occasionally the complex-variable in a Laplace transform
$\mathbf{q}$	velocity vector
q	velocity magnitude; rarely as the parameter of a theta function (see §4.4)
s	distance measured along a surface or a streamline
t	time; in part II to denote value of z on the boundary of a domain
$\mathcal{t}$	reduced time (see §6.8)
$\mathrm{U}(x)$	unit function (see §3.6)
$U = q_a$	speed of fluid at average flow conditions
$w = \phi + i\psi$	complex stream-function
$z = x + iy$	complex variable—the physical (Argand) plane
$Z = x + i\beta_a y$	modified Argand plane for linear perturbation theory
$Z_1(x), Z_2(x), \ldots$	zeta functions (see §4.5)
β_a	$(1 - M_a^2)^{\frac{1}{2}}$, M_a being average Mach number
Γ	circulation
$\boldsymbol{\delta}(x)$	delta function or derivative of unit function
$\zeta = \gamma + i\eta$	an independent complex variable related analytically to w
θ	angle between $\mathbf{q}$ and Ox-axis
ρ	density
$\tau = \Omega + i\theta$	dependent complex variable (see §2.14)

PRINCIPAL SYMBOLS

ϕ	velocity potential
ψ	stream-function
$\Omega = -\int_{U}^{q} \beta\, dq/q$	a speed variable $(= \ln(U/q)$ in incompressible flow)
$\Omega^* = \tanh^{-1}\beta_a$	a critical value for $-\Omega$ (see §2.13)
ϖ	leap in Ω across a vortex sheet

Subscripts

a	values under 'average' flow conditions (often taken to be conditions upstream at infinity)
s	surface values; sometimes steady flow values
$\infty, -\infty$	values at $\phi = \infty$ (downstream at infinity) and $\phi = -\infty$ (upstream at infinity)
$0, h$	values on the boundary streamlines $\psi = 0$ and $\psi = h$.

PART I
FLUID MOTION THEORY

CHAPTER 1

INTRODUCTORY THEORY

1.1 Introduction

In the account of the theory of two-dimensional fluid flow given in this book it will be assumed that the reader is familiar with the elements of fluid dynamics, although the demands made on his knowledge are not extensive. In this chapter we provide a brief review of the basic theory of the subject, in which those aspects of special significance to the theory and applications of later chapters are stressed.

With few exceptions we shall be dealing with the two-dimensional or plane flow of an inviscid compressible fluid, moving past aerofoils, through channels, etc., at speeds which are everywhere subsonic. Our aim is to give a systematic and unified account of most of the important subsonic flow problems that can be treated by linear methods, and complex variable theory will be fully exploited for this purpose. However, this means that we cannot attempt an exact theory of compressibility effects since to do so would lead immediately to non-linearities of one kind or another.

Compressibility is therefore allowed for by approximating the ideal gas by a Chaplygin–Kármán–Tsien tangent gas, i.e. an imaginary gas whose isentrope is a straight line tangential to the ideal gas isentrope. This form of linearization is more general than the linear perturbation theory usually adopted, for it does not impose the restriction that the deviations from some basic uniform flow be small—although such a restriction is implicit at high Mach numbers. This means that, unlike linear perturbation theory, tangent gas theory contains *all* of the classical incompressible flow theory as a special case (see § 1.2).

Similarly, in order to preserve linearity, our treatment of unsteady motions, e.g. aerofoil flutter, must be restricted to the case of small unsteady perturbations about the mean motion. Compressible unsteady flow which is rapidly varying is found to depend on solutions of the wave equation with mixed boundary conditions. This yields a very complicated analysis (see the review by van Spiegel and van de Voren, 1953) not fitting into the main theme of this text, and it is therefore omitted.

The restriction of compressible unsteady flow to slowly varying flows, and the tangent gas approximation, have the consequence that every problem we consider is reducible to that of solving Laplace's equation subject to various boundary conditions. The very powerful complex variable techniques of Cauchy integrals and conformal mapping are therefore available.

An account of these techniques, adequate for engineers and applied mathematicians, and not too detailed with excessive rigour, has not yet been published. Muskhelishvili's (1953) brilliant book on singular integral equations deals with the Cauchy integral more generally and rigorously than we need, but does not give sufficient details of the kind of application in which we are interested. Similarly, texts on conformal mapping (e.g. Nehari, 1952) are usually more general than we need, and also fail to provide the kind of detail and practical application required by the engineer. Part II of this book is aimed at filling this gap: and it is hoped that it will be of interest to a wider class of readers than will the other two parts of the book.

Part III is devoted to applications of the theory of parts I and II to aero and fluid motion problems. It does not pretend to provide a complete cover of the subject as a whole, but it does go much further in this direction than existing texts. The aim in part III is to show the reader how to apply the theory rather than to exhibit all the details of any particular application, although complete details are given for a few selected problems.

By separating the basic mathematics in part II from the fluid dynamics of parts I and III we have been able to condense considerably our treatment of various topics, with, it is hoped, a consequent gain in clarity.

Summarizing, we can say that the scope of our studies will be those inviscid flow problems governed by the two-dimensional form of Laplace's equation. The difficulties of our subject-matter will lie in the character of the boundary conditions, which, for example, may be mixed, unsteady, not defined explicitly, and so on. The principal mathematical techniques will be those of complex variable theory.

THE BASIC EQUATIONS

1.2   The isentropic equation of state

The fluid motion is assumed to be such that viscosity and heat conduction can be neglected, which means that changes of state at a fluid particle are *adiabatic*. In this case the pressure p is a function only of the fluid density ρ:

$$p = p(\rho).$$

This is often called the *adiabatic* equation, but an equally suitable name is the *isentropic* equation, for neglect of viscosity implies thermodynamic reversibility, so that adiabatic changes are always isentropic.

For the cases in which we shall be interested the isentropic equation can be written in the form

$$p = k\rho^{-n} + b \quad (p \geqslant 0), \tag{1}$$

where k, b and n are constants, and the impossibility of negative pressures is indicated. With a perfect or *ideal* gas $b = 0$ and $n = -\gamma$, where γ is the ratio of the specific heats of the gas. For our purposes air can be regarded as being an ideal gas with $\gamma = 7/5$. The isentrope of the ideal gas

$$p = k\rho^\gamma \tag{2}$$

is shown in figure 1.2.

The non-linearity of (1) adds considerable complexity to the theory of gas flow, and it is necessary in most problems to use an approximation in

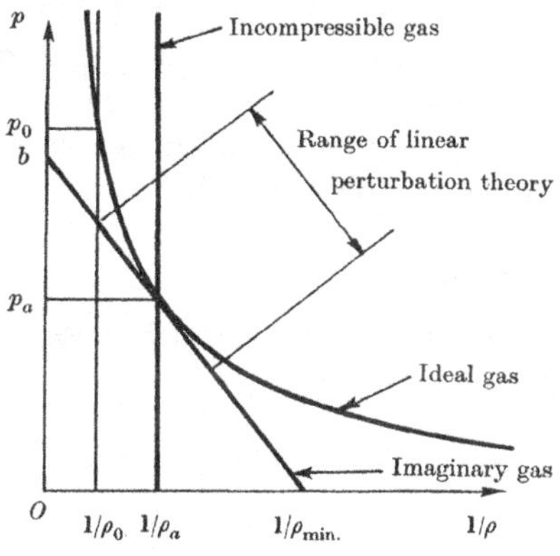

Fig. 1.2

its place. The most obvious step is to linearize the equation in the $(p, 1/\rho)$-plane by putting $n = 1$. Consider a point $(p_a, 1/\rho_a)$ on the ideal gas isentrope I; the most suitable straight line approximation to I near $(p_a, 1/\rho_a)$ will clearly be the *tangent* to I at $(p_a, 1/\rho_a)$. Instead of regarding this tangent as being an approximation to the curved isentrope, we may postulate an *imaginary* gas which has exactly this tangent for its isentrope. Such a gas can be termed a Chaplygin–Kármán–Tsien tangent gas after the authors (Chaplygin, 1902; Kármán, 1941; Tsien, 1939) who have made the most significant contributions to its theory, but for brevity we shall refer to it simply as a *tangent gas*.

The question of the selection of the most appropriate tangency point $(p_a, 1/\rho_a)$ will be discussed later (§ 2.10); it is sufficient to note here that $(p_a, 1/\rho_a)$ will be a point corresponding to some suitably averaged flow condition.

It is apparent from Fig. 1.2 that the tangent gas will be a reasonable approximation to an ideal gas provided either (i) p and $1/\rho$ do not vary markedly from p_a, $1/\rho_a$, or (ii) the point $(p_a, 1/\rho_a)$ lies on the ideal gas isentrope at a point of small curvature, e.g. near the point $(p_0, 1/\rho_0)$ shown. When (i) applies we have a linear perturbation of the flow from the state $(p_a, 1/\rho_a)$, and the resulting theory is termed *linear perturbation theory*. With (ii) it is only the case of high pressure and density with which we shall be concerned in this book. When $(p_a, 1/\rho_a)$ is at a point of relatively high density the tangent to the ideal gas isentrope becomes nearly vertical, and very little error is introduced by assuming that it is the vertical line $\rho = \rho_a$. This is the very widely used *incompressible flow* approximation, which is reasonably good provided very low pressures are not attained in the flow. Except for this restriction (often ignored in applications of the theory), quite large pressure variations are permissible.

Summarizing, we have that the tangent gas theory contains both linear perturbation theory (small variations in p and ρ) and incompressible flow theory (large variations in p and none in ρ) as special cases. The relation between tangent gas theory and linear perturbation theory will be discussed in some detail below (§ 2.12).

1.3   The speed of sound

If a is the speed of small pressure disturbances in the gas, then it is shown in elementary texts that

$$a^2 = \frac{dp}{d\rho}.$$

Differentiation of (1) therefore gives

$$a^2 = -nk\rho^{-(n+1)} = \frac{n(b-p)}{\rho}, \tag{3}$$

so that

$$\frac{a}{a_a} = \left(\frac{\rho_a}{\rho}\right)^{\frac{1}{2}(n+1)}. \tag{4}$$

With a tangent gas, $n = 1$, and (4) gives

$$a\rho = a_a\rho_a, \tag{5}$$

whereas with an ideal gas, $n = -\gamma$, and so

$$\frac{a}{a_a} = \left(\frac{\rho}{\rho_a}\right)^{\frac{1}{2}(\gamma-1)}. \tag{6}$$

At the tangency point $dp/d\rho$ is the same for both the ideal and tangent gas. Hence from (3)

$$a_a^2 = \frac{\gamma p_a}{\rho_a}. \tag{7}$$

The slope of the isentrope of the tangent gas is $\{dp/d(1/\rho)\}_a = -\rho_a^2 a_a^2$; hence this isentrope has the equation

$$p - p_a = \rho_a^2 a_a^2 \left(\frac{1}{\rho_a} - \frac{1}{\rho}\right). \tag{8}$$

As p cannot be negative the minimum value of ρ in a tangent gas is given by

$$\frac{1}{\rho_{\text{min.}}} = \frac{1}{\rho_a} + \frac{p_a}{\rho_a^2 a_a^2} = \frac{\gamma+1}{\gamma}\frac{1}{\rho_a},$$

by (7). Thus

$$\rho_{\text{min.}} = \frac{\gamma}{\gamma+1}\rho_a = \tfrac{7}{12}\rho_a. \tag{9}$$

1.4. The continuity equation

Assume the fluid to be moving without discontinuities except those occurring at the boundaries. (By 'boundaries' here we are not referring merely to solid boundaries: vortex sheets, i.e. surfaces of discontinuity in the velocity vector $\mathbf{q}$ may exist in the flow, and in this case we regard the sheet as a type of boundary.)

In Cartesian coordinates x, y, z let ∇ be the vector operator

$$\nabla \equiv \mathbf{i}\frac{\partial}{\partial x} + \mathbf{j}\frac{\partial}{\partial y} + \mathbf{k}\frac{\partial}{\partial z},$$

where $\mathbf{i}, \mathbf{j}, \mathbf{k}$ are unit vectors in the Ox, Oy, and Oz directions respectively. Then the time rate of change following the motion of a particle is

$$\frac{d}{dt} = \frac{\partial}{\partial t} + \mathbf{q}.\nabla, \tag{10}$$

where the dot denotes the scalar product between two vectors.

In this notation the equation of continuity can be written†

$$\frac{\partial\rho}{\partial t} + \nabla.(\rho\mathbf{q}) = \frac{d\rho}{dt} + \rho\nabla.\mathbf{q} = 0. \tag{11}$$

These equations apply at all interior points of the fluid, except where fluid is being 'created' or 'destroyed', i.e. except at sources or sinks.

1.5 The equation of motion

Euler's equation of motion is†

$$\frac{d\mathbf{q}}{dt} = \mathbf{F} - \frac{1}{\rho}\nabla p, \tag{12}$$

where $\mathbf{F}$ is any external body force, such as gravity, which may be acting on the fluid, per unit mass. The vector transformation

$$\frac{d\mathbf{q}}{dt} = \frac{\partial\mathbf{q}}{\partial t} + (\mathbf{q}.\nabla)\mathbf{q} = \frac{\partial\mathbf{q}}{\partial t} + \tfrac{1}{2}\nabla q^2 - \mathbf{q}\wedge\boldsymbol{\zeta},$$

† The reader is referred to a standard text such as *Theoretical Hydrodynamics* by Milne-Thomson (1949) for more details of the theory outlined in §§ 1.4–1.9, 1.12 and 1.13.

where $\boldsymbol{\zeta}$ is the vorticity vector $\nabla \wedge \mathbf{q}$, and $\wedge$ indicates a vector product, enables us to write (12) in the form

$$\frac{\partial \mathbf{q}}{\partial t} - \mathbf{q} \wedge \boldsymbol{\zeta} = \mathbf{F} - \frac{1}{\rho} \nabla p - \tfrac{1}{2} \nabla q^2.$$

Let $\mathbf{r} = x\mathbf{i} + y\mathbf{j} + z\mathbf{k}$, then

$$d\mathbf{r} \cdot \frac{1}{\rho} \nabla p = \frac{1}{\rho} (d\mathbf{r} \cdot \nabla) \, p = \frac{1}{\rho} dp = d \int \frac{dp}{\rho} = d\mathbf{r} \cdot \nabla \int \frac{dp}{\rho},$$

i.e. $\dfrac{1}{\rho} \nabla p = \nabla \int \dfrac{dp}{\rho}$, and further if we take $\mathbf{F}$ to be a conservative force, and therefore expressible in the form $\mathbf{F} = -\nabla \chi$, where χ is a scalar function, the equation of motion can be written

$$\frac{\partial \mathbf{q}}{\partial t} - \mathbf{q} \wedge \boldsymbol{\zeta} = -\nabla \left(\chi + \int \frac{dp}{\rho} + \tfrac{1}{2} q^2 \right). \tag{13}$$

1.6 The vorticity equation

By operating on (13) with $\nabla \wedge$, and recalling that $\nabla \wedge \nabla = 0$, we obtain

$$\frac{\partial \boldsymbol{\zeta}}{\partial t} - \nabla \wedge (\mathbf{q} \wedge \boldsymbol{\zeta}) = 0,$$

or

$$\frac{\partial \boldsymbol{\zeta}}{\partial t} + (\mathbf{q} \cdot \nabla) \boldsymbol{\zeta} = (\boldsymbol{\zeta} \cdot \nabla) \mathbf{q} - \boldsymbol{\zeta}(\nabla \cdot \mathbf{q}) + \mathbf{q}(\nabla \cdot \boldsymbol{\zeta}),$$

the last term of which vanishes as $\nabla \cdot \boldsymbol{\zeta} = \nabla \cdot (\nabla \wedge \mathbf{q}) = 0$. Equations (10) and (11) now enable us to write the equation in the form

$$\frac{d(\boldsymbol{\zeta}/\rho)}{dt} = \left(\frac{\boldsymbol{\zeta}}{\rho} \cdot \nabla \right) \mathbf{q}. \tag{14}$$

Two important conclusions can be drawn from this equation. The first is that in two-dimensional motions

$$\frac{d(\boldsymbol{\zeta}/\rho)}{dt} = 0, \tag{15}$$

for in this case $\boldsymbol{\zeta}$ is necessarily orthogonal to $\mathbf{q}$, so $(\boldsymbol{\zeta} \cdot \nabla) \mathbf{q}$ is a derivative along a direction normal to the plane of the motion, and is therefore zero. The second result, which is not restricted to plane flows, is that if a fluid particle has no vorticity initially, it never acquires any. One proof of this (due in essence to Stokes) depends on the uniqueness of the solution of (14) with specified initial conditions—$\boldsymbol{\zeta} = 0$ is obviously *a* solution of (14) satisfying $\boldsymbol{\zeta} = 0$ at $t = 0$, and so by the uniqueness theorem it is *the* solution. An alternative proof due to Kelvin is based on the theorem on the conservation of circulation given below in § 1.8.

1.7   Irrotational motion

A consequence of the results of the previous section is that in a motion which has commenced from rest, the vorticity is everywhere zero, since this is initially so. This does not exclude the possibility of vortex sheets springing from the boundaries of solid bodies, for in this case the velocity vector is no longer continuous, as assumed tacitly in the proof of (14).

The assumption is now introduced that the fluid motions in which we shall be mainly interested can all be generated from rest (e.g. by an impulsive start). Hence

$$\zeta = 0, \tag{16}$$

throughout the fluid, and we have irrotational flow.

1.8   Conservation of circulation (Kelvin)

The circulation Γ about any closed curve C is defined by

$$\Gamma = \int_C \mathbf{q}.d\mathbf{r}, \tag{17}$$

$d\mathbf{r}$ being a vector element tangential to C. Kelvin's theorem states that when the external forces are conservative, the circulation about any closed circuit which moves with the fluid is constant.

Proof. As C moves with the fluid

$$\mathbf{q}.\frac{d}{dt}(d\mathbf{r}) = \mathbf{q}.d\mathbf{q} = d(\tfrac{1}{2}q^2) = \nabla(\tfrac{1}{2}q^2).d\mathbf{r},$$

so

$$\frac{d\Gamma}{dt} = \int_C \left\{ \frac{d\mathbf{q}}{dt}.d\mathbf{r} + \mathbf{q}.\frac{d}{dt}(d\mathbf{r}) \right\}$$

$$= -\int_C \left\{ \nabla\chi + \frac{1}{\rho}\nabla p - \nabla(\tfrac{1}{2}q^2) \right\}.d\mathbf{r},$$

by (12) and $\mathbf{F} = -\nabla\chi$ (conservative external forces). Therefore

$$\frac{d\Gamma}{dt} = -\int_C d\left(\chi + \int\frac{dp}{\rho} - \tfrac{1}{2}q^2\right) = -\left(\chi + \int\frac{dp}{\rho} - \tfrac{1}{2}q^2\right)_C = 0,$$

as each quantity in the bracket is single valued. Thus Γ is independent of the time.

This theorem applies whether the flow is irrotational or not.

1.9   Velocity potential. Pressure equation

If the flow is irrotational $\nabla \wedge \mathbf{q} = 0$, and as the operator $\nabla \wedge \nabla$ reduces any scalar to zero, it follows that $\mathbf{q}$ can be derived from a *potential* function ϕ, i.e.

$$\mathbf{q} = \nabla\phi, \tag{18}$$

In this case (13) can be written

$$\nabla\left(\frac{\partial\phi}{\partial t}+\chi+\int\frac{dp}{\rho}+\tfrac{1}{2}q^2\right)=0,$$

or

$$\frac{\partial\phi}{\partial t}+\chi+\int\frac{dp}{\rho}+\tfrac{1}{2}q^2=C(t), \tag{19}$$

where $C(t)$ depends only on t. Equation (19) is the *pressure equation*, or *Bernoulli's* equation for irrotational flow.

In steady flow $\partial\phi/\partial t=0$, and (19) reduces to

$$\chi+\int\frac{dp}{\rho}+\tfrac{1}{2}q^2=C, \tag{20}$$

where C is now a constant, sometimes called the *Bernoulli* constant.

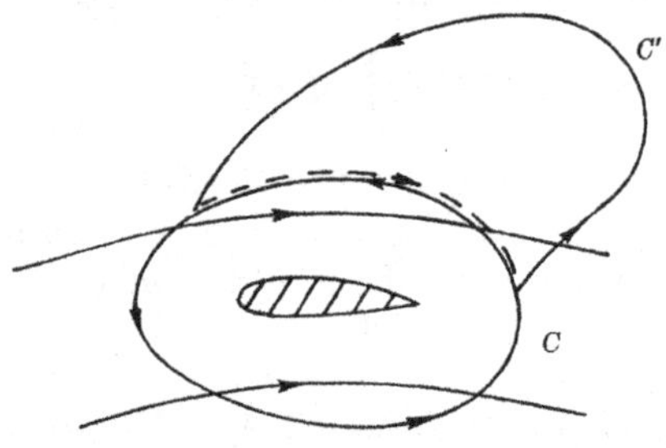

Fig. 1.9

In irrotational flow the circulation around a closed contour C equals the change in the potential function ϕ about C, for by (17) and (18)

$$\Gamma=\int_C\nabla\phi.d\mathbf{r}=\int_C d\phi=[\phi]_C. \tag{21}$$

The contour C can be deformed in any way without altering the value of the circulation, provided it does not cross any singularity in the fluid and it remains within the fluid. For if C' is a contour partially coincident with a part of C (Fig. 1.9), ϕ will be a single-valued function on and within C', and clearly

$$\int_{C+C'}d\phi=[\phi]_C+[\phi]_{C'}=[\phi]_C=\Gamma.$$

1.10   Isentropic relations in steady flow. Critical values

Suppose we have a steady flow for which the external forces are negligible; the flow may be rotational or irrotational. Let $\hat{\mathbf{s}}$ be a unit vector parallel to $\mathbf{q}$, and let s be the distance measured along a streamline. Then by multiplying (13) by $\hat{\mathbf{s}}$ we find for the case under consideration that

$$\hat{\mathbf{s}} \cdot \nabla \left(\int \frac{dp}{\rho} + \tfrac{1}{2} q^2 \right) = 0,$$

i.e.
$$2 \int \frac{dp}{\rho} + q^2 = 2C, \tag{22}$$

where C is, in general, a different constant for each streamline. From (1)

$$\int \frac{dp}{\rho} = \frac{nk\rho^{-(n+1)}}{n+1} = -\left(\frac{dp/d\rho}{n+1} \right) = -\frac{a^2}{n+1},$$

so (22) can be written

$$q^2 - \frac{2}{n+1} a^2 = 2C.$$

Conditions at the tangency point $(p_a, 1/\rho_a)$ are the same in both the ideal and tangent gases (see §1.3), so the constant $2C$ can be replaced by $U^2 - 2a_a^2/(n+1)$, where $U = q_a$ is the velocity corresponding to tangency point conditions. Hence

$$q^2 - \frac{2}{n+1} a^2 = U^2 - \frac{2}{n+1} a_a^2, \tag{23}$$

or alternatively
$$a^2 \left(M^2 - \frac{2}{n+1} \right) = a_a^2 \left(M_a^2 - \frac{2}{n+1} \right), \tag{24}$$

where M is the ratio of the fluid speed q to the corresponding sound speed a, i.e. the local *Mach* number.

By combining (4) and (24) we obtain

$$\frac{\rho_a}{\rho} = \left\{ \frac{1 - \tfrac{1}{2}(n+1) M_a^2}{1 - \tfrac{1}{2}(n+1) M^2} \right\}^{1/(n+1)}. \tag{25}$$

With a tangent gas $n = 1$, so

$$\frac{\rho_a}{\rho} = \frac{\beta_a}{\beta}, \tag{26}$$

where
$$\beta = \sqrt{(1 - M^2)}. \tag{27}$$

For this case (24) yields
$$a\beta = a_a \beta_a. \tag{28}$$

If (9) for the minimum value of ρ in a tangent gas is combined with (26) to (28) and (23) in turn, the following limiting values in a tangent gas are obtained

$$\beta_{\text{min.}} = \frac{\gamma}{\gamma+1} \beta_a, \quad M_{\text{max.}} = \sqrt{\left\{ 1 - \frac{\gamma^2}{(\gamma+1)^2} \beta_a^2 \right\}}, \quad a_{\text{max.}} = \frac{\gamma+1}{\gamma} a_a, \tag{29}$$

and
$$q_{\text{max.}} = U \sqrt{\left\{ 1 + \frac{2\gamma+1}{\gamma^2} \frac{1}{M_a^2} \right\}}. \tag{30}$$

The number β defined in (27) is an important speed parameter, which occurs throughout the theory developed in later chapters. By writing (28) in the form $q\beta/\sqrt{(1-\beta^2)} = U\beta_a/\sqrt{(1-\beta_a^2)}$ and solving for β we get the tangent gas relation

$$\beta = \frac{\beta_a}{M_a}\left\{\left(\frac{q}{U}\right)^2 + \left(\frac{\beta_a}{M_a}\right)^2\right\}^{-\frac{1}{2}}, \tag{31}$$

which will be required in the next chapter.

1.11   The equation for the velocity potential

From (11), (18) and $a^2 = dp/d\rho$ it follows that

$$a^2\nabla^2\phi + \frac{1}{\rho}\frac{dp}{dt} = 0,$$

or
$$a^2\nabla^2\phi + \frac{1}{\rho}\frac{\partial p}{\partial t} + \frac{1}{\rho}\mathbf{q}.\nabla p = 0. \tag{32}$$

If the constant in (19) is made independent of t by incorporating any time variable terms in $\partial\phi/\partial t$, the equation can be differentiated to give

$$\frac{\partial(\nabla\phi)}{\partial t} + \frac{1}{\rho}\nabla p + \nabla(\tfrac{1}{2}q^2) = 0,$$

and
$$\frac{\partial^2\phi}{\partial t^2} + \frac{1}{\rho}\frac{\partial p}{\partial t} + \mathbf{q}.\frac{\partial(\nabla\phi)}{\partial t} = 0,$$

where the external forces have been assumed negligible. Using these two equations to eliminate ρ and p from (32) we arrive at

$$a^2\nabla^2\phi - \tfrac{1}{2}\nabla\phi.\nabla(\nabla\phi.\nabla\phi) = 2\nabla\phi.\frac{\partial(\nabla\phi)}{\partial t} + \frac{\partial^2\phi}{\partial t^2}, \tag{33}$$

which is the most general equation for the velocity potential.

In incompressible flow $a = \infty$, and (33) reduces to

$$\nabla^2\phi = 0, \tag{34}$$

in both steady and unsteady flow.

FORCES ON OBSTACLES

1.12   The momentum theorems

Equations (11) and (12) can be combined to give

$$\frac{\partial(\rho\mathbf{q})}{\partial t} + \nabla.(\rho\mathbf{q}\mathbf{q}) + \nabla p - \rho\mathbf{F} = 0. \tag{35}$$

On integrating this equation over a region τ of the fluid, containing no singularities, and making use of the divergence theorem, we arrive at

$$\int_S (\rho \mathbf{q}\mathbf{n}.\mathbf{q} + p\mathbf{n})\, dS + \int_\tau \left\{ \frac{\partial(\rho\mathbf{q})}{\partial t} - \rho\mathbf{F} \right\} d\tau = 0, \tag{36}$$

where $\mathbf{n}$ is a unit vector along the outwards normal to the surface S enclosing τ. Equation (36) is known as Euler's *momentum theorem.*

Let $\mathbf{r}$ be a vector drawn from the origin into τ, then it follows from (35) that

$$\int_\tau \mathbf{r} \wedge \{\nabla.(\rho\mathbf{q}\mathbf{q}) + \nabla p\}\, d\tau + \int_\tau \mathbf{r} \wedge \left\{ \frac{\partial(\rho\mathbf{q})}{\partial t} - \rho\mathbf{F} \right\} d\tau = 0.$$

Use of the result $\nabla \wedge \mathbf{r} = 0$, and the divergence theorem gives

$$\int_\tau \mathbf{r} \wedge \nabla.(\rho\mathbf{q}\mathbf{q})\, d\tau = \int_\tau \nabla.(\mathbf{r} \wedge \rho\mathbf{q}\mathbf{q})\, d\tau$$

$$= -\int_S \mathbf{n}.(\mathbf{r} \wedge \rho\mathbf{q}\mathbf{q})\, dS = \int_S \mathbf{r} \wedge (\rho\mathbf{q}\mathbf{n}.\mathbf{q})\, dS.$$

Similarly

$$\int_\tau \mathbf{r} \wedge \nabla p\, d\tau = \int_S \mathbf{r} \wedge p\mathbf{n}\, dS;$$

hence

$$\int_S \mathbf{r} \wedge (\rho\mathbf{q}\mathbf{n}.\mathbf{q} + p\mathbf{n})\, dS + \int_\tau \mathbf{r} \wedge \left\{ \frac{\partial(\rho\mathbf{q})}{\partial t} - \rho\mathbf{F} \right\} d\tau = 0. \tag{37}$$

This is the *angular momentum theorem.*

1.13   Forces acting on a rigid body

These two momentum theorems find an important application in the calculation of the forces acting on a fixed rigid body immersed in the steady flow of a fluid not subject to external forces.

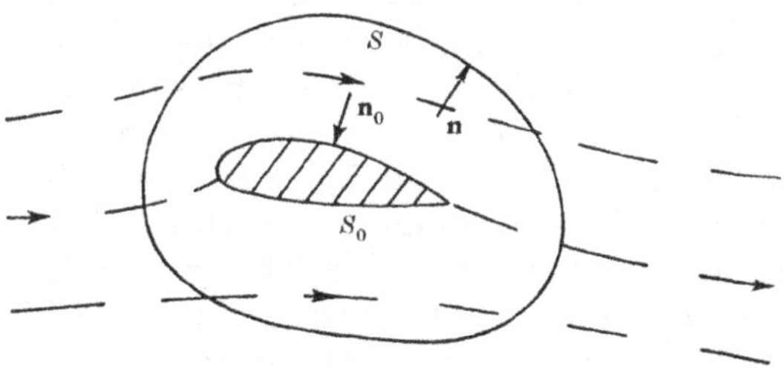

Fig. 1.13

Let S_0 be the surface of the fixed solid body, and S any other closed surface entirely surrounding the obstacle, then in the present application (36) reduces to

$$\int_{S_0} (\rho \mathbf{q}\mathbf{n}_0.\mathbf{q} + p\mathbf{n}_0)\, dS_0 + \int_S (\rho\mathbf{q}\mathbf{n}.\mathbf{q} + p\mathbf{n})\, dS = 0.$$

Now the thrust $\mathbf{T}$ on the body due to the fluid pressure is

$$\mathbf{T} = \int_{S_0} p\mathbf{n_0}\,dS_0,$$

and, as the surface is solid, $\mathbf{n_0}.\mathbf{q} = 0$ on S_0; hence

$$\mathbf{T} = -\int_S (\rho\mathbf{q}\mathbf{n}.\mathbf{q} + p\mathbf{n})\,dS, \tag{38}$$

provided, of course, that no singularities lie between S and S_0.

It is similarly found from (37) that the moment about the origin of the pressures acting on the body is

$$\mathbf{M} = -\int_S \mathbf{r} \wedge (\rho\mathbf{q}\mathbf{n}.\mathbf{q} + p\mathbf{n})\,dS. \tag{39}$$

The virtue of these equations for $\mathbf{T}$ and $\mathbf{M}$ is that we are free to choose the surface S, for in most problems involving the calculating of forces on bodies it is possible to find a surface which is more convenient than any other. This point is illustrated in the next section.

1.14   The jet-thrust paradox

Euler's momentum theorem can be used to establish a surprising result in steady potential flow concerning the *ideal* forward thrust or upstream force acting on a body, which emits a jet of air in any direction into the main stream flowing past it. It is the result for an infinite stream we wish to establish here, but first it is convenient to consider a body-jet combination in a cylindrical wind tunnel of arbitrary cross-section. This facilitates the application of the momentum theorem to the problem.

To generalize the problem slightly we shall suppose that the body also draws in some air from the main stream: the situation is shown in Fig.1.14. Air is drawn into the body at BB', energy and mass are added to the stream at a section EE' within the body, and it is emitted at DD' as a high-speed jet. It is assumed that the jet and main-stream fluids do not mix, and that the flow in the main stream is isentropic. Thus our model is only a crude approximation to a real flow, but it does yield the *ideal* thrust, which, when compared with the actual thrust obtained in practice, gives a useful estimate of the loss of jet efficiency due to the action of viscosity and turbulence.

The flow can be regarded as two distinct duct flows through the regions Σ and Σ' shown in the figure. The inner duct is bounded by the surface S', and the planes $L_{-\infty}L'_{-\infty}$, $M_\infty M'_\infty$ at right angles to the tunnel axis, and at infinity, while the outer duct is bounded by S, $L_{-\infty}L'_{-\infty}$, $M_\infty M'_\infty$ and the tunnel wall. The pressure is continuous across S and S'

where they coincide, for upstream the flow is homogeneous, and downstream the surface S, S' is a vortex sheet across which the pressure is continuous (see § 1.21).

Let $\mathbf{n}$, $\mathbf{n}'$ be unit vectors along the outward normals to the regions Σ, Σ', and let $\mathbf{i}$ be a unit vector directed downstream along the tunnel axis. Then the upstream thrust T acting on the body is given by

$$T = -\int_S p\mathbf{n}.\mathbf{i}\,dS - \int_{S'} p\mathbf{n}'.\mathbf{i}\,dS', \tag{40}$$

for where S and S' coincide $p(S) = p(S')$ and $\mathbf{n} = -\mathbf{n}'$.

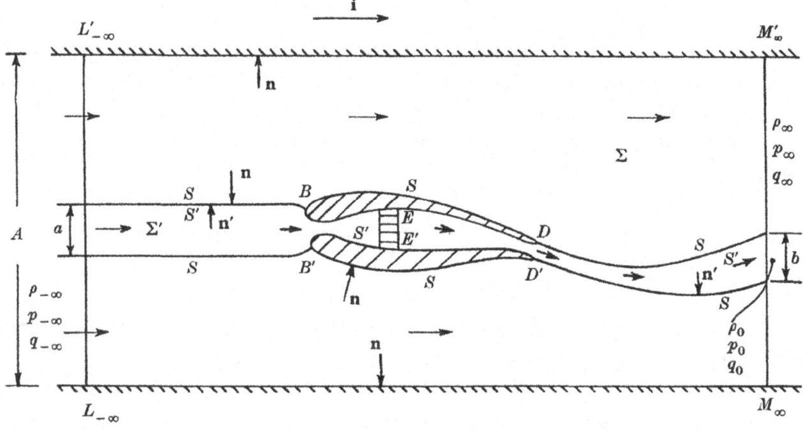

Fig. 1.14

We are dealing here with the steady flow of a fluid not subject to external body forces, so it follows from (36) that for each of the regions Σ and Σ'

$$\int p\mathbf{n}.\mathbf{i}\,dS = -\int p\mathbf{q}.\mathbf{in}.\mathbf{q}\,dS. \tag{41}$$

Let the subscripts $-\infty$, ∞ and 0 denote stream values at infinity (i) upstream, (ii) downstream in the external flow, and (iii) downstream in the jet flow, respectively, and let Σ' have cross-sectional areas of a and b at infinity upstream and downstream respectively. Let A be the cross-sectional area of the tunnel. Then applying (41) to the region Σ we find

$$\int_S p\mathbf{n}.\mathbf{i}\,dS - (A-a)\,p_{-\infty} + (A-b)\,p_\infty = (A-a)\,\rho_{-\infty}q_{-\infty}^2 - (A-b)\,\rho_\infty q_\infty^2,$$

as $\mathbf{n}.\mathbf{i} = 0$ on the tunnel walls, and $\mathbf{n}.\mathbf{q} = 0$ except on the planes $L_{-\infty}L'_{-\infty}$ and $M_\infty M'_\infty$. A similar calculation for Σ' gives

$$\int_{S'} p\mathbf{n}'.\mathbf{i}\,dS' - ap_{-\infty} + bp_0 = a\rho_{-\infty}q_{-\infty}^2 - b\rho_0 q_0^2.$$

Substitution of these values in (40) gives

$$T = b(\rho_0 q_0^2 - \rho_\infty q_\infty^2) - A(p_{-\infty} - p_{-\infty} + \rho_{-\infty} q_{-\infty}^2 - \rho_\infty q_\infty^2), \tag{42}$$

on making use of the fact that p_0 and p_∞ are equal.

The next step is to deduce results for the infinite stream by taking the limit $A \to \infty$; in this case $p_{-\infty}, \rho_{-\infty}, q_{-\infty} \to p_\infty, \rho_\infty, q_\infty$, and the last term of (42) is indefinite. It can be calculated as follows. Mass conservation for Σ gives $(A-a)\rho_{-\infty} q_{-\infty} = (A-b)\rho_\infty q_\infty$, hence on eliminating A we can write the term in question as

$$(a\rho_{-\infty} q_{-\infty} - b\rho_\infty q_\infty)\left(\frac{\delta p + \delta(\rho q^2)}{\delta(\rho q)}\right),$$

where δB is the small quantity $B_{-\infty} - B_\infty$. We are dealing here with the potential flow in the main stream, hence from (20) $\delta p + \rho q \delta q = 0$. Therefore to first order in small quantities $\delta p + \delta(\rho q^2) = q\delta(\rho q)$, so in the limit $A \to \infty$ the last term of (42) reduces to $\rho_\infty q_\infty^2 (a-b)$. Thus in an infinite stream

$$T = b\rho_0 q_0^2 - a\rho_\infty q_\infty^2. \tag{43}$$

This result can be expressed in the non-dimensional form

$$C_T = C_J - 2C_Q, \tag{44}$$

where C_T is the *thrust* coefficient, $C_T \equiv T/\frac{1}{2}c\rho_\infty q_\infty^2$, C_J is the jet *momentum* coefficient, $C_J \equiv b\rho_0 q_0^2/\frac{1}{2}c\rho_\infty q_\infty^2$, C_Q is the *mass* coefficient of the air drawn into the body, $C_Q \equiv a\rho_\infty q_\infty/c\rho_\infty q_\infty$, and c is an area typical of the body.

The paradox in (44) lies in the fact that C_T is apparently independent of the jet exit angle, so that, taking an extreme example, the air could be ejected in the *upstream* direction, and provided only that its *final* flow direction is downstream, it would still give a forward thrust! While C_Q must be constant along the jet, C_J will vary, so that a given value of C_J at the trailing edge will result in a value of C_J at infinity (the value appearing in (44)) which varies to some extent with the exit angle. From this point of view the exit angle will have *some* effect on C_T, but by no means the effect that our engineering intuition might at first suggest.

1.15   D'Alembert's paradox: the force on semi-infinite bodies

There are three special cases of (44) of some interest. These are as follows.

(1) When the jet is derived from a source within the body alone (e.g. as in a rocket), $C_Q = 0$, and (44) becomes $C_T = C_J$.

(2) When there is no jet, but air intakes are operating (e.g. as in suction for boundary-layer control on wings), (44) gives $C_T = -2C_Q$, so there is a 'sink-drag' of $2C_Q$ on the body.

(3) With no jet and no air intake, $C_T = 0$, and we have D'Alembert's

well-known paradox, namely that there is no drag force on a body moving with uniform velocity through an unbounded inviscid fluid.

These results are all quite independent of the particular form of the isentropic equation of state, so apply to both ideal and tangent gases. However, they do depend on one important assumption about the character of the flow that we have not yet stated explicitly. This is that the flow is completely attached to the body, that is, it does not separate to form an 'open' wake, or dead-air region extending to infinity behind the body (cf. Fig. 1.16a). Inviscid potential flows of the type just excluded do give rise to a drag force acting on the body, these flows are considered at length in ch. 11.

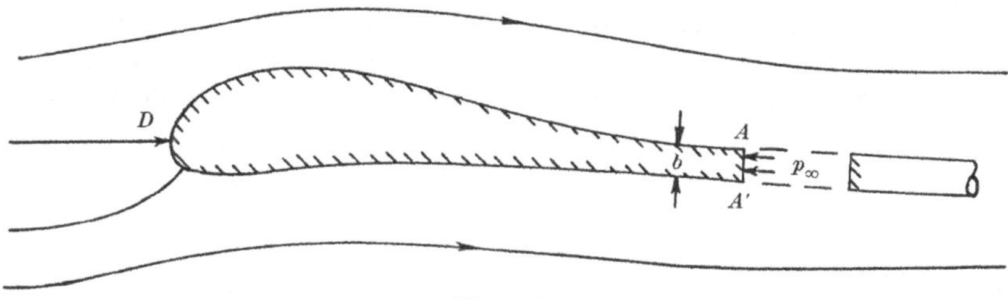

Fig. 1.15

Equation (44) can also be used to determine the drag force D acting on the curved surface of a body which extends downstream to infinity. Let the cross-section of this semi-infinite body be constant, and equal to b well downstream (Fig. 1.15). For this case there is no intake of fluid, and no outflow along the 'jet', i.e. $a = 0$ and $q_0 = 0$. Hence (44) gives

$$T = 0. \tag{45}$$

But in obtaining (44) we took the pressure and momentum integrals across the surface of the jet downstream, and consequently the left-hand side of (45) contains a thrust term $p_\infty b$ in addition to the force $-D$ acting on the curved surface. Hence

$$D = bp_\infty. \tag{46}$$

Equation (45) can be termed D'Alembert's Paradox for a semi-infinite body. It means that if that part of the body to the rear of some section AA' well downstream is removed, then the static pressure of the stationary fluid replacing it is sufficient to support the pressure forces on the curved surface upstream of AA'.

1.16   The drag in a real fluid: inviscid flow models

In the flow of a real fluid past a body a drag force is a matter of common experience. This force has three main components, namely the *induced* drag, the *form* drag and the *skin-friction* drag. Induced drag is a three-dimensional phenomenon, properly accounted for by potential theory (see Robinson & Laurmann, 1956), so we shall not mention it further, but profile drag—the sum of form and skin friction—is a viscosity effect occurring in both two- and three-dimensional flows. A detailed study of profile drag obviously lies outside the scope of this text (the reader can refer to vol. II of Goldstein (1938) for these details), but it is sometimes necessary to take the main effects of this drag into consideration in certain potential flows, therefore we shall give a brief outline of the main features.

Form drag is associated with thick bluff bodies, and is due to flow separation from the surface of the body in the manner shown in Fig. 1.16a. This results in a large wake of slowly eddying fluid extending downstream of the body. An inviscid flow model is easily made—the edges of AB, $A'B'$ of the wake are assumed to form the boundaries of

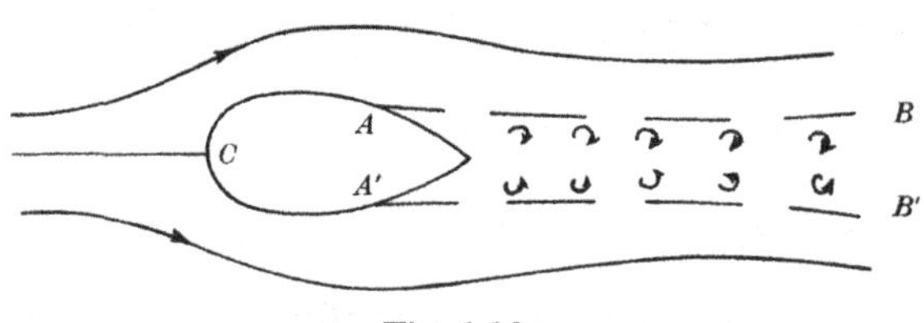

Fig. 1.16a

a potential flow in the fluid outside the wake. Some indeterminancy is introduced here in that the points of flow separation A and A' and the wake pressure behind the body are initially unknown, but in that the model (i) gives a drag force, and (ii) displaces the streamlines outwards from the wake it is a considerable improvement over the non-separating, or *Dirichlet* type, flow used in § 1.14. A study of separating potential flows—usually termed *Helmholtz* or *Kirchhoff* flows—will be made in ch. 11.

Skin-friction drag, on the other hand, is predominant with slender bodies, such as aerofoils and wings. This drag is associated with a loss in fluid momentum in the boundary layer over the aerofoil surface. The resulting velocity deficiency is carried downstream, and produces a narrow region of high vorticity in the aerofoil's wake. The velocity reduction means that the streamlines will be displaced outwards from the wake centre-line, just as with form drag, but of course to nothing like the same extent.

An important formula for the drag D of a body can be deduced from (43). This equation was obtained on the assumption that the fluid passes through an internal duct in the body (Fig. 1.14), but this is not an essential part of the argument. Hence the fluid in a stream-tube element of area da at $x = -\infty$, which passes over the exterior surface of a body (see Fig. 1.16b) and loses momentum in the process, and which flows downstream at infinity with a velocity q through an element of area db, contributes by (43)

$$dD = \rho_\infty q_\infty^2 \, da - \rho_\infty q^2 \, db \tag{47}$$

to the drag. We have put $\rho_0 = \rho_\infty$ here because it is the same fluid throughout the stream tube. Some pressure terms are omitted from the right-hand side of (47) but they vanish on integration, in just the same way as the pressure terms vanish in the summation used in equation (40). By conservation of mass $\rho_\infty q_\infty \, da = \rho_\infty q \, db$, and so the integral of (47) over the plane S_∞ at $x = \infty$ can be written

$$D = \rho_\infty q_\infty^2 \vartheta, \tag{48}$$

where

$$\vartheta \equiv \int_{S_\infty} \frac{q}{q_\infty}\left(1 - \frac{q}{q_\infty}\right) db \tag{49}$$

is termed the *momentum thickness* of the wake. These equations are the basis of the wake traverse method of estimating the drag of a body (see Howarth 1953, vol. II).

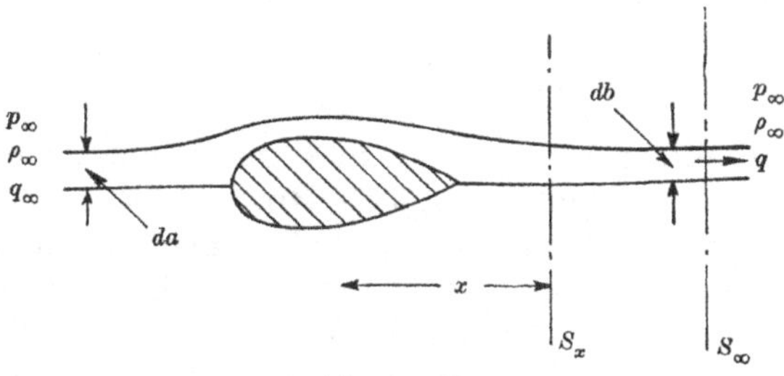

Fig. 1.16b

Equation (48) can be applied at any section S_x (Fig. 1.16b) sufficiently far downstream of the body for the pressure p over S_x to be sensibly the same as p_∞, as this is all that was assumed in proving (43). Consequently sufficiently far downstream ϑ is independent of x. Another important characteristic of wakes is their widening with increasing downstream distance x by viscous and turbulent diffusion. The constancy of ϑ means that as the wake spreads the velocity deficiency is reduced, until eventually q tends to q_∞, and the effective range of the integral in (49) tends to infinity in such a way as to keep ϑ constant. This effect allows us to replace ϑ in (48) by the wake *displacement thickness* δ defined by

$$\delta \equiv \int_{S_x} \left(1 - \frac{q}{q_\infty}\right) db, \tag{50}$$

for this is equivalent to neglecting $(q - q_\infty)^2$ compared with $(q - q_\infty)$. In two-dimensional flow this displacement thickness is clearly the additional distance separating the streamlines in the potential flow on one side of the wake from those in the potential flow on the other side of the wake.

Summarizing the above we have that the main features of real wakes are (i) the loss in fluid momentum, (ii) the outwards displacement of the streamlines from the wake, and (iii) the spreading of the wake with downstream distance, until eventually—well downstream—uniform conditions are re-established.

It is difficult to see how an inviscid flow model of a wake could be devised to reproduce (iii), but fortunately in many cases this is not important because of the relatively slow rate at which the wake widens with downstream distance. From equations for the turbulent wake given in vol. II of Goldstein (1938) it is found that the edge of the wake satisfies the equation

$$\left(\frac{y}{c}\right)^2 \simeq 0{\cdot}43 \, C_D\left(\frac{x}{c}\right), \tag{51}$$

where c is a typical dimension of the cylinder (e.g. the chord length for an aerofoil, or the diameter of a circular cylinder), and C_D is the drag coefficient,

$$C_D \equiv \frac{D}{\tfrac{1}{2}c\rho_\infty q_\infty^2}. \tag{52}$$

Thus in the case of aerofoils moving at subsonic speeds, even if C_D has the rather large value of $0{\cdot}03$ (see ch. 12 of Howarth. 1953) (51) gives a total wake width of less than $\tfrac{1}{2}c$ at a distance of $4c$ downstream of the trailing edge. With bluff bodies, having a high form drag, the position is somewhat poorer. For example a circular cylinder has a C_D of about unity for a considerable Reynolds number range (see p. 419 of Goldstein, 1938), and (51) gives a wake width of about $2\tfrac{1}{2}$ diameters 4 diameters downstream.

The displacement property of wakes can be represented in two distinct ways in potential flow. First when the body is bluff, and the flow separates as shown in Fig. 1.16a, we adopt the Helmholtz model—or rather an adaptation of it—which prescribes a pressure distribution along the separated or free streamlines AB, $A'B'$. In the limit as $x \to \infty$, if our model is to give an accurate representation of the displacement thickness, the normal distance between AB and $A'B'$ should tend to δ, where in accordance with (48) to (50)

$$\delta = \frac{D}{\rho_\infty q_\infty^2} = \tfrac{1}{2}cC_D. \tag{53}$$

The asymptotic form of the pressure distribution selected for AB, $A'B'$ must be such to secure this result, but this still leaves a considerable degree of freedom, permitting a fairly accurate model to be devised.

With slender bodies a less accurate but much simpler representation is usually sufficient. This is simply to use (53) over the whole of the wake, i.e. to add a semi-infinite solid sting of thickness δ to the rear of the body as shown in Fig. 1.16c. The sting is faired into the body smoothly at the trailing edge so as to avoid discontinuities.

Instead of regarding the sting as a solid body one could alternatively place a *source* of fluid at the trailing edge of sufficient strength to displace the flow by an amount δ at $x = \infty$. (The appropriate theory appears in §6.11.) This representation, due to Prandtl[†] and Thom (1943), gives a continuous fluid flow over the wake region, and would clearly have an advantage over the solid sting in cases when the wake position is initially unknown—as, for example, in the case of the lifting aerofoil. Two minor difficulties arise with this model, namely (i) the flow at the trailing edge is

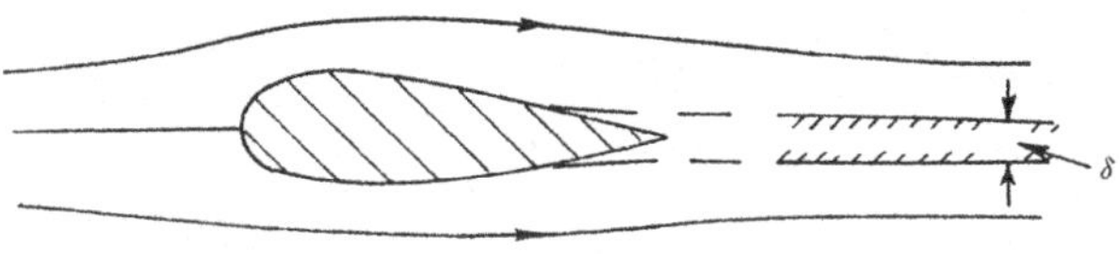

Fig. 1.16c

not smooth, and (ii) the source—if regarded as being attached to the aerofoil—gives a *thrust* of $\delta\rho_\infty q_\infty^2$ instead of a drag of this magnitude. This last result follows by putting b, a, ρ_0, q_0 equal to δ, 0, ρ_∞ and q_∞ respectively in (43). Of course as far as wake displacement is concerned this thrust is really immaterial.

Turning finally to the question of drag, we see from the remark just made about the thrust of a source, and from equation (46) that neither the source model nor the solid sting model yields the right value for the drag. (There is some confusion in the literature on this point.) The more elaborate Helmholtz model, however, does give a drag force, which can be quite close to that obtained in a real flow (see §11.8). The model used will therefore depend on the purpose of the investigation.

Summarizing the position we have: (1) if it is desired to predict the *form* drag of bluff bodies we employ Helmholtz or similar models; (2) no inviscid model can predict the skin-friction drag predominate with aerofoil shapes, but in this case the effect of the wake displacement thickness on the external flow can be allowed for by adding either a solid sting or a source to the rear of the profile, employing in this the experimental value of the drag (or a theoretical value deduced from boundary-layer theory, see Robinson & Laurmann, 1956, p. 167).

[†] *See Aerodynamic Theory*, vol. III, p. 195 (Durand, 1935).

1.17   The lift in a real fluid: Joukowski's hypothesis

Now we have shown in §1.15 that there is no force *parallel* to the stream direction acting on a closed obstacle in Dirichlet (i.e. inviscid and non-separating) flow, and hence if there is a force at all, it must act at right angles to the stream direction. Such a force is called a *lift*, and a great deal of the theory given in part III will be concerned with the calculation of this force in a variety of circumstances.

It is sufficient here to consider the two-dimensional flow about a slender body set at an incidence α to the stream. The cross-section or profile C of this body is shown in Fig. 1.17. If the incidence is large the body will be

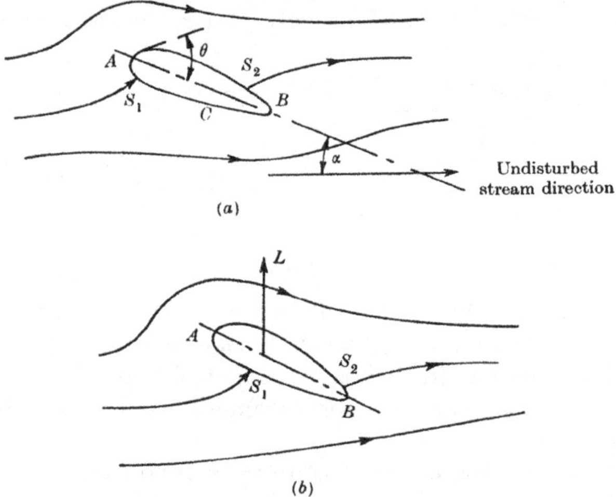

Fig. 1.17. The lift force. (*a*), no lift. (*b*), lift.

relatively 'bluff' to the stream, and flow separation will be likely. This case is excluded by requiring α to be small (e.g. $|\alpha| < \alpha_{\text{crit.}}$, where $\alpha_{\text{crit.}}$ lies between about 10 and 20 degrees).

The main features of the flow are the lift force L, the incidence α, and the locations, S_1 and S_2, of the two points on C, where the streamline of which C is a part divides and rejoins. In general these are points at which q vanishes—*stagnation points*.

It will be established later (§8.4), that the Dirichlet flow about a body is uniquely determined if the flow direction, θ say, is completely known on the given profile surface C. Let θ be measured from the chord line direction $\overrightarrow{AB}$—a convenient line fixed in C—and let θ_s be the value of θ on C. Then if S_1 and S_2 are fixed in position, θ_s is completely known, and the flow direction at all points in the flow is fixed. In particular θ_∞, the flow direction at infinity, is fixed. As the incidence α is the angle between AB and the stream direction at infinity, $\alpha = \theta_\infty$. Thus corresponding to each pair of locations S_1, S_2 there is a unique value of α. The pressures over C are also uniquely determined, and therefore the lift L is fixed in value. Conversely, if L and α are prescribed the positions S_1 and S_2 will be uniquely determined. More generally we need to prescribe any *pair* of the quantities α, L_1, S_1, S_2 to obtain a unique Dirichlet flow.

The qualitative dependence of L and α on S_1 and S_2 is easily decided by considering a completely symmetrical body having an elliptical (or similar) cross-section. If S_1 and S_2 are symmetrically situated so that $S_1 S_2$ passes through the centre of the ellipse, the 'symmetry' of the flow about AB (see Fig. 1.17 a) must make L zero. An increase in α will move S_1 and S_2 about C in the *same* direction if L remains zero. On the other hand, if S_1 and S_2 move in *opposite* directions from symmetrical positions, the flow becomes unsymmetrical, and a lift is established.

Now suppose we set the profile at a fixed incidence α. Then it is a matter of experience that in a real fluid the flow is completely determined. However, our inviscid flow is indeterminate if only α is given. *Joukowski's hypothesis* removes this indeterminacy by postulating that the *rear* stagnation point S_2 always be at a fixed point on C, regardless of the value of α. This hypothesis was originally made only for those profiles possessing a sharp trailing edge Σ, and S_2 was fixed at Σ. As very high velocities are theoretically possible at sharp corners, this selection of S_2 removes the singularity in the flow at Σ.

The very small viscosity in a real fluid establishes a state of affairs close to that postulated by Joukowski, by considerably reducing the high velocities around Σ initially obtained in setting up the motion from rest. Continuity then causes S_2 to move towards Σ, and a lift force develops. When equilibrium is achieved the flow is a complicated viscous flow, and no stagnation point S_2 really exists. A full discussion of the phenomenon is to be found in Goldstein (1938, pp. 40–1, 66–9). It is found that the Joukowski lift L_J (the lift obtained when S_2 coincides with Σ) exceeds the real lift L_R for the same incidence by 10–20 % or more. In terms of a (fictitious) stagnation point S_2, this means that S_2 does not quite reach Σ.

Joukowski's hypothesis can be applied to rounded trailing edges by fixing S_2 at the point of minimum radius of curvature in the trailing edge region, but it is clear that in this case the viscosity effect will be less than for a sharp trailing edge, and therefore the lift deficiency $L_J - L_R$ will be greater. Inviscid theory provides no means of establishing $L_J - L_R$; we are forced to use either empirical methods or boundary-layer theory (Spence, 1954).

BOUNDARY CONDITIONS

1.18 Solid boundaries

Consider the Dirichlet flow past a solid obstacle or wall. Let $\mathbf{v}$ be the velocity of a point P on the boundary, and let $\mathbf{q}$ be the fluid velocity at P. Then continuity of contact between the solid boundary and the fluid requires that the fluid velocity relative to the surface be tangential to it. Hence

$$(\mathbf{q}-\mathbf{v}).\mathbf{n} = 0, \tag{54}$$

where $\mathbf{n}$ is a unit vector normal to the boundary.

In two-dimensional flow this result can be expressed in a form more convenient for later applications as follows. Let $\mathbf{t}$ be a unit vector lying along the tangent to the surface at P, and let θ_s be the angle between $\mathbf{t}$ and some reference line or axis AB fixed in the obstacle (Fig. 1.18). Let θ be

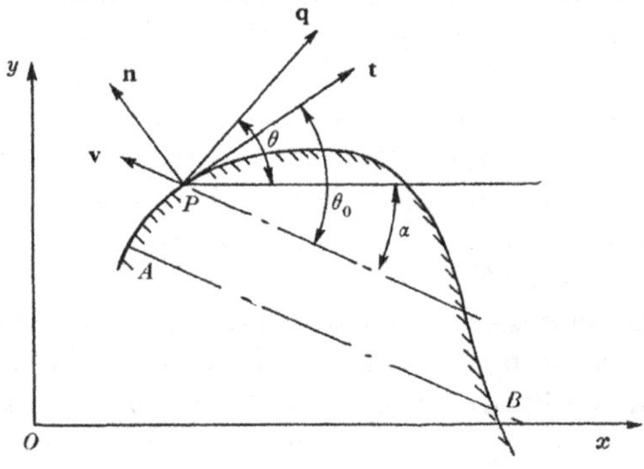

Fig. 1.18

the angle between $\mathbf{q}$ and the Ox axis, and α that between Ox and AB, then it is clear from the figure that

$$\theta = \tan^{-1}\!\left(\frac{\mathbf{n}.\mathbf{q}}{\mathbf{t}.\mathbf{q}}\right) + \theta_s - \alpha,$$

or by (54)
$$\theta = \tan^{-1}\!\left(\frac{\mathbf{n}.\mathbf{v}}{\mathbf{t}.\mathbf{q}}\right) + \theta_s - \alpha. \tag{55}$$

In general, while θ_s depends only on the shape of the body—assuming a rigidly connected boundary—both α and $\mathbf{v}$ will be functions of time. As stated in § 1.1 it is necessary to restrict our treatment of unsteady flows to the case of small perturbations in the displacement of the boundaries from their mean positions. If the axes xOy are fixed relative to the obstacle's mean position, this implies that, except in the immediate neighbourhood of stagnation points, where $\mathbf{q} = 0$, the unsteady perturbation speed $|\mathbf{v}|$ is much smaller than the fluid speed $|\mathbf{q}|$. By (54) this means that the angle between $\mathbf{q}$ and $\mathbf{t}$ must be small, and that (55) can be replaced by the approximation

$$\theta = \theta_s + \frac{\mathbf{n}.\mathbf{v}}{|\mathbf{q}|} - \alpha. \tag{56}$$

The role of stagnation points as boundary conditions is discussed in § 1.23.

1.19   Design problems: given boundary pressures

With the given solid boundaries discussed in the preceding section, the principal task is usually to determine the fluid pressures acting on these boundaries, and then by integration to find the forces acting. The direct inverse of this is given the boundary pressures on a solid obstacle find the shape of this obstacle. This is called the *design* problem, or sometimes the *indirect* problem of fluid and aeromechanics. The boundary condition is simply

$$p = p_s(s), \tag{57}$$

where p_s is a given function of the distance s along the boundary in question.

If s is measured in the direction of flow, and the gradient $(\partial p/\partial s)$ on the boundary is *positive*, it is said to be an *adverse* pressure gradient. This term refers to two well-known effects of such gradients on boundary layers, namely (i) they increase the likelihood of separation from the surface, and (ii) they promote the transition from laminar to turbulent flow within the boundary layer (see Goldstein, 1938).

As far as closed bodies such as aerofoils are concerned, (i) means an increase in form drag, and (ii) an increase in skin-friction drag. Aerofoils are therefore often designed so as to reduce adverse pressure gradients to a minimum, subject of course to a number of constraints arising from structural and other considerations. Broadly, the method here is to distribute the unavoidable adverse gradient fairly uniformly so that high peaks in this gradient do not occur. Aerofoils designed in this way are termed *laminar flow* aerofoils. Piercy (1947) devotes a whole chapter to this topic, which the reader can study with profit.

An alternative technique, which is due to A. A. Griffith (see Goldstein, 1948), is to concentrate the unavoidable pressure gradients over a very narrow region and then to remove the boundary layer arriving at this region by suction into the aerofoil. This method can be employed to prevent flow separation from quite thick profiles—profiles which would not normally qualify for the term 'aerofoil'. If gradients are made zero, except at the suction region the pressure distribution will be almost a step-function as shown in Fig. 1.19. Lighthill (1945b) has designed aerofoils on this principle. For a general account of the subject readers can refer to Goldstein's (1948) paper.

In some cases the shape of *part* of the boundary surface may be initially prescribed, the problem being that of designing the remaining surface, so that it satisfies some pressure condition. In these cases the boundary conditions are a mixture of those considered in § 1.18 and the present section; they are termed *mixed* boundary conditions. The Helmholtz flow shown in Fig. 1.16a is an example of mixed boundary conditions.

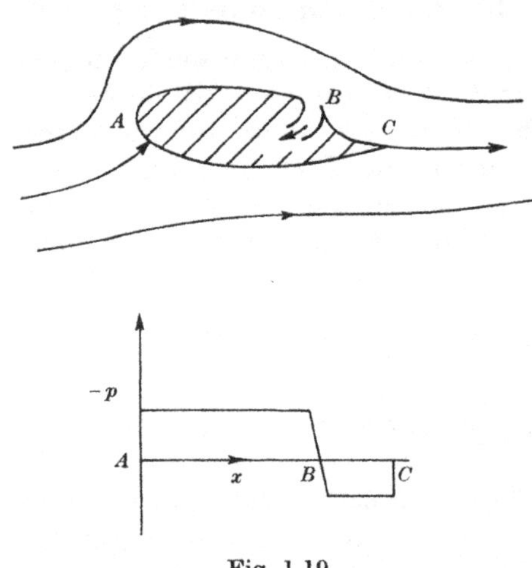

Fig. 1.19

1.20 Porous and perforated surfaces

We shall now consider a boundary condition which lies mid-way in character between that of §§ 1.18 and 1.19. It is the boundary condition appropriate to the potential flow past one side of a porous wall S.

Let p be the pressure in the potential flow at a point P on S, and let p_s be the pressure at P on the opposite side of S. If $\mathbf{n}$ is a unit vector normal to S in the direction away from the potential flow, then experiment shows that there is a functional relation $\mathbf{n} \cdot \mathbf{q} = f(p - p_s)$ between the normal velocity through the surface and the pressure difference $p - p_s$. The experimental data usually satisfies a power law

$$f(p - p_s) = \tfrac{1}{2}\lambda_s q_a \left(\frac{p - p_s}{\tfrac{1}{2}\rho_a q_a}\right)^n, \tag{58}$$

fairly closely (see, for example, Preston & Rawcliffe, 1950). Here λ_s and n are non-dimensional constants depending on the character of the porous wall, and ρ_a, q_a are reference values of the density and speed.

Let the angle between the tangent to S at P and the x-axis be θ_s, then the angle θ between $\mathbf{q}$ and Ox is given by (see Fig. 1.20a) $\theta = \theta_s + \sin^{-1}\{(\mathbf{n}.\mathbf{q})/|\mathbf{q}|\}$. In the applications of interest to us the velocity normal to S will be very much smaller than $|\mathbf{q}|$, so we can write $\theta = \theta_s + (\mathbf{n}.\mathbf{q})/|\mathbf{q}|$. On combining this with (58) we arrive at the porous wall boundary condition

$$\theta = \theta_s + \frac{\lambda_s q_a}{2|\mathbf{q}|}\left(\frac{p-p_s}{\tfrac{1}{2}\rho_a q_a^2}\right)^n. \tag{59}$$

Notice that in the limit $\lambda_s \to 0$ (59) reduces to the steady ($\mathbf{v} = 0$) form of (56) (incorporating α in θ_s), while in the limit $\lambda_s \to \infty$ it reduces to (57) as $\theta - \theta_s$ must remain finite. The mixed boundary conditions referred to in the last paragraph of §1.19 are obtained by letting $\lambda_s = 0$ in some intervals, and $\lambda_s = \infty$ in the remainder.

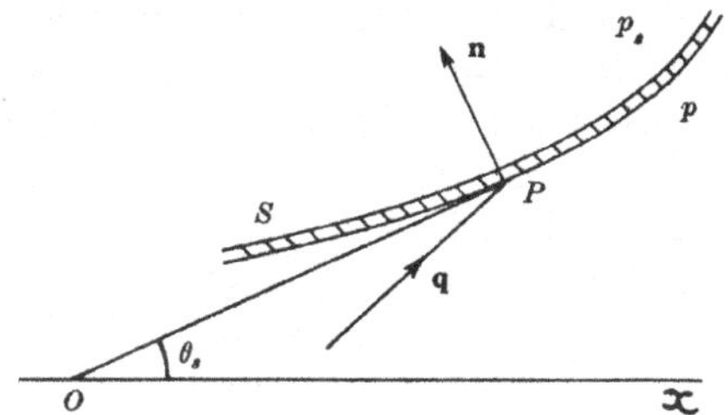

Fig. 1.20a. Porous surface.

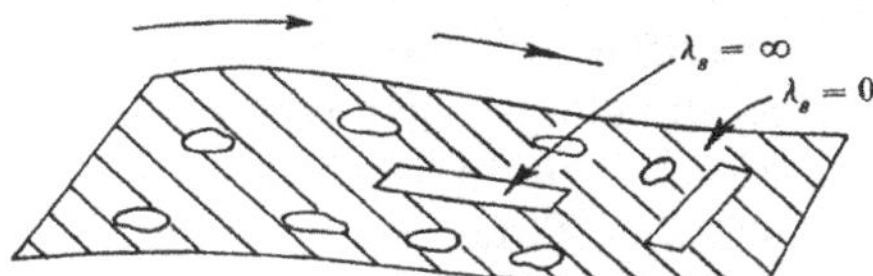

Fig. 1.20b. Perforated surface.

Perforated surfaces are surfaces which are solid in parts and open or 'free' in the remainder, i.e. θ is prescribed in some parts, and p is prescribed in the remainder, i.e. in the gaps or perforations (see Fig. 1.20b). From the remark of the last paragraph such a surface can be regarded as being a special case of a porous surface, with λ_s having the values 0 and ∞. When the perforations are small in size and distributed uniformly, at some distance from the surface the effect on the flow is just like that of a uniformly porous wall for which the parameter n in (59) has the convenient value of unity. This important result is established in §7.29.

1.21 Surfaces of separation: vortex sheets

It is easily verified that, in the absence of surface tension, the equilibrium of a non-rigid surface S separating two inviscid fluids, or separating different regions of the same fluid, demands continuity of pressure across S. Such surfaces are often described as being 'free', by which no more is meant than that their positions are initially unknown. Usually the velocity vector is discontinuous across these surfaces, and if in addition the surface separates different regions of the *same* fluid, it is called a *vortex sheet*.

Vortex sheets are convenient idealizations in cases where concentrated vorticity makes an appearance (see discussion in Goldstein, 1938, pp. 40–6). In three-dimensional problems the velocity vector can be discontinuous in both magnitude and direction across a vortex sheet S, but it follows from (19) and (20), and the continuity of p and ρ across S (continuity of ρ follows from the isentropic equation of state) that it can be discontinuous in *magnitude* only if either the flow is unsteady or S separates regions having different Bernoulli constants. An example of the latter case is shown in Fig. 1.14.

In two-dimensional flows the velocity vector can not be discontinuous in direction, so $|\mathbf{q}|$ jumps in value across a vortex sheet. Let this jump be ω, then the element of circulation $\delta\Gamma$ about a small, narrow, rectangular circuit threading a length δs of the surface S, as shown in Fig. 1.21, is $(q_+ - q_-)\,\delta s = \omega\,\delta s$. Let this circuit be fixed in the fluid particles of the vortex sheet, which we shall assume have a velocity $\bar{q}$ at the point in question, then by Kelvin's theorem ($\S 1.8$), $d(\delta\Gamma)/dt = 0$. Thus

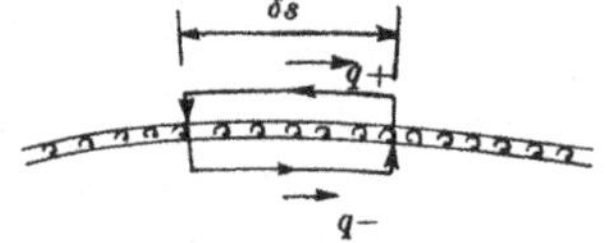

Fig. 1.21

$$\frac{d}{dt}(\omega\,\delta s) = \frac{d\omega}{dt}\,\delta s + \omega\,\delta\bar{q} = 0,$$

as $ds/dt = \bar{q}$. We have now proved that the strength of the vortex sheet, ω, satisfies the equation

$$\frac{d\omega}{dt} + \omega\frac{\partial\bar{q}}{\partial s} = 0, \tag{60}$$

or as $(d\omega/dt) = (\partial\omega/\partial t) + \bar{q}(\partial\omega/\partial s)$,

$$\frac{\partial\omega}{\partial t} + \frac{\partial(\bar{q}\omega)}{\partial s} = 0. \tag{61}$$

This result can also be derived directly from (13), a derivation which has the added virtue of establishing the value of $\bar{q}$. Let $\mathbf{s}$ be a unit vector parallel to $\mathbf{q}$, then the scalar product of $\mathbf{s}$ and (13) gives

$$\frac{\partial q}{\partial t} + \frac{\partial \chi}{\partial s} + \frac{1}{\rho}\frac{\partial p}{\partial s} + \frac{1}{2}\frac{\partial q^2}{\partial s} = 0. \tag{62}$$

Now χ, p and ρ are continuous across the sheet, so by applying this equation to opposite points on each side of the sheet, and subtracting the results we obtain an equation identical with (61), except that $\bar{q}$ is replaced by $\frac{1}{2}(q_+ + q_-)$. From this we conclude that the sheet strength ω is convected downstream at a speed equal to the average of the speeds on each side of the sheet.

If $\bar{q}$ is constant, or changes very slowly, (61) can be written

$$\frac{\partial \omega}{\partial t} + \bar{q}\frac{\partial \omega}{\partial s} = 0, \tag{63}$$

which has the general solution

$$\omega = f(s - \bar{q}t), \tag{64}$$

where f is an arbitrary function. In this case (64) shows that ω is convected downstream unchanged in magnitude, a result which proves very useful in unsteady aerofoil theory (§ 9.1).

1.22   Boundary conditions at the free surface of a heavy incompressible liquid

With a heavy incompressible liquid the external body force per unit mass is g downwards. Hence if the Oy axis is vertical the potential χ (see § 1.5) is gy, and (62) becomes

$$\frac{\partial q}{\partial t} + g\frac{\partial y}{\partial s} + \frac{1}{\rho}\frac{\partial p}{\partial s} + q\frac{\partial q}{\partial s} = 0. \tag{65}$$

Let $y = \eta(x, z)$ be the free surface, S, of this liquid, i.e. the surface which separates the liquid from a gas or vapour, then there are two boundary conditions to be applied on S, namely (i) the pressure condition, and (ii) the kinematic condition.

The pressure condition is simply that—neglecting surface tension—the pressure at S just inside the liquid equals the pressure on S in the gas. This means that as the pressure in the gas can be taken as constant, $\partial p/\partial s = 0$ along S, and so from (65) we have the boundary condition

$$\frac{\partial q}{\partial t} + g\frac{\partial \eta}{\partial s} + q\frac{\partial q}{\partial s} = 0. \tag{66}$$

The kinematic condition is derived from the fact that fluid particles initially on S remain on S. The vertical velocity, v, of a fluid particle moving along a path on S will therefore be $d\eta/dt = \partial\eta/\partial t + q\,\partial\eta/\partial s$. Hence

$$\frac{\partial\eta}{\partial t} + q\frac{\partial\eta}{\partial s} = v. \tag{67}$$

Now consider the linear perturbation of a two-dimensional flow in the (x, y)-plane (this is the only case that can be treated by the methods of this book). Let the velocity components be $U + u$, v in the Ox, Oy directions, and let θ_s, θ be the slopes of the free surface and the velocity

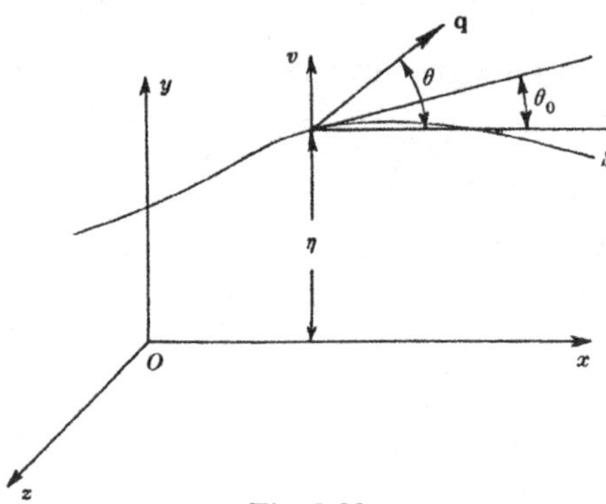

Fig. 1.22

vector respectively (see Fig. 1.22). Then θ_s, θ are small, and to first order $\theta_s = \partial\eta/\partial s$, $\theta = v/U$. Therefore neglecting second-order terms in (66) and (67) we have

$$\frac{\partial u}{\partial t} + g\theta_s + U\frac{\partial u}{\partial s} = 0, \tag{68}$$

and

$$\frac{\partial\eta}{\partial t} + U\theta_s = U\theta, \tag{69}$$

or, equivalently,

$$\frac{\partial\theta_s}{\partial t} + U\frac{\partial\theta_s}{\partial s} = U\frac{\partial\theta}{\partial s}. \tag{70}$$

In steady motion $\theta = \theta_s$, and we have the boundary condition

$$g\theta + U\frac{\partial u}{\partial s} = 0, \tag{71}$$

but in unsteady motion this is replaced by the somewhat complicated condition

$$\frac{\partial^2 u}{\partial t^2} + 2U\frac{\partial^2 u}{\partial t\,\partial s} + U^2\frac{\partial^2 u}{\partial s^2} + gU\frac{\partial\theta}{\partial s} = 0, \tag{72}$$

which follows by eliminating θ_s from (68) and (70).

1.23   Boundary conditions at infinity

Consider a fluid flow which extends to unlimited distances, i.e. which extends to 'infinity'. Such a flow is not determined uniquely unless restrictions on the asymptotic form of the flow near infinity are prescribed. For example the velocity magnitude and direction at infinity must be prescribed:

$$\lim_{|\mathbf{r}|\to\infty} |\mathbf{q}| = q_\infty, \quad \lim_{|\mathbf{r}|\to\infty} \cos^{-1}(\mathbf{q}.\mathbf{i}/|\mathbf{q}|) = \theta_\infty, \tag{73}$$

where $\mathbf{i}$ is unit vector parallel to Ox, and θ is the angle between $\mathbf{q}$ and Ox. In some cases (e.g. flow in channels) the values of q_∞, θ_∞ in (73) will depend on the *direction* from which infinity is approached. Usually further restrictions, such as

$$|\mathbf{q}| - q_\infty \sim O\{|\mathbf{r}|^{-n}\} \quad (n > 0), \tag{74}$$

where n is given, are required.

Restrictions of this type are termed the 'boundary conditions at infinity'; they are no less important than ordinary or 'local' boundary conditions, and in some cases prove to be more significant than local conditions.

In two-dimensional potential flow about a closed obstacle in an infinite stream Kelvin's equation (§1.8) provides a particularly important boundary condition at infinity. From (17) and (18) this is that

$$\Gamma = \int_C \mathbf{q}.d\mathbf{r} = [\phi]_C = k, \tag{75}$$

where k is a finite constant, and C is a contour enclosing the obstacle. As shown in §1.9 C may be extended until it is the large circle near infinity. Comparison of (74) and (75) reveals that if Γ is finite and non-zero, $n = 1$, and if Γ is zero $n > 1$. Equation (75) is thus equivalent to a restriction on the form of $|\mathbf{q}|$ near infinity, and so can be regarded as a boundary condition at infinity.

Other restrictions similar to (75) are the self-evident '*closure*' conditions of two-dimensional flow:

$$\int_C dx = 0, \quad \int_C dy = 0, \tag{76}$$

where C is any contour enclosing the obstacle. Further consideration of (76) is best deferred until the concept of *stream-function* is introduced.

1.24 Miscellaneous boundary conditions

(a) *Positions of stagnation points*

The significance of stagnation points as boundary conditions is made clear in § 1.17. They determine the *sense* of the velocity vector over the surface. Unfortunately they usually occur in positions initially unknown.

If only one stagnation point exists, its position is fixed by one of the boundary conditions at infinity, but when *two* stagnation points occur on a solid boundary, such as an aerofoil, an additional boundary condition has to be imposed to fix the positions of the two points. The Joukowski condition (§ 1.17) is the best example of this.

(b) *Positions of points of flow separation*

As described in § 1.16 the fluid sometimes leaves the solid boundary at some point A, and forms a free vortex sheet. The positions of such points are obviously boundary conditions of considerable importance, but they can be determined only by employing additional hypothesis derived from viscous flow considerations. However, it is found (§§ 11.6, 11.7) that potential theory and the physical reality requirement (see § 1.25 below) considerably restrict the range of possible positions for these points.

(c) *Positions of points of flow re-attachment*

Under certain circumstances, which are not yet fully understood, the separated flow can re-attach to the solid boundary to form a closed cavity or bubble. As with (b) potential theory is again inadequate to determine the positions of the points of flow re-attachment, and viscous theory must be employed to complete the boundary conditions.

Boundary conditions (a), (b) and (c) are all due to viscosity, and illustrate the role that viscosity plays in potential flow theory.

In the *interior* of the fluid, i.e. away from solid boundaries, dead-air regions, vortex sheets, etc., viscosity is negligible, so we ignore it completely, and employ the relatively simple potential flow theory. However on the *boundaries* viscosity effects cannot, in general, be ignored, but must be allowed for in the boundary conditions themselves—otherwise the flows obtained may be indeterminant, unrealistic, or just simply inaccurate as models of real flows.

It should be mentioned here that it is by no means always clear how the boundary conditions ought to be adjusted to make proper allowance for viscosity effects. It is really the task of boundary-layer theory to provide the answer here. While this theory can predict separation points, it is unfortunately not yet sufficiently developed to cope with separated 'boundary' layers, and therefore to predict re-attachment points.

1.25 Existence and uniqueness theorems

The existence and uniqueness of solutions of Laplace's equation depends entirely on the nature of the boundary conditions imposed. For example if the position of both stagnation points *and* the incidence for an aerofoil in Dirichlet flow are prescribed, the boundary conditions are 'over-determined' (see §1.17), and in general no solution exists. On the other hand, if the incidence alone is prescribed the boundary conditions are incomplete, and an infinite number of solutions exist, i.e. any particular solution is not unique. Thus the complementary requirements of existence and uniqueness can be regarded as providing restrictions on the boundary conditions themselves. Hence we have the 'boundary condition'

$$\left.\begin{array}{l}\text{The boundary conditions must be chosen in such} \\ \text{a way as to ensure that (i) a solution exists, and} \\ \text{(ii) it is unique.}\end{array}\right\} \qquad (77)$$

There are two methods of ensuring that (77) is satisfied, viz. (*a*) the establishment of general theorems covering the type of boundary condition in question, and (*b*) trial-and-error methods, i.e. attempt to find a solution in a given case, and if this succeeds then try to establish that it is unique. In general (*a*) is much more difficult than (*b*), but of course the results obtained by (*a*) are far more general.

Few texts on fluid mechanics give any attention to existence theorems: on uniqueness theorems the easily established results for Dirichlet flows (see, for example, Milne-Thomson, 1949, pp. 92–4) are usually all that are offered. There are three good reasons for this; first the mathematics required is often difficult and tedious, secondly the theorems that have been established for separated flows are seldom general enough to meet the applied mathematician's or engineer's needs, and thirdly the trial-and-error method is much more fruitful in that it yields 'existence' as a by-product of a practical solution. Indeed the trial-and-error method for specific cases has sometimes led to the discovery of more general existence theorems (see, for example, Beckenback (ed.) 1952, pp. 149–63). Two recent texts that do consider existence and uniqueness theorems at some length are Bergman & Shiffer (1953) and Birkhoff & Zarantonello (1957).

The trial-and-error method often yields mathematical solutions which satisfy the boundary conditions up to a point, but which fail to make sense physically. An example of such a physically unrealistic solution is shown in Fig. 1.25. Here we have a potential flow separating at A, and then passing through a region occupied by part of the solid boundary downstream of A. When this feature of the trial-and-error solution is discovered the correct conclusion is that no (sensible) solution exists for the given boundary conditions, i.e. for the given obstacle shape, and the type of separated flow assumed. Clearly the next step in the trial-and-error method is to modify the boundary condition by assuming a re-attachment point somewhere on the boundary downstream of A (near B), and a final separation farther downstream (near C).

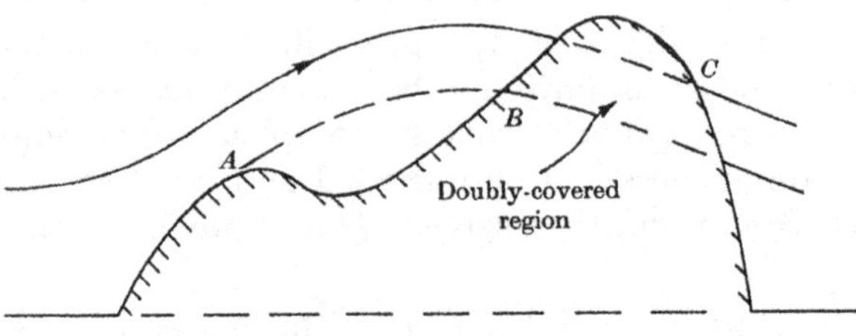

Fig. 1.25

Except with Dirichlet flows, for which the kind of difficulty just discussed cannot occur, our solutions must always be investigated to ensure that they do make physical sense. But even if they do, this is of course no guarantee that they are sensible representations of real flows.

CHAPTER 2

GENERAL THEORY OF TWO-DIMENSIONAL FLOW

INTRINSIC EQUATIONS: THE STREAM FUNCTION

2.1 Summary of basic equations

Two-dimensional or plane flow is characterized by the condition that there is a fixed direction, Oz, say, along which all rates of change of the flow variables vanish. Let $\hat{\mathbf{k}}$ be a unit vector along Oz then the flow takes place in planes parallel to the Oxy-plane that have $\hat{\mathbf{k}}$ for their unit normal.

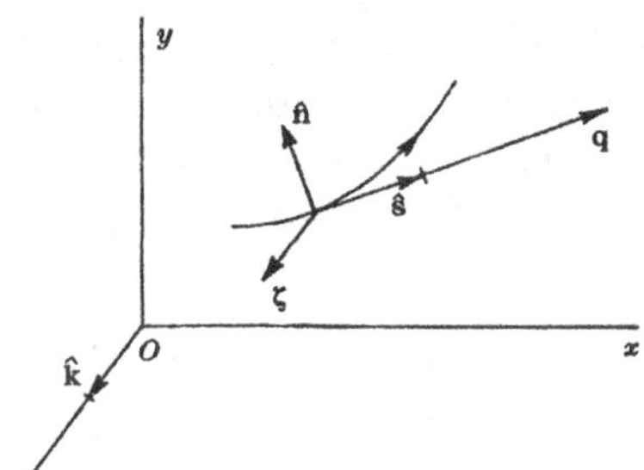

Fig. 2.1

Let $\hat{\mathbf{s}}$ and $\hat{\mathbf{n}}$ be unit vectors lying in the flow plane such that $\hat{\mathbf{s}}$ is parallel to the velocity $\mathbf{q}$, and $\hat{\mathbf{s}}$, $\hat{\mathbf{n}}$, $\hat{\mathbf{k}}$ form a right-handed orthogonal triad (see Fig. 2.1).

The vorticity vector $\boldsymbol{\zeta}$ is necessarily at right-angles to the flow plane, so $\boldsymbol{\zeta} = \zeta\hat{\mathbf{k}}$: and as $\mathbf{q} = q\,\hat{\mathbf{s}}$ it follows that $\boldsymbol{\zeta} \wedge \mathbf{q} = \zeta q\,\hat{\mathbf{k}} \wedge \hat{\mathbf{s}} = \zeta q\,\hat{\mathbf{n}}$, and

$$\zeta\hat{\mathbf{k}} = \nabla \wedge (q\,\hat{\mathbf{s}}) = q\nabla \wedge \hat{\mathbf{s}} - \hat{\mathbf{s}} \wedge \nabla q. \tag{1}$$

Equations (1.11) and (1.13)† can be written

$$\frac{\partial \rho}{\partial t} + \rho q\nabla\cdot\hat{\mathbf{s}} + \hat{\mathbf{s}}\cdot\nabla(\rho q) = 0, \tag{2}$$

and

$$q\frac{\partial \hat{\mathbf{s}}}{\partial t} + \frac{\partial q}{\partial t}\hat{\mathbf{s}} + q\zeta\hat{\mathbf{n}} = -\nabla C, \tag{3}$$

where

$$C = \chi + \int\frac{dp}{\rho} + \tfrac{1}{2}q^2. \tag{4}$$

In addition to these equations there is the isentropic equation of state

$$\left.\begin{aligned}
p &= p_a\left(\frac{\rho}{\rho_a}\right)^{\gamma} \quad \text{(ideal gas)}, \\
p &= p_a + \rho_a^2 a_a^2\left(\frac{1}{\rho_a} - \frac{1}{\rho}\right) \quad \text{(tangent gas)}
\end{aligned}\right\} \tag{5}$$

† This indicates (13) of ch. 1. This notation will be used only when referring to equations in other chapters.

(§§ 1.2, 1.3) and the relations

$$a^2 = \frac{dp}{d\rho} \left\{ \begin{array}{ll} \dfrac{a}{a_a} = \left(\dfrac{\rho}{\rho_a}\right)^{\frac{1}{2}(\gamma-1)} & \text{(ideal gas),} \\[2ex] \dfrac{a}{a_a} = \dfrac{\rho_a}{\rho}, & \text{(tangent gas),} \end{array} \right\} \tag{6}$$

for the sound speed.

Equations (1)–(6) contain seven scalar equations for the seven unknowns ρ, q, $\hat{\mathbf{s}}$, ζ, C, p and a.

2.2 Intrinsic equations: natural coordinates

Let θ be the angle between the vector $\mathbf{q}$ and the x-axis at a fixed point P in the flow field, and at a particular instant of time. Also let s, n be distances measured in the directions, $\hat{\mathbf{s}}$, $\hat{\mathbf{n}}$ at this instant. Then as $\hat{\mathbf{s}}$ is a unit vector it can change only as a result of a change in its direction, i.e. a change in the angle θ. From Fig. 2.2 $d\hat{\mathbf{s}} = \hat{\mathbf{n}}\,d\theta$, and $dn = -\hat{\mathbf{s}}\,d\theta$. Hence

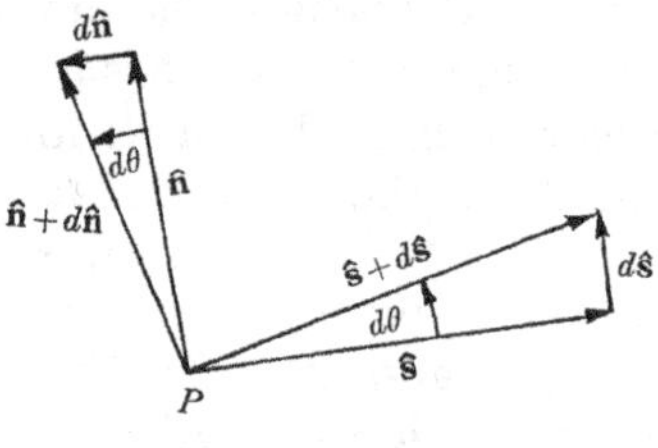

Fig. 2.2

$$\frac{\partial \hat{\mathbf{s}}}{\partial t} = \hat{\mathbf{n}}\frac{\partial \theta}{\partial t}, \quad \frac{\partial \hat{\mathbf{s}}}{\partial s} = \hat{\mathbf{n}}\frac{\partial \theta}{\partial s}, \quad \frac{\partial \hat{\mathbf{s}}}{\partial n} = \hat{\mathbf{n}}\frac{\partial \theta}{\partial n},$$

$$\frac{\partial \hat{\mathbf{n}}}{\partial s} = -\hat{\mathbf{s}}\frac{\partial \theta}{\partial s}, \quad \frac{\partial \hat{\mathbf{n}}}{\partial n} = -\hat{\mathbf{s}}\frac{\partial \theta}{\partial n}. \tag{7}$$

By taking Cartesian axes instantaneously coincident with the directions $\hat{\mathbf{s}}$ and $\hat{\mathbf{n}}$ at a given point P, we can write the operator ∇ in the form

$$\nabla = \hat{\mathbf{s}}\frac{\partial}{\partial s} + \hat{\mathbf{n}}\frac{\partial}{\partial n}. \tag{8}$$

From (7) and (8)

$$\nabla.\hat{\mathbf{s}} = \hat{\mathbf{n}}.\hat{\mathbf{n}}\frac{\partial \theta}{\partial n} = \frac{\partial \theta}{\partial n}, \quad \nabla.\hat{\mathbf{n}} = -\frac{\partial \theta}{\partial s}, \quad \hat{\mathbf{s}}.\nabla = \frac{\partial \theta}{\partial s}, \quad \hat{\mathbf{s}}\wedge\nabla = \hat{\mathbf{s}}\wedge\hat{\mathbf{n}}\frac{\partial}{\partial n} = \hat{\mathbf{k}}\frac{\partial}{\partial n},$$

$$\nabla\wedge\hat{\mathbf{s}} = \nabla\wedge(\hat{\mathbf{n}}\wedge\hat{\mathbf{k}}) = (\hat{\mathbf{k}}.\nabla)\hat{\mathbf{n}} - \hat{\mathbf{k}}(\nabla.\hat{\mathbf{n}}) = -\hat{\mathbf{k}}(\nabla.\hat{\mathbf{n}}) = \hat{\mathbf{k}}\frac{\partial \theta}{\partial s}.$$

These results enable us to write (1), (2) and (3)

$$q\frac{\partial \theta}{\partial s} - \frac{\partial q}{\partial n} = \zeta, \tag{9}$$

$$\frac{\partial \rho}{\partial t} + \rho q\frac{\partial \theta}{\partial n} + \frac{\partial}{\partial s}(\rho q) = 0, \tag{10}$$

$$\frac{\partial q}{\partial t} = -\frac{\partial C}{\partial s}, \tag{11}$$

and
$$q\frac{\partial\theta}{\partial t}+q\zeta = -\frac{\partial C}{\partial n}. \tag{12}$$

These are the intrinsic equations for the flow, and the variables s, n are sometimes termed 'natural coordinates'.

Equation (11) shows that in steady flow C is constant along each streamline; further, if the flow is irrotational, $\zeta = 0$, and it follows from (12) that C is constant throughout the fluid.

2.3   The stream function

If the flow is either steady or incompressible, (2) can be written $\nabla.(\rho q\hat{s}) = 0$. As the operator $\nabla.\nabla\wedge$ reduces any vector to zero, this equation will be satisfied if
$$\rho q\,\hat{s} = \rho_s \nabla\wedge\psi, \tag{13}$$

where ψ is some vector to be determined, and ρ_s is a constant—often taken to be equal to ρ_0, the fluid density at a stagnation point in an ideal fluid.

Clearly ψ can have no components in the plane of the flow, for if this were so, $\nabla\wedge\psi$ would have components in the $\hat{k}$ direction, which is impossible by (13). So $\psi = \psi\hat{k}$, and $\nabla\wedge\hat{k}\psi = -\hat{k}\wedge\nabla\psi$, as $\hat{k}$ is a constant vector. Hence from (8) we deduce that

$$\nabla\wedge\psi = -\hat{k}\wedge\hat{s}\frac{\partial\psi}{\partial s}-\hat{k}\wedge\hat{n}\frac{\partial\psi}{\partial n} = \frac{\partial\psi}{\partial n}\hat{s}-\frac{\partial\psi}{\partial s}\hat{n}.$$

Substituting this in (13) we arrive at

$$\frac{\partial\psi}{\partial s} = 0, \quad \frac{\partial\psi}{\partial n} = \frac{\rho q}{\rho_s}. \tag{14}$$

The first of these shows that ψ is constant on each line traced out by a fluid particle. In steady flow these lines are the streamlines, and ψ is termed the *stream-function*.

Integration of (14) from P_0 on the streamline $\psi = \psi_0$ to any point P_1 on another streamline $\psi = \psi$, gives

$$\rho_s\int_{P_0}^{P_1} d\psi = \rho_s\int_{P_0}^{P_1}\left(\frac{\partial\psi}{\partial n}\right)dn = \int_{P_0}^{P_1}\rho q\,dn = \rho_s\psi_1-\rho_s\psi_0. \tag{15}$$

The third integral in (15) is clearly the mass flow per unit time between ψ_0 and ψ_1, and it follows that in steady flow ψ has the physical significance of being proportional to the mass of fluid flowing per unit time between a given datum streamline on which ψ is assigned the value zero, and the streamline on which it has the (constant) value ψ. This is, of course, on the assumption that (2) is valid at all points between the streamlines in question, i.e. that there are no sources or sinks of fluid between these streamlines.

By combining (1) and (13), and using the facts that $\hat{\mathbf{k}}$ is a constant vector, and that derivatives in the $\hat{\mathbf{k}}$ direction are zero, we can readily show that the vorticity and stream-function are related by

$$\zeta = -\frac{\rho_s}{\rho}\nabla^2\psi - (\nabla\psi).\nabla\left(\frac{\rho_s}{\rho}\right). \tag{16}$$

2.4 Linear shear flow

Linear shear flow is defined by the condition that $\zeta = $ constant. In a region of the stream having negligible curvature (9) shows that $\partial q/\partial n$ is constant in such a flow. This type of flow, or rather something approaching it, occurs in a number of practical problems, an important example of which is the effect of the wind gradient near the ground on the lift of wings (Fig. 2.4).

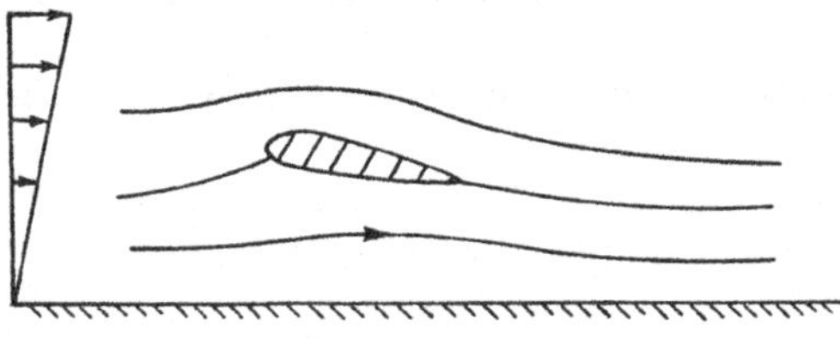

Fig. 2.4

A glance at (16) shows that compressible shear flow will not be amenable to linear methods, except perhaps in some special cases.

In incompressible flow ρ is constant. Giving it the value ρ_s reduces (16) to

$$\zeta = -\nabla^2\psi. \tag{17}$$

Thus incompressible linear shear flow satisfies the Poisson equation

$$\nabla^2\psi + k = 0, \tag{18}$$

where k is the constant value of ζ.

Let ∇^2 be expressed in Cartesian coordinates x, y, i.e.

$$\nabla^2 = \frac{\partial^2}{\partial x^2} + \frac{\partial^2}{\partial y^2},$$

then the introduction of the pseudo stream-function $\bar{\psi}$ defined by

$$\bar{\psi} = \psi - \tfrac{1}{2}ky^2, \tag{19}$$

reduces (18) to $\qquad\qquad\qquad \nabla^2\bar{\psi} = 0. \tag{20}$

IRROTATIONAL STEADY FLOW

2.5   The equations for q and θ

In irrotational steady flow, $\zeta = 0$ and time derivatives vanish, so (9) to (11) can be written

$$q\frac{\partial\theta}{\partial s} - \frac{\partial q}{\partial n} = 0, \tag{21}$$

$$q\frac{\partial\theta}{\partial n} + \frac{\partial q}{\partial s} + \frac{q}{\rho}\frac{\partial\rho}{\partial s} = 0, \tag{22}$$

and

$$\frac{\partial\chi}{\partial s} + \frac{1}{\rho}\frac{\partial p}{\partial s} + q\frac{\partial q}{\partial s} = 0. \tag{23}$$

From (23) and the first of (6)

$$\frac{1}{\rho}\frac{\partial p}{\partial s} = \frac{a^2}{\rho}\frac{\partial\rho}{\partial s} = -q\frac{\partial q}{\partial s} - \frac{\partial\chi}{\partial s},$$

therefore we can eliminate $\partial\rho/\partial s$ from (22) to find

$$q\frac{\partial\theta}{\partial n} + (1 - M^2)\frac{\partial q}{\partial s} - \frac{1}{q}M^2\frac{\partial\chi}{\partial s} = 0, \tag{24}$$

where M is the local Mach number, q/a. The pair of differential equations (21) and (24), together with the relation between q and M, determines the velocity (q, θ).

With a gravitational field parallel to the Oy-axis, $\chi = gy$ (see § 1.22) and the last term of (24) can be written $-(gc/a^2)\,(q/c)\,(\partial y/\partial s)$, where c is a typical length in the flow (e.g. aerofoil chord). In all the problems studied in this book gc/a^2 will be very small (e.g. less than $0{\cdot}001$) compared with unity, so the last term of (24) is negligible. The equation can therefore be replaced by

$$q\frac{\partial\theta}{\partial n} + (1 - M^2)\frac{\partial q}{\partial s} = 0. \tag{25}$$

The independent variables n, s in (21) and (25) for q and θ have the defect of not forming an orthogonal net over the *whole* of the flow plane, so the powerful technique of conformal mapping is not applicable with these variables. This is rectified in the next section.

2.6   Transformation to the (ϕ, ψ)-plane

In irrotational steady flow both the potential and stream-functions exist. Equation (1.18) can be written

$$q\hat{\mathbf{s}} = \nabla\phi = \hat{\mathbf{s}}\frac{\partial\phi}{\partial s} + \hat{\mathbf{n}}\frac{\partial\phi}{\partial n}, \tag{26}$$

i.e.
$$\frac{\partial \phi}{\partial s} = q, \quad \frac{\partial \phi}{\partial n} = 0. \tag{27}$$

From (14) and (27)
$$\nabla \phi . \nabla \psi = \frac{\partial \phi}{\partial s} \frac{\partial \psi}{\partial s} + \frac{\partial \phi}{\partial n} \frac{\partial \psi}{\partial n} = 0,$$

which shows that the equipotential lines, $\phi = $ constant, cut the stream-lines, $\psi = $ constant, orthogonally over the whole of the flow plane. In view of the last remark of § 2.5 the next logical step in our development of the theory of plane flow is to replace the variables s and n by ϕ and ψ.

Equations (14) and (27) enable us to write (21) and (25) in the forms
$$\frac{\partial \theta}{\partial \phi} - \frac{\rho}{\rho_s q} \frac{\partial q}{\partial \psi} = 0, \quad \frac{\partial \theta}{\partial \psi} + (1 - M^2)\frac{\rho_s}{\rho q} \frac{\partial q}{\partial \phi} = 0.$$

We shall now restrict the theory to *subsonic* flow, i.e. to flows for which $M < 1$ everywhere. In this case our equations can be simplified by introducing a new independent 'speed' variable Ω defined by
$$d\Omega = -\frac{\sqrt{(1 - M^2)}}{q} dq = -\frac{\beta}{q} dq, \tag{28}$$

and a non-dimensional number m defined by
$$m = \sqrt{(1 - M^2)}\frac{\rho_s}{\rho} = \frac{\beta \rho_s}{\rho}. \tag{29}$$

These changes reduce our equations to the compact forms
$$\frac{\partial \theta}{\partial \phi} + \frac{1}{m} \frac{\partial \Omega}{\partial \psi} = 0, \quad \frac{\partial \theta}{\partial \psi} - m\frac{\partial \Omega}{\partial \phi} = 0. \tag{30}$$

As M and ρ depend only on q it follows from (28) and (29) that m can be expressed as a function of Ω. Equations (30) are therefore non-linear, but they could be much worse, for they do have the virtue that the coefficient m is independent of the remaining variables ϕ, ψ and θ. This independence suggests that the role of the dependent and independent variables in (30) might be profitably interchanged.

2.7   The hodograph transformation

The interchange just mentioned is known as the *hodograph transformation*, and the plane of the new independent variables Ω, θ can be termed the *pseudo-hodograph plane*. The *true* hodograph plane is the (u, v)-plane, where u, v are the x, y components of the velocity vector, i.e. $u = q\cos\theta$, $v = q\sin\theta$, but the (q, θ)-plane is so much more convenient that many authors now loosely call this the hodograph plane. The (Ω, θ)-plane is derived from the (q, θ)-plane by simply altering the ordinate scale, since by (1.24), (28) and the definition of M it follows that Ω is a single-valued function of q.

If J is the Jacobian $\partial(\phi, \psi)/\partial(\Omega, \theta)$, then provided J is finite and non-zero,

$$\frac{\partial\Omega}{\partial\phi} = \frac{1}{J}\frac{\partial\psi}{\partial\theta}, \quad \frac{\partial\theta}{\partial\phi} = -\frac{1}{J}\frac{\partial\psi}{\partial\Omega}, \quad \frac{\partial\Omega}{\partial\psi} = -\frac{1}{J}\frac{\partial\phi}{\partial\theta}, \quad \frac{\partial\theta}{\partial\psi} = \frac{1}{J}\frac{\partial\phi}{\partial\Omega},$$

by which (30) are transformed into the linear forms

$$\frac{\partial\psi}{\partial\Omega} + \frac{1}{m}\frac{\partial\phi}{\partial\theta} = 0, \quad \frac{\partial\phi}{\partial\Omega} - m\frac{\partial\psi}{\partial\theta} = 0. \tag{31}$$

These equations are called the *hodograph equations*; they were applied to problems of gaseous jets by the Russian Chaplygin over fifty years ago.

The hodograph equations have the great virtue of being linear, that is to say if $\phi_1, \phi_2, \dots, \phi_n, \dots$, are solutions then so are all linear combinations of $\phi_1, \phi_2, \dots, \phi_n, \dots$, and similarly for ψ. This means there is no difficulty in finding *solutions* to the hodograph equations, but this gain is at the expense of very complicated boundary conditions in the hodograph plane, so solutions to *given* flow problems are, in general, very difficult to calculate. In fact, except in special cases, the exact boundary position in the hodograph plane corresponding to a given boundary shape in the physical plane is unknown until the problem in question is completely solved!

Of course the inverse method of finding solutions to the linear hodograph equations, and then transforming these back to the physical plane in order to discover what particular flow problems they solve, is possible, and has been developed by a number of authors. Notable amongst these are Chaplygin (1902), Cherry (1947) and Lighthill (1947). However, this method leads to comparatively difficult mathematics even for relatively simple boundary conditions such as occur in the lifting aerofoil problem (Lighthill, 1947). It would be quite out of the question for more involved boundary conditions, such as those provided by lifting aerofoils in wind tunnels, unsteady aerofoil motion, separated flows, etc. For this reason we cannot adopt the hodograph method in the present text, the main aim of which is to provide practical solutions to a large number of boundary value problems of varying complexity.†

The proper use of the hodograph method lies in providing a number of exact solutions, which (i) may reveal characteristics of the flow not exposed by approximate treatments—as, for example, in Ringleb's (1940) solution of the compressible flow past a half plane—and (ii) can act as 'datum' solutions to check approximate theories. Furthermore, with a different speed variable from that defined in (28), the hodograph

† A recent account of the hodograph method is to be found in Sears (1954).

method can be used to deal with transonic flows, i.e. flows consisting of both subsonic and supersonic regions.

An alternative definition of the speed variable which permits supersonic flow is

$$d\Omega_1 = -\frac{\rho}{\rho_s q}\,dq, \tag{32}$$

and if this is used instead of (28) it will be found that (30) and (31) are replaced by

$$\frac{\partial\theta}{\partial\phi} + \frac{\partial\Omega_1}{\partial\psi} = 0, \quad \frac{\partial\theta}{\partial\psi} - m^2\frac{\partial\Omega_1}{\partial\phi} = 0, \tag{33}$$

and

$$\frac{\partial\psi}{\partial\Omega_1} + \frac{\partial\phi}{\partial\theta} = 0, \quad \frac{\partial\phi}{\partial\Omega_1} - m^2\frac{\partial\psi}{\partial\theta} = 0, \tag{34}$$

respectively. For completely subsonic flows these are not as mathematically convenient as equations (31), but they avoid the number $\sqrt{(1-M^2)}$. By replacing the rather complicated exact functional relation found from (29) and (32) to hold between m^2 and Ω_1 by a simpler expression, which agrees closely with this exact relation in the transonic range, Tomotika & Tamada (1950) have developed a comparatively simple (but still indirect) method of obtaining transonic flows past symmetrical aerofoils.

Incidentally the simplicity of the form of (33) make them very suitable for solution by the method of finite differences, which has been developed by Thom (1933), Southwell (1948) and others for difficult linear and non-linear partial differential equations.

2.8 Non-linear boundary conditions

Differential equations, such as (30), are described as *non-linear* because coefficients of the derivatives are functions of the *dependent* variables, i.e. of the required solution, and so must be unknown initially. Similarly, boundary conditions that are initially unknown because they depend on the solution of the boundary value problem in question, can be called 'non-linear'.

A boundary condition has two distinct components; these are (i) the boundary *position* or shape, and (ii) the boundary *relation*. Examples of boundary relations have been given in §§ 1.18–1.25. The boundary shape is no less important than the boundary relation—in fact from the practical point of view of calculation the boundary shape is generally much more important. Important types of non-linear boundary conditions are thus those in which (i) the boundary *shape* is known exactly, but the boundary relation depends on the final solution, and conversely (ii) the boundary *relation* is known exactly, but the shape is not. The first type of non-linear boundary condition occurs in the (ϕ, ψ)-plane, for in this plane the boundaries are known straight lines parallel to $\psi = 0$ (porous walls are an exception to this). But as the relation between the (x, y)-plane (the physical plane) and the (ϕ, ψ)-plane is initially unknown, the actual boundary values (or relations) to be employed on these boundary lines are unknown. The second type of non-linear boundary condition occurs in the hodograph plane for, as described in § 2.7, the boundary shape is generally unknown in this plane, while ψ and $\partial\phi/\partial n$ have known values.

The two most powerful methods of solving linear partial differential equations, namely the separation of variables technique, and the kernel-function method (of which the use of Green's function for harmonic functions subject to given boundary values is an example), both depend for their success in practical applications on an initial knowledge of the boundary *shape*. This explains why the first type of non-linear boundary condition is very much to be preferred, and why solutions to given flow problems are so difficult to calculate by the hodograph method.

The only way of escaping from the hodograph plane with its linear equations, but difficult non-linear boundary conditions, is to return to the (ϕ, ψ)-plane, and use linear approximations to (30). The non-linearity of the (ϕ, ψ)-plane boundary conditions remains a source of difficulty, but the fact that at least formal solutions, satisfying boundary conditions similar to those prescribed, can be written down by kernel-function methods in this plane, is a very big gain. It will be seen in part III that this approach reduces the problem of solving differential equations in two independent variables to the (usually) easier one of solving an integral equation involving only one independent variable. In many cases this integral equation degenerates to an ordinary algebraic equation, and the solution to the problem is obtained immediately.

It may seem from the foregoing discussion that if an approximation is to be used to linearize the differential equations in the (ϕ, ψ)-plane, then one might as well return right back to the (x, y)-plane in which the flow is usually first defined, and use an approximation to linearize the differential equations in this plane. This method is in fact widely used; it is the *linear perturbation theory* referred to in § 1.2, and it has the advantage of completely linearizing the boundary conditions in certain cases. However, it does have several disadvantages compared with the (ϕ, ψ)-plane linearization, disadvantages which are best discussed after the two theories have been described.

LINEAR THEORIES

2.9   Linearization in the (ϕ, ψ)-plane

In the (ϕ, ψ)-plane the equations for Ω and θ are

$$\frac{\partial \theta}{\partial \phi} + \frac{1}{m}\frac{\partial \Omega}{\partial \psi} = 0, \quad \frac{\partial \theta}{\partial \psi} - m\frac{\partial \Omega}{\partial \phi} = 0. \tag{30, bis}$$

These will be linearized if we can write $m = k$, where k is a constant. The only possible alternative to $m = k$ is $m = f(\phi, \psi)$, but as m is, in fact, a function of Ω alone, there is no way of selecting f before the final solution $\Omega(\phi, \psi)$ is known.

The 'best' value of the constant k for our purposes would be that which makes the average error in Ω and θ resulting from this approximation a minimum. It seems reasonable to assume that the average value of m over the (ϕ, ψ)-plane is the best for this purpose. Adopting the notation for average values introduced in § 1.2 we shall denote this value of m by m_a. Hence from (29) our approximation reads

$$m \simeq m_a = \frac{\beta_a \rho_s}{\rho_a}. \tag{35}$$

Now the constant ρ_s—first introduced quite arbitrarily in (13)—is still at our disposal. Let

$$\rho_s = \frac{\rho_a}{\beta_a}, \tag{36}$$

then (35) assumes the very convenient form

$$m \simeq m_a = 1, \tag{37}$$

and equations (30) are reduced to

$$\frac{\partial \theta}{\partial \phi} + \frac{\partial \Omega}{\partial \psi} = 0, \quad \frac{\partial \theta}{\partial \psi} - \frac{\partial \Omega}{\partial \phi} = 0. \tag{38}$$

Instead of regarding (35) as an approximation for an ideal gas we can regard it as an *exact* equation defining some imaginary gas. Equation (35) will be exact if

$$\frac{\beta}{\rho} = \frac{\beta_a}{\rho_a},$$

and comparing this with (1.26) for the tangent gas we draw the following important conclusion:

The equations for Ω and θ in the (ϕ, ψ)-plane are linearized exactly by replacing the ideal gas by a tangent gas.

This result is, of course, quite independent of the choice of the tangency point.

2.10   The tangency point

In an ideal gas

$$\frac{\rho_a}{\rho} = \left\{\frac{1 + \tfrac{1}{2}(\gamma - 1)\,M^2}{1 + \tfrac{1}{2}(\gamma - 1)\,M_a^2}\right\}^{1/(\gamma-1)}, \tag{39}$$

which follows from (1.25) on replacing n by $-\gamma$ (see §1.2). Hence from (29), (36) and $\beta \equiv (1 - M^2)^{\frac{1}{2}}$ we find that

$$m = \frac{\beta}{\beta_a}\frac{\rho_a}{\rho} = \left\{\frac{1 - M^2}{1 - M_a^2}\right\}^{\frac{1}{2}}\left\{\frac{1 + \tfrac{1}{2}(\gamma - 1)\,M^2}{1 + \tfrac{1}{2}(\gamma - 1)\,M_a^2}\right\}^{1/(\gamma-1)} \tag{40}$$

in an ideal gas. Expansion in series gives

$$m = 1 - \tfrac{1}{8}(\gamma + 1)\,(M^4 - M_a^4) + O(M^6 - M_a^6). \tag{41}$$

Hence regarding the equation $m = 1$ as an approximation for an ideal gas, we conclude that this is a good approximation if

$$\tfrac{1}{8}(\gamma + 1)|M^4 - M_a^4| \ll 1 \quad (\gamma = 1\cdot4) \tag{42}$$

almost everywhere in the flow. 'Almost' is used here because it is reasonable to assume that the existence of a finite number of singular points at which (42) is violated (e.g. sharp corners and stagnation points) would not introduce significant errors, except in the immediate neighbourhoods of the points in question. We shall return to this aspect shortly. Fig. 2.10 shows the region in which the error in $m = 1$ is less than 10%. Table 2.10 shows the variation of m with M_a and M.

Clearly the best tangency point will be that which makes the left-hand side of (42) as small as possible over most of the flow field. For example, in the flow of an infinite stream past an aerofoil the selec

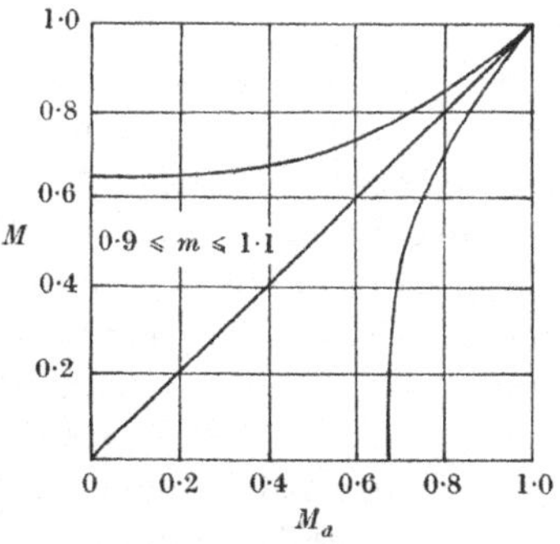

Fig. 2.10

tion $M_a = M_\infty$ (and therefore $U = q_a = q_\infty$, $p_a = p_\infty$, $\rho_a = \rho_\infty$, etc.), where the subscript ∞ denotes values in the (undisturbed) stream at infinity, will make the left-hand side of (42) almost zero everywhere, except in the close vicinity of the aerofoil. In all problems involving infinite streams, therefore, we shall adopt this choice of tangency conditions—a choice first employed by von Kármán (1941) and Tsien (1939).

In flow problems in which q varies considerably over a large area (e.g. flow in channels having large variations in width) the best choice for M_a, or equivalently for the corresponding speed U, is not easily decided, and no definite rule can be given. The difficulty is that the best value to take for U is probably $\bar{q}$, the average of $q(\phi, \psi)$ over the whole field, however $\bar{q}$ cannot be calculated until the solution to the flow problem is known. One must therefore rely on a rough estimate of $\bar{q}$.

Chaplygin (1902), and later Busemann (1937) and Demtchenko (1932), took the tangency point to be at stagnation conditions, i.e. they set $M_a = 0$. This introduces a small but constant error throughout the flow; in addition, it is not a good approximation if $M > 0\cdot6$ (cf. Fig. 2.10). The Kármán–Tsien method is clearly preferable.

Table 2.10. *Values of m for an ideal gas*

	M								
M_a	0	0·4	0·6	0·7	0·8	0·85	0·9	0·95	1·00
0	1·000	0·992	0·952	0·902	0·811	0·738	0·635	0·473	0
0·4	1·008	1·000	0·960	0·910	9·818	0·745	0·640	0·477	0
0·6	1·051	1·042	1·000	0·947	0·852	0·776	0·666	0·497	0
0·7	1·109	1·099	1·055	1·000	0·899	0·818	0·703	0·524	0
0·8	1·233	1·233	1·174	1·113	1·000	0·910	0·783	0·583	0
0·85	1·355	1·343	1·289	1·222	1·098	1·000	0·859	0·640	0
0·9	1·576	1·563	1·400	1·322	1·278	1·164	1·000	0·745	0
0·95	2·115	2·098	2·014	1·908	1·715	1·562	1·342	1·000	0
1·00	∞	∞	∞	∞	∞	∞	∞	∞	1·000

Stagnation points are isolated points at which $m - 1$ is larger than usual. For example, if $M_a = 0\cdot8$, $m - 1 = 0\cdot23$ at a stagnation point. But such errors occur in only small regions of flow and are not serious.

'Corners' projecting into the fluid cause more serious errors, for in the neighbourhood of these corners the speed can be very large—in fact with *sharp* corners the speed is theoretically unbounded (§ 7.4). However, as the theory cannot permit fluid speeds exceeding q_{max}. (§ 1.10), sharp corners are not really permissible, and there must be some restriction on the smallness of the curvature of these projections. But even if the curvature is selected to limit M so that it does not exceed $0\cdot95$, (40) reveals that the error in (37) can be as high as 54%. However, just as in the case of stagnation points, these comparatively large errors occur in only one or two small, isolated regions, if at all, and it is reasonable to assume that their effects *outside* their regions of occurrence are negligible.

In spite of the remark just made about *sharp* corners, we shall occasionally employ them in part III for mathematical convenience. A solution so obtained will have no meaning in the immediate neighbourhood of a corner, but at some distance from the corner the solution can be quite properly interpreted as being that appropriate to a *rounded* projection on which $q < q_{\text{max}}$, and which has the same deflexion angle as the corner employed to represent it. Thus our corners are mathematical devices very similar to the sources, sinks and vortices commonly employed in fluid mechanics.

2.11   Linearization in the (x, y)-plane

Equations (21), (25) and (28) give

$$\frac{\partial\theta}{\partial s}+\frac{1}{\beta}\frac{\partial\Omega}{\partial n}=0, \quad \frac{\partial\theta}{\partial n}-\beta\frac{\partial\Omega}{\partial s}=0.$$

Transforming these into the (x, y)-plane by (cf. Fig. 2.11a)

$$\frac{\partial}{\partial s}=\cos\theta\frac{\partial}{\partial x}+\sin\theta\frac{\partial}{\partial y}, \quad \frac{\partial}{\partial n}=\cos\theta\frac{\partial}{\partial y}-\sin\theta\frac{\partial}{\partial x},$$

then combining the resulting equations we arrive at

$$\left.\begin{aligned}
\left(\beta\sin^2\theta+\frac{1}{\beta}\cos^2\theta\right)\frac{\partial\theta}{\partial y}-\left(\frac{1}{\beta}-\beta\right)\sin\theta\cos\theta\frac{\partial\theta}{\partial x}-\frac{\partial\Omega}{\partial x}=0,\\
\left(\beta\cos^2\theta+\frac{1}{\beta}\sin^2\theta\right)\frac{\partial\theta}{\partial x}-\left(\frac{1}{\beta}-\beta\right)\sin\theta\cos\theta\frac{\partial\theta}{\partial y}+\frac{\partial\Omega}{\partial y}=0.
\end{aligned}\right\} \quad (43)$$

These are the equations we must linearize; they should be compared with the corresponding, and much simpler, equations in the (ϕ, ψ)-plane, equations (30). The linearization of (43) clearly requires a more restrictive approximation than replacing the ideal gas by a tangent gas, which was sufficient in the (ϕ, ψ)-plane.

Equations (43) are partially linearized by requiring that θ and its derivatives be of order ϵ, where ϵ is a small first-order quantity, almost everywhere in the flow, and then ignoring terms of order ϵ^2 and higher. This reduces (43) to

$$\frac{1}{\beta}\frac{\partial\theta}{\partial y}-\frac{\partial\Omega}{\partial x}=0, \quad \beta\frac{\partial\theta}{\partial x}+\frac{\partial\Omega}{\partial y}=0. \quad (44)$$

These are still non-linear as β is a function of q, and hence depends on Ω. Now

$$\beta=\beta_a+(\Omega-\Omega_a)\left(\frac{d\beta}{d\Omega}\right)_{\Omega=\Omega_a}+O(\Omega-\Omega_a)^2,$$

so if we assume $\Omega-\Omega_a=O(\epsilon)$, and ignore terms $O(\epsilon^2)$, (44) becomes

$$\frac{\partial\theta}{\partial(\beta_a y)}-\frac{\partial\Omega}{\partial x}=0, \quad \frac{\partial\theta}{\partial x}+\frac{\partial\Omega}{\partial(\beta_a y)}=0, \quad (45)$$

which are the required linear equations.

A convenient origin for Ω is that which makes Ω_a vanish. From (28) this means that

$$\Omega=-\int_U^q\frac{\beta}{q}dq. \quad (46)$$

The approximations leading to (45) can now be expressed

$$\theta, \; \frac{\partial\theta}{\partial x}, \; \frac{\partial\theta}{\partial y}, \; \Omega=O(\epsilon), \quad O(\epsilon^2)\simeq 0. \quad (47)$$

These mean, in particular, that the boundary slopes and curvatures must be small. In infinite streams the restrictions on θ are sufficient to ensure that Ω is $O(\epsilon)$, but there are cases, e.g. flow in a slowly converging channel, when this is not so.

The restriction that Ω be $O(\epsilon)$ enables us to deduce from (46) that

$$\Omega = \frac{\beta_a}{U} \int_U^q dq = \beta_a\left(1 - \frac{q}{U}\right), \tag{48}$$

to first order. Similarly,

$$\theta = \tan^{-1}\left(\frac{dy}{dx}\right) = \frac{dy}{dx}, \tag{49}$$

to first order.

In incompressible flow $\beta = \beta_a = 1$, and (43) reduces immediately to (45) (with $\beta_a = 1$). Thus the approximations in (47) are necessary only in compressible flow.

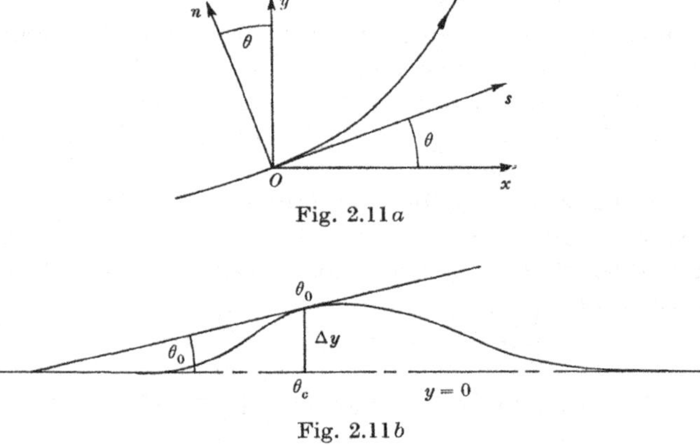

Fig. 2.11 a

Fig. 2.11 b

The usual method of solving (45) requires yet a further restriction; this is that on the boundaries

$$\Delta y = O(\epsilon), \tag{50}$$

where Δy is the maximum deviation of the boundary in the y-direction. If (50) holds, the boundary conditions can be applied on straight lines parallel to $y = 0$, instead of on the actual curved boundaries. These straight lines are selected so as to pass through the 'average' positions of the real boundaries. For example, if the boundary condition is $\theta = \theta_0$ at $y = \Delta y$, then the boundary condition at $y = 0$ is

$$\theta_c \simeq \theta_0 - \Delta y (d\theta/dy)_0 = \theta_0\{1 + 0(\epsilon)\}$$

by (47) and (50). Hence we can apply $\theta = \theta_0$ at $y = 0$ and incur only a second-order error term.

A less restrictive condition than (50) is sometimes useful, e.g. in the example just considered (50) can be replaced by

$$\Delta y \left(\frac{d\theta}{dy}\right)_0 = O(\epsilon^2), \tag{51}$$

which permits Δy to be $O(\epsilon^{-n})$ provided $(d\theta/dy)_0$ is $O(\epsilon^{n+2})$.

Equations (50) and (51) apply to both compressible and incompressible flow.

2.12   Comparison of the two theories

Linear perturbation theory has several obvious disadvantages compared with the linear theory for the (ϕ, ψ)-plane given in § 2.9; these are as follows.

(1) Perturbation theory can be applied to only a limited class of problems, namely those in which the boundaries are inclined at small angles to the x-axis. As θ is usually large ($\approx \frac{1}{2}\pi$) near stagnation points, the theory must *fail* in the neighbourhoods of these points, whereas the (ϕ, ψ)-plane theory is merely a little *inaccurate* near such points (see § 2.10).

(2) The boundary shape in the (x, y)-plane is not as simple as in the (ϕ, ψ)-plane, in which the boundaries are, in general, straight lines parallel to $\psi = 0$. This difficulty is obviated only if (50) or (51) is valid— a restriction which must be used even in incompressible flow.

(3) Finally, whereas the (ϕ, ψ)-plane theory can be made the basis of a more accurate theory with little difficulty, this is not so with linear perturbation theory. The difficulty of a second-order theory based on linear perturbation theory occurs because of the failure of this theory at stagnation points. Lighthill (Sears, 1954) discusses this point at some length in an interesting account of the various approximations that have been used in subsonic flow theory.

On the credit side the linear perturbation theory is algebraically a little simpler than the (ϕ, ψ)-plane theory, which is naturally so as the latter contains the former as a special case (see § 2.17). Where the two theories yield identical results we shall adopt linear perturbation theory for this gain in simplicity.

2.13   Relations for the velocity, pressure and density

(1) *Tangent gas theory*

From (46) and (1.31)

$$\Omega = -\frac{\beta_a}{M_a} \int_U^q \frac{1}{q} \left\{ \left(\frac{q}{U}\right)^2 + \left(\frac{\beta_a}{M_a}\right)^2 \right\}^{-\frac{1}{2}} dq,$$

which can be integrated to give the relation

$$\frac{q}{U} = \sinh \Omega^* \operatorname{cosech} (\Omega + \Omega^*), \qquad (52)$$

where
$$\Omega^* = \sinh^{-1}\left(\frac{\beta_a}{M_a}\right) = \tanh^{-1}(\beta_a)$$

$$= \ln\left(\frac{1+\beta_a}{M_a}\right) = \ln\left(\frac{M_a}{1-\beta_a}\right). \qquad (53)$$

(These equivalent forms for Ω^* will prove useful later.)

The relation between β and Ω is found from (1.31) and (52) to be

$$\beta = \tanh(\Omega + \Omega^*), \qquad (54)$$

so
$$M = \operatorname{sech}(\Omega + \Omega^*). \qquad (55)$$

The density is thus
$$\rho = \frac{\rho_a}{\beta_a} \tanh(\Omega + \Omega^*), \qquad (56)$$

by (1.26) and (54).

Equation (1.8) for the pressure in a tangent gas reads

$$p = p_a + \rho_a^2 \frac{U^2}{M_a^2}\left(\frac{1}{\rho_a} - \frac{1}{\rho}\right),$$

on replacing a_a by U/M_a. By (53) and (56) this can be written

$$p = p_a + \rho_a U^2 \sinh \Omega \cosh \Omega^* \operatorname{cosech}(\Omega + \Omega^*). \qquad (57)$$

The pressure coefficient, C_p, defined by

$$C_p = \frac{p - p_a}{\frac{1}{2}\rho_a U^2}, \qquad (58)$$

is therefore
$$C_p = 2 \sinh \Omega \cosh \Omega^* \operatorname{cosech}(\Omega + \Omega^*). \qquad (59)$$

The minimum value of Ω, corresponding to zero pressure, is given by

$$\Omega_{\min.} = -\tanh^{-1}\left(\frac{\beta_a}{1 + \gamma M_a^2}\right), \qquad (60)$$

which follows from the first of (1.29), (53) and (54). From (52)

$$q_{\max.} = U \sinh \Omega^* \operatorname{cosech}(\Omega_{\min.} + \Omega^*). \qquad (61)$$

From (52) $q = 0$ if $\Omega = \infty$, $q = \infty$ if $\Omega = -\Omega^*$, and q is negative (i.e impossible) when $\Omega < -\Omega^*$. As $0 \leqslant q \leqslant q_{\max.} < \infty$, it follows that the speed variable Ω must satisfy

$$-\Omega^* < \Omega_{\min.} \leqslant \Omega \leqslant \infty. \qquad (62)$$

Notice from (56) that $M = 1$ at $\Omega = -\Omega^*$, so (62) shows that supersonic patches are impossible in a tangent gas.

Table 2.13 of special values, which can be deduced from (1.29) and the equations given above, will be useful. In this table γ has been given the value $1\cdot4$.

Table 2.13. *Special values for a tangent gas*

Variable ...	q	a	M	β	ρ	p	C_p	Ω
Stagnation conditions	0	$a_a\beta_a$	0	1	$\dfrac{\rho_a}{\beta_a}$	$p_a+\dfrac{p_aU^2}{1+\beta_u}$	$\dfrac{2}{1+\beta_u}$	∞
Average flow conditions	U	a_a	M_a	β_a	ρ_a	p_a	0	0
Zero pressure conditions	$q_{\text{max.}}$	$\tfrac{12}{7}a_a$	$M_{\text{max.}}$	$\tfrac{7}{12}\beta_a$	$\tfrac{7}{12}\rho_a$	0	$\dfrac{-p_a}{\frac{1}{2}\rho_aU^2}$	$\Omega_{\text{min.}}$
Sonic conditions	∞	∞	1	0	0	$-\infty$	$-\infty$	$-\Omega^*$

A relation for the pressure gradient, $\partial p/\partial s$, along a streamline, which will be of use later, can be deduced from (27), (52) and (57):

$$\frac{\partial p}{\partial s} = q\,\frac{\partial p}{\partial \phi} = q\,\frac{dp}{d\Omega}\,\frac{\partial \Omega}{\partial \phi},$$

so

$$\frac{\partial p}{\partial s} = \frac{\rho_a U^3 \sinh^2 \Omega^* \cosh \Omega^*}{\sinh^3 (\Omega + \Omega^*)}\left(\frac{\partial \Omega}{\partial \phi}\right). \tag{63}$$

From (53), (62) and (63) we conclude that $\partial p/\partial s$ and $\partial\Omega/\partial\phi$ are of the same sign, in particular an adverse (i.e. positive) pressure gradient means that

$$\frac{\partial \Omega}{\partial \phi} > 0. \tag{64}$$

(2) *Linear perturbation theory*

In linear perturbation theory Ω is a small first-order quantity, and second-order quantities are negligible. This permits us to replace (52), (54), (56) and (59) by the linear forms

$$\frac{q}{U} = 1 - \frac{\Omega}{\beta_a}, \tag{65}$$

$$\beta = \beta_a\left(1 + \frac{M_a^2}{\beta_a}\,\Omega\right), \tag{66}$$

$$\rho = \rho_a\left(1 + \frac{M_a^2}{\beta_a}\,\Omega\right), \tag{67}$$

and

$$C_p = \frac{2\Omega}{\beta_a}. \tag{68}$$

(3) *Incompressible flow*

If the flow is incompressible $M_a = 0$, $\beta_a = 1$, so from (53) $\Omega^* = \infty$. In this limit (52), (56) and (59) reduce to

$$\frac{q}{U} = e^{-\Omega}, \tag{69}$$

$$\rho = \rho_a, \tag{70}$$

and
$$C_p = 1 - e^{-2\Omega}. \tag{71}$$

COMPLEX VARIABLE FORMULATION

2.14 Complex variable equations for the tangent gas

If $f(z)$ is a one-valued and differentiable function of $z = x + iy$ in a domain D of the Argand or z-plane, except possibly at a finite number of exceptional or *singular* points, we say that it is *analytic* in D. If no singular points occur in D, we shall term $f(z)$ *regular* in D. Now any text on complex variable theory (e.g. Nehari, 1952) contains the fundamental result that the necessary and sufficient conditions for $f(z) \equiv u + iv$ to be regular in D is that the four partial derivatives $\partial u/\partial x$, $\partial v/\partial y$, $\partial u/\partial y$, $\partial v/\partial x$ exist (u and v are real functions), are continuous and satisfy the Cauchy–Riemann equations

$$\frac{\partial u}{\partial x} = \frac{\partial v}{\partial y}, \quad \frac{\partial u}{\partial y} = -\frac{\partial v}{\partial x}, \tag{72}$$

at each point of D.

Equations (38) for the tangent gas can be written

$$\frac{\partial \Omega}{\partial \phi} = \frac{\partial \theta}{\partial \psi}, \quad \frac{\partial \Omega}{\partial \psi} = -\frac{\partial \theta}{\partial \phi}, \tag{73}$$

and comparing these with (72) we conclude that in a domain D of the plane
$$w = \phi + i\psi, \tag{74}$$

in which the derivatives in (73) exist and are continuous, the complex function τ defined by
$$\tau = \Omega + i\theta, \tag{75}$$

is a regular function of w. This is expressed by writing

$$\tau = \tau(w), \tag{76}$$

in D. We shall also use the functional notation (76) even when exceptional points occur in D, and τ is only an analytic function of w.

Notice that further differentiation in (73) gives

$$\nabla^2 \Omega = 0, \quad \nabla^2 \theta = 0, \tag{77}$$

where ∇^2 is the Laplacian operator,

$$\nabla^2 \equiv \frac{\partial^2}{\partial \phi^2} + \frac{\partial^2}{\partial \psi^2}. \tag{78}$$

Functions like Ω and θ, which satisfy Laplace's equation (77), are *harmonic* functions. If a pair of harmonic functions are related by Cauchy–Riemann equations, they are said to be the harmonic *conjugates* of each other. Thus Ω and θ are harmonic conjugate functions in the (ϕ, ψ)-plane.

From the above theory it follows that if $f(w)$ is any analytic function of w, then

$$\Omega = \tfrac{1}{2}\{f(w) + \overline{f(w)}\}, \quad \theta = \tfrac{1}{2}i\{\overline{f(w)} - f(w)\}, \tag{79}$$

where the bar denotes a *complex* conjugate, are solutions of equations (77) compatible with (73).

2.15 The relation between the w- and z-planes

When the solution $\tau(w)$ to a given flow problem has been found it remains to transform this into the physical or z-plane. We therefore need the relation connecting w and z.

From Fig. 2.11a it is apparent that

$$dx = \cos\theta\, ds - \sin\theta\, dn, \quad dy = \sin\theta\, ds + \cos\theta\, dn,$$

and therefore

$$dz = e^{i\theta}(ds + i\, dn). \tag{80}$$

By (14), (27) and (36)

$$d\phi = q\, ds, \quad d\psi = \left(\frac{\beta_a}{\rho_a}\right)\rho q\, dn, \tag{81}$$

so

$$dz = \frac{e^{i\theta}}{q}\left\{d\phi + i\left(\frac{\rho_a}{\beta_a}\right)\frac{1}{\rho}\, d\psi\right\}$$

$$= \frac{e^{i\theta}}{2q}\left\{\left(1 + \frac{\rho_a}{\beta_a \rho}\right)dw + \left(1 - \frac{\rho_a}{\beta_a \rho}\right)d\overline{w}\right\},$$

as $d\phi = \tfrac{1}{2}(dw + d\overline{w})$, $i\,d\psi = \tfrac{1}{2}(dw - d\overline{w})$ from (74). Equations (52), (53) and (56) yield

$$\frac{1}{q}\left(1 \pm \frac{\rho_a}{\beta_a \rho}\right) = \frac{\sinh(\Omega + \Omega^*)}{U \sinh \Omega^*}\{1 + \coth(\Omega + \Omega^*)\}$$

$$= \pm \frac{1}{U}\frac{1 \pm \beta_a}{\beta_a} e^{\pm\Omega},$$

so that the dz, dw relation can be written in the form

$$dz = \frac{1}{2U\beta_a}\{(1+\beta_a)\,e^\tau\,dw - (1-\beta_a)\,e^{-\bar\tau}\,d\bar w\}, \tag{82}$$

where we have used (75). Finally we have the required relation

$$z = z_c + \frac{1+\beta_a}{2U\beta_a}\int_{w_c}^{w} e^{\tau(w)}\,dw - \frac{1-\beta_a}{2U\beta_a}\overline{\int_{w_c}^{w} e^{-\tau(w)}\,dw}, \tag{83}$$

where z_c and w_c are corresponding points in the z- and w-planes.

Thus when $\tau = \tau(w)$ is known it can be substituted in (83) to give a relation between the (x, y) and (ϕ, ψ)-planes in the form

$$z = z(w, \overline{w}). \tag{84}$$

2.16 Incompressible flow

In incompressible flow $\beta_a = 1$, and (82) reduces to

$$dz = \frac{1}{U}e^\tau\,dw, \tag{85}$$

i.e. instead of (84) we have

$$z = z(w). \tag{86}$$

Equation (86)—or rather its inverse—expresses that in incompressible flow, as distinct from compressible flow, w is an analytic function of z, and it also follows that τ is an analytic function of z. This is the really important mathematical difference between compressible and incompressible flows, and it means that in some cases incompressible flow is analytically easier to calculate.

For example suppose we have an aerofoil at zero incidence, and that the slope of the surface is $\theta_s(s)$, where s is the perimeter distance measured around the surface from some convenient origin. If the incidence is now increased to α, i.e. the chord line is inclined at an angle $-\alpha$ to Ox, the new slope, θ_α say, will be

$$\theta_\alpha(s) = \theta_s(s) - \alpha + \Delta, \tag{87}$$

where Δ is a term arising from the displacement of the stagnation points. Thus if the increase in incidence displaces a stagnation point from $s = s_0$ to $s = s_1$, the sense of the vector $\mathbf{q}$ is changed in (s_0, s_1), and so $\Delta = \pm\pi$ in $s_0 \leqslant s \leqslant s_1$, and vanishes outside this interval. The important conclusion from (87) is that on the aerofoil $\mathscr{I}\tau$ is the sum of a term due to aerofoil thickness alone and a term due to incidence alone. Consequently, the contributions to $\tau(z)$ due to thickness and incidence can be calculated independently and then linearly superimposed. This important simplification is not possible in compressible flow, for the linearity of (87) holds only in the z-plane, whereas—except at $\beta_a = 1$—the equation for τ is linear only in the w-plane.

From (85)

$$\tau = \ln\left(\frac{U\,dz}{dw}\right) = \ln\left(U\left|\frac{dz}{dw}\right|\right) + i\arg\left(\frac{dz}{dw}\right),\tag{88}$$

so by (69) and (75)

$$\left|\frac{dz}{dw}\right| = \frac{1}{q}, \quad \arg\left(\frac{dz}{dw}\right) = \theta.\tag{89}$$

Equations (88) and (89) have been widely used in theoretical fluid mechanics for nearly a century.

2.17　Linear perturbation theory

Comparison of equations (45) and (72) leads to the conclusion that in linear perturbation theory

$$\tau = \tau(Z),\tag{90}$$

where

$$Z = x + i\beta_a y.\tag{91}$$

Equations (76) and (90) imply that there must be a relation between w and Z in linear perturbation theory. Equation (91) can be written

$$z = \frac{1+\beta_a}{2U\beta_a}UZ - \frac{1-\beta_a}{2U\beta_a}U\bar{Z},\tag{92}$$

and comparing this with (82) we conclude that the relation in question must be contained in $e^\tau\,dw = U\,dZ$ and $e^{-\bar\tau}\,d\bar{w} = U\,d\bar{Z}$. But these are incompatible unless it is valid to put $\tau = 0$ in the w, Z relation, when

$$w = UZ.\tag{93}$$

It must be emphasized here that (93) is to be regarded only as a relation between the *independent* variables w and Z, and not to imply that τ is zero. Hence its real meaning is contained in the equation

$$\tau(w) = \tau(UZ).\tag{94}$$

The following alternative derivation of (94), which does not use (90), may be clearer. From (47) $\tau = \Omega + i\theta$ is a small first-order quantity, while from (83) and (92) the exact relation between w and Z is $w = UZ + O\,|\tau|$. Hence

$$\tau(w) = \tau(UZ) + O\,|\tau|\left(\frac{d\tau}{dw}\right) + \ldots = \tau(UZ)$$

to first order.

It follows from (49), (65), (68), (91) and (94) that any 'exact' solution

$\tau = \tau(w)$ obtained for a tangent gas can be converted into the corresponding linear perturbation theory by the equations

$$\tau(\phi, \psi) = \tau(Ux, U\beta_a y), \tag{95}$$

$$\Omega = \beta_a\left(1 - \frac{q}{U}\right) = \tfrac{1}{2}\beta_a C_p, \tag{96}$$

and
$$\theta = \frac{dy}{dx}. \tag{97}$$

2.18   The force and moment acting on a streamline

We shall now calculate expressions for the force and moment acting on an interval AB of a streamline due to the pressures acting on one side of AB (see Fig. 2.18).

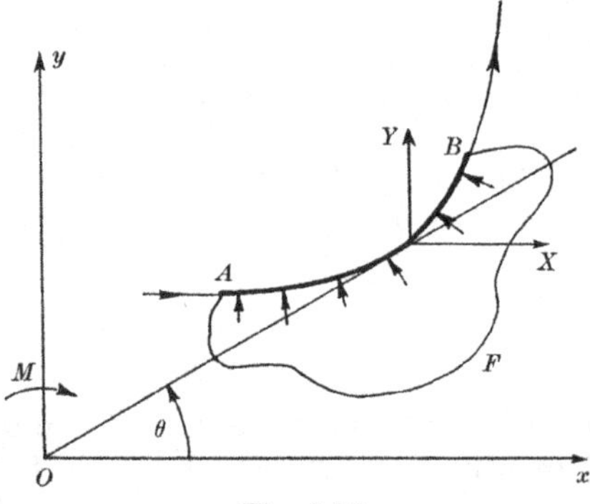

Fig. 2.18

Let X, Y be the components of this force in the Ox-, Oy-directions and let M be the *clockwise* moment about the origin O. Then

$$X = -\int_{AB} p\,dy, \quad Y = \int_{AB} p\,dx,$$

or
$$X + iY = i\int_{AB} (p - p_a)\,dz + ip_a\int_{AB} dz.$$

Therefore from (58)

$$X + iY = \frac{i}{2}\rho_a U^2 \int_{AB} C_p\,dz + ip_a(z_B - z_A). \tag{98}$$

Also
$$M = -\int_{AB} p(x\,dx + y\,dy) = -\mathscr{R}\int_{AB} p(x - iy)\,(dx + i\,dy)$$

$$= -\mathscr{R}\int_{AB} (p - p_a)\bar{z}\,dz - p_a\mathscr{R}\int_{AB} \bar{z}\,dz,$$

where '$\mathscr{R}$' denotes 'the real part of'. Hence

$$M = -\tfrac{1}{2}\rho_a U^2 \mathscr{R} \int_{AB} C_p \bar{z}\, dz - \tfrac{1}{2} p_a (z_B \bar{z}_B - z_A \bar{z}_A). \tag{99}$$

On AB $\psi = \text{constant}$, so (80), (81) and (52) give

$$U\, dz = e^{i\theta} \sinh(\Omega + \Omega^*)\, \text{cosech}\, \Omega^*\, d\phi.$$

Hence $UC_p\, dz = 2\coth \Omega^* \sinh \Omega e^{i\theta}\, d\phi$, by (53) and (59). This can be written

$$C_p\, dz = \frac{1}{U\beta_a}(e^\tau\, dw - e^{-\bar{\tau}}\, d\overline{w}), \tag{100}$$

as $\tau = \Omega + i\theta$ and $d\phi = dw = d\overline{w}$ on AB. If (82) is now used to eliminate $e^\tau\, dw$ from (100), there is obtained

$$C_p\, dz = \frac{2}{1+\beta_a}\left(dz - \frac{1}{U}e^{-\bar{\tau}}\, d\overline{w}\right). \tag{101}$$

On substituting this in (98), and taking the complex conjugate, we arrive at

$$X - iY = i\frac{\rho_a U}{1+\beta_a}\int_{AB} e^{-\tau}\, dw - i(\bar{z}_B - \bar{z}_A)\left(p_a + \frac{\rho_a U^2}{1+\beta_a}\right), \tag{102}$$

which enables the force on AB to be calculated directly from the solution $\tau = \tau(w)$.

Similarly it follows from (101) and (99) that

$$M = \frac{\rho_a U}{1+\beta_a}\mathscr{R}\int_{AB} e^{-\tau} z\, dw - \tfrac{1}{2}(z_B \bar{z}_B - z_A \bar{z}_A)\left(p_a + \frac{\rho_a U^2}{1+\beta_a}\right), \tag{103}$$

but as z is a function of both w and $\overline{w}$ this is not a very convenient expression to evaluate. A more convenient form can be obtained as follows.

Let $z = z_0$ at $w = 0$, then from (83)

$$z = z_0 + \frac{1+\beta_a}{2U\beta_a}\int_0^w e^\tau\, dw - \frac{1-\beta_a}{2U\beta_a}\overline{\int_0^w e^{-\tau}\, dw}.$$

Therefore

$$\mathscr{R}\int_{AB} e^{-\tau} z\, dw = \mathscr{R} z_0 \int_{AB} e^{-\tau}\, dw + \frac{1+\beta_a}{2U\beta_a}\mathscr{R}\int_{AB} e^{-\tau}\left(\int_0^w e^\tau\, dw\right)$$

$$-\frac{1-\beta_a}{2U\beta_a}\mathscr{R}\int_{AB} e^{-\tau}\overline{\left(\int_0^w e^{-\tau}\, dw\right)}\, dw.$$

An integration by parts now gives

$$\int_{AB} e^{-\tau}\left(\overline{\int_0^w e^{-\tau}\, dw}\right) dw = \left[\overline{\int_0^w e^{-\tau}\, dw}\int_0^w e^{-\tau}\, dw\right]_{AB}$$

$$-\int_{AB} e^{-\bar{\tau}}\left(\int_0^w e^{-\tau}\, dw\right) d\overline{w},$$

and noting that the last term of this equation is the conjugate of the first, we conclude that

$$\mathscr{R}\int_{AB} e^{-\tau}\left(\overline{\int_0^w e^{-\tau}\,dw}\right)dw = \tfrac{1}{2}\left|\int_0^w e^{-\tau}\,dw\right|^2_{AB},$$

where $|\ |^2$ denotes the 'square of the modulus'. Therefore

$$\frac{\rho_a U}{1+\beta_a}\mathscr{R}\int_{AB} e^{-\tau}z\,dw = \mathscr{R}z_0\frac{\rho_a U}{1+\beta_a}\int_{AB} e^{-\tau}\,dw + \frac{\rho_a}{2\beta_a}\mathscr{R}\int_{AB} e^{-\tau}$$

$$\times\left(\int_0^w e^{\tau}\,dw\right)dw - \frac{\rho_a}{4\beta_a}\frac{1-\beta_a}{1+\beta_a}\left|\int_{AB} e^{-\tau}\,dw\right|^2.$$

Substituting this in (103), and making use of (102) we find that

$$M = -\mathscr{R}z_0(Y+iX) + \frac{\rho_a}{2\beta_a}\mathscr{R}\int_{AB} e^{-\tau}\left(\int_0^w e^{\tau}\,dw\right)dw - \frac{\rho_a}{4\beta_a}\frac{1-\beta_a}{1+\beta_a}$$

$$\times\left|\int_{AB} e^{-\tau}\,dw\right|^2 + \left(p_a + \frac{\rho_a U^2}{1+\beta_a}\right)\mathscr{R}\{z_0(\bar z_B - \bar z_A) - \tfrac{1}{2}(z_B\bar z_B - z_A\bar z_A)\};\quad (104)$$

we now have an expression permitting M to be calculated directly from the solution $\tau = \tau(w)$.

Equations (102) and (104) have the advantage over other expressions for the force and moment of having analytic (in the complex variable sense) integrands, and this means (see § 3.2) that the streamline path AB can be replaced by any other path, such as AFB (see Fig. 2.18), which has A and B as its terminal points, provided that the closed contour $AFBA$ contains no singular points.

The force and moment acting on a solid closed cylinder, such as an aerofoil, can be derived from (102) and (104), for the pressures acting on the cylinder are pressures acting on one side of a (divided) streamline. As the cylinder is closed $\bar z_A = \bar z_B$, and the last term of each of (102) and (104) vanishes. Hence if C is the cylinder contour, these equations reduce to

$$X - iY = i\frac{\rho_a U}{1+\beta_a}\int_C e^{-\tau}\,dw, \qquad (105)$$

and

$$M = -\mathscr{R}z_0(Y+iX) + \frac{\rho_a}{2\beta_a}\mathscr{R}\int_C e^{-\tau}\left(\int_0^w e^{\tau}\,dw\right)dw$$

$$-\frac{\rho_a}{4\beta_a}\frac{1-\beta_a}{1+\beta_a}\left|\int_C e^{-\tau}\,dw\right|^2. \qquad (106)$$

2.19 Blasius's formulae

Equations (105) and (106) are generalizations of the well-known formulae of Blasius (1910) for the forces on a solid closed body in incompressible flow.

In incompressible flow $\beta_a = 1$, and (105) and (106) reduce to

$$X - iY = \frac{i}{2}\rho_a U \int_C e^{-\tau}\, dw, \tag{107}$$

and

$$M = -\mathscr{R}z_0(Y + iX) + \tfrac{1}{2}\rho_a\mathscr{R}\int_C e^{-\tau}\left(\int_0^w e^{\tau}\, dw\right) dw. \tag{108}$$

Equations (85) and (86) permit us to write (107) in the alternative form

$$X - iY = \frac{i}{2}\rho_a\int_C \left(\frac{dw}{dz}\right) dw = \frac{i}{2}\rho_a\int_C \left(\frac{dw}{dz}\right)^2 dz. \tag{109}$$

Similarly,

$$M = \tfrac{1}{2}\rho_a\mathscr{R}\int_C z\left(\frac{dw}{dz}\right)^2 dz. \tag{110}$$

These are Blasius's formulae.

PERTURBATIONS OF STEADY IRROTATIONAL FLOW

2.20   The possible perturbations

The greater part of this book will be concerned with the solution of a variety of *steady-*, *irrotational-*, *plane*-flow problems. But some problems of practical interest involve weaker conditions, in which the flow can be: (1) irrotational and plane, but not *quite* steady; or (2) steady and plane, but not quite irrotational; or (3) steady and irrotational, but not quite plane. The modifications resulting from these weaker conditions to a *basic*, steady, irrotational, plane flow (hereafter referred to simply as the 'basic' flow) are called *perturbations* of the basic flow. Mathematical progress with the modified theories requires that in general these perturbations be *small*, which is why the word 'quite' is used in our list of the three main types of perturbation possible on the classical two-dimensional flow.

Of course *two* conditions could be weakened simultaneously, e.g. as in unsteady rotational flow, but it is unnecessary to consider these together, for a consequence of assuming that the perturbations are small is that they are *linear*. This means that the perturbations can be calculated separately, and then linearly superimposed in any order on the basic flow. The effect of one type of perturbation on another is thus assumed to be a (negligible) second-order effect compared with the effect of the perturbations on the basic flow.

A further consequence of making the perturbations small is that the perturbation boundary conditions can be applied at the boundaries of the basic flow. (This point is pertinent only with unsteady flows, for the boundaries of the perturbed flow and the basic flow obviously coincide with the other types of perturbation.) This consequence follows by an argument similar to that given at the end of § 2.11, in which it is shown that the errors introduced in this way are second order, except possibly at one or two singular points.

We must expect the possible existence of singular points (or even regions), at which the approximations that yield our linear perturbation theories break down. In other words, our approximations may not be *uniformly* valid over the whole of the (ϕ, ψ)-plane. This point has already been discussed in § 2.12 in connexion with the linear perturbation theory of the *basic* flow. As the errors introduced by this lack of uniform validity are not, in general, of the first order, we can ignore them: however, it should be noted that the terms producing these errors can play a significant role in a more accurate theory allowing for second-order terms. (See the papers by Lighthill (1949b) and Whitham (1950).)

In the following sections we shall derive the three perturbation equations: in later chapters these will be applied to specific problems.

2.21 Unsteady irrotational flow

In unsteady irrotational flow (4), (6), (9) to (12) and (28) yield

$$\left.\begin{aligned}
\frac{\partial\theta}{\partial s}+\frac{1}{\beta}\frac{\partial\Omega}{\partial n}=0, \qquad \frac{\partial\theta}{\partial n}-\beta\frac{\partial\Omega}{\partial s}+\frac{1}{\rho q}\frac{\partial\rho}{\partial t}+\frac{q}{a^2\beta}\frac{\partial\Omega}{\partial t}-\frac{1}{a^2}\frac{\partial\chi}{\partial s}=0, \\
\frac{1}{\rho}\frac{\partial p}{\partial s}=\frac{q^2}{a^2\beta}\frac{\partial\Omega}{\partial s}+\frac{q}{a^2\beta}\frac{\partial\Omega}{\partial t}-\frac{1}{a^2}\frac{\partial\chi}{\partial s}, \qquad \frac{\partial\theta}{\partial t}+\frac{1}{q}\frac{\partial\chi}{\partial n}+\frac{a^2}{\rho q}\frac{\partial\rho}{\partial n}-\frac{q}{\beta}\frac{\partial\Omega}{\partial n}=0.
\end{aligned}\right\} \quad (111)$$

It is easily verified that these equations cannot be completely linearized by the use of a tangent gas. We shall therefore consider only the two important cases of (1) an incompressible fluid in which the external force potential, χ, is not necessarily zero; and (2), a linear perturbation of a steady compressible flow in which χ is negligible.

(1) *Incompressible flow*

In this case $\rho = \text{constant}$, $a = \infty$, $\beta = 1$, so the first pair of (111) reduce to

$$\frac{\partial\theta}{\partial s}+\frac{\partial\Omega}{\partial n}=0, \qquad \frac{\partial\theta}{\partial n}-\frac{\partial\Omega}{\partial s}=0.$$

Transforming these back to the (x, y)-plane by the equations preceding (43) we find

$$\frac{\partial\theta}{\partial y}-\frac{\partial\Omega}{\partial x}=0, \qquad \frac{\partial\theta}{\partial x}+\frac{\partial\Omega}{\partial y}=0.$$

Hence $$\tau = \tau(z), \qquad (112)$$

where $\tau = \Omega + i\theta$ and $z = x + iy$. Thus unsteady and steady incompressible flows satisfy the same equation.

Let the subscript 0 indicate the basic *steady* flow, then (see §2.16) $z = z(w_0)$, and (112) can be replaced by

$$\tau = \tau(w_0) = \tau_0(w_0) + \Delta\tau(w_0),\qquad(113)$$

where $\Delta\tau$ is the unsteady component of τ.

Notice that this theory is not restricted to small perturbations, and consequently can be applied to flows for which no basic steady flow exists —e.g. the uniform acceleration of an aerofoil at a fixed incidence.

(2) *Perturbation theory*

Now consider a small perturbation of steady compressible flow, in which χ is negligible. In this case the coefficients of the derivatives in (111) can be replaced by the values they have in the basic steady flow, values which we shall distinguish by the subscript 0. Furthermore, the operators $\partial/\partial s$ and $\partial/\partial n$ can be replaced by $q_0\,\partial/\partial\phi_0$ and $(\rho_0 q_0/\rho_s)\,\partial/\partial\psi_0$ to the same order of accuracy. The resulting equations are almost completely linearized by assuming the fluid to be a tangent gas; (6), (35) to (37) are then applicable to the coefficients of the derivatives. By these transformations and approximations (111) become

$$\frac{\partial\theta}{\partial\phi_0} + \frac{\partial\Omega}{\partial\psi_0} = 0, \qquad \frac{\partial\theta}{\partial\psi_0} - \frac{\partial\Omega}{\partial\phi_0} + \frac{1}{\beta_0\rho_0 q_0^2}\frac{\partial\rho}{\partial t} + \frac{1}{(a_a\beta_a)^2}\frac{\partial\Omega}{\partial t} = 0,$$

$$\frac{1}{\beta_0\rho_0 q_0^2}\frac{\partial\rho}{\partial\phi_0} = \frac{1}{(a_a\beta_a)^2}\frac{\partial\Omega}{\partial\phi_0} + \frac{1}{(a_a\beta_a q_0)^2}\frac{\partial\Omega}{\partial t},$$

$$\frac{1}{\beta_0\rho_0 q_0^2}\frac{\partial\rho}{\partial\psi_0} = \frac{1}{(a_a\beta_a)^2}\frac{\partial\Omega}{\partial\psi_0} - \frac{1}{(a_a\beta_a q_0)^2}\frac{\partial\theta}{\partial t}.$$

If the first, second and third of these equations are differentiated with respect to ψ_0, ϕ_0 and t respectively, ρ and θ can be eliminated from the resulting equations to give

$$\frac{\partial^2\Omega}{\partial\phi_0^2} + \frac{\partial^2\Omega}{\partial\psi_0^2} = \frac{2}{(a_a\beta_a)^2}\frac{\partial^2\Omega}{\partial\phi_0\,\partial t} + \frac{1}{(a_a\beta_a q_0)^2}\frac{\partial^2\Omega}{\partial t^2},\qquad(114)$$

in which we have neglected a term containing the product $(\partial\Omega_0/\partial\phi_0)(\partial\rho/\partial t)$. This term is negligible compared to those retained if: (i) the flow is only a small perturbation from a uniform flow along the x-axis, i.e. $\partial\Omega_0/\partial\phi_0$ is relatively small; or if (ii) the Mach number is low, so that the possible variations in ρ are small; or if (iii) time rates of changes are small, so that $\partial\rho/\partial t$ is quite small.

The last term of (114) is linear, but inconvenient in that the coefficient contains q_0, which is a function of ϕ_0 and ψ_0. Assumption (i) would enable us to replace q_0 by U in this coefficient. With assumption (iii) the term in question could be ignored compared with that containing $\partial^2\Omega/(\partial\phi_0\,\partial t)$; in fact if the time rates of change were really small, the *whole* of the right-hand side of (114) could be ignored compared with the left-hand side, and Ω would satisfy Laplace's equation. Laplace's equation also results if in assumption (ii) the Mach number is small enough to make a_a relatively large, and the right-hand side of (114) negligible. It is therefore consistent to write (114) in the form

$$\frac{\partial^2\Omega}{\partial\phi_0^2}+\frac{\partial^2\Omega}{\partial\psi_0}=\frac{2}{(a_a\beta_a)^2}\frac{\partial^2\Omega}{\partial\phi_0\,\partial t}+\frac{1}{(a_a\beta_a U)^2}\frac{\partial^2\Omega}{\partial t^2}. \tag{115}$$

A similar calculation shows that θ also satisfies (115), and hence as $\tau=\Omega+i\theta$,

$$\frac{\partial^2\tau}{\partial\phi_0^2}+\frac{\partial^2\tau}{\partial\psi_0^2}=\frac{2}{(a_a\beta_a)^2}\frac{\partial^2\tau}{\partial\phi_0\,\partial t}+\frac{1}{(a_a\beta_a U)^2}\frac{\partial^2\tau}{\partial t^2}. \tag{116}$$

This wave equation cannot be treated by the complex variable methods developed in part II, and we shall therefore restrict our attention to the cases of: (1) incompressible flow, when τ is exactly harmonic in the w_0- $(=\phi_0+i\psi_0)$ plane; and (2) compressible flow varying slowly enough for *second*-order time derivatives to be ignored. In the latter case (116) can be written

$$\nabla^2\tau=\frac{2}{(a_a\beta_a)^2}\frac{\partial^2\tau}{\partial\phi_0\,\partial t}, \tag{117}$$

operating on this with $\nabla^2\left(\equiv\dfrac{\partial^2}{\partial\phi_0^2}+\dfrac{\partial^2}{\partial\psi_0^2}\right)$ we get

$$\nabla^4\tau=\frac{4}{(a_a\beta_a)^4}\frac{\partial^4\tau}{\partial\phi_0^2\,\partial t^2}\simeq 0,$$

on ignoring the second-order time derivative. This is equivalent to replacing (117) by the pair of equations

$$\nabla^2\tau_1=\frac{2}{(a_a\beta_a)^2}\frac{\partial^2\tau}{\partial\phi_0\,\partial t}, \tag{118}$$

$$\nabla^2\tau=0. \tag{119}$$

The first of these is a Poisson type equation, which is amenable to complex variable methods (see § 3.20).

If τ is made to satisfy the boundary conditions for τ_1 exactly, then (119) and (118) can be regarded as being successive steps of an iterative process of solving (117), the third step of which is to put τ_1 into the right-hand side of (117); this yields τ_2 and so on.

2.22 Small vorticity flow

In steady rotational flow (4), (6), (9) to (12) and (28) yield

$$\frac{\partial\theta}{\partial s}+\frac{1}{\beta}\frac{\partial\Omega}{\partial n}=\frac{\zeta}{q}, \quad \frac{\partial\theta}{\partial n}-\beta\frac{\partial\Omega}{\partial s}=0, \\[2mm] \frac{1}{\rho}\frac{\partial\rho}{\partial s}=\frac{q^2}{a^2\beta}\frac{\partial\Omega}{\partial s}, \quad \frac{1}{\rho}\frac{\partial\rho}{\partial n}=\frac{q^2}{a^2\beta}\frac{\partial\Omega}{\partial n}-\frac{q}{a^2}\zeta. \tag{120}$$

where the external forces acting on the fluid have been neglected. Let the subscript 0 denote the flow that results when the boundary conditions of the flow are unchanged, but ζ is zero. This flow we term the 'basic irrotational flow'; it will have a potential function ϕ_0 and a stream function ψ_0. If ζ is assumed to be small, the coefficients in (120) can be replaced by their values in the basic irrotational flow, and the operators $\partial/\partial s$, $\partial/\partial n$ can be changed to $q_0\,\partial/\partial\phi_0$ and $(\rho_0 q_0/\rho_s)\,\partial/\partial\psi_0$. This transforms (120) into

$$\frac{\partial\theta}{\partial\phi_0}+\frac{\partial\Omega}{\partial\psi_0}=\frac{\zeta}{q_0^2}, \quad \frac{\partial\theta}{\partial\psi_0}-\frac{\partial\Omega}{\partial\phi_0}=0, \tag{121}$$

$$\frac{1}{\rho_s M_0^2}\frac{\partial\rho}{\partial\phi_0}=\frac{\partial\Omega}{\partial\phi_0}, \quad \frac{\zeta}{q_0^2}=\frac{\partial\Omega}{\partial\psi_0}-\frac{1}{\rho_s M_0^2}\frac{\partial\rho}{\partial\psi_0}, \tag{122}$$

on assuming, as usual, that the fluid is a tangent gas.

At a first glance it appears that the substitution

$$d\kappa=-\frac{d\rho}{\rho_s M_0^2}=-\frac{dp}{\rho_s q_0^2}, \tag{123}$$

can be used to reduce these equations to the Cauchy–Riemann form. However (123) is not generally valid as, unlike irrotational flow, q_0 does not depend uniquely on p owing to the variation of the Bernoulli constant from streamline to streamline (see (12)). However, in a linear perturbation from a uniform flow along the x-axis (123) can be approximated by

$$d\kappa=-\frac{1}{\rho_s U^2}dp, \tag{124}$$

and in this case $\kappa+i\theta$ is an analytic function of $w_0=\phi_0+i\psi_0$, or by (94),

$$J\equiv\kappa+i\theta=J(Z) \quad (Z=x+i\beta_a y). \tag{125}$$

Lighthill (1950) has used this method to deal with the weak rotational field behind shock waves on aerofoils.

In addition to (121) and (122) there is the result

$$\frac{d}{dt}\left(\frac{\zeta}{\rho}\right)=q\frac{\partial}{\partial s}\left(\frac{\zeta}{\rho}\right)\simeq q_0^2\frac{\partial}{\partial\phi_0}\left(\frac{\zeta}{\rho}\right)=0,$$

which follows from (1.15), the steadiness of the flow, and the approximations introduced above. Hence

$$\frac{\partial \zeta}{\partial \phi_0} \simeq \frac{\zeta}{\rho_0}\frac{\partial \rho}{\partial \phi_0} = \zeta \frac{M_0^2}{\beta_0}\frac{\partial \Omega}{\partial \phi_0}, \tag{126}$$

where the last form follows from (35) to (37) and (122). The right hand of (126) can be replaced by zero if (i) the flow is nearly incompressible, or if (ii) ζ is very small, or if (iii) the flow is only a small perturbation from a uniform flow along the x-axis. We shall assume that one of these restrictions—or a suitable combination of them—applies, then

$$\frac{\partial \zeta}{\partial \phi_0} = 0. \tag{127}$$

It now follows from (121) and (127) that

$$\nabla^2 \Omega = \zeta \frac{\partial}{\partial \psi_0}\left(\frac{1}{q_0^2}\right), \quad \nabla^2 \theta = \zeta \frac{\partial}{\partial \phi_0}\left(\frac{1}{q_0^2}\right),$$

i.e.
$$\nabla^2 \tau = \zeta \left(\frac{\partial}{\partial \psi_0} + i\frac{\partial}{\partial \phi_0}\right)\left(\frac{1}{q_0^2}\right), \tag{128}$$

where $\tau = \Omega + i\theta$. This reduces the problem to the solution of a Poisson type of equation.

2.23   Quasi-plane flows

By *quasi*-plane flows we mean three-dimensional flows which are reducible to plane flows. (Axially-symmetric flows are excluded because although they depend only on two independent variables, they are not 'two-dimensional' in the sense defined in the first sentence of §2.1.) Quasi-plane flows are characterized by having a relatively small or zero rate of change in some special direction $\hat{\mathbf{k}}$ say. When $\hat{\mathbf{k}}$ is at right angles to the stream direction and the rate of change is exactly zero we have real plane flow (see §2.1).

Important examples of quasi-plane flows are: (i) the flow past infinite-yawed wings; (ii) the cross-flow in the slender body theory of Munk (1924) and Jones (1946); and (iii) the flow in slotted wind tunnels (see §7.30).

Take Oz to lie along the special direction $\hat{\mathbf{k}}$, and let u, v, w be the components of velocity along Ox, Oy, Oz, then with quasi-plane flow

$$\frac{\partial u}{\partial z}, \frac{\partial v}{\partial z}, \frac{\partial w}{\partial z} = O(\epsilon) \simeq 0, \tag{129}$$

where ϵ is small compared with the derivatives of u, v, w with respect to x and y.

In steady irrotational flow (§§ 1.4, 1.7)

$$\frac{\partial v}{\partial x} - \frac{\partial u}{\partial y} = 0, \tag{130}$$

$$\frac{\partial w}{\partial x} = \frac{\partial u}{\partial z}, \quad \frac{\partial w}{\partial y} = \frac{\partial v}{\partial z}, \tag{131}$$

and

$$\frac{\partial(\rho u)}{\partial x} + \frac{\partial(\rho v)}{\partial y} + \frac{\partial(\rho w)}{\partial z} = 0. \tag{132}$$

It follows from (129) and (131) that

$$dw = \frac{\partial w}{\partial x}dx + \frac{\partial w}{\partial y}dy + \frac{\partial w}{\partial z}dz = \frac{\partial u}{\partial z}dx + \frac{\partial v}{\partial z}dy + \frac{\partial w}{\partial z}dz = O(\epsilon) \simeq 0, \tag{133}$$

so that

$$dq^2 = d(u^2 + v^2 + w^2) \simeq dq_0^2, \tag{134}$$

where

$$q_0 = \sqrt{(u^2 + v^2)}. \tag{135}$$

Equations (4) and (6) and the assumption that gc/a^2 is very small (see § 2.5) now give

$$d\rho = -\frac{\rho}{2a^2}dq^2 \simeq -\rho\frac{q_0}{a^2}dq_0$$

by (134). Hence

$$q_0 d\rho = -\rho M_0^2 dq_0, \tag{136}$$

where

$$M_0 \equiv \frac{q_0}{a}. \tag{137}$$

Equations (129), (135) and (136) now permit us to replace (132) by

$$\frac{\partial(\rho u)}{\partial x} + \frac{\partial(\rho v)}{\partial y} = 0, \tag{138}$$

to first order. Equation (138) is exact in two cases of quasi-plane flow, namely: (i) flows for which $dw = 0$ exactly (e.g. an infinite yawed wing, see Fig. 2.23), and (ii) flows having u, v, w periodic in the Oz-direction (see § 7.30). Case (i) is obvious, in (ii) we average (132) over one period to get (138) exactly, except that ρu, ρv are replaced by their averages $\rho\bar{u}$, $\rho\bar{v}$. Equation (130) is similarly averaged.

Equations (130) and (138) permit the introduction of a velocity potential ϕ and a stream function ψ defined by

$$u = \frac{\partial\phi}{\partial x}, \quad v = \frac{\partial\phi}{\partial y}, \quad \rho u = \rho_s\frac{\partial\psi}{\partial y}, \quad \rho v = -\rho_s\frac{\partial\psi}{\partial x}, \tag{139}$$

where ρ_s is an arbitrary constant. If these are used to change the

independent variables in (130) and (138) from x, y to ϕ, ψ, and if the
dependent variables are changed to q_0, θ_0, where q_0 is defined in (135) and

$$\theta_0 \equiv \tan^{-1}\left(\frac{v}{u}\right), \tag{140}$$

there results
$$\frac{\partial \theta_0}{\partial \phi} - \frac{\rho}{\rho_s q_0} \frac{\partial q_0}{\partial \psi} = 0, \tag{141}$$

and
$$\frac{\partial(\rho q_0)}{\partial \phi} + \frac{\rho^2 q_0}{\rho_s} \frac{\partial \theta_0}{\partial \psi} = 0.$$

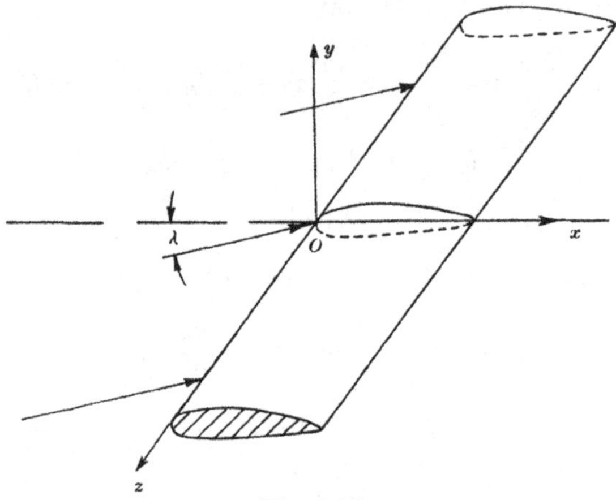

Fig. 2.23

By (136) this last equation can be written

$$\frac{\partial \theta_0}{\partial \psi} + \frac{\rho_s \beta_0^2}{\rho q_0} \frac{\partial q_0}{\partial \phi} = 0, \tag{142}$$

where
$$\beta_0 = \sqrt{(1 - M_0^2)}. \tag{143}$$

On introducing
$$d\Omega_0 \equiv -\frac{\beta_0}{q_0} dq_0, \tag{144}$$

(141) and (142) become

$$\frac{\partial \theta_0}{\partial \phi} + \frac{\rho}{\rho_s \beta_0} \frac{\partial \Omega_0}{\partial \psi} = 0, \quad \frac{\partial \theta_0}{\partial \psi} - \frac{\rho_s \beta_0}{\rho} \frac{\partial \Omega_0}{\partial \phi} = 0, \tag{145}$$

which should be compared with the corresponding equations for plane
flow—(30).

It follows from (133) that w is approximately constant, and therefore
we can write
$$w = U \sin \lambda, \tag{146}$$

where λ is a constant, and $U = q_a$, the average value of q. As $q_0^2 = q^2 - w^2$,

$$q_{0a} = U_0 = \sqrt{(U^2 - U^2 \sin^2 \lambda)} = U \cos \lambda. \tag{147}$$

Hence

$$\frac{\beta_0^2}{\beta_{0a}^2} = \frac{1 - M_0^2}{1 - M_{0a}^2} = \frac{a^2 - q^2 + U^2 \sin^2 \lambda}{a_a^2 - U^2 + U^2 \sin^2 \lambda} \frac{a_a^2}{a^2},$$

so

$$\frac{a^2 \beta_0^2}{a_a^2 \beta_{0a}^2} = \frac{a^2 \beta^2 + U^2 \sin^2 \lambda}{a_a^2 \beta_a^2 + U^2 \sin^2 \lambda}. \tag{148}$$

In a tangent gas the relations (see (6) and (1.26))

$$a\rho = a_a \rho_a, \quad a\beta = a_a \beta_a, \quad \frac{\rho}{\beta} = \frac{\rho_a}{\beta_a}, \tag{149}$$

hold, and so it follows from (148) and (149) that

$$a\rho = a_a \rho_a, \quad a\beta_0 = a_a \beta_{0a}, \quad \frac{\rho}{\beta_0} = \frac{\rho_a}{\beta_{0a}}, \tag{150}$$

apply in the quasi-plane flow of a tangent gas.

Consequently, if we limit the theory to that for a tangent gas, and set

$$\rho_s = \frac{\rho_a}{\beta_{0a}}, \tag{151}$$

(cf. (36)), (145) reduce to

$$\frac{\partial \theta_0}{\partial \phi} + \frac{\partial \Omega_0}{\partial \psi} = 0, \quad \frac{\partial \theta_0}{\partial \psi} - \frac{\partial \Omega_0}{\partial \phi} = 0. \tag{152}$$

Thus if

$$\tau_0 = \Omega_0 + i\theta_0, \tag{153}$$

and

$$w_0 = \phi + i\psi,$$

then (152) imply

$$\tau_0 = \tau_0(w_0). \tag{154}$$

If (144), (151) and (154) are compared with the corresponding equations for the plane flow of a tangent gas, namely (28), (36) and (76), it is immediately apparent that we have reduced our quasi-plane flow problem to that of finding the plane flow in Oxy of a stream whose average velocity is $U_0 = U \cos \lambda$. There is thus no need to pursue this theory further. All of the equations in §§ 2.13–2·19 are applicable to quasi-plane flow—it is only necessary to replace U, q, β, M, Ω, Ω^* and similar quantities by U_0, q_0, β_0, M_0, Ω_0, Ω_0^*, etc.

In an ideal gas it follows from (1.24) and $M = q/a$ that

$$\left(\frac{q}{U}\right)^2 = \frac{1 + \frac{1}{2}(\gamma - 1) M_a^2}{1 + \frac{1}{2}(\gamma - 1) M^2} \frac{M^2}{M_a^2}. \tag{155}$$

From (137) and (146)

$$\left(\frac{q}{U}\right)^2 = \left(\frac{q_0}{U}\right)^2 + \sin^2\lambda = \frac{M_0^2}{M^2}\left(\frac{q}{U}\right)^2 + \sin^2\lambda,$$

and so using (155) to eliminate q/U, and rearranging we find

$$M_0^2 = M^2 - \sin^2\lambda \, M_a^2 \frac{1 + \frac{1}{2}(\gamma-1)\,M^2}{1 + \frac{1}{2}(\gamma-1)\,M_a^2}. \tag{156}$$

This equation can be written

$$\frac{1 + \frac{1}{2}(\gamma-1)\,M_0^2}{1 + \frac{1}{2}(\gamma-1)\,M_{0a}^2} = \frac{1 + \frac{1}{2}(\gamma-1)\,M^2}{1 + \frac{1}{2}(\gamma-1)\,M_a^2}, \tag{157}$$

on using $M_{0a} = M_a \cos\lambda$. It follows from (39) and (157) that

$$\frac{\beta_0}{\beta_{0a}}\frac{\rho_a}{\rho} = \left\{\frac{1 - M_0^2}{1 - M_{0a}^2}\right\}^{\frac{1}{2}}\left\{\frac{1 + \frac{1}{2}(\gamma-1)\,M_0^2}{1 + \frac{1}{2}(\gamma-1)\,M_{0a}^2}\right\}^{1/(\gamma-1)}. \tag{158}$$

In tangent gas theory the left-hand side of (158) is exactly unity (see (150)), and from § 2.10 this will be a reasonable approximation provided the point (M_0, M_{0a}) lies within the region shown in Fig. 2.10. This restriction does not apply to (M, M_a), which can lie in the supersonic region if λ is large enough: for example if $M_a = 1\cdot5$, $M = 1\cdot6$ and $\lambda = 66°$ we find that $M_{0a} = 0\cdot598$, $M_0 = 0\cdot766$ and $\beta_0\rho_a/(\beta_{0a}\rho) = 0\cdot890$.

Slender body theory is obtained by putting $\lambda = \frac{1}{2}\pi - \alpha$ where α is small.

PART II
COMPLEX VARIABLE THEORY

CHAPTER 3

CAUCHY INTEGRALS AND THEIR APPLICATION TO HARMONIC FUNCTION BOUNDARY VALUE PROBLEMS

CAUCHY'S THEOREM

3.1 Introduction

It will be assumed that the reader has some acquaintance with the theory of functions of a complex variable; most of §§ 3.1–3.3 is therefore a revisionary outline provided for reference purposes. The material in §§ 3.4–3·20 is reasonably complete, as it covers important ground very often neglected in courses in applied mathematics and theoretical fluid mechanics.

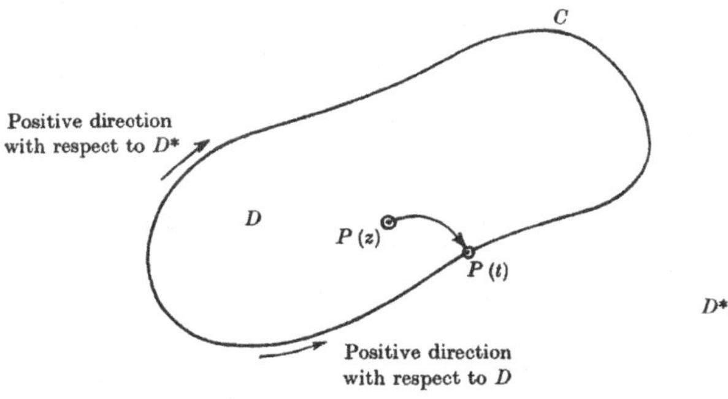

Fig. 3.1 a

Let C be a continuous arc, not cutting itself, and possessing a continuously turning tangent, then it is termed a *simple smooth arc*. A continuous chain of a finite number of such arcs forming a closed non-intersecting curve will be termed a closed, piecewise-smooth curve, or simply a *contour*. We shall accept as intuitively obvious that a contour C divides the plane into two distinct *domains D and $D*$*. If C is a contour of finite length as shown in Fig. 3.1.a, one of these domains (D) will be *bounded*, while the other ($D*$) will be *unbounded*. More generally, when some points on C lie at infinity, both D and its complement $D*$ will be unbounded.

The positive direction along C with respect to D is that which puts D on the left hand when facing this direction. If D is bounded as shown in the figure, the positive direction with respect to D will be in an anticlockwise direction about C, while the positive direction with respect to the un-bounded D^* is clockwise. We shall usually indicate the positive sense of a contour by an arrow, and a contour C will be defined completely only when this sense is specified. Points are said to lie *within* C, or *outside* C, depending whether or not they lie in that domain with respect to which the sense of C has been fixed.

A domain D is described as being *singly connected* if the boundary contour C separating it from its complement D^* is in one piece. Thus each of the domains D and D^* shown in Fig. 3.1a are singly connected. A *doubly connected* domain is a contiguous or *connected* set of points

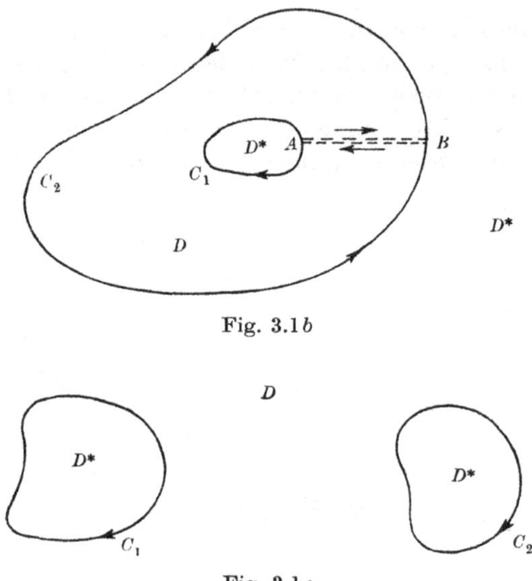

Fig. 3.1b

Fig. 3.1c

separated from its complement by *two* distinct non-intersecting contours C_1 and C_2. The domains labelled D in Figs. 3.1b and 3.1c are doubly connected. A connected domain defined by n non-intersecting contours is said to be of connectivity n.

If the contours C_1 and C_2 in Fig. 3.1b are connected by a *barrier* AB, the contour defining D becomes the continuous curve $C_2 + \overrightarrow{BA} + C_1 + \overrightarrow{AB}$, where $\overrightarrow{BA}$ and $\overrightarrow{AB}$ are opposite sides of the barrier AB. The connectivity of D is then reduced from 2 to 1. More generally if D is of connectivity n the introduction of m ($m < n$) barriers reduces the connectivity to $n-m$. In some cases one or more of the contours $C_1, C_2, \ldots$ will be simply *points* $P_1, P_2, \ldots$ not belonging to D, but such degeneracy will clearly not affect the connectivity.

If a function $F(z)$ $(z = x + iy)$ is one-valued and differentiable at every point of a domain D, save possibly a finite number of exceptional points, it is said to be *analytic* in D; if there are no exceptional points then $F(z)$ is *regular* in D.

Let D be a singly-connected domain bounded by a contour C of finite length. Let $F(z)$ be an analytic function, regular within C, and continuous as C is approached along a path lying entirely in D (see Fig. 3.1 a). Then if z lies on such a path terminating at a point t on C we can write $\lim_{z \to t} F(z) = F(t)$. Cauchy's theorem states that

$$\int_C F(t)\, dt = 0. \tag{1}$$

This is the fundamental theorem of the complex integral calculus. For its proof, which is rather difficult, the reader is referred to Nehari (1952) or Copson (1935).

This result, and others similar to it given below, can be extended to domains of connectivity n very easily. Consider the case shown in Fig. 3.1 b. With the barrier AB inserted D becomes singly connected, and (1) can be written

$$\int_C F(t)\, dt = \int_{C_1} F(t)\, dt + \int_{AB} F(t)\, dt + \int_{C_2} F(t)\, dt + \int_{BA} F(t)\, dt = 0.$$

The continuity of $F(z)$ across AB gives $\displaystyle\int_{AB} + \int_{BA} = 0$, therefore Cauchy's theorem reads

$$\int_{C_1} F(t)\, dt + \int_{C_2} F(t)\, dt = 0,$$

and this is the same as (1) if C is interpreted to be $C_1 + C_2$. Notice from Fig. 3.1 b that C_1 and C_2 are taken in opposite senses.

3.2　The residue theorem

An immediate and obvious consequence of Cauchy's theorem is that the value of the contour integral of an analytic function is unaltered by deformation of the contour provided that the contour crosses no singularity of the integrand during deformation. This freedom to deform contours is very important, and enables us to establish a valuable generalization of Cauchy's theorem known as the residue theorem.

First we recall that if $F(z)$ has an isolated singularity at $z = a$, it can be represented in the neighbourhood of a by the Laurent series

$$F(z) = \sum_0^\infty b_{-n}(z - a)^n + \sum_1^\infty b_n(z - a)^{-n}. \tag{2}$$

The second sum in (2) is called the *principal* part of $F(z)$ at $z = a$. If the principal part is a terminating series, of which b_m is the last non-zero coefficient, the singularity is termed a *pole* of order m. The coefficient b_1 is called the *residue* of $F(z)$ at $z = a$. If the series comprising the principal part is not terminating, a is called an *isolated essential singularity*. If the principal part vanishes, i.e. if

$$F(z) = \sum_0^\infty b_{-n}(z-a)^n, \tag{3}$$

the function is regular at a.

The residue theorem states that if $F(z)$ is continuous within and on a closed contour C and regular, save for a finite number of poles, within C, then

$$\int_C F(t)\,dt = 2\pi i \Sigma R, \tag{4}$$

where ΣR is the sum of the residues of $F(z)$ at its poles within C.

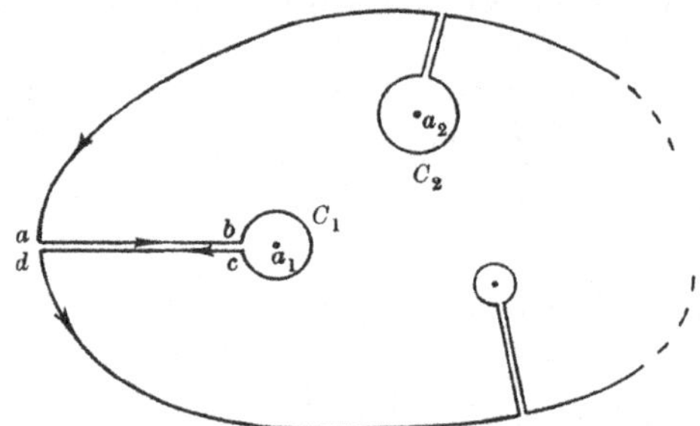

Fig. 3.2

Consider the contour shown in Fig. 3.2, which is comprised of C plus small circles, $C_1, C_2, \ldots$, centred on the singularities at $a_1, a_2, \ldots$, together with paths such as ab, cd, connecting the circles to C. This contour contains no singular points of $F(z)$, so by (1)

$$\int_C F(t)\,dt + \Sigma \int_{C_i} F(z)\,dz = 0, \tag{5}$$

the contributions from paths like ab and cd cancelling out. Near a_i we can write

$$\int_{C_i} F(z)\,dz = \sum_{-m}^\infty b_{-n} \int_{C_i} (z-a_i)^n\,dz = -\sum_{-m}^\infty b_{-n}\,ir^{n+1} \int_0^{2\pi} e^{(n+1)i\theta}\,d\theta,$$

where $r = |z-a_i|$, $\theta = \arg(z-a_i)$, and the minus sign arises because $C_1, C_2, \ldots$ are traversed in the negative direction. The integral vanishes unless $n = -1$, in which case it has the value 2π. Thus

$$\int_{C_i} F(z)\,dz = -2\pi i b_1 = -2\pi i \times \text{residue of } F(z) \text{ at } a_1.$$

Equation (4) now follows by substituting this result in (5).

It is useful to note that at a simple pole, i.e. a pole of order unity, at $z = a$,

$$b_1 = \lim_{z \to a} \{(z - a)\, F(z)\}; \tag{6}$$

more generally at a pole of order m,

$$b_1 = \lim_{z \to a} \left\{ \frac{d^{m-1}}{dz^{m-1}} \frac{(z-a)^m}{(m-1)!} F(z) \right\}, \tag{7}$$

but it is usually better to write $\zeta = z - a$ and expand in powers of ζ.

3.3   The point at infinity

It will be assumed that the reader is familiar with the concept of the *point at infinity*, namely that 'point' whose neighbourhood is the exterior of a circle of arbitrarily large radius. The situation is best visualized by means of the well-known stereographic projection of a sphere from a point of itself on to its equatorial plane (see Copson, 1935, p. 8). The point at infinity can be defined as that point which corresponds to the origin in the transformation $\zeta = 1/z$, and a function $F(z)$ is said to have a zero, pole, or essential singularity at infinity as the function $F(1/z)$ has a zero, pole or essential singularity at the origin.

Let $F(z)$ have an isolated singularity at infinity, and let C be a contour large enough to enclose the singularities of $F(z)$ in the finite part of the plane. Then the residue of $F(z)$ at infinity is defined to be

$$-\frac{1}{2\pi i} \int_C F(z)\, dz,$$

where C is taken in the usual anticlockwise direction. The minus sign is introduced here to allow for the fact that anticlockwise contours about the origin are equivalent in sense to *clockwise* contours about the point at infinity. Thus the sum of the residues of a function for *all* its singularities (assumed to be isolated and finite in number) including infinity is zero.

By replacing $z - a$ by $1/z$ in (2) we find that near infinity a function $F(z)$ having an isolated singularity at infinity can be expanded

$$F(z) = \sum_0^\infty b_{-n}\, z^{-n} + \sum_1^\infty b_n\, z^n. \tag{8}$$

If the second term of (8)—the principal part at infinity—terminates at $b_m z^m$, then $F(z)$ has a pole of order m at infinity; if m is zero $F(z)$ is regular at infinity, and

$$F(z) = \sum_0^\infty b_{-n}\, z^{-n}. \tag{9}$$

By an argument similar to that leading to (4) it is easily verified that only the term b_{-1}/z in (8) contributes to $\int_C F(z)\, dz$, and therefore the residue at infinity must be $-b_{-1}$. From this we deduce the very useful result:

If $F(z)$ is regular in the whole of the z-plane, except at a finite number of points where $F(z)$ has poles, then

$$\int_C F(z)\,dz = 2\pi i$$

$$\times \text{(coefficient of } 1/z \text{ in the expansion of } F(z) \text{ near infinity),} \quad (10)$$

where C is a contour enclosing all the poles in the finite part of the plane.

CAUCHY INTEGRALS

3.4 Cauchy's integral formula

Let $F(z)$ be continuous within and on a contour C and regular within C, then for any point z lying *within* C,

$$F(z) = \frac{1}{2\pi i}\int_C \frac{F(t)}{t-z}\,dt. \quad (11)$$

This result, known as Cauchy's integral formula, follows immediately from (4) and (6) and the regularity of $F(z)$ within C.

If z lies *outside* C it is a direct consequence of Cauchy's theorem that

$$0 = \frac{1}{2\pi i}\int_C \frac{F(t)}{t-z}\,dt, \quad (12)$$

as the only pole of the integrand now lies outside C.

Let t_0 be a point on C, and consider the limits

$$\frac{1}{2\pi i}\lim_{\substack{z\to t_0 \\ (z\,\epsilon\,D)}}\int_C \frac{F(t)}{t-z}\,dt, \quad \frac{1}{2\pi i}\lim_{\substack{z\to t_0 \\ (z\,\epsilon\,D^*)}}\int_C \frac{F(t)}{t-z}\,dt, \quad (13)$$

where z approaches t_0 along a path which lies either (i) entirely within C ($z\,\epsilon\,D$† see Fig. 3.1a), or (ii) entirely outside C ($z\,\epsilon\,D^*$). As (11) and (12) hold, however small $|z-t_0|$ may be, it is clear that the limits in question are (i) $F(t_0)$, and (ii) zero.

The value of the integral in (11) and (12) *at* $z = t_0$, i.e.

$$\mathscr{F}(t_0) = \frac{1}{2\pi i}\int_C \frac{F(t)}{t-t_0}\,dt, \quad (14)$$

is as yet undefined, for if $F(t_0) \neq 0$ the integrand becomes infinite like $|t-t_0|^{-1}$ as $t\to t_0$, and so the integral is not convergent in the ordinary

---

† The symbol ϵ introduced here means 'belongs to' or 'contained within'.

sense. Clearly any meaning we give $\mathscr{F}(t_0)$ must differ from at least one of the limits in (13); it is expedient to attach a meaning which is quite independent of the limits in (13). The interpretation we shall give the improper integral $\mathscr{F}(t_0)$ is the well-known *Cauchy principal value*, and following this we shall establish important relations—the Plemelj formulae—between $\mathscr{F}(t_0)$ and the limits in (13).

3.5   Cauchy principal values

First consider a generalization of the Cauchy integral defined in (11), viz.

$$F(z) \equiv \frac{1}{2\pi i} \int_C \frac{f(t)\,dt}{t-z}, \tag{15}$$

where $f(t)$ is (in general) a complex function of position defined on C, and z is any point in the plane not on C. (Restrictions will be imposed on the 'density' function $f(t)$ below.) The symbol t in $f(t)$ denotes the *point* $t(x, y)$ rather than the corresponding complex number $t = x + iy$, for $f(t)$ is usually defined *only* on C and cannot be regarded as an analytic function of a complex variable. In general $f(t)$ does not equal $F(t)$, and only in exceptional circumstances (e.g. as in § 3.4) can an analytic continuation of $f(t)$ off C in D or D^* be found.

We shall employ the notation

$$\lim_{\substack{z\to t \\ (z\,\in\, D)}} F(z) = F^+(t_0), \quad \lim_{\substack{z\to t \\ (z\,\in\, D^*)}} F(z) = F^-(t_0), \tag{16}$$

for the two limit values of $F(z)$ as z tends to an ordinary point $P(t_0)$ on C. The value of $F(z)$ of the integral (15) at $z = t_0$ shall be denoted by $\mathscr{F}(t_0)$, i.e.

$$\mathscr{F}(t_0) \equiv \frac{1}{2\pi i} \int_C \frac{f(t)\,dt}{t-t_0}. \tag{17}$$

A meaning can be given to this improper integral as follows.

Let t_0 be a point lying on a smooth arc $t_1 t_2$ of C where t_1, t_2 are the points of intersection of C and the small circle $|z - t_0| = \epsilon$ (see Fig. 3.5a). Designate this arc by γ. If the limit

$$\lim_{\epsilon\to 0} \frac{1}{2\pi i} \int_{C-\gamma} \frac{f(t)\,dt}{t-t_0}$$

exists it is called the '*principal*' *value* of the improper integral $\mathscr{F}(t_0)$. Obviously if the integral in (17) exists in the ordinary sense (i.e. exists when no restriction is placed on the way t_1 and t_2 tend to t_0) then the principal value also exists and has the same value, but not conversely. Therefore the same symbol will be used throughout this text to denote both ordinary integrals and principal values, with the understanding that if the integral has no meaning in the ordinary sense its principal value is to be taken.

We shall now give a condition on $f(t)$ which is *sufficient* for the existence of a principal value at the point t_0. This is that $f(t)$ satisfies the *Hölder* condition at t_0, i.e. that for all t of C sufficiently close to t_0 the inequality

$$|f(t)-f(t_0)| \leqslant A\,|t-t_0|^\mu \quad (A,\mu\,\text{real}, 0 < A < \infty, 0 < \mu < 1),\quad (18)$$

holds. This condition is clearly stronger than 'continuity', and weaker than 'differentiability'. It is not the most general condition, but it will suffice for our purposes.

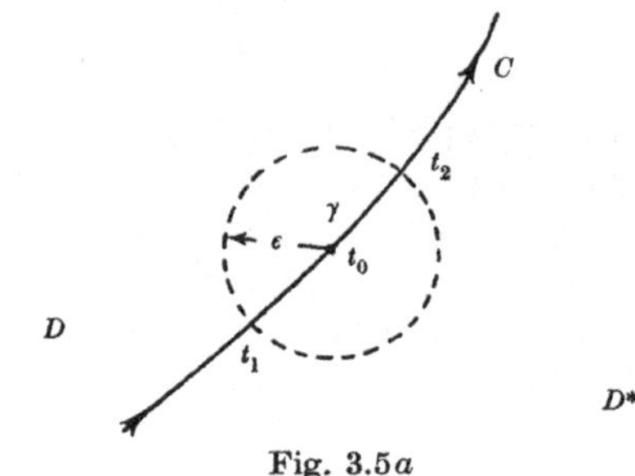

Fig. 3.5*a*

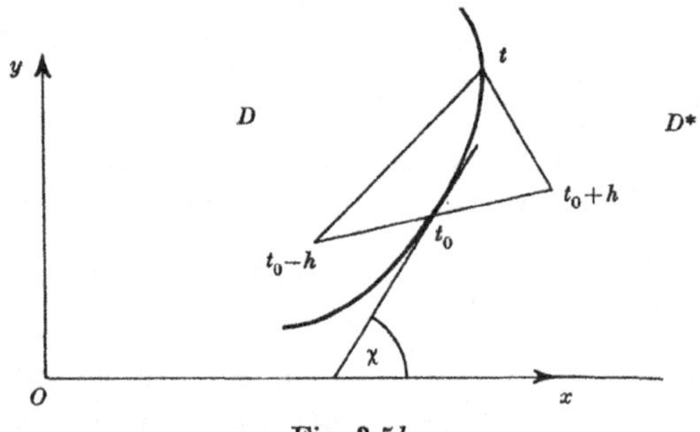

Fig. 3.5*b*

The proof is as follows:

$$\int_{C-\gamma}\frac{f(t)\,dt}{t-t_0} = f(t_0)\int_{C-\gamma}\frac{dt}{t-t_0} + \int_{C-\gamma}\frac{f(t)-f(t_0)}{t-t_0}\,dt$$

$$= f(t_0)\ln\left(\frac{t_1-t_0}{t_2-t_0}\right) + \int_{C-\gamma}\frac{f(t)-f(t_0)}{t-t_0}\,dt. \quad (19)$$

Now $|t_2-t_0| = |t_1-t_0| = \epsilon$, so $\lim_{\epsilon\to 0}\{\ln(t_1-t_0)-\ln(t_2-t_0)\} = i\pi$. Also (18) ensures that the limit of the second integral on the right-hand side of (19) exists in the ordinary sense. Therefore the principal value of the integral in (17) also exists, and is given by

$$\mathscr{F}(t_0) = \tfrac{1}{2}f(t_0) + \frac{1}{2\pi i}\int_C \frac{f(t)-f(t_0)}{t-t_0}\,dt. \tag{20}$$

Let h be a small complex number such that its vector in the Argand plane is not parallel to the tangent to C at t_0 (Fig. 3.5b), then an alternative definition of the principal value of the integral in (17) is

$$\mathscr{F}(t_0) = \lim_{|h|\to 0}\frac{1}{2\pi i}\int_C f(t)\left\{\frac{t-t_0}{(t-t_0)^2-h^2}\right\}dt. \tag{21}$$

The equivalence of this definition to that first given can be shown as follows:

$$\frac{1}{2\pi i}\int_C f(t)\left\{\frac{t-t_0}{(t-t_0)^2-h^2}\right\}dt = \frac{1}{2\pi i}\int_C \{f(t)-f(t_0)\}\left\{\frac{t-t_0}{(t-t_0)^2-h^2}\right\}dt$$

$$+\frac{1}{4\pi i}f(t_0)\left\{\int_C \frac{dt}{t-t_0-h}+\int_C \frac{dt}{t-t_0+h}\right\}. \tag{22}$$

The restriction on arg h means that one of the points $z = t_0+h$, $z = t_0-h$ must lie in D and the other must lie in D^* (see figure). Therefore by Cauchy's theorem and Cauchy's integral formula the last term of (22) reduces to $\tfrac{1}{2}f(t_0)$. Furthermore, (18) ensures the uniform convergence of the second integral in (22) with respect to h in a range which contains $h = 0$, for at $h = 0$ the integrand has the form $|t-t_0|^{\mu-1}$ near $t = t_0$, and the integral converges in the ordinary sense. This means that we can take the limit $|h| \to 0$ past the integral sign of the integral in question. Hence taking the limit of (22), and using (21) as the definition of $\mathscr{F}(t_0)$ we arrive back at (20), and the equivalence of the two definitions is established.

3.6   The Riemann–Stieltjes integral: the delta function

It is convenient to interrupt the development of the theory of Cauchy integrals at this point with a brief outline of the theory of the Riemann–Stieltjes integral—which will be used throughout the following chapters —for this integral greatly facilitates the derivation of the formulae relating $F^+(t_0)$, $F^-(t_0)$ and $\mathscr{F}(t_0)$.

Let $f(x)$ and $g(x)$ be bounded functions of x defined in (a, b), and let $x_1, x_2, \dots, x_{r+i}, \dots, x_n$ be a set of increasing values of x between a and b, such that each interval $x_{r+1}-x_r$ is less than a given ϵ. If the sum

$$S_n = f(\xi_0)\{g(x_1)-g(a)\}+f(\xi_1)\{g(x_2)-g(x_1)\}+ \dots +f(\xi_n)\{g(b)-g(x_n)\}, \tag{23}$$

where $x_r \leqslant \xi_r \leqslant x_{r+1}$, tends to a unique limit S as ϵ tends to zero, which is independent of the choice of the x_r and ξ_r, this limit is called a Riemann–Stieltjes integral. The usual notation is

$$S = \int_a^b f(x)\,dg(x),$$

where the limits b and a are understood to apply to x, not g.

A sufficient condition for the existence of the R–S integral is that over (a, b) $f(x)$ (or $g(x)$) is of bounded variation,† and $g(x)$ (or $f(x)$) is continuous. It follows from this, and a subdivision of the range of integration, that if $f(x)$ and $g(x)$ are continuous save for a finite number of simple discontinuities, the R–S integral exists provided that none of the discontinuities in $f(x)$ and $g(x)$ coincide. Where $g(x)$ is differentiable the R–S integral reduces to an ordinary Riemann integral, viz.

$$\int_a^b f(x)\,dg(x) = \int_a^b f(x)\,g'(x)\,dx. \tag{24}$$

The formula for integration by parts of R–S integrals is

$$\int_a^b f(x)\,dg(x) = \left[f(x)\,g(x) \right]_a^b - \int_a^b g(x)\,df(x), \tag{25}$$

which is more general than the corresponding formula for Riemann integrals in not requiring that $f(x)$ and $g(x)$ be differentiable functions.

Suppose $g(x)$ is differentiable, save at $x = c$ $(a < c < b)$, where it has a simple discontinuity of magnitude β, then from the definition of an R–S integral

$$S = \lim_{\epsilon \to 0} \left\{ \int_{c+\frac{1}{2}\epsilon}^b f(x)\,g'(x)\,dx + \int_a^{c-\frac{1}{2}\epsilon} f(x)\,g'(x)\,dx \right\}$$

$$+ f\left(c + \tfrac{1}{2}\delta\epsilon\right)\left\{ g\left(c + \tfrac{1}{2}\epsilon\right) - g\left(c - \tfrac{1}{2}\epsilon\right) \right\} \quad (|\delta| < 1),$$

i.e.
$$S = \int_{c+0}^b f(x)\,g'(x)\,dx + \int_a^{c-0} f(x)\,g'(x)\,dx + \beta f(c).$$

More generally if there are n discontinuities β_i at $x = c_i$ $(i = 1, 2, ..., n)$,

$$S = \sum_{i=0}^n \int_{c_i+0}^{c_{i+1}-0} f(x)\,g'(x)\,dx + \sum_{i=1}^n \beta_i f(c_i) \quad (c_0 = a,\ c_{n+1} = b). \tag{26}$$

The complexity of the right-hand side of this equation illustrates one of the advantages of the R–S integral notation, namely its concise handling of both continuous and discrete distributions in the one integral form.

---

† The *total variation* V of $f(x)$ in (a, b) is defined to be the upper bound of the sums $v = |f(x_1) - f(a)| + |f(x_2) - f(x_1)| + ... + |f(b) - f(x_n)|$, for all modes of subdivision of (a, b). If V is finite $f(x)$ is said to be of *bounded variation*.

In terms of the *unit function* $U(x)$, i.e. a step-function defined by

$$U(x) = \begin{cases} 1 & (x > 0), \\ \tfrac{1}{2} & (x = 0), \\ 0 & (x < 0), \end{cases} \tag{27}$$

the function $g(x)$ employed in (26) could be expressed

$$g(x) = h(x) + \sum_{i=1}^{n} \beta_i\, U(x - c_i), \tag{28}$$

where $h(x)$ is continuous, having the same derivative as $g(x)$ except at the points $x = c_i$. Hence

$$\int_a^b f(x)\, dg(x) = \int_a^b f(x)\, h'(x)\, dx + \sum_{i=1}^{n} \beta_i \int_a^b f(x)\, d\mathrm{U}(x - c_i),$$

and as

$$\int_a^b f(x)\, d\mathrm{U}(x - c_i) = f(c_i), \tag{29}$$

we recover (26).

The *delta* function $\delta(x)$ is an 'improper' function, which can be defined in several ways (see ch. 4 of van der Pol & Bremmer, 1950), the most convenient of which for our purposes is

$$\delta(x) = \mathrm{U}'(x). \tag{30}$$

From (27) it is clear that $\delta(x)$ vanishes everywhere except at $x = 0$, where it becomes infinite. From (29) and (30) there follows the fundamental property of the delta function, viz.

$$\int_a^b f(x)\, \delta(x - c)\, dx = f(c) \quad (a < c < b). \tag{31}$$

Extensions of the Riemann integral to improper integrals such as

$$\int_0 x^m \ln |x|\, dx \quad (m > -1), \quad \int_0^\infty f(x)\, dx,$$

and the integral defined in (17), also hold for the R–S integral. For example we can write

$$\mathscr{R}\int_a^b \frac{f(x)}{x - c}\, dx = \int_a^b f(x)\, d\{\ln |x - c|\}$$
$$= f(b) \ln (b - c) - f(a) \ln (c - a) - \int_a^b \ln |x - c|\, df(x)$$
$$(a < c < b), \tag{32}$$

by (25). It is easily verified that it is sufficient for the existence of the R–S integrals in this equation that $f(x)$ satisfies the Hölder condition at $x = c$.

3.7 The Plemelj Formulae

We can now derive two very important formulae due to Plemelj (1908).

Let $f(t)$ satisfy the Hölder condition (henceforth called the H condition) at an ordinary point t_0 of C, then in the notation of (16) and (17) the first of Plemelj's formulae can be written

$$F^+(t_0) + F^-(t_0) = 2\mathscr{F}(t_0), \tag{33}$$

one proof of which is as follows.

Let z and z' be the points $t_0 - h$, $t_0 + h$ shown in Fig. 3.5b, then (15) gives

$$F(z) + F(z') = \frac{1}{2\pi i} \int_C f(t) \left\{ \frac{1}{t - t_0 + h} + \frac{1}{t - t_0 - h} \right\} dt$$

$$= \frac{1}{2\pi i} \int_C f(t) \left\{ \frac{t - t_0}{(t - t_0)^2 - h^2} \right\} dt.$$

Assuming for the moment that F^+, F^- do exist, we now obtain (33) immediately by taking the limit $|h| \to 0$, and making use of (21).

The second Plemelj formula is

$$F^+(t_0) - F^-(t_0) = f(t_0), \tag{34}$$

which can be proved as follows. In the notation just introduced

$$F(z) - F(z') = \frac{1}{2\pi i} \int_C f(t) \left\{ \frac{1}{t - t_0 + h} - \frac{1}{t - t_0 - h} \right\} dt$$

$$= \frac{1}{\pi i} \int_C \{ f(t) - f(t_0) \} \left\{ \frac{h}{(t - t_0)^2 - h^2} \right\} dt$$

$$+ \frac{1}{2\pi i} f(t_0) \int_C \left\{ \frac{1}{t - t_0 + h} - \frac{1}{t - t_0 - h} \right\} dt. \tag{35}$$

By the same argument as used for the last term of (22), the last term of (35) reduces to $f(t_0)$. The penultimate term of (35) is uniformly convergent with respect to h provided only that $f(t)$ is continuous at $t = t_0$—the (stronger) H condition is not needed here—and it vanishes in the limit $|h| \to 0$. It remains only to take the limit $|h| \to 0$ of (35) to obtain (34). It now follows from (17), (33) and (34) that $F^+(t_0)$ and $F^-(t_0)$ both exist, and have the values

$$\left. \begin{aligned} F^+(t_0) &= \tfrac{1}{2} f(t_0) + \frac{1}{2\pi i} \int_C \frac{f(t)}{t - t_0} dt, \\[2mm] F^-(t_0) &= -\tfrac{1}{2} f(t_0) + \frac{1}{2\pi i} \int_C \frac{f(t)}{t - t_0} dt. \end{aligned} \right\} \tag{36}$$

These are the forms of the very important Plemelj formulae to which we shall make frequent reference in later pages. They give the limits of the analytic function

$$F(z) = \frac{1}{2\pi i} \int_C \frac{f(t)}{t-z}\, dt, \tag{37}$$

as $z \to t_0$ from one side of C or the other.

In the special case $f(t) = F^+(t)$ (36) and (37) reduce to

$$F^+(t_0) = \frac{1}{\pi i} \int_C \frac{F^+(t)}{t-t_0}\, dt, \quad F^-(t_0) = 0, \quad F(z) = \frac{1}{2\pi i} \int_C \frac{F^+(t)}{t-z}\, dt \tag{38}$$

(cf. §3.4).

3.8   Derivatives of Cauchy integrals

Derivatives of any order of the function $F(z)$ defined in (37) can be obtained by differentiating the integral with respect to z: thus

$$F'(z) = \frac{1}{2\pi i} \int_C \frac{f(t)}{(t-z)^2}\, dt, \tag{39}$$

and in general

$$F^{(n)}(z) = \frac{n!}{2\pi i} \int_C \frac{f(t)\, dt}{(t-z)^{n+1}}. \tag{40}$$

Repeated integration of (40) by parts yields the alternative formula

$$F^{(n)}(z) = \frac{1}{2\pi i} \int_C \frac{f^{(n)}(t)}{t-z}\, dt. \tag{41}$$

Comparing this with (37) we conclude that a sufficient condition for the existence of the limits $F^{(n)+}(t_0)$ and $F^{(n)-}(t_0)$ is that $f^{(n)}(t)$ exists and satisfies the H condition at t_0.

Equations (39) and (40) suggest that the notion of the principal value of a Cauchy integral should be extended to cover improper integrals of the form

$$n! \int_C \frac{f(t)}{(t-t_0)^{n+1}}\, dt.$$

A suitable interpretation of this integral is contained in

$$n! \int_C \frac{f(t)}{(t-t_0)^{n+1}}\, dt \equiv \frac{\partial^n}{\partial t_0^n} \int_C \frac{f(t)\, dt}{t-t_0}. \tag{42}$$

These generalized principal parts are sometimes indicated by rendering the integral sign 'not negotiable' thus:

$$\fint_C \{f(t)/(t-t_0)^{n+1}\}\, dt,$$

but we shall make no distinction. The reader is asked to take principal values of all integrals not existing in the ordinary sense. These improper integrals can always be avoided by an integration by parts (cf. (32)), but there is no particular virtue in this. They can always be evaluated from the definition (42), but a much simpler method can be adopted if the indefinite integral is known. For if

$$\int^t \frac{f(t)\,dt}{(t-t_0)^{n+1}} = \psi(t_0, t) + C,$$

then it is easily proved by mathematical induction that

$$\int_a^b \frac{f(t)\,dt}{(t-t_0)^{n+1}} = \psi(t_0, b) - \psi(t_0, a) \tag{43}$$

(see pp. 229 to 235 of Sears (1954)).

3.9   Inversion formulae: the Poincaré–Bertrand formula

Let $\phi(t)$, $\psi(t)$ be functions satisfying the H condition on C then each of the equations

$$\psi(t_0) = \frac{1}{\pi i} \int_C \frac{\phi(t)\,dt}{t-t_0}, \quad \phi(t_0) = \frac{1}{\pi i} \int_C \frac{\psi(t)\,dt}{t-t_0}, \tag{44}$$

follows from the other. Thus if the first of (44) is regarded as an integral equation for an unknown function $\phi(t)$, the second of (44) gives its solution.   One proof of the inversion formulae (44) is as follows.

Consider the analytic function

$$\chi(z) = \frac{1}{\pi i} \int_C \frac{\phi(t)\,dt}{t-z}.$$

By (36) it has the boundary value

$$\chi^+(t) = \phi(t) + \frac{1}{\pi i} \int_C \frac{\phi(t_1)\,dt_1}{t_1-t}.$$

Applying the first of (38) to this function we have

$$\chi^+(t_0) = \frac{1}{\pi i} \int_C \frac{\chi^+(t)\,dt}{t-t_0}.$$

Eliminating χ^+ from the last two equations we find

$$\frac{1}{\pi i} \int_C \frac{dt}{t-t_0} \frac{1}{\pi i} \int_C \frac{\phi(t_1)\,dt_1}{t_1-t} = \phi(t_0), \tag{45}$$

which holds for any function $\phi(t)$ satisfying the H condition on C. Equations (44) follow immediately from (45).

Inversion formulae of great importance in aerofoil theory can be derived by taking C to be the unit circle $|z| = 1$. If γ is the angular displacement of a point on C from OX, then $t = e^{i\gamma}$, $t_0 = e^{i\gamma_0}$ and

$$\frac{dt}{t-t_0} = \frac{i e^{i\gamma} d\gamma}{e^{i\gamma} - e^{i\gamma_0}} = \tfrac{1}{2} \cot \tfrac{1}{2}(\gamma - \gamma_0)\,d\gamma + \tfrac{1}{2} i\,d\gamma.$$

Let $\Omega(\gamma)$, $\theta(\gamma)$ be periodic functions of period 2π satisfying the H condition on C, then $\phi(t)$ and $\psi(t)$ in (44) can be replaced by $\Omega(\gamma)$ and $i\theta(\gamma)$. It is found that

$$\frac{1}{2\pi}\int_{-\pi}^{\pi}\Omega(\gamma)\cot\tfrac{1}{2}(\gamma-\gamma_0)\,d\gamma+\frac{i}{2\pi}\int_{-\pi}^{\pi}\Omega(\gamma)\,d\gamma = -\theta(\gamma_0), \qquad (46)$$

$$\frac{1}{2\pi}\int_{-\pi}^{\pi}\theta(\gamma)\cot\tfrac{1}{2}(\gamma-\gamma_0)\,d\gamma+\frac{i}{2\pi}\int_{-\pi}^{\pi}\theta(\gamma)\,d\gamma = \Omega(\gamma). \qquad (47)$$

Suppose $\theta(\gamma)$ is a known function, and $\Omega(\gamma)$ is defined by the integral equation (cf. §8.6)

$$\theta(\gamma_0) = -\frac{1}{2\pi}\int_{-\pi}^{\pi}\Omega(\gamma)\cot\tfrac{1}{2}(\gamma-\gamma_0)\,d\gamma. \qquad (48)$$

Integrating (48) over $(-\pi, \pi)$ we get

$$\int_{-\pi}^{\pi}\theta(\gamma_0)\,d\gamma_0 = -\frac{1}{2\pi}\int_{-\pi}^{\pi}d\gamma_0\int_{-\pi}^{\pi}\Omega(\gamma)\cot\tfrac{1}{2}(\gamma-\gamma_0)\,d\gamma$$

$$= -\frac{1}{2\pi}\int_{-\pi}^{\pi}\Omega(\gamma)\,d\gamma\int_{-\pi}^{\pi}\cot\tfrac{1}{2}(\gamma-\gamma_0)\,d\gamma_0 = 0;$$

the interchange of the order of integration here will be justified at the end of this section. Hence for (48) to be capable of solution $\theta(\gamma)$ must satisfy the restriction

$$\int_{-\pi}^{\pi}\theta(\gamma)\,d\gamma = 0. \qquad (49)$$

If a like restriction is now imposed on the unknown function, i.e.

$$\int_{-\pi}^{\pi}\Omega(\gamma)\,d\gamma = 0,$$

(46) and (48) become identical. By (47) and (49) the required solution of (48) is therefore

$$\Omega(\gamma_0) = \frac{1}{2\pi}\int_{-\pi}^{\pi}\theta(\gamma)\cot\tfrac{1}{2}(\gamma-\gamma_0)\,d\gamma. \qquad (50)$$

Equations (48) and (50) are known as Hilbert's inversion formulae (see Hilbert, 1924, p. 75).

Equation (45) is actually a special case of the Poincaré–Bertrand formula, viz.

$$\frac{1}{\pi i}\int_C\frac{dt}{t-t_0}\frac{1}{\pi i}\int_C\frac{\phi(t,t_1)}{t_1-t}\,dt_1 = \phi(t_0,t_0)+\frac{1}{\pi i}\int_C dt_1\frac{1}{\pi i}\int_C\frac{\phi(t,t_1)\,dt}{(t-t_0)(t_1-t)}, \qquad (51)$$

which can be established as follows. Let z be a point not on C, and let

$$\chi(t_1,z) \equiv \frac{t_1-z}{\pi i}\int_C\frac{\phi(t,t_1)\,dt}{(t-z)(t_1-t)},$$

then
$$\frac{1}{\pi i}\int_C \frac{\chi(t_1,z)}{t_1-z}\,dt_1 = \frac{1}{\pi i}\int_C \frac{dt}{t-z}\,\frac{1}{\pi i}\int_C \frac{\phi(t,t_1)}{t_1-t}\,dt_1,$$

assuming that it is legitimate to invert the order of integration of an ordinary integral and a Cauchy integral (see final paragraph of this section). Applying the first of (36) to each side of this equation we obtain

$$\chi^+(t_0,t_0)+\frac{1}{\pi i}\int_C \frac{\chi^+(t_1,t_0)}{t_1-t_0}\,dt_1 = \int_C \frac{\phi(t_0,t_1)}{t_1-t_0}\,dt_0 + \frac{1}{\pi i}\int_C \frac{dt}{t-t_0}\,\frac{1}{\pi i}\int_C \frac{\phi(t,t_1)}{t_1-t}\,dt,$$

and similarly from the definition of $\chi(t_1,z)$,

$$\chi^+(t_1,t_0) = \phi(t_0,t_1)+\frac{t_1-t_0}{\pi i}\int_C \frac{\phi(t,t_1)\,dt}{(t-t_0)(t_1-t)}.$$

Elimination of χ^+ now yields (51), which holds for any function $\phi(t,t_1)$ satisfying the H condition with respect to both t and t_1 on C.

When C passes through the point at infinity, and t_0 is at this point, (51) becomes

$$\int_C \frac{dt}{t}\int_C \frac{\phi(t,t_1)}{t_1-t}\,dt_1 = \pi^2\phi(\infty,\infty) + \int_C dt_1\int_C \frac{\phi(t,t_1)}{t(t-t_1)}\,dt \tag{52}$$

(put $t_0 = 0$ and then apply the inversion $t' = 1/t$).

Equations (51) and (52) reveal the need for some care when inverting the order of integration of repeated Cauchy integrals.

Let a function $K(t,t_1)$ given on C, be of the form

$$K(t,t_1) = \frac{\psi(t,t_1)}{|t_1-t|^\alpha}\quad (0\leqslant \alpha < 1),$$

where $\psi(t,t_1)$ satisfies the H condition. Then it follows directly from (51) that

$$\frac{1}{\pi i}\int_C \frac{dt}{t-t_0}\,\frac{1}{\pi i}\int_C K(t,t_1)\,dt_1 = \frac{1}{\pi i}\int_C dt_1\int_C \frac{K(t,t_1)}{t-t_0}\,dt_1,$$

for here

$$\phi(t,t_1) = (t_1-t)\,K(t,t_1) = |t_1-t|^{1-\alpha}\exp\{i\arg(t_1-t)\}\,\psi(t,t_1),$$

whence $\phi(t,t) = 0$. This justifies the interchange of the order of integration of Cauchy and ordinary integrals.

BOUNDARY SINGULARITIES

3.10 Singularities in the density function

The Plemelj and other formulae established in §§ 3.7 and 3.8 hold at points t_0 on C at which $f(t)$ satisfies the H condition. However, it would be too much of a restriction on $f(t)$ to require that it satisfies the H condition at *all* points on C. We shall therefore admit the possibility of a finite number of exceptional points $\{t^*\}$ on C at which the H condition is not satisfied. Of course we cannot expect the Plemelj formulae to hold at such points.

The left-hand side of equation (37), i.e. of

$$F(z) = \frac{1}{2\pi i} \int_C \frac{f(t)\,dt}{t-z},$$

will remain finite provided the singularity in $f(t)$ at $t = t^*$ is integrable in the ordinary sense. This admits the following important types of singularity:

(1) *Simple discontinuity*:

$$f(t) = A\,\mathrm{U}(t-t^*), \tag{53}$$

where A is a finite constant and U is the unit function defined in (27). This type of singularity (see Fig. 3.10a) is important in two-dimensional flow problems for it occurs in the flow direction θ at stagnation points and sharp projections on solid boundaries.

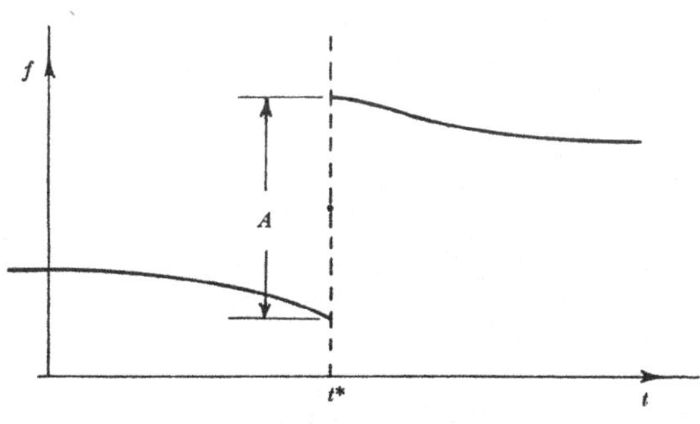

Fig. 3.10a

(2) *Logarithmic infinity*:

$$f(t) = A \ln |t-t^*|$$

This singularity, shown in Fig. 3.10b, is also important in fluid motion problems. The harmonic function Ω, which is the harmonic conjugate of θ (see § 2.14) becomes logarithmically infinite at stagnation points, i.e. at points where θ has simple discontinuities. This suggests combining the logarithmic singularity and (53) into the single complex logarithmic singularity

$$f(t) = A \ln |t-t^*| + iA \arg (t-t^*),$$

or

$$f(t) = A \ln (t-t^*), \tag{54}$$

as $\arg (t-t^*)$ behaves like $-\pi\mathrm{U}(t-t^*)$ near $t = t^*$.

(3) *Power-law infinity*:

$$f(t) = A\,|t-t^*|^{-\alpha} \quad (0 \leqslant \alpha < 1). \tag{55}$$

A typical example of this occurs in the pressure gradient at a point of flow separation (see §11.6).

Notice that if $g(t) \propto |t-t^*|^{1-\alpha}$, and therefore satisfies the H condition at t^*, its first derivative has a power-law infinity at $t = t^*$, and its second derivative is not integrable in the ordinary sense. It could however be given the interpretation embodied in (42).

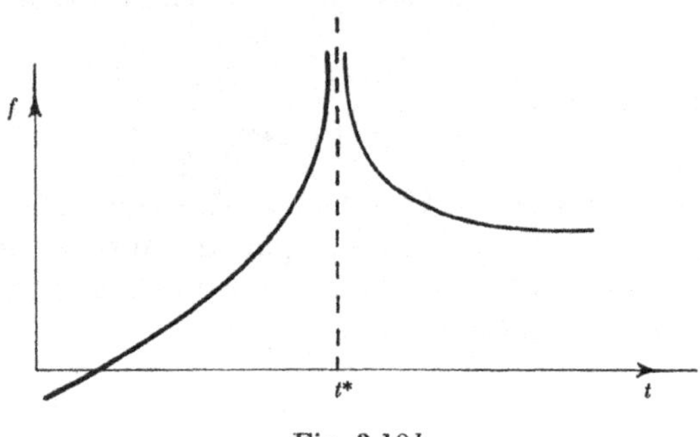

Fig. 3.10*b*

(4) *Cauchy-type singularity*:

$$f(t) = A(t-t^*)^{-1}. \tag{56}$$

This is the derivative of a logarithmic infinity; the Cauchy principal value of the integral must be taken in this case.

(5) *Impulse or delta function*:

$$f(t) = A\delta(t-t^*). \tag{57}$$

From (30) and (53) it follows that this type of singularity will occur in the derivative of the flow direction θ at a stagnation point.

3.11   Extension of the residue theorem

The residue theorem given in §3.2 can be extended to cover the case of a function $F(z)$ having not only a finite number of poles *within* a given contour C, but also a finite number of *simple* poles *on* the contour C. In this case the contour integral $\int_C F(t)\,dt$ is improper in the ordinary sense, and must be given its Cauchy principal value.

Suppose, for example, the pole has a residue b, and lies at a point a on C, at which the tangent slope has a simple discontinuity of $\pi - \alpha$ (see

Fig. 3.11). When $\alpha = \pi$, a becomes an ordinary point on C, which is thus included in the theory given below as a particular case.

In the neighbourhood of a we can write

$$F(z) = \frac{b}{z-a} + G(z),$$

where $G(z)$ is regular at a. Now indent the contour around a by a small circular arc γ of radius ϵ, then on γ $z = a + \epsilon\, e^{i\theta}$, so

$$\int_\gamma F(z)\, dz = ib\Delta\theta + i\epsilon \int_\gamma G(\theta)\, e^{i\theta}\, d\theta,$$

where $\Delta\theta$ is the change in θ over the circular arc γ. Clearly $\lim\limits_{\epsilon \to 0} \Delta\theta = -\alpha$,

and as $\displaystyle\int_\gamma G(\theta)\, e^{i\theta}\, d\theta$ is bounded as $\epsilon \to 0$, then

$$\lim_{\epsilon \to 0} \int_\gamma F(z)\, dz = -ib\alpha. \tag{58}$$

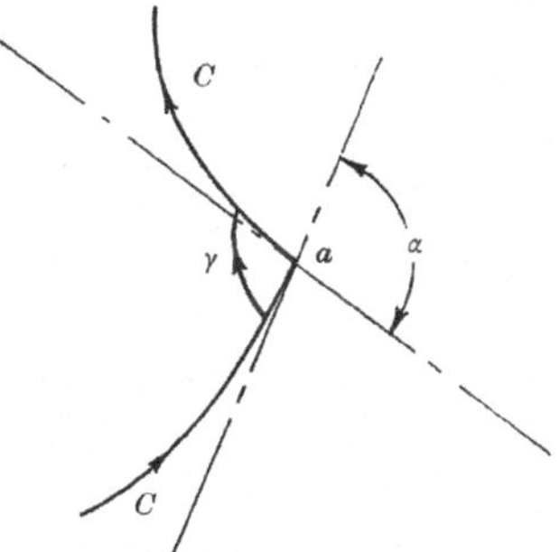

Fig. 3.11

Now by the ordinary form of the residue theorem

$$\frac{1}{2\pi i}\left\{\int_{C-\gamma} F(t)\, dt + \int_\gamma F(z)\, dz\right\} = \Sigma R,$$

where ΣR is the sum of the residues of $F(z)$ at the poles *within* C. In the limit as $\epsilon \to 0$ this equation and (58) give

$$\frac{1}{2\pi i}\int_C F(t)\, dt = \Sigma R + \frac{\alpha}{2\pi} b.$$

More generally if there are simple poles of residues $b_1, b_2, \ldots, b_n$ on C at points of discontinuity in the tangent direction of $\pi - \alpha_1, \pi - \alpha_2, \ldots, \pi - \alpha_n$, then

$$\frac{1}{2\pi i}\int_C F(t)\, dt = \Sigma R + \frac{1}{2\pi} \sum_{i=1}^{n} \alpha_i b_i, \tag{59}$$

where the integral is to be interpreted as a Cauchy principal value.

Poles of higher order than unity on C render the contour integral infinite, unless of course the interpretation given in (42) can be used.

3.12   Boundary passing through the point at infinity

Another type of boundary singularity occurs when the boundary passes through the point at infinity. The simplest example occurs when C is the infinite straight line, say $y = 0$. We shall extend the definition of Cauchy integrals to this case.

Consider the function $F(z)$ defined by

$$F(z) = \lim_{R \to \infty} \frac{1}{2\pi i} \int_{-R}^{R} \frac{f(x)\,dx}{x-z} = \frac{1}{2\pi i} \int_{-\infty}^{\infty} \frac{f(x)\,dx}{x-z} \quad (\mathscr{I}z \neq 0), \qquad (60)$$

where $f(x)$ satisfies the H condition on $-\infty < x < \infty$ and where for sufficiently large x

$$|f(x) - f(\infty)| \leqslant A\,|x|^{-\mu} \quad (0 < A < \infty,\ \mu > 0). \qquad (61)$$

Now $\lim\limits_{R \to \infty} \int_{a}^{R} \ldots$ and $\lim\limits_{R \to \infty} \int_{-R}^{a} \ldots$ will not exist separately unless $f(\infty) = 0$, and consequently $\lim\limits_{R \to \infty} \int_{-R}^{R} \ldots$ defines a kind of 'principal' value at infinity. At first sight this seems to be a new kind of principal value, but as the transformation $\xi = 1/z$ clearly changes this particular value of the integral into an 'ordinary' Cauchy principal value at the origin, no new concept is involved. Thus $\lim\limits_{R \to \infty} \int_{-R}^{R} \ldots$ is simply the *Cauchy principal value at infinity*. Equation (61) is the H condition at infinity (cf. (18)).

The integral can be transformed into an ordinary type of infinite integral by using the result

$$\lim_{R \to \infty} \int_{-R}^{R} \frac{dx}{x-z} = \lim_{R \to \infty} \ln\left(\frac{z-R}{z+R}\right) = \left\{ \begin{array}{l} i\pi\,(\mathscr{I}z > 0), \\ -i\pi\,(\mathscr{I}z < 0). \end{array} \right\} \qquad (62)$$

This enables us to write (cf. (20))

$$F(z) = \frac{1}{2\pi i} \int_{-\infty}^{\infty} \frac{f(x) - f(\infty)}{x-z} \left\{ \begin{array}{l} +\tfrac{1}{2}f(\infty)\,(\mathscr{I}z > 0), \\ -\tfrac{1}{2}f(\infty)\,(\mathscr{I}z < 0). \end{array} \right\} \qquad (63)$$

An application of the Plemelj formulae to (60) gives

$$\left. \begin{array}{l} F^{+}(x_0) = \tfrac{1}{2}f(x_0) + \dfrac{1}{2\pi i} \displaystyle\int_{-\infty}^{\infty} \dfrac{f(x)\,dx}{x-x_0}, \\[3mm] F^{-}(x_0) = -\tfrac{1}{2}f(x_0) + \dfrac{1}{2\pi i} \displaystyle\int_{-\infty}^{\infty} \dfrac{f(x)\,dx}{x-x_0}, \end{array} \right\} \qquad (64)$$

where F^+, F^- are the limiting values of $F(z)$ as $z \to x_0$ from $\mathscr{I}z > 0$ and $\mathscr{I}z < 0$ respectively (the domains D and D^* with respect to the real axis, see §3.1). The values of the integrals in (64) are defined by

$$\int_{-\infty}^{\infty} \frac{f(x)\,dx}{x-x_0} = \lim_{\substack{R\to\infty \\ \epsilon\to 0}} \left[\left\{ \int_{-R}^{x_0-\epsilon} + \int_{x_0+\epsilon}^{R} \right\} \frac{f(x)\,dx}{x-x_0} \right]. \tag{65}$$

In order to study the behaviour of $F(z)$ near $z = \infty$ apply the inversion $z = -1/\zeta$ to (60). This maps the upper half of the z-plane on to the upper half of the ζ-plane, and the point $z = \infty$ on to $\zeta = 0$. With $x = -1/\gamma$, $F(z) = F_1(\zeta)$ and $f(x) = f_1(\gamma)$, we find that (60) transforms into

$$F(z) = F_1(\zeta) = \frac{\zeta}{2\pi i} \int_{-\infty}^{\infty} \frac{f_1(\gamma)\,d\gamma}{\gamma(\gamma-\zeta)} = \frac{1}{2\pi i} \int_{-\infty}^{\infty} \frac{f_1(\gamma)\,d\gamma}{\gamma-\zeta} - \frac{1}{2\pi i} \int_{-\infty}^{\infty} \frac{f_1(\gamma)\,d\gamma}{\gamma}.$$

The existence of those integrals as Cauchy principal values is ensured because the H conditions on $-\infty \leqslant x \leqslant \infty$ transform into H conditions on $-\infty \leqslant \gamma \leqslant \infty$; in particular the H condition at $x = \infty$—(61)—is transformed in the H condition at $\gamma = 0$.

It remains only to take the limit $\zeta \to 0$ to find by the Plemelj formula that

$$F^+(\infty) = F_1^+(0) = \tfrac{1}{2}f_1(0) + \frac{1}{2\pi i} \int_{-\infty}^{\infty} \frac{f_1(\gamma)\,d\gamma}{\gamma} - \frac{1}{2\pi i} \int_{-\infty}^{\infty} \frac{f_1(\gamma)\,d\gamma}{\gamma},$$

whence

$$F^+(\infty) = \tfrac{1}{2}f(\infty); \qquad \left. \begin{array}{l} \\ \\ \end{array} \right.$$

similarly

$$F^-(\infty) = -\tfrac{1}{2}f(\infty). \tag{66}$$

Comparison of (66) with (63) and (64) reveals that

$$\lim_{z\to\infty} \int_{-\infty}^{\infty} \frac{f(x)-f(\infty)}{x-z}\,dx = 0, \quad \lim_{x_0\to\infty} \int_{-\infty}^{\infty} \frac{f(x)\,dx}{x-x_0} = 0, \tag{67}$$

of which we shall make frequent use in later chapters. Finally consider the special case when $F^-(x_0)$ is constant. From (64) and (66) $F^-(x_0) = -F^+(\infty)$, and $f(x_0) = F^+(x_0) + F^+(\infty)$, hence (60) gives

$$F(z) = \frac{1}{2\pi i} \int_{-\infty}^{\infty} \frac{F^+(x)+F^+(\infty)}{x-z}\,dx$$

$$= \frac{1}{2\pi i} \int_{-\infty}^{\infty} \frac{F^+(x)}{x-z}\,dx \left\{ \begin{array}{l} +\tfrac{1}{2}F^+(\infty)\,(\mathscr{I}z > 0), \\ -\tfrac{1}{2}F^+(\infty)\,(\mathscr{I}z < 0). \end{array} \right. \tag{68}$$

For $\mathscr{I}z < 0$, $F(z) = F^-(x_0) = -F^+(\infty)$, hence (68) can be written in the form

$$\frac{1}{2\pi i} \int_{-\infty}^{\infty} \frac{F^+(x)\,dx}{x-z} + \tfrac{1}{2}F^+(\infty) = \left\{ \begin{array}{l} F(z)\,(\mathscr{I}z > 0), \\ 0\ \ (\mathscr{I}z < 0). \end{array} \right. \tag{69}$$

These may be called Cauchy's integral and Cauchy's theorem respectively (cf. (11) and (12)) for the regular function $F(z)$ in $\mathscr{I}z > 0$.

We must leave our treatment of the general theory of Cauchy integrals at this point; the reader who wishes to pursue this field further should study chs. 1–4 of *Singular Integral Equations* by Muskhelishvili (1953).

THE FUNDAMENTAL PROBLEMS
OF POTENTIAL THEORY

3.13   The Dirichlet and Neumann problems

The Dirichlet problem for a given contour C in the z- ($= x + iy$) plane is that of finding a function $\Omega(x, y)$ harmonic (i.e. satisfying Laplace's equation, see §2.14) within C, and having prescribed values of Ω on C. The Neumann problem is similar, except that the normal derivatives of Ω are given on C.

First consider the Dirichlet problem. Let

$$\Omega = \Omega_0(t), \tag{70}$$

on C, where $\Omega(t)$ satisfies the H condition except at a finite number of points where any of the singularities listed in §3.10 may occur. (We shall henceforth refer to this modified H condition as the 'M.H.' condition on C.) The Dirichlet problem can be transformed into one in the theory of analytic functions as follows. Let $\theta(x, y)$ be the harmonic conjugate (see §2.14) of $\Omega(x, y)$, and let $\tau = \Omega + i\theta$, then τ is an analytic function of z, regular within C. The boundary condition (70) can now be written $\mathscr{R}\tau^+(t) = \Omega_0(t)$, i.e.

$$\tau^+(t) + \overline{\tau^+(t)} = 2\Omega_0(t), \tag{71}$$

where $\overline{\tau^+(t)}$ is the complex conjugate of $\tau^+(t)$.† When the function $\tau(z)$ satisfying (71) is found, the real part of this function is the required solution to the Dirichlet problem.

The boundary condition for the Neumann problem can be expressed in a similar form. Let the *inward* normal derivative of a harmonic function $\theta(x, y)$ have a boundary value on C given by

$$\frac{\partial \theta}{\partial n} = g(s), \tag{72}$$

where s is distance measured along C from some convenient reference

---

† The reader is reminded that if $\tau(z) = \Omega(x, y) + i\theta(x, y)$, where Ω and θ are real functions, $\overline{\tau(z)} = \overline{\tau}(\bar{z}) = \Omega(x, y) - i\theta(x, y)$, $\tau(\bar{z}) = \Omega(x, -y) + i\theta(x, -y)$, and
$$\overline{\tau(\bar{z})} = \overline{\tau}(z) = \Omega(x, -y) - i\theta(x, -y).$$

point at $t = t_0$. If χ is the angle between the tangent at a typical point $P(t)$ on C and Ox, then (72) can be written (see Fig. 3.13)

$$\cos\chi\frac{\partial\theta}{\partial y} - \sin\chi\frac{\partial\theta}{\partial x} = g(s), \text{ on } C. \tag{73}$$

As before let $\Omega(x,y)$ be the harmonic conjugate of $\theta(x,y)$, and $\tau(z) = \Omega(x,y) + i\theta(x,y)$, then the Cauchy–Riemann equations, namely

$$\frac{\partial\Omega}{\partial x} = \frac{\partial\theta}{\partial y}, \quad \frac{\partial\Omega}{\partial y} = -\frac{\partial\theta}{\partial x}, \tag{74}$$

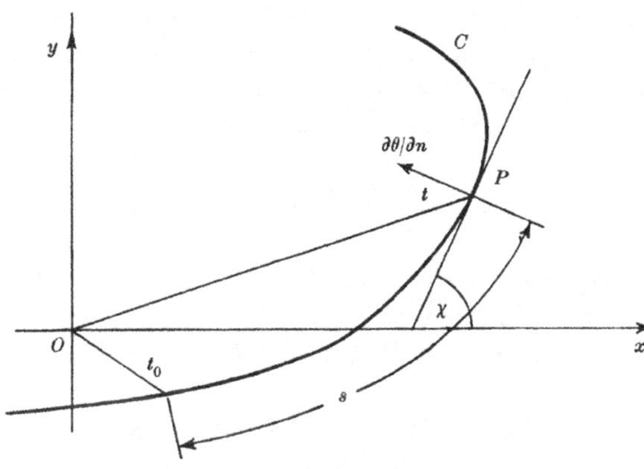

Fig. 3.13

permit us to write the derivative

$$\frac{d\tau}{dz} = \tau'(z) = \frac{\partial\Omega}{\partial x} + i\frac{\partial\theta}{\partial x},$$

in the form

$$\tau'(z) = \frac{\partial\theta}{\partial y} + i\frac{\partial\theta}{\partial x}. \tag{75}$$

Also

$$\frac{dt}{ds} = \frac{dx}{ds} + i\frac{dy}{ds} = \cos\chi + i\sin\chi. \tag{76}$$

Equations (75) and (76) now permit (73) to be expressed

$$\mathscr{R}[\tau'(t)\,dt] = g(s)\,ds, \quad \text{i.e.} \quad d[\tau^+(t) + \overline{\tau^+(t)}] = 2g(s)\,ds,$$

on introducing the usual notation for (inner) boundary values. Therefore corresponding to (71) we have

$$\tau^+(t) + \overline{\tau^+(t)} = 2\int^s g(s)\,ds, \tag{77}$$

which reduces the Neumann problem to the same problem in analytic function theory as already deduced for the Dirichlet problem, except that we now take the *imaginary* part of $\tau(z)$ instead of the real part.

If (75) is replaced by the alternative form

$$\tau'(z) = \frac{\partial \Omega}{\partial x} - i\frac{\partial \Omega}{\partial y}, \quad \text{then} \quad \mathscr{R}\tau'(z)\,dt = \frac{\partial \Omega}{\partial x}\,dx + \frac{\partial \Omega}{\partial y}\,dy = d\Omega.$$

Hence
$$\Omega(s) = \int^{s} g(s)\,ds, \tag{78}$$

which is otherwise obvious by comparison of (71) and (77).

There is an important restriction that $g(s)$ must satisfy if a solution to the Neumann problem is to exist. This is that after one circuit of the contour C, $\Omega(s)$, as defined by (78), must return to its original value (assuming that the circuit does not start at one of the exceptional points discussed in § 3.10). And this is possible only if

$$\int_{C} g(s)\,ds = 0. \tag{79}$$

The Dirichlet and Neumann problems are sometimes called the *first* and *second fundamental problems of potential theory*. Further progress in their solution requires that the shape of the contour C be given.

3.14 Solutions of the fundamental problems for a circle

As a special and simple example of the foregoing theory we shall now seek an analytic function $\tau(z)$ within unit circle $|z| = 1$, whose boundary values on this circle satisfy equation (71) (or equation (77)). There are two distinct but essentially equivalent methods of solution.

(1) *The continuation method*

In the 'continuation' method we try to continue $\tau(z)$ *outside* of the unit circle γ, say, in such a way that it is regular outside γ, and has a boundary value $\tau^{-}(t)$ on γ given by

$$\tau^{-}(t) = -\overline{\tau^{+}(t)}. \tag{80}$$

This choice of boundary value for $\tau^{-}(t)$ is suggested by the Plemelj formula (34), for it permits (71) to be written in the form

$$\tau^{+}(t) - \tau^{-}(t) = 2\Omega_0(t). \tag{81}$$

We shall now show that the required continuation to $|z| > 1$ is given by

$$\tau(z) = -\overline{\tau(1/\bar{z})}, \tag{82}$$

i.e. in words the value of τ at a point z outside γ equals minus the complex conjugate of its value at the inverse point $\{1/\bar{z} = x/(x^2+y^2) + (iy/(x^2+y^2))\}$ with respect to γ. First (82) satisfies (80) as on γ $t = 1/\bar{t}$; secondly, $\overline{\tau(1/\bar{z})}$

is an analytic function of z, being in fact equal to $\bar{\tau}(1/z)$ (see footnote on p. 92), and thirdly if $\tau(z)$ is regular within γ, its continuation outside γ must also be regular.

In the general case (82) does not give an *analytic* continuation of $\tau(z)$ outside γ, because of the discontinuity in $\tau(z)$ across γ, but the whole function now defined inside and outside of γ can be called a *sectionally regular function* (Muskhelishvili, 1953). However, the real and imaginary parts of (80), namely $\Omega^+(t) = -\Omega^-(t)$, $\theta^+(t) = \theta^-(t)$, suggest that for the special case in which the real part (Ω) of $\tau(z)$ vanishes in any interval AB on the circle γ, (80) may give a proper analytic continuation of $\tau(z)$ through this interval. This conjecture is true, and is known as the *Schwarzian inversion* or *reflexion principle*. We shall not prove it here, for it is not essential to the present calculation, but as it is a most useful technique in our subject, and does follow on naturally from the considerations of this section, we shall return to it shortly (§3.15).

Returning now to the problem of finding $\tau(z)$, we first observe from (82) that if at the centre of the circle, $z = 0$, τ has the value $\alpha + i\beta$, then $\tau(0) = -\bar{\tau}(\infty)$, i.e.

$$\tau(\infty) = -\bar{\tau}(0) = -\alpha + i\beta. \tag{83}$$

Our problem can now be stated as follows. It is required to find a sectionally regular function $\tau(z)$, satisfying (83) at infinity, and (81) on the unit circle γ. A solution to this problem is given immediately by Plemelj's formula for by (34) and (37) it is clear that the function

$$\tau(z) = \frac{1}{\pi i} \int_\gamma \frac{\Omega_0(t)\,dt}{t-z} - \alpha + i\beta \tag{84}$$

satisfies all the conditions. No other solutions are possible, for otherwise the difference $\sigma(z)$ of two solutions would vanish at infinity and satisfy the condition $\sigma^+(t) = \sigma^-(t)$. Thus $\sigma(z)$ would be regular over the whole plane, and by Liouville's theorem on such functions (see Copson, 1935, p. 70) must be constant, i.e. in this case be zero everywhere.

As $\tau(0) = \alpha + i\beta$, (84) gives

$$2\alpha = \frac{1}{\pi i} \int_\gamma \frac{\Omega_0(t)}{t}\,dt, \tag{85}$$

so that eliminating α from (84)

$$\tau(z) = \frac{1}{2\pi i} \int_\gamma \Omega_0(t) \left(\frac{2}{t-z} - \frac{1}{t} \right) dt + i\beta,$$

i.e.

$$\tau(z) = \frac{1}{2\pi i} \int_\gamma \Omega_0(t) \frac{t+z}{t-z} \frac{dt}{t} + i\beta. \tag{86}$$

This result is known as the *Schwarz formula*.

An alternative form of (86) is obtained by putting $z = r\,e^{i\gamma}$, $0 \leqslant r < 1$, $t = e^{i\gamma_0}$. It will be found that

$$\tau(r\,e^{i\gamma}) = \frac{1}{2\pi} \int_{-\pi}^{\pi} \Omega_0(\gamma_0) \frac{1 + 2ir\sin(\gamma - \gamma_0) - r^2}{1 - 2r\cos(\gamma - \gamma_0) + r^2} \, d\gamma_0 + i\beta. \qquad (87)$$

The solution to Dirichlet's problem now follows by taking the real part of (87). This is

$$\Omega(r, \gamma) = \frac{1}{2\pi} \int_{-\pi}^{\pi} \Omega_0(\gamma_0) \frac{(1 - r^2)\, d\gamma_0}{1 - 2r\cos(\gamma - \gamma_0) + r^2}, \qquad (88)$$

a well-known result known as *Poisson's formula*.

The imaginary part of (87) can be written

$$\theta(r, \gamma) = \frac{1}{2\pi} \int_{-\pi}^{\pi} \Omega_0(\gamma_0)\, d\left\{\ln\left|1 - 2r\cos(\gamma - \gamma_0) + r^2\right|\right\} + \beta.$$

Hence by integrating by parts, and using the periodic character of $\Omega_0(\gamma_0)$ we obtain

$$\theta(r, \gamma) = \beta - \frac{1}{2\pi} \int_{-\pi}^{\pi} \frac{d\Omega_0}{d\gamma_0} \ln\left|1 - 2r\cos(\gamma - \gamma_0) + r^2\right| d\gamma_0. \qquad (89)$$

But on unit circle $\gamma = s$, so by (78) $d\Omega_0/d\gamma = g(\gamma)$. Also β is clearly the value of θ at the origin. Therefore (89) can be written

$$\theta(r, \gamma) = \theta(0, \gamma) - \frac{1}{2\pi} \int_{-\pi}^{\pi} g(\gamma) \ln\left|1 - 2r\cos(\gamma - \gamma_0) + r^2\right| d\gamma_0, \qquad (90)$$

which solves the Neumann problem for the circle.

(2) *The Image method*

While the continuation method of solving the fundamental problems is very instructive, it is rather more complicated than is strictly necessary, and unless there is a special reason for wanting $\tau(z)$ to be defined outside the circle, the 'image' method described below is preferable.

We start with Cauchy's *integral* for $\tau(z)$, viz.

$$\tau(z) = \frac{1}{2\pi i} \int_{\gamma} \frac{\tau^+(t)}{t - z} \, dt, \qquad (91)$$

where z is a point lying within the unit circle γ. As the point $1/\bar{z}$ must lie outside γ, Cauchy's *theorem* gives that

$$0 = \frac{1}{2\pi i} \int_{\gamma} \frac{\tau^+(t)\, dt}{t - 1/\bar{z}}. \qquad (92)$$

The complex conjugate of (92) is

$$0 = -\frac{1}{2\pi i} \int_{\gamma} \frac{\overline{\tau^+(t)\, dt}}{\bar{t} - 1/z},$$

or

$$0 = \frac{1}{2\pi i}\int_\gamma \frac{z\,\overline{\tau^+(t)}\,dt}{(z-t)\,t},$$

as $t\bar{t} = 1$. The last equation can be written

$$0 = \frac{1}{2\pi i}\int_\gamma \frac{\overline{\tau^+(t)}\,dt}{t-z} - \frac{1}{2\pi i}\int_\gamma \frac{\overline{\tau^+(t)}\,dt}{t}. \tag{93}$$

We now add (93) to (91) and obtain

$$\tau(z) = \frac{1}{\pi i}\int_\gamma \frac{\Omega_0(t)}{t-z}\,dt - \frac{1}{2\pi i}\int_\gamma \{\Omega_0(t) - i\theta_0(t)\}\frac{dt}{t}, \tag{94}$$

with the aid of (71) and the fact that $\Omega_0(t)$, $\theta_0(t)$ are the real and imaginary parts of $\tau^+(t)$. Hence

$$\tau(z) = \frac{1}{2\pi i}\int_\gamma \Omega_0(t)\frac{t+z}{t-z}\frac{dt}{t} + \frac{1}{2\pi}\int_\gamma \frac{\theta_0(t)}{t}\,dt, \tag{95}$$

which is the Schwarz formula, (86), so the calculation of the solutions to the Dirichlet and Neumann problems now follows as above.

At the centre of the circle, $z = 0$, and (95) yields

$$\Omega(0) + i\theta(0) = \frac{1}{2\pi}\int_\gamma \Omega_0(t)\frac{dt}{it} + \frac{i}{2\pi}\int_\gamma \theta_0(t)\frac{dt}{it}.$$

Let $t = e^{i\gamma}$, then it is found that

$$\Omega(0) = \frac{1}{2\pi}\int_{-\pi}^{\pi} \Omega_0(\gamma)\,d\gamma, \quad \theta(0) = \frac{1}{2\pi}\int_{-\pi}^{\pi} \theta_0(\gamma)\,d\gamma. \tag{96}$$

These equations express a very important result in harmonic-function theory due to Gauss, namely that the mean value of a harmonic function over a circle, which encloses no singularities of the function, is equal to the value of the function at the centre of the circle. (We are of course assuming a *plane* harmonic field here.)

Notice that if equation (93) is *subtracted* from (91), then as $\tau^+(t) - \overline{\tau^+(t)} = 2i\theta_0(t)$, we arrive at

$$\tau(z) = \frac{1}{2\pi}\int_\gamma \theta_0(t)\frac{t+z}{t-z}\frac{dt}{t} + \frac{1}{2\pi i}\int_\gamma \frac{\Omega_0(t)}{t}\,dt. \tag{97}$$

This may be described as the *harmonic conjugate equation* to (95). It also follows directly from (95) on replacing $\tau(t)$ by $-i\tau(z)$. A simple, and obvious rule in the derivation of conjugate equations is '*interchange the roles of Ω and $i\theta$*'.

In the limit $z \to t_0$, a point on the unit circle, (95) yields

$$\Omega(t_0) + i\theta(t_0) = \Omega(t_0) + \frac{1}{2\pi i}\int_\gamma \Omega_0(t)\frac{t+t_0}{t-t_0}\frac{dt}{t} + \frac{1}{2\pi}\int_\gamma \theta_0(t)\frac{dt}{t}, \tag{98}$$

where the first term on the right-hand side corresponds to the similar term in the Plemelj formulae (36) (cf. (37) and (95)).

Both the methods given above of solving the fundamental problems for a circle can be applied to a number of other singly-connected domains, e.g. infinite strips (§ 3.16) and rectangles (§ 4.14); we shall always employ the simpler image method.

As we shall see later (ch. 5) it is really unnecessary to solve the fundamental problems again for singly-connected regions other than the circle, for these regions can be mapped on to the interior of a circle by a conformal transformation. As the harmonic character of the functions is unchanged by the conformal mapping, our problem is reduced to the one solved above. However, this is not always the most convenient or instructive procedure, and at the risk of being a little uneconomical in the development of our subject, we shall generate several solutions of the fundamental problems *ab initio*.

3.15   The 'Schwarzian inversion' or 'reflexion' principle

We occasionally find it very convenient to be able to make an analytic continuation of our complex function $\tau(z)$ over portion of a boundary Γ into a region in which it is not naturally defined (e.g. see § 11.8). This is easily effected in the cases when (*a*) Γ is a straight line or a circular arc; and when (*b*) either the real or imaginary part of $\tau(z)$ is constant over Γ. There is no essential loss of generality if this constant is taken to be zero, for suppose $\Omega = k$ on Γ, then the real part of the analytic function $\tau - k$ is zero on Γ, and it is necessary only to add k to the analytic continuation of $\tau - k$ to find the correct continuation of τ.

First suppose the boundary Γ is a circular arc, and select the origin in the z-plane to be at the centre of the circle of which the arc is a part. It has been shown in § 3.14 that the relation

$$\tau(z) = -\overline{\tau(1/\bar{z})} \qquad\qquad (82\ bis)$$

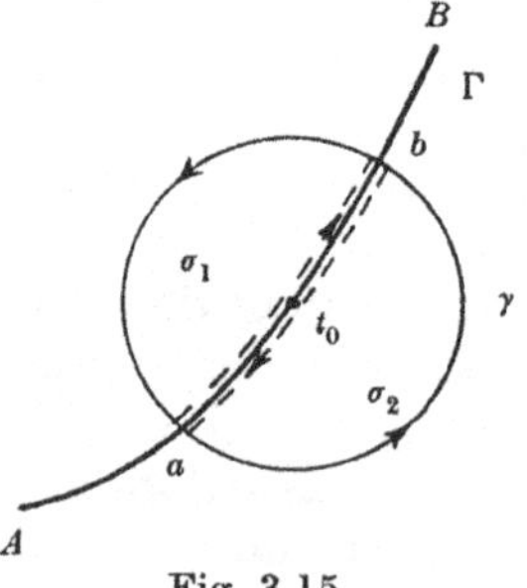

Fig. 3.15

provides at least a *sectionally* analytic continuation of $\tau(z)$ across Γ, and further if Ω vanishes on Γ the function $\tau(z)$ so continued is continuous across Γ. We shall now prove that in this latter case (i.e. Ω vanishing on Γ) (82) provides a proper analytic continuation of $\tau(z)$ across Γ.

For the proof it is obviously sufficient to show that the function $\tau(z)$ is analytic in the neighbourhood of any point t_0 on Γ not coinciding with its ends A, B. Describe with t_0 as centre a circle γ small enough to cut Γ in just two points a and b (see Fig. 3.15). Let σ_1 be the part of this circular region in which $\tau(z)$ is defined initially, σ_2 the part where $\tau(z)$ is defined by (82), and $\sigma = \sigma_1 + \sigma_2$. Now define a function $\Phi(z)$ by†

$$\Phi(z) = \tau(z) \quad (z \in \sigma_1): \quad \Phi(z) = -\overline{\tau(1/\bar z)} \quad (z \in \sigma_2):$$

$$\Phi(t) = \tau(t) \quad (t \in \Gamma).$$

Cauchy's theorem gives

$$\Phi(z) = \frac{1}{2\pi i} \int_{\gamma_1} \frac{\tau(z)}{t-z} \, dt + \frac{1}{2\pi i} \int_{\gamma_2} \frac{\{-\overline{\tau(1/\bar z)}\}}{t-z} \, dt \quad (z \in \sigma),$$

where γ_1, γ_2 are the boundary contours of σ_1 and σ_2, for if $z \in \sigma_1$ the second integral vanishes, and the first yields $\Phi = \tau(z)$, whereas if $z \in \sigma_2$ the first integral vanishes and the second gives $\Phi = -\overline{\tau(1/\bar z)}$. Because of the continuity of $\tau(z)$ across Γ, and the resultant cancelling of the contributions of the arcs ab and ba, the sum of the two integrals reduces to a single integral taken along γ, i.e.

$$\Phi(z) = \frac{1}{2\pi i} \int_{\gamma} \frac{\Phi(t) \, dt}{t-z}.$$

This completes the proof, as the right-hand side of this equation clearly represents a function analytic within γ.

It should be obvious that the role of the sign of the right-hand side of (82) is simply to ensure the continuity of $\tau(z)$ across Γ, so if it is the *imaginary* part (θ) of τ which vanishes on Γ the analytic continuation of $\tau(z)$ is given by

$$\tau(z) = \overline{\tau(1/\bar z)}. \tag{99}$$

Now suppose the boundary Γ is part of a straight line, and select the origin and orientation of the z-plane so that this line is the Ox-axis. Then a function $\tau(z) = \Omega(x, y) + i\theta(x, y)$ initially defined in some region bounded by Γ and some curve lying in the upper half-plane can be continued to the lower half-plane across Γ by either of

$$\tau(z) = \overline{\tau(\bar z)}, \quad \tau(z) = -\overline{\tau(\bar z)}. \tag{100}$$

We use the first of these when θ vanishes on Γ, and the second when it is Ω that vanishes, then these continuations are analytic. The proof is similar to that given for circular arcs, so we shall omit it.

3.16   Solution of the fundamental problems for an infinite strip

Let $\tau(z)$ be an analytic function, regular in the infinite strip $-\infty < x < \infty$, $0 < y < h$, whose real part $\Omega(x, y)$ has the boundary values

$$\Omega_0(x) = \lim_{\epsilon \to 0} \Omega(x, \epsilon), \quad \Omega_h(x) = \lim_{\epsilon \to 0} \Omega(x, h - \epsilon) \quad (\epsilon > 0). \tag{101}$$

† See footnote on p. 76.

These functions will be assumed to satisfy the modified Hölder conditions (§ 3.13). Also let the imaginary part of $\tau(z)$, $\theta(x, y)$, have finite limits θ_∞, $\theta_{-\infty}$ defined by

$$\theta_\infty = \lim_{x \to \infty} \theta(x, y), \quad \theta_{-\infty} = \lim_{x \to -\infty} \theta(x, y). \tag{102}$$

We shall now show that $\tau(z)$ is related to these boundary values by

$$\tau(z) = \frac{1}{2hi} \int_{-\infty}^{\infty} \left\{ \Omega_0(x) \coth \frac{\pi}{2h}(x - z) - \Omega_h(x) \tanh \frac{\pi}{2h}(x - z) \right\} dx$$
$$+ \tfrac{1}{2} i(\theta_\infty + \theta_{-\infty}). \tag{103}$$

Proof. Let z be a point lying within the rectangle $-R \leqslant x \leqslant R$, $0 \leqslant y \leqslant h$ (see Fig. 3.16), then the points of affix $z - 2mih$, $m = \pm 1, \pm 2, \pm 3, \ldots$, all lie outside this rectangle. It therefore follows from Cauchy's *theorem* when $m \neq 0$, and Cauchy's *integral* when $m = 0$ that

$$\begin{aligned}(m = 0) \, \tau(z) \\ (m \neq 0) \quad 0\end{aligned} \Bigg\} = \frac{1}{2\pi i} \int_{-R}^{R} \left\{ \frac{\tau_0(x)}{x + 2mih - z} - \frac{\tau_h(x)}{x + (2m+1)ih - z} \right\} dx$$
$$+ \frac{1}{2\pi} \int_0^h \left\{ \frac{\tau(R, y)}{R + 2mih + iy - z} - \frac{\tau(-R, y)}{-R + 2mih + iy - z} \right\} dy, \tag{104}$$

where $\tau_0(x)$ and $\tau_h(x)$ are defined by equations similar to (101). Adding (104) from $m = -M$ to $m = M$ we find that

$$\tau(z) = \frac{1}{2\pi i} \int_{-R}^{R} \sum_{-M}^{M} \left\{ \frac{\tau_0(x)}{x + 2mih - z} - \frac{\tau_h(x)}{x + (2m+1)ih - z} \right\} dx$$
$$+ \frac{1}{2\pi} \int_0^h \sum_{-M}^{M} \left\{ \frac{\tau(R, y)}{R + 2mih + iy - z} - \frac{\tau(-R, y)}{-R + 2mih + iy - z} \right\} dy. \tag{105}$$

Now it is well known that (see Whittaker & Watson, 1952, p. 136)

$$\lim_{M \to \infty} \sum_{m=-M}^{M} \left(\frac{1}{a + mib} \right) = \frac{\pi}{b} \coth \left(\frac{\pi a}{b} \right), \tag{106}$$

so taking this limit in (105) we obtain

$$\tau(z) = \frac{1}{4ih} \int_{-R}^{R} \left\{ \tau_0(x) \coth \frac{\pi}{2h}(x - z) - \tau_h(x) \coth \frac{\pi}{2h}(x + ih - z) \right\} dx$$
$$+ \frac{1}{4h} \int_0^h \left\{ \tau(R, y) \coth \frac{\pi}{2h}(R + iy - z) \right.$$
$$\left. - \tau(-R, y) \coth \frac{\pi}{2h}(-R + iy - z) \right\} dy.$$

Now let $R \to \infty$. Then

$$\tau(z) = \frac{1}{4ih} \int_{-\infty}^{\infty} \left\{ \tau_0(x) \coth \frac{\pi}{2h}(x-z) - \tau_h(x) \tanh \frac{\pi}{2h}(x-z) \right\} dx$$
$$+ \tfrac{1}{4}(\tau_\infty + \tau_{-\infty}), \quad (107)$$

where τ_∞ and $\tau_{-\infty}$ are defined by equations similar to (102).†

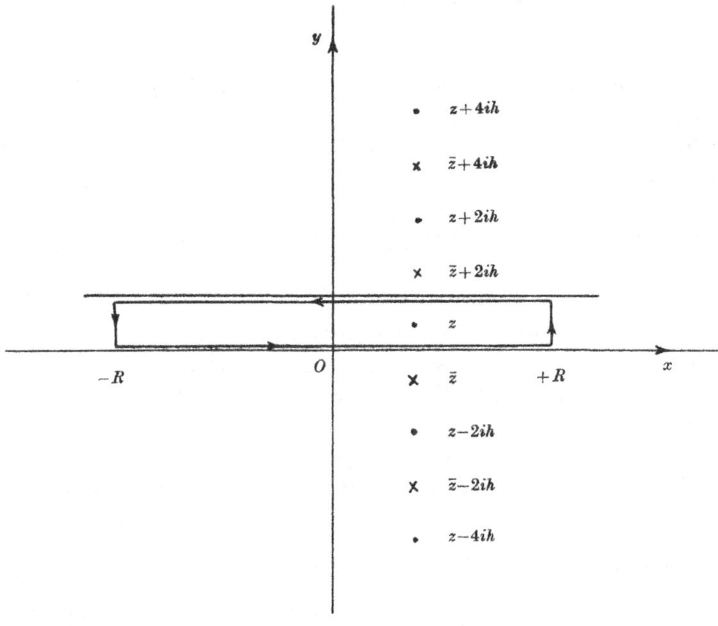

Fig. 3.16

If the same argument is developed for the conjugate point $\bar{z}$ instead of for z, an equation similar to (107) is obtained, but with z replaced by $\bar{z}$, and, since the points of affix $\bar{z} - 2mih$, $m = 0, \pm 1, \pm 2, \dots$ *all* lie outside

† In writing (107) we have tacitly assumed that $\lim\limits_{R\to\infty} \tau(\pm R, y)$ exist and are independent of y. This is certainly the case in almost all of the applications in which we shall be interested. However, exceptions do arise and to allow for these we should define $\tau_\infty, \tau_{-\infty}$ to be

$$\tau_{\pm\infty} = \lim_{R\to\infty} \frac{1}{h} \int_0^h \tau(\pm R, y)\, dy. \qquad (A)$$

This leaves the form of (107) unchanged.

For example let the boundary values $\theta(R, 0)$, $\theta(R, y)$ have the constant values $\theta_{\infty 0}$, $\theta_{\infty h}$ near infinity, then as θ is harmonic it must be given by $\theta(R, y) = \theta_{\infty 0} + (y/h)(\theta_{\infty h} - \theta_{\infty 0})$. Therefore (A) gives

$$\theta_\infty = \tfrac{1}{2}(\theta_{\infty 0} + \theta_{\infty h}), \qquad (B)$$

as the appropriate limit value.

the rectangle $-R \leqslant x \leqslant R, 0 \leqslant y \leqslant h$ (see Fig. 3.16) with the left-hand side of the equation replaced by zero. Hence

$$0 = \frac{1}{4ih} \int_{-\infty}^{\infty} \left\{ \tau_0(x) \coth \frac{\pi}{2h}(x - \bar{z}) - \tau_h(x) \tanh \frac{\pi}{2h}(x - \bar{z}) \right\} dx + \tfrac{1}{4}(\tau_\infty + \tau_{-\infty}).$$

The complex conjugate to the last equation is

$$0 = \frac{-1}{4ih} \int_{-\infty}^{\infty} \left\{ \bar{\tau}_0(x) \coth \frac{\pi}{2h}(x - z) - \bar{\tau}_h(x) \tanh \frac{\pi}{2h}(x - z) \right\} dx + \tfrac{1}{4}(\bar{\tau}_\infty + \bar{\tau}_{-\infty}). \tag{108}$$

It now remains only to subtract (108) from (107), and to use the results $\tau + \bar{\tau} = 2\Omega$, $\tau - \bar{\tau} = 2i\theta$, in order to arrive at the result we wish to establish, viz. (103). Addition of (107) and (108) yields the harmonic conjugate to (103), namely

$$\tau(z) = \frac{1}{2h} \int_{-\infty}^{\infty} \left\{ \theta_0(x) \coth \frac{\pi}{2h}(x - z) - \theta_h(x) \tanh \frac{\pi}{2h}(x - z) \right\} dx$$
$$+ \tfrac{1}{2}(\Omega_\infty + \Omega_{-\infty}). \tag{109}$$

As in §3.14 the real part of (103) gives the solution to the Dirichlet problem, whereas the imaginary part, together with integration by parts, yields the solution to the Neumann problem. These straightforward calculations are left to the reader.

Now consider the limit $z \to x_0$, where x_0 is a point on the boundary $y = 0$. We shall apply this limit to (109), which can be written

$$\tau(z) = \frac{1}{\pi} \int_{-\infty}^{\infty} \frac{\theta_0(x)\,dx}{x - z} + \frac{1}{2h} \int_{-\infty}^{\infty} \theta_0(x) \left\{ \coth \frac{\pi}{2h}(x - z) - \frac{2h}{\pi} \frac{1}{(x - z)} \right\} dx$$
$$- \frac{1}{2h} \int_{-\infty}^{\infty} \theta_h(x) \tanh \frac{\pi}{2h}(x - z)\,dx + \tfrac{1}{2}(\Omega_\infty + \Omega_{-\infty}).$$

Each of the last two integrals in this equation exists in the ordinary sense in the limit $z \to x_0$, for it is easily verified that

$$\coth \frac{\pi}{2h}(x - z) - \frac{2h}{\pi} \frac{1}{(x - z)} = \frac{\pi}{6h}(x - z) + O(x - z)^3,$$

while the first integral tends to the limit (see §3.12)

$$i\theta_0(x_0) + \frac{1}{\pi} \int_{-\infty}^{\infty} \frac{\theta_0(x)\,dx}{x - x_0}.$$

The term $i\theta_0(x_0)$ cancels out with the imaginary part of $\tau(x_0)$ so we are left with

$$\Omega(x_0) = \frac{1}{2h} \int_{-\infty}^{\infty} \left\{ \theta_0(x) \coth \frac{\pi}{2h}(x - x_0) - \theta_h(x) \tanh \frac{\pi}{2h}(x - x_0) \right\} dx$$
$$+ \tfrac{1}{2}(\Omega_\infty + \Omega_{-\infty}), \tag{110}$$

A similar result holds on $z = x_0 + ih$.

3.17   The two-sided character of the point at infinity

It may puzzle the reader that we have been able to distinguish *two* values for Ω at infinity in (109) and (110) (and similarly for θ in equation (103)). Certainly there are *not* two points at infinity, even though in an infinite strip this appears to be so. The reader should think of the strip as a lune just encompassing the surface of a sphere (§ 3.3.), so that its points just touch at the antipode of the origin (Fig. 3.17) at S. S is the point at infinity. It will be noticed that S does not lie *within* the area of the lune, but is a boundary point which acts as a *barrier* and prevents the existence of contours within the strip which completely encircle the sphere. If such contours were possible the region would not be singly connected. As S is a barrier it is necessary to distinguish between the two 'sides' of S, sides which we have labelled S_{∞} and $S_{-\infty}$. Then to traverse the boundary of our region in the positive direction we must travel along $OS_{\infty}BS_{-\infty}O$ (see Fig. 3.17).

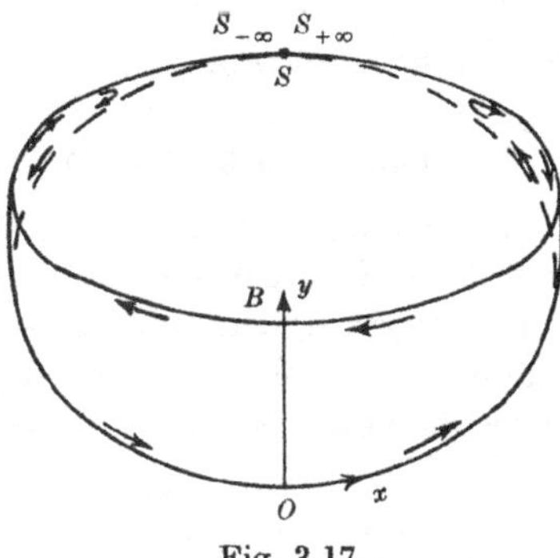

Fig. 3.17

Summarizing, we have that in the case of the infinite strip the point at infinity, S, is a (two-sided) barrier point, and therefore any function analytic within the strip can have *two* distinct values at S, depending on which *side* of S is approached. Thus, despite the fact there is only one point at infinity, this point does have the characteristics of two points, and we shall usually talk of the point 'upstream at infinity' ($S_{-\infty}$) and the 'point downstream at infinity' (S_{∞}) as if they were distinct.

We can now consider the effect on (109) of taking the limits $\mathscr{R}z \to \pm\infty$. As $S_{\pm\infty}$ are boundary points, it is simplest to move first to a point on the boundary, e.g. take the limit $z \to x_0$, then to take the limits $x_0 \to \pm\infty$, for otherwise both limiting processes occur at once and complicate the calculation (cf. derivation of (66)). Equation (110) is the result of taking the limit $z \to x_0$. In the limits $x_0 \to \pm\infty$, (110) yields (formally)

$$\Omega_{\pm\infty} = \mp \frac{1}{2h} \int_{-\infty}^{\infty} \{\theta_0(x) - \theta_h(x)\}\, dx + \tfrac{1}{2}(\Omega_{\infty} + \Omega_{-\infty}),$$

i.e.
$$\Omega_{\infty} - \Omega_{-\infty} = -\frac{1}{h} \int_{-\infty}^{\infty} \{\theta_0(x) - \theta_h(x)\}\, dx. \tag{111}$$

This step can be justified by writing

$$\coth \frac{\pi}{2h}(x - x_0) = \left\{ \coth \frac{\pi}{2h}(x - x_0) - \frac{2h}{\pi(x - x_0)} \right\} + \frac{2h}{\pi(x - x_0)},$$

and then using the second of (67) to deal with the singular term. Thus the boundary values Ω_∞, $\Omega_{-\infty}$, $\theta_0(x)$, $\theta_h(x)$ cannot all be assigned arbitrarily, but must satisfy the condition expressed in (111), e.g. if we prescribe $\Omega_{-\infty}$, $\theta_0(x)$ and $\theta_h(x)$ freely, (111) fixes the value of Ω_∞.

Similarly, we can deduce from (102) that

$$\theta_\infty - \theta_{-\infty} = \frac{1}{h}\int_{-\infty}^{\infty} \{\Omega_0(x) - \Omega_h(x)\}\,dx, \tag{112}$$

which is the harmonic conjugate of (111).

3.18   Basic solution for the upper half-plane

The basic solution for the upper half-plane can be derived from (69), but it is of interest to derive it from the results given above for the infinite strip by taking the limit $h \to \infty$. In this limit the two-sided character of the point at infinity clearly disappears, and so we must have $\tau_{-\infty} \to \tau_\infty$—a result which also follows from (111) and (112). As

$$\lim_{h\to\infty}\frac{\pi}{2h}\coth\frac{\pi}{2h}(x-z) = \frac{1}{x-z}, \quad \lim_{h\to\infty}\frac{\pi}{2h}\tanh\frac{\pi}{2h}(x-z) = 0,$$

(109) yields in this limit

$$\tau(z) = \frac{1}{\pi}\int_{-\infty}^{\infty}\frac{\theta_0(x)\,dx}{x-z} + \Omega_\infty. \tag{113}$$

On $y = 0$ at $z = x_0$ (113) gives †

$$\Omega(x_0) = \frac{1}{\pi}\int_{-\infty}^{\infty}\frac{\theta_0(x)\,dx}{x-x_0} + \Omega_\infty. \tag{114}$$

GREEN'S FUNCTION METHODS

3.19   The complex variable form of Green's formulae

The method of finding solutions to the fundamental problems illustrated in §§ 3.14 and 3.16 (also see § 4.9) clearly depends on the fact that the *shapes* of the contours enclosing the domains in question are simple enough to permit the use of image points. Solutions for regions defined

† We shall make no further reference to the limiting process $z \to t_0$ discussed at some length in § 3.7, and which yields the Plemelj's formulae. By the type of argument given on p. 102 the reader can easily check that the following rule always succeeds.

' The limiting form taken by the solution of a fundamental problem of potential theory (expressed in its complex form) on the boundary contour can be obtained by putting $z = t_0$ in the solution—which will thus be rendered either wholly real or wholly imaginary —and then discarding either $\Omega(t_0)$ or $i\theta(t_0)$ from the left-hand side, whichever makes the equation completely real or completely imaginary.'

by contours of more complicated shape can be determined by a combination of the basic solutions already found and the conformal mapping technique to be outlined in ch. 5. There is, however, an alternative method of dealing with contours of general shape, which is of considerable importance. This is the well-known Green's function method, a method which is applicable both to more general forms of elliptic differential equations than Laplace's equation, and also to domains in 3, 4 or more dimensioned spaces. Clearly there will be a close relation between the two-dimensional Green's function for a given contour C, and a harmonic function θ, and the conformal mapping method applied to C and θ. This connexion is established in § 5.17.

The Green's function method is usually expressed in real variable theory—good accounts occur in many texts—but we shall depart from the usual treatment by giving a complex variable form to the theory, a form which has some advantages from the point of view of this book. Nehari (1952) and Bergman & Schiffer (1953) have also given accounts of the complex variable treatment of Green's function methods.

First observe that the equations

$$x = \tfrac{1}{2}(z+\bar{z}), \quad y = \frac{1}{2i}(z-\bar{z}), \tag{115}$$

permit any function $u(x,y)$ to be expressed as a function of z and $\bar{z}$, i.e. $u = u(z,\bar{z})$. By (115) the partial derivative operators transform according to

$$\left.\begin{aligned}
\frac{\partial}{\partial x} &= \frac{\partial}{\partial z}+\frac{\partial}{\partial \bar{z}}, \quad \frac{\partial}{\partial y} = i\frac{\partial}{\partial z}-i\frac{\partial}{\partial \bar{z}}, \\
\frac{\partial}{\partial z} &= \frac{1}{2}\left(\frac{\partial}{\partial x}-i\frac{\partial}{\partial y}\right), \quad \frac{\partial}{\partial \bar{z}} = \frac{1}{2}\left(\frac{\partial}{\partial x}+i\frac{\partial}{\partial y}\right).
\end{aligned}\right\} \tag{116}$$

We shall denote partial derivatives by subscripts, e.g. $u_x = \partial u/\partial x$, etc., then by (116)

$$u_x = u_z + u_{\bar{z}}, \quad u_{xx} = (u_z + u_{\bar{z}})_z + (u_z + u_{\bar{z}})_{\bar{z}} = u_{zz} + 2u_{z\bar{z}} + u_{\bar{z}\bar{z}},$$

$$u_y = iu_z - iu_{\bar{z}}, \quad u_{yy} = i(iu_z - iu_{\bar{z}})_z - i(iu_z - iu_{\bar{z}})_{\bar{z}} = 2u_{z\bar{z}} - u_{zz} - u_{\bar{z}\bar{z}},$$

so
$$\nabla^2 u = u_{xx} + u_{yy} = 4u_{z\bar{z}}. \tag{117}$$

Thus the transformation (115) reduces Laplace's equation to

$$u_{z\bar{z}} = 0, \tag{118}$$

the general solution of which is

$$u(z,\bar{z}) = f_1(z) + f_2(\bar{z}), \tag{119}$$

where f_1, f_2 are arbitrary functions.

It is often convenient to replace the function notation $u(z,\bar{z})$ by $u(\bar{z})$, which thus means that u is a function of the *position* of the point of affix z, and not necessarily a function of the complex variable z (or $\bar{z}$) alone (see also first paragraph of § 3.5). With this notation it is apparent from (119) that for a *harmonic* function $u(\tilde{z})$

$$\frac{\partial u(\tilde{z})}{\partial z} = f_3(z), \quad \frac{\partial u(\tilde{z})}{\partial \bar{z}} = f_4(\bar{z}), \tag{120}$$

where f_3 and f_4 are *analytic* functions of z and $\bar{z}$.

Let $f(x, y)$, $g(x, y)$ be functions such that f, g, f_y, g_x are continuous over a domain D bounded by a contour C, then the two-dimensional form of Green's theorem (see, for example, Nehari, 1952, p. 240) can be expressed

$$\int_C f\,dx = -\int_D f_y\,dA, \quad \int_C g\,dy = \int_D g_x\,dA,$$

where dA is an element of area in D. Let $g = if$, and then add the equations to get

$$\int_C f(dx + i\,dy) = \int_C f\,dz = -\int_D (f_y - if_x)\,dA,$$

or regarding f as a function of z and $\bar{z}$, and using (116)

$$\int_C f(\tilde{z})\,dz = 2i\int_D f_{\bar{z}}(\tilde{z})\,dA. \tag{121}$$

Similarly, by subtracting the equations we get

$$\int_C f(\tilde{z})\,d\bar{z} = -2i\int_D f_z(\tilde{z})\,dA. \tag{122}$$

Now let $u(\tilde{z})$, $v(\tilde{z})$ be two real functions of z and $\bar{z}$, sufficiently differentiable for the following argument. Set $f = uv_z$ in (121), and $f = vu_{\bar{z}}$ in (122), then subtract the resulting equations to find

$$2i\int_D (uv_{z\bar{z}} - vu_{z\bar{z}})\,dA = \int_C (uv_z\,dz + vu_{\bar{z}}\,d\bar{z}), \tag{123}$$

a result which is generally known as 'Green's second formula'.

Employing the notation of § 3.13—$dx/ds = \cos\chi$, $dy/ds = \sin\chi$—and (116), we have

$$2v_z\,dz = (v_x - iv_y)(dx + i\,dy) = v_x\,dx + v_y\,dy - i\left(v_y\frac{dx}{ds} - v_x\frac{dy}{ds}\right)ds$$

$$= dv - i(v_y\cos\chi - v_x\sin\chi)\,ds,$$

i.e.
$$2v_z\,dz = dv - i\left(\frac{\partial v}{\partial n}\right)ds, \tag{124}$$

where $(\partial v/\partial n)$ is the inward normal derivative (see Fig. 3.13), and we have made use of (73).

Similarly,

$$2u_z\,d\bar{z} = du + i\left(\frac{\partial u}{\partial n}\right)ds. \tag{125}$$

With (117), (124) and (125) we can write (123) in the form

$$\int_D (u\,\nabla^2 v - v\,\nabla^2 u)\,dA = \int_C \left(v\frac{\partial u}{\partial n} - u\frac{\partial v}{\partial n}\right)ds - i\int_C d(uv),$$

the last term of which vanishes, leaving us with the usual form of Green's second formula.

3.20 Green's function

The Green's function $g(\tilde{z}, \tilde{\zeta})$ of a domain D with respect to a fixed point $P(\zeta)$ is defined by the conditions: (i) within C, the boundary of D, g is of the form

$$g(\tilde{z}, \tilde{\zeta}) = -\ln|z - \zeta| + f(\tilde{z}, \tilde{\zeta}), \tag{126}$$

where f is harmonic at all points of D; and (ii) g tends to zero as $P(z)$ tends to any point $Q(t)$ of C, i.e.

$$g(\tilde{t}, \tilde{\zeta}) = 0. \tag{127}$$

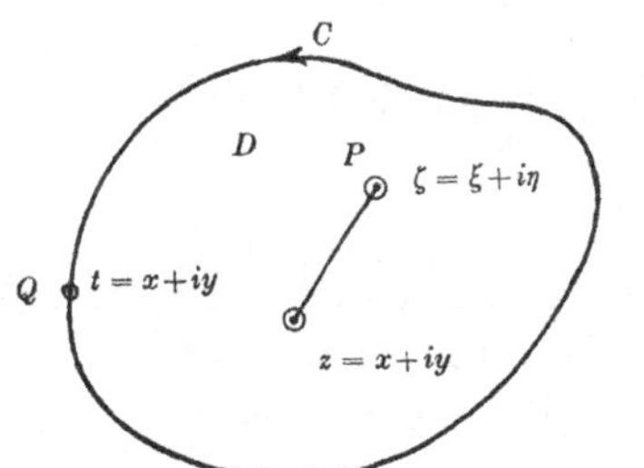

Fig. 3.20a

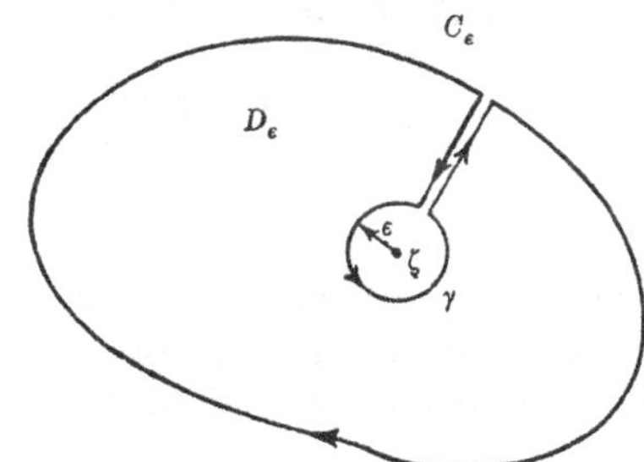

Fig. 3.20b

Writing

$$\ln|z - \zeta| = \tfrac{1}{2}\ln(z - \zeta)(\bar{z} - \bar{\zeta}),$$

we see that

$$2\frac{\partial}{\partial z}\ln|z - \zeta| = \frac{1}{z - \zeta}, \tag{128}$$

and

$$4\frac{\partial^2}{\partial z\,\partial\bar{z}}\ln|z - \zeta| = 0.$$

Therefore by differentiating (126) we find that $g(\tilde{z}, \tilde{\zeta})$ is harmonic, except at $z = \zeta$.

Now consider the indented contour C_ϵ shown in Fig. 3.20b, which excludes the singularity in g at ζ by a small circle γ of radius ϵ. At all points of the remaining domain D_ϵ g is harmonic, i.e. $g_{z\bar{z}} = 0$. Identifying the function v in (123) with g we get

$$\int_{C_\epsilon} (ug_z\,dz + gu_z\,d\bar{z}) = -2i\int_{D_\epsilon} gu_{z\bar{z}}\,dA.$$

The contributions to $\int_{C_\epsilon}$ from the straight portions of the indentation connecting C and γ cancel out, and so $\int_{C_\epsilon} = \int_C - \int_\gamma$ (taking γ in the positive sense). Thus by (126), (127) and (128) we find that

$$\int_\gamma u\left(-\frac{1}{2}\frac{1}{z-\zeta}+f_z\right)dz + u_{\bar z}(-\ln|z-\zeta|+f)\,d\bar z = \int_C ug_t\,dt + 2i\int_{D_\epsilon} gu_{z\bar z}\,dA.$$
$$(129)$$

Put $z-\zeta = \epsilon\,e^{i\delta}$, then the left-hand side of this equation reads

$$-\frac{i}{2}\int_0^{2\pi} u\,d\delta + i\epsilon\int_0^{2\pi}\{uf_z e^{i\delta} - (-\ln\epsilon + f)\,u_{\bar z}e^{-i\delta}\}\,d\delta,$$

and as u, f and their derivatives are continuous at $z = \zeta$, it follows that this expression tends to $-\pi i u(\zeta)$ in the limit $\epsilon \to 0$. Hence (129) becomes

$$u(\zeta) = \frac{i}{\pi}\int_C ug_t\,dt - \frac{2}{\pi}\int_D gu_{z\bar z}\,dA,\qquad(130)$$

when $\epsilon \to 0$. As g vanishes on C, $2g_t\,dt = -i\,(\partial g/\partial n)\,ds$, by (124) so

$$u(\zeta) = \frac{1}{2\pi}\int_C u(t)\frac{\partial g}{\partial n}(t, \zeta)\,ds - \frac{1}{2\pi}\int_D g(\bar z, \zeta)\,h(\bar z)\,dA,\qquad(131)$$

where
$$4u_{z\bar z} = \nabla^2 u = h(\bar z).\qquad(132)$$

If $h(\bar z)$ is a known function (132) is called *Poisson's* equation, and (131) is thus the solution of Poisson's equation for a given domain D.

In particular if u is harmonic, $h = 0$, and (131) becomes

$$u(\zeta) = \frac{1}{2\pi}\int_C u(t)\frac{\partial g}{\partial n}(t, \zeta)\,ds,\qquad(133)$$

or
$$u(\zeta) = \frac{i}{\pi}\int_C u(t)\frac{\partial g}{\partial t}(t, \zeta)\,dt.\qquad(134)$$

Thus knowledge of the Green's function for a domain enables us to solve the Dirichlet problem for that domain.

Equation (134) is a generalized form of several equations derived earlier in this chapter. For example if D is the semi-infinite strip $-\infty < x < \infty$, $0 \leqslant y \leqslant h$, then by comparing (134) and (107) we conclude that

$$\frac{\partial g}{\partial x}(x, \bar z) = \frac{\pi}{2hi}\,\mathscr{I}\left\{\coth\frac{\pi}{2h}(x-z)\right\}.$$

It is not easy to infer the form of g from this. An explicit formula for g for domains of general shape is given in §5.17.

The second fundamental problem of potential theory—the Neumann problem—can be solved in a similar way with the aid of the *Neumann* function $N(\tilde{z}, \tilde{\zeta})$ of the domain D. This function is similar to Green's function except that the condition given in equation (127) is replaced by

$$\frac{\partial N}{\partial n}(\tilde{t}, \tilde{\zeta})\,ds = k\,ds, \quad \int_C N(\tilde{t}, \tilde{\zeta})\,ds = 0, \tag{135}$$

where k is a constant. The second condition in (135) is added in order to define N uniquely. In view of the fact that the second fundamental problem can always be converted into the form of the first fundamental problem (§ 3.13) we do not need to consider the Neumann function any further.

CHAPTER 4

MIXED AND PERIODIC BOUNDARY CONDITIONS

4.1 The various types of mixed boundary conditions

There are many types of mixed boundary conditions of importance in fluid dynamics. We shall not consider them all, but confine our attention to a few of the most important.

Let C be a contour comprised of $2N$ ($N = 1, 2, \ldots$) continuous sections $C_1, C_2, \ldots, C_{2N}$, i.e. $C = \sum_{n=1}^{2N} C_n$, and let $\tau(z)\,(= \Omega + i\theta)$ be analytic within C. Denote by $g(t)$ and $(h)t$ known functions on C, and let $\Omega(t)$, $\theta(t)$ denote the limit values $\Omega^+(t)$, $\theta^+(t')$ (see §3.5) on C. Then in order of difficulty the mixed boundary conditions in which we shall be interested are:

(a) *Simply mixed*. This is the case

$$\left.\begin{aligned}
\Omega(t) &= h(t) \quad \text{for} \quad t \in \sum_{r=0}^{N-1} C_{2r+1}, \\
\theta(t) &= g(t) \quad \text{for} \quad t \in \sum_{r=1}^{N} C_{2r},
\end{aligned}\right\} \tag{1}$$

where $N = 1$.

(b) *Doubly mixed*. The boundary conditions are given by (1) with $N = 2$.

(c) *Periodic*. Suppose that as a point $P(t_r)$ moves over C_r a corresponding point $P(t_s)$ moves over $C_s(r \neq s)$, the correspondence between the points being one to one, then we shall describe the boundary conditions as being 'periodic' if

$$\left.\begin{aligned}
\theta(t) &= g(t) \quad \text{for} \quad t \in \sum_{\substack{n=1 \\ n\neq r,s}}^{2N} C_n, \\
\Omega(t_r) - \Omega(t_s) &= h(t), \quad \theta(t_r) = \theta(t_s).
\end{aligned}\right\} \tag{2}$$

(The roles of Ω and θ can of course be interchanged here to yield the conjugate problem.)

The word 'periodic' is used here for want of a better term. It is appropriate in the particular and important case in which C_s and C_r are identical in shape, and $t_r = t_s + a$, where a is a constant. For in this case the periodic relation

$$\tau(z) = \tau(z + na) \quad (n = \pm 1, \pm 2, \ldots (z \in D_0)), \tag{3}$$

clearly provides a *sectional* analytic continuation (see §3.14) of τ into an infinite domain $D_\infty = \sum\limits_{-\infty}^{\infty} D_i$ (see Fig. 4.1b). If $h = 0$ in (2) τ is continuous in D_∞ and the original boundary conditions on C can be replaced by periodic boundary conditions on the boundary of D_∞.

(d) *Simply mixed and periodic.* In this case C is divided into four intervals, and (1) applied on one pair of intervals, and (2) applied on the other pair.

(e) *Multiply mixed.* The boundary conditions are given by (1) with no restriction on N. With $N > 2$ untabulated functions (hyperelliptic integrals) are generally unavoidable in the theory.

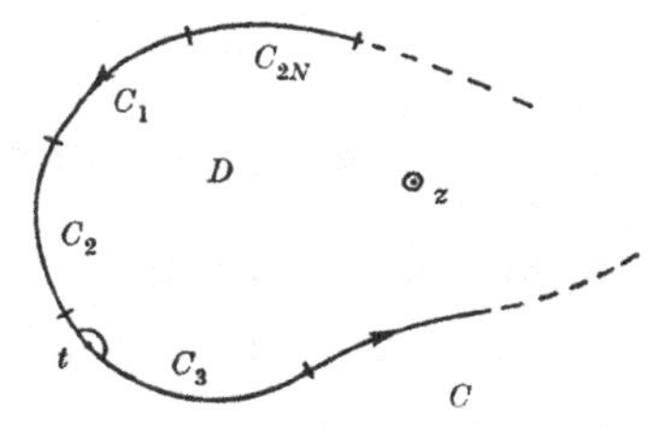

Fig. 4.1a

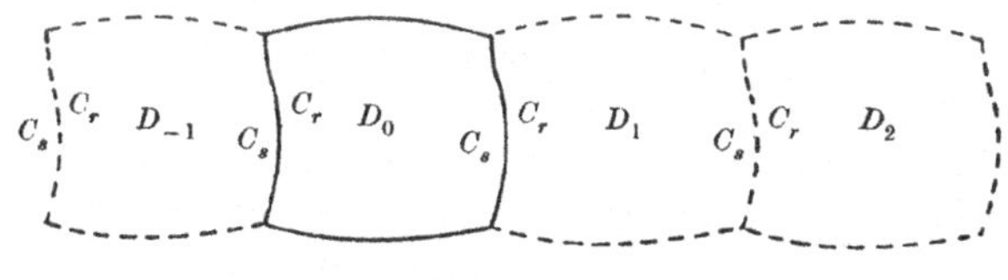

Fig. 4.1b

(f) *Continuously mixed.* The boundary condition on C is of the form

$$\theta(t) = \lambda(t)\,\Omega(t) + g(t), \tag{4}$$

where $\lambda(t)$ and $g(t)$ are known functions. When $\lambda = 0$, $\theta = g(t)$, and when $\lambda = \infty$, $\Omega = \lim\limits_{\lambda \to \infty} (\theta - g)/\lambda = h(t)$, say. Thus (4) yields the boundary conditions of (1) if $\lambda = 0$ for $t \in \sum\limits_{r=1}^{N} C_{2r}$, and $\lambda = \infty$ for $t \in \sum\limits_{r=0}^{N-1} C_{2r+1}$. Equation (4) may be termed the *Riemann–Hilbert* boundary condition (see Musk-helishvili, 1953, p. 99).

(g) *Continuously mixed and periodic.* Here C is divided into four intervals with (2) applying on one pair and (4) on the other pair.

(h) *Relation between a function and its derivatives.* Let s be distance measured around C, and n distance measured along the inwards normal to C, then a type of relation frequently met in fluid dynamics (e.g. in problems involving surface waves) can be expressed in the form

$$A(t)\frac{\partial\theta}{\partial n}+B(t)\frac{\partial\theta}{\partial s}+c(t)\,\theta = g(t), \tag{5}$$

where A, B, c and g are known functions of t. The problem defined by (5) is often called the *Poincaré* Problem, as it was investigated by Poincaré (1910) in a study of the theory of tides.

In the remainder of this chapter we shall consider the problems defined by the boundary conditions listed above in some detail.

INFINITE AND SEMI-INFINITE STRIPS

4.2   Simply mixed boundary conditions

With simply mixed boundary conditions C is divided into just two intervals with $\Omega(t)$ prescribed on one and $\theta(t)$ on the other. The first step towards the solution of the problem so defined is to map the interior of C into the infinite strip $-\infty < x < \infty$, $0 < y < h$ (see theory given in §5.5) in such a way that the interval in which θ is known is mapped on to $y = 0$, $-\infty < x < \infty$, and the interval in which Ω is known on to $y = h$, $-\infty < x < \infty$.

The calculation now proceeds along almost identical lines to that given in §3.16, the only difference being that before (3.104) are added together over $n = -N$ to $n = N$, they are multiplied by the factor $(-1)^n$. Thus in place of the limit in (3.106) we have to use (see Whittaker & Watson, 1952, p. 136)

$$\lim_{N\to\infty}\sum_{n=-N}^{N}\left(\frac{(-1)^n}{a+nib}\right) = \frac{\pi}{b}\operatorname{cosech}\frac{\pi a}{b}, \tag{6}$$

and consequently instead of (3.107) we arrive at

$$\tau(z) = \frac{1}{4ih}\int_{-\infty}^{\infty}\left\{\tau_0(x)\operatorname{cosech}\frac{\pi}{2h}(x-z) - \tau_h(x)\operatorname{cosech}\frac{\pi}{2h}(x+ih-z)\right\}dx,$$

$$\text{or}\quad \tau(z) = \frac{1}{4ih}\int_{-\infty}^{\infty}\left\{\tau_0(x)\operatorname{cosech}\frac{\pi}{2h}(x-z) + i\tau_h(x)\operatorname{sech}\frac{\pi}{2h}(x-z)\right\}dx. \tag{7}$$

The same type of argument by which (3.108) was derived in the present case yields

$$0 = \frac{-1}{4ih}\int_{-\infty}^{\infty}\left\{\bar{\tau}_0(x)\operatorname{cosech}\frac{\pi}{2h}(x-z) - i\bar{\tau}_h(x)\operatorname{sech}\frac{\pi}{2h}(x-z)\right\}dx. \tag{8}$$

The addition of (7) and (8) gives

$$\tau(z) = \frac{1}{2h} \int_{-\infty}^{\infty} \left\{ \theta_0(x) \operatorname{cosech} \frac{\pi}{2h}(x-z) + \Omega_h \operatorname{sech} \frac{\pi}{2h}(x-z) \right\} dx, \qquad (9)$$

as $\bar{\tau} + \tau = 2\Omega$, and $\tau - \bar{\tau} = 2i\theta$. This is the solution to the simply mixed boundary value problem defined above. On $y = 0$ (9) becomes

$$\Omega(x_0) = \frac{1}{2h} \int_{-\infty}^{\infty} \left\{ \theta_0(x) \operatorname{cosech} \frac{\pi}{2h}(x-x_0) + \Omega_h(x) \operatorname{sech} \frac{\pi}{2h}(x-x_0) \right\} dx, \quad (10)$$

while on $y = h$ it becomes

$$\theta(x_0) = \frac{1}{2h} \int_{-\infty}^{\infty} \left\{ \theta_0(x) \operatorname{sech} \frac{\pi}{2h}(x-x_0) + \Omega_h(x) \operatorname{cosech} \frac{\pi}{2h}(x-x_0) \right\} dx. \quad (11)$$

At first sight these equations appear to be incomplete, for there is no term corresponding to $\frac{1}{2}(\Omega_\infty + \Omega_{-\infty})$, which occurs in (3.109). Furthermore, in the limits $x_0 \to \pm\infty$, (10) and (11) apparently give $\theta_{\pm\infty} = 0$, $\Omega_{\pm\infty} = 0$, but these particular values are nowhere demanded in the derivation of the equations in question. The fact is that $\theta_{\pm\infty}$, $\Omega_{\pm\infty}$ are *boundary* values, which must be prescribed before the integrals in (10) and (11) can be properly evaluated, and unless these values happen to be zero, it is not valid to interchange $\lim_{x_0\to\infty}$ and $\int_{-\infty}^{\infty}$ in these equations.

Suppose, for example, $\Omega_h(x)$ has the constant value a over $(-\infty, \infty)$, then as

$$\int_{-\infty}^{\infty} \operatorname{sech} \frac{\pi}{2h}(x-z)\, dx = \frac{4h}{\pi} \left[\tan^{-1} \left\{ \tanh \frac{\pi}{4h}(x-z) \right\} \right]_{-\infty}^{\infty} = 2h,$$

(9) becomes

$$\tau(z) = a + \frac{1}{2h} \int_{-\infty}^{\infty} \theta_0(x) \operatorname{cosech} \frac{\pi}{2h}(x-z)\, dx. \qquad (12)$$

The limits $z \to x_0$, then $x_0 \to \pm\infty$, now correctly give $\Omega_{\pm\infty} = a$. As a further check notice that in the limit $h \to \infty$ (12) properly reduces to (3.113).

4.3 Periodic boundary conditions for the upper half-plane

First method : $\tau(z)$ continuous. Let $\theta_0(x)$ be prescribed on $y = 0$ so that it satisfies the periodic relation $\theta_0(x) = \theta_0(x + 2l)$, then by (3.113) we have that in the upper half-plane

$$\tau(z) = \frac{1}{\pi} \int_{-\infty}^{\infty} \frac{\theta_0(x)\, dx}{x-z} + \Omega_\infty = \frac{1}{\pi} \lim_{N\to\infty} \sum_{n=-N}^{N} \int_{(2n-1)l}^{(2n+1)l} \frac{\theta_0(x)\, dx}{x-z} + \Omega_\infty$$

$$= \frac{1}{\pi} \int_{-l}^{l} \theta_0(x) \lim_{N\to\infty} \sum_{n=-N}^{N} \left(\frac{1}{x+2nl-z} \right) dx + \Omega_\infty,$$

on making use of the periodic character of $\theta_0(x)$ and the fact that $\int_{-\infty}^{\infty}$ means $\lim_{R\to\infty}\int_{-R}^{R}$ here (see § 3.12). Now

$$\lim_{N\to\infty}\sum_{n=-N}^{N}\left(\frac{1}{x+2nl-z}\right)=\frac{\pi}{2l}\cot\frac{\pi}{2l}(x-z) \tag{13}$$

(cf. (3.106)), so our solution can be written

$$\tau(z)=\frac{1}{2l}\int_{-l}^{l}\theta_0(x)\cot\frac{\pi}{2l}(x-z)\,dx+\Omega_\infty. \tag{14}$$

This may be regarded as the solution to the problem: find an analytic function $\tau(z)$ in the semi-infinite strip $-l<x<l$, $0<y<\infty$, having a given imaginary part on $y=0$, and satisfying the condition $\tau(-l+iy)=\tau(l+iy)$. Equation (14) is fundamental in classical aerofoil theory.

Second method: $\tau(z)$ *discontinuous*. We shall now obtain an important generalization of (14), which proves to be fundamental in *unsteady* aerofoil theory. This is the case in which $\tau(-l+iy)$ is *not* equal to $\tau(l+iy)$. Instead we shall put

$$\theta^+(y)=\theta^-(y),\quad \omega(y)=\Omega^+(y)-\Omega^-(y), \tag{15}$$

where
$$\left.\begin{aligned}\Omega^+(y)+i\theta^+(y)=\tau^+(y)&\equiv\lim_{\substack{\epsilon\to 0\\(\epsilon>0)}}\tau(l-\epsilon,y),\\[2mm]\Omega^-(y)+i\theta^-(y)=\tau^-(y)&\equiv\lim_{\substack{\epsilon\to 0\\(\epsilon>0)}}\tau(-l+\epsilon,y),\end{aligned}\right\} \tag{16}$$

so that the real part of τ is discontinuous across $x=(2n+1)l$, $n=\pm 1,\pm 2,\ldots$. There is no way of modifying the derivation of (14) to deal with this case—we must calculate afresh using a method similar to that employed in §§ 3.16 and 4.2.

Let z be a point lying within the rectangle $-l<x<l$, $0<y<Y$, then the points $z-2nl$, $n=\pm 1,\pm 2,\ldots$, all lie outside this rectangle (see Fig. 4.3). Therefore, using Cauchy's theorem and integral and adding the results over $n=-N$ to $n=N$, just as in § 3.16, we arrive at

$$\tau(z)=\frac{1}{2\pi i}\int_{-l}^{l}\sum_{n=-N}^{N}\left(\frac{\tau_0(x)}{x+2nl-z}-\frac{\tau(x,Y)}{x+iY+2nl-z}\right)dx$$

$$+\frac{1}{2\pi i}\int_{0}^{Y}\sum_{n=-N}^{N}\left(\frac{\tau^+(y)}{iy+(2n+1)l-z}-\frac{\tau^-(y)}{iy+(2n-1)l-z}\right)i\,dy. \tag{17}$$

Now take the limit $N \to \infty$, and make use of (13) to obtain

$$\tau(z) = \frac{1}{4il} \int_{-l}^{l} \tau_0(x) \cot \frac{\pi}{2l}(x-z)\,dx - \frac{1}{4il} \int_{-l}^{l} \tau(x, Y) \cot \frac{\pi}{2l}(x+iY-z)\,dx$$

$$+ \frac{1}{4l} \int_0^Y \tau^+(y) \cot \frac{\pi}{2l}(iy+l-z)\,dy - \frac{1}{4l} \int_0^Y \tau^-(y) \cot \frac{\pi}{2l}(iy-l-z)\,dy.$$

Next we take the limit $Y \to \infty$, then as $\cot(a \pm \tfrac{1}{2}\pi) = -\tan a$, and $\cot(\pi/2l)(x+iY-z) = -i \coth(\pi/2l)(Y - ix + iz) \to -i$, our equation yields the limit form

$$\tau(z) = \tfrac{1}{2}\tau_\infty + \frac{1}{4il} \int_{-l}^{l} \tau_0(x) \cot \frac{\pi}{2l}(x-z)\,dx$$

$$- \frac{1}{4l} \int_0^\infty \{\tau^+(y) - \tau^-(y)\} \tan \frac{\pi}{2l}(iy-z)\,dy. \qquad (18)$$

From (15) it follows that $\tau^+(y) - \tau^-(y) = \varpi(y)$, which can be used to simplify the last integral of (18).

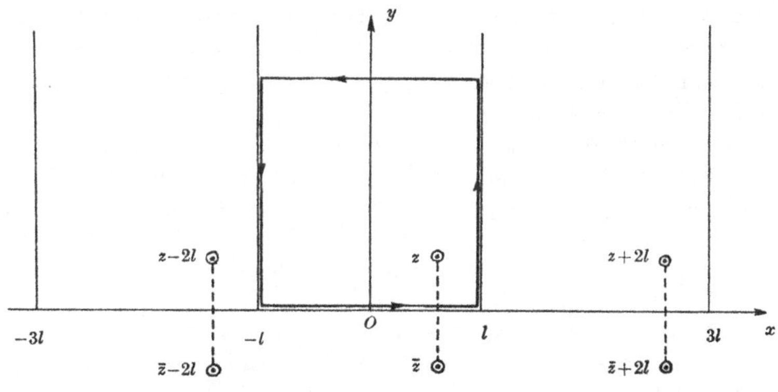

Fig. 4.3

The conjugate points $\bar{z} - 2nl$, $n = 0, \pm 1 \pm 2 \ldots$, all lie outside the basic semi-infinite rectangle (see Fig. 4.3), so that if z is replaced by $\bar{z}$ in the right-hand side of (18), the left-hand side vanishes. The complex conjugate to the equation just described is

$$0 = \tfrac{1}{2}\bar{\tau}_\infty - \frac{1}{4il} \int_{-l}^{l} \bar{\tau}_0(x) \cot \frac{\pi}{2l}(x-z)\,dx + \frac{1}{4l} \int_0^\infty \varpi(y) \tan \frac{\pi}{2l}(iy+z)\,dy.$$

Addition of this to (18) gives

$$\tau(z) = \Omega_\infty + \frac{1}{2l} \int_{-l}^{l} \theta_0(x) \cot \frac{\pi}{2l}(x-z)\,dx$$

$$+ \frac{1}{4l} \int_0^\infty \varpi(y) \left\{ \tan \frac{\pi}{2l}(iy+z) - \tan \frac{\pi}{2l}(iy-z) \right\} dy, \qquad (19)$$

or

$$\tau(z) = \Omega_\infty + \frac{1}{2l}\int_{-l}^{l}\theta_0\cot\frac{\pi}{2l}(x-z)\,dx + \frac{1}{2l}\sin\frac{\pi z}{l}\int_0^\infty \frac{\varpi(y)\,dy}{\cosh(\pi y/l)+\cos(\pi z/l)},$$

$$(20)$$

which is the required generalization of (14).

In the limit $z \to i\infty$ (20) yields

$$\Omega_\infty + i\theta_\infty = \Omega_\infty + \frac{i}{2l}\int_{-l}^{l}\theta_0(x)\,dx + \frac{i}{2l}\int_0^\infty \varpi(y)\,dy,$$

i.e.
$$\theta_\infty = \frac{1}{2l}\int_{-l}^{l}\theta_0(x)\,dx + \frac{1}{2l}\int_0^\infty \varpi(y)\,dy, \qquad (21)$$

so it is not possible to assign θ_∞ arbitrarily if $\varpi(y)$ is given. Because of the first of (15) the harmonic conjugate of (21) (see paragraph following (3.97)) reduces to

$$\Omega_\infty = \frac{1}{2l}\int_{-l}^{l}\Omega_0(x)\,dx. \qquad (22)$$

Equation (20) is the solution of the problem: find $\tau(z)$ in a semi-infinite strip given (i) $\mathscr{I}\tau\,(=\theta)$ on the *end* of the strip, and (ii) the *difference* in $\mathscr{R}\tau\,(=\Omega)$ across the width of the strip. The reader will see the relevance of this problem to unsteady aerofoil theory if he glances at §1.21. The theory is fully developed in ch. 9.

The first method given above of dealing with periodic boundary conditions can be applied to infinite strips. We start with (3.103) instead of (3.113), and find that in place of (13) we have to evaluate the sum

$$\lim_{N\to\infty}\sum_{n=-N}^{N}\coth\frac{\pi}{2h}(x+2nl-z),$$

and a similar one involving tanh.... Similar summations arise if the second method is adopted, and as the background theory of the functions emerging from these summations is rather complicated we shall defer dealing with periodic boundary conditions in infinite strips until we have sketched out some of the more important characteristics of the functions in question. Readers acquainted with the theory of theta, zeta and Jacobian elliptic functions can turn immediately to §4.9.

THE THETA, ZETA AND JACOBIAN ELLIPTIC FUNCTIONS

4.4   The theta functions

In the following pages we shall frequently refer to the account of the theta and closely related Jacobian elliptic functions given by Whittaker & Watson (1952) (henceforth referred to as 'W. & W.'), whose notation we shall adopt.

The theta functions are defined by the series

$$\begin{aligned}
\vartheta_1(z,q) &= 2\sum_{n=0}^{\infty}(-1)^n q^{(n+\frac{1}{2})^2}\sin(2n+1)z, \\
\vartheta_2(z,q) &= 2\sum_{n=0}^{\infty} q^{(n+\frac{1}{2})^2}\cos(2+1)z, \\
\vartheta_3(z,q) &= 1+2\sum_{n=1}^{\infty} q^{n^2}\cos 2nz, \\
\vartheta_4(z,q) &= 1+2\sum_{n=1}^{\infty}(-1)^n q^{n^2}\cos 2nz,
\end{aligned} \tag{23}$$

where $|q|<1$. The parameter q is often replaced by a parameter μ defined by

$$q = e^{i\pi\mu}\quad(\mathscr{I}\mu>0), \tag{24}$$

in which case the functional symbol is written $\vartheta(z\,|\,\mu)$, or sometimes simply $\vartheta(z)$ when there is no ambiguity about the value of μ.

By virtue of the relations

$$\begin{aligned}
&\vartheta_1(z+\pi)=-\vartheta_1(z),\quad \vartheta_2(z+\pi)=-\vartheta_2(z),\quad \vartheta_3(z+\pi)=\vartheta_3(z), \\
&\hspace{7cm}\vartheta_4(z+\pi)=\vartheta_4(z), \\
&\vartheta_1(z+\pi\mu)=\frac{-\vartheta_1(z)}{q\,e^{2iz}},\quad \vartheta_2(z+\pi\mu)=\frac{\vartheta_2(z)}{q\,e^{2iz}},\quad \vartheta_3(z+\pi\mu)=\frac{\vartheta_3(z)}{q\,e^{2iz}}, \\
&\hspace{4cm}\vartheta_4(z+\pi\mu)=\frac{-\vartheta_4(z)}{q\,e^{2iz}},
\end{aligned} \tag{25}$$

which follow from (23), the theta functions are said to be *quasi* doubly-periodic functions. They are evidently regular throughout the finite part of the z-plane.

It can also be shown from (23) that

$$\begin{aligned}
\{\vartheta_1,\vartheta_2,\vartheta_3,\vartheta_4\}(z) &= \{-\vartheta_2,\vartheta_1,\vartheta_4,\vartheta_3\}(z+\tfrac{1}{2}\pi) \\
&= M\{-i\vartheta_4,\vartheta_3,\vartheta_2,-i\vartheta_1\}(z+\tfrac{1}{2}\pi\mu) \\
&= M\{-i\vartheta_3,\vartheta_4,\vartheta_1,-i\vartheta_2\}(z+\tfrac{1}{2}\pi+\tfrac{1}{2}\pi\mu),
\end{aligned} \tag{26}$$

where $M=q^{\frac{1}{4}}e^{iz}$, and $\{\vartheta_1,\vartheta_2,\vartheta_2,\vartheta_4\}(z)$ is the column vector $\{\vartheta_1(z),\vartheta_2(z),\vartheta_3(z),\vartheta_4(z)\}$, etc. (We shall frequently save space by using a condensed notation of this type.)

For brevity $\vartheta_2(0),\vartheta_3(0),\vartheta_4(0)$ will be denoted by $\vartheta_2,\vartheta_3,\vartheta_4\,(\vartheta_1(0)=0)$, and similarly $[\vartheta_1'(z)]_{z=0}$ by ϑ_1'. These numbers are related by (W. & W. p. 472)

$$\vartheta_1' = \vartheta_2\vartheta_3\vartheta_4. \tag{27}$$

From (26) we observe that as $\vartheta_1(z)$ has a zero at $z=0$, $\vartheta_2(z),\vartheta_3(z)$ and $\vartheta_4(z)$ have corresponding zeros at $\tfrac{1}{2}\pi$, $\tfrac{1}{2}\pi+\tfrac{1}{2}\pi\mu$ and $\tfrac{1}{2}\pi\mu$, and it is easily shown (W. & W. p. 465) that these are the only zeros of the zeta functions within the cell of corners, a, $a+\pi$, $a+\pi+\pi\mu$, $a+\pi\mu$.

Landen's transformations, namely (W. & W. p. 476)

$$\frac{\vartheta_3(z\mid\mu)\,\vartheta_4(z\mid\mu)}{\vartheta_4(2z\mid 2\mu)} = \frac{\vartheta_2(z\mid\mu)\,\vartheta_1(z\mid\mu)}{\vartheta_1(2z\mid 2\mu)} = \frac{\vartheta_3(0\mid\mu)\,\vartheta_4(0\mid\mu)}{\vartheta_4(0\mid 2\mu)}, \tag{28}$$

prove important in our work, a remark also true of Jacobi's imaginary transformations (W. & W. p. 475), which can be written in the vector form

$$\{\vartheta_1,\vartheta_2,\vartheta_3,\vartheta_4\}(\mu z\mid\mu) = (-i\mu)^{-\frac{1}{2}}\exp\left(\frac{\mu z^2}{\pi i}\right)\{-i\vartheta_1,\vartheta_4,\vartheta_3,\vartheta_2\}\left(z\,\middle|\,-\frac{1}{\mu}\right). \tag{29}$$

4.5   The zeta functions

We shall define the zeta functions by the 'logarithmic derivatives'

$$Z_i(z\mid\mu) \equiv \frac{d}{dz}[\ln\{\vartheta_i(z\vartheta_3^{-2}\mid\mu)\}] = \vartheta_3^{-2}\frac{\vartheta_i'(z\vartheta_3^{-2}\mid\mu)}{\vartheta_i(z\vartheta_3^{-2}\mid\mu)}, \tag{30}$$

$i = 1, 2, 3, 4$. It will be noticed (cf. W. & W. p. 518) that $Z_4(z)$ is identical to Jacobi's zeta function, $Z(z)$.

It is convenient to define real numbers K and K' by

$$K = \tfrac{1}{2}\pi\,\vartheta_3^2, \quad iK' = \tfrac{1}{2}\pi\mu\vartheta_3^2, \tag{31}$$

then
$$\mu = iK'/K, \tag{32}$$

and
$$Z_i(z\mid\mu) = \frac{\pi}{2K}\frac{\vartheta_i'\left(\dfrac{\pi z}{2K}\,\middle|\,\dfrac{iK'}{K}\right)}{\vartheta_i\left(\dfrac{\pi z}{2K}\,\middle|\,\dfrac{iK'}{K}\right)}. \tag{33}$$

Using these definitions in (25) and (26) we obtain the following relations between the zeta functions

$$Z_i(z+2K) = Z_i(z), \quad Z_i(z+2iK') = -\frac{i\pi}{K}+Z_i(z), \tag{34}$$

$$\begin{aligned}
\{Z_1,Z_2,Z_3,Z_4\}(z) &= \frac{i\pi}{2K}+\{Z_4,Z_3,Z_2,Z_1\}(z+iK') \\
&= \{Z_2,Z_1,Z_4,Z_3\}(z+K) \\
&= \frac{i\pi}{2K}+\{Z_3,Z_4,Z_1,Z_2\}(z+K+iK').
\end{aligned} \tag{35}$$

From (33) it follows that $Z_i(z)$ has poles at the zeros of $\vartheta_i(\pi z/2K)$, so the poles of $Z_1(z)$, $Z_2(z)$, $Z_3(z)$ and $Z_4(z)$ are at the points 0, K, $K+iK'$ and iK' respectively, and further these are the only poles of the zeta functions within a rectangle with corners a, $a+2K$, $a+2K+2iK'$, $a+2iK'$. Near

$z = 0$, $\vartheta_1(\pi z/2K) \approx \vartheta_1' \pi z/2K$, so $Z_1(z) \approx 1/z$, i.e. $Z_1(z)$ has a simple pole of residue unity at $z = 0$. Equation (35) shows that the other zeta functions also have simple poles of residue unity at their singular points.

The two most important relations involving change of parameter can be derived from (28) and (29) by taking their logarithmic derivatives. Equation (28) gives

$$\left\{\frac{\vartheta_1'}{\vartheta_1}, \frac{\vartheta_3'}{\vartheta_2}\right\}(u \mid \mu) + \left\{\frac{\vartheta_2'}{\vartheta_2}, \frac{\vartheta_4'}{\vartheta_4}\right\}(u \mid \mu) = 2\left\{\frac{\vartheta_1'}{\vartheta_1}, \frac{\vartheta_4'}{\vartheta_4}\right\}(2u \mid 2\mu).$$

Put $u = \pi z/2K$ and use (33) to rewrite this in the form

$$K\{Z_1, Z_3\}(z \mid \mu) + K\{Z_2, Z_4\}(z \mid \mu) = 2K_1\{Z_1, Z_4\}\left(\frac{2K_1}{K}z \mid 2\mu\right), \quad (36)$$

where $K_1 = \tfrac{1}{2}\pi\,\vartheta_3^2(0 \mid 2\mu)$, i.e. corresponds to a parameter of $\mu_1 = 2\mu$. The relation between K and K_1 is given below in (54).

The logarithmic derivative of (29) is

$$\tau\left\{\frac{\vartheta_1'}{\vartheta_1}, \frac{\vartheta_2'}{\vartheta_2}, \frac{\vartheta_3'}{\vartheta_3}, \frac{\vartheta_4'}{\vartheta_4}\right\}(\mu u \mid \mu) = \frac{2u\mu}{\pi i} + \left\{\frac{\vartheta_1'}{\vartheta_1}, \frac{\vartheta_4'}{\vartheta_4}, \frac{\vartheta_3'}{\vartheta_3}, \frac{\vartheta_2'}{\vartheta_2}\right\}\left(u \mid -\frac{1}{\mu}\right). \quad (37)$$

We observe from (31) and (32) that K and K' interchange roles when μ is replaced by $-1/\mu$. Bearing this in mind we can transform (37) into a relation between zeta functions. We find

$$K\mu\{Z_1, Z_2, Z_3, Z_4\}\left(\frac{2K\mu}{\pi}u \mid \mu\right) = \frac{u\mu}{i} + K'\{Z_1, Z_4, Z_3, Z_2\}\left(\frac{2K'u}{\pi} \mid -\frac{1}{\mu}\right).$$

Thus putting $z = 2K'u/\pi$, and using (32) we arrive at

$$i\{Z_1, Z_2, Z_3, Z_4\}(iz \mid \mu) = \frac{\pi z}{2KK'} + \{Z_1, Z_4, Z_3, Z_2\}\left(z \mid -\frac{1}{\mu}\right). \quad (38)$$

From the expansions (W. & W. p. 489)

$$Z_1(z \mid \mu) = \frac{\pi}{2K}\left\{\cot\frac{\pi z}{2K} + 4\sum_{n=1}^{\infty}\left(\frac{q^{2n}\sin(\pi z/K)}{1 - 2q^{2n}\cos(\pi z/K) + q^{4n}}\right)\right\}, \quad (39)$$

$$Z_3(z \mid \mu) = \frac{\pi}{2K}\left\{-4\sum_{n=1}^{\infty}\left(\frac{q^{2n-1}\sin(\pi z/K)}{1 + 2q^{2n-1}\cos(\pi z/K) + q^{4n-2}}\right)\right\}, \quad (40)$$

where
$$q = e^{-\pi K/K'} = e^{\pi i\mu}, \quad (41)$$

(see (24) and (32)), and $Z_2(z) = Z_1(z + K)$, $Z_4(z) = Z_3(z + K)$, it follows that

$$\lim_{\substack{\mu \to i\infty \\ (q \to 0) \\ (K' \to \infty)}} \{Z_1, Z_2, Z_3, Z_4\}(z \mid \mu) = \frac{\pi}{2K}\left\{\cot\frac{\pi z}{2K}, -\tan\frac{\pi z}{2K}, 0, 0\right\}. \quad (42)$$

Combining this result with (38) we obtain the complementary limit

$$\lim_{\substack{\mu\to i0\\(K\to\infty)}} \{Z_1, Z_2, Z_3, Z_4\}(z\,|\,\mu) = i \lim_{\mu_1\to i\infty} \{Z_1, Z_4, Z_3, Z_2\}(iz\,|\,\mu_1)$$

$$= \frac{i\pi}{2K'}\left\{\cot\frac{\pi iz}{2K'}, 0, 0, -\tan\frac{\pi iz}{2K'}\right\},$$

i.e.
$$\lim_{\substack{\mu\to i0\\(K\to\infty)}} \{Z_1, Z_2, Z_3, Z_4\}(z\,|\,\mu) = \frac{\pi}{2K'}\left\{\coth\frac{\pi z}{2K'}, 0, 0, \tanh\frac{\pi z}{2K'}\right\}. \tag{43}$$

The special values

$$Z_2(0), Z_3(0), Z_4(0), Z_1(K), Z_3(K), Z_4(K) = 0, \tag{44}$$

follow from (23) and (33).

4.6   Jacobian elliptic functions

Important functions closely related to the theta and zeta functions are the Jacobian elliptic functions, which may be defined by (W. & W. p. 492)

$$\left.\begin{aligned}
\operatorname{sn}(z,k) &\equiv \frac{1}{k^{\frac{1}{2}}}\frac{\vartheta_1\!\left(\dfrac{\pi z}{2K}\,\middle|\,\dfrac{iK'}{K}\right)}{\vartheta_4\!\left(\dfrac{\pi z}{2K}\,\middle|\,\dfrac{iK'}{K}\right)},\\[2ex]
\operatorname{cn}(z,k) &\equiv \left(\frac{k'}{k}\right)^{\frac{1}{2}}\frac{\vartheta_2\!\left(\dfrac{\pi z}{2K}\,\middle|\,\dfrac{iK'}{K}\right)}{\vartheta_4\!\left(\dfrac{\pi z}{2K}\,\middle|\,\dfrac{iK'}{K}\right)},\\[2ex]
\operatorname{dn}(z,k) &\equiv (k')^{\frac{1}{2}}\frac{\vartheta_3\!\left(\dfrac{\pi z}{2K}\,\middle|\,\dfrac{iK'}{K}\right)}{\vartheta_4\!\left(\dfrac{\pi z}{2K}\,\middle|\,\dfrac{iK'}{K}\right)},
\end{aligned}\right\} \tag{45}$$

$\operatorname{ns}(z,k) = 1/\operatorname{sn}(z,k)$, $\operatorname{dc}(z,k) = \operatorname{dn}(z,k)/\operatorname{cn}(z,k)$, ..., etc. (twelve distinct functions), where

$$k = \frac{\vartheta_2^2\!\left(0\,\middle|\,\dfrac{iK'}{K}\right)}{\vartheta_3^2\!\left(0\,\middle|\,\dfrac{iK'}{K}\right)}, \tag{46}$$

and
$$k' = \sqrt{\{1 - k^2\}}. \tag{47}$$

We cannot list here all the properties of these functions required in this text, but for the reader's convenience we shall give, in a condensed form, some of the principal results. Whittaker & Watson give an account of elliptic functions sufficient for all our needs. A convenient summary of the main properties, together with tables of the functions $\operatorname{sn}(z, k)$, $\operatorname{cn}(z, k)$, $\operatorname{dn}(z, k)$, K', K, E', E (see § 4.8) and $Z_4(z, k)$, are to be found in a text by Milne-Thomson (1950). For an exhaustive list of elliptic and other related integrals the reader is referred to Byrd & Friedmann (1954).

The periodic relations between the elliptic functions can be written (omitting the modulus k for brevity)

$$
\begin{aligned}
\{\operatorname{sn}, \operatorname{cn}, \operatorname{dn}\}(u) &= \{-\operatorname{cd}, k'\operatorname{sd}, k'\operatorname{nd}\}(u + K) \\
&= \left\{\frac{1}{k}\operatorname{ns}, \frac{i}{k}\operatorname{ds}, i\operatorname{cs}\right\}(u + iK') \\
&= -\left\{\frac{1}{k}\operatorname{dc}, \frac{ik'}{k}\operatorname{nc}, ik'\operatorname{sc}\right\}(u + K + iK'),
\end{aligned}
\tag{48}
$$

$$
\{\operatorname{sn}, \operatorname{cn}, \operatorname{dn}\}(u + 2K) = \{-\operatorname{sn}, -\operatorname{cn}, \operatorname{dn}\}u,
\tag{49}
$$

and
$$
\{\operatorname{sn}, \operatorname{cn}, \operatorname{dn}\}(u + 2iK') = \{\operatorname{sn}, -\operatorname{cn}, -\operatorname{dn}\}(u).
\tag{50}
$$

Also
$$
\{\operatorname{sn}, \operatorname{cn}, \operatorname{dn}\}(-u) = \{-\operatorname{sn}, \operatorname{cn}, \operatorname{dn}\}(u).
\tag{51}
$$

Special values of the argument are:

u	0	$\tfrac{1}{2}K$	K	$2K$	$\tfrac{1}{2}iK'$	iK'	$2iK'$	$K+iK'$	$2K+2iK'$
$\operatorname{sn} u$	0	$1/(1+k')^{\frac{1}{2}}$	1	0	$i/k^{\frac{1}{2}}$	∞	0	$1/k$	0
$\operatorname{cn} u$	1	$k'^{\frac{1}{2}}/(1+k')^{\frac{1}{2}}$	0	-1	$(1+k)^{\frac{1}{2}}/k^{\frac{1}{2}}$	∞	-1	$-ik'/k$	1
$\operatorname{dn} u$	1	$k'^{\frac{1}{2}}$	k'	1	$(1+k)^{\frac{1}{2}}$	∞	-1	0	-1

The periods, zeros, poles and residues are

	periods		zeros		poles		residues	
$\operatorname{sn} u$	$4K,$	$2iK'$	$0,$	$2K$	$iK',$	$2K+iK'$	$1/k,$	$-1/k$
$\operatorname{cn} u$	$4K,$	$2K+2iK'$	$K,$	$3K$	$iK',$	$2K+iK'$	$-i/k,$	i/k
$\operatorname{dn} u$	$2K,$	$4iK'$	$K+iK,$	$K+3iK'$	$iK',$	$3iK'$	$-i,$	i

The squares of the functions are related as follows:

$$
\begin{aligned}
\operatorname{sn}^2 u + \operatorname{cn}^2 u = 1, \quad \operatorname{dn}^2 u + k^2 \operatorname{sn}^2 u &= 1, \\
\operatorname{dn}^2 u - k^2 \operatorname{cn}^2 u &= k'^2,
\end{aligned}
\tag{52}
$$

and the derivatives are given by

$$
\begin{aligned}
\frac{d}{du}\operatorname{sn} u = \operatorname{cn} u\, \operatorname{dn} u, \quad \frac{d}{du}\operatorname{cn} u &= -\operatorname{sn} u\, \operatorname{dn} u, \\
\frac{d}{du}\operatorname{dn} u &= -k^2 \operatorname{sn} u\, \operatorname{cn} u.
\end{aligned}
\tag{53}
$$

These equations are easily extended to the other nine functions by use of (48).

Landen's transformation for elliptic functions follows directly from (28) and (45). It is

$$\text{sn}\left(\frac{2K_1}{K}u, k_1\right) = \frac{k}{k_1^{\frac{1}{2}}}\,\text{sn}\,(u, k)\,\text{cd}\,(u, k).$$

From (53) and the result $\text{cn}\,0 \times \text{dn}\,0 = 1$ it follows that

$$\lim_{u\to 0}\frac{\text{sn}\,(2K_1 u/K)}{u} = 2K_1/K,$$

and therefore $2K_1/K = k/k_1^{\frac{1}{2}}$. The particular value $u = \frac{1}{2}K$, and the table of special values given above yields $k/k_1^{\frac{1}{2}} = (1+k')$, which can be re-arranged to give $k/k_1^{\frac{1}{2}} = 2/(1+k_1)$. Finally, as the parameter of the theta functions is doubled by the Landen transformation, it follows from (32) that $K_1/K = K_1'/2K'$. Combining these results we have

$$\left.\begin{aligned}
\text{sn}\left(\frac{2K_1}{K}u, k_1\right) &= \frac{2K_1}{K}\,\text{sn}\,(u, k)\,\text{cd}\,(u, k), \\[2mm]
\frac{K_1}{K} = \frac{K_1'}{2K'} &= \frac{1}{1+k_1} = \tfrac{1}{2}(1+k') = \frac{k}{2k_1^{\frac{1}{2}}}.
\end{aligned}\right\} \tag{54}$$

where

The inverse transformation readily found to be

$$\text{sn}\,(u, k) = \frac{K}{K_1}\frac{\text{sn}\left(\dfrac{K_1 u}{K}, k_1\right)}{1 + k_1\,\text{sn}^2\left(\dfrac{K_1 u}{K}, k_1\right)}. \tag{55}$$

Jacobi's imaginary transformation is

$$\text{sn}\,(iu, k) = i\,\text{sc}\,(y, k'), \quad \text{cn}\,(iy, k) = \text{nc}\,(y, k'), \quad \text{dn}\,(iy, k) = \text{dc}\,(y, k'). \tag{56}$$

The following limits will be useful:

$$\lim_{\substack{k\to 0 \\ k'\to 1}}\{\text{sn}\,u, \text{cn}\,u, \text{dn}\,u\} = \{\sin u, \cos u, 1\}, \tag{57}$$

and

$$\lim_{\substack{k'\to 0 \\ k\to 1}}\{\text{sn}\,u, \text{cn}\,u, \text{dn}\,u\} = \{\tanh u, \text{sech}\,u, \text{sech}\,u\}. \tag{58}$$

The addition theorems are

$$\text{sn}\,(u \pm v) = \frac{\text{sn}\,u\,\text{cn}\,v\,\text{dn}\,v \pm \text{sn}\,v\,\text{cn}\,u\,\text{dn}\,u}{1 - k^2\,\text{sn}^2 u\,\text{sn}^2 v}, \tag{59}$$

$$\text{cn}\,(u \pm v) = \frac{\text{cn}\,u\,\text{cn}\,v \mp \text{sn}\,u\,\text{dn}\,u\,\text{sn}\,v\,\text{dn}\,v}{1 - k^2\,\text{sn}^2 u\,\text{sn}^2 v}, \tag{60}$$

$$\text{dn}\,(u \pm v) = \frac{\text{dn}\,u\,\text{dn}\,v \mp k^2\,\text{sn}\,u\,\text{cn}\,u\,\text{sn}\,v\,\text{cn}\,v}{1 - k^2\,\text{sn}^2 u\,\text{sn}^2 v}. \tag{61}$$

From these it can be established that

$$\frac{\operatorname{sn}^2 u \,\operatorname{dn}^2 u}{\operatorname{cn}^2 u} = \frac{1-\operatorname{cn} 2u}{1+\operatorname{cn} 2u}, \quad \frac{k^2 \operatorname{sn}^2 u \,\operatorname{cn}^2 u}{\operatorname{dn}^2 u} = \frac{1-\operatorname{dn} 2u}{1+\operatorname{dn} 2u}, \tag{62}$$

$$\operatorname{sn}^2 \tfrac{1}{2}u = \frac{1-\operatorname{cn} u}{1+\operatorname{dn} u}, \quad \operatorname{cn}^2 \tfrac{1}{2}u = \frac{\operatorname{dn} u + \operatorname{cn} u}{1+\operatorname{dn} u},$$

and

$$\operatorname{dn}^2 \tfrac{1}{2}u = \frac{k'^2 + \operatorname{dn} u + k^2 \operatorname{cn} u}{1+\operatorname{dn} u}. \tag{63}$$

Finally we note the following expansions

$$\operatorname{sn} u = \frac{2\pi}{kK} \sum_{n=0}^{\infty} \frac{q^{n+\frac{1}{2}}}{1-q^{2n+1}} \sin\left\{\frac{(2n+1)\pi u}{2K}\right\},$$

$$\operatorname{cn} u = \frac{2\pi}{kK} \sum_{n=0}^{\infty} \frac{q^{n+\frac{1}{2}}}{1+q^{2n+1}} \cos\left\{\frac{(2n+1)\pi u}{2K}\right\}, \tag{64}$$

$$\operatorname{dn} u = \frac{\pi}{2K} + \frac{2\pi}{K} \sum_{n=1}^{\infty} \frac{q^n}{1+q^{2n}} \cos\frac{n\pi u}{K},$$

$$\left(\frac{2kK}{\pi}\right)^{\frac{1}{2}} = \vartheta_2 = 2q^{\frac{1}{4}}(1+q^2+q^6+q^{12}+q^{20}+\ldots),$$

$$\left(\frac{2K}{\pi}\right)^{\frac{1}{2}} = \vartheta_3 = 1+2q+2q^4+2q^9+\ldots, \tag{65}$$

$$\left(\frac{2k'K}{\pi}\right)^{\frac{1}{2}} = \vartheta_4 = 1-2q+2q^4+2q^9-\ldots,$$

where from (24) and (32) $\qquad q = e^{-\pi K'/K}, \tag{66}$

and the relations involving the theta functions in (65) have been derived from (31), (46) and (47).

4.7　Some useful relations between the zeta and Jacobian functions

From (33) and (45)

$$Z_1(z,k) - Z_4(z,k) = \frac{d}{dz}\ln\left\{\frac{\vartheta_1\left(\frac{\pi z}{2K}\bigg|\frac{iK'}{K}\right)}{\vartheta_4\left(\frac{\pi z}{2K}\bigg|\frac{iK'}{K}\right)}\right\} = \frac{d}{dz}\ln\{\operatorname{sn}(z,k)\}.$$

Therefore by the first of (53)

$$\frac{\operatorname{sn}'(z,k)}{\operatorname{sn}(z,k)} = \frac{\operatorname{cn}(z,k)\,\operatorname{dn}(z,k)}{\operatorname{sn}(z,k)} = Z_1(z,k) - Z_4(z,k), \tag{67}$$

where the parameter μ has been replaced in the zeta functions by the parameter k defined in (46).

A particularly useful relation which can be derived from the definitions of the zeta and Jacobian functions is (W. & W. p. 519)

$$Z_4(y,k) + Z_4(z,k) - Z_4(y+z,k) = k^2 \operatorname{sn}(y,k)\operatorname{sn}(z,k)\operatorname{sn}(y+z,k). \quad (68)$$

Substitution of values $y = x + iK'$, $z = K$ in (68) and use of (35), (44) and (48) gives

$$Z_1(x,k) - Z_2(x,k) = \operatorname{ns}(x,k)\operatorname{dc}(x,k)$$
$$= \frac{2K_1}{K}\operatorname{ns}\left(\frac{2K_1 x}{K}, k_1\right), \quad (69)$$

by (54). The first of (36) can be written

$$Z_1(x,k) + Z_2(x,k) = \frac{2K_1}{K}Z_1\left(\frac{2K_1 x}{K}, k_1\right); \quad (70)$$

adding this to (69) and putting $x = Kz$ we arrive at

$$KZ_1(Kz,k) = K_1 Z_1(2K_1 z, k_1) + K_1 \operatorname{ns}(2K_1 z, k_1), \quad (71)$$

which is a useful 'change of parameter' rule.

One further relation required later is

$$KZ_1\left\{\frac{K}{l}(x-z), k\right\} = K_1 Z_4\left\{\frac{2K_1}{l}x, k_1\right\} - K_1 Z_4\left\{\frac{2K_1}{l}z, k_1\right\}$$
$$+ K_1 \frac{\operatorname{cn}\dfrac{2K_1}{l}z\,\operatorname{dn}\dfrac{2K_1}{l}z + \operatorname{cn}\dfrac{2K_1}{l}x\,\operatorname{dn}\dfrac{2K_1}{l}x}{\operatorname{sn}\dfrac{2K_1}{l}x - \operatorname{sn}\dfrac{2K_1}{l}z}, \quad (72)$$

where the modulus of the elliptic functions is k_1.

To prove this we start with (68); the particular values $y = \beta$, $z = iK' - \gamma$ give

$$Z_4(\beta) - Z_1(\gamma) - Z_1(\beta - \gamma) = \frac{\operatorname{sn}\beta}{\operatorname{sn}\gamma\operatorname{sn}(\beta-\gamma)},$$

on using (35) and (48). Hence eliminating $Z_1(\gamma)$ by (67) we find

$$Z_1(\beta - \gamma) = Z_4(\beta) - Z_4(\gamma) + \operatorname{ns}\gamma\left\{\frac{\operatorname{sn}\beta}{\operatorname{sn}(\beta-\gamma)} - \operatorname{cn}\gamma\operatorname{dn}\gamma\right\},$$

i.e.

$$Z_1(\beta - \gamma) + \operatorname{ns}(\beta-\gamma) = Z_4(\beta) - Z_4(\gamma) + \operatorname{ns}\gamma\left\{\frac{\operatorname{sn}\beta + \operatorname{sn}\gamma}{\operatorname{sn}(\beta-\gamma)} - \operatorname{cn}\gamma\operatorname{dn}\gamma\right\}. \quad (73)$$

Now it can be shown from (59) that (see example 3, p. 496, W. & W.)

$$\frac{\operatorname{sn}\beta + \operatorname{sn}\gamma}{\operatorname{sn}(\beta-\gamma)} = \frac{\operatorname{sn}\beta\operatorname{cn}\gamma\operatorname{dn}\gamma + \operatorname{sn}\gamma\operatorname{cn}\beta\operatorname{dn}\beta}{\operatorname{sn}\beta - \operatorname{sn}\gamma}, \quad (74)$$

therefore (73) can be written

$$Z_1(\beta-\gamma)+\operatorname{ns}(\beta-\gamma) = Z_4(\beta)-Z_4(\gamma)+\frac{\operatorname{cn}\beta\,\operatorname{dn}\beta+\operatorname{cn}\gamma\,\operatorname{dn}\gamma}{\operatorname{sn}\beta-\operatorname{sn}\gamma}.$$

Let k_1 be the modulus in this equation, and let $\beta = (2K_1/l)\,x$, $\gamma = (2K_1/l)z$, then (72) follows immediately by an application of the change of parameter rule expressed in (71).

The second of (36) enables us to write (72) in the alternative form

$$K\left[Z_1\!\left\{\frac{K}{l}(x-z),k\right\}-\tfrac12 Z_3\!\left\{\frac{Kx}{l},k\right\}-\tfrac12 Z_4\!\left\{\frac{Kx}{l},k\right\}\right]$$

$$= K_1\frac{\operatorname{cn}\dfrac{2K_1}{l}z\,\operatorname{dn}\dfrac{2K_1}{l}z+\operatorname{cn}\dfrac{2K_1}{l}x\,\operatorname{dn}\dfrac{2K_1}{l}x}{\operatorname{sn}\dfrac{2K_1}{l}x-\operatorname{sn}\dfrac{2K_1}{l}z}-K_1Z_4\!\left\{\frac{2K_1}{l}z,k_1\right\}. \quad (75)$$

4.8 Jacobi's incomplete elliptic functions of the second and third kinds

In certain fluid flow problems (see §13.1) elliptic functions of the second and third kinds arise. The definition we shall adopt for the third kind is

$$\Pi(z,y) \equiv \int_0^z \frac{k^2\operatorname{sn}y\,\operatorname{cn}y\,\operatorname{dn}y\,\operatorname{sn}^2z}{1-k^2\operatorname{sn}^2y\,\operatorname{sn}^2z}\,dz, \quad (76)$$

$\Pi(z,y)$ being the function due to Jacobi (W. & W. pp. 522–3). It can be deduced from (59) and (68) that

$$\tfrac12\{Z_4(z-y)-Z_4(z+y)+2Z_4(y)\} = \frac{k^2\operatorname{sn}y\,\operatorname{cn}y\,\operatorname{dn}y\,\operatorname{sn}^2z}{1-k^2\operatorname{sn}^2y\,\operatorname{sn}^2z}, \quad (77)$$

consequently it follows from (33) and (76) that

$$\Pi(z,y) = \tfrac12\ln\left[\frac{\vartheta_4\!\left\{\dfrac{\pi}{2K}(z-y)\,\middle|\,\dfrac{iK'}{K}\right\}}{\vartheta_4\!\left\{\dfrac{\pi}{2K}(z+y)\,\middle|\,\dfrac{iK'}{K}\right\}}\right]+zZ_4(y). \quad (78)$$

It is obvious from this form that

$$\Pi(z,y) = \Pi(y,z)+zZ_4(y)-yZ_4(z). \quad (79)$$

Equations (26) and (78) enable the particular values

$$\Pi(K,y) = KZ_4(y), \quad \Pi(iK',y) = iK'Z_4(y)-\frac{i\pi}{2}+\frac{i\pi}{2K}y,\left.\vphantom{\frac{i\pi}{2K}}\right\}$$

and

$$\Pi(K+iK',y) = (K+iK')\,Z_4(y)+\frac{i\pi}{2K}y \qquad\qquad\quad (80)$$

to be obtained.

The functions defined in this section frequently arise in the integration of expressions involving elliptic functions. An important example is

$$\operatorname{cn}\zeta\,\operatorname{dn}\zeta\int\frac{d\gamma}{\operatorname{sn}\gamma-\operatorname{sn}\zeta}=\tfrac{1}{2}\ln\left[\frac{\operatorname{cd}\gamma-\operatorname{cd}\zeta}{\operatorname{cd}\gamma+\operatorname{cd}\zeta}\right]+\Pi(\gamma+iK',\zeta),\qquad(81)$$

which is easily verified by differentiation.

The elliptic integral of the second kind can be defined by

$$E(u,k)\equiv\int_0^u\operatorname{dn}^2(u,k)\,du.\qquad(82)$$

The function $E(k)\equiv E(K,k)$ is usually termed the *complete* elliptic integral of the second kind. It is not difficult to show that (W. & W. p. 518)

$$E(u,k)=Z_4(u,k)+\frac{u}{K}E(k).\qquad(83)$$

The notation $E'\equiv E(K',k')$ is also used in elliptic integral theory.

As the squares of the twelve Jacobian elliptic functions can each be expressed in terms of $\operatorname{dn}^2 u$, (82) enables us to integrate $\operatorname{sn}^2 u$, $\operatorname{cn}^2 u$, ..., etc.

RECTANGLES

4.9   Periodic boundary conditions for the infinite strip

The results to be derived below are important in the theory of fluid motions through doubly-connected domains; they include the theory of §§ 3.16 and 4.3 as special cases.

We shall solve two similar boundary value problems for the rectangle $-l < x < l$, $0 < y < h$, labelled B in Fig. 4.9. Let $\tau^+(y)$, $\tau^-(y)$ be defined as in (16), and also let

$$\tau_0(x)=\lim_{\epsilon\to 0}\tau(x,\epsilon),\quad \tau_h(x)=\lim_{\epsilon\to 0}\tau(x,y-\epsilon)\quad(\epsilon\geqslant 0),\qquad(84)$$

as in § 3.16. Then the first of the boundary value problems can be stated as follows.

Determine an analytic function $\tau(z)=\Omega(x,y)+i\theta(x,y)$, regular within B, and satisfying the boundary conditions

$$\left.\begin{array}{ll}(a)\ \theta_0(x)\text{ prescribed,}&(b)\ \theta_h(x)\text{ prescribed,}\\(c)\ \theta^+(y)=\theta^-(y),&(d)\ \Omega^+(y)-\Omega^-(y)=\varpi(y),\end{array}\right\}\qquad(85)$$

where $\varpi(y)$ is a given function.

Notice that if $\varpi(y)=0$, the discontinuities of $\tau(z)$ in $-\infty < x < \infty$, $0 < y < h$ vanish, and we have the case of periodic boundary conditions for the infinite strip.

The method of calculating $\tau(z)$ is simply a combination of the image methods already employed in §§ 3.16 and 4.3, and it is an excellent example of this powerful technique.

Let z be a point within the base rectangle B, then all the points of affix $z + 2nl + 2ihm$, n, $m = \pm 1$, ± 2, ± 3, ..., be outside B. Hence by using

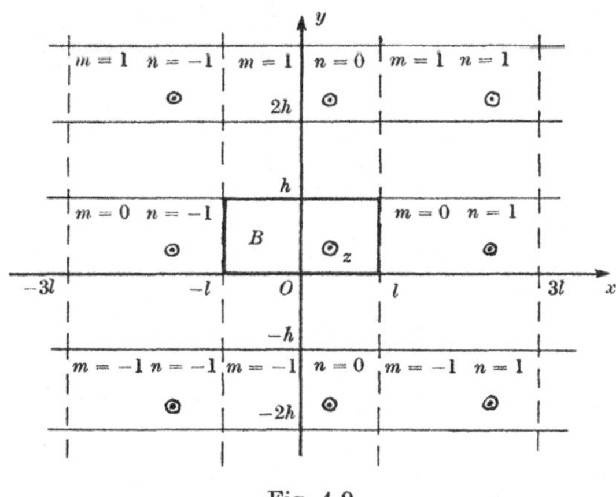

Fig. 4.9

Cauchy's integral and theorem, and simple addition—just as in the earlier calculations—we arrive at

$$\tau(z) = \frac{1}{2\pi i} \int_{-l}^{l} \sum_{m=-M}^{M} \sum_{n=-N}^{N}$$

$$\times \left\{ \frac{\tau_0(x)}{x + 2nl + 2mih - z} - \frac{\tau_h(x)}{x + 2nl + (2m+1)ih - z} \right\} dx$$

$$+ \frac{1}{2\pi} \int_{0}^{h} \sum_{m=-M}^{M} \sum_{n=-N}^{N}$$

$$\times \left\{ \frac{\tau^+(y)}{iy + (2n+1)l + 2mih - z} - \frac{\tau^-(y)}{iy + (2n-1)l + 2mih - z} \right\} dy \quad (86)$$

(cf. (3.105) and (17)).

The next step is to take the limits $M \to \infty$ and $N \to \infty$, and for this purpose we need the value of repeated series of the type

$$S(a, b, c) \equiv \lim_{M \to \infty} \lim_{N \to \infty} \sum_{m=-M}^{M} \sum_{n=-N}^{N} \left(\frac{1}{a + bn + icm} \right). \quad (87)$$

First applying (13) we have

$$
\begin{aligned}
S &= \frac{\pi}{b} \lim_{M\to\infty} \sum_{m=-M}^{M} \cot\frac{\pi}{b}(a+icm) \\
&= \frac{\pi}{b}\left\{ \cot\frac{\pi a}{b} + \sum_{m=1}^{\infty}\left[\cot\frac{\pi}{b}(a+icm) + \cot\frac{\pi}{b}(a-icm)\right]\right\} \\
&= \frac{\pi}{b}\left\{ \cot\frac{\pi a}{b} + 4\sum_{m=1}^{\infty}\left(\frac{q^{2m}\sin(2\pi a/b)}{1-2q^{2m}\cos(2\pi a/b)+q^{4m}}\right)\right\},
\end{aligned}
\tag{88}
$$

where $q = \exp(-\pi|c/b|)$. Comparison with (39) and (41) now yields

$$
S = \frac{2K}{b} Z_1\!\left(\frac{2Ka}{b}\,\bigg|\,\frac{ic}{b}\right),
\tag{89}
$$

where c and b have been taken to be positive without any essential loss in generality.

If (89) is now applied to the limit form of (86) there is obtained

$$
\begin{aligned}
\tau(z) = \frac{K}{2\pi i l}\int_{-l}^{l}&\left(\tau_0(x)\,Z_1\!\left\{\frac{K}{l}(x-z)\,\bigg|\,\frac{ih}{l}\right\} - \tau_h(x)\,Z_1\!\left\{\frac{K}{l}(x+ih-z)\,\bigg|\,\frac{ih}{l}\right\}\right)dx \\
+ \frac{K}{2\pi l}\int_{0}^{h}&\left(\tau^{+}(y)\,Z_1\!\left\{\frac{K}{l}(iy+l-z)\,\bigg|\,\frac{ih}{l}\right\} - \tau^{-}(y)\,Z_1\!\left\{\frac{K}{l}(iy-l-z)\,\bigg|\,\frac{ih}{l}\right\}\right)dy,
\end{aligned}
\tag{90}
$$

the uniform convergence of the series involved permitting the interchange $\lim\lim\int = \int\lim\lim$. Equations (35) and (85) enable (90) to be written

$$
\begin{aligned}
\tau(z) = \frac{1}{4l}\int_{-l}^{l} \tau_h(x)\,dx + \frac{K}{2\pi i l}\int_{-l}^{l}&\left(\tau_0(x)\,Z_1\!\left\{\frac{K}{l}(x-z)\right\} \right. \\
&\left. - \tau_h(x)\,Z_4\!\left\{\frac{K}{l}(x-z)\right\}\right)dx + \frac{K}{2\pi l}\int_{0}^{h}\varpi(y)\,Z_2\!\left\{\frac{K}{l}(iy-z)\right\}dy,
\end{aligned}
\tag{91}
$$

where the parameter of the zeta functions here, viz.

$$
\mu = i\frac{h}{l} = i\frac{K'}{K},
\tag{92}
$$

is omitted for brevity.

Now as *all* the points of affix $\bar{z} + 2nl + 2mih$, $n, m = 0, \pm 1, \pm 2, \ldots$, lie outside the basic rectangle B, it follows that if z is replaced by $\bar{z}$ in (91) the left-hand side vanishes. The complex conjugate of the equation just described is

$$
\begin{aligned}
0 = \frac{1}{4l}\int_{-l}^{l} \bar{\tau}_h(x)\,dx - \frac{K}{2\pi i l}\int_{-l}^{l}&\left(\bar{\tau}_0(x)\,Z_1\!\left\{\frac{K}{l}(x-z)\right\} \right. \\
&\left. - \bar{\tau}_h(x)\,Z_4\!\left\{\frac{K}{l}(x-z)\right\}\right)dx - \frac{K}{2\pi l}\int_{0}^{h}\varpi(y)\,Z_2\!\left\{\frac{K}{l}(iy+z)\right\}dy,
\end{aligned}
\tag{93}
$$

using the fact that all the zeta functions are odd functions of their argument. Addition of (91) and (93) now gives

$$\tau(z) = \frac{1}{2l}\int_{-l}^{l}\Omega_h(x)\,dx + \frac{K}{\pi l}\int_{-l}^{l}\left(\theta_0(x)\,Z_1\left\{\frac{K}{l}(x-z)\right\} - \theta_h(x)\,Z_4\left\{\frac{K}{l}(x-z)\right\}\right)dx$$

$$+ \frac{K}{2\pi l}\int_{0}^{h}\varpi(y)\left[Z_2\left\{\frac{K}{l}(iy-z)\right\} - Z_2\left\{\frac{K}{l}(iy+z)\right\}\right]dy, \quad (94)$$

which is one form of the solution to the boundary value problem, defined by (85).

There is an important restriction on the boundary values of θ_0, θ_h, and ϖ, which can be deduced as follows. As $\tau(z)$ is an analytic function it must satisfy $\int_C \tau(z)\,dz = 0$, taken around the basic rectangle, thus

$$\int_{-l}^{l}\{\tau_0(x) - \tau_h(x)\}\,dx + i\int_{0}^{h}\{\tau^+(y) - \tau^-(y)\}\,dy = 0, \quad (95)$$

or

$$\int_{-l}^{l}\{\Omega_0(x) - \Omega_h(x)\}\,dx = 0, \quad (96)$$

and

$$\int_{-l}^{l}\{\theta_0(x) - \theta_h(x)\}\,dx + \int_{0}^{h}\varpi(y)\,dy = 0, \quad (97)$$

by (85). Equation (96) permits us to replace the first right-hand term of (94) by the symmetrical form

$$\frac{1}{4l}\int_{-l}^{l}\{\Omega_0(x) + \Omega_h(x)\}\,dx.$$

4.10   Alternative form for the solution

The following alternative form of (94), due to Rosenblat (1957), is sometimes more convenient:

$$\tau(z) = \tfrac{1}{2}\Omega(l,h) + \tfrac{1}{2}\Omega(0,h)$$

$$+ \frac{K_1}{\pi l}\int_{-l}^{l}\theta_0(x)\;\frac{\operatorname{cn}\dfrac{2K_1}{l}z\,\operatorname{dn}\dfrac{2K_1}{l}z + \operatorname{cn}\dfrac{2K_1}{l}x\,\operatorname{dn}\dfrac{2K_1}{l}x}{\operatorname{sn}\dfrac{2K_1}{l}x - \operatorname{sn}\dfrac{2K_1}{l}z}\;dx$$

$$- \frac{K_1}{\pi l}\int_{-l}^{l}\theta_h(x)\;\frac{\operatorname{cn}\dfrac{2K_1}{l}z\,\operatorname{dn}\dfrac{2K_1}{l}z - k_1^{-1}\,\operatorname{ds}\dfrac{2K_1}{l}x\,\operatorname{cs}\dfrac{2K_1}{l}x}{k_1^{-1}\operatorname{ns}\dfrac{2K_1}{l}x - \operatorname{sn}\dfrac{2K_1}{l}z}\;dx$$

$$+ \frac{K_1}{\pi l}\operatorname{sn}\frac{2K_1}{l}z\int_{0}^{h}\varpi(y)\;\frac{\operatorname{cn}\dfrac{2K_1}{l}z\,\operatorname{dn}\dfrac{2K_1}{l}z\;\operatorname{cn}\dfrac{2K_1}{l}iy\,\operatorname{dn}\dfrac{2K_1}{l}iy}{\operatorname{sn}^2\dfrac{2K_1}{l}iy - \operatorname{sn}^2\dfrac{2K_1}{l}z}\;dy, \quad (98)$$

where the modulus of the elliptic functions is k_1, implicitly determined by the ratio

$$\frac{K_1'}{K_1} = \frac{2h}{l}.$$ (99)

The proof is as follows. Put $z = \pm l + ih$ in (94), and use (35) and (97) to find

$$\Omega(\pm l, h) = \frac{1}{4l}\int_{-l}^{l}\{\Omega_0(x) + \Omega_h(x)\}\,dx + \frac{K}{\pi l}\int_{-l}^{l}\left\{\theta_0(x)\,Z_3\!\left(\frac{Kx}{l}\right)\right.$$
$$\left. -\,\theta_h(x)\,Z_2\!\left(\frac{Kx}{l}\right)\right\}dx.$$ (100)

Similarly,

$$\Omega(0, h) = \frac{1}{4l}\int_{-l}^{l}\{\Omega_0(x) + \Omega_h(x)\}\,dx + \frac{K}{\pi l}\int_{-l}^{l}\left\{\theta_0(x)\,Z_4\!\left(\frac{Kx}{l}\right)\right.$$
$$\left. -\,\theta_h(x)\,Z_1\!\left(\frac{Kx}{l}\right)\right\}dx.$$ (101)

We now add half of each of these equations to (94). The three resulting integrands are converted to elliptic-function form by means of (75), in which x is given the values x, $x + ih$ and $x + l$ in turn, and in which (35) and (49) are used to obtain the particular forms given in (98). The term involving $Z_4\{(2K_1/l)\,z, k_1\}$ vanishes by (97). Equation (99) follows from (54) and (92), and the proof is complete.

The only virtue of (98) as compared with (94) is that the Jacobian elliptic functions are sometimes easier to manipulate algebraically than are the zeta functions. However the integrals in (94) are much more readily evaluated for simple distributions of θ_0, θ_h, ..., e.g. when these are step-functions.

4.11   Limiting cases

In the limits $l \to \infty$ and $h \to \infty$ equation (94) yields (3.109) and (20) as special cases. This can be verified as follows.

First consider the limit $l \to \infty$, which by (92) is equivalent to $\mu \to i0$. From (43) and (92)

$$\lim_{l\to\infty}\frac{K}{2l}\{Z_1, Z_4\}\left\{\frac{K}{l}\,(x-z)\left|\frac{ih}{l}\right.\right\} = \frac{K'}{2h}\lim_{\mu\to i0}\{Z_1, Z_4\}\left\{\frac{K'}{h}\,(x-z)\,\mu\,|\,\mu\right\}$$
$$= \frac{\pi}{2h}\left\{\coth\frac{\pi}{2h}\,(x-z), \quad \tanh\frac{\pi}{2h}\,(x-z)\right\}.$$

Similarly, the limit form of the last integral in (94) vanishes, so the limit of this equation reads

$$\tau(z) = \lim_{l\to\infty}\frac{1}{2l}\int_{-l}^{l}\Omega_h(x)\,dx + \frac{l}{2h}\int_{-\infty}^{\infty}\left\{\theta_0(x)\coth\frac{\pi}{2h}\,(x-z)\right.$$
$$\left. -\,\theta_h\tanh\frac{\pi}{2h}\,(x-z)\right\}dx.$$

It remains only to take the limits $z \to \pm\infty$ in this equation to establish that

$$\lim_{l\to\infty} \frac{1}{2l} \int_{-l}^{l} \Omega_h(x)\,dx = \tfrac{1}{2}(\Omega_\infty + \Omega_{-\infty}), \tag{102}$$

and we have recovered (3.109).

The limit $h \to \infty$, i.e. $\mu \to i\infty$ is easily derived with the aid of (42)—we recover (19) from (94) immediately.

Before leaving this problem we should show that the *order* of limit taking in (87) is immaterial—an obvious requirement for the uniqueness of our solution. Let $\tilde{S}$ be the value of the double sum in (87) when the limits are reversed in order, then $\tilde{S}$ can be found merely by interchanging b and ic, and therefore K and K', in (89). Thus

$$\tilde{S} = \frac{2K'}{ic} Z_1\!\left(\frac{2K'a}{ic}\,\middle|\,\frac{ib}{c}\right) = \frac{2Ki}{b} Z_1\!\left(\frac{2Ka}{b}\,i\,\middle|\,\frac{ib}{c}\right),$$

as $K'/K = c/b$. Hence from (38) and (89)

$$\tilde{S} = \frac{2K}{b} Z_1\!\left(\frac{2Ka}{b}\,\middle|\,\frac{ic}{b}\right) + \frac{2\pi a}{bc} = S + \frac{2\pi a}{bc}. \tag{103}$$

This result means that the *double* series in (87) is not convergent, while the repeated series are, their sums differing by $2\pi a/bc$. The use of $\tilde{S}$ in place of S in the limit form of (86) clearly results in an additional term on the right-hand side of (91). From (103) this term is found to be

$$\frac{1}{4hli} \int_{-l}^{l} [\tau_0(x)\,\{x-z\} - \tau_h(x)\,\{x+ih-z\}]\,dx + \int_{0}^{h} [\tau^+(y)\,\{iy+l-z\}$$
$$- \tau^-(y)\,\{iy-l-z\}]\,d(iy) = \frac{1}{4hli} \int_{B} \tau(z)\,(t-z)\,dt,$$

where B is the basic rectangle (see Fig. 4.9). The integrand of this contour integral is regular within B, so the integral must vanish by Cauchy's theorem. And therefore it makes no difference to (91), and the subsequent theory, which of S and $\tilde{S}$ are used, i.e. in which order the limits in (87) are taken.

4.12 Mixed and periodic boundary conditions for the infinite strip

Suppose now the boundary conditions (85) are modified to read

$$(a)\ \ \theta_0(x) \text{ prescribed}, \quad (b)\ \ \Omega_h(x) \text{ prescribed},$$
$$(c)\ \ \theta^+(y) = \theta^-(y), \quad (d)\ \ \Omega^+(y) - \Omega^-(y) = \varpi(y), \tag{104}$$

then we shall have a generalization of the mixed boundary value problem treated in §4.2. Just as the solution to this later problem, viz. (9) was readily derived by a simple modification to the previous theory for the Dirichlet problem in an infinite strip, so the solution to (104) can be derived by making a similar modification to the theory of § 4.9.

The modification referred to is that before adding the Cauchy integrals comprising the terms in (86) they are multiplied by $(-1)^m$. This makes no difference to the left-hand side of (86), which arises only from the term $m = 0$, $n = 0$ in the double sum, but it does mean that instead of the repeated series (87) we now need the value of

$$S_1(a,b,c) = \lim_{M\to\infty}\lim_{N\to\infty}\sum_{m=-M}^{M}\sum_{n=-N}^{N}\left(\frac{(-1)^m}{a+bn+icn}\right). \tag{105}$$

This is easily calculated with the aid of (88) and (89) as follows. Instead of (88) we obtain for the present case

$$S_1 = \frac{\pi}{b}\left[\cot\frac{\pi a}{b}+\sum_{m=1}^{\infty}(-1)^m\left\{\cot\frac{\pi}{b}(a+icm)+\cot\frac{\pi}{b}(a-icm)\right\}\right].$$

The absolute convergence of this series permits its rearrangement into

$$S_1 = \frac{\pi}{b}\left[\cot\frac{\pi a}{b}+\sum_{r=1}^{\infty}\left\{\cot\frac{\pi}{b}(a+2icr)+\cot\frac{\pi}{b}(a-2icr)\right\}\right]$$

$$-\frac{\pi}{b}\sum_{s=1}^{\infty}\left\{\cot\frac{\pi}{b}(a+2ics-ic)+\cot\frac{\pi}{b}(a-2ics+ic)\right\}$$

$$=\frac{\pi}{b}\left[\cot\frac{\pi a}{b}+\sum_{m=1}^{\infty}\left\{\cot\frac{\pi}{b}(a+2icm)+\cot\frac{\pi}{b}(a-2icm)\right\}\right]$$

$$-\frac{\pi}{b}\left[\cot\frac{\pi}{b}(a+ic)+\lim_{M\to\infty}\sum_{m=1}^{M}\left\{\cot\frac{\pi}{b}(a+ic+2icm)\right.\right.$$

$$\left.\left.+\cot\frac{\pi}{b}(a+ic-2icm)\right\}+\lim_{M\to\infty}\cot\frac{\pi}{b}(a+ic-2icM)\right]$$

$$=\frac{2K_1}{b}\left\{Z_1\left(\frac{2K_1a}{b}\Big|\mu_1\right)-Z_1\left(\frac{2K_1a}{b}+iK_1'\,|\,\mu_1\right)-\frac{\pi i}{2K_1}\right\},$$

$$\mu_1=\frac{iK_1'}{K_1}=\frac{2ic}{b}, \tag{106}$$

by (88) and (89). Hence from (35) and (67)

$$S_1 = \frac{2K_1}{b}\left\{Z_1\left(\frac{2K_1a}{b}\Big|\mu_1\right)-Z_4\left(\frac{2K_1a}{b}\Big|\mu_1\right)\right\}=\frac{2K_1}{b}\frac{\mathrm{sn}'}{\mathrm{sn}}\left(\frac{2K_1a}{b},k_1\right), \tag{107}$$

where
$$k_1 = \vartheta_2^2(0\,|\,\mu_1)/\vartheta_3^2(0\,|\,\mu_1), \tag{108}$$

and
$$\frac{\mathrm{sn}'}{\mathrm{sn}}(u,k)\equiv\frac{\mathrm{sn}'(u,k)}{\mathrm{sn}(u,k)}.$$

The subscript 1 has been introduced here (for K_1, K_1', and k_1) in conformity with the notation introduced in (36), for we are now dealing with a parameter μ_1 twice the value of that appearing in §4.9.

If $\tilde{S}_1$ is the value of the double sum in (105) when the limits are reversed in order then, by an application of (6),

$$
\begin{aligned}
\tilde{S}_1 &= \frac{\pi}{c} \lim_{N\to\infty} \sum_{n=-N}^{N} \operatorname{cosech} \frac{\pi}{c}(a+bn) \\[2mm]
&= \frac{\pi i}{2c} \lim_{N\to\infty} \sum_{n=-N}^{N} \left\{ \cot\frac{\pi i}{2c}(a+bn) + \tan\frac{\pi i}{2c}(a+bn) \right\} \\[2mm]
&= \frac{\pi i}{2c}\left[\cot\frac{\pi i a}{2c} + \sum_{n=1}^{\infty}\left\{ \cot\frac{\pi i}{2c}(a+bn) + \cot\frac{\pi i}{2c}(a-bn) \right\} \right] \\[2mm]
&\quad - \frac{\pi i}{2c}\left[\cot\left(\frac{\pi i a}{2c} + \frac{\pi}{2}\right) + \sum_{n=1}^{\infty}\left\{ \cot\frac{\pi i}{2c}(a-ic+bn) + \cot\frac{\pi i}{2c}(a-ic-bn) \right\} \right] \\[2mm]
&= \frac{K_1' i}{c}\left\{ Z_1\left(\frac{K_1' i a}{c} \,\middle|\, \frac{ib}{2c}\right) - Z_1\left(\frac{K_1' i a}{c} - K_1' \,\middle|\, \frac{ib}{2c}\right) \right\},
\end{aligned}
$$

by an application of (88) and (89), and using the K-notation already introduced in (106). It now follows from the second of (35), (38) and (106) that

$$
\tilde{S}_1 = \frac{2K_1}{b}\left\{ Z_1\left(\frac{2K_1 a}{b} \,\middle|\, \tau_1\right) - Z_4\left(\frac{2K_1 a}{b} \,\middle|\, \tau_1\right) \right\} = S_1,
$$

by (107). Thus the double series in (105) is convergent, the order of limit taking being immaterial.

Applying the result contained in (105) and (107) to the equation corresponding to (86), i.e. an equation containing the factor $(-1)^m$, now gives

$$
\begin{aligned}
\tau(z) &= \frac{K_1}{2\pi i l}\int_{-l}^{l}\left[\tau_0(x)\frac{\operatorname{sn}'}{\operatorname{sn}}\left\{\frac{K_1}{l}(x-z)\right\} - \tau_h(x)\frac{\operatorname{sn}'}{\operatorname{sn}}\left\{\frac{K_1}{l}(x+ih-z)\right\} \right]dx \\[2mm]
&\quad + \frac{K_1}{2\pi l}\int_{0}^{h}\left[\tau^+(y)\frac{\operatorname{sn}'}{\operatorname{sn}}\left\{\frac{K_1}{l}(iy+l-z)\right\} - \tau^-(y)\frac{\operatorname{sn}'}{\operatorname{sn}}\left\{\frac{K_1}{l}(iy-l-z)\right\} \right]dy,
\end{aligned}
\tag{109}
$$

where
$$
\frac{K_1'}{K_1} = \frac{\mu_1}{i} = \frac{2h}{l}.
\tag{110}
$$

Now (48) gives

$$
\operatorname{sn}\left\{\frac{K_1}{l}(iy\pm l-z)\right\} = \operatorname{sn}\left\{\frac{K_1}{l}(iy-z)\pm K_1\right\} = \pm\operatorname{cd}\left\{\frac{K_1}{l}(iy-z)\right\},
$$

so
$$
\frac{\operatorname{sn}'}{\operatorname{sn}}\left\{\frac{K_1}{l}(iy\pm l-z)\right\} = \frac{\operatorname{cd}'}{\operatorname{cd}}\left\{\frac{K_1}{l}(iy-z)\right\},
$$

and (109) can be written

$$\tau(z) = \frac{K_1}{2\pi i l} \int_{-l}^{l} \left[\tau_0(x) \frac{\mathrm{sn}'}{\mathrm{sn}} \left\{ \frac{K_1}{l}(x-z) \right\} - \tau_h(x) \frac{\mathrm{sn}'}{\mathrm{sn}} \left\{ \frac{K_1}{l}(x-z) + \frac{iK_1'}{2} \right\} \right] dx$$
$$+ \frac{K_1}{2\pi l} \int_0^h \varpi(y) \frac{\mathrm{cd}'}{\mathrm{cd}} \left\{ \frac{K_1}{l}(iy-z) \right\} dy, \quad (111)$$

on making use of (104) and (110).

The same type of argument as used to derive (93) yields in the present case that

$$0 = \frac{-K_1}{2\pi i l} \int_{-l}^{l} \left[\bar{\tau}_0(x) \frac{\mathrm{sn}'}{\mathrm{sn}} \left\{ \frac{K_1}{l}(x-z) \right\} - \bar{\tau}_h(x) \frac{\mathrm{sn}'}{\mathrm{sn}} \left\{ \frac{K_1}{l}(x-z) - \frac{iK_1'}{2} \right\} \right] dx$$
$$- \frac{K_1}{2\pi l} \int_0^h \varpi(y) \frac{\mathrm{cd}'}{\mathrm{cd}} \left\{ \frac{K_1}{l}(iy+z) \right\} dy, \quad (112)$$

where advantage has been taken of the fact that $(\mathrm{cd}'/\mathrm{cd})(u)$ is an odd function of u. Now from (48)

$$\mathrm{sn}\left\{ \frac{K_1}{l}(x-z) - \frac{iK_1'}{2} \right\} = \frac{1}{k}\,\mathrm{ns}\left\{ \frac{K_1}{l}(x-z) + \frac{iK_1'}{2} \right\},$$

so

$$\frac{\mathrm{sn}'}{\mathrm{sn}}\left\{ \frac{K_1}{l}(x-z) - \frac{iK_1'}{2} \right\} = -\frac{\mathrm{sn}'}{\mathrm{sn}}\left\{ \frac{K_1}{l}(x-z) + \frac{iK_1'}{2} \right\}. \quad (113)$$

If therefore we add (111) and (112) we find with the aid of (113) that

$$\tau(z) = \frac{K_1}{\pi l} \int_{-l}^{l} \left[\theta_0(x) \frac{\mathrm{sn}'}{\mathrm{sn}} \left\{ \frac{K_1}{l}(x-z) \right\} + i\Omega_h(x) \frac{\mathrm{sn}'}{\mathrm{sn}} \left\{ \frac{K_1}{l}(x-z) + \frac{iK_1'}{2} \right\} \right] dx$$
$$+ \frac{K_1}{2\pi l} \int_0^h \varpi(y) \left[\frac{\mathrm{cd}'}{\mathrm{cd}} \left\{ \frac{K_1}{l}(iy-z) \right\} - \frac{\mathrm{cd}'}{\mathrm{cd}} \left\{ \frac{K_1}{l}(iy+z) \right\} \right] dy, \quad (114)$$

and this is one form of the solution of the boundary value problem specified in (104).

It contains (9) and (19) as the special cases $l \to \infty$ and $h \to \infty$, which can be verified by a straightforward calculation based on (53), (57) and (58).

The alternative form,

$$\tau(z) = \frac{K_1}{\pi l} \int_{-l}^{l} \theta_0(x) \frac{\mathrm{cn}\frac{2K_1}{l}(x-z) + \mathrm{dn}\frac{2K_1}{l}(x-z)}{\mathrm{sn}\frac{2K_1}{l}(x-z)} dx$$

$$+ \frac{K_1}{\pi l} \int_{-l}^{l} \Omega_h(x) \frac{\mathrm{ds}\frac{2K_1}{l}(x-z) + k\,\mathrm{cs}\frac{2K_1}{l}(x-z)}{\mathrm{ns}\frac{2K_1}{l}(x-z)} dx$$

$$+ \frac{K_1 k_1'^2}{\pi l} \int_0^h \frac{\mathrm{sn}\frac{2K_1}{l}z \; \varpi(y)\,dy}{\mathrm{cn}\frac{2K_1}{l}iy\,\mathrm{dn}\frac{2K_1}{l}z + \mathrm{dn}\frac{2K_1}{l}iy\,\mathrm{cn}\frac{2K_1}{l}z}, \quad (115)$$

where the modulus of the elliptic functions is still k_1, can be obtained from (114) as follows.

Making use of several of the relations given in §4.6 we get

$$\frac{\mathrm{sn}'}{\mathrm{sn}}\{\tfrac{1}{2}(x-y)\} = \frac{1}{2}\frac{d}{dx}\ln\{\mathrm{sn}^2\,\tfrac{1}{2}(x-y)\} = \frac{1}{2}\frac{d}{dx}\ln\left(\frac{1-\mathrm{cn}\,(x-y)}{1+\mathrm{dn}\,(x-y)}\right)$$

$$= \frac{\mathrm{sn}\,(x-y)\,\mathrm{dn}\,(x-y)}{1-\mathrm{cn}\,(x-y)} + \frac{k_1^2\,\mathrm{sn}\,(x-y)\,\mathrm{cn}\,(x-y)}{1+\mathrm{dn}\,(x-y)}$$

$$= [\{1+\mathrm{cn}(x-y)\}\,\mathrm{dn}\,(x-y)$$

$$+\{1-\mathrm{dn}\,(x-y)\}\,\mathrm{cn}\,(x-y)]/\mathrm{sn}\,(x-y).$$

Therefore
$$\frac{\mathrm{sn}'}{\mathrm{sn}}\{\tfrac{1}{2}(x-y)\} = \frac{\mathrm{cn}\,(x-y)+\mathrm{dn}\,(x-y)}{\mathrm{sn}\,(x-y)}. \tag{116}$$

Also
$$\frac{\mathrm{cd}'}{\mathrm{cd}}\{\tfrac{1}{2}(x-y)\} - \frac{\mathrm{cd}'}{\mathrm{cd}}\{\tfrac{1}{2}(x+y)\} = \frac{d}{dx}\ln\frac{\mathrm{cd}\,\tfrac{1}{2}(x-y)}{\mathrm{cd}\,\tfrac{1}{2}(x+y)}$$

$$= \frac{d}{dx}\ln\left\{\frac{\mathrm{cn}\,\tfrac{1}{2}x\,\mathrm{dn}\,\tfrac{1}{2}x\,\mathrm{cn}\,\tfrac{1}{2}y\,\mathrm{dn}\,\tfrac{1}{2}y + (1-k_1^2)\,\mathrm{sn}\,\tfrac{1}{2}x\,\mathrm{sn}\,\tfrac{1}{2}y}{\mathrm{cn}\,\tfrac{1}{2}x\,\mathrm{dn}\,\tfrac{1}{2}x\,\mathrm{cn}\,\tfrac{1}{2}y\,\mathrm{dn}\,\tfrac{1}{2}y - (1-k_1^2)\,\mathrm{sn}\,\tfrac{1}{2}x\,\mathrm{sn}\,\tfrac{1}{2}y}\right\},$$

which after differentiation and some straightforward algebra based on (62) and (63) reduces to

$$\frac{\mathrm{cd}'}{\mathrm{cd}}\{\tfrac{1}{2}(x-y)\} - \frac{\mathrm{cd}'}{\mathrm{cd}}\{\tfrac{1}{2}(x+y)\} = \frac{2(1-k_1^2)\,\mathrm{sn}\,y}{\mathrm{cn}\,x\,\mathrm{dn}\,y + \mathrm{dn}\,x\,\mathrm{cn}\,y}. \tag{117}$$

Equations (116) and (117) enable us to change (114) into the form given in (115).

4.13   The various boundary value problems for the rectangle

Fig. 4.13a shows the basic rectangle $-l \leqslant x \leqslant l$, $0 \leqslant y \leqslant h$, labelled $(0,0)$, together with the doubly infinite lattice of alternate rectangles $(2n-1)l \leqslant x \leqslant (2n+1)l, rh \leqslant y \leqslant (r+1)h$, where $n, r = \pm 2, \pm 4, \pm 6, \ldots$. The value of (n, r) is shown in the figure for some of these rectangles in the vicinity of the basic rectangle. Notice that r is related to m of §4.9 by $r = 2m$.

Referring to Fig. 4.9 and to the theory of §§4.9 and 4.12 we see that:

(i) By adding Cauchy integrals for points in the set of all rectangles having n values, $0, \pm 1, \pm 2, \pm 3, \ldots$ (i.e. rectangles *contiguous* in the x-direction) we were able to generate *periodic* boundary conditions in the x-direction—or more generally boundary conditions involving the difference of Ω (or θ) over one period.

(ii) By similarly using points in the rectangles $r = 0, \pm 2, \pm 4, \ldots$, together with a type of conjugate Cauchy integral (93), which in fact applies to the set of points in the alternate rectangles $r = \pm 1, \pm 3, \pm 5, \ldots,$ we obtained *Dirichlet*-type boundary conditions in the y-direction, i.e. on $y = 0, y = h$; and

(iii) By *adding* the integrals corresponding to the set of values

$$r = 0, \pm 4, \pm 8, \ldots,$$

and *subtracting* the integrals appropriate to

$$r = \pm 2, \pm 6, \pm 10, \ldots,$$

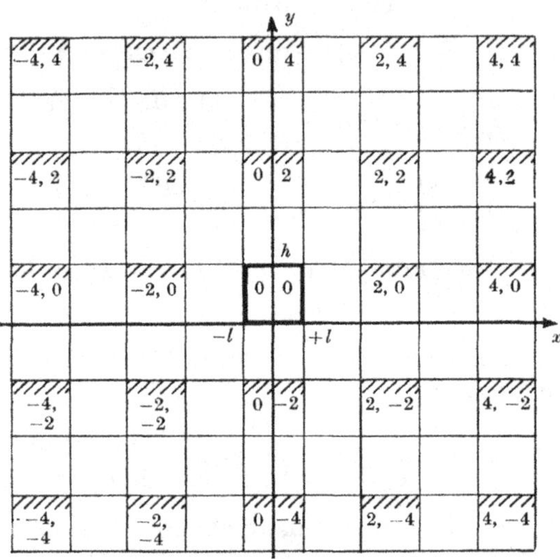

Fig. 4.13 a

then using the alternate rectangles

$$r = \pm 1, \pm 5, \pm 9, \ldots, \quad r = \pm 3, \pm 7, \pm 11, \ldots,$$

as in (ii), we generated *mixed* boundary conditions in the y-direction.

This analysis of our earlier methods into three basic operations, viz. (i) add for contiguous rectangles, (ii) add for every second rectangle and (iii) of the rectangles in (ii) add and subtract alternatively, immediately suggests how a variety of solutions for boundary value problems in a rectangle may be obtained. Each of operations (i), (ii) and (iii) can be carried out in both the x- and y-directions, and this would generate nine distinct solutions, not counting harmonic conjugate solutions. These latter cannot really be counted as distinct so we shall ignore them. Let the nine solutions be labelled (i, ii), (ii, ii), (iii, ii), etc., where the first operation indicated in the bracket applies to the x-direction, and the second to the y-direction.

Of the nine solutions three are immediately seen to be redundant, for there is no essential difference between (i, ii) and (ii, i), and other similar pairs—only a rotation of axes being involved. We shall examine the remaining six solutions in turn.

(1) *Solution* (i, i)

If we add for contiguous rectangles in *both* directions we obtain an equation like (86), except that $2m$ is replaced by r, and the sum extends now over all integer values of r. This is equivalent to replacing the ratio h/l by $h/2l$, i.e. to dividing the parameter μ of the theta functions in (90) by 2. Let $\mu_0 = \tfrac{1}{2}\mu = ih/2l$, $K_0 = (\tfrac{1}{2}\pi)\,\vartheta_3^2(0\,|\,\mu_0)$, $iK_0' = K_0\mu_0$, then changing the parameter in (90), and making use of (34), we find for the present problem

$$\tau(z) = \frac{1}{2l}\int_{-l}^{l}\tau_h(x)\,dx + \frac{K_0}{2\pi i l}\int_{-l}^{l}\{\tau_0(x)-\tau_h(x)\}\,Z_1\!\left\{\frac{K_0}{l}(x-z)\,\big|\,\mu_0\right\}dx$$

$$+\frac{K_0}{2\pi l}\int_{0}^{h}\{\tau^+(y)-\tau^-(y)\}\,Z_2\!\left\{\frac{K_0}{l}(iy-z)\,\big|\,\mu_0\right\}dy. \quad (118)$$

The first term on the right-hand side can be expressed in a variety of forms with the aid of (95). Thus we have the solution to the problem 'find a harmonic function given the differences $\tau_0-\tau_h$, $\tau^+-\tau^-$', but practical problems in aerodynamics rarely, if ever, give rise to such boundary value problems.

(2) *Solution* (i, ii)

This has been calculated in §4.9. It is important in unsteady aerofoil theory (see §14.1). A slight generalization of the solution, (94), can be obtained by replacing (c) of (85) by $\theta^+(y)-\theta^-(y)=f(y)$, the effect of which is to replace $\varpi(y)$ in (94) by $\varpi(y)+if(y)$.

(3) *Solution* (i, iii)

This is given in §4.12, and the same remarks as given for solution (i, ii) apply to it.

(4) *Solution* (ii, ii)

This solution, given below in §4.14, yields two distinct cases, namely (a) the solution to the Dirichlet problem in a rectangle, e.g. $\theta_0(x)$, $\theta_h(x)$, $\theta^+(y)$, $\theta^-(y)$ given functions—not important in fluid mechanics—and (b) the solution to a mixed boundary value problem of the type $\theta_0(x)$, $\theta_h(x)$, $\Omega^+(y)$, $\Omega^-(y)$ given functions—of importance in cavitating flows (see §7.24).

(5) *Solution (ii, iii)*

We have not found this solution, of which a typical case is shown in Fig. 4.13c, of value in fluid mechanics. The reader will have no difficulty in deriving the solution by an obvious modification of the theory given in § 4.12, if it is of interest to him.

Solution (ii, ii) a

Solution (ii, ii) b

Fig. 4.13b

(6) *Solution (iii, iii)*

A typical case is shown in Fig. 4.13d; it is not important in fluid dynamics, and a remark similar to that given for (ii, iii) applies.

Solution (ii, iii)

Solution (iii, iii)

Fig. 4.13c

Fig. 4.13d

4.14   The Dirichlet and mixed problems for the rectangle

The first step in the calculation of solution (ii, ii) is to replace n by $2s$ in (86), and then sum over integer values of s instead of n. This is clearly equivalent to replacing l by $2l$. Thus corresponding to (90) we obtain

$$\tau(z) = \frac{K_0}{4\pi i l} \int_{-l}^{l} \left[\tau_0(x) Z_1\left\{\frac{K_0}{2l}(x-z)\middle| \frac{ih}{2l}\right\} - \tau_h(x) Z_1\left\{\frac{K_0}{2l}(x+ih-z)\middle| \frac{ih}{2l}\right\} \right] dx$$

$$+ \frac{K_0}{4\pi l} \int_0^h \left[\tau^+(y) Z_1\left\{\frac{K_0}{2l}(iy+l-z)\middle| \frac{ih}{2l}\right\} \right.$$

$$\left. - \tau^-(y) Z_1\left\{\frac{K_0}{2l}(iy-l-z)\middle| \frac{ih}{2l}\right\} \right] dy, \quad (119)$$

where the parameter of the zeta functions is

$$\mu_0 = \frac{iK_0'}{K_0} = \frac{ih}{2l} \tag{120}$$

(cf. (92) and (110)).

A glance at Fig. 4.13a shows that for each rectangle occupied by one of the points used in deriving (119), three other rectangles remain unoccupied. Fig. 4.14 shows the basic rectangle B, and three adjacent rectangles containing the image points $\bar{z}$, $2l-\bar{z}$ and $2l-z$. As the points $\bar{z}+4sl+2mih$, $2l-\bar{z}+4sl+2mih$, $2l-z+4sl+2mih$, $s, n = 0, \pm 1, \pm 2, \ldots$, all lie outside B, it follows from Cauchy's theorem that if z in (119) is replaced by $\bar{z}$, $2l-\bar{z}$, and $2l-z$ in turn, the left-hand sides of the three resulting equations will be zero. The complex conjugates of the equations derived from the points $\bar{z}$ and $2l-\bar{z}$ are

$$0 = \frac{-K_0}{4\pi i l} \int_{-l}^{l} \left[\bar{\tau}_0(x) Z_1\left\{\frac{K_0}{2l}(x-z) \,|\, \mu_0\right\} - \bar{\tau}_h(x) Z_1\left\{\frac{K_0}{2l}(x-ih-z) \,|\, \mu_0\right\} \right] dx$$

$$- \frac{K_0}{4\pi l} \int_0^h \left[\bar{\tau}^+(y) Z_1\left\{\frac{K_0}{2l}(iy-l+z) \,|\, \mu_0\right\} \right.$$

$$\left. - \bar{\tau}^-(y) Z_1\left\{\frac{K_0}{2l}(iy+l+z) \,|\, \mu_0\right\} \right] dy, \quad (121)$$

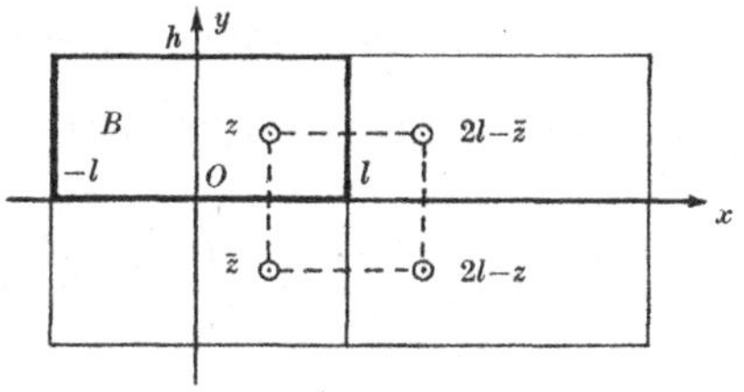

Fig. 4.14

and

$$0 = -\frac{K_0}{4\pi i l} \int_{-l}^{l} \left[\bar{\tau}_0(x) Z_2\left\{\frac{K_0}{2l}(x+z) \,|\, \mu_0\right\} - \bar{\tau}_h(x) Z_2\left\{\frac{K_0}{2l}(x-ih+z) \,|\, \mu_0\right\} \right] dx$$

$$- \frac{K_0}{4\pi l} \int_0^h \left[\bar{\tau}^+(y) Z_2\left\{\frac{K_0}{2l}(iy-l-z) \,|\, \mu_0\right\} \right.$$

$$\left. - \bar{\tau}^-(y) Z_2\left\{\frac{K_0}{2l}(iy+l-z) \,|\, \mu_0\right\} \right] dy, \quad (122)$$

where we have made use of (35) and the fact that the zeta functions are odd functions of their arguments. The equation derived from the point $2l-z$ can be written

$$0 = \frac{K_0}{4\pi i l} \int_{-l}^{l} \left[\tau_0(x) Z_2\left\{\frac{K_0}{2l}(x+z) \,|\, \mu_0\right\} - \tau_h(x) Z_2\left\{\frac{K_0}{2l}(x+ih+z) \,|\, \mu_0\right\} \right] dx$$

$$+ \frac{K_0}{4\pi l} \int_0^h \left[\tau^+(y) Z_2\left\{\frac{K_0}{2l}(iy+l+z) \,|\, \mu_0\right\} \right.$$

$$\left. - \tau^-(y) Z_2\left\{\frac{K_0}{2l}(iy-l+z) \,|\, \mu_0\right\} \right] dy, \quad (123)$$

on again using (35). It remains only to combine (119), (121) to (123) in the two distinct ways possible in order to reach the solutions we desire.

First combining (119) and (123), and making further use of (35) we obtain

$$\tau(z) = \mathscr{A} + \frac{K_0}{4\pi i l}\int_{-l}^{l}\left(\tau_0(x)\left[Z_1\Big\{\frac{K_0}{2l}(x-z)\Big\} \pm Z_2\Big\{\frac{K_0}{2l}(x+z)\Big\}\right]\right.$$

$$-\tau_h(x)\left[Z_4\Big\{\frac{K_0}{2l}(x-z)\Big\} \pm Z_3\Big\{\frac{K_0}{2l}(x+z)\Big\}\right]\bigg)\,dx$$

$$+\frac{K_0}{4\pi l}\int_{0}^{h}\left(\tau^+(y)\left[Z_1\Big\{\frac{K_0}{2l}(iy+l-z)\Big\} \pm Z_2\Big\{\frac{K_0}{2l}(iy+l+z)\Big\}\right]\right.$$

$$-\tau^-(y)\left[Z_2\Big\{\frac{K_0}{2l}(iy+l-z)\Big\} \pm Z_1\Big\{\frac{K_0}{2l}(iy+l+z)\Big\}\right]\bigg)\,dy,\quad (124)$$

where $\mathscr{A}$ is $\dfrac{1}{4l}\displaystyle\int_{-l}^{l}\tau_h(x)\,dx$ or zero depending on whether the upper or lower sign is taken. Similarly, it follows from (121) and (122) that

$$0 = \bar{\mathscr{A}} - \frac{K_0}{4\pi i l}\int_{-l}^{l}\left(\bar{\tau}_0(x)\left[Z_1\Big\{\frac{K_0}{2l}(x-z)\Big\} \pm Z_2\Big\{\frac{K_0}{2l}(x+z)\Big\}\right]\right.$$

$$-\bar{\tau}_h(x)\left[Z_4\Big\{\frac{K_0}{2l}(x-z)\Big\} \pm Z_3\Big\{\frac{K_0}{2l}(x+z)\Big\}\right]\bigg)\,dx$$

$$-\frac{K_0}{4\pi l}\int_{0}^{h}\left(\bar{\tau}^+(y)\left[Z_2\Big\{\frac{K_0}{2l}(iy+l+z)\Big\} \pm Z_1\Big\{\frac{K_0}{2l}(iy+l-z)\Big\}\right]\right.$$

$$-\bar{\tau}^-(y)\left[Z_1\Big\{\frac{K_0}{2l}(iy+l+z)\Big\} \pm Z_2\Big\{\frac{K_0}{2l}(iy+l-z)\Big\}\right]\bigg)\,dy.\quad (125)$$

If (124) and (125) are now added, and the two distinct solutions separated, we obtain

$$\tau(z) = \frac{1}{2l}\int_{-l}^{l}\Omega_h(x)\,dx + \frac{K_0}{2\pi l}\int_{-l}^{l}\left(\theta_0(x)\left[Z_1\Big\{\frac{K_0}{2l}(x-z)\Big\} + Z_2\Big\{\frac{K_0}{2l}(x+z)\Big\}\right]\right.$$

$$-\theta_h(x)\left[Z_4\Big\{\frac{K_0}{2l}(x-z)\Big\} + Z_3\Big\{\frac{K_0}{2l}(x+z)\Big\}\right]\bigg)\,dx$$

$$+\frac{iK_0}{2\pi l}\int_{0}^{h}\left(\theta^+(y)\left[Z_1\Big\{\frac{K_0}{2l}(iy+l-z)\Big\} + Z_2\Big\{\frac{K_0}{2l}(iy+l+z)\Big\}\right]\right.$$

$$-\theta^-(y)\left[Z_1\Big\{\frac{K_0}{2l}(iy+l+z)\Big\} + Z_2\Big\{\frac{K_0}{2l}(iy+l-z)\Big\}\right]\bigg)\,dy,\quad (126)$$

and

$$\tau(z) = \frac{K_0}{2\pi l} \int_{-l}^{l} \left(\theta_0(x) \left[Z_1\left\{\frac{K_0}{2l}(x-z)\right\} - Z_2\left\{\frac{K_0}{2l}(x+z)\right\} \right] \right.$$

$$- \theta_h(x) \left[Z_4\left\{\frac{K_0}{2l}(x-z)\right\} - Z_3\left\{\frac{K_0}{2l}(x+z)\right\} \right] \right) dx$$

$$+ \frac{K_0}{2\pi l} \int_{0}^{h} \left(\Omega^+(y) \left[Z_1\left\{\frac{K_0}{2l}(iy+l-z)\right\} - Z_2\left\{\frac{K_0}{2l}(iy+l+z)\right\} \right] \right.$$

$$\left. - \Omega^-(y) \left[Z_2\left\{\frac{K_0}{2l}(iy+l-z)\right\} - Z_1\left\{\frac{K_0}{2l}(iy+l+z)\right\} \right] \right) dy, \quad (127)$$

which are the solutions to the problems defined in Fig. 4.13b.

More convenient forms of these two equations can be derived with the aid of (75). If we replace the parameters k and k_1 in (75) by k_0 and k respectively, i.e. replace l by $2l$ (see (92) and (120)) then we shall have

$$K_0 Z_1\left\{\frac{K_0}{2l}(x-z), k_0\right\} = KZ_4\left\{\frac{K}{l}x, k\right\} - KZ_4\left\{\frac{Kz}{l}, k\right\}$$

$$+ K \frac{\operatorname{cn}\dfrac{Kz}{l}\operatorname{dn}\dfrac{Kz}{l} + \operatorname{cn}\dfrac{Kx}{l}\operatorname{dn}\dfrac{Kx}{l}}{\operatorname{sn}\dfrac{Kx}{l} - \operatorname{sn}\dfrac{Kz}{l}}. \quad (128)$$

Now replace z by $2l-z$ in this expression, and use (34) and (35) to find

$$K_0 Z_2\left\{\frac{K_0}{2l}(x+z), k_0\right\} = KZ_4\left\{\frac{Kx}{l}, k\right\} + KZ_4\left\{\frac{Kz}{l}, k\right\}$$

$$- K \frac{\operatorname{cn}\dfrac{Kz}{l}\operatorname{dn}\dfrac{Kz}{l} - \operatorname{cn}\dfrac{Kx}{l}\operatorname{dn}\dfrac{Kx}{l}}{\operatorname{sn}\dfrac{Kx}{l} - \operatorname{sn}\dfrac{Kz}{l}}. \quad (129)$$

Therefore, by addition and subtraction of (128) and (129) we get

$$K_0\left[Z_1\left\{\frac{K_0}{2l}(x-z)\right\} + Z_2\left\{\frac{K_0}{2l}(x+z)\right\} \right] = 2KZ_4\left(\frac{Kx}{l}, k\right) + \frac{2K\operatorname{cn}\dfrac{Kx}{l}\operatorname{dn}\dfrac{Kx}{l}}{\operatorname{sn}\dfrac{Kx}{l} - \operatorname{sn}\dfrac{Kz}{l}},$$

$$(130)$$

and

$$K_0\left[Z_1\left\{\frac{K_0}{2l}(x-z)\right\} - Z_2\left\{\frac{K_0}{2l}(x+z)\right\} \right] = -2KZ_4\left(\frac{Kz}{l}, k\right)$$

$$+ \frac{2K\operatorname{cn}\dfrac{Kz}{l}\operatorname{dn}\dfrac{Kz}{l}}{\operatorname{sn}\dfrac{Kx}{l} - \operatorname{sn}\dfrac{Kz}{l}}. \quad (131)$$

The substitution of $x+ih$ for x in these equations gives

$$K_0\left[Z_4\left\{\frac{K_0}{2l}(x-z)\right\}+Z_3\left\{\frac{K_0}{2l}(x+z)\right\}\right]=2KZ_1\left(\frac{Kx}{l},k\right)-\frac{2K\,\mathrm{ds}\dfrac{Kx}{l}\,\mathrm{cs}\dfrac{Kx}{l}}{\mathrm{ns}\dfrac{Kx}{l}-k\,\mathrm{sn}\dfrac{Kz}{l}},$$

$$\tag{132}$$

and

$$K_0\left[Z_4\left\{\frac{K_0}{2l}(x-z)\right\}-Z_3\left\{\frac{K_0}{2l}(x+z)\right\}\right]=-2KZ_4\left(\frac{Kz}{l},k\right)$$

$$+\frac{2Kk\,\mathrm{cn}\dfrac{Kz}{l}\,\mathrm{dn}\dfrac{Kz}{l}}{\mathrm{ns}\dfrac{Kx}{l}-k\,\mathrm{sn}\dfrac{Kz}{l}},\quad\tag{133}$$

by (35) and (48). Finally the substitution $x=iy\pm l$ in (130) and (131) yields similarly that

$$K_0\left[Z_1\left\{\frac{K_0}{2l}(iy+l\pm z)\right\}+Z_2\left\{\frac{K_0}{2l}(iy+l\mp z)\right\}\right]=2KZ_3\left(\frac{Kiy}{l},k\right)$$

$$-\frac{2K}{k'^2}\frac{\mathrm{sd}\dfrac{Kiy}{l}\,\mathrm{nd}\dfrac{Kiy}{l}}{\mathrm{cd}\dfrac{Kiy}{l}\pm\mathrm{sn}\dfrac{Kz}{l}},\quad\tag{134}$$

and

$$K_0\left[Z_1\left\{\frac{K_0}{2l}(iy+l\pm z)\right\}-Z_2\left\{\frac{K_0}{2l}(iy+l\mp z)\right\}\right]=\pm2KZ_4\left(\frac{Kz}{l},k\right)$$

$$+\frac{2K\,\mathrm{cn}\dfrac{Kz}{l}\,\mathrm{dn}\dfrac{Kz}{l}}{\mathrm{cd}\dfrac{Kiy}{l}\pm\mathrm{sn}\dfrac{Kz}{l}}.\quad\tag{135}$$

The contour integral about the basic rectangle B,

$$\int_B\tau(t)Z_4\left(\frac{Kt}{l}\right)dt=\int_{-l}^{l}\tau_0(x)Z_4\left(\frac{Kx}{l}\right)dx-\int_{-l}^{l}\tau_h(x)Z_1\left(\frac{Kx}{l}\right)dx$$

$$+\frac{i\pi}{2K}\int_{-l}^{l}\tau_h(x)\,dx+i\int_0^h\tau^+(y)Z_3\left(\frac{Kiy}{l}\right)dy$$

$$-i\int_0^h\tau^-(y)Z_3\left(\frac{Kiy}{l}\right)dy=\frac{l\pi i}{K}\tau(ih),\tag{136}$$

by the extension of the Residue Theorem given in §3.11. For $\tau(z)$ is regular within B and has no simple poles on B, while $Z_4(Kz/l)$ is regular within B, but has a simple pole of residue l/K (see discussion following (35)) at $z=iK'l/K=ih$.

Equations (130), (132), (134) and the imaginary part of (136) now permit us to write (126) in the form

$$\tau(z) = \Omega(ih) + \frac{K}{\pi l}\int_{-l}^{l}\left(\theta_0(x)\frac{\operatorname{cn}\dfrac{Kx}{l}\operatorname{dn}\dfrac{Kx}{l}}{\operatorname{sn}\dfrac{Kx}{l}-\operatorname{sn}\dfrac{Kz}{l}}+\theta_h(x)\frac{\operatorname{ds}\dfrac{Kx}{l}\operatorname{cs}\dfrac{Kx}{l}}{\operatorname{ns}\dfrac{Kx}{l}-k\operatorname{sn}\dfrac{Kz}{l}}\right)dx$$

$$-\frac{iK}{\pi lk'^2}\int_{0}^{h}\left(\theta^+(y)\frac{\operatorname{sd}\dfrac{Kiy}{l}\operatorname{nd}\dfrac{Kiy}{l}}{\operatorname{cd}\dfrac{Kiy}{l}-\operatorname{sn}\dfrac{Kz}{l}}-\theta^-(y)\frac{\operatorname{sd}\dfrac{Kiy}{l}\operatorname{nd}\dfrac{Kiy}{l}}{\operatorname{cd}\dfrac{Kiy}{l}+\operatorname{sn}\dfrac{Kz}{l}}\right)dy,\tag{137}$$

while (131), (133), (135) and the imaginary part of (95) enable us to rewrite (127) as

$$\tau(z) = \frac{K}{\pi l}\operatorname{cn}\frac{Kz}{l}\operatorname{dn}\frac{Kz}{l}\left[\int_{-l}^{l}\left(\frac{\theta_0(x)}{\operatorname{sn}\dfrac{Kx}{l}-\operatorname{sn}\dfrac{Kz}{l}}-\frac{k\theta_h(x)}{\operatorname{ns}\dfrac{Kx}{l}-k\operatorname{sn}\dfrac{Kz}{l}}\right)dx\right.$$

$$\left.+\int_{0}^{h}\left(\frac{\Omega^+(y)}{\operatorname{cd}\dfrac{Kiy}{l}-\operatorname{sn}\dfrac{Kz}{l}}+\frac{\Omega^-(y)}{\operatorname{cd}\dfrac{Kiy}{l}+\operatorname{sn}\dfrac{Kz}{l}}\right)dy\right].\tag{138}$$

THE RIEMANN–HILBERT PROBLEM

4.15 The boundary value problem

The Riemann–Hilbert problem for a singly-connected domain D is that of determining an analytic function $\tau(z) = \Omega(x,y) + i\theta(x,y)$ within D, which satisfies on the boundary C the linear relation

$$\theta(t) = \lambda(t)\,\Omega(t) + g(t)\quad (t\in C),\tag{139}$$

where $\lambda(t)$ and $g(t)$ are given functions. This is the boundary condition described in § 4.1 as being 'continuously mixed'. When $\lambda = 0$, $\theta(t) = g(t)$. Let $\lim_{\lambda\to\infty}\{(\theta-g)/\lambda\} = h(t)$, then (139) yields $\Omega(t) = h(t)$ in the limit $\lambda = \infty$. Thus the boundary conditions studied in §§ 4.2 and 4.14 can be regarded as being special cases of (139).

Of course the function $g(t)$ must be of such a form that $\lim_{\lambda\to\infty}(g/y)$ does exist. A suitable form is $g = \theta_s(t) - \lambda(t)\,\Omega_0(t)$, where θ_s, Ω_s are independent of λ. Then (139) becomes

$$\theta = \lambda(\Omega - \Omega_s) + \theta_s,\tag{140}$$

and we have $\theta = \theta_s$ when $\lambda = 0$, and $\Omega = \Omega_s$ when $\lambda = \pm\infty$.

This type of boundary condition arises in fluid dynamics in the subsonic flow of a fluid past certain kinds of porous walls and also closely perforated walls (see §§ 1.20, 6.4). The porous walls must be of such a construction that the normal velocity through them is proportional to the pressure difference across them, which requires that the index n in (1.59) equals unity. It is also necessary that the velocity normal to the walls be small compared with the tangential component. The theory is given in detail in § 6.4; we shall restrict our attention in this chapter to the general mathematical theory of analytic functions satisfying (140).

4.16   The reduction of the boundary value problem to two Dirichlet problems

The first step towards the solution of our problem is that of finding an analytic function $\tau_0(z)$ $(\tau_0 = \Omega_0 + i\theta_0)$, which satisfies the homogeneous boundary condition

$$\theta_0 = \lambda\Omega_0. \tag{141}$$

This is easily reduced to a Dirichlet-type boundary condition—we simply take the logarithm of $\tau_0(z)$. Thus†

$$f(z) \equiv \ln\tau_0(z) = \ln|\Omega_0 + i\theta_0| + i\tan^{-1}(\theta_0/\Omega_0) \tag{142}$$

is an analytic function satisfying the boundary condition $\mathscr{I}f(t) = \tan^{-1}\lambda$ by (141).

It is convenient here to introduce a new function $\epsilon(t)$ defined by

$$\epsilon \equiv \frac{1}{\pi}\tan^{-1}\lambda; \tag{143}$$

as λ varies over $(-\infty, \infty)$, ϵ ranges over $(-\tfrac{1}{2}, \tfrac{1}{2})$. Our boundary condition for $f(t)$ can now be written

$$\mathscr{I}f(t) = \pi\epsilon. \tag{144}$$

Solutions appropriate to this type of boundary condition have been given for a variety of boundary contours C in §§ 3.14, 3.16, 4.2, 4.9 and 4.14.

The second step towards the solution of (140) is to calculate the value of the function Φ defined by

$$\Phi(z) \equiv \frac{\tau(z)}{\tau_0(z)}, \tag{145}$$

which we shall find also satisfies a Dirichlet type of boundary condition. Now

$$\Phi = \frac{\Omega + i\theta}{\Omega_0 + i\theta_0} = \frac{i+\lambda}{i+\lambda}\frac{\Omega+i\theta}{\Omega_0+i\theta_0} = \frac{\lambda\Omega - \theta + i(\Omega + \lambda\theta)}{\lambda\Omega_0 - \theta_0 + i(\Omega_0 + \lambda\theta_0)},$$

---

† 'ln' denotes the principal value of the logarithm.

and therefore by (140) and (141), on C

$$\Phi = \frac{\Omega + \lambda\theta}{\Omega_0 + \lambda\theta_0} + i\frac{\theta_s - \lambda\Omega_s}{\Omega_0 + \lambda\theta_0}.$$ (146)

Hence from (141), (143) and (146)

$$\mathscr{I}\Phi = \frac{\theta_s - \Omega_s \tan \pi\epsilon}{\Omega_0(1 + \tan^2 \pi\epsilon)},$$

i.e.

$$\mathscr{I}\Phi = \frac{\theta_s \cos \pi\epsilon - \Omega_s \sin \pi\epsilon}{\Omega_0 \sec \pi\epsilon}.$$ (147)

The right-hand side of this equation will be completely known when the first Dirichlet problem (for $\ln \tau_0(z)$) has been solved.

The final step in finding the solution $\tau(z)$ of (140) is simply to combine the solutions for $\tau_0(z)$ and $\Phi(z)$ in accordance with (145).

4.17　General solution for a singly-connected region

If $\mathscr{F}(z)$ is an analytic function in a singly-connected domain D, then it can be expressed in terms of its imaginary part $\mathscr{I}\mathscr{F}(t)$ on the boundary C of D by an equation of the type

$$\mathscr{F}(z) = \frac{1}{2\pi} \int_C \mathscr{I}\mathscr{F}(t)\, Q(z,t)\, dt + \ln A,$$ (148)

where A is a real positive constant depending on $\mathscr{R}\mathscr{F}(t)$, and $Q(z,t)$ is a function which depends on the shape of C (cf. (3.97), (3.109), (3.113) and, more generally, the harmonic conjugate of (5.116)). In applications of (148) the constant A is usually determined from a knowledge of the real part of $\mathscr{F}$ at a particular point z_0 within D.

Applying (148) to the function $f(z)$ defined in (142), and making use of (144) we get

$$\tau_0(z) = A \exp\left\{ \int_C \tfrac{1}{2}\epsilon(t)\, Q(z,t)\, dt \right\}.$$ (149)

Let t^* be a point on C, then from (149)

$$|\tau_0(t^*)| = A \exp\left\{ \mathscr{R} \int_C \tfrac{1}{2}\epsilon(t)\, Q(t^*,t)\, dt \right\},$$

but by (141) and (143)

$$|\tau_0| = |\Omega_0 + i\theta_0| = \sqrt{(\Omega_0^2 + \theta_0^2)} = \Omega_0\sqrt{(1 + \lambda^2)} = \Omega_0 \sec \pi\epsilon.$$ (150)

Hence (147) can now be written

$$\mathscr{I}\Phi = \frac{1}{A}(\theta_s \cos \pi\epsilon - \Omega_s \sin \pi\epsilon) \exp\left\{ -\mathscr{R}\int_C \tfrac{1}{2}\epsilon(t)\, Q(t^*,t)\, dt \right\}.$$ (151)

Applying (148) to this function Φ therefore,

$$\Phi(z) = \frac{1}{2\pi A}\int_C (\theta_s \cos\pi\epsilon - \Omega_s \sin\pi\epsilon)\exp\left\{-\mathscr{R}\int_C \tfrac{1}{2}\epsilon(t^*)\,Q(t,t^*)\,dt^*\right\}$$
$$\times Q(z,t)\,dt + B, \quad (152)$$

where B is a new real constant. It now remains only to combine (145) and (149) with this last result to find the required general solution of the boundary value problem defined in (140). We get

$$\tau(z) = \exp\left\{\int_C \tfrac{1}{2}\epsilon(t)\,Q(z,t)\,dt\right\}\left[K + \frac{1}{2\pi}\int_C (\theta_s \cos\pi\epsilon - \Omega_s \sin\pi\epsilon)\right.$$
$$\left.\times\exp\left\{-\mathscr{R}\int_C \tfrac{1}{2}\epsilon(t^*)\,Q(t,t^*)\,dt^*\right\}Q(z,t)\,dt\right], \quad (153)$$

where K is a new real constant ($= AB$), fixed in value by a prescribed value of τ at a particular point of D.

Notice that when θ_s and Ω_s vanish on C, (153) correctly reduces to (149) for $\tau_0(z)$.

A particular case of (153) of some practical importance occurs when ϵ is a step-function of the form

$$\epsilon = \tfrac{1}{2}[\ \pm\{\mathbf{U}(t-t_1)-\mathbf{U}(t-t_2)\}\pm\{\mathbf{U}(t-t_3)-\mathbf{U}(t-t_4)\}\pm\{\ldots\}]$$

(see §3.6 for a definition of the unit function $\mathbf{U}(t)$), i.e. ϵ has its limit values $\tfrac{1}{2}$ or $-\tfrac{1}{2}$ in the intervals (t_1,t_2), (t_3,t_4), ..., and vanishes in the alternative intervals (t_2,t_3), (t_4,t_5), ... This is the case mentioned in (a) of §4.1, viz. θ is specified over some sections of C, and Ω is specified over the remainder.

4.18   Examples

(1) *The infinite strip*

Let D be the infinite strip $-\infty < x < \infty$, $0 < y < h$, then (3.109) for this domain reads

$$\tau(z) = \frac{1}{2h}\int_{-\infty}^{\infty}\left\{\theta_0(x)\coth\frac{\pi}{2h}(x-z) - \theta_h(x)\tanh\frac{\pi}{2h}(x-z)\right\}dx$$
$$+ \tfrac{1}{2}(\Omega_\infty + \Omega_{-\infty}). \quad (154)$$

Hence for any analytic function $\mathscr{F}(z)$ in this strip

$$\mathscr{F}(z) = \frac{1}{2\pi}\int_C \mathscr{I}\{\mathscr{F}(t)\}\frac{\pi}{h}\coth\frac{\pi}{2h}(t-z)\,dt + \tfrac{1}{2}(\mathscr{R}\mathscr{F}_\infty + \mathscr{R}\mathscr{F}_{-\infty}). \quad (155)$$

Similarly, (3.111) yields

$$\mathscr{R}\mathscr{F}_\infty - \mathscr{R}\mathscr{F}_{-\infty} = -\frac{1}{h}\int_C \mathscr{I}\{\mathscr{F}(t)\}\,dt. \quad (156)$$

Comparison of (148) and (155) gives

$$Q(z, t) = \frac{\pi}{h} \coth \frac{\pi}{2h} (t - z), \tag{157}$$

and the general solution for the infinite strip follows on the substitution of this value of Q in (153).

The constant K in (153) can be evaluated in terms of known quantities as follows. From (148), (149) and (155), in which $\mathscr{F} = \ln \tau_0$, it is clear that the number A satisfies

$$\ln A = \tfrac{1}{2} \{ \mathscr{R}(\ln \tau_0)_\infty + \mathscr{R} \ln (\tau_0)_{-\infty} \} = \tfrac{1}{2} \ln |\tau_{0\infty} \tau_{-0\infty}|,$$

using an obvious notation. Thus

$$A = |\tau_{0\infty} \tau_{-0\infty}|^{\frac{1}{2}}. \tag{158}$$

Similarly, putting $\mathscr{F} = \ln \tau_0$ in (156) and using (142) and (144) we get

$$|\tau_{-0\infty}/\tau_{0\infty}|^{\frac{1}{2}} = \exp \left\{ \frac{\pi}{2h} \int_C \epsilon(t) \, dt \right\}. \tag{159}$$

If next we put $\mathscr{F} = \Phi$ in (155) and use (146) and (150) we find that the constant B in (152) is

$$B = \tfrac{1}{2} \mathscr{R} (\Phi_\infty + \Phi_{-\infty}) = \frac{(\Omega \cos \pi\epsilon + \theta \sin \pi\epsilon)_\infty}{2 |\tau_{0\infty}|} + \frac{(\Omega \cos \pi\epsilon + \theta \sin \pi\epsilon)_{-\infty}}{2 |\tau_{-0\infty}|}. \tag{160}$$

Hence as $K = AB$ it only remains to combine (158) to (160) to find

$$K = \frac{1}{2} \left[(\Omega \cos \pi\epsilon + \theta \sin \pi\epsilon)_\infty \exp \left\{ \frac{\pi}{2h} \int_C \epsilon(t) \, dt \right\} \right.$$
$$\left. + (\Omega \cos \pi\epsilon + \theta \sin \pi\epsilon)_{-\infty} \exp \left\{ \frac{-\pi}{2h} \int_C \epsilon(t) \, dt \right\} \right], \tag{161}$$

which expresses K in terms of the distribution $\epsilon(t)$ and the values at the points $x = \infty, x = -\infty$. It is actually unnecessary to know the values of Ω and θ at *both* of these points, for by putting $\mathscr{F} = \Phi$ in (156) and making use of (151), (157) to (159) we obtain

$$(\Omega \cos \pi\epsilon + \theta \sin \pi\epsilon)_\infty \exp \left\{ \frac{\pi}{2h} \int_C \epsilon(t) \, dt \right\} = (\Omega \cos \pi\epsilon + \theta \sin \pi\epsilon)_{-\infty}$$

$$\times \exp \left\{ -\frac{\pi}{2h} \int_C \epsilon(t) \, dt \right\} - \frac{1}{h} \int_C (\theta_s \cos \pi\epsilon - \Omega_s \sin \pi\epsilon)$$

$$\times \exp \left\{ -\mathscr{R} \frac{\pi}{2h} \int_C \epsilon(t^*) \coth \frac{\pi}{2h} (t^* - t) \, dt^* \right\} dt, \tag{162}$$

which enables us to express K in terms of the values at either $x = \infty$ or $x = -\infty$.

A special and very simple example of the above theory is contained in §4.2. It is $\epsilon = 0$ on $y = 0$ (θ_s specified) and $\epsilon = \frac{1}{2}$ on $y = h$ (Ω_s specified). The value $\epsilon = \frac{1}{2}$ makes $\lambda = \infty$ by (143), and so reduces (140) to $\Omega = \Omega_s$. The limit $\epsilon = -\frac{1}{2}$ has the same result, but clearly gives a different solution, so we must be careful to take the correct limit. This is easily decided from (140), for if $\Omega > \Omega_s$ implies $\theta > \theta_s$ for finite λ, we must take the limit $\epsilon \to \frac{1}{2}$, whereas if $\Omega > \Omega_s$ implies that $\theta < \theta_s$ the limit to take is $\epsilon \to -\frac{1}{2}$.

(2) *The upper half-plane*

The strip domain just considered becomes the upper half-plane in the limit $h \to \infty$. In this limit (161) and (162) can be combined to yield

$$K = (\Omega \cos \pi\epsilon + \theta \sin \pi\epsilon)_\infty. \tag{163}$$

the distinction between $x = \infty$ and $x = -\infty$ now vanishing. Equation (157) gives $Q(z,t) = 2/(t-z)$. Substitution of this value and (163) in (153) now yields

$$\tau(z) = \exp\left\{\int_{-\infty}^{\infty} \frac{\epsilon(t)}{t-z}\,dt\right\}\left[(\Omega \cos \pi\epsilon + \theta \sin \pi\epsilon)_\infty\right.$$
$$\left. + \frac{1}{\pi}\int_{-\infty}^{\infty} (\theta_s \cos \pi\epsilon - \Omega_s \sin \pi\epsilon)\exp\left\{-\mathscr{R}\int_{-\infty}^{\infty} \frac{\epsilon(t^*)\,dt^*}{t-t^*}\right\}\frac{dt}{t-z}\right]. \tag{164}$$

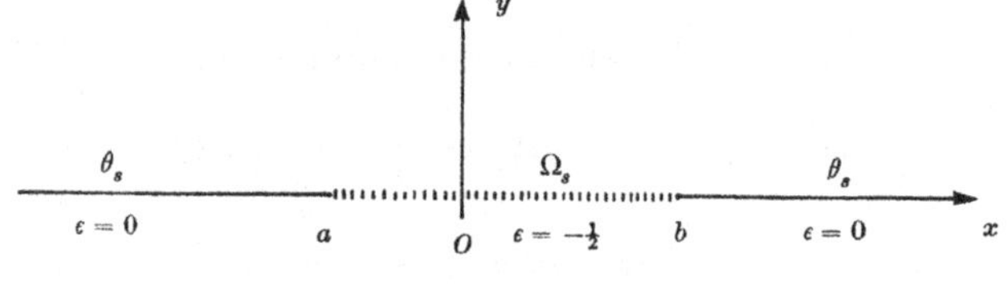

Fig. 4.18

A particular case of (164) of which some use will be made in §10.12 is

$$\epsilon = \begin{cases} 0\,(-\infty < t < a, \quad b < t < \infty), \\ -\tfrac{1}{2}(a < t < b) \end{cases} \tag{165}$$

(see Fig. 4.18). With this distribution of ϵ (164) becomes

$$\tau(z) = \left(\frac{z-a}{z-b}\right)^{\frac{1}{2}}\left\{\Omega_\infty + \frac{1}{\pi}\left(\int_{-\infty}^{a} + \int_{b}^{\infty}\right)\theta_s\left(\frac{t-b}{t-a}\right)^{\frac{1}{2}}\frac{dt}{t-z} + \frac{1}{\pi}\int_{a}^{b}\Omega_s\left(\frac{t-b}{a-t}\right)^{\frac{1}{2}}\frac{dt}{t-z}\right\}. \tag{166}$$

4.19   Mixed periodic boundary conditions

We shall now make an extension of the mixed boundary value problem defined in §4.15—and solved in §4.17 for a singly-connected region—

which will enable us to deal with doubly-connected regions. First suppose the doubly-connected region in question is mapped by the method of §5.9 into the rectangle $-l < x < l$, $0 < y < h$ (see Fig. 5.9). Then the boundary value problem solved below is: find a function $\tau = \Omega + i\theta$, analytic in the rectangle, given that equation (140), viz. $\theta = \lambda(\Omega - \Omega_s) + \theta_s$, is satisfied on $y = 0$ and on $y = h$, and that

$$\lim_{\epsilon \to +0} \{\tau(l-\epsilon, y) - \tau(-l+\epsilon, y)\} \equiv [\tau(y)] = \varpi(y) \tag{167}$$

is a known real function. In the particular case $\varpi = 0$, both the real and imaginary parts of τ have identical values on $x = \pm l$ for a given y, and so $\tau(z)$ can be continued outside the basic rectangle to a periodic function, $\tau(z) = \tau(z + 2l)$, in the infinite strip $-\infty < x < \infty$, $0 < y < h$ (cf. §4.9).

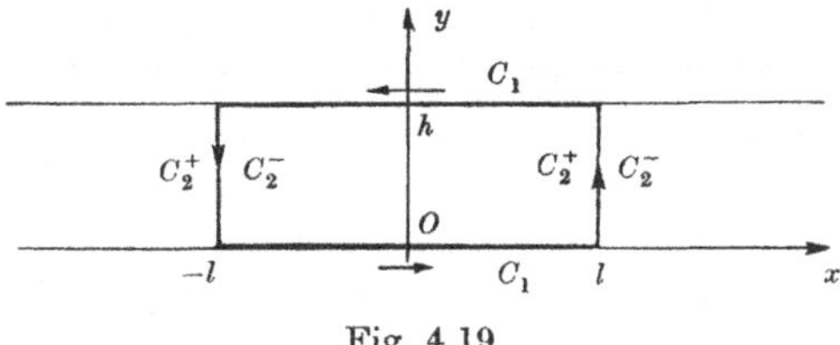

Fig. 4.19

The periodic and mixed boundary value problem just defined can be solved by a method similar to that used in §4.17. The solution is based on an equation given in §4.9 for the particular case $\lambda = 0$. The equation in question is (4.94), which can be rewritten in the form

$$\mathscr{F}(z) = \frac{1}{2\pi} \int_{C_1} \mathscr{I}\{\mathscr{F}(t)\} Q(z, t)\, dt + \frac{1}{2\pi} \int_{C_2} [\mathscr{R}\{\mathscr{F}(t)\}] P(z, t)\, dt + B, \tag{168}$$

where

$$Q(z, t) = \frac{2K}{l} Z_1 \left\{ \frac{K}{l}(x - z) \left| \frac{ih}{l} \right. \right\}, \tag{169}$$

$$P(z, t) = \frac{K}{il} \left[Z_2 \left\{ \frac{K}{l}(iy - z) \left| \frac{ih}{l} \right. \right\} - Z_2 \left\{ \frac{K}{l}(iy + z) \left| \frac{ih}{l} \right. \right\} \right], \tag{170}$$

$[\mathscr{R}\{\mathscr{F}(t)\}]$ is the jump in $\mathscr{R}\{\mathscr{F}(t)\}$ from $x = l$ to $x = -l$, C_1 is the sides of the rectangle $y = 0$, $y = h$, C_2 is the side $x = l - 0$ (Fig 4.19), and B is a real constant given by

$$B = \frac{1}{4l} \int_{-l}^{l} [\mathscr{R}\{\mathscr{F}(x, 0)\} + \mathscr{R}\{\mathscr{F}(x, h)\}]\, dx. \tag{171}$$

It is a simple matter to extend (168) to cover the case in which $\mathscr{I}\{\mathscr{F}(t)\}$, as well as $\mathscr{R}\{\mathscr{F}(t)\}$, is discontinuous across C_2. A glance at (85) and (91) shows that in this case (168) will be replaced by

$$\mathscr{F}(z) = \frac{1}{2\pi}\int_{C_1}\mathscr{I}\{\mathscr{F}(t)\}\,Q(z,t)\,dt + \frac{1}{2\pi}\int_{C_2}([\mathscr{R}\{\mathscr{F}(t)\}]\,P(z,t)$$
$$+\,[\mathscr{I}\{\mathscr{F}(t)\}]\,R(z,t))\,dt + B, \quad (172)$$

where B is now (from (95))

$$B = \frac{1}{4l}\int_{-l}^{l}[\mathscr{R}\{\mathscr{F}(x,0)\} + \mathscr{R}\{\mathscr{F}(x,h)\}]\,dx + \frac{1}{4l}\int_{0}^{h}[\mathscr{I}\{\mathscr{F}(y)\}]\,dy, \quad (173)$$

and

$$R(z,t) \equiv \frac{K}{l}\left[Z_2\!\left\{\frac{K}{l}(iy-z)\,\Big|\,\frac{ih}{l}\right\} + Z_2\!\left\{\frac{K}{l}(iy+z)\,\Big|\,\frac{ih}{l}\right\}\right]. \quad (174)$$

If we were to transform the rectangle back into the original doubly-connected region the functions P, Q and R would be modified, but (172) would retain the same general form, and so it can be regarded as being applicable to any doubly-connected region of a general shape.

We apply (172) to the present problem as follows. First let $\mathscr{F}(z)$ be the function $\ln\tau_0(z)$, where $\tau_0(z)$ satisfies the boundary condition expressed in (141) on C_1, and is continuous across C_2, i.e. $[\tau_0(y)] = 0$. Then from (144) and (172)

$$\tau_0(z) = A\,\exp\left\{\int_{C_1}\tfrac{1}{2}\epsilon(t)\,Q(z,t)\,dt\right\}, \quad (175)$$

where A is a real constant.

Secondly, let $\mathscr{F}(z)$ be the function $\Phi(z)$ defined in (145). Comparison of (149) and (175) shows that $\mathscr{I}\Phi$ on C_1 will be given by an equation similar in form to (151), so it only remains to evaluate the jump in $\Phi(t)$ across C_2. From (145), (167) and the continuity of $\tau_0(z)$ this is

$$[\Phi] = \frac{1}{A}\,\varpi(y)\exp\left\{-\int_{C_1}\tfrac{1}{2}\epsilon(t^*)\,Q(t,t^*)\,dt^*\right\},$$

where $P(t)$ is now a point on C_2. Substituting the values of $\mathscr{I}\Phi$ and $[\Phi]$ we have just calculated into (172) we derive an equation for $\Phi(z)$; then we find the required equation for $\tau(z)$ on using (145) and (175). The final result is

$$\tau(z) = \exp\left\{\int_{C_1}\tfrac{1}{2}\epsilon(t)\,Q(z,t)\,dt\right\}\left[\frac{1}{2\pi}\int_{C_1}(\theta_s\cos\pi\epsilon - \Omega_s\sin\pi\epsilon)\right.$$
$$\times\exp\left\{-\mathscr{R}\int_{C_1}\tfrac{1}{2}\epsilon(t^*)\,Q(t,t^*)\,dt^*\right\}Q(z,t)\,dt$$
$$+\frac{1}{2\pi}\int_{C_2}\varpi(y)\left(\mathscr{R}\left\{\exp\left[-\int_{C_1}\tfrac{1}{2}\epsilon(t^*)\,Q(t,t^*)\,dt^*\right]\right\}P(z,t)\right.$$
$$\left.\left.+\,\mathscr{I}\left\{\exp\left[-\int_{C_1}\tfrac{1}{2}\epsilon(t^*)\,Q(t,t^*)\,dt^*\right]\right\}R(z,t)\right)dt + K\right], \quad (176)$$

where K is a new real constant. It can be evaluated in terms of known quantities in much the same way as in the first example of §4.18.

4.20   An example: the upper half-plane

A relatively simple example of the above theory, of which we shall make later use, is obtained by putting $\varpi(y) = 0$, and letting $h \to \infty$. In this limit (169) and (42) yield

$$Q(z, t) = \frac{\pi}{l} \cot \frac{\pi}{2l} (x - z),$$

so (176) becomes

$$\tau(z) = \exp\left\{ \frac{\pi}{2l} \int_{-l}^{l} \epsilon(x) \cot \frac{\pi}{2l} (x - z) \, dx \right\} \left[K + \frac{1}{2l} \int_{-l}^{l} (\theta_s \cos \pi\epsilon - \Omega_s \sin \pi\epsilon) \right.$$

$$\left. \times \exp\left\{ -\mathscr{R} \frac{\pi}{2l} \int_{-l}^{l} \epsilon(x^*) \cot \frac{\pi}{2l} (x^* - x) \, dx^* \right\} \cot \frac{\pi}{2l} (x - z) \, dx \right]. \quad (177)$$

The (real) constant K can be evaluated in terms of Ω_∞ and θ_∞ by simply taking the limit $z \to i\infty$. In this limit $\cot\{\pi(x - z)/2l\} \to i$, so the real and imaginary parts of (177) give

$$K = \mathscr{R}\left\{ (\Omega_\infty + i\theta_\infty) \exp\left[-\left\{ \frac{\pi i}{2l} \int_{-l}^{l} \epsilon(x) \, dx \right\} \right] \right\}, \quad (178)$$

and

$$\frac{1}{2l} \int_{-l}^{l} (\theta_s \cos \pi\epsilon - \Omega_s \sin \pi\epsilon) \exp\left\{ -\mathscr{R} \frac{\pi}{2l} \int_{-l}^{l} \epsilon(x^*) \cot \frac{\pi}{2l} (x^* - x) \, dx^* \right\} dx$$

$$= \mathscr{I}\left[(\Omega_\infty + i\theta_\infty) \exp\left\{ -\frac{\pi i}{2l} \int_{-l}^{l} \epsilon(x) \, dx \right\} \right]. \quad (179)$$

For a given value Ω_∞ this pair of equations determines the values of K and θ_∞ in terms of the known distributions θ_s, Ω_s and ϵ.

Equation (177) could have been derived directly by using (14) in place of (172). Comparing (14) and (172) we observe that the correct form of the constant B in the present example is $\mathscr{R}\{\mathscr{F}_\infty\}$, where $\mathscr{F}_\infty$ is the value of $\mathscr{F}$ at $z \equiv i\infty$. This shows that the constant A in (175) must be $|\tau_{0\infty}|$, and that the constant K in (178) is therefore $\mathscr{R}\{\Phi_\infty\} |\tau_{0\infty}|$.

Thus by (145)

$$K = \mathscr{R}\left\{ \Omega_\infty + i\theta_\infty) \frac{|\tau_{0\infty}|}{\tau_{0\infty}} \right\}.$$

The limit $z \to i\infty$ in (175) gives

$$\tau_{0\infty} = |\tau_{0\infty}| \exp\left\{ \frac{\pi i}{2h} \int_{-l}^{l} \epsilon(x) \, dx \right\},$$

and combining this with the previous equation gives us an alternative derivation of (178).

In the first example of § 4.18 the reader can readily verify that (161) for K *cannot* be derived from (153) and (157) by simply taking the limits $z \to \pm \infty$. And yet in the example just given we have verified that the limit-taking method is perfectly valid. The distinction between the two examples in this connexion is that in §4.18 the point at infinity is on the *boundary* of the domain (see discussion in §3.17), and this is not so in the example of this section. It is particularly important to take care with limiting processes when dealing with boundary conditions as complicated as those we have been considering in this chapter.

THE POINCARÉ PROBLEM

4.21   Reduction of the problem to an integral equation

The Poincaré problem has been defined at the end of §4.1 in (5). Our attention will be confined to the particular case of (5) in which $B = 0$, for this case is by far the most frequently experienced in fluid dynamics and aerodynamics. In this case the boundary relation for the harmonic function $\theta(x, y)$ can be written in the form

$$k(t)\frac{\partial\theta}{\partial n}+\theta = f(t), \tag{180}$$

where $k(t), f(t)$ are known functions and $(\partial/\partial n)$ is the derivative along the inwards normal to the boundary C. The problem can be reduced to that of solving a quasi-regular Fredholm integral equation as follows.

Let $\tau(z) = \Omega(x, y) + i\theta(x, y)$, where $\Omega(x, y)$ is thus the harmonic conjugate of θ, then the harmonic conjugate of (148) enables us to write

$$\tau(z) = -\frac{i}{2\pi}\int_C \Omega(t)\,Q(z, t)\,dt + i\ln A, \tag{181}$$

where A is a real positive constant depending on θ and $Q(z, t)$ is a known function depending on the shape of C. On C at $t = t_0$ (181) gives

$$\theta(t_0) = -\frac{1}{2\pi}\mathscr{R}\int_C \Omega(t)\,Q(t_0, t)\,dt + \ln A, \tag{182}$$

where $Q(t_0, t)$ behaves like $2/(t - t_0)$ near $t = t_0$.† Integrating by parts we get

$$\theta(t_0) = \frac{1}{2\pi}\mathscr{R}\int_C F(t_0, t)\frac{\partial\Omega}{\partial s}\frac{ds}{dt}\,dt + \ln A, \tag{183}$$

where

$$F(t_0, t) \equiv \int^t Q(t_0, t)\,dt, \tag{184}$$

and behaves like $2\ln|t - t_0|$ near $t = t_0$.

† The existence of this Cauchy-type singularity in Q should now be obvious to the reader from the various examples of (182) given in chs. 3 and 4; a rigorous proof is easily constructed from the integrand of (5.116) on p. 196.

Let χ be the angle between Ox and the tangent to C (see Fig. 3.13), then

$$\frac{\partial \theta}{\partial n} = \cos\chi \frac{\partial \theta}{\partial y} - \sin\chi \frac{\partial \theta}{\partial x} = \cos\chi \frac{\partial \Omega}{\partial x} + \sin\chi \frac{\partial \Omega}{\partial y} = \frac{\partial \Omega}{\partial s},$$

by the Cauchy–Riemann equations, and therefore (180) can be written

$$\theta = f(t) - k(t)\frac{\partial \Omega}{\partial s}. \tag{185}$$

Eliminating θ from (183) and (185), and using $ds/dt = \cos\chi + i\sin\chi$, we arrive at the quasi-regular Fredholm integral equation for $\partial\Omega/\partial s$

$$f(t_0) - k(t_0)\frac{\partial \Omega}{\partial s} = \frac{1}{2\pi}\mathscr{R}\int_C F(t_0, t)\, e^{i\chi}\frac{\partial \Omega}{\partial s}\, dt + \ln A. \tag{186}$$

We term this 'quasi-regular' because the kernel $F(t_0, t)\,e^{i\chi}$ has a logarithmic singularity at $t = t_0$. A *singular* integral equation has a Cauchy-type singularity in its kernel, whereas a (regular) Fredholm equation has no singularity. The quasi-regular form is thus intermediate between the singular and Fredholm forms. Many of the iterative techniques for solving Fredholm equations are applicable to quasi-regular equations, but some care is necessary to avoid the occurrence of improper or divergent integrals.

It very seldom occurs that (186) passes a solution which can be expressed in a closed analytical form, and so iterative or other approximate methods are usually necessary.

When (186) has been solved for $\partial\Omega/\partial s$, $\theta(t)$ is given by (185), and $\theta(x, y)$ follows from (183) on replacing t_0 by $z = x + iy$.

Another form of integral equation can be deduced from the harmonic conjugate of (182). This is obtained by replacing $\theta(t_0)$ and $-\Omega(t)$ by $\Omega(t_0)$ and $\theta(t)$ respectively. If $\theta(t)$ is then eliminated by (185) the result is the integro-differential equation

$$\Omega(t_0) = \frac{1}{2\pi}\mathscr{R}\int_C \left\{ f(t) - k(t)\frac{\partial \Omega}{\partial s}\right\} Q(t_0, t)\, dt + \ln B, \tag{187}$$

where $\ln B$ is a real constant. An integration by parts will convert this into an integral equation for Ω, but its kernel will behave like $(t - t_0)^{-2}$ near $t = t_0$, and consequently the integral will be divergent, unless the interpretation of (3.42) is adopted. This strong singularity will usually cause the failure of iterative methods of solution, and therefore (186) is to be preferred over (187).

4.22   The Poincaré problem in an infinite strip

A boundary condition like (185) occurs at the surface of a heavy fluid (see (1.71)), and therefore a domain of some importance is the infinite strip $-\infty < x < \infty$, $0 < y < h$. Furthermore, any singly-connected domain D can be transformed into an infinite strip (e.g. by the method to be described in §5.5), which can thus be taken as a standard domain in which to solve the Poincaré problem. There is therefore no essential loss in generality in the following form of the theory.

From (3.103) and (3.112)

$$\tau(z) = -\frac{i}{2h} \int_{-\infty}^{\infty} \left\{ \Omega_0(x) \coth \frac{\pi}{2h} (x-z) - \Omega_h(x) \tanh \frac{\pi}{2h} (x-z) \right\} dx$$
$$+ \tfrac{1}{2} i(\theta_\infty + \theta_{-\infty}), \quad (188)$$

and
$$\theta_\infty - \theta_{-\infty} = \frac{1}{h} \int_{-\infty}^{\infty} \{ \Omega_0(x) - \Omega_h(x) \} \, dx. \quad (189)$$

From these we find that corresponding to (186) we have the pair of equations

$$f_0(x_0) - k_0(x_0) \, \Omega_0'(x_0) = \frac{1}{\pi} \int_{-\infty}^{\infty} \ln \left| \sinh \frac{\pi}{2h} (x-x_0) \right| \Omega_0'(x) \, dx$$

$$-\frac{1}{\pi} \int_{-\infty}^{\infty} \ln \left\{ \cosh \frac{\pi}{2h} (x-x_0) \right\} \Omega_h'(x) \, dx + \tfrac{1}{2} (\theta_\infty + \theta_{-\infty}), \quad (190)$$

and
$$f_h(x_0) - k_h(x_0) \, \Omega_h'(x_0) = \frac{1}{\pi} \int_{-\infty}^{\infty} \ln \left\{ \cosh \frac{\pi}{2h} (x-x_0) \right\} \Omega_0'(x) \, dx$$

$$-\frac{1}{\pi} \int_{-\infty}^{\infty} \ln \left| \sinh \frac{\pi}{2h} (x-x_0) \right| \Omega_h'(x) \, dx + \tfrac{1}{2} (\theta_\infty + \theta_{-\infty}), \quad (191)$$

where f_0, k_0 apply on $y = 0$, and f_h, k_h on $y = h$. In the integration by parts employed here we have made use of $\lim_{R \to \pm\infty} R\{\Omega_0(R) - \Omega_h(R)\} = 0$, which is implicit in (189), to eliminate the integrated terms. Equations (190) and (191) must be solved as a pair of simultaneous equations for $\Omega_0'(x)$ and $\Omega_h'(x)$.

If $k_0 \to \infty$ and $f_0 \to \infty$ so that $f_0/k_0 \to g_0$, g_0 being finite, then on $y = 0$ (180) becomes $\partial\theta_0/\partial y = \Omega_0'(x) = g_0(x)$, and either of (190) or (191) serves to define $\Omega_h'(x)$. Notice that in this case (190) becomes a *regular* Fredholm integral equation of the first kind for $\Omega_h'(x)$.

The complementary case occurs when $k_0 \to 0$, and (180) reduces to $\theta_0(x) = f_0(x)$, but this only reduces (190) and (191) from integral equations of the second kind to ones of the first kind—a relatively trivial change.

A better formulation can be obtained by starting with (11) and integrating only the term involving $\Omega_h(x)$ by parts. The result is

$$\theta_h(x_0) = f_h(x_0) - k_h(x_0)\,\Omega_h'(x_0) = \frac{1}{2h}\int_{-\infty}^{\infty} f_0(x)\,\mathrm{sech}\,\frac{\pi}{2h}\,(x - x_0)\,dx$$

$$-\frac{1}{\pi}\int_{-\infty}^{\infty}\ln\left|\tanh\frac{\pi}{4h}\,(x - x_0)\right|\Omega_h'(x)\,dx. \quad (192)$$

Except in the special case considered in the next section the integral equations given above cannot be solved in closed forms.

4.23　Explicit solution of a special case by the use of Laplace transforms

Equations (190), (191) and (192) can be solved without difficulty in the event that k_0 and k_h are constants. This occurs in the steady flow of a heavy liquid (1.71), and also in several other problems in fluid dynamics (see § 7.32). We shall illustrate the method by solving (192); a similar method can be applied to the other equations.

The Laplace transform technique is especially suitable for equations like (192) (with k_h = constant) which have 'difference' kernels, i.e. kernels which depend only on the difference $(x - x_0)$.

We shall adopt the two-sided Laplace transform of van der Pol & Bremmer (1950), and use the following notation for the transform and its inverse:

$$\mathscr{L}\{f(t); p\} \equiv p\int_{-\infty}^{\infty} e^{-pt}f(t)\,dt, \quad (\alpha < \mathscr{R}p < \beta), \quad (193)$$

$$\mathscr{L}^{-1}\{h(p); t\} \equiv \frac{1}{2\pi i}\int_{c-i\infty}^{c+i\infty} e^{pt}\frac{h(p)}{p}\,dp, \quad (\alpha < c < \beta), \quad (194)$$

where the restrictions on the real numbers c and $\mathscr{R}p$ are chosen so as to secure convergence of the integrals.

The principal transform rules we shall need are

$$\mathscr{L}\{f(\pm\lambda t); p\} = \pm\mathscr{L}\left\{f(t); \pm\frac{p}{\lambda}\right\} \quad (\lambda > 0), \quad (195)$$

$$\mathscr{L}\{f(t+\lambda); p\} = e^{\lambda p}\mathscr{L}\{f(t); p\}, \quad (196)$$

$$\mathscr{L}\{e^{-\lambda t}f(t); p\} = \frac{p}{p+\lambda}\mathscr{L}\{f(t); p+\lambda\}, \quad (197)$$

$$\mathscr{L}\left\{\frac{d^n f(t)}{dt^n}; p\right\} = p^n\mathscr{L}\{f(t); p\}, \quad (198)$$

$$\mathscr{L}\left\{\int_{-\infty}^{\infty} f(\tau)\,h(t-\tau)\,d\tau; p\right\} = \frac{1}{p}\mathscr{L}\{f(t); p\}\,\mathscr{L}\{h(t); p\}, \quad (199)$$

and $\qquad\qquad \mathscr{L}\left\{\int_{-\infty}^{t} f(\tau)\,d\tau; p\right\} = \frac{1}{p}\mathscr{L}\{f(t); p\}. \quad (200)$

In addition to these results we shall also need the transforms of a number of functions, but these will not be given as they can all be obtained from a long list given at the end of van der Pol & Bremmer's text (henceforth P.B.).

Returning to our problem it is apparent from (192) and (199) that in finding the transform of (192) we shall need the transforms of sech $\pi x/2h$ and $\ln |\tanh \pi x/4h|$. P.B. give

$$\mathscr{L}\{\ln |\coth \tfrac{1}{2}t| ; p\} = \pi \tan \left(\frac{\pi p}{2}\right); \quad \mathscr{L}\left\{\frac{1}{e^t + 1}; p\right\} = -\frac{\pi p}{\sin \pi p},$$

provided $-1 < \mathscr{R}p < 0$, so by (195) and (197) we get

$$\mathscr{L}\left\{\ln \left|\tanh \frac{\pi x}{4h}\right|; p\right\} = -\pi \tan hp, \quad \mathscr{L}\left\{\frac{2e^{\pi x/2h}}{e^{\pi x/h} + 1}; p\right\} = \frac{2ph}{\cos ph},$$

provided $-\pi/2h < \mathscr{R}p < 0$. Therefore if the transform of (192) is taken, and use is made of (198) and (199), there results

$$\mathscr{L}\{f_h(x_0); p\} - k_h p \, \mathscr{L}\{\Omega_h(x_0); p\} = \sec hp \, \mathscr{L}\{f_0(x_0); p\}$$
$$+ \tan hp \, \mathscr{L}\{\Omega_h(x_0); p\}.$$

Solving this for $\mathscr{L}\{\Omega_h\}$, and inverting by (194) we get

$$\Omega_h = \frac{1}{2\pi i} \int_{c-i\infty}^{c+i\infty} \frac{e^{pt}}{p} \frac{\mathscr{L}\{f_h(x_0); p\} \cos hp - \mathscr{L}\{f_0(x_0); p\}}{\sin hp + kp \cos hp} dp; \quad (201)$$

the derivative of this is the required solution of the special case of (192) under consideration.

CHAPTER 5

CONFORMAL MAPPING

BASIC MAPPING FORMULAE

5.1 General principles

It is not our purpose to give a detailed account of conformal mapping in fluid mechanics, but simply to revise for the reader the main concepts, and, more important, to provide a list, with brief derivations, of the mappings or transformations of importance in classical fluid mechanics. For an introductory account of conformal mapping, and its use in fluid motion problems the reader is referred to ch. 6 of Glauert's (1948) classic text. Milne-Thomson (1949) also gives a useful account of this topic, which goes somewhat further than that of Glauert. A really extensive account of conformal mapping—but without any reference to applications of the theory—has been written by Nehari (1952). However, none of these accounts provide the reader with a list of all the transformations he is likely to need, and therefore he must acquire some skill in calculating the transformations required in particular flow problems. Fortunately this requires experience in the application of only one theorem, viz. the Schwarz–Christoffel mapping theorem (§ 5.3), which is the key theorem of the whole subject.

If z and w are the affixes of points in two distinct Argand planes, then a functional relation

$$w = f(z), \tag{1}$$

may be regarded as a *mapping* of points in the z-plane on to points in the w-plane. If z lies within a domain D bounded by a contour C, then w will lie within an *image* domain D^* bounded by an image contour C^*. (That D^* is a *domain* really requires proof, see Nehari, p. 153, but we can accept this as intuitively obvious.) We say that D is mapped by (1) on to D^*; the correspondence between D and D^* is not necessarily *one-to-one*, i.e. to each point P of D there is not necessarily only one corresponding point P^* of D^* (see Fig. 5.1). The ambiguity likely to arise from this situation, which does occur in some fluid motion applications (see § 11.25) can be overcome by the *Riemann-surface* concept, i.e. the concept of different overlapping sheets connected together so as to make a continuous domain. A point P now has to be regarded as belonging to only one particular sheet of the domain D (e.g. in the case shown in Fig. 5.1 P belongs either to the upper sheet or to the lower sheet, but not to both).

The derivative of (1) is

$$dw = f'(z)\,dz,$$

which can be written

$$|dw| = |f'(z)|\,|dz|, \quad \arg(dw) = \arg\{f'(z)\} + \arg(dz),$$

      (2)

by the usual rules for taking the modulus and argument of a complex equation. These equations show that the vector interval dw is derived from the vector interval dz by (i) increasing the magnitude of dz by a factor $|f'(z)|$, and then by (ii) rotating the resulting vector through an angle $\arg\{f'(z)\}$. Points at which $f'(z)$ is zero or infinite are called *critical*

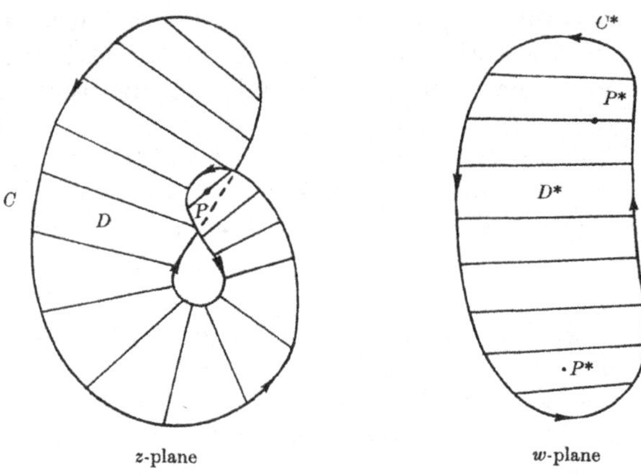

Fig. 5.1

or *singular* points of the transformation. Suppose, for example, that $f'(z)$ has a zero of order n at the point z_0, then in the neighbourhood of this point

$$f(z) = f(z_0) + a(z - z_0)^{n+1} + \dots.$$

Hence

$$w - w_0 = a(z - z_0)^{n+1} + \dots,$$

and

$$\arg(w - w_0) = \arg a + (n+1)\arg(z - z_0) + \dots.$$

Thus

$$\lim_{w \to w_0} \arg(w - w_0) = \arg a + (n+1)\lim_{z \to z_0}\arg(z - z_0). \qquad (3)$$

First suppose that $P(z)$ is not a singular point of the transformation, and let dw, dw_1 be two distinct vectors passing through $P^*(w)$. Then from the last of (2), $\arg(dw) - \arg(dw_1) = \arg(dz) - \arg(dz_1)$, as $f'(z)$ depends only on the position of P. This result means that the image curves to two intersecting curves in D intersect at the *same* angle in D^* as do the original curves in D. This is the important *conformal* property of the transformations being considered.

At a singular point of the transformation, on the other hand, it follows by a similar argument based on (3) that the angle between the image curves in D^* is not equal to the angle between the original curves in D.

Let $z = x + iy$, $w = \phi + i\psi$, and suppose the curves $\phi = \text{constant}$, $\psi = \text{constant}$ are drawn in the w-plane. The Cauchy–Riemann equations connecting ϕ, ψ, x and y are

$$\frac{\partial \phi}{\partial x} = \frac{\partial \psi}{\partial y}, \quad \frac{\partial \phi}{\partial y} = -\frac{\partial \psi}{\partial x},$$

so

$$\frac{\partial \phi}{\partial x}\frac{\partial \psi}{\partial x} + \frac{\partial \phi}{\partial y}\frac{\partial \psi}{\partial y} = 0,$$

or

$$\nabla \phi_0 . \nabla \psi = 0.$$

Hence the curves $\psi = \text{constant}$, $\phi = \text{constant}$ form an orthogonal net in the z-plane. The transformation to the w-plane maps these curves into straight lines parallel to the real and imaginary axes, so this orthogonal property is preserved in the transformation. This is, of course, simply a special case of the conformal property of the transformation.

Another important property is that a conformal mapping which is *one-to-one* preserves the connectivity of a domain—a result we can accept here as obvious.

The following theorem provides a method of checking that a given transformation claimed to map D into D^* does in fact do this.

Let C and C^* be two (closed) contours in the z-plane and w-plane, respectively, and let $w = f(z)$ be regular on and within C. If $w = f(z)$ maps C on to C^* in such a way that C^* is traversed by w exactly once in the positive sense, then $w = f(z)$ maps the domain D bounded by C on to the domain D^* bounded by C^*.

The proof is quite simple—it depends on the *argument principle* (see Nehari, 1952, p. 154)—but we shall omit it for brevity, and in any case the result is almost obvious.

A much more important and fundamental theorem is that known as the *Riemann mapping theorem*, which may be stated:

Any simply-connected schlicht† domain D whose boundary consists of more than one point can be mapped conformally on to the interior of the unit circle, in such a way that an arbitrary point of D and a direction through this point correspond, respectively, to the origin and the positive direction of the real axis.

This theorem assures us from the outset that there *exists* a function $f(z)$, which will map a given schlicht, simply-connected domain D via the unit circle, into another given schlicht, simply-connected domain D^*. The

---

† Non-overlapping, e.g. the w-plane of Fig. 5.1 is 'schlicht', but the z-plane is not.

explicit *calculation* of this function is another matter, and excepting special cases can only be carried out approximately. The proof of the Riemann mapping theorem is rather difficult; the reader may refer here to Nehari, p. 173.

5.2 The elementary transformations

(1) We shall assume the reader to be familiar with the *linear transformation* $w = az + b$, which involves (i) a shift of origin, (ii) a rotation of $\arg a$, and (iii) a stretching of $|a|$.

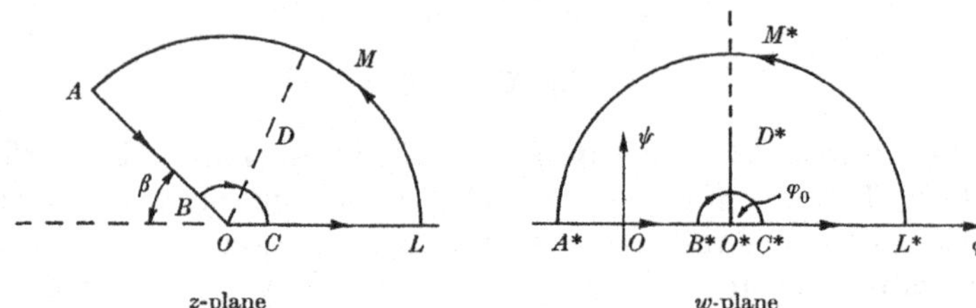

Fig. 5.2 *a*

(2) The next most important transformation is the '*corner-removing*' *transformation*, which can be defined as follows. Let β be an angle in the range $-\pi \leqslant \beta < \pi$, then the sector $0 \leqslant \arg z \leqslant (\pi - \beta)$, $0 < |z| < \infty$ is mapped on to the upper half of the w-plane, i.e. on to the sector $0 \leqslant \arg w \leqslant \pi$, $0 < |w| < \infty$, by

$$w = z^n, \tag{4}$$

where

$$n = 1 \Big/ \left(1 - \frac{\beta}{\pi}\right). \tag{5}$$

This is easily verified, for (4) and (5) yield $\arg w / \arg z = \pi/(\pi - \beta)$, so that $\arg z$ is increased by the factor $\pi/(\pi - \beta)$ in the transformation to the w-plane (see Fig. 5.2 *a*).

Combining transformations 1 and 2 we have that

$$w = A^n(z - z_0)^n + \phi_0, \tag{6}$$

maps the sector

$$-\frac{\arg A}{n} \leqslant \arg(z - z_0) \leqslant \pi - \beta - \frac{\arg A}{n}, \quad 0 < |z - z_0| < \infty,$$

on to the upper half of the w-plane, such that the corner at z_0 maps on to $w = \phi_0$. To fix the value of A we need to assign another pair of corresponding points z_1, w_1.

A particular case of (6) we shall need below is

$$w = a\left(1 + \frac{z}{b}\right)^n + \phi_0. \tag{7}$$

If a and b are positive real numbers then the points $w = \phi_0$, $a + \phi_0$ will correspond to the points $z = -b$ and $z = 0$ respectively (Fig. 5.2b).

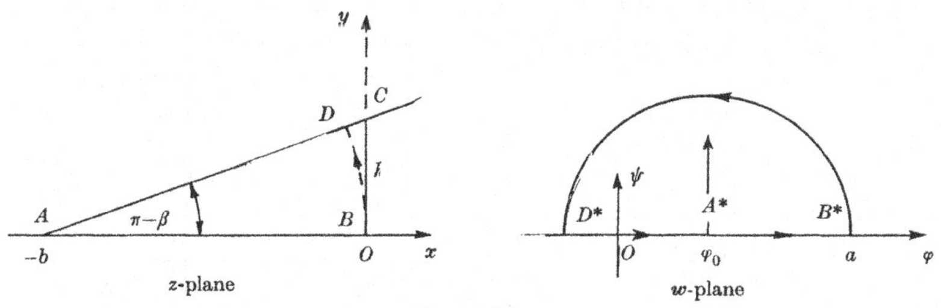

Fig. 5.2b

(3) The '*infinite-strip*' transformation is the special case of the corner-removing transformation obtained by letting $\beta \to \pi$, or by (5), $n \to \infty$. If the point A in Fig. 5.2b is moved to infinity by letting $b \to \infty$, and the intercept p of AC on the y-axis is kept fixed, the z-plane becomes the infinite strip $-\infty \leqslant x \leqslant \infty$, $0 \leqslant y \leqslant h$, shown in Fig. 5.2$c$. The w-plane remains as in Fig. 5.2b, being unaffected by this limit.

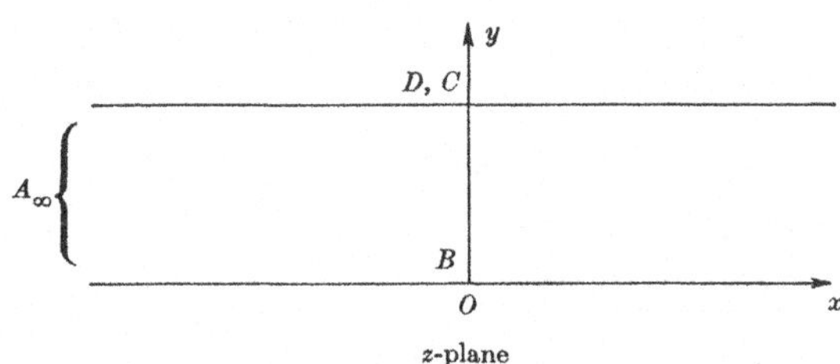

Fig. 5.2c

From (5) and Fig. 5.2b

$$h/b = \tan(\pi - \beta) = \tan(\pi/n),$$

so (7) can be written

$$w = a\left(1 + \frac{z}{h}\tan\frac{\pi}{n}\right)^n + \phi_0.$$

Taking the limit $n \to \infty$, and using the well-known result

$$\lim_{n \to \pm\infty}\left(1 + \frac{cz}{n}\right)^n = e^{cz}, \tag{8}$$

we obtain
$$w = a\,e^{\pi z/h} + \phi_0, \tag{9}$$

for the infinite-strip transformation.

(4) The *inversion* transformation

$$w = -1/z, \tag{10}$$

transforms the upper half of the z-plane into the upper half of the w-plane, the relation between the planes being that shown in Fig. $5.2d$. The point at infinity in the z-plane becomes the origin in the w-plane.

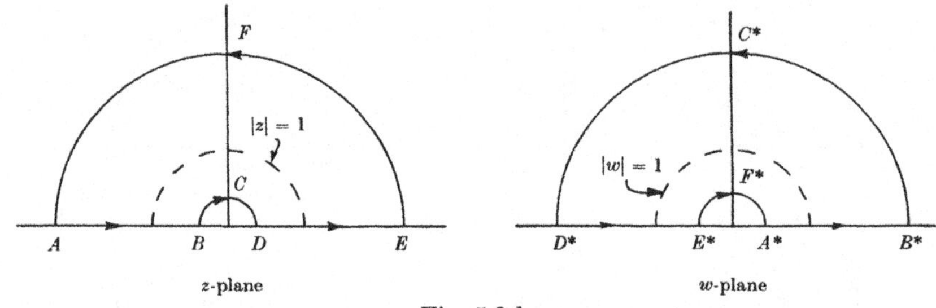

z-plane   w-plane

Fig. $5.2d$

Notice from (4) and (5) that, apart from the rotation factor $e^{i\pi} = -1$, (10) can be regarded as the special case of (5), $\beta = 2\pi$. The 'corner' of 2π is at $z = \infty$, as can be seen by tracing the curve $DEFAB$ in Fig. $5.2d$. This corner is 'straightened' by (10), and mapped on to $w = 0$.

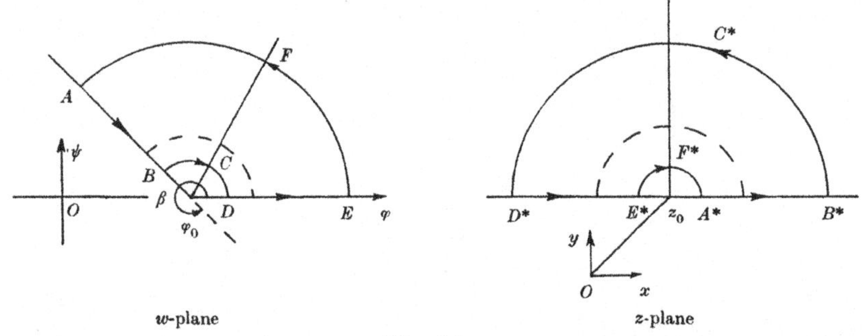

w-plane   z-plane

Fig. $5.2e$

(5) Equation (10) is only a special case of the *generalized inversion* (cf. (6))

$$w = A(z - z_0)^n + \phi_0, \tag{11}$$

where n is given by (5) and β lies in the range $\pi < \beta \leqslant 2\pi$. When $A = -1$, the transformation is that shown in Fig. $5.2e$. Equation (11) transforms the point $z = \infty$ on to $w = \phi_0$, and at the same time 'straightens' out the angle β at $z = \infty$.

Summarizing the above transformations we have that sectors in the z-plane (counting the infinite strip as a 'sector') are mapped on to the upper half of the w-plane by transformations of the type

$$
w = \begin{cases} A(z-z_0)^{\pi/(\pi-\beta)} + \phi_0 & (-\pi \leqslant \beta < \pi), \\ Ae^{\pi z/h} + \phi_0 & (\beta = \pi), \\ A(z-z_0)^{\pi/(\pi-\beta)} + \phi_0 & (\pi < \beta \leqslant 2\pi), \end{cases} \tag{12}
$$

which therefore remove simple discontinuities of amount β at those corners of these sectors that are mapped on to $w = \phi_0$.

5.3   The Schwarz–Christoffel mapping theorem

We shall map the singly-connected region bounded by the contour $C = PABGD_\infty E_\infty FP$ shown in Fig. 5.3 on to the upper half of the w-plane, in such a way that C maps on to the real axis $\psi = 0$, $-\infty < \phi < \infty$, and a selected point P of C maps on to the point at infinity. The contour C is assumed to have a continuous turning tangent save for a finite number of exceptional points, where simple discontinuities in the slope θ of C may exist (e.g. points B and G in Fig. 5.3). Also a finite number of points of C are assumed to be at infinity (e.g. points D_∞ and E_∞). The point P which will be moved to infinity by the transformation may be an ordinary point of C, or one of the exceptional points (e.g. B, G, D_∞ or E_∞) just described. In the figure P is shown to be at a point where the tangent jumps through an angle β; an ordinary point is obtained if $\beta = 0$, while P is a point at infinity if $\beta = \pi$ (as at D_∞), or if $\pi < \beta \leqslant 2\pi$ (as at E_∞). The particular cases $\beta = 0$, π and 2π are especially important in practical applications of the mapping theorem.

The contour C is indented about P by the curve $\gamma = FHA$ of such a shape that it maps into the large semicircle $|w| = R$. We shall need the relation between w and z on this indentation, i.e. near $w = \infty$. The mapping $z \to w$ must (i) map C on to $\psi = 0$, (ii) map P on to $w = \infty$, and (iii) remove the discontinuity β in the slope at P. An elementary transformation $z \to w$, which does (ii) and (iii) can be deduced from (12), by putting $\phi_0 = 0$, $z_0 = z_p$ (the value of z at P), and replacing w by $1/w$ (this is equivalent to putting $\phi_0 = \infty$). We find that

$$
Aw_1 = (z-z_p)^{-\pi/(\pi-\beta)} \quad (-\pi \leqslant \beta < \pi, \quad \pi < \beta \leqslant 2\pi),
$$

$$
Aw_1 = e^{-\pi z/h} \quad (\beta = \pi).
$$

The remaining transformation $w_1 \to w$ must be regular at infinity, and therefore must be of the form (cf. equation (3.9))

$$
Aw_1 = kw\left(1 + \frac{b_0}{w} + \frac{b_1}{w^2} + \dots\right),
$$

where A, k, b_0, b_1, ... are constants. Combining the transformations we have

$$z = z_p + (kw)^{(\beta-\pi)/\pi}\left\{1 + \frac{\beta-\pi}{\pi}\frac{b_0}{w} + O\left(\frac{1}{w^2}\right)\right\} \quad (\beta \neq \pi),$$

and

$$z = -\frac{h}{\pi}\ln(kw) - \frac{h}{\pi}\frac{b_0}{w} + O\left(\frac{1}{w^2}\right). \tag{13}$$

These relations give some idea of how complicated the relation between the z- and w-planes must be.

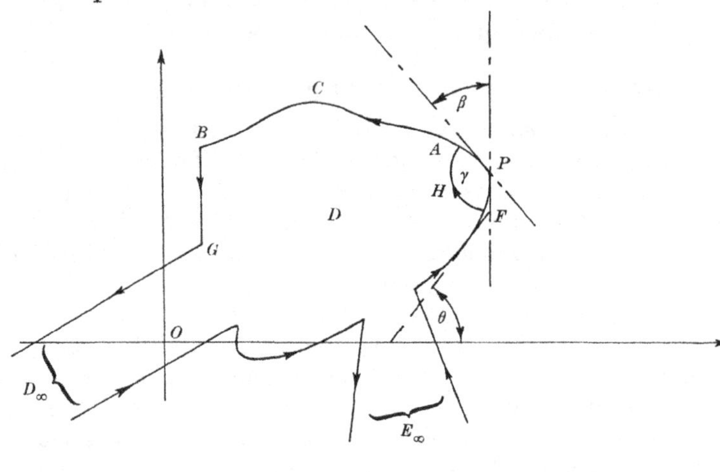

z-plane

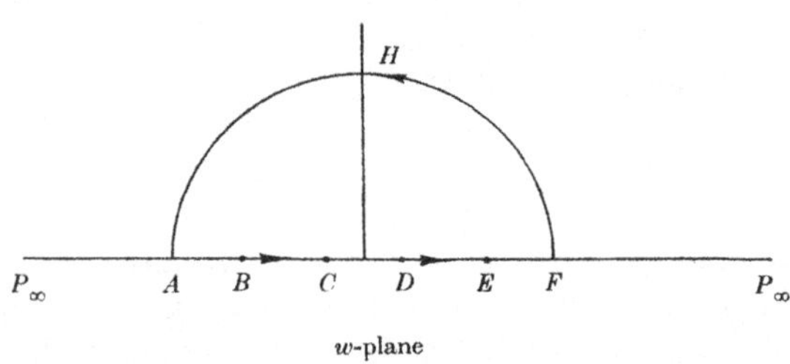

w-plane

Fig. 5.3

Instead of seeking a direct relation between z and w we could look for a derivative of z with respect to w which simplifies (13) into a 'regular' relation, then find the relation between this derivative and w, and finally integrate back to the (z, w)-relation. A derivative which does reduce the right-hand side of (13) to a regular function is

$$\delta = \frac{d}{dw}\left(\ln\frac{dz}{dw}\right), \tag{14}$$

for from (13) and (14) we find that for all values of β

$$\delta = -\left(2-\frac{\beta}{\pi}\right)\frac{1}{w}+O\left(\frac{1}{w^2}\right) \quad (-\pi \leqslant \beta \leqslant 2\pi). \tag{15}$$

From (15) we have

$$\lim_{w \to \infty} \delta = 0. \tag{16}$$

We have still to investigate the form of δ in the neighbourhood of the other singular points on C, i.e. the points B, G, D_∞ and E_∞ shown in Fig. 5.3. This can be done in just the same way as for P. Suppose there is a singular point on C at $z = z_0$ which maps on to $w = \phi_0$, and which has a leap in the slope θ of C of magnitude β_0. Then by (12) the elementary transformation $z \to w_1$ which removes this discontinuity in θ, and which maps z_0 on to ϕ_0, is

$$\frac{w_1-\phi_0}{A} = \begin{cases} (z-z_0)^{\pi/(\pi-\beta_0)} & (\beta \neq \pi), \\ e^{\pi z/h} & (\beta = \pi). \end{cases}$$

The remaining transformation $w_1 \to w$ is regular at $w_1 = w = \phi_0$, so (cf. §3.2)

$$\frac{w_1-\phi_0}{A} = k(w-\phi_0)\{1+b_0(w-\phi_0)+b_1(w-\phi_0)^2+\ldots\},$$

where A, k, b_0, b_1, ... are constants. Eliminating w_1 we find

$$z = \begin{cases} z_0 + [k(w-\phi_0)]^{(\pi-\beta_0)/\pi}\{1+0(w-\phi_0)\} & (\beta \neq \pi), \\ \dfrac{h}{\pi}\{\ln k(w-\phi_0) + 0(w-\phi_0)\} & (\beta = \pi). \end{cases} \tag{17}$$

Equations (14) and (17) now give

$$\delta = -\frac{\beta_0}{\pi}\frac{1}{w-\phi_0}+O(1) \quad (-\pi \leqslant \beta < 2\pi), \tag{18}$$

i.e. we have a Cauchy-type singularity in δ at ϕ_0 (see §3.10).

The slope θ of the curves $\psi = $ constant in the z-plane (C is the curve $\psi = 0$) is given by $\arg (dz/dw)$, for on such curves

$$\arg\left(\frac{dz}{dw}\right) = \arg\left(\frac{dz}{d\phi}\right) = \arg\left(\frac{dx}{d\phi}+i\frac{dy}{d\phi}\right) = \tan^{-1}\left(\frac{dy}{dx}\right) = \theta.$$

Then on $\psi = $ constant,

$$\delta = \frac{d}{dw}\left\{\ln\left|\frac{dz}{dw}\right| + i\arg\left(\frac{dz}{dw}\right)\right\}$$

$$= \frac{d}{d\phi}\left\{\ln\left|\frac{dz}{d\phi}\right| + i\theta\right\},$$

so

$$\mathscr{I}\delta = \frac{d\theta}{d\phi} = \theta'. \tag{19}$$

Comparing (18) and (19) with the theory of §3.10 we conclude that $\theta'(\phi)$ has a delta-function singularity at $\psi = 0$, $\phi = \phi_0$.

The function $\delta(w)$ is analytic in the upper half-plane $\mathscr{I}w > 0$, vanishes at infinity by (16), and has integrable singularities in its imaginary part $\theta'(\phi)$ on $\psi = 0$. From the theory given in §3.18, and in particular from (3.113) applied to the function δ in the w-plane, we can therefore write

$$\delta(w) = \frac{1}{\pi} \int_{-\infty}^{\infty} \frac{\theta'(\phi)\, d\phi}{\phi - w},$$

or
$$\delta(w) = \frac{1}{\pi} \int_{-\infty}^{\infty} \frac{d\theta(\phi)}{\phi - w}, \tag{20}$$

in Stieltjes's integral notation.

It only remains to integrate (20) twice in accordance with (14) to obtain the $z(w)$ relation. We find

$$\ln \frac{dz}{dw} = \ln K - \frac{1}{\pi} \int_{-\infty}^{\infty} \ln (w - \phi)\, d\theta(\phi), \tag{21}$$

where $\ln K$ is the constant of integration. Thus

$$\frac{dz}{dw} = K \exp \left\{ -\frac{1}{\pi} \int_{-\infty}^{\infty} \ln (w - \phi)\, d\theta(\phi) \right\}, \tag{22}$$

and finally
$$z = z_0 + K \int_{w_0}^{w} \exp \left\{ -\frac{1}{\pi} \int_{-\infty}^{\infty} \ln (w - \phi)\, d\theta(\phi) \right\} dw, \tag{23}$$

where z_0 and w_0 are corresponding points.

Thus in order to make the mapping unique it is necessary to assign *two* corresponding points (z_p, ∞) and (z_0, w_0). Then the constant K follows from

$$z_p = z_0 + K \int_{w_0}^{\infty} \exp \left\{ \frac{1}{\pi} \int_{-\infty}^{\infty} \ln (w - \phi)\, d\theta(\phi) \right\} dw. \tag{24}$$

Equation (22) is a generalization of the Schwarz–Christoffel mapping formula for polygons. When C is a polygon $\theta(\phi)$ is a step-function, and the R-S integral in (22) degenerates to a sum (see §3.6). Suppose the discontinuities in $\theta(\phi)$ in this case are of magnitude α_n at $\phi = \phi_n$, $n = 1, 2, \ldots, N$, then (22) becomes

$$\frac{dz}{dw} = K \exp \left\{ -\sum_{n=1}^{N} \frac{\alpha_n}{\pi} \ln (w - \phi_n) \right\},$$

or
$$\frac{dz}{dw} = K \prod_{n=1}^{N} (w - \phi_n)^{-\alpha_n/\pi}, \tag{25}$$

and this is the usual form of the well-known mapping equation due to Schwarz and Christoffel. It is one of the most important equations in the whole of classical two-dimensional fluid mechanics. We shall make very considerable use of it in the following pages.

Returning to the more general form, (22), we note that the distribution of the slope, $\theta(\phi)$, is subject to the obvious restriction

$$\int_{-\infty}^{\infty} d\theta(\phi) = 2\pi - \beta, \tag{26}$$

$2\pi - \beta$ being the total change in θ in one circuit of C. When $\theta(\phi)$ is a step-function (26) degenerates to

$$\sum_{n=1}^{N} \alpha_n = 2\pi - \beta. \tag{27}$$

For a given distribution $\theta(\phi)$ (23) and (24) enable us to calculate the mapping relation $z = z(w)$, but in practical problems it will be the *shape* of C that is known, and the distribution $\theta(\phi)$ will not be known until the mapping relation has been determined. This situation leads to a complicated integral equation when C is a curved boundary; this case is discussed in the next section.

5.4 The basic integral equation of mapping theory

Knowledge of the shape of C enables us to compute its intrinsic equation, $s = s(\theta)$, where s is the distance measured around C from some suitable origin. On $\psi = 0$, i.e. on C,

$$\left|\frac{dz}{dw}\right| = \left|\frac{dx}{d\phi} + i\frac{dy}{d\phi}\right| = \sqrt{\left\{\left(\frac{dx}{d\phi}\right)^2 + \left(\frac{dy}{d\phi}\right)^2\right\}} = \frac{ds}{d\phi} = \frac{ds}{d\theta}\frac{d\theta}{d\phi},$$

so
$$\left|\frac{dz}{dw}\right| = R\theta'(\phi), \tag{28}$$

where $R = s'(\theta)$ is the radius of curvature of C.

On $\psi = 0$, at $w = \phi^*$, the real part of (21) gives

$$\ln\left|\frac{dz}{dw}\right| = \ln|K| - \frac{1}{\pi}\int_{-\infty}^{\infty} \ln|\phi^* - \phi|\, d\theta(\phi), \tag{29}$$

and combining this with (28) we arrive at

$$\ln\theta'(\phi^*) = \ln|K| - \ln R\{\theta(\phi^*)\} - \frac{1}{\pi}\int_{-\infty}^{\infty} \ln|\phi^* - \phi|\, d\theta(\phi). \tag{30}$$

The constant $\ln|K|$ occurring in (30) can be computed in the following way. By (28) and (29), on $\psi = 0$

$$\left|\frac{dz}{dw}\right| = \frac{ds}{d\phi} = |K|\exp\left\{-\frac{1}{\pi}\int_{-\infty}^{\infty} \ln|\phi - \phi^*|\, d\theta(\phi^*)\right\},$$

so
$$s(\phi) = |K|\int_{\phi_0}^{\phi} \exp\left\{-\frac{1}{\pi}\int_{-\infty}^{\infty} \ln|\phi - \phi^*|\, d\theta(\phi^*)\right\} d\phi, \tag{31}$$

where ϕ_0 is the value of ϕ at the origin of s. Let $s(\phi_1) = \ell$, where ϕ_1 and ℓ are conveniently selected values of ϕ and s (this fixes the *scale* of the mapping of C on to $\psi = 0$), then $|K|$ is determined as a function of $\theta(\phi)$ by

$$\ell = |K| \int_{\phi_0}^{\phi_1} \exp\left\{ -\frac{1}{\pi} \int_{-\infty}^{\infty} \ln|\phi - \phi^*|\, d\theta(\phi^*) \right\} d\phi. \tag{32}$$

Substituting this in (30) we arrive at the basic integro-differential equation for $\theta(\phi)$, namely

$$\ln \theta'(\phi^*) = -\ln \int_{\phi_0}^{\phi_1} \exp\left\{ -\frac{1}{\pi} \int_{-\infty}^{\infty} \ln|\phi - \phi^*|\, d\theta(\phi^*) \right\} d\phi$$

$$-\ln \frac{R}{\ell}\{\theta(\phi^*)\} - \frac{1}{\pi} \int_{-\infty}^{\infty} \ln|\phi^* - \phi|\, d\theta(\phi), \tag{33}$$

where ϕ_1, ϕ_0 and ℓ are known numbers, and $R\{\theta\}$ is a known function. This non-linear equation is well beyond the standard methods of solution; numerical or iterative methods must be employed. When $\theta(\phi)$ has been found from (33) it is substituted in (23) to give the required mapping relation.

If C is finite the most convenient selection of ϕ_0 and ϕ_1 is $-\infty$ and $+\infty$ respectively. In this case ℓ becomes the perimeter length of C.

Probably the best method of solving (33) is the following. First C is approximated to by a figure comprising of sections over each of which $\theta'(\phi)$ has a constant value, i.e. $\theta'(\phi)$ is a step-function. When the constant value of $\theta'(\phi)$ is zero, the corresponding section will be a straight line, otherwise it will be curved having a slowly varying curvature. Convergence of the integrals in (33) requires that we choose $\theta'(\phi)$ to vanish in the intervals bounded by $\phi = -\infty$ and $\phi = +\infty$. In practice we can start with an approximating polygon (see Fig. 5.4), and then round off those corners not corresponding to corners on C. For the approximating polygon it follows from (31) that

$$s(\phi) = |K| \int_{\phi_0}^{\phi} \prod_{n=1}^{N} |\phi - \phi_n|^{-\alpha_n/\pi}\, d\phi, \tag{34}$$

where the notation is the same as for (25). The perimeter distance to the rth corner will be

$$s_r = s(\phi_r) = |K| \int_{\phi_0}^{\phi_r} \prod_{n=1}^{N} |\phi - \phi_n|^{-\alpha_n/\pi}\, d\phi \quad (r = 1, 2, ..., N). \tag{35}$$

The s_r will be known, so the N equations (35) can be used to find the values of ϕ_r. Obviously numerical methods will be necessary in most cases to evaluate the integrals in (35).

When this stage of the calculation is complete the corners not corresponding to corners on C can be rounded off as follows. Suppose the approximating polygon has an interior angle of $\pi - \alpha_m$ at $\phi = \phi_m$, i.e. there is a discontinuity of α_m in the slope θ, then we can round this off by putting $\theta'(\phi) = \alpha_m/a$ in $\phi_m - \frac{1}{2}a < \phi < \phi_m + \frac{1}{2}a$. The range a is selected to give the closest agreement between the modified 'polygon' and C.

In this, or other ways, we finally arrive at an approximating curve, C_1 say, having a known distribution $\theta_1(\phi)$. Equations (31) and (32) give the corresponding function $s_1(\phi)$. We now connect C and C_1 by making the approximation† $s(\phi)/\ell$ on C equals $s_1(\phi)/\ell_1$, i.e. that the $s(\phi)/\ell$ relation on C is the same as that on C_1. Then from the known relations $R(s/\ell)/\ell$, $\theta(s/\ell)$ on C we find the approximate functions $R_2(\phi)/\ell_2$, $\theta_2(\phi)$ and $\theta_2'(\phi)$ for C. Substituting $R_2(\phi)/\ell_2$ and $\theta_2'(\phi)$ in the right-hand side

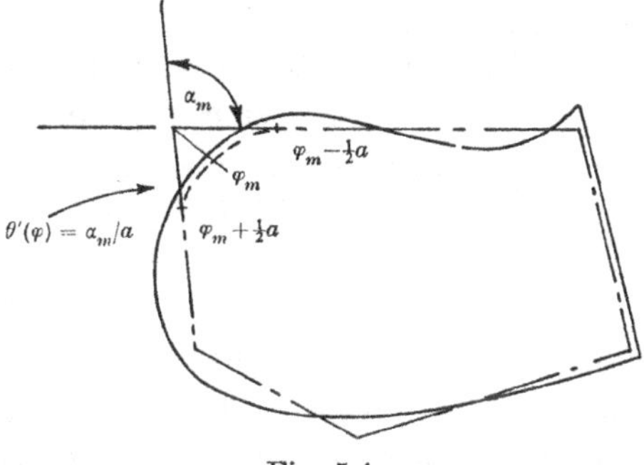

Fig. 5.4

of (33) we obtain a new result for $\theta'(\phi)$, say $\theta_3'(\phi)$ from the left-hand side of (33). This is now inserted in (31) and (32) to yield the corresponding function $s_3(\phi)/\ell_3$. We now set $s(\phi)/l = s_3(\phi)/\ell_3$, i.e. make an improved mapping approximation, and go through the process just outlined again. This is repeated until no change in successive iterations shows that the method has converged.

Proof of the theoretical convergence of this iterative process would be difficult, and, in any case, rather an academic exercise, for the important question is 'is convergence sufficiently rapid for the method to be practicable?' The answer depends to some extent on the accuracy of the numerical methods adopted to evaluate the integrals in (31), (32) and (33). Regions of high curvature on C can upset the convergence considerably, for small errors in integration will cause errors in $s(\phi)$ sufficient to 'move' a peak in R a distance $\Delta\phi$ which exceeds the width (in ϕ) of this peak. Such a movement in a peak of R can, in the next iteration, cause greater local errors in $s(\phi)$, and so on   the process diverging in the neighbourhood of the peak. However, it is not difficult to devise practical techniques to

† This important type of approximation occurs in many places in this text, and will be referred to as the 'mapping approximation'.

overcome this. On the whole convergence is found to be quite rapid, provided there are no marked peaks in R (excepting sharp corners), and experience shows that the contour resulting from the distribution $\theta_3'(\phi)$ is indistinguishable from C for practical purposes.

An example given below in § 7.6 illustrates the details of the method.

The method outlined above of mapping a singly-connected domain D of general shape on to the upper half-plane $-\infty \leqslant \phi \leqslant \infty$, $0 \leqslant \psi \leqslant \infty$ enables us to map any two such domains D_1 and D_2 on to each other, for if each of D_1 and D_2 are mapped on to the same upper half-plane, then we will have $z_1 = z_1(w)$, $z_2 = z_2(w)$, i.e. a parametric relation between z_1 and z_2. The upper half-plane is here in the role of a standard or *canonical* domain. Other canonical domains, e.g. that within unit circle, may be more suitable for certain applications. In §§ 5.5 and 5.6 we shall introduce two more canonical domains for singly-connected regions, of great value in fluid mechanics. These are the semi-infinite strip $-a \leqslant \phi \leqslant a$, $0 \leqslant \psi \leqslant \infty$, and the infinite strip $-\infty \leqslant \phi \leqslant \infty$, $0 \leqslant \psi \leqslant h$. These can be regarded as generalizations of the upper half-plane domain, which is obtained in each of the limits $a \to \infty$, or $h \to \infty$.

5.5　A first generalization of the mapping theorem

Suppose that instead of just *one* point P, we map *two* points P and Q on to the point at infinity in the w-plane. The two sections of C lying between P and Q can be mapped on to the parallel lines $\psi = 0$, $\psi = h$, so that P lies on one 'side' of the point at infinity at $\phi = -\infty$, and Q lies on the other side at $\phi = \infty$. The domain within C then maps on to the infinite strip $-\infty \leqslant \phi \leqslant \infty$, $0 \leqslant \psi \leqslant h$.

The method of calculating the mapping formula is very similar to that used in § 5.3. We introduce the function $\delta(w)$ defined by (14), and investigate its behaviour at the singular points of C. Equations (18) and (19) clearly apply in the present case, and it only remains to investigate the behaviour of δ near P and Q.

Consider first the point P; suppose the discontinuity in the slope of C at P is β_p. This discontinuity is removed and P(at z_p) is mapped on to the origin of a w_1-plane by the elementary transformation (cf. (12))

$$w_1 = A(z - z_p)^{\pi/(\pi - \beta_p)},$$

where A is a constant. If P is now removed to the side $\phi = -\infty$ of the point at infinity in the w-plane by the infinite strip transformation (see (3) of § 5.2), then

$$w_1 = B e^{\pi w/h},$$

where B is a constant. Hence near $z = z_p$, $\phi = -\infty$, the overall transformation of the domain within C on to the infinite strip in the w-plane is of the form

$$(z - z_p)\{1 + O(z - z_p)\} = K\, e^{(\pi - \beta_p)w/h}, \tag{36}$$

where K is a new constant, and the term $\{1 + O(z - z_p)\}$ represents the regular (at $z = z_p$) transformation remaining after the two elementary transformations have been applied. Therefore

$$z = z_p + K\, e^{(\pi - \beta_p)w/h}\{1 + O(e^{(\pi - \beta_p)w/h})\}, \tag{37}$$

and by (14)

$$\delta = \frac{\pi - \beta_p}{h} + O(e^{(\pi - \beta_p)w/h}). \tag{38}$$

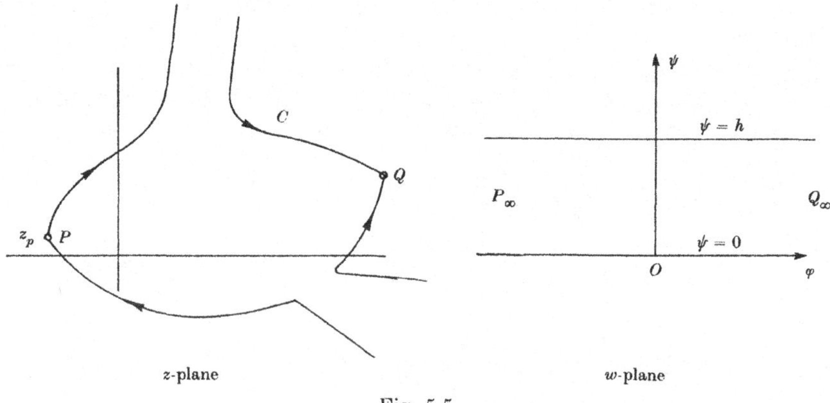

Fig. 5.5

A similar argument for the point Q (rotate the w-plane through $180°$) yields

$$z = z_q + K\, e^{-(\pi - \beta_q)w/h}\{1 + O(e^{-(\pi - \beta_q)w/h})\}, \tag{39}$$

and

$$\delta = -\frac{\pi - \beta_q}{h} + O(e^{-(\pi - \beta_q)w/h}). \tag{40}$$

From (38) and (49) we see that

$$\delta_\infty + \delta_{-\infty} = (\beta_q - \beta_p)/h. \tag{41}$$

Let δ have imaginary parts $\theta_0'(\phi)$ and $\theta_h'(\phi)$ on $\psi = 0$ and $\psi = h$ respectively (cf. (19)), then the conditions satisfied by $\delta(w)$ are similar to those satisfied by $\tau(z)$ in §3.16. And therefore corresponding to (3.109) we can write

$$\delta(w) - \frac{1}{2h} \int_{-\infty}^{\infty} \left\{ \theta_0'(\phi) \coth\frac{\pi}{2h}(\phi - w) - \theta_h'(\phi)\tanh\frac{\pi}{2h}(\phi - w) \right\} d\phi$$

$$+ \frac{1}{2h}(\beta_q - \beta_p),$$

on taking (41) into account. Integrating this with respect to w:

$$\ln\frac{dz}{dw} = \ln K - \frac{1}{\pi}\int_{-\infty}^{\infty}\left[\ln\sinh\frac{\pi}{2h}(w-\phi)\,d\theta_0(\phi)\right.$$

$$\left. - \ln\cosh\frac{\pi}{2h}(w-\phi)\,d\theta_h(\phi)\right] + \frac{w}{2h}(\beta_q - \beta_p), \quad (42)$$

where $\ln K$ is the constant of integration, and we have introduced the Stieltjes's notation.

A pair of simultaneous integral equations can be derived from (42) by putting $w = \phi^*$, and $w = \phi^* + ih$. The theory is an exact parallel to that given in §5.4, and the same sort of iterative method of solution can be employed.

When C is a polygon both $\theta_0(\phi)$ and $\theta_h(\phi)$ are step-functions. Let $\theta_0(\phi)$ have leaps of α_n at $\phi = \phi_n$, $n = 1, 2, ..., N$, and $\theta_h(\phi)$ have leaps of α_m at $\phi = \phi_m$, $m = 1, 2, ..., M$, then (42) reduces to

$$\ln\frac{dz}{dw} = \ln K + \frac{w}{2h}(\beta_q - \beta_p) - \frac{1}{\pi}\sum_{n=1}^{N}\alpha_n\ln\sinh\frac{\pi}{2h}(w-\phi_n)$$

$$+ \frac{1}{\pi}\sum_{m=1}^{M}\alpha_m\ln\cosh\frac{\pi}{2h}(w-\phi_m),$$

i.e.

$$\frac{dz}{dw} = K\exp\left[\frac{w}{2h}(\beta_q - \beta_p)\right]\prod_{m=1}^{M}\left[\cosh\frac{\pi}{2h}(w-\phi_m)\right]^{\alpha_m/\pi}\Bigg/$$

$$\prod_{n=1}^{N}\left[\sinh\frac{\pi}{2h}(w-\phi_n)\right]^{\alpha_n/\pi}. \quad (43)$$

Some use will be made of this result in §5.12.

5.6   A second generalization of the mapping theorem

Instead of mapping a point P lying *on* the contour C on to $w = \infty$, it is often more convenient to take P to be a point lying *within* C. Then as no point of C lies at infinity in the w-plane, the complete contour C must be mapped on to a finite interval, say $-a < \phi < a$, of $\psi = 0$.

The points $\phi = a - 0$, $\phi = -a + 0$ on $\psi = 0$ will correspond to the two 'sides' of some point A on C (see Fig. 5.6), while that 'side' of the point at infinity attained by keeping ϕ within $(-a, a)$ and letting ψ tend to infinity must correspond to the point P. The curves $\phi = \pm a$, $0 \leqslant \psi \leqslant \infty$, in the z-plane will be on opposite sides of a single curve joining A to P. The family of curves $\phi = $ constant will all begin on C and terminate at P, while their orthogonal trajectories, $\psi = $ constant, will be closed curves, all of which enclose P. Thus the domain D within C maps on to the semi-infinite strip $0 \leqslant \psi \leqslant \infty$, $-a \leqslant \phi \leqslant a$.

The function δ defined in (14) has an imaginary part $\theta'(\phi)$ on $\psi = 0$, and satisfies the condition

$$\delta(-a+i\psi) = \delta(a+i\psi),$$

there being no discontinuity across PA.

The real part of δ vanishes at infinity; the proof of this result is similar to that used in deriving (40). The point P can be transformed to $\psi = \infty$ by two elementary transformations. The first of these is

$$w_1 = A(z-z_p)^{\frac{1}{2}},$$

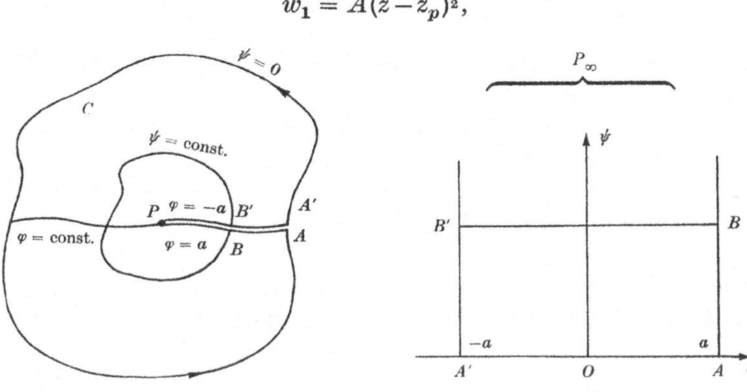

Fig. 5.6

which eliminates the angle $\beta = -\pi$ on the indentation $A'PA$ at P. The second transformation, namely the infinite-strip transformation

$$w_1 = B\,e^{i\pi w/2a},$$

then removes P to $i\infty$ in the w-plane, the width of the strip being $h = -2ia$ in this case (cf. Figs. 5.2c and 5.6). Thus the overall transformation can be written (cf. (36))

$$(z-z_p)\{1+O(z-z_p)\} = K\,e^{i\pi w/a}, \tag{44}$$

where K is a constant. Hence

$$z = z_p + K\,e^{i\pi w/a}\{1+O(e^{i\pi w/a})\}. \tag{45}$$

From this equation and the definition of δ it now follows that

$$\lim_{w\to i\infty} \delta = i\pi/a, \tag{46}$$

and our remark about the real part of this limit is established.

The conditions satisfied by $\delta(w)$ are similar to those satisfied by the function $\tau(z)$ in § 4.3, and it therefore follows by comparison with (4.14) that

$$\delta(w) = \frac{1}{2a} \int_{-a}^{a} \theta'(\phi) \cot \frac{\pi}{2a} (\phi - w) \, d\phi.$$

Integration now gives

$$\ln \frac{dz}{dw} = \ln K - \frac{1}{\pi} \int_{-a}^{a} \ln \sin \frac{\pi}{2a} (w - \phi) \, d\theta(\phi), \tag{47}$$

i.e.

$$\frac{dz}{dw} = K \exp\left\{ -\frac{1}{\pi} \int_{-a}^{a} \ln \sin \frac{\pi}{2a} (w - \phi) \, d\theta(\phi) \right\}, \tag{48}$$

$\ln K$ being the integration constant. These equations reduce to (21) and (22) in the limit $a \to \infty$.

In particular if C is a polygon, so that $\theta(\phi)$ is a step-function having leaps of α_n, $n = 1, 2, \ldots, N$, (48) reduces to

$$\frac{dz}{dw} = K \prod_{n=1}^{N} \left[\sin \frac{\pi}{2a} (w - \phi_n) \right]^{-\alpha_n/\pi}. \tag{49}$$

An integral equation for $\theta'(\phi)$ can be derived from (47) by a method similar to that given in § 5.4 for (22). Corresponding to (33) we find

$$\ln \theta'(\phi^*) = -\ln \int_{\phi_0}^{\phi_1} \exp\left\{ -\frac{1}{\pi} \int_{-a}^{a} \theta'(\phi^*) \ln \left| \sin \frac{\pi}{2a} (\phi - \phi^*) \right| d\phi^* \right\} d\phi$$

$$-\ln \frac{R}{\ell} \{ \theta(\phi^*) \} - \frac{1}{\pi} \int_{-a}^{a} \theta'(\phi) \ln \left| \sin \frac{\pi}{2a} (\phi - \phi^*) \right| d\phi. \tag{50}$$

One advantage of (50) over (33) is that the finite range of integration facilitates the use of numerical methods. Another important advantage occurs in the mapping of nearly circular domains, which occur frequently in practical problems. This is treated in the next section.

When (50) has been solved for $\theta'(\phi)$, the relation $s = s(\phi)$ will be known exactly, and therefore the value of z on C will be known as a function of ϕ. Then the position of P in the z-plane, z_p, can be determined from

$$z_p = \frac{1}{2a} \int_{-a}^{a} z(\phi) \, d\phi. \tag{51}$$

This equation can be established as follows.

With $\mathscr{I}z > 0$,

$$\cot z = -i \left[1 + 2 \sum_{n=1}^{\infty} e^{2inz} \right], \tag{52}$$

and hence by integration

$$\ln \sin z = -i \left[z - i \sum_{n=1}^{\infty} \frac{1}{n} e^{2inz} \right]. \tag{53}$$

Therefore

$$\ln \sin \frac{\pi}{2a}(w-\phi) = -i\left[\frac{\pi}{2a}(w-\phi) - i\sum_{n=1}^{\infty}\frac{1}{n}e^{\pi i n(w-\phi)/a}\right], \tag{54}$$

provided $\psi > 0$. Substituting this in (48) we obtain

$$\frac{dz}{dw} = K\,e^{i\pi w/a}\exp\left\{\frac{1}{\pi}\sum_{n=1}^{\infty}\frac{1}{n}e^{in\pi w/a}\int_{-a}^{a}e^{-in\pi\phi/a}\,d\theta(\phi)\right\}, \tag{55}$$

where we have used the obvious result (see Fig. 5.6),

$$\int_{-a}^{a}d\theta(\phi) = 2\pi,$$

and absorbed a term not dependent on w into the constant K.

Integration now gives an expression of the form

$$z(w) = z_p + \frac{aK}{i\pi}e^{i\pi w/a}\left\{1 + \sum_{n=1}^{\infty}b_n e^{in\pi w/a}\right\}, \tag{56}$$

where the b_n are functions of $\theta'(\phi)$, and independent of w. Now integrate this along $\psi = $ constant from $\phi = -a$ to $\phi = a$. The exponential series vanishes and we are left with

$$\int_{-a}^{a}z(w)\,d\phi = 2az_p.$$

This is true for all $\psi > 0$, and in the limit $\psi \to 0$ we have (51).

For a given position of P (51) is an implicit restriction on the distribution $\theta'(\phi)$ which must be taken into account in any method of solving (50). Equations (50) and (51) can be solved by an iterative method similar to that described in § 5.4.

5.7   The mapping of nearly circular domains

First consider the circle $|z-z_p| = R$, of radius R and centre at z_p. Let z_p be mapped on to $w = \infty$, then the curves $\phi = $ constant will be radial lines uniformly distributed around the circle. This symmetry means that $\theta'(\phi)$ is constant (see Fig. 5.7), and as the total change in θ is 2π over $(-a, a)$,

$$\theta'(\phi) = \pi/a. \tag{57}$$

Hence by (47) $\quad \ln\dfrac{dz}{dw} = \ln K - \dfrac{1}{a}\int_{-a}^{a}\ln\sin\dfrac{\pi}{2a}(w-\phi)\,d\phi. \tag{58}$

Integrating (54) over $-a \leqslant \phi \leqslant a$ we find that the integral in (58) equals $i\pi w/a$, so

$$\frac{dz}{dw} = K\,e^{i\pi w/a};$$

hence
$$z = z_p + \frac{aK}{i\pi}\, e^{i\pi w/a}. \tag{59}$$

To determine K we have to assign a second pair of corresponding points z_1, w_1. For example let the point

$$z_1 = z_p - R$$

(Q in Fig. 5.7) map on to $w = 0$, then (59) gives $R = iaK/\pi$, and

$$z = z_p - R\, e^{i\pi w/a}. \tag{60}$$

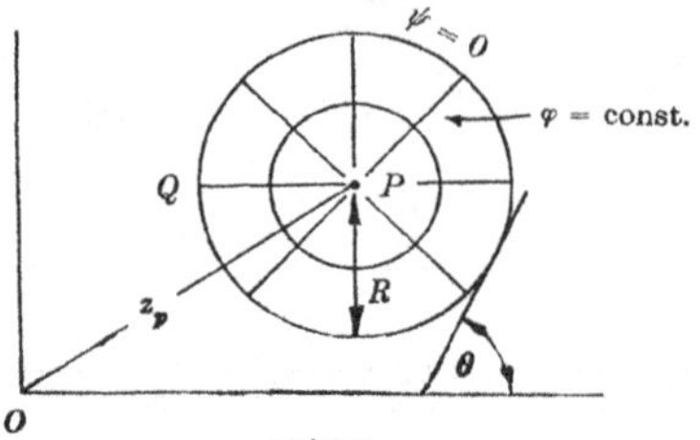

z-plane

Fig. 5.7

Differentiation gives
$$\ln\frac{dz}{dw} = \ln\frac{R\pi}{ia} + \frac{i\pi w}{a}. \tag{61}$$

Now suppose we have a nearly circular contour C to which the circle $|z - z_p| = R$ is a close approximation. Combining (47), (58) and (61) we obtain

$$\ln\frac{dz}{dw} = \ln\frac{iR\pi}{a} + \frac{i\pi w}{a} - \frac{1}{\pi}\int_{-a}^{a}\ln\sin\frac{\pi}{2a}(w-\phi)\left\{\theta'(\phi) - \frac{\pi}{a}\right\}d\phi,$$

or
$$\ln\frac{dz}{dw} = \ln\frac{iR\pi}{a} + \frac{i\pi w}{a} + \frac{1}{2a}\int_{-a}^{a}\left\{\theta(\phi) - \frac{\pi\phi}{a}\right\}\cot\frac{\pi}{2a}(\phi-w)\,d\phi, \tag{62}$$

on integrating by parts. Equation (62) is exact, but because C is almost a circle $\theta(\phi) - \pi\phi/a$ will be small compared to π, and the equation can be written

$$\frac{dz}{dw} = \frac{iR\pi}{a}\, e^{i\pi w/a}\left[1 + \frac{1}{2a}\int_{-a}^{a}\left\{\theta(\phi) - \frac{\pi\phi}{a}\right\}\cot\frac{\pi}{2a}(\phi-w)\,d\phi\right], \tag{63}$$

where the relation between ϕ and s can be taken to be

$$s = R\theta = \frac{\pi R\phi}{a}, \tag{64}$$

to the same order of accuracy. Equation (63) is now an explicit formula for dz/dw, which is reasonably accurate for nearly circular domains.

5.8 The Theodorsen–Garrick method for 'star-shaped' contours

While it suits the method of this book to employ the Schwarz–Christoffel function $\ln(dz/dw)$ in all conformal mapping problems, an account of mapping, especially in relation to fluid motion problems, which omitted the Theodorsen–Garrick method (1933) would be quite incomplete. In this method the conjugate harmonic functions $\ln|z - z_p|$,

$\arg(z - z_p)$, where P is the point within C mapped on to $w = i\infty$, replace $\ln|dz/dw|$ and $\arg(dz/dw)$. However C must be 'star-shaped' with respect to P, i.e. the radius vector from P must intersect C only once, for otherwise the mapping on to the auxiliary plane will not be one-to-one.

Suppose C is to be mapped on to the semi-infinite strip in the w-plane (Fig. 5.8) such that the points P, Q at z_p, $z_p + b$ map on to $w = i\infty$, 0. Consider the function

$$\chi(w) = \ln\left(\frac{z - z_p}{R\,e^{i\pi w/a}}\right). \tag{65}$$

From (60) we observe that $\chi = 0$ if C is a circle of radius R. Also (56) shows that χ vanishes at $z = z_p$, $w = i\infty$, and is analytic near this point provided a barrier AP is introduced (see figure) to make $\chi(w)$ single valued.

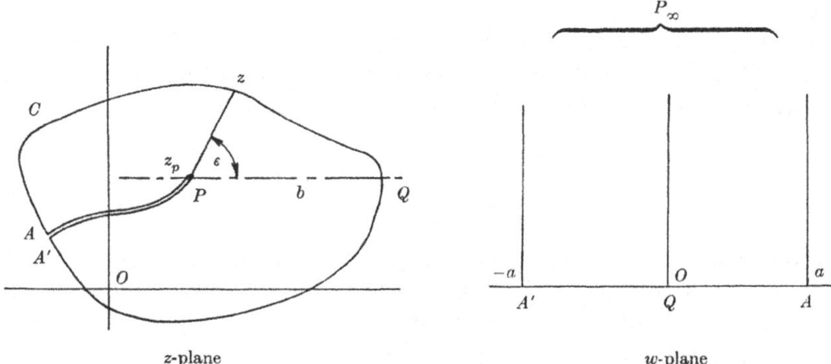

z-plane   w-plane

Fig. 5.8

Let C have the polar equation with respect to P,

$$|z - z_p| = R\,e^{P(\epsilon)}, \tag{66}$$

where

$$\epsilon = \arg(z - z_p). \tag{67}$$

Then

$$\mathscr{R}\chi(w) = P\{\epsilon(\phi)\}, \tag{68}$$

on C, as ϵ can be regarded as a function of ϕ.

An application of the conjugate of the formula in (4.14) to $\chi(w)$ now gives

$$\ln\left\{\frac{z - z_p}{R\,e^{i\pi w/a}}\right\} = -\frac{i}{2a}\int_{-a}^{a} P\{\epsilon(\phi)\}\cot\frac{\pi}{2a}(\phi - w)\,d\phi. \tag{69}$$

On $\psi = 0$ at $\phi = \phi^*$ this yields the integral equation

$$\epsilon(\phi^*) - \frac{\pi\phi^*}{a} = -\frac{1}{2a}\int_{-a}^{a} P\{\epsilon(\phi)\}\cot\frac{\pi}{2a}(\phi - \phi^*)\,d\phi. \tag{70}$$

Equation (69) has the harmonic conjugate

$$\ln\left(\frac{z-z_p}{R}\right) = \frac{i\pi w}{a} + \frac{1}{2a}\int_{-a}^{a}\left\{\epsilon(\phi)-\frac{\pi\phi}{a}\right\}\cot\frac{\pi}{2a}(\phi-w)\,d\phi. \tag{71}$$

When C is almost a circle it is apparent from Fig. (5.8) that $\epsilon(\phi)-\dfrac{\pi\phi}{a}$ will be small, so we have the approximation

$$z = z_p + R\,e^{i\pi w/a}\left[1+\frac{1}{2a}\int_{-a}^{a}\left\{\epsilon(\phi)-\frac{\pi\phi}{a}\right\}\cot\frac{\pi}{2a}(\phi-w)\,d\phi\right], \tag{72}$$

corresponding to (63) for dz/dw. Nehari (p. 265) obtains the conjugate of (72), i.e. the approximate form of (69), by a somewhat different approach.

Theodorsen & Garrick solved (70) by 'the method of successive conjugates' (Garrick, 1952); in this they start from an initial value $\epsilon_0(\phi)$, and then derive a sequence of functions by the recurrence formula

$$\epsilon_{n+1}(\phi^*) - \frac{\pi\phi^*}{a} = -\frac{1}{2a}\int_{-a}^{a} P\{\epsilon_n(\phi)\}\cot\frac{\pi}{2a}(\phi-\phi^*)\,d\phi. \tag{73}$$

A suitable starting function is $\epsilon_0 = \pi\phi/a$ if nothing better is known, but convergence is more rapid if a better choice of ϵ_0 is made. This iterative process is very similar to that described in § 5.4. Ostrowski (1952) shows that it converges provided $|P''(\epsilon)|$ is uniformly bounded.

5.9 The mapping of doubly-connected regions

The next logical extension of the mapping theorem is to the case when the point P lying within C at $\psi = \infty$ is replaced by a contour C' lying entirely within C. The domain D lying between C and C' is *doubly connected*, and can be mapped on to a singly-connected domain only by inserting a suitable barrier AB connecting C and C' (see § 3.1).

We shall follow Woods (1958a), and map D on to the rectangle with sides $\phi = -a$, $\psi = 0$, $\phi = a$, $\psi = h$ in the w-plane, in such a way that C and C' fall on $\psi = 0$ and $\psi = h$ respectively. Let the sides of AB correspond to $\phi = a$ and $\phi = -a$, as shown in Fig. 5.9, then by continuity across AB the analytic function δ defined in (14) satisfies

$$\delta(-a+i\psi) = \delta(a+i\psi).$$

Let δ have imaginary parts $\theta_0'(\phi)$ and $\theta_h'(\phi)$ on $\psi = 0$ and $\psi = h$ respectively. Then the conditions satisfied by $\delta(w)$ are similar to those satisfied by the function $\tau(z)$ in § 4.9, and corresponding to (4.94) we are therefore able to write

$$\delta(w) = \frac{1}{4a}\int_{-a}^{a}\left\{\frac{d}{d\phi}\left(\ln\left|\frac{dz}{d\phi}\right|\right)_0 + \frac{d}{d\phi}\left(\ln\left|\frac{dz}{d\phi}\right|\right)_h\right\}d\phi$$
$$+ \frac{K}{\pi a}\int_{-a}^{a}\left[\theta_0'(\phi)\,Z_1\left\{\frac{K}{a}(\phi-w)\right\} - \theta_h'(\phi)\,Z_4\left\{\frac{K}{a}(\phi-w)\right\}\right]d\phi, \quad (74)$$

where the form of the first term follows from the real part of (14) and the last remark of § 4.9, the 'ϖ' term in (4.94) is omitted as δ is continuous across AB, and the parameter of the zeta functions is (see (4.92))

$$\mu = i\frac{h}{a} = i\frac{K'}{K}. \quad (75)$$

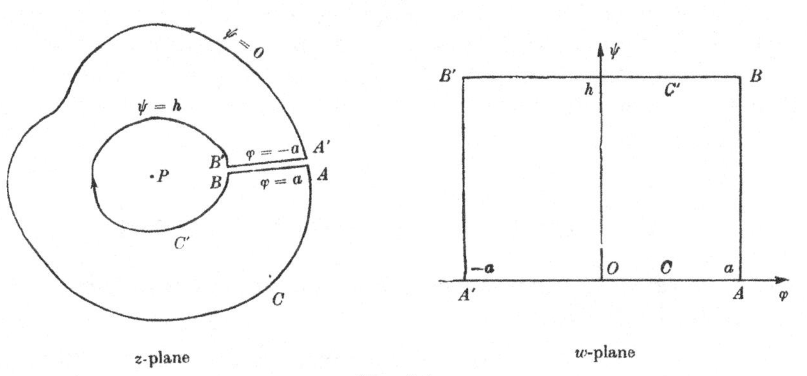

Fig. 5.9

The first term of (74) vanishes by the continuity of $\ln|dz/d\phi|$ across AB. It now remains only to integrate (74) with respect to w to obtain

$$\ln\frac{dz}{dw} = \ln K - \frac{1}{\pi}\int_{-a}^{a}\left[\ln\vartheta_1\left\{\frac{\pi}{2a}(w-\phi)\right\}d\theta_0(\phi) - \ln\vartheta_4\left\{\frac{\pi}{2a}(w-\phi)\right\}d\theta_h(\phi)\right], \quad (76)$$

where we have made use of (4.33) and (14), and have expressed the integrals in the convenient Riemann–Stieltjes form.

Equation (76) can be regarded as a somewhat generalized form of the classical Schwarz–Christoffel mapping formula for polygons. In the limit $C' \to P$, i.e. $h \to \infty$ it yields (47), as is readily verified by a calculation similar to that given in § 4.11.

A particular form of (76), which will be of some use in later chapters, is obtained when both C and C' are polygon shaped, i.e. when $\theta_0(\phi)$ and $\theta_h(\phi)$ are step-functions. Let the steps or leaps in $\theta_0(\phi)$ be α_n at $\phi = \phi_n$

$(n = 1, 2, ..., N)$, and those in $\theta_h(\phi)$ be α_m at $\phi = \phi_m$ $(m = 1, 2, ..., M)$. Then for this case (76) yields

$$\ln\frac{dz}{dw} = \ln K - \frac{1}{\pi}\sum_{n=1}^{N}\alpha_n \ln\vartheta_1\left\{\frac{\pi}{2a}(w - \phi_n)\right\} + \frac{1}{\pi}\sum_{m=1}^{M}\alpha_m \ln\vartheta_4\left\{\frac{\pi}{2a}(w - \phi_m)\right\},$$

$$\text{or}\quad \frac{dz}{dw} = K\prod_{m=1}^{M}\left[\vartheta_4\left\{\frac{\pi}{2a}(w - \phi_m)\right\}\right]^{\alpha_m/\pi}\bigg/\prod_{n=1}^{N}\left[\vartheta_1\left\{\frac{\pi}{2a}(w - \phi_n)\right\}\right]^{\alpha_n/\pi}. \quad (77)$$

This formula will be used in § 5.15 below.

Returning to the general formula (76) there are two comments we should make. First when the exponential of (76) is integrated to give a formula for $z(w)$ there will be four unknown constants appearing in the equation, namely K, a, h, and one additional constant of integration. In general this will require the specification of four corresponding points in the z- and w-plane for a unique mapping. Secondly, by a calculation similar to that given in § 5.4 we can derive a pair of simultaneous integro-differential equations from (76) for $\theta_0'(\phi)$ and $\theta_h'(\phi)$, and these can be solved by the same type of iterative process as already outlined for singly-connected regions.

MAPPING AND FLUID MOTIONS

5.10   The relation between conformal mapping and incompressible fluid motion

In the account of conformal mapping given above we have deliberately used the symbol w to denote the plane of the (canonical) region into which the z-plane domain is mapped. The reason for this will be apparent to the reader if he turns back to § 2.16. Equations (2.85) to (2.89) of this section suggest that the 'w' of the present chapter could be interpreted as the complex variable of some incompressible flow. We shall now look at the mappings given in §§ 5.3, 5.5, 5.6 and 5.9 from this point of view.

First consider the incompressible flow implicit in Fig. 5.3. The surface C is the streamline $\psi = 0$. Hence the fluid flows out from P towards A, around C in a clockwise direction, and then past F back into P. In the channel terminated by D_∞ the fluid near the right-hand side flows some distance towards D_∞, and then returns along the opposite side, and only the streamline $\psi = 0$ actually reaches D_∞; similarly, for the flow up divergent channel to E_∞.

The behaviour of the flow at the point P is highly singular. It is clear from the infinite extent of the w-plane in the ψ-direction, and the mass flow interpretation given to ψ in § 2.3, that an infinite mass of fluid is emitted from one side of P and simultaneously a similar mass is absorbed on the other side of P. Conservation of mass clearly prevents there being any net 'creation' or 'destruction' of mass at P. The flow in the neighbourhood of P is said to be due to a *doublet* at P—'doublet' because there is a *source* on one side of P, and a *sink* of equal strength on the other side. We shall study the nature of the flow near the doublet with the aid of equations (13), which apply in the neighbourhood of P.

The discontinuity in the slope at $P(\beta)$ is not an essential part of the doublet singularity, for the discussion given above is applicable whatever the value of β, and in particular for $\beta = 0$. For non-zero values of β we have in effect two singularities imposed at P—the 'corner-removing' singularity, which is simply the flow past a corner, in along one wall, and out along the other—and the *inversion* singularity. It is this latter singularity—or rather its fluid flow interpretation—in which we are at present interested.

With $\beta = 0$ in (13) we get

$$z - z_p = \frac{1}{(kw)}\left\{1 - \frac{b_0}{w} + O\left(\frac{1}{w^2}\right)\right\} \quad (w \text{ near } \infty).$$

This expansion can be reversed to give

$$w = \frac{-\mu\, e^{i\sigma}}{z - z_p}\{1 + O(z - z_p)\} \quad (z \text{ near } z_p), \tag{78}$$

where μ and σ are (new) real constants. Equation (78) shows that $w(z)$ has a simple pole at a doublet.

Using the equations of § 2.16 we have

$$\frac{dw}{dz} = U\, e^{-\tau} = U\, e^{-\Omega} e^{-i\theta},$$

i.e.
$$\frac{dw}{dz} = q\, e^{-i\theta}. \tag{79}$$

Thus from (78) it follows that the velocity vector (q, θ) due to the doublet alone is

$$(q, \theta) = \left(\frac{\mu}{|z - z_p|^2},\, 2\chi - \sigma\right), \tag{80}$$

where $\chi = \arg(z - z_p)$. The flow field due to this doublet is shown in Fig. 5.10.

If we replace $z - z_p$ in (78) by $1/z$, the pole at $z = z_p$ will be removed to the point at infinity (see § 3.3). Thus for a doublet at infinity we have

$$w = U\, e^{i\sigma} z\left\{1 + O\left(\frac{1}{z}\right)\right\}, \tag{81}$$

where we have introduced a new constant $U = -\mu$. In this case

$$\lim_{z \to \infty} q = U, \quad \lim_{z \to \infty} \theta = -\sigma, \tag{82}$$

so we have a parallel stream at infinity making an angle $-\sigma$ with the x-axis.

Next consider the transformation shown in Fig. 5.5. This clearly corresponds to the flow of a finite mass of fluid per second from a *source* at P, across the domain D, and into a *sink* at Q. The character of the flow near the source at P can be studied with the help of (36), which can be reversed to give

$$w = \frac{h}{\pi - \beta_p} \ln \frac{z - z_p}{K} + O(z - z_p). \tag{83}$$

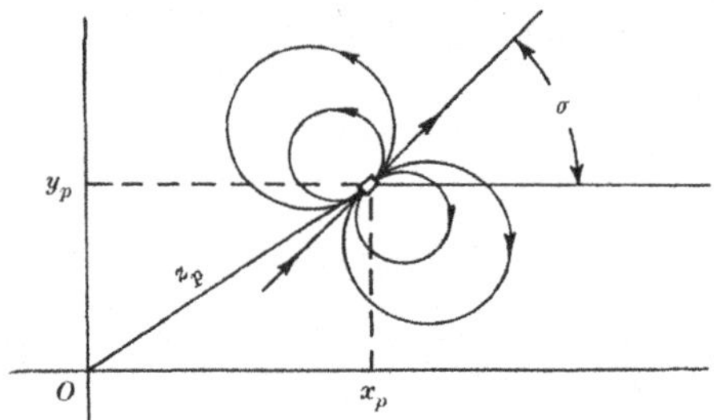

Fig. 5.10

Now h is proportional to the mass flow from P (§ 2.3), and as this flow is emitted into an angle $\pi - \beta_P$, $m \equiv h/(\pi - \beta_P)$ is proportional to the mass flow into unit angle at P.

Because

$$\frac{dw}{dz} = \frac{m}{z - z_p},$$

the velocity vector due to a source of *strength* m is

$$(q, \theta) = \left(\frac{m}{|z - z_p|}, \quad \arg(z - z_p) \right). \tag{84}$$

Equation (39) gives similar formulae for sinks, the only difference being a change in the sign of m.

The transformation shown in Fig. 5.6 corresponds to a *circulating* flow about the point P, the circulation being (see § 1.9)

$$\int_C \mathbf{q} \cdot d\mathbf{r} = \int_C \nabla\phi \cdot d\mathbf{r} = \int_{-a}^{a} d\phi = 2a,$$

for all streamlines. The singular point P of this flow is termed a *vortex*. Near P (44) gives

$$w = -i\kappa \ln\left(\frac{z-z_p}{K}\right) + O(z-z_p), \tag{85}$$

where $\kappa = 2a/2\pi$ is the circulation per unit angle about P. For this vortex

$$\frac{dw}{dz} = \frac{-i\kappa}{(z-z_p)},$$

so the corresponding velocity vector is

$$(q,\theta) = \left(\frac{\kappa}{|z-z_p|}, \quad \arg(z-z_p) + \tfrac{1}{2}\pi\right). \tag{86}$$

Finally Fig. 5.9 shows another circulating flow, this time between the cylinders C and C'. There are no doublets, sources, sinks or vortices to consider in this flow.

5.11   The use of mapping methods in fluid motion problems

The direct correspondence between incompressible flow and conformal mapping discussed above is only a small part of the role played by conformal mapping in fluid motions. In each of the incompressible flows considered the appropriate region of the w-plane happened to be one of the canonical shapes for which the various boundary value problems considered in chs. 3 and 4 have been solved, e.g. upper half-planes, infinite strips and rectangles.

But much more general and complicated types of w-planes are encountered in fluid dynamics, and before the solutions to the corresponding fluid motion problems can be written down these w-planes have to be transformed into one of the canonical regions for which the appropriate solution is known. This means we have to introduce an auxiliary variable ζ, where ζ is the plane of the selected canonical region, and then find the relation between ζ and w with the aid of one of the mapping formulae of this chapter.

Now the region in the w-plane—where, in general, w is the complex potential function defined by (2.74) for subsonic compressible flow—appropriate to some fluid motion, will be bounded by straight lines parallel to $\psi = 0$, i.e. by streamlines (an exception to this occurs with porous walls, see § 7.27). In other words, the flow region in the w-plane will be a *polygon*, which is quite easily mapped on to the canonical ζ-plane by means of one of (25), (43), (49) or (77). The solution can be written down in the ζ-plane by using the theory given in chs. 3 and 4. For example if τ is the variable defined in § 2.14, we find $\tau(\zeta)$, and as we know $w(\zeta)$, we can eliminate ζ to arrive at $\tau(w)$. Then (2.83) yields the relation

$z = z(w, \overline{w})$, and the problem is (formally) solved. It is, of course, not essential to eliminate ζ from the theory; it is often more convenient to retain it as a parameter.

We shall not elaborate the method of solving fluid motion problems just described further at this stage, for many examples are given in part III of this book, which make the method quite clear.

In the following sections we prepare some of the material necessary to part III by obtaining some $w(\zeta)$ relations of importance in practical applications.

SOME IMPORTANT CONFORMAL TRANSFORMATIONS

5.12 Flow in channels with mixed boundary conditions

With channel flows through single branches (branched channels are considered in § 7.33) the w-plane is simply the infinite strip $-\infty \leqslant \phi \leqslant \infty$, $0 \leqslant \psi \leqslant h$. Now suppose the boundary condition for such a flow is of the type: the imaginary (or real) part of an analytic function $\tau(w)$ prescribed over an interval AB of $\psi = 0$, while the real (or imaginary) part of $\tau(w)$ is prescribed over the rest of $\psi = 0$, and over the whole of $\psi = h$. Then it is clear from the discussion on this type of mixed boundary condition given in § 4.2 that it will be convenient to map the w-plane on to an auxiliary ζ- ($\zeta = \gamma + i\eta$) plane, in such a way that AB maps on to the whole of $\eta = 0$, and the rest of the w-plane boundary maps on to $\eta = \frac{1}{2}\pi$. The two planes are shown in Fig. 5.12. The origin of the w-plane has been chosen so that AB is the interval $-\phi_0 \leqslant \phi \leqslant \phi_0$.

Let the points $\phi = \infty$, $\phi = -\infty$ map on to $\eta = \frac{1}{2}\pi$; $\gamma = \gamma_0$, $\gamma = -\gamma_0$. Travelling around the w-plane contour in the positive direction we observe that the slope of this contour jumps by an amount π at C_∞, and an amount π at E_∞. As this is the negative direction along $\eta = \frac{1}{2}\pi$ we conclude that $\theta(\gamma)$ is a step-function having leaps of $-\pi$ at $\gamma = \gamma_0$ and $\gamma = -\gamma_0$ on $\eta = \frac{1}{2}\pi$. An application of (43), in which z, w are replaced by w, ζ respectively, and h has the value $\frac{1}{2}\pi$, now yields

$$\frac{dw}{d\zeta} = K[\cosh(\zeta - \gamma_0)]^{-1}[\cosh(\zeta + \gamma_0)]^{-1}.$$

Hence
$$dw = \frac{K\,d\zeta}{\cosh^2\zeta\cosh^2\gamma_0 - \sinh^2\zeta\sinh^2\gamma_0}$$

$$= \frac{2K}{\sinh 2\gamma_0}\frac{d(\tanh\gamma_0\tanh\zeta)}{(1 - \tanh^2\gamma_0\tanh^2\zeta)}.$$

Therefore
$$w = \frac{2K}{\sinh 2\gamma_0}\tanh^{-1}(\tanh\gamma_0\tanh\zeta), \tag{87}$$

the integration constant vanishing as $w = 0$ and $\zeta = 0$ are corresponding points.

At the point D in Fig. 5.12 $w = ih$ and $\zeta = i\pi/2$. Thus from (87)

$$ih = \frac{2K}{\sinh 2\gamma_0} \tanh^{-1} (\tanh \gamma_0 \tanh \tfrac{1}{2} i\pi).$$

Now $\tanh \tfrac{1}{2} i\pi = i \tan \tfrac{1}{2}\pi = i\infty$, and $\tanh^{-1} (\tanh \gamma_0 \, i\infty) = i \tan^{-1} \infty = \tfrac{1}{2} i\pi$ so

$$h = \frac{2K}{\sinh 2\gamma_0} \tfrac{1}{2}\pi.$$

Hence (87) can be written

$$\tanh \frac{\pi w}{2h} = \tanh \gamma_0 \tanh \zeta. \tag{88}$$

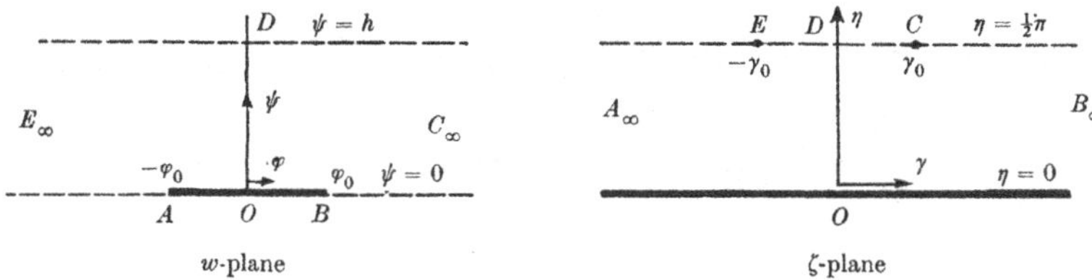

Fig. 5.12

The point $w = \phi_0$ maps on to $\zeta = \infty$, so from (88) $\gamma_0 = \pi\phi_0/2h$. Thus (88) can be written

$$\tanh \zeta = \coth \frac{\pi\phi_0}{2h} \tanh \frac{\pi w}{2h}, \tag{89}$$

and this is the required mapping formula. We shall make some use of (89) in §§ 7.13 to 7.19.

5.13   Non-circulating flow about a cylinder

Consider an isolated cylinder C of general shape lying in the z-plane in an infinite stream of fluid, and let the fluid be passing smoothly over the whole of C, so that we have a Dirichlet-type flow (see § 1.16). If there is no circulation about C the potential function ϕ will be a continuous single-valued function in the z-plane (see § 1.9). The boundary of the w-plane in this case is simply a straight line of finite length lying parallel to $\psi = 0$. Let the origin be chosen so that this slit, corresponding to the aerofoil, is $\psi = 0$, $-2a \leqslant \phi \leqslant 2a$, as shown in Fig. 5.13.

Now the most convenient canonical domain for the region 'inside' the boundary slit $BDACB$ (see § 3.1) is the ζ-plane shown in Fig. 5.13, for the continuity of the flow across BD_∞ means that the boundary conditions in the ζ-plane are similar to those studied in § 4.3. Moving around the slit in the positive direction we see that the slope of this 'polygon' has leaps of $-\pi$ at the front and rear stagnation points A and B, i.e. at the points $\gamma = 0$ and $\gamma = \pi$ in the ζ-plane. The relation between the w- and ζ-planes

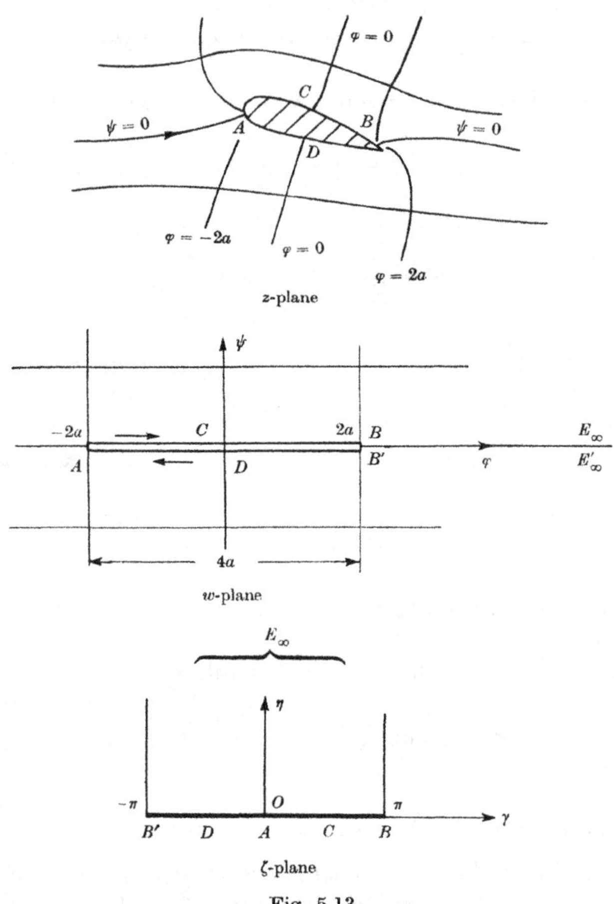

Fig. 5.13

can now be written down directly from (49), in which z, w, a are replaced in the present application by w, ζ, π. This yields

$$\frac{dw}{d\zeta} = K \sin \tfrac{1}{2}(\zeta - 0) \sin \tfrac{1}{2}(\zeta - \pi) = -\tfrac{1}{2}K \sin \zeta. \tag{90}$$

Thus
$$w = \tfrac{1}{2}K \cos \zeta,$$

the constant of integration vanishing as $w = 0$ at $\zeta = \pm \tfrac{1}{2}\pi$ (points C and D in Fig. 5.13). Now $w = 2a$ at $\zeta = \pm \pi$ (point B), so $K = -4a$, and

$$w = -2a \cos \zeta. \tag{91}$$

This is the required mapping formula. Notice that in this example of the theory given in § 5.6 it is the point at infinity in the w-plane that is mapped on to $\zeta = i\infty$.

5.14   Circulating flow about a cylinder

Next consider the same problem as above (Dirichlet flow about a cylinder) with the difference that now there is a circulation Γ about the cylinder. By § 1.9 it follows that ϕ increases by Γ on each complete circulation of the cylinder, and so it is now necessary to introduce some sort of discontinuity which allows for this in the representation of the w-plane. Two apparently distinct methods of doing this are shown in Fig. 5.14.

In the first w-plane we have inserted a semi-infinite strip $E'_\infty D'DE_\infty$ of width Γ in the ϕ-direction; then any contour $F'HF$ which is closed in the physical or z-plane is open in the w-plane, with a net gain of Γ in ϕ. Clearly the flow conditions on $D'E'_\infty$ are identical with those on DE_∞. We shall map this w-plane on to the ζ-plane shown in the same figure with the aid of (49).

It is not immediately apparent that the open 'polygon' $D'ACBB'D$ can be mapped into the ζ-plane by the method of § 5.6. However, a glance through § 5.6 reveals that at no stage was it assumed that the two sides AP, $A'P$ of the curve joining P to C were *adjacent*, but only that the functions involved were continuous across this curve. Of course in the case shown in Fig. 5.6 the two sides of the curve in question cannot be other than adjacent, but if P is at infinity separation of these sides is clearly possible. We conclude from this that it is valid to apply (49) to the mapping of the first w-plane on to the ζ-plane shown in Fig. 5.14.

As in the case of non-circulating flow, there are leaps of $-\pi$ in the slope of the w-plane polygon at the front and rear stagnation points (points A and B in the figure). Thus taking the rear stagnation point to be at $\gamma = -\pi$, and the front stagnation point to be at $\gamma = -2\alpha_0$, we find from (49) that

$$\frac{dw}{d\zeta} = K \sin \tfrac{1}{2}(\zeta + 2\alpha_0) \sin \tfrac{1}{2}(\zeta - \pi),$$

on replacing z, w, a in (49) by w, ζ, π. If K is given the value $-4a$ as in the non-circulating case, then

$$\frac{dw}{d\zeta} = 2a\{\sin(\zeta + \alpha_0) + \sin \alpha_0\}. \tag{92}$$

Hence $\qquad w = -2a \cos(\zeta + \alpha_0) + (\zeta + \alpha_0 - \tfrac{1}{2}\pi)\, 2a \sin \alpha_0, \tag{93}$

selecting the constant of integration so that $w = 0$ at $\zeta = \tfrac{1}{2}\pi - \alpha_0$. Equation (93) is the required generalization of (91).

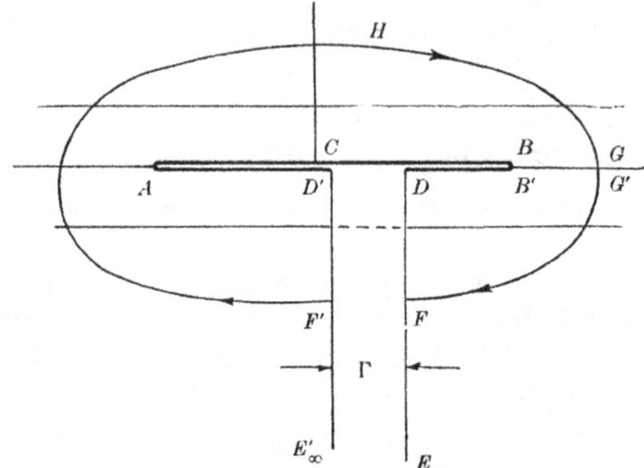

First w-plane

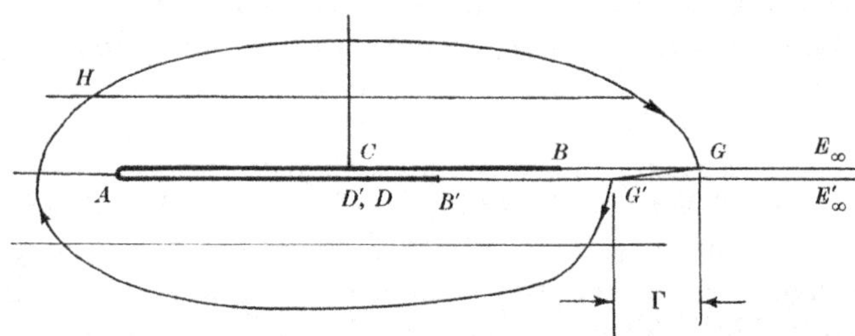

Second w-plane

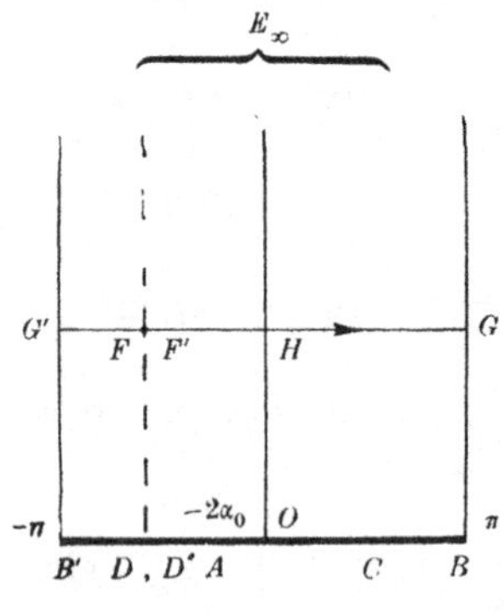

ζ-plane

Fig. 5.14

One circuit of the cylinder, by a path such as $G'FF'HG$ in Fig. 5.14, increases ζ by 2π, and so by (93) the corresponding increase in ϕ, the real part of w, is

$$\Gamma = 4\pi a \sin \alpha_0. \tag{94}$$

Turning now to the second w-plane shown in Fig. 5.14, we easily verify that (93) also maps this on to the ζ-plane, and therefore there is no essential difference between the two w-planes. The second representation does have the slight advantage that the leap in ϕ occurs across the trailing edge streamline, where in unsteady flow there will also be a leap in the derivative $(q = \partial\phi/\partial s)$ of ϕ. It is mathematically convenient to keep these discontinuities to the same curve.

5.15 Circulating flow about a cylinder in a stream of finite width

Our next example of the methods of this chapter is an obvious generalization of the example just completed.

Let the cylinder lie on $\psi = 0$ in $-2b \leqslant \phi \leqslant 2b$, the boundary stream-lines of the stream be $\psi = h_1$, $\psi = -h_2$ $(h_1, h_2 > 0)$, and the circulation be Γ. Then the w-plane can be represented as in Fig. 5.15. The symmetrical arrangement of the sides $D'F''$, DF (about $\phi = 0$) of the curve of discontinuity in ϕ simplifies the algebra of the transformation without any real loss in generality.

The most suitable ζ-plane on which to map the w-plane is the rectangle $-K \leqslant \gamma \leqslant K$, $0 \leqslant \eta \leqslant K'$ shown in Fig. 5.15, for in this plane we shall have the same type of boundary conditions as those studied in §4.9, i.e. known values on $\eta = 0$ (the cylinder surface) and $\eta = K'$ (the boundaries of the finite stream) and periodic conditions in the γ-direction.

We have therefore to map the region between the two open 'polygons' $D'ACBD$ and $F'I_\infty HE_\infty F$ on to the ζ-plane rectangle; the appropriate equation for this is (77). By an argument similar to that given for the open polygon in Fig. 5.14 the fact that these polygons are open does not affect the application of (77) at all. The only essential requirement is the continuity of derivatives of ϕ and ψ across the 'gap' $F'D'DF$, and this is obviously met in the present application.

The following table of corresponding values taken from Fig. 5.15 will be useful:

Point ...	F'	I	H	E	F	D'	A	C	B	D
w-plane	$-\tfrac{1}{2}\Gamma - ih_2$	$-\infty$	ih_1	∞	$\tfrac{1}{2}\Gamma - ih_2$	$-\tfrac{1}{2}\Gamma$	$-2b$	0	$2b$	$\tfrac{1}{2}\Gamma$
ζ-plane	$-K + iK'$	$-\gamma_1 + iK'$	iK'	$\gamma_1 + iK'$	$K + iK'$	$-K$	$-\gamma_0$	0	γ_0	K

The polygon $D'ACBD$ has leaps of $-\pi$ in its slope at points A and B, where $\gamma = -\gamma_0$ and γ_0; similarly, the polygon $FEHIF'$ has leaps of π at $\gamma = -\gamma_1$ and γ_1, but this is in the direction of decreasing γ, and therefore

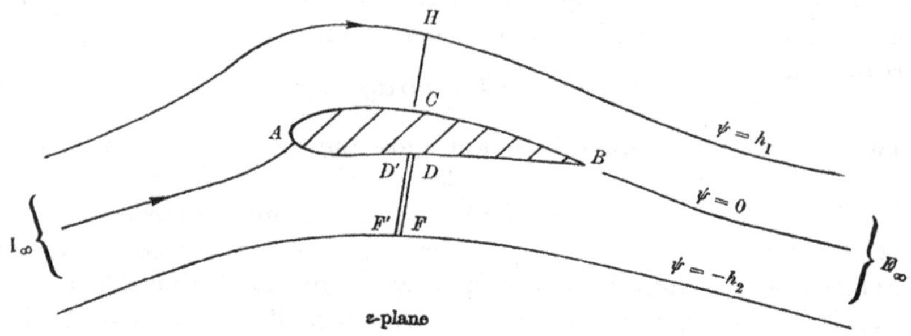

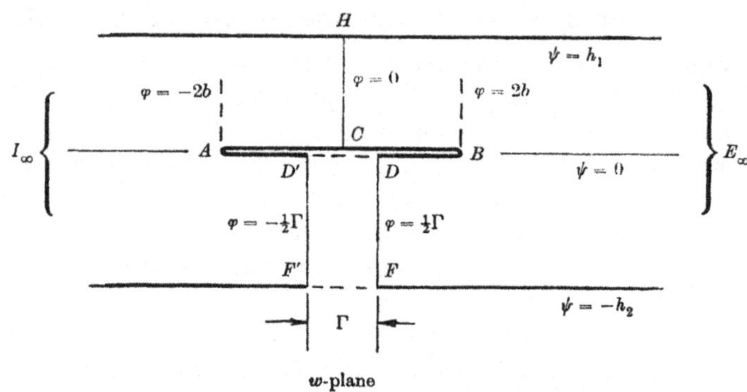

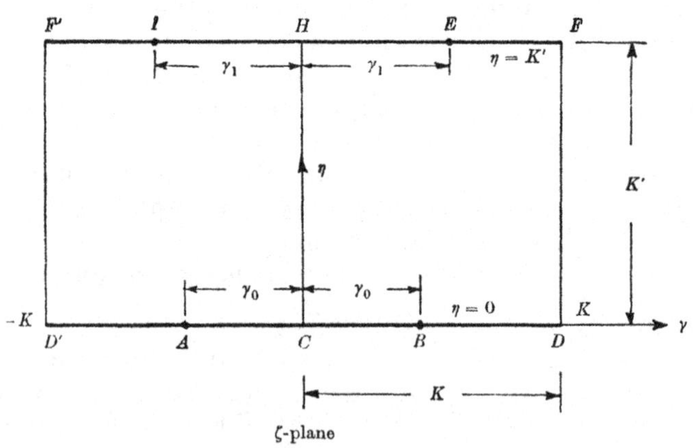

H
C
A
B
D' D
F' F
$\psi = h_1$
$\psi = 0$
$\psi = -h_2$
I_∞
E_∞
z-plane

H
$\psi = h_1$
$\varphi = -2b$
$\varphi = 0$
$\varphi = 2b$
C
I_∞
A
D'
D
B
$\psi = 0$
E_∞
$\varphi = -\frac{1}{2}\Gamma$
$\varphi = \frac{1}{2}\Gamma$
F'
F
Γ
$\psi = -h_2$
w-plane

F'
I
H
E
F
γ_1
γ_1
$\eta = K'$
η
K'
γ_0
γ_0
$\eta = 0$
K
-K
D'
A
C
B
D
γ
K
ζ-plane

Fig. 5.15

regarded as a function of γ the slope has leaps of $-\pi$ at $\gamma = -\gamma_1$ and γ_1. In the notation of (77) $\alpha_n = -\pi$ at $\gamma_n = \pm\gamma_0$, and $\alpha_m = -\pi$ at $\gamma_m = \pm\gamma_1$. Thus replacing z, w, a and h in (75) and (77) by w, ζ, K and K', and using these values of α_n and α_m we arrive at

$$\frac{dw}{d\zeta} = K_0 \frac{\vartheta_1\left\{\frac{\pi}{2K}(\zeta-\gamma_0)\left|\frac{iK'}{K}\right.\right\}\vartheta_1\left\{\frac{\pi}{2K}(\zeta+\gamma_0)\left|\frac{iK'}{K}\right.\right\}}{\vartheta_4\left\{\frac{\pi}{2K}(\zeta-\gamma_1)\left|\frac{iK'}{K}\right.\right\}\vartheta_4\left\{\frac{\pi}{2K}(\zeta+\gamma_1)\left|\frac{iK'}{K}\right.\right\}}. \tag{95}$$

The addition formulae (Whittaker & Watson, 1952, p. 487)

$$\vartheta_1(y+z)\,\vartheta_1(y-z)\,\vartheta_4^2 = \vartheta_1^2(y)\,\vartheta_4^2(z) - \vartheta_4^2(y)\,\vartheta_1^2(z), \tag{96}$$

$$\vartheta_4(y+z)\,\vartheta_4(y-z)\,\vartheta_4^2 = \vartheta_4^2(y)\,\vartheta_4^2(z) - \vartheta_1^2(y)\,\vartheta_1^2(z), \tag{97}$$

applied to (95) yield

$$\frac{dw}{d\zeta} = K_0 \frac{\vartheta_1^2\left(\frac{\pi\zeta}{2K}\right)\vartheta_4^2\left(\frac{\pi\gamma_0}{2K}\right) - \vartheta_4^2\left(\frac{\pi\zeta}{2K}\right)\vartheta_1^2\left(\frac{\pi\gamma_0}{2K}\right)}{\vartheta_4^2\left(\frac{\pi\zeta}{2K}\right)\vartheta_4^2\left(\frac{\pi\gamma_1}{2K}\right) - \vartheta_1^2\left(\frac{\pi\zeta}{2K}\right)\vartheta_1^2\left(\frac{\pi\gamma_1}{2K}\right)}.$$

Dividing the numerator by $\vartheta_4^2\left(\frac{\pi\gamma_0}{2K}\right)\vartheta_4^2\left(\frac{\pi\zeta}{2K}\right)$ and the denominator by $\vartheta_4^2\left(\frac{\pi\gamma_1}{2K}\right)\vartheta_4^2\left(\frac{\pi\zeta}{2K}\right)$, using the first of (4.45) and modifying the constant K_0 we arrive at

$$\frac{dw}{d\zeta} = A\frac{\operatorname{sn}^2\zeta - \operatorname{sn}^2\gamma_0}{1 - k^2\operatorname{sn}^2\gamma_1\operatorname{sn}^2\zeta},$$

A being the new constant. With a further change of constant this can be expressed

$$\frac{dw}{d\zeta} = B\left\{\frac{k^2\operatorname{sn}\gamma_1\operatorname{cn}\gamma_1\operatorname{dn}\gamma_1\operatorname{sn}^2\gamma_0}{1-k^2\operatorname{sn}^2\gamma_0\operatorname{sn}^2\gamma_1} - \frac{k^2\operatorname{sn}\gamma_1\operatorname{cn}\gamma_1\operatorname{dn}\gamma_1\operatorname{sn}^2\zeta}{1-k^2\operatorname{sn}^2\gamma_1\operatorname{sn}^2\zeta}\right\}. \tag{98}$$

The terms in the curly brackets can be expressed in terms of the zeta functions defined in § 4.5. From (4.68) and (4.59) it follows that

$$Z_4(z-y) - Z_4(z+y) + 2Z_4(y) = k^2\operatorname{sn}y\operatorname{sn}z\{\operatorname{sn}(y+z) + \operatorname{sn}(z-y)\}$$

$$= \frac{2k^2\operatorname{sn}y\operatorname{cn}y\operatorname{dn}y\operatorname{sn}^2z}{1-k^2\operatorname{sn}^2z\operatorname{sn}^2y}. \tag{99}$$

Hence (98) can be written

$$\frac{dw}{d\zeta} = B\{\tfrac{1}{2}[Z_4(\gamma_0-\gamma_1) - Z_4(\gamma_0+\gamma_1) + 2Z_4(\gamma_1)]$$

$$- \tfrac{1}{2}[Z_4(\zeta-\gamma_1) - Z_4(\zeta+\gamma_1) + 2Z_4(\gamma_1)]\}. \tag{100}$$

We can now use the theory given in § 4.8 to integrate (100) and find that

$$w = \tfrac{1}{2}B\{[Z_4(\gamma_0-\gamma_1)-Z_4(\gamma_0+\gamma_1)+2Z_4(\gamma_1)]\zeta-2\Pi(\zeta,\gamma_1)\}, \quad (101)$$

taking into account that the origins of the w- and ζ-planes correspond.

The constant B can be found from the fact that the imaginary part of w jumps from $\psi = h_1$ to $\psi = -h_2$ as ζ passes through the point E on $\eta = K'$ (cf. Fig. 5.15). The second and third of (4.80) show that $\mathscr{I}\Pi(\zeta,\gamma_1)$ must leap $\tfrac{1}{2}\pi$ at E, and therefore the imaginary part of the right-hand side of (101) leaps $-\tfrac{1}{2}B\pi$ at E. Hence $-\tfrac{1}{2}B\pi = -(h_1+h_2)$, or $B = 2(h_1+h_2)/\pi$. Our transformation can now be written

$$w = \frac{h_1+h_2}{\pi}[\{Z_4(\gamma_0-\gamma_1)-Z_4(\gamma_0+\gamma_1)+2Z_4(\gamma_1)\}\zeta-2\Pi(\zeta,\gamma_1)], \quad (102)$$

and this is the desired generalization of (93).

If the numbers b, Γ, h_1 and h_2 are assumed to be known then the mapping into the ζ-plane introduces three unknown constants, viz. γ_0, γ_1 and the modulus k (K' and K are uniquely determined by the value of k). Three relations defining these constants can be found by substituting the corresponding values of w and ζ at points B and F into (102). At B $w = 2b$ and $\zeta = \gamma_0$, so

$$2b = \frac{h_1+h_2}{\pi}[\{Z_4(\gamma_0-\gamma_1)-Z_4(\gamma_0+\gamma_1)+2Z_4(\gamma_1)\}\gamma_0-2\Pi(\gamma_0,\gamma_1)]; \quad (103)$$

at F $w = \tfrac{1}{2}\Gamma-ih_2$, $\zeta = K+iK'$ so with the help of (4.80) we find

$$\tfrac{1}{2}\Gamma-ih_2 = \frac{h_1+h_2}{\pi}\left[\{Z_4(\gamma_0-\gamma_1)-Z_4(\gamma_0+\gamma_1)\}(K+iK')-\frac{i\pi}{K}\gamma_1\right].$$

The real part of this equation is

$$\frac{\pi\Gamma}{2(h_1+h_2)} = \{Z_4(\gamma_0-\gamma_1)-Z_4(\gamma_0+\gamma_1)\}K, \quad (104)$$

while the imaginary part and (104) can be combined to yield

$$\frac{2\gamma_1}{K} = 1+\frac{h_2-h_1}{h_2+h_1}+\frac{\Gamma K'}{(h_1+h_2)K}. \quad (105)$$

The transformation of the w-plane on to the ζ-plane is now completely defined by (102) to (105).

5.16   The use of simultaneous Schwarz–Christoffel transformations

It will be obvious to the reader that the mapping formulae for the examples given in §§ 5.10–5.15 above could have been obtained by using the Schwarz–Christoffel transformation to map each of the w- and ζ-planes on to the upper half of some intermediate t-plane, and then eliminating t from the resulting pair of equations.

To illustrate this method we shall consider the problem of designing a bend in a channel. The various planes are shown in Fig. 5.16. The mixed boundary value problem is: θ given on $CE_\infty A$, $BF_\infty D$ and Ω given on AB, CD. The most convenient ζ-plane to use is the rectangle $-K \leqslant \gamma \leqslant K$, $0 \leqslant \eta \leqslant K'$, such that the two intervals in which θ are prescribed cover opposite sides of this rectangle. The solution in this ζ-plane can be written down immediately from the harmonic conjugate of (4.138), as a glance at Fig. 4.13b will make clear. This problem is considered further in §7.26; here we are concerned only with the mapping of the w-plane on to the ζ-plane.

We have assumed the w-plane to be symmetrical about $\phi = 0$; in this case it is clear that if the origins of the w- and t-planes coincide, and the points D, F, B map on to $t = 1$, t_0, $1/k$, $(k < 1)$, then A, E, C map on to $t = -1$, $-t_0$, $-1/k$. In the w-plane there are angles $\alpha_1 = \pi$, $\alpha_2 = \pi$ at E and F. Hence by an application of (25) $dw/dt = K(t-t_0)^{-1}(t+t_0)^{-1}$, so

$$w = \frac{h}{\pi}\ln\left(\frac{t_0+t}{t_0-t}\right), \tag{106}$$

on using the facts that $w = 0$ at $t = 0$, and $\mathscr{I}w = \psi$ leaps by an amount h at $t = t_0$. Solving (106) for t we obtain

$$t = t_0 \tanh\frac{\pi w}{2h}. \tag{107}$$

As $t = 1$ at $w = a$, and $t = 1/k$ at $w = b+ih$ (see figure), it follows from (107) that $t_0 = \coth\dfrac{\pi a}{2h}$ and

$$k = \tanh\frac{\pi a}{2h}\tanh\frac{\pi b}{2h}. \tag{108}$$

Next we map the ζ-plane into the t-plane. Angles of $\frac{1}{2}\pi$ occur at each of the points A, B, C and D. Equation (25) applied to this case therefore gives $d\zeta/dt = K(t-1)^{-\frac{1}{2}}(t+1)^{-\frac{1}{2}}(t-1/k)^{-\frac{1}{2}}(t+1/k)^{-\frac{1}{2}}$, so

$$\zeta = \frac{1}{\alpha}\int\frac{dt}{\sqrt{\{(1-t^2)(1-k^2t^2)\}}},$$

where α is a constant. This can be integrated by introducing Jacobian elliptic functions of modulus k, such that $t = \operatorname{sn}(x, k)$. Then $1-t^2 = \operatorname{cn}^2 x$, $1-k^2t^2 = \operatorname{dn}^2 x$, $dt = \operatorname{cn}x\operatorname{dn}x\,dx$ (see §4.6). Therefore $\zeta = x/\alpha$, i.e. $t = \operatorname{sn}(\alpha\zeta, k)$, as $\zeta = 0$ at $t = 0$. Let the numbers K and K', defined in Fig. 5.16, be the real and imaginary quarter-periods of the elliptic

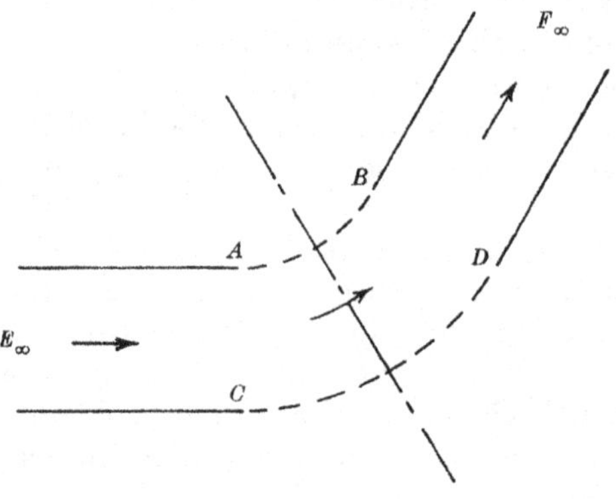

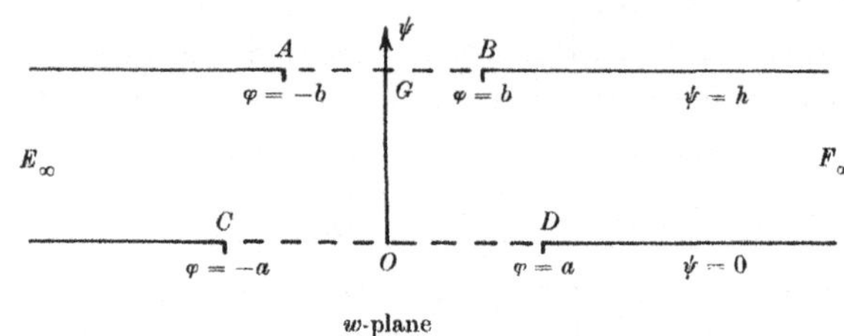

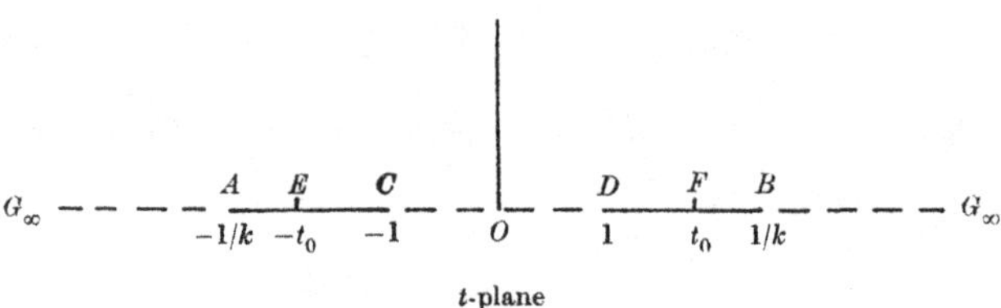

F∞
B
A
D
E∞
C
z-plane

ψ
A
B
G
φ = −b
φ = b
ψ = h
E∞
F∞
C
D
O
φ = −a
φ = a
ψ = 0
w-plane

A E C
D F B
G∞
O
G∞
−1/k −t₀ −1
1 t₀ 1/k
t-plane

−K + iK'
G
K + iK'
A
B
E
F
η
C
γ
D
−K
O
K
ζ-plane

Fig. 5.16

functions, then as $t = 1$ at $\zeta = K$ it follows that $\alpha = 1$ ($\operatorname{sn} K = 1$). We now have

$$t = \operatorname{sn}(\zeta, k). \tag{109}$$

Finally eliminating t from (107) and (109) we arrive at

$$\tanh\frac{\pi w}{2h} = \tanh\frac{\pi a}{2h}\operatorname{sn}(\zeta, k), \tag{110}$$

where k is given by (108). This is the required mapping equation.

GREEN'S FUNCTION

5.17 The relation between Green's function and conformal mapping

There is a very important relation between the Green's function for a singly-connected domain D in the z-plane, and the mapping function $w = f(z)$ which maps D on to the interior of unit circle (with origin for centre) in the w-plane. The methods we have given in ch. 3 of dealing with Dirichlet problems make unnecessary knowledge of this relation for such problems, but it is very useful if Poisson's equation has to be solved (see (3.131)). In the notation of § 3.20 the relation in question is

$$g(\tilde{z}, \tilde{\zeta}) = \ln\left|\frac{1 - f(z)\,\overline{f(\zeta)}}{f(z) - f(\zeta)}\right|. \tag{111}$$

This will be proved if we can show that the right-hand side of (111) has the characteristic properties of the Green's function of D, viz. is a harmonic function of $\tilde{z}$, and satisfies (3.126) and (3.127) of § 3.20. First, g is clearly a harmonic function of $\tilde{z}$—except at $z = \zeta$—being the real part of an analytic function of z. Secondly, near $z = \zeta$

$$g(\tilde{z}, \tilde{\zeta}) = -\ln|z - \zeta| + \ln\left|\frac{\{1 - f(z)\,\overline{f(\zeta)}\}\,(z - \zeta)}{f(z) - f(\zeta)}\right|$$

and the last term is harmonic even at $z = \zeta$. Finally on the boundary of D $|w| = |f(z)| = 1$, so $f(z)\,\overline{f(z)} = 1$. Thus on this boundary (111) can be written

$$g(\tilde{z}, \tilde{\zeta}) = \ln\left|\frac{\overline{f(z)} - \overline{f(\zeta)}}{f(z) - f(\zeta)}\right| + \ln|f(z)|,$$

which obviously vanishes. Thus the right-hand side of (111) is the Green's function of D.

On writing g in the form

$$g(\tilde{z}, \tilde{\zeta}) = \tfrac{1}{2}\ln\left\{\frac{\{1 - f(z)\,\overline{f(\zeta)}\}\{1 - \overline{f(z)}\,f(\zeta)\}}{\{f(z) - f(\zeta)\}\{\overline{f(z)} - \overline{f(\zeta)}\}}\right\}, \tag{112}$$

we see that the symmetry property

$$g(\tilde{z}, \xi) = g(\xi, \tilde{z}) \tag{113}$$

holds.

If Ω is harmonic within D then by (3.133)

$$\Omega(\xi) = \frac{1}{2\pi} \int_C \Omega(\tilde{t}) \frac{\partial g}{\partial n}(\tilde{t}, \xi)\, ds.$$

Equation (111) enables this to be written

$$\Omega(\xi) = \mathscr{R}\left[\frac{1}{2\pi} \int_C \Omega(\tilde{t}) \frac{\partial}{\partial n} \ln\left\{ \frac{1 - f(\zeta)\,\overline{f(t)}}{f(\zeta) - f(t)} \right\} ds \right].$$

The imaginary part of the term in the square brackets is—apart from an additive constant—the harmonic conjugate function θ of Ω. Hence if $\tau = \Omega + i\theta$,

$$\tau(\zeta) = \frac{1}{2\pi} \int_C \Omega(\tilde{t}) \frac{\partial}{\partial n} \ln\left\{ \frac{1 - f(\zeta)\,\overline{f(t)}}{f(\zeta) - f(t)} \right\} ds + iA, \tag{114}$$

where A is a real constant.

If we put $u = v$ in (3.125) then eliminate dv by using (3.124) we find

$$\frac{\partial v}{\partial n}\, ds = i\left(\frac{\partial v}{\partial t}\, dt - \frac{\partial v}{\partial \tilde{t}}\, d\tilde{t} \right). \tag{115}$$

Hence
$$\frac{\partial}{\partial n} \ln\left\{ \frac{1 - f(\zeta)\,\overline{f(t)}}{f(\zeta) - f(t)} \right\} ds = i\left\{ \frac{f'(t)\, dt}{f(\zeta) - f(t)} + \frac{f(\zeta)\,\overline{f'(t)}\, d\tilde{t}}{1 - f(\zeta)\,\overline{f(t)}} \right\}.$$

But on C, the boundary of D, $f(t)\,\overline{f(t)} = 1$, and the right-hand side of this equation can be reduced to

$$i\frac{f(\zeta) + f(t)}{f(\zeta) - f(t)} \frac{f'(t)\, dt}{f(t)}.$$

Substituting this value in (114) we have

$$\tau(\zeta) = \frac{1}{2\pi i} \int_C \Omega(\tilde{t}) \frac{f(\zeta) + f(t)}{-f(\zeta) + f(t)} \frac{f'(t)\, dt}{f(t)} + iA. \tag{116}$$

It is a simple calculation to show from (116) that the constant A is given by

$$A = \theta(\zeta_0) = \frac{1}{2\pi i} \int_C \theta(\tilde{t}) \frac{f'(t)\, dt}{f(t)}, \tag{117}$$

where ζ_0 is that point which maps on to the origin of the w-plane, i.e. such that $w = f(\zeta_0) = 0$.

The result contained in (116) and (117) could have also been derived directly from (3.86) of which it is an obvious generalization.

As an example consider the semi-infinite strip

$$-\pi \leqslant \gamma \leqslant \pi, \quad 0 \leqslant \eta \leqslant \infty \quad (\zeta = \gamma + i\eta),$$

shown in Fig. 5.13. This domain is mapped into the unit circle in the w-plane by

$$w = f(\zeta) = -e^{i\zeta}, \tag{118}$$

a result which follows from (60) on replacing z, z_p, $-R$, w and a in that equation by w, 0, 1, ζ and π. If this mapping function is substituted in (111) there results

$$g(\tilde{\zeta}, \tilde{t}) = \ln \left| \frac{1 - e^{i\zeta}\,e^{-i\bar{t}}}{e^{i\zeta} - e^{it}} \right|, \tag{119}$$

which, after some elementary algebra, can be written in the real variable form

$$g(\gamma, \eta; \gamma_0, \eta_0) = \ln \left\{ \frac{\cosh(\eta + \eta_0) - \cos(\gamma - \gamma_0)}{\cosh(\eta - \eta_0) - \cos(\gamma - \gamma_0)} \right\}, \tag{120}$$

where $t = \gamma_0 + i\eta_0$.

Substituting (118) in (116) we obtain

$$\begin{aligned}
\tau(\zeta) &= \frac{1}{2\pi i} \int_C \Omega(\tilde{t}) \cot \tfrac{1}{2}(t - \zeta)\,dt + iA \\
&= \frac{1}{2\pi i} \int_{-\pi}^{\pi} \Omega(\gamma_0) \cot \tfrac{1}{2}(\gamma_0 - \zeta)\,d\gamma_0 + iA \\
&\quad + \frac{1}{2\pi} \int_0^{\infty} \{\Omega(i\eta_0 + \pi) - \Omega(i\eta_0 - \pi)\} \tan \tfrac{1}{2}(i\eta_0 - \zeta)\,d\eta_0,
\end{aligned} \tag{121}$$

and this result is a generalization of that given (in conjugate form) in (4.14).

If instead of being harmonic, Ω satisfies Poisson's equation in the semi-infinite strip, i.e. satisfies

$$\nabla^2 \Omega = h(\gamma, \eta),$$

then from (3.131), (120) and (121) it follows that

$$\begin{aligned}
\Omega(\gamma, \eta) = \mathscr{R} &\left[\frac{1}{2\pi i} \int_{-\pi}^{\pi} \Omega(\gamma_0) \cot \tfrac{1}{2}(\gamma_0 - \gamma - i\eta)\,d\gamma_0 \right. \\
&\left. + \frac{1}{2\pi} \int_0^{\infty} \{\Omega(\pi, \eta_0) - \Omega(-\pi, \eta_0)\} \tan \tfrac{1}{2}(i\eta_0 - i\eta - \gamma)\,d\eta_0 \right] \\
&- \frac{1}{2\pi} \int_0^{\infty} \int_{-\pi}^{\pi} h(\gamma_0, \eta_0) \ln \left\{ \frac{\cosh(\eta + \eta_0) - \cos(\gamma - \gamma_0)}{\cosh(\eta - \eta_0) - \cos(\gamma - \gamma_0)} \right\} d\gamma_0\,d\eta_0.
\end{aligned} \tag{122}$$

PART III
APPLICATIONS

CHAPTER 6

GENERAL ACCOUNT OF METHODS: SOURCES, DOUBLET AND VORTICES

6.1 Recapitulation

Before entering into the details of the application of the theory developed in parts I and II to problems in fluid and aerodynamics, we shall give a general account of the methods adopted. To this end it will be convenient to have a brief summary of the salient equations and ideas so far developed.

The most important equations are those of § 2.14, which show that the complex function

$$\tau = \Omega + i\theta, \tag{1}$$

is an analytic function of

$$w = \phi + i\psi. \tag{2}$$

In (1) θ is the slope of the streamline, while Ω is a speed parameter depending only on q. In § 2.13 it is shown that

$$\frac{q}{U} = \sinh \Omega^* \operatorname{cosech} (\Omega + \Omega^*) \quad (-\Omega^* < \Omega_{\text{min.}} \leqslant \Omega < \infty), \tag{3}$$

where U is the velocity corresponding to 'average' or 'tangency point' conditions ($U \equiv q_a$),

$$\Omega^* = \sinh^{-1}\left(\frac{\beta_a}{M_a}\right) = \tanh^{-1}\beta_a = \ln\frac{1+\beta_a}{M_a} = \ln\frac{M_a}{1-\beta_a}, \tag{4}$$

$$\Omega_{\text{min.}} = -\tanh^{-1}\left\{\frac{\beta_a}{1+\gamma M_a^2}\right\}, \tag{5}$$

and β_a is related to the tangency point Mach number M_a by $\beta_a = (1 - M_a^2)^{\frac{1}{2}}$. From (3) and (4) we find that

$$\frac{U}{q} = \frac{1}{2\beta_a}\{(1+\beta_a)\,e^{\Omega} - (1-\beta_a)\,e^{-\Omega}\}. \tag{6}$$

In problems involving infinite streams the tangency point is usually taken to correspond to the flow conditions at infinity in the upstream direction. Thus, employing the notation $\lim_{\phi \to -\infty} A = A_{-\infty}$, we would have for this case $A_a = A_{-\infty}$, i.e. the subscript 'a' and '$-\infty$' would be equivalent.

In (2) ϕ and ψ are the usual potential and stream-functions, viz.

$$d\phi = q\,ds, \quad d\psi = \frac{\rho q}{\rho_s}\,dn \quad (\rho_s = \rho_a/\beta_a), \tag{7}$$

where s, n are distances along the streamline and its normal respectively, the value of ρ_s comes from (2.36), and ρ is the fluid density.

The pressure coefficient and density are related to Ω by (§ 2.13)

$$C_p \equiv \frac{p - p_a}{\frac{1}{2}\rho_a U^2} = 2 \sinh \Omega \cosh \Omega^* \operatorname{cosech} (\Omega + \Omega^*), \tag{8}$$

and

$$\rho = \frac{\rho_a}{\beta_a} \tanh (\Omega + \Omega^*). \tag{9}$$

Equations (3), (4), (6) and (9) enable us to write (7) in the forms

$$ds = \frac{1}{2\beta_a U} \{(1 + \beta_a) e^{\Omega} - (1 - \beta_a) e^{-\Omega}\} d\phi, \tag{10}$$

$$dn = \frac{1}{2\beta_a U} \{(1 + \beta_a) e^{\Omega} + (1 - \beta_a) e^{-\Omega}\} d\psi. \tag{11}$$

The physical ($z = x + iy$)-plane is given by integrating

$$dz = \frac{1}{2\beta_a U} \{(1 + \beta_a) e^{\tau} dw - (1 - \beta_a) e^{-\tau} d\overline{w}\}. \tag{12}$$

Thus if τ is expressed in terms of some parameter ζ,

$$z(\zeta, \bar{\zeta}) = \frac{1}{2\beta_a U} \left[(1 + \beta_a) \int^{\zeta} e^{\tau(\zeta)} \frac{dw}{d\zeta} d\zeta - (1 - \beta_a) \overline{\int^{\zeta} e^{-\tau(\zeta)} \frac{dw}{d\zeta} d\zeta} \right]. \tag{13}$$

The force and moment acting on a streamline AB (see Fig. 2.18) are given by

$$X - iY = i \frac{\rho_a U}{1 + \beta_a} \int_{AB} e^{-\tau} dw - i(\bar{z}_B - \bar{z}_A) \left(p_a + \frac{p_a U^2}{1 + \beta_a} \right), \tag{14}$$

and

$$M = -\mathscr{R} z_0 (Y + iX) + \frac{\rho_a}{2\beta_a} \mathscr{R} \int_{AB} e^{-\tau} \left(\int_0^w e^{\tau} dw \right) dw$$

$$- \frac{\rho_a}{4\beta_a} \frac{1 - \beta_a}{1 + \beta_a} \left| \int_0^w e^{-\tau} dw \right|_{AB}^2 + \left(p_a + \frac{\rho_a U^2}{1 + \beta_a} \right) \mathscr{R} \{z_0 (\bar{z}_B - \bar{z}_A)$$

$$- \tfrac{1}{2}(z_B \bar{z}_B - z_A \bar{z}_A)\}, \tag{15}$$

where the moment is in the clockwise sense about the origin in the z-plane, and where z_0 is the value of z corresponding to $w = 0$.

Equations (1)–(15) are the principal equations for the steady flow of a tangent gas. In the limits $M_a \to 0$ ($\beta_a \to 1$) they reduce to the equations of classical incompressible flow theory. In taking this limit we must remember that the specific heat ratio γ tends to infinity, and that the sound speed, $a_a = \sqrt{(\gamma p_a/\rho_a)}$ (see (1.7)), does likewise. From this equation for a_a it follows that the product $\gamma M_a^2 = \gamma U^2/a_a^2$, which appears in (5) and in several other places in the theory, tends to $U^2 \rho_a/\rho_a$.

As pointed out in §2.16 the principal difference between the theories for compressible and incompressible flow is that $z = z(w, \overline{w})$ in the former and $z = z(w)$ in the latter. This means that τ is an analytic function of w *and* z, a result which at times somewhat simplifies the theory. The principal equations in incompressible flow are (§2.16)

$$\tau = \ln\left(U dz/dw\right), \tag{16}$$

$$\frac{q}{U} = e^{-\Omega}, \tag{17}$$

$$C_p = 1 - e^{-2\Omega}, \tag{18}$$

$$ds = d\phi/q, \tag{19}$$

and the forms of (14) and (15) obtained by putting $\beta_a = 1$.

In incompressible flow two other dependent variables are sometimes used instead of τ. These are:

(1) *The 'displacement' function,*

$$\delta(w) \equiv Uz - w = \int (e^\tau - 1)\, dw, \tag{20}$$

and

(2) *the complex velocity,*

$$\chi(w) = \frac{dw}{dz} = q\, e^{-i\theta} = u - iv, \tag{21}$$

where u and v are the velocity components in the Ox and Oy directions respectively. We exclude w itself because—for reasons given in 2 of §2.12—it is preferable to z as an *independent* variable in most problems, and naturally it cannot play both roles. We call δ the displacement function because it gives a measure of the displacement of the streamlines and equipotentials from the positions they would have in a uniform parallel stream.

The computational difficulties of compressible flow can be reduced when it is valid to use *linear perturbation theory.* For this case (see §2.17) we can replace w in $\tau(w)$ by $Z = x + i\beta_a y$, i.e.

$$\tau = \tau(Z) \quad (Z = x + i\beta_a y), \tag{22}$$

where to the same order of accuracy

$$\tau = \beta_a\left(1 - \frac{q}{U}\right) + i\theta = \tfrac{1}{2}\beta_a C_p + i\frac{dy}{dx}. \tag{23}$$

Terms $O\,|\tau|^2$ are neglected in this theory, consequently (12) can be written

$$Udz = \frac{1}{2\beta_a}\{(1 + \beta_a)(1 + \tau)\, dw - (1 - \beta_a)(1 - \overline{\tau})\, d\overline{w}\},$$

which when rearranged and integrated assumes the form

$$\frac{1}{2}\left(1+\frac{1}{\beta_a^2}\right)(UZ-w)+\frac{1}{2}\left(1-\frac{1}{\beta_a^2}\right)(U\bar{Z}-\bar{w})=\frac{1}{\beta_a}\int\tau(w)\,dw. \tag{24}$$

Comparing this with (20) we conclude that the function

$$\delta(w)=\frac{1}{\beta_a}\int(e^\tau-1)\,dw\simeq\frac{1}{\beta_a}\left(\int\tau\,dw\right), \tag{25}$$

where terms $O\,|\tau|^2$ are ignored, is a suitable generalization of the displacement function of incompressible flow. From the left-hand side of (24) it is found that

$$\delta=(Ux-\phi)+i\left(\frac{Uy}{\beta_a}-\frac{\psi}{\beta_a^2}\right). \tag{26}$$

If the first equation in (25) is adopted as the definition of the displacement function for both linear perturbation theory and incompressible flow theory, then it should be remembered that this function is restricted to flows about slender bodies, except in the limit $\beta_a=1$, when there is no such restriction.

Another useful equation, valid in linear perturbation theory, is

$$\rho=\rho_a\left(1+\frac{M_a^2}{\beta_a}\Omega\right), \tag{27}$$

which follows from (4), (9) and the disregard of terms $O(\Omega^2)$.

We will not summarize the perturbation theories given in §§ 2.20–2.23, for by far the greater number of the applications given in the following chapters are based on the equations given above.

BOUNDARY CONDITIONS

6.2   Unmixed boundary conditions

In steady flow the boundary conditions most commonly met are:

(1) *Solid boundaries*:

$$\theta=\theta_s-\alpha, \tag{28}$$

(1.56), where θ_s is the slope of the surface measured from some datum direction, which is at an angle $-\alpha$ to the x-axis.

(2) *Free Boundaries*—or boundaries on which the pressure distribution is known:

$$\Omega=\Omega_s, \tag{29}$$

where Ω_s is related to the known pressure p_s by (8).

6.3   Porous boundaries

The boundary condition, appropriate to porous boundaries is the very complicated relation between θ and Ω obtained by eliminating p from (1.59) and (8). Significant algebraic progress is possible only if this is a linear relation, i.e. only if $n = 1$. This means that the velocity normal to the wall must be directly proportional to the pressure difference $(p - p_s)$ across the wall (see Fig. 6.3 a).

Walls having approximately this character have been designed (e.g. see Preston & Rawcliffe, 1950), but even if this were not so the theory which results from the assumption $n = 1$ would still be valuable, owing to its relevance to more than just porous boundaries (see §§ 6.4 and 6.5).

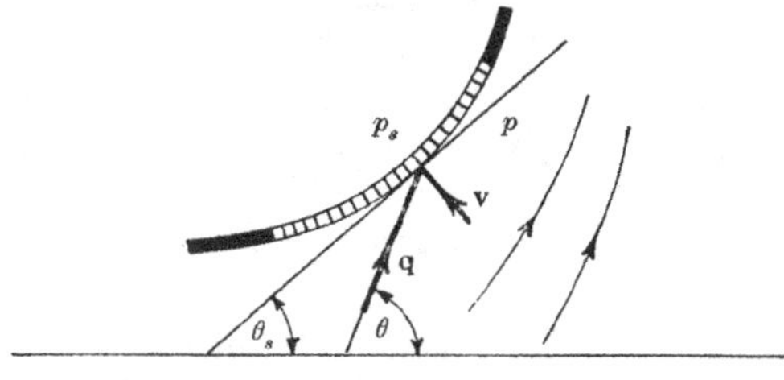

Fig. 6.3 a

With $n = 1$, and taking the datum values ρ_a and q_a in (1.59) to be the average conditions introduced above, we have

$$\theta - \theta_s = \frac{\lambda_s U}{2q} \left\{ \frac{p - p_a}{\frac{1}{2}\rho_a U^2} - \frac{p_s - p_a}{\frac{1}{2}\rho_a U^2} \right\},$$

i.e. $\quad \theta - \theta_s = \frac{\lambda_s}{\beta_a} \sinh(\Omega + \Omega^*) \left\{ \frac{\sinh \Omega}{\sinh(\Omega + \Omega^*)} - \frac{\sinh \Omega_s}{\sinh(\Omega_s + \Omega^*)} \right\}, \quad (30)$

by (3), (4) and (8). It appears from this relation that a further assumption is necessary in order that the desired linear relation between θ and Ω be obtained. But this is not so, for it is only necessary to make further use of an assumption already employed in the derivation of (1.59). This is that $\theta - \theta_s$ is a small angle, i.e. the velocity component normal to the surface is small compared with the tangential component. This means that the right-hand side of (30) is small, and so neglecting terms of order $(\Omega - \Omega_s)^2$ and higher, we can replace (30) by the linear relation

$$\theta - \lambda\Omega = \theta_s - \lambda\Omega_s, \tag{31}$$

where $\quad\quad\quad\quad \lambda = \frac{\lambda_s}{\beta_a} \frac{\sinh \Omega^*}{\sinh(\Omega^* + \Omega_s)}. \tag{32}$

The mathematical theory for this boundary condition has been given in §§ 4.15–4.20.

Notice that the number λ in (31) will be positive if an increase in $(p - p_s)$ increases $(\theta - \theta_s)$ (as in the situation shown in Fig. 6.3a), and negative if an increase in $(p - p_s)$ reduces $(\theta - \theta_s)$ (as in Fig. 6.3b). This could be formulated into a rule involving the sense in which the boundary of the flow domain is traversed, but it is easier to deduce the sign of λ from physical considerations in each case.

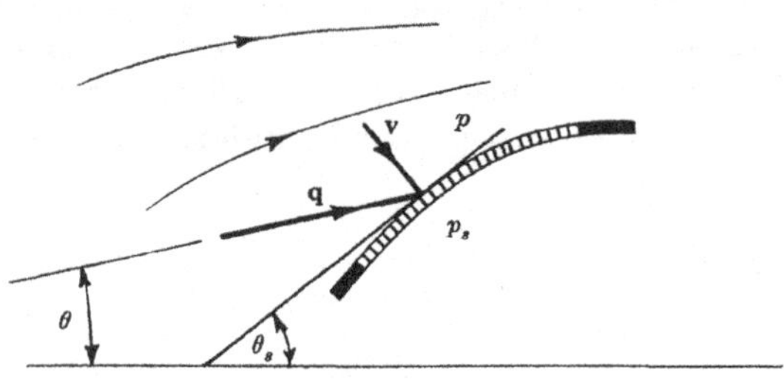

Fig. 6.3b

The quantities θ_s, λ and Ω_s are known functions of the distance s along the porous wall. We note from (32) that even if the 'porosity' λ_s is constant, λ will vary along the wall unless either (i) $p_s - p_a$, and therefore Ω_s is small, and (32) can be written

$$\lambda = \lambda_s/\beta_a, \tag{33}$$

or (ii) p_s is constant. Equation (33) is therefore applicable in linear perturbation theory.

An alternative formulation of the boundary conditions for flow past porous walls has been given by Baldwin, Turner & Knechtel (1954), but their method is valid only when *both* θ and θ_s are small, and would not be applicable, for example, to the flow past a porous circular cylinder.

6.4 Perforated boundaries

As shown in § 4.16 it is convenient to replace λ in (31) by a variable ϵ defined by

$$\epsilon = \frac{1}{\pi} \tan^{-1} \lambda \quad (-\tfrac{1}{2} \leqslant \epsilon \leqslant \tfrac{1}{2}), \tag{34}$$

so that (31) can be expressed

$$\theta \cos \pi\epsilon - \Omega \sin \pi\epsilon = \theta_s \cos \pi\epsilon - \Omega_s \sin \pi\epsilon. \tag{35}$$

Then if $\epsilon = 0$, $\theta = \theta_s$, and if $\epsilon = \pm\tfrac{1}{2}$, $\Omega = \Omega_s$; i.e. $\epsilon = 0$ gives solid boundaries and $\epsilon = \pm\tfrac{1}{2}$ gives free boundaries. With free boundaries

$\epsilon = \frac{1}{2}$ if the flow is 'below' the free surface (cf. Fig. 6.3a), and $\epsilon = -\frac{1}{2}$ if the flow is 'above' the free surface (cf. Fig. 6.3b).

The relevance of this to perforated boundaries can be seen from Fig. 6.4a, which shows a small outflow through a section of a perforated surface. If it is assumed that free jets of fluid emerge from each hole as shown, and that the angles they make with the main stream is small, the boundary condition over the holes is approximately the same as that over the free surfaces of the jets, i.e. Ω = constant. On the solid parts of the wall θ is known, and consequently we have the distribution of ϵ shown in the figure.

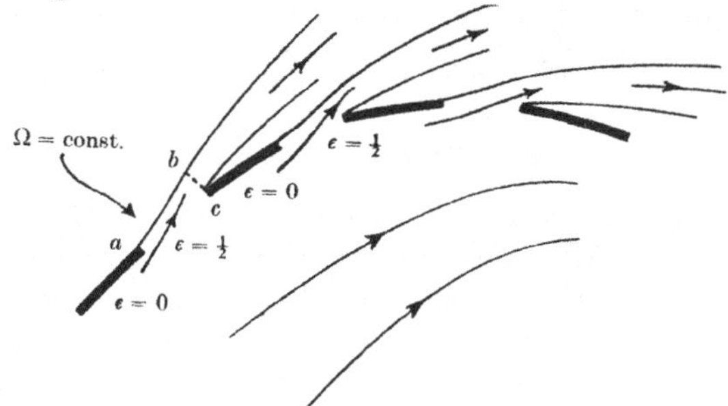

Fig. 6.4a

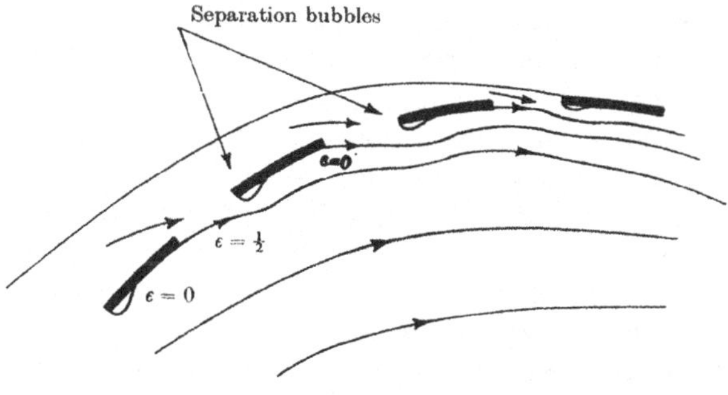

Fig. 6.4b

If the perforations are small and relatively close we can replace the distribution of ϵ by an average or 'effective' ϵ defined by

$$\bar{\epsilon} = \frac{O \times S_0 + \frac{1}{2}S_{\frac{1}{2}}}{S_0 + S_{\frac{1}{2}}} = \frac{1}{2}\frac{S_{\frac{1}{2}}}{S_0 + S_{\frac{1}{2}}}, \tag{36}$$

where $S_{\frac{1}{2}}$ is the area of the perforations and S_0 is the solid surface area. Thus in a given (small) region of the surface $2\bar{\epsilon}$ is the ratio of the area of the holes in the region to the total area of the region. That (36) does define the porous surface equivalent in its effect on the flow to a given perforated surface remains to be established, for it is not obvious that a composite ϵ should be a linear function of its components. The theory is verified for the case of a wind tunnel in § 7.29.

Inwards flow through a porous surface is more complicated than outwards flow; an ideal flow pattern which may be a rough approximation to the real flow is shown in Fig. 6.4b. The lower pressure on the inside of the solid sections will facilitate the flow re-attachment shown, and it may be reasonable to ignore the distortion in the flow pattern caused by these separations. The pressure over the gaps will again be approximately the same as that outside the wall, and so we shall again use (36) as the definition of the equivalent porous wall for inwards flow. The theory will obviously be rather better for outwards flow than for inwards flow.

6.5   Mixed boundary conditions

Many flow problems involve boundaries which are partly free and partly constrained to lie along given solid surfaces. Thus some parts of the boundaries have prescribed pressures ($\epsilon = \pm \frac{1}{2}$) and other parts prescribed shapes ($\epsilon = 0$). It is obvious that this type of boundary condition can be regarded as a special case of the porous wall boundary condition (see (35)).

6.6   Boundary conditions at the point at infinity

When the point at infinity in the z-plane lies in the domain of the fluid motion there will always be some restrictions on the behaviour of the function $\tau(w)$ in the neighbourhood of this point. The origin of the w-plane is usually selected so that z is finite at $w = 0$, and hence $z \to \infty$ always implies $w \to \infty$. The reverse of this, i.e. $w \to \infty$ implies $z \to \infty$, is not always true, e.g. at sources and sinks in the *finite* part of the z-plane (see § 5.10) $w \to \infty$ but z does not. The behaviour of $\tau(w)$ at such points, corresponding to sources and sinks, will also be known (see following section). By the 'boundary conditions at the point at infinity' in this book we shall mean all the restrictions on $\tau(w)$ in the neighbourhood of $w = \infty$. In most of our applications, however, there will be no need to distinguish between the points $z = \infty$ and $w = \infty$.

The most important boundary condition at infinity is a specification of the value or values assumed by τ at $z = \infty$. With channel flow τ will have a distinct value for each channel branch of infinite length; whereas in streams of infinite width τ will have only the single value τ_∞. In applying this boundary condition it is often useful to bear in mind that any changes in boundary shape or pressure distribution occurring in the *finite* part of the plane cannot alter the value (or values) of τ_∞.

Another important type of boundary condition at infinity depends on the conservation of mass principle. As an example consider the steady flow through an unbranched channel. Let the channel width, fluid density and velocity at $\phi = -\infty$, be $H_{-\infty}$, $\rho_{-\infty}$ and $q_{-\infty}$, and the corresponding values downstream be H_∞, ρ_∞ and q_∞. Then conservation of mass gives

$$H_\infty \rho_\infty q_\infty = H_{-\infty} \rho_{-\infty} q_{-\infty}. \tag{37}$$

Combining (3), (4) and (9) we get

$$\rho q = \rho_a U \frac{\cosh \Omega^*}{\cosh (\Omega + \Omega^*)}, \tag{38}$$

so (37) can be written

$$\frac{H_\infty}{H_{-\infty}} = \frac{\cosh (\Omega_\infty + \Omega^*)}{\cosh (\Omega_{-\infty} + \Omega^*)}. \tag{39}$$

By squaring this equation and making use of (3) and (4) we obtain the alternative relation

$$\left(\frac{H_\infty}{H_{-\infty}}\right)^2 = \frac{M_a^2 + (1 - M_a^2)\,(U/q_\infty)^2}{M_a^2 + (1 - M_a^2)\,(U/q_{-\infty})^2}. \tag{40}$$

Equation (39) restricts the choice of values for Ω_∞ and $\Omega_{-\infty}$.

A further type of boundary condition at infinity can often be derived from some geometrical consideration. For example if a closed body lies in an infinite stream, then we can write down the so-called *closure condition*

$$\int_C dz = 0, \tag{41}$$

where C is any closed contour enclosing the body. Let C transform into C^* in the w-plane, then by (12)

$$(1 + \beta_a) \int_{C^*} e^\tau \, dw = (1 - \beta_a) \overline{\int_{C^*} e^{-\tau} \, dw}. \tag{42}$$

Each integral here can usually be evaluated by finding a series expansion for $\tau(w)$ near $w = \infty$, and then applying the residue theorem. As C^* is not necessarily closed (cf. Fig. 5.14) it may be necessary first to find a convenient transformation $w = w(\zeta)$ such that C^* maps into a closed curve in the ζ-plane.

6.7   Circulation about and mass flow from an isolated body

Let C be a closed contour about a body, then the circulation Γ is given by (see (1.21))

$$\Gamma = \int_C d\phi.$$

Also let Q be the net mass flow out of the body (e.g. through a porous surface or from point sources on the surface), then from (7)

$$Q = \rho_s \int_C d\psi.$$

Hence

$$\int_C dw = \int_C d\phi + i\int_C d\psi = \Gamma + iQ/\rho_s. \tag{43}$$

If Γ and Q are given quantities this result gives a restriction on the shape of the w-plane, and it therefore indirectly affects the function $\tau(w)$.

In incompressible flow (43) can be regarded as a boundary condition at infinity in the z-plane, for in this case $w = w(z)$, and we may write

$$\int_C \frac{dw}{dz} dz = \Gamma + iQ/\rho_s. \tag{44}$$

The left-hand side of (44) can be evaluated by the residue theorem.

6.8   Boundary conditions in unsteady flow

Equation (1.56) for the flow direction in unsteady flow past a moving solid boundary is

$$\theta = \theta_s + \frac{\mathbf{n}.\mathbf{v}}{|\mathbf{q}|} - \alpha, \tag{45}$$

where $\mathbf{q}$ is the fluid velocity, $\mathbf{v}$ is the small unsteady perturbation velocity of the surface, $\mathbf{n}$ is the unit normal to the surface, and θ_s, α are defined in (28) above. These velocities and directions are measured in an xOy frame which may itself be moving non-uniformly, provided of course this motion is such that the perturbation $\mathbf{v}$ remains small. We shall select xOy so that the mean position of the solid boundary, with respect to the perturbation $\mathbf{v}$, is stationary in this frame. This divides the unsteadiness into two components, namely (i) small perturbations about (ii) a more slowly varying mean motion. An example of (i) alone is the transverse flutter of an aerofoil in a uniform stream, and of (ii) alone the steady acceleration of an aerofoil at constant incidence.

If the boundary is rigid the slope θ_s is independent of time t, whereas $\mathbf{n}$, $\mathbf{v}$, $\mathbf{q}$ and α are, in general, functions of t. The unsteady motion adds a further time-dependent term to (45) which arises from the motion of stagnation points not fixed in position by Joukowski's condition. Fig. 6.8 shows two stagnation points s_0 and s_1 at the nose of a body in a stream; if s_0 and s_1 refer to the mean and perturbed motions respectively, then it is clear from the figure that a term

$$\pi\{\mathbf{U}(s-s_1) - \mathbf{U}(s-s_0)\}$$

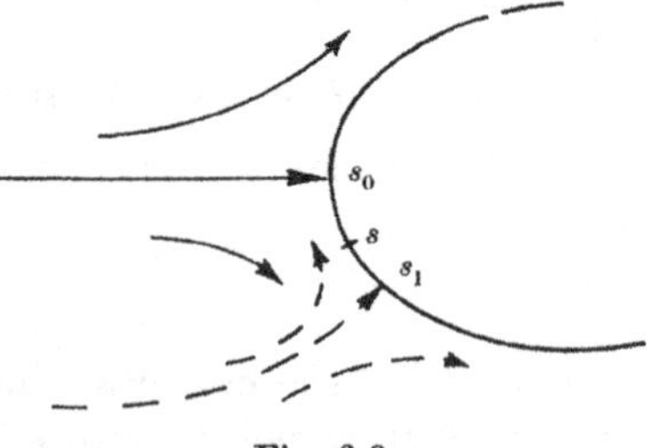

Fig. 6.8

must be added to the right-hand side of (45) to account for the perturbations of the stagnation point. As

$U'(x) = \delta(x)$ (see §3.6), this term can be written $\pi(s_0 - s_1)\,\delta(s - s_0)$ for small perturbations. Hence we have

$$\theta = \theta_s + \frac{\mathbf{n}.\mathbf{v}}{|\mathbf{q}|} - \alpha + \pi(s_0 - s_1)\,\delta(s - s_0). \tag{46}$$

In the general case both s_0 and s_1 will be functions of time.

With an isolated body in unsteady motion it is found necessary to introduce a vortex into the flow in order to satisfy Kelvin's circulation theorem (§1.8). This sheet extends downstream from the rear of the body—the arguments supporting this model of the flow will be given in §9.1. The equation governing the strength of such a vortex is (§1.61)

$$\frac{\partial \omega}{\partial t} + \frac{\partial(\bar{q}\omega)}{\partial s} = 0, \tag{47}$$

where ω is the leap in the velocity magnitude across the sheet and $\bar{q}$ is the average of the velocities on each side of the sheet.

If ϖ is the leap in Ω across the sheet it follows from the derivative of (3) that to first order in ω and ϖ,

$$\omega = - U \sinh \Omega^* \, \frac{\cosh(\overline{\Omega} + \Omega^*)}{\sinh^2(\overline{\Omega} + \Omega^*)}, \quad \varpi = - \frac{\bar{q}}{\bar{\rho}}\frac{\rho_a}{\beta_a}\varpi,$$

the last form following from (3) and (9). The unsteady flow theory given in §2.21 applies either to incompressible flow or to slowly varying compressible flow. In each case derivatives of $\bar{\rho}$ can be ignored, and (47) can be written

$$\frac{\partial(\bar{q}\varpi)}{\partial t} + \frac{\partial}{\partial s}(\bar{q}^2\varpi) = 0.$$

We shall define the 'reduced time' $\mathfrak{t}$ by

$$\mathfrak{t} \equiv \frac{1}{c}\int_0^t \bar{q}\,dt, \tag{48}$$

where c is a reference length of the body, e.g. the chord length when the body is an aerofoil. The equation for ϖ can now be written

$$\frac{1}{c}\frac{\partial(\bar{q}\varpi)}{\partial \mathfrak{t}} + \frac{\partial(\bar{q}\varpi)}{\partial s} + \varpi\frac{\partial \bar{q}}{\partial s} = 0. \tag{49}$$

In most problems of interest to us $\partial\bar{q}/\partial s$ is very small or zero, and the last term of (49) can be ignored. The most general solution of (49) then reads

$$\bar{q}\varpi(s,\mathfrak{t}) = f(s - c\mathfrak{t}), \tag{50}$$

where f is an arbitrary function.

Finally we have Kelvin's circulation theorem, $d\Gamma/dt = 0$, where Γ is the circulation about any contour *completely* enclosing the body and any trailing vortex sheets.

<h1 style="text-align:center">SOURCES, SINKS AND VORTICES</h1>

6.9   Sources and sinks

It is now necessary to discuss the point singularities which may occur
in fluid motion problems. One such singularity—the stagnation point—
has already received attention, and there remain those singularities
known as sources (or sinks), vortices and doublets.

Suppose that in the immediate neighbourhood of a point z_0 of the
z-plane the direction of the fluid motion is radially outwards from z_0, and
symmetrical in all directions from z_0, then there is said to be a *fluid* source
at z_0. As we are dealing with two-dimensional motions this source should
strictly be called a line source, but we shall take this as understood.

If the flow direction is reversed, i.e. is radially *inwards* towards z_0, we
have a negative source, or a *sink* at z_0.

The *strength* of a source is measured by Q, its mass *output* per unit time:
a negative value of Q means that there is a sink of strength $-Q$ at the
point z_0. From (7) and (43) we see that a source at z_0 causes a jump in ψ
at z_0 of amount

$$h = \frac{Q}{\rho_s} = \frac{\beta_a Q}{\rho_a}. \tag{51}$$

Conservation of mass requires that in the neighbourhood of an isolated
source

$$\rho q = \frac{Q}{2\pi r}, \tag{52}$$

where $r = |z - z_0|$. Hence ρq must be infinite at $z = z_0$, and a *line* source
is not physically realistic. This difficulty can be overcome by replacing
the line by a small circular cylinder of such a radius r_0 that the velocity
and fluid density at $r = r_0$ are physically attainable. The maximum
possible fluid speed, $q_{\text{max.}}$, corresponds to the minimum possible density,
$\rho_{\text{min.}}$ (see §§ 1.3 and 2.13) and hence the minimum value r_0 can have is

$$r_0 = \frac{Q}{2\pi} \frac{1}{q_{\text{max.}} \rho_{\text{min.}}} = \frac{Q}{2\pi} \frac{1}{\left(\dfrac{\gamma}{\gamma+1}\right)\left(1 + \dfrac{2\gamma+1}{\gamma^2}\dfrac{1}{M_a^2}\right)\rho_a U} \tag{53}$$

by equations (1.9) and (1.30). The unrealistic flow in $|z - z_0| < r_0$ is
simply ignored.

The z- and w-planes for the flow due to an isolated source are shown in
Fig. 6.9. If the flow direction θ is assumed to be zero on BA_∞, then
continuity of τ within the region $A_\infty B_\infty DB'A'_\infty$ requires that θ have the
value 2π on $B'A'_\infty$.

We now have to find an analytic function $\tau(w)$ in an infinite strip given
its imaginary part on $\psi = 0$ and $\psi = h$. The solution to this type of

problem is contained in (3.109). Replacing z by w, and putting $\theta_0 = 0$, $\theta_h = 2\pi$ in (3.109) we get

$$\tau(w) = -\frac{\pi}{h}\int_{-\infty}^{\infty} \tanh \frac{\pi}{2h}(\phi - w)\, d\phi + K,$$

where K is a real constant to be determined (the form for K given in (3.109) obviously has little value in the present application as $\Omega_\infty = \infty$

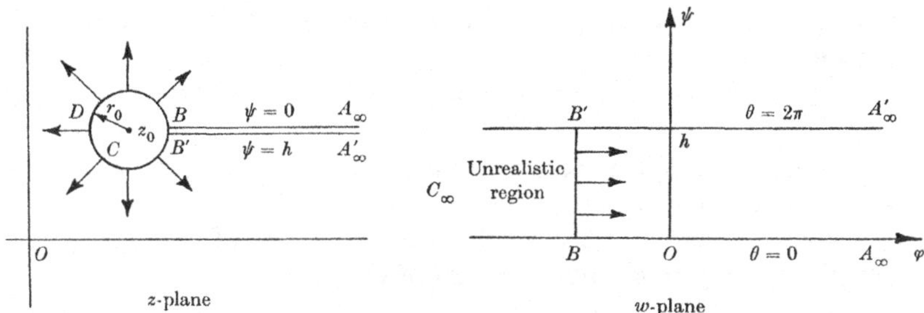

Fig. 6.9

and $\Omega_{-\infty} = -\infty$). Therefore, remembering that $\displaystyle\int_{-\infty}^{\infty}$ is to be interpreted as $\displaystyle\lim_{R\to\infty}\int_{-R}^{R}$ (§ 3.12), we get

$$\tau(w) = -2 \lim_{R\to\infty} \ln \left(\frac{\cosh \dfrac{\pi}{2h}(R - w)}{\cosh \dfrac{\pi}{2h}(R + w)} \right) + K,$$

i.e.
$$\tau(w) = \frac{2\pi}{h} w + K. \tag{54}$$

The next step is to substitute this in (12), and to integrate the resulting expression to find the relation between z and w:

$$z - z_0 = \frac{h}{4\pi U}\left\{ \left(\frac{1+\beta_a}{\beta_a}\right) e^{(2\pi w/h)+K} + \left(\frac{1-\beta_a}{\beta_a}\right) e^{-(2\pi \overline{w}/h)-K} \right\}. \tag{55}$$

That z_0 is the correct constant of integration here is verified below.

Replacing $(2\pi w/h) + K$ in (55) by $\tau = \Omega + i\theta$, and making use of (4) we get

$$r = |z - z_0| = \frac{h}{4\pi U \sinh \Omega^*}\{e^{\Omega+\Omega^*} + e^{-(\Omega+\Omega^*)}\}$$

$$= \frac{h}{2\pi\beta_a U}\frac{\cosh(\Omega+\Omega^*)}{\cosh \Omega^*}$$

$$= \frac{Q}{2\pi\beta_a U\rho_s}\frac{\rho_a U}{\rho q},$$

by (38) and (51). As $\rho_s = \rho_a/\beta_a$ it follows that

$$r = \frac{Q}{2\pi}\frac{1}{\rho q},$$

and the agreement of this with (52) yields the required verification of (55).

Notice from (55) that the limit $\phi \to -\infty$ (cf. Fig. 6.9) does *not* give $z \to z_0$, but of course neither (54) nor (55) have any validity for $|z-z_0| < r_0$, where r_0 is defined in (53). From the real part of (54) this restriction can be written

$$\phi > \frac{h}{2\pi}(\Omega_{\min.} - K). \tag{56}$$

To fix the constant K we need to assign a value to q at a given w, e.g. if $q = U$ at $\phi = 0$, then $\Omega = 0$ at $\phi = 0$, and K vanishes.

Equations (54) and (55) also apply if there is a *sink* of strength Q at z_0— it is only necessary to change the sign of Q in (51).

6.10 Vortices

A line *vortex* is a singularity similar to a line source, except that the streamlines and equipotentials are interchanged, i.e. the fluid *circulates* about z_0 instead of radiating from it. The strength of the vortex is measured by the circulation Γ about it:

$$\Gamma = \int_C q\,ds = \int_0^{2\pi} qr\,d\theta = 2\pi qr.$$

Hence
$$q = \Gamma/2\pi r, \tag{57}$$

and the motion is therefore physically impossible for $|z-z_0| < r_0$, where

$$r_0 = \frac{\Gamma}{2\pi q_{\max.}} = \frac{\Gamma}{2\pi U}\frac{1}{\left[1 + \dfrac{2\gamma+1}{\gamma^2}\dfrac{1}{M_a^2}\right]}. \tag{58}$$

This can be overcome by replacing the line vortex by a rigid core of radius r_0, in which q will obey the linear law, $q = \Gamma r/2\pi r_0$.

The w-plane for this flow is the infinite strip $0 \leqslant \phi \leqslant \Gamma$, $-\infty < \psi < \infty$, and by almost the same algebra as used for the source we can deduce that

$$\tau(w) = \frac{2\pi i}{\Gamma}w + K, \tag{59}$$

and
$$z - z_0 = \frac{\Gamma}{4\pi i U}\left(\frac{1+\beta_a}{\beta_a}e^{(2\pi i w/\Gamma)+K} - \frac{1-\beta_a}{\beta_a}e^{(2\pi i \bar{w}/\Gamma)-K}\right), \tag{60}$$

where K is a real constant. With the aid of (3) it is easily verified that the real part of (60) is the same as (57).

In incompressible flow $\beta_a = 1$, when (55) and (60) can be written in the form

$$w = \frac{h}{2\pi} \ln (z - z_0) \pm \text{constant}, \tag{61}$$

where $h = Q/\rho_a$ for a source, and $h = -i\Gamma$ for a vortex. This result was derived in §5.10 by a somewhat different approach.

6.11   Source in a uniform stream

Fig. 6.11a shows the upper halves of the z- and w-planes for a source at D in a uniform stream. As the flow is obviously symmetrical we need consider only half the flow pattern. Alternatively we could regard the flow shown in the figure as being due to a 'half-source' emitting fluid into a stream flowing past a rigid wall $A_\infty E_\infty$.

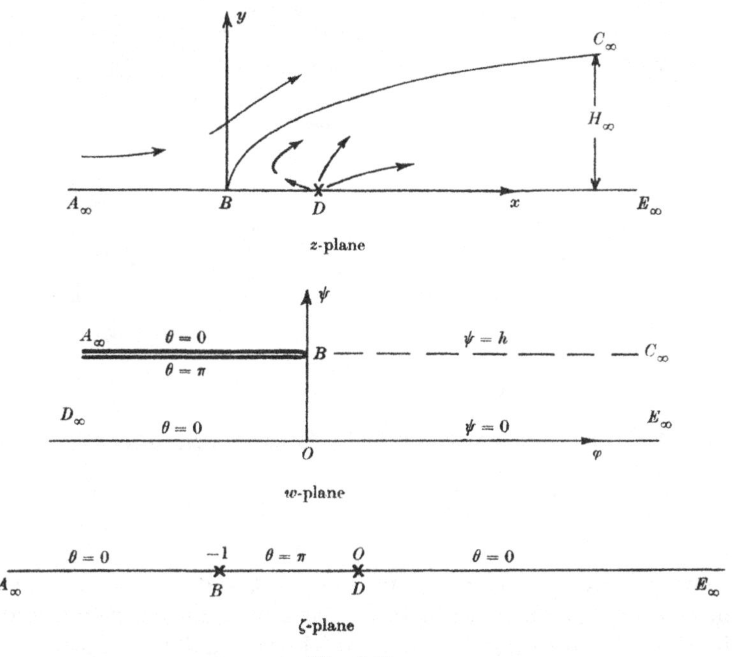

Fig. 6.11a

The discontinuity of $-h$ in ψ at D_∞ will be related to the strength Q of the complete source by (cf. (51))

$$h = \frac{\beta_a Q}{2\rho_a}. \tag{62}$$

The first step is to transform the w-plane into a convenient 'ζ-plane'— the most convenient in the present problem is the upper half-plane shown in the figure. Let the source D map on to $\zeta = 0$, and the stagnation point B at $w = ih$ map on to $\zeta = -1$. Then the w-plane 'polygon' has exterior angles of $-\pi$ at $\zeta = -1$, and π at $\zeta = 0$. An application of (5.25) (replace z, w in (5.25) by w, ζ) therefore gives $dw/d\zeta = K(\zeta+1)\zeta^{-1}$. Hence $w = K(\zeta + \ln \zeta) + B$. The discontinuity in ψ at $\zeta = 0$ requires that $K = h/\pi$, while $w = ih$, $\zeta = -1$ at B yields $A = h/\pi$. Therefore the transformation is

$$w = \frac{h}{\pi}(\zeta+1) + \frac{h}{\pi}\ln \zeta. \tag{63}$$

Let $\zeta = \gamma + i\eta$, then on $\eta = 0$ we have the boundary values $\theta = \pi$ in $-1 \leqslant \gamma \leqslant 0$, and $\theta = 0$ elsewhere. Let $q = U$ at $z = \infty$, then from (3) $\Omega_\infty = 0$. We can now apply (3.113) in the ζ-plane to find

$$\tau(\zeta) = \frac{1}{\pi}\int_{-1}^{0} \frac{\pi\,d\gamma}{\gamma-\zeta} + 0 = \ln\frac{\zeta}{\zeta+1}. \tag{64}$$

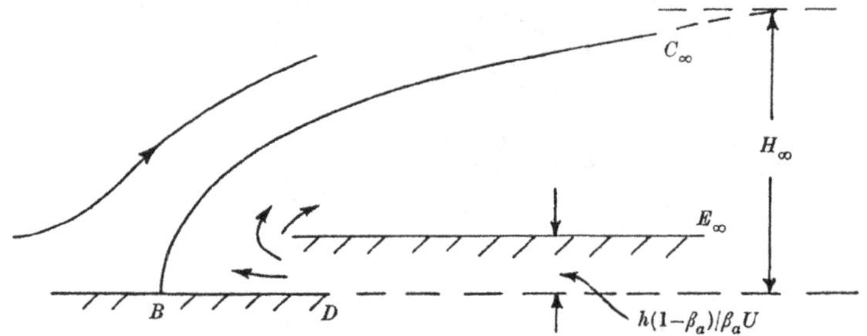

Fig. 6.11*b*

From (13), (63) and (64) the relation between the z- and ζ-planes is

$$z = \frac{h}{2\pi U}\left\{\frac{1+\beta_a}{\beta_a}(\zeta+1) - \frac{1-\beta_a}{\beta_a}\left(\bar{\zeta} + 2\ln\bar{\zeta} - \frac{1}{\bar{\zeta}}\right)\right\}, \tag{65}$$

where we have assumed the stagnation point to be at $z = 0$. The solution of our flow problem is contained in a parametric form in (63) and (65).

The limit $\zeta \to 0$ cannot be taken in those equations because—for the reason given in §6.9—the solution is not valid in the immediate neighbourhood of the source. A more serious difficulty is that (65) shows that the value of y on $\gamma > 0$, $\eta = 0$, is not equal to its value on $\gamma < 0$, $\eta = 0$, exceeding it by $h(1-\beta_a)/\beta_a U$. We have therefore not solved the problem shown in the z-plane of Fig. 6.11*a*, but instead have solved the flow problem shown in Fig. 6.11*b*. In incompressible flow $\beta_a = 1$, and the difficulty disappears. Notice from (65) that in this case the z- and ζ-planes are identical save for a scale factor and a shift of origin.

Fig. 6 11b suggests an alternative and more important interpretation of the flow we have been considering: this is that it is the flow of a uniform stream over a cliff BC_∞. Let the height of this cliff tend to H_∞ downstream of the source. At $w = \infty + ih$ (63) gives $\eta = \pi$, $\gamma = \infty$, and these values in (65) give $H_\infty = h/U$, i.e. from (62)

$$H_\infty = Q/2U\rho_s. \tag{66}$$

It is easily verified from (14), (64) and (66) that the horizontal force acting on this cliff is $X = p_a H_\infty$ (cf. (1.46)).

6.12   Doublet in a uniform stream

Suppose there is a source of strength Q at $z = -\tfrac{1}{2}H$, and a sink of equal strength at $z = \tfrac{1}{2}H$ in a uniform stream moving along the x-axis in the positive direction, then the z- and w-planes for this flow will be as shown in Fig. 6.12. The boundary displacement, corresponding to that shown in Fig. 6.11b will produce a strip (in which no flow occurs) connecting the source and sink.

The fluid emitted at D will be completely removed at E, and the main-stream fluid will be deflected around the curve BCF, which could be replaced by a solid surface. Hence the source-sink pair can be regarded as being equivalent to a solid obstacle as far as the main-stream flow is concerned.

Referring to Fig. 6.12 we see that there are exterior angles $-\pi, \pi, \pi$ and $-\pi$ at $\gamma = -1, -k, k$ and 1 respectively, hence the Schwarz–Christoffel mapping formula (5.25) gives $dw/d\zeta = K(\zeta^2 - 1)/(\zeta^2 - k^2)$. Integrating this and using the fact that the w-plane origin is at C, and $\psi = -h$ in $\eta = 0$, $-k < \gamma < k$ $(\zeta = \gamma + i\eta)$ we arrive at

$$w = \frac{2hk\zeta}{\pi(1-k^2)} + \frac{h}{\pi}\ln\left(\frac{\zeta+k}{\zeta-k}\right), \tag{67}$$

where k is related to the values a, $-a$ of ϕ at the stagnation points F, B by

$$a = \frac{2hk}{\pi(1-k^2)} + \frac{h}{\pi}\ln\frac{1+k}{1-k}, \tag{68}$$

and h is given by (62).

A glance at Fig. 6.12 shows that $\theta = \pi$ in $-1 \leqslant \gamma \leqslant -k$, $\theta = -\pi$ in $k \leqslant \gamma \leqslant 1$, and elsewhere $\theta = 0$. The value $\theta = -\pi$ may puzzle the reader, but on tracing out the streamline $DBCFE$, it becomes apparent that the flow on FE is at an angle of 2π greater than that on DB. Similar care in distinguishing between $\theta = \pi$ and $\theta = -\pi$ is required in many other flow problems.

With $q = U$ at $z = \infty$ we get $\Omega_\infty = 0$ from (3), and (3.113) can now be applied to the ζ-plane to give

$$\tau = \frac{1}{\pi}\int_{-1}^{-k}\frac{\pi\,d\gamma}{\gamma-\zeta}+\frac{1}{\pi}\int_{k}^{1}\frac{(-\pi)\,d\gamma}{\gamma-\zeta} = \ln\frac{\zeta^2-k^2}{\zeta^2-1}. \tag{69}$$

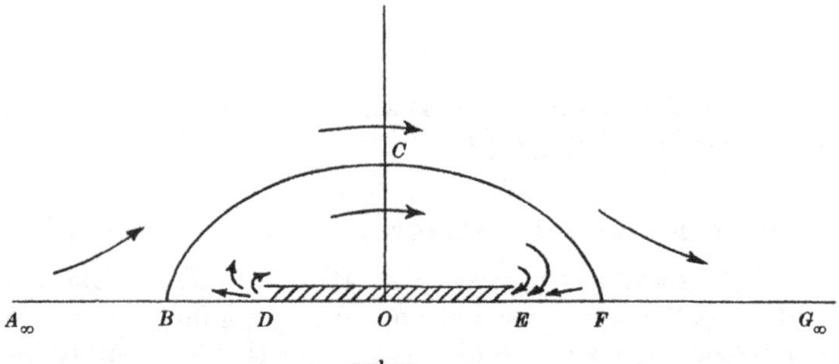

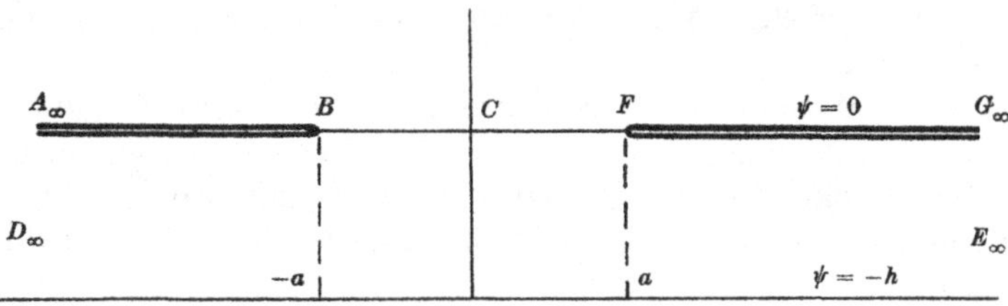

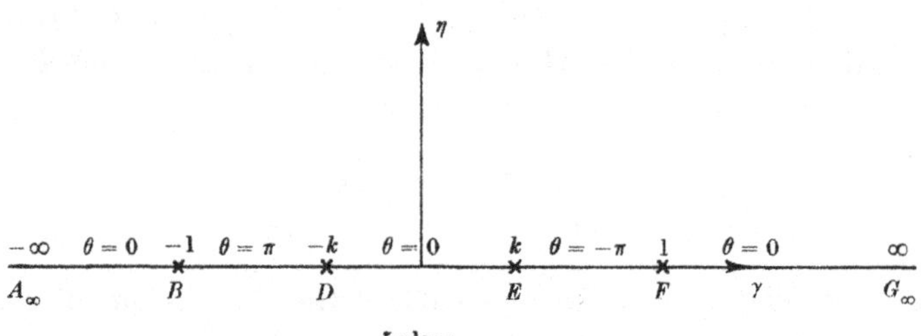

Fig. 6.12

Substituting (69) and the derivative of (67) into (13), and taking the origin of the z-plane at the point O shown in Fig. 6.12, we get

$$z = \frac{hk}{\pi U\beta_a(1-k^2)}\left[(1+\beta_a)\,\zeta-(1-\beta_a)\left\{\bar{\zeta}\left(1-\frac{(1-k^2)^2}{2k^2(\bar{\zeta}^2-k^2)}\right)\right.\right.$$
$$\left.\left.+\frac{1-k^2}{k}\left(1+\frac{1-k^2}{4k^2}\right)\ln\frac{\bar{\zeta}+k}{\bar{\zeta}-k}\right\}\right]. \tag{70}$$

The basic equations for a source-sink pair are contained in (67)–(70). This pair can be converted into the singularity known as a *doublet* (see §5.10), by allowing the source and sink to coalesce into a single point, and at the same time increasing their strength infinitely.

From (62) we see that h is a measure of the strength of the source-sink pair, while from Fig. 6.12 it is clear that k is a measure of their separation. The doublet is obtained in the limit $k \to 0$, $h \to \infty$, $kh \to$ a finite non-zero number. From (68) it follows that

$$hk = \frac{\pi a}{4}, \tag{71}$$

in the limit. Similarly, (67), (69) and (70) reduce to

$$w = \frac{a}{2}\left(\zeta + \frac{1}{\zeta}\right), \tag{72}$$

$$\tau = \ln\frac{\zeta^2}{\zeta^2 - 1}, \tag{73}$$

and
$$z = \frac{a}{4U\beta_a}\left\{(1+\beta_a)\zeta - (1-\beta_a)\left(\bar{\zeta} + \frac{2}{\bar{\zeta}} - \frac{1}{3\bar{\zeta}^2}\right)\right\}. \tag{74}$$

If $\beta_a = 1$, we find that $|z|$ is constant on the streamline $\psi = 0$ in $-a < \phi < a$, i.e. the curve BCF in the z-plane of Fig. 6.12 is a semicircle; but this is not true in compressible flow as is readily verified from (74).

6.13   Vortex in a uniform stream

Consider a vortex of strength $-\Gamma$ at the origin A in the z-plane, and let there be a uniform stream moving in the positive direction of the x-axis. If Γ is positive (i.e. the *circulation* about the vortex is negative), then the vortex will cause a clockwise rotation of the fluid in its immediate neighbourhood, while the fluid beyond a certain critical streamline will simply be deflected around the vortex in the way shown in Fig. 6.13.

The discontinuity in ϕ about any contour enclosing A will equal Γ (see §1.9), and this discontinuity can be introduced across any convenient curve joining A and the point at infinity; in Fig. 6.13 we have selected the negative half of the imaginary axis. Thus there is a jump in ϕ of amount Γ across AC_∞ in the z-plane.

The critical streamline separating the main-stream flow from the vortex flow is labelled $\psi = 0$, but further progress from B to A reduces ψ, the stream direction being discontinuous at B, and at A ψ will have returned to a value of $-\infty$. These facts enable us to draw the w-plane shown in the figure.

It is easily verified that the w-plane is mapped on to the ζ-plane shown in Fig. 6.13 by

$$w = \frac{i\Gamma}{2\pi}(1-\zeta^2) + \frac{i\Gamma}{\pi}\ln\zeta + \tfrac{1}{2}\Gamma. \tag{75}$$

A glance at the z-plane shows that on $\eta = 0$ θ will have the distribution

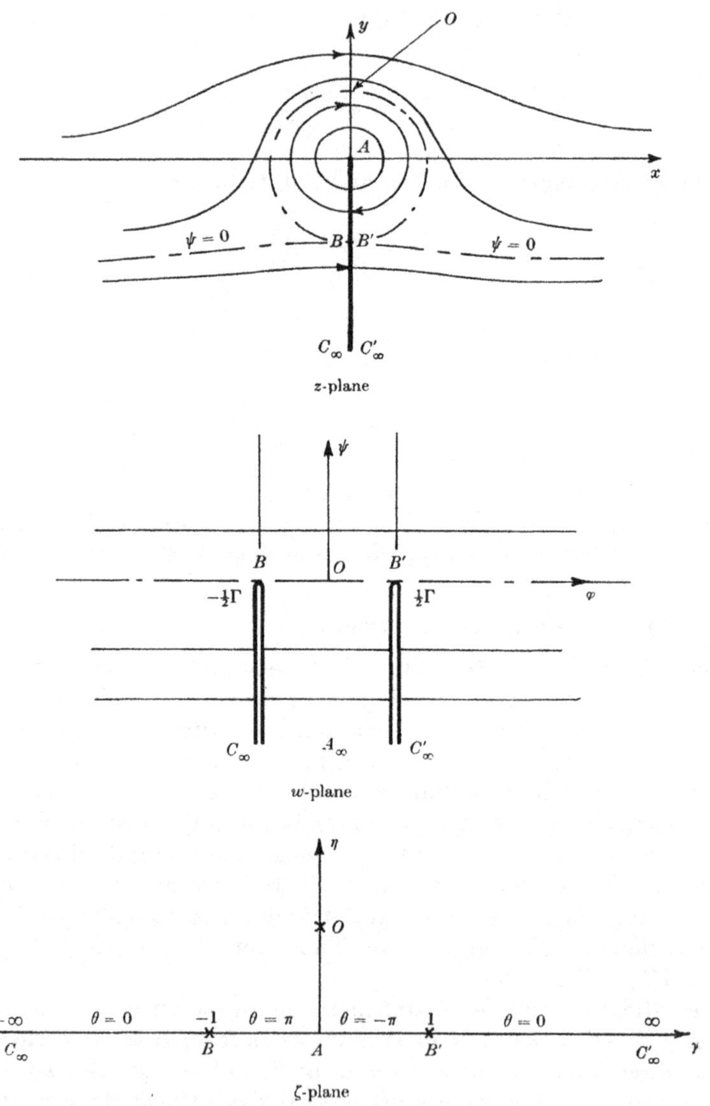

Fig. 6.13

$\theta = \pi, -1 < \gamma < 0, \theta = -\pi, 0 < \gamma < 1, \theta = 0$ elsewhere. And if $q = U$ at $z = \infty$, Ω_∞ will vanish, so (3.113) gives

$$\tau = \frac{1}{\pi} \int_{-1}^{0} \frac{\pi\,d\gamma}{\gamma - \zeta} + \frac{1}{\pi} \int_{0}^{1} \frac{(-\pi)\,d\gamma}{\gamma - \zeta} = \ln \frac{\zeta^2}{1 - \zeta^2}. \tag{76}$$

Substituting (76) and the derivative of (75) into (13), and making the w- and z-plane origins coincide, we obtain

$$z = \frac{i\Gamma}{2\pi U \beta_a} \{(1 + \beta_a)\tfrac{1}{2}\zeta^2 - (1 - \beta_a)(\tfrac{1}{2}\bar{\zeta}^2 - 2\ln\bar{\zeta} - \tfrac{1}{2}\bar{\zeta}^{-2})\}. \tag{77}$$

The logarithmic term in this result immediately shows that unless the flow is incompressible, x is *not* continuous across $C_\infty A$ as implied in Fig. 6.13, and so (75)–(77) do not correspond to a physically realistic flow. Exactly the same type of difficulty was met with a source in a uniform stream (cf. Fig. 6.11b), but in this latter case we were able to find a physically realistic interpretation of our equations, likewise with a doublet in a uniform stream. However, with an isolated vortex no such interpretation suggests itself.

The failure of the solution for an isolated vortex to satisfy the *closure* condition (42)—for this is what the difficulty amounts to—does not cause any embarrassment in any of the flow problems we consider in later chapters, because the vortices we introduce are always associated with bodies, such as aerofoils, and, except in the degenerate case of a flat plate (see § 8.7), these provide sufficient disposable parameters to ensure that the closure condition is satisfied.

GENERAL METHODS

6.14 Outline of the general procedure of solving fluid motion problems

The simple examples of fluid flows considered in §§ 6.9–6.13 illustrate quite well the method of dealing with more general fluid motion problems. The steps are as follows.

(1) *Define the problem clearly in the z-plane*

This is not always as simple as it may at first seem, especially if the problem requires that the physically most likely flow be determined. For in this case flow separation and re-attachment points may have to be estimated (see § 1.24), and in general this is not a simple task. It is frequently necessary to find a family of solutions, each member of which corresponds to a particular choice of separation or re-attachment point, and then to use some hypothesis to select the most likely flow from these members. Examples of this will be given in ch. 11.

(2) *Construct the w-plane*

From a knowledge of the physical nature of the problem, and its definition in the z-plane, it is usually a simple matter to settle the character of the w-plane. Sources, sinks and vortices may sometimes require a little thought, and it was for this reason that simple cases involving these singularities were considered in some detail above.

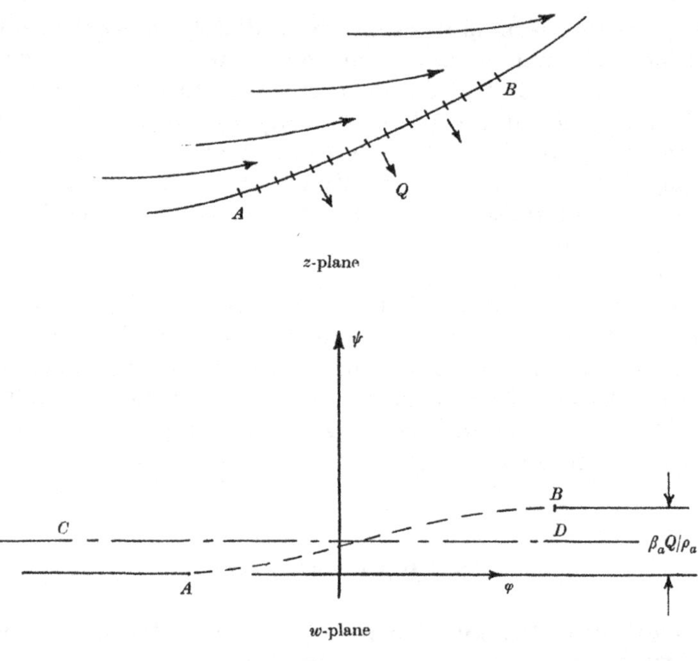

Fig. 6.14

In general, the w-plane will be bounded by straight lines parallel to $\psi = 0$. Porous walls provide an exception to this, for if an amount of fluid Q is absorbed between points A and B on the wall then ψ will change by $\beta_a Q/\rho_a$ (see (43)) over AB. If the fluid mass per unit length sucked through the porous wall is known, then the curve AB in the w-plane can be drawn. In many problems it is sufficient to assume that the displacement $\beta_a Q/\rho_a$ is small enough compared with $\phi_B - \phi_A$ to be ignored, and that the boundary condition (see (31)) can be applied on some average 'streamline', such as CD in Fig. 6.14, which intersects the porous section.

(3) *Map the w-plane on to an auxiliary ζ-plane*

An auxiliary plane is selected such that the solution to the boundary value problem (for $\tau = \Omega + i\theta$) is either known or readily deduced for this plane. As the boundary shape will be known in the w-plane, the mapping $w = w(\zeta)$ is easily found; usually the Schwarz–Christoffel formula suffices for this purpose, for curved boundaries are rarely unavoidable in the w-plane.

(4) *Calculate $\tau(\zeta)$*

If the ζ-plane has been properly selected it will be possible to write down $\tau(\zeta)$ in terms of the boundary conditions in this plane by using an appropriate result from chs. 3 or 4.

(5) *Calculate $z(\zeta, \bar{\zeta})$*

By substituting $\tau(\zeta)$ in (13) we obtain $z = z(\zeta, \bar{\zeta})$; then the desired relation $w = w(z, \bar{z})$ is obtained on eliminating ζ and $\bar{\zeta}$ from $z(\zeta, \bar{\zeta})$ and $w(\zeta)$.

(6) *Check the solution for physical realism*

The final step is to check that the solution does represent a physically attainable flow; this is especially important when the problem involves free streamlines (see discussion in § 1.24).

The integrations involved in steps 4 and 5 frequently demand approximate or numerical methods. Furthermore, for certain types of problem it becomes necessary to solve an integral equation in order to find the function $\tau(\zeta)$ (step 4) appropriate to a given boundary value problem in the z-plane. An example of this has already been given in § 5.4, and we shall not enlarge on it further at this stage.

CHAPTER 7

FLOW IN CHANNELS

7.1 Introduction

The expression 'channel flow' is used in this chapter in a broader sense
than is usually the case. It will be used to refer to any simply-connected
region of flow regardless of the character of the boundaries. In the case
of 'single-channel flow' the flow is bounded by two streamlines, as shown
in Fig. 7.1 a. The simplest case of channel flow is obtained if one of these
streamlines is at infinity, and we have the flow of a stream of infinite

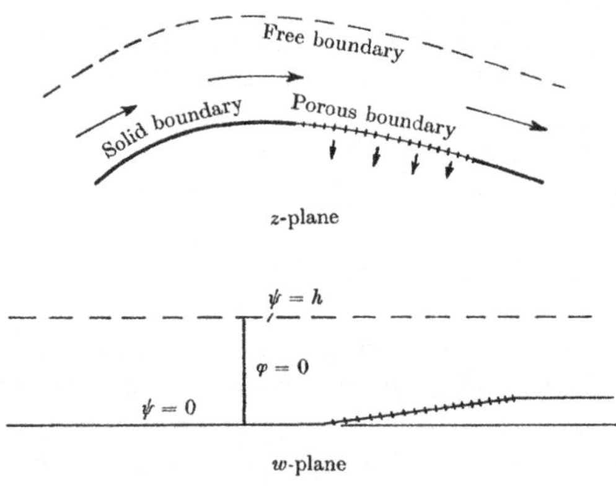

Fig. 7.1 a

width along a wall. Most of our attention will be given to single-channel
flow, the w-plane for which is just the infinite strip $-\infty < \phi < \infty$,
$0 \leqslant \psi \leqslant h$. The boundary conditions are applied on the bounding streamlines $\psi = 0$, $\psi = h$, and at $\phi = -\infty$, $\phi = \infty$. A notation similar to that
introduced in § 3.16 will be employed, viz.

$$\Omega_0(\phi) = \lim_{\psi \to 0} \Omega(\phi, \psi), \quad \Omega_h(\phi) = \lim_{\psi \to h} \Omega(\phi, \psi), \tag{1}$$

and similarly for θ_0, θ_h, τ_0 and τ_h.

With branched channels (see Fig. 7.1 b) the boundary conditions must
not only be applied on the two extreme boundaries $\psi = 0$, $\psi = h$, but also
on a number of intermediate semi-infinite slits $\psi = h_1$, $\phi_1 \leqslant \phi \leqslant \infty$,
$\psi = h_2$, $-\infty \leqslant \phi \leqslant \phi_2$, ... These channels will be considered in § 7.33.

The three principal types of boundary condition studied below are
(i) θ prescribed on sections of $\psi = 0$, $\psi = h$ (i.e. solid boundaries), (ii) Ω

similarly prescribed (free boundaries), and (iii) a linear relation holding between θ and Ω over sections of the bounding streamlines (porous walls). To distinguish these in our diagrams we shall use continuous curves for (i), broken curves for (ii), and a cross-etched curve for (iii) (see Fig. 7.1a). Steady flow only will be studied in this chapter.

Let values at $\phi = -\infty$ and $\phi = \infty$ be denoted as usual by the subscripts $-\infty$ and ∞. If Q is the total mass of fluid removed from the

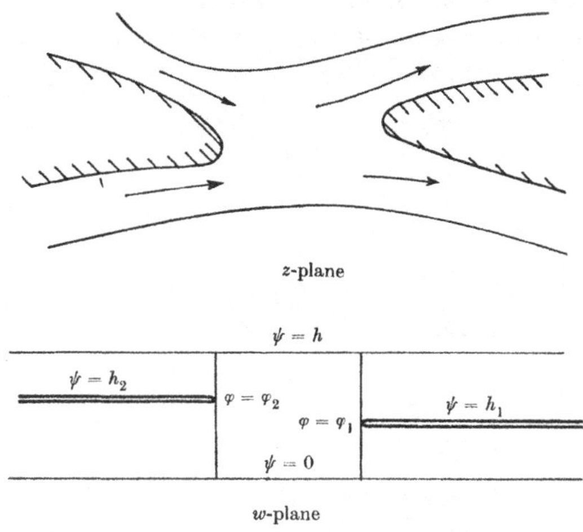

Fig. 7.1b

channel through porous sections or sinks on the walls, then conservation of mass gives $H_{-\infty}\rho_{-\infty}q_{-\infty} = H_{\infty}\rho_{\infty}q_{\infty} + Q$. Hence by (6.38)

$$\frac{H_{-\infty}}{H_{\infty}} = \frac{\cosh\left(\Omega_{-\infty}+\Omega^*\right)}{\cosh\left(\Omega_{\infty}+\Omega^*\right)} + C_m, \tag{2}$$

where
$$C_m \equiv Q/(\rho_{-\infty}q_{-\infty}H_{-\infty}). \tag{3}$$

If $\Omega_{-\infty}$ and C_m are prescribed, (2) enables us to calculate Ω_{∞} for a given channel.

Let the channel be parallel to the x-axis at $\phi = -\infty$, then the integral of (6.12) across the channel upstream at infinity gives

$$iH_{-\infty} = \frac{1}{2U\beta_a}\{(1+\beta_a)\,e^{\Omega_{-\infty}}\,(ih) - (1-\beta_a)\,e^{-\Omega_{-\infty}}\,(-ih)\}.$$

With the help of (6.4), viz. $\beta_a = \tanh\Omega^*$ this yields

$$h = \beta_a H_{-\infty} U\,\frac{\cosh\Omega^*}{\cosh\left(\Omega^* + \Omega_{-\infty}\right)}. \tag{4}$$

A similar result holds at $\phi = \infty$.

FLOW THROUGH GIVEN CHANNELS

7.2 General theory

The first channel flow problem we consider is that of calculating the flow through a channel having solid walls of a prescribed shape. This is often termed the *direct* problem of channel flow. From the shape of the channel walls we can determine the wall slopes θ_0 and θ_h as functions of the distance s along the walls. Thus we start with the boundary conditions

$$\theta_0 = \theta_0(s), \quad \theta_h = \theta_h(s), \tag{5}$$

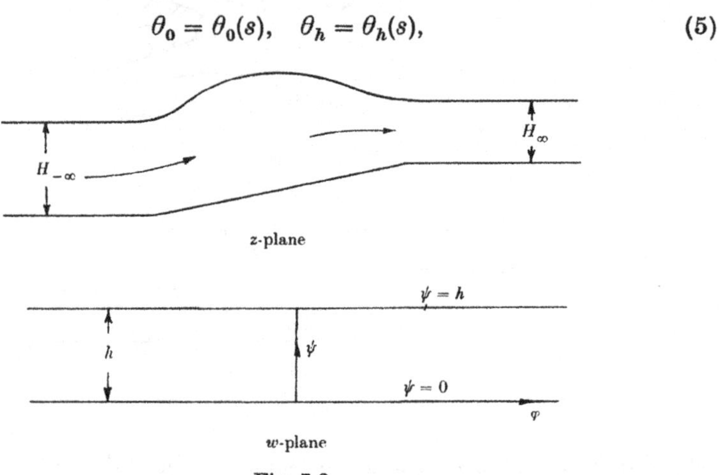

Fig. 7.2

together with a prescribed value of $q_{-\infty}$, and hence by (6.3) of $\Omega_{-\infty}$. By (2) and the fact that $C_m = 0$,

$$\Omega_\infty = -\Omega^* + \cosh^{-1}\left\{\frac{H_\infty}{H_{-\infty}}\cosh(\Omega^* + \Omega_{-\infty})\right\}. \tag{6}$$

Let $b = H_{-\infty} - H_\infty$ be the overall change in the channel width, then from (4), (6) and $\beta_a = \tanh\Omega^*$,

$$b = H_{-\infty}\left\{1 - \frac{H_\infty}{H_{-\infty}}\right\} = \frac{h}{\beta_a U \cosh\Omega^*}\{\cosh(\Omega^* + \Omega_{-\infty}) - \cosh(\Omega^* + \Omega_\infty)\},$$

i.e.

$$b = \frac{2h}{\beta_a U}\sinh\tfrac{1}{2}(\Omega_{-\infty} - \Omega_\infty)\{\sinh\tfrac{1}{2}(\Omega_\infty + \Omega_{-\infty}) + \beta_a\cosh\tfrac{1}{2}(\Omega_\infty + \Omega_{-\infty})\}. \tag{7}$$

For the moment we shall proceed as if $\theta_0(\phi)$ and $\theta_h(\phi)$ are the known boundary relations, and later we shall show how these functions can be deduced from (5). The w-plane is so simple (Fig. 7.2) that it is unnecessary to map it on an auxilary ζ-plane. The solution to our boundary value problem for $\tau(w)$ follows from (3.109). It is

$$\tau(w) = \tfrac{1}{2}(\Omega_\infty + \Omega_{-\infty}) + \frac{1}{2h} \int_{-\infty}^{\infty} \Big\{ \theta_0(\phi) \coth \frac{\pi}{2h}(\phi - w)$$
$$- \theta_h(\phi) \tanh \frac{\pi}{2h}(\phi - w) \Big\} d\phi, \quad (8)$$

where from (3.111) $\theta_0(\phi)$ and $\theta_h(\phi)$ must satisfy the restriction

$$\Omega_\infty - \Omega_{-\infty} = -\frac{1}{h} \int_{-\infty}^{\infty} \{\theta_0(\phi) - \theta_h(\phi)\} \, d\phi. \quad (9)$$

The relation between the distances s_0 and s_h along the boundary streamlines measured from $\phi = 0$ is obtained by integrating $ds = d\phi/q$. With the aid of (6.3) we get

$$\left. \begin{aligned} s_0(\phi) &= \int_0^\phi \frac{d\phi}{q_0} = \frac{\operatorname{cosech} \Omega^*}{U} \int_0^\phi \sinh(\Omega^* + \Omega_0) \, d\phi, \\ s_h(\phi) &= \int_0^\phi \frac{d\phi}{q_h} = \frac{\operatorname{cosech} \Omega^*}{U} \int_0^\phi \sinh(\Omega^* + \Omega_h) \, d\phi. \end{aligned} \right\} \quad (10)$$

For the right-hand side of (9) to be finite $(\theta_0 - \theta_h)$ must vanish faster than ϕ^{-1} near infinity. Using this fact we can integrate (8) by parts to find

$$\tau(w) = -\frac{1}{\pi} \int_{-\infty}^{\infty} \ln \sinh \frac{\pi}{2h}(\phi - w) \quad \theta_0'(\phi) \, d\phi$$
$$+ \frac{1}{\pi} \int_{-\infty}^{\infty} \ln \cosh \frac{\pi}{2h}(\phi - w) \quad \theta_h'(\phi) \, d\phi + \tfrac{1}{2}(\Omega_\infty + \Omega_{-\infty}), \quad (11)$$

where, in accordance with §3.10 (also see (5.19)), $\theta_0'(\phi)$ and $\theta_h'(\phi)$ may behave like delta functions at a finite number of points. (Alternatively we could separate out from the integral the sum of terms arising from the jumps in θ_0 and θ_h, but this makes the algebra cumbersome.)

On $\psi = 0$ at $\phi = \phi^*$ (11) yields (see footnote on p. 104)

$$\Omega_0(\phi^*) = -\frac{1}{\pi} \int_{-\infty}^{\infty} \ln \left| \sinh \frac{\pi}{2h}(\phi - \phi^*) \right| \theta_0'(\phi) \, d\phi$$
$$+ \frac{1}{\pi} \int_{-\infty}^{\infty} \ln \left| \cosh \frac{\pi}{2h}(\phi - \phi^*) \right| \theta_h'(\phi) \, d\phi + \tfrac{1}{2}(\Omega_\infty + \Omega_{-\infty}). \quad (12)$$

Let $R_0 = s_0'(\theta)$ and $R_h = s_h'(\theta)$ be the radius of curvature of the channel walls, then

$$d\theta_0(\phi) = \theta_0'(\phi) \, d\phi = \frac{d\theta_0}{ds_0} \frac{ds_0}{d\phi} d\phi = \frac{d\phi}{R_0 q_0}, \quad (13)$$

and similarly for $\theta_h'(\phi) \, d\phi$. At corners of discontinuity $\Delta\theta_i$ in θ at $\phi = \phi_i$

we must put $1/R_0 q_0 = \Delta\theta_i \delta(\phi - \phi_i)$. These expressions and (6.3) permit us to write (12) as

$$\sinh^{-1}\left\{\frac{U}{q_0}(\phi^*)\sinh\Omega^*\right\} = \Omega^* - \frac{1}{\pi}\int_{-\infty}^{\infty}\frac{1}{R_0 q_0}(\phi)\ln\left|\sinh\frac{\pi}{2h}(\phi-\phi^*)\right|d\phi$$

$$+\frac{1}{\pi}\int_{-\infty}^{\infty}\frac{1}{R_h q_h}(\phi)\ln\left|\cosh\frac{\pi}{2h}(\phi-\phi^*)\right|d\phi + \tfrac{1}{2}(\Omega_\infty + \Omega_{-\infty}). \quad (14)$$

Similarly, by putting $w = \phi^* + ih$ in (11) we arrive at

$$\sinh^{-1}\left\{\frac{U}{q_h}(\phi^*)\sinh\Omega^*\right\} = \Omega^* - \frac{1}{\pi}\int_{-\infty}^{\infty}\frac{1}{R_0 q_0}(\phi)\ln\left|\cosh\frac{\pi}{2h}(\phi-\phi^*)\right|d\phi$$

$$+\frac{1}{\pi}\int_{-\infty}^{\infty}\frac{1}{R_h q_h}(\phi)\ln\left|\sinh\frac{\pi}{2h}(\phi-\phi^*)\right|d\phi + \tfrac{1}{2}(\Omega_\infty + \Omega_{-\infty}). \quad (15)$$

In (14) and (15) we now have a pair of simultaneous integral equations for the velocities $q_0(\phi)$ and $q_h(\phi)$.

These non-linear equations are somewhat complicated by the unknown functions $R_0(\phi)$, $R_h(\phi)$ under the integral signs. These functions will depend on q_0 and q_h through (10) and the known functions $R_0(s_0)$ and $R_h(s_h)$, which follow from the given coordinates of the channel walls. Thus from the point of view of an exact solution (14) and (15) are hopelessly complicated, even for quite simple functions $R_0(s_0)$ and $R_h(s_h)$; numerical iterative methods must be employed. One complete iteration will consist of the following steps:

(1) We have approximations $q_0^{(n)}(\phi), q_h^{(n)}(\phi)$ from the previous iteration. (To start it is usually sufficient to take $q_0^{(1)}(\phi), q_h^{(1)}(\phi)$ to have the values which would be given by a one-dimensional channel flow theory, i.e. take ρq to be inversely proportional to channel width.)

(2) Deduce $s_0^{(n)}(\phi), s_h^{(n)}(\phi)$ from (10).

(3) Eliminate $s_0^{(n)}, s_h^{(n)}$ from $s_0^{(n)}(\phi), s_h^{(n)}(\phi)$ and $R_0(s_0), R_h(s_h)$ to obtain $R_0^{(n)}(\phi), R_h^{(n)}(\phi)$.

(4) Compute $\dfrac{1}{R_0^{(n)} q_0^{(n)}}, \dfrac{1}{R_h^{(n)} q_h^{(n)}}$.

(5) Use the functions computed in step 4 in the integrals in (14) and (15), and from the left-hand sides of these equations obtain the new approximations $q_0^{(n+1)}(\phi), q_h^{(n+1)}(\phi)$.

This iterative process is repeated until it makes no appreciable change in the functions $q_0(\phi)$ and $q_h(\phi)$. Convergence is usually rapid, except where R_0 and R_h are very small. A rigorous proof of convergence will not be given here, but experience with particular cases leaves little doubt that the method does converge quite quickly in all practical cases.

7.3   The displacement function method

In incompressible flow there is an alternative approach of some value based on the displacement function $\delta(w)$ defined in (6.20), viz.

$$\delta(w) = Uz - w = \int (e^{\tau(w)} - 1)\, dw. \tag{16}$$

But this method is restricted to channels which do not have an overall deflexion and which are the same width upstream as downstream, for otherwise one of δ_∞ and $\delta_{-\infty}$ will be infinite. We shall set $H_\infty = H_{-\infty} = H$, and $q_\infty = q_{-\infty} = U$, then $\Omega_\infty = \Omega_{-\infty} = 0$ and $h = HU$ from (4). The boundary values of $\mathscr{I}\delta(w)$ are Uy_0 and $U(y_h - H)$, and corresponding to (8) we have

$$\delta(w) \equiv Uz - w = K + \frac{1}{H} \int_{-\infty}^{\infty} \left\{ y_0(\phi) \coth \frac{\pi}{HU}(\phi - w) \right.$$

$$\left. - (y_h(\phi) - H) \tanh \frac{\pi}{HU}(\phi - w) \right\} d\phi, \quad (17)$$

where
$$K = \tfrac{1}{2}\mathscr{R}(\delta_\infty + \delta_{-\infty}) = \tfrac{1}{2} U \lim_{\phi \to \infty} \{ x(\phi) + x(-\phi) \}. \tag{18}$$

Corresponding to (9) there is

$$\lim_{\phi \to \infty} \{ Ux(\phi) - Ux(-\phi) - 2\phi \} = -\frac{1}{H} \int_{-\infty}^{\infty} \{ y_0(\phi) + H - y_h(\phi) \}\, d\phi. \tag{19}$$

Suppose that $z = z_0$ at $w = 0$, then K can be calculated from

$$K = Uz_0 - \frac{1}{H} \int_{-\infty}^{\infty} \left\{ y_0(\phi) \coth \frac{\pi\phi}{HU} - (y_h(\phi) - H) \tanh \frac{\pi\phi}{HU} \right\} d\phi, \quad (20)$$

which follows from (17).

In the limit $H \to \infty$ (17) to (20) reduce to

$$\delta(w) = K + \frac{1}{\pi} \int_{-\infty}^{\infty} \frac{y_0(\phi)\, d\phi}{\phi - w}, \tag{21}$$

and
$$K = \mathscr{R}\delta_\infty = \mathscr{R}\delta_{-\infty} = Uz_0 - \frac{1}{\pi} \int_{-\infty}^{\infty} \frac{y_0(\phi)}{\phi}\, d\phi. \tag{22}$$

We shall make no use of these results, other than to establish one interesting result on the displacement of the equipotentials at infinity in channel flow, a result due to Thom (1947). Thom's result was obtained for the case of the flow about a symmetrical aerofoil on the centre-line of a wind tunnel. Half of the flow pattern is shown in Fig. 7.3. In this case $y_h = H$, and (17) reduces to

$$\delta(w) = K + \frac{1}{H} \int_{-\infty}^{\infty} y_0(\phi) \coth \frac{\pi}{HU}(\phi - w)\, d\phi.$$

The displacements upstream and downstream at infinity are thus

$$\delta_{\infty} = K - \frac{1}{H}\int_{-\infty}^{\infty} y_0(\phi)\,d\phi, \quad \delta_{-\infty} = K + \frac{1}{H}\int_{-\infty}^{\infty} y_0(\phi)\,d\phi.$$

For an aerofoil of moderate thickness : chord ratio we have the linear approximation

$$\int_{-\infty}^{\infty} y_0(\phi)\,d\phi = \int_{-\infty}^{\infty} y_0 q\,ds \simeq U\int_{-\infty}^{\infty} y_0\,dx = \tfrac{1}{2}UA,$$

where A is the aerofoil's cross-sectional area. Hence we have Thom's result, namely

$$\lim_{\phi\to\infty}\left(x - \frac{\phi}{U}\right) = \frac{K}{U} - \frac{A}{\mathscr{H}}, \quad \lim_{\phi\to-\infty}\left(x + \frac{\phi}{U}\right) = \frac{K}{U} + \frac{A}{\mathscr{H}}, \tag{23}$$

where $\mathscr{H} = 2H$ is the wind-tunnel width.

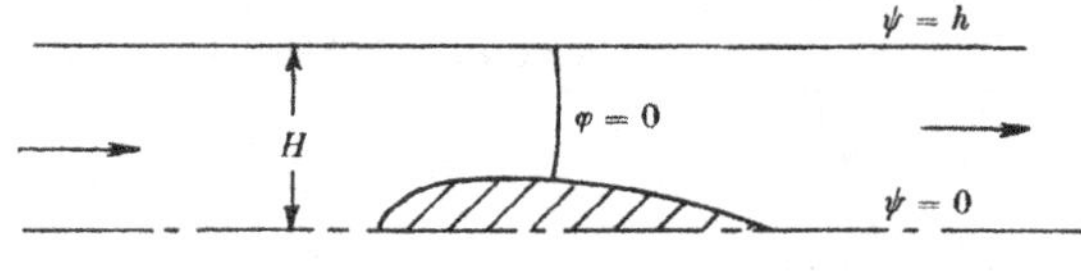

Fig. 7.3

Goldstein & Lighthill (1943) have used equations equivalent to (17) and (21) in calculations of compressible flow through channels by the Janzen–Rayleigh method.

7.4 Rectilinear walls

With rectilinear walls θ_0 and θ_h will be step-functions, and the mathematical difficulties of solving (14) and (15) vanish. The following theory is not strictly valid because of the existence of sharp projecting corners in the flow (see last paragraph of § 2.10), but outside the neighbourhoods of these corners the theory—to the extent that it can represent a real flow—will yield fairly accurate results.

Let θ_0 be constant except for jumps α_j at $\phi = \phi_j, j = 1, 2, \ldots, m$, and θ_h be constant except for jumps α_i at $\phi = \phi_i$ ($i = 1, 2, \ldots, n$). Then returning to the Riemann–Stieltjes integral form by replacing $\theta_0' d\phi$ by $d\theta_0(\phi)$ and $\theta_h' d\phi$ by $d\theta_h(\phi)$ in (11), and evaluating these in the usual way (see § 3.6), we obtain

$$\tau(w) = -\frac{1}{\pi}\sum_{j=1}^{m}\alpha_j\ln\sinh\frac{\pi}{2h}(\phi_j - w) + \frac{1}{\pi}\sum_{i=1}^{n}\alpha_i\ln\cosh\frac{\pi}{2h}(\phi_i - w)$$

$$+ \tfrac{1}{2}(\Omega_{\infty} + \Omega_{-\infty}). \tag{24}$$

Now the reference velocity U should be chosen so as to be as near as possible to the average value in the channel (see § 2.10), but it is not easy to decide on this reference value in a channel for which $H_\infty \neq H_{-\infty}$. A rough average can be obtained by making $q = U$ when Ω has the average of its upstream and downstream limiting values, $\frac{1}{2}(\Omega_\infty + \Omega_{-\infty})$. From (6.3) $q = U$ at $\Omega = 0$, hence with this choice

$$\Omega_\infty + \Omega_{-\infty} = 0. \tag{25}$$

A further restriction on Ω_∞ and $\Omega_{-\infty}$ comes from equation (9), which for the rectilinear channel reduces to

$$\Omega_{-\infty} - \Omega_\infty = \frac{1}{h} \sum_{j=1}^{m} \alpha_j \phi_j - \frac{1}{h} \sum_{i=1}^{n} \alpha_i \phi_i. \tag{26}$$

Equations (6), (25) and (26) define the values of Ω_∞ and $\Omega_{-\infty}$, and impose *one* restriction on the $(n+m)$ numbers ϕ_i, ϕ_j. A further $(n+m-1)$ restrictions can be determined from the channel geometry.

To obtain these $(n+m-1)$ equations we must first substitute (24) in (6.12), and then integrate to find

$$z = z_0 + \frac{1+\beta_a}{2\beta_a U} \int_0^w \frac{\prod_{i=1}^{n} \left\{ \cosh \frac{\pi}{2h}(\phi_i - w) \right\}^{\alpha_i/\pi}}{\prod_{j=1}^{m} \left\{ \sinh \frac{\pi}{2h}(\phi_j - w) \right\}^{\alpha_j/\pi}} dw$$

$$- \frac{1-\beta_a}{2\beta_a U} \int_0^{\bar{w}} \frac{\prod_{j=1}^{m} \left\{ \sinh \frac{\pi}{2h}(\phi_j - \bar{w}) \right\}^{\alpha_j/\pi}}{\prod_{i=1}^{n} \left\{ \cosh \frac{\pi}{2h}(\phi_i - \bar{w}) \right\}^{\alpha_i/\pi}} d\bar{w}, \tag{27}$$

where z_0 is the value of z at $w = 0$, and (25) has been taken into account. The points z_i, z_j corresponding to $w = \phi_i + ih$, $w = \phi_j$ yield $2(m+n)$ relations from the real and imaginary parts of (27), but as the angles α_i, α_j are given, there are $(n+m)$ restrictions like $\arg(z_{i+1} - z_i) = \sum_{r=1}^{i} \alpha_r$ on the choice of z_i, z_j. Furthermore, prescribing the ratio $H_\infty/H_{-\infty}$ (see (6)) removes one more degree of freedom in the choice of z_i, z_j. Thus the net result from this use of (27) is the $(n+m-1)$ equations we need to define completely the numbers ϕ_i, ϕ_j. Alternatively, we could use an equation like (5.35) to determine the ϕ_i, ϕ_j. A glance at (27) shows that numerical methods are unavoidable, even with rectilinear channels, unless they are very simple in shape.

7.5   A simple example: a step in a channel

The simplest non-trivial example of rectilinear-channel flow is that shown in Fig. 7.5. In this case $\alpha_j = 0$, and $\alpha_i = \tfrac{1}{2}\pi$ at $\phi = -a$, and $-\tfrac{1}{2}\pi$ at $\phi = a$. Substitution of these values in (24), (26) and (27) gives

$$\tau(w) = \tfrac{1}{2}\ln\left(\frac{\sinh\dfrac{\pi}{2h}(w-a)}{\sinh\dfrac{\pi}{2h}(w+a)}\right), \tag{28}$$

$$\Omega_{-\infty} - \Omega_{\infty} = \frac{\pi a}{h}, \tag{29}$$

and

$$z = z_0 + \frac{1+\beta_a}{2\beta_a U}\int_0^w \left\{\frac{\sinh\dfrac{\pi}{2h}(w-a)}{\sinh\dfrac{\pi}{2h}(w+a)}\right\}^{\frac{1}{2}} dw - \frac{1-\beta_a}{2\beta_a U}\int_0^{\overline{w}} \left\{\frac{\sinh\dfrac{\pi}{2h}(\overline{w}+a)}{\sinh\dfrac{\pi}{2h}(\overline{w}-a)}\right\}^{\frac{1}{2}} d\overline{w}. \tag{30}$$

The substitution
$$p^2 = \frac{\sinh\dfrac{\pi}{2h}(w-a)}{\sinh\dfrac{\pi}{2h}(w+a)},$$

i.e.
$$w = \frac{h}{\pi}\ln\left\{\frac{e^{-\pi a/h}p^2 - 1}{p^2 - e^{-\pi a/h}}\right\},$$

enables the first integral in (30) to be evaluated, and a similar substitution serves for the second.

The height of the step, b, follows from (30), for

$$ib = z(w = a) - z(w = -a).$$

Integrating (30) and using (29) we obtain

$$b = \frac{2h}{U}\sinh\frac{\pi a}{2h}, \tag{31}$$

which also follows directly from (7), (25) and (29). From (25) and (29) $\Omega_{-\infty} = \pi a/2h$. Substituting this in (4) and using $\beta_a = \tanh\Omega^*$ we get

$$H_{-\infty} = \frac{h}{\beta_2 U}\left\{\cosh\frac{\pi a}{2h} + \beta_a\sinh\frac{\pi a}{2h}\right\}. \tag{32}$$

Thus for a given channel (i.e. b and $H_{-\infty}$ given), h/U and a/h are calculated from (31) and (32), then substituted in (30) to complete the solution of the flow problem.

It is interesting to note that (30) does not fail at the singular points $w = \pm a$, where τ becomes infinite; z remains finite at both points and the value of b found from (30) agrees exactly on that based on the more general considerations yielding (7). In view of our experience with sources and vortices (§ 6.9 and 6.10), failure of the theory to give sensible z-values at sharp corners and stagnation points might well have been expected. This is especially so at sharp corners for, as noted in § 2.10, the solution for τ has no physical significance in the immediate neighbourhood of a sharp corner. The difference between the two types of singularity is that $w \to \infty$ at a source or vortex, but remains finite at a stagnation point or sharp corner, and it is this limit $\overline{w} \to \infty$ in the second term of the integration for z that causes the difficulty.

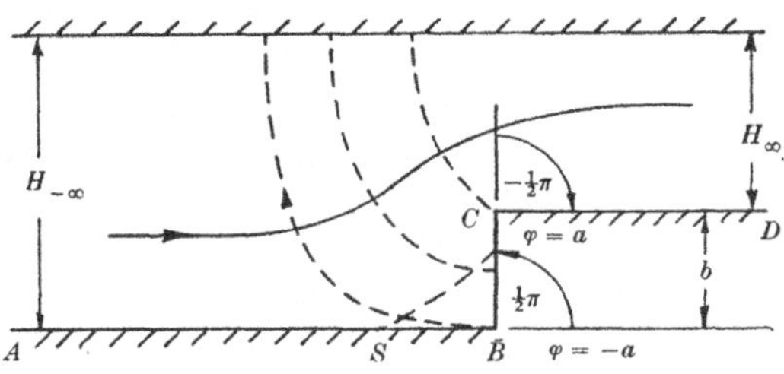

Fig. 7.5

The flow shown in Fig. 7.5 is not physically realistic, for the adverse pressure gradient along AB would produce flow separation at some point S on AB and would prevent the fluid reaching B. A 'dead-air' bubble would form in the corner. Further the flow would separate again at the sharp corner C (assuming the flow to re-attach on BC) and produce a downstream wake lying along CD. This more realistic model of the flow is considered in § 7.16.

7.6　A symmetrical cylinder in a channel

In order to illustrate the method of dealing with curved boundaries we shall now consider the flow about a cylinder placed centrally between straight parallel walls a distance $2H$ apart, with the cylinder having a line of symmetry coinciding with the tunnel axis (see Fig. 7.6). These restrictions on the position and shape of the cylinder mean that the flow is symmetrical about the stagnation streamline $\psi = 0$, and so only the half of the flow lying between $\psi = 0$ and $\psi = h$ needs to be considered. This flow can be regarded as being a 'channel' flow.

Let the front and rear stagnation points of the cylinder be at $\phi = -2a$, $\phi = 2a$, then the boundary conditions are that $\theta_h = 0$, and $\theta_0 = 0$, except in $-2a \leqslant \phi \leqslant 2a$. If we take U to be $q_{-\infty}$, then $q_\infty = U$ by continuity, and by (6.3) $\Omega_{-\infty} = \Omega_{\infty} = 0$. Equations (4), (9) and (12) now reduce to

$$h = \beta_a HU, \tag{33}$$

$$0 = \int_{-2a}^{2a} \theta_0(\phi)\,d\phi, \tag{34}$$

and

$$\Omega_0(\phi^*) = -\frac{1}{\pi}\int_{-2a}^{2a} \ln\left|\sinh\frac{\pi}{2h}(\phi-\phi^*)\right| d\theta_0(\phi). \tag{35}$$

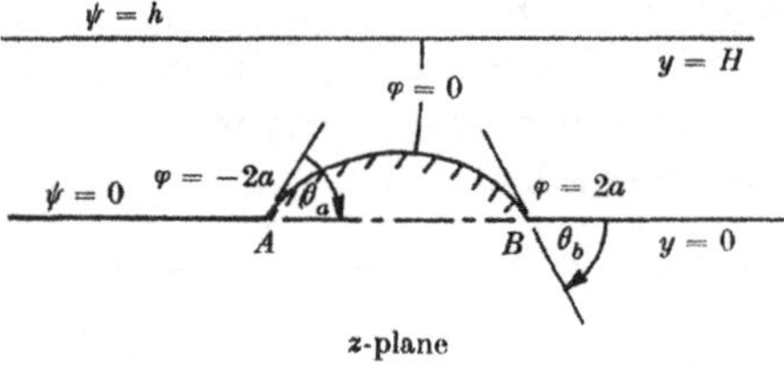

z-plane

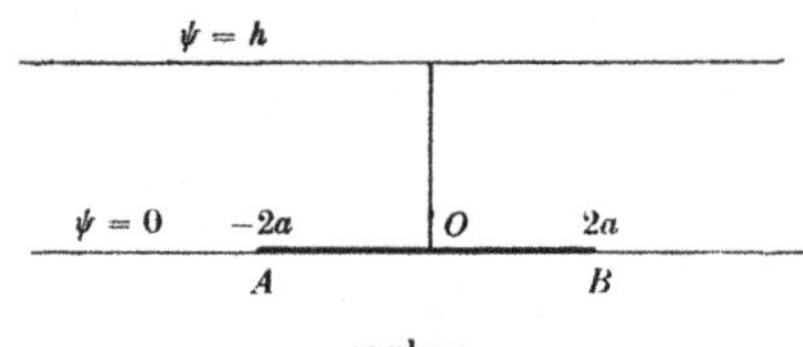

w-plane

Fig. 7.6

The overall deflexion of the channel is zero, hence

$$0 = \int_{-2a}^{2a} d\theta_0(\phi), \tag{36}$$

which will prove very useful in the numerical solution of (35). The integrals in (34)–(36) must be regarded as being over the range $(2a+0, -2a-0)$, i.e. as *including* the end-points $\phi = -2a$, $\phi = 2a$, for at these points there are discontinuities in θ, which make important contributions to the integrals.

Let θ_a and θ_b be the discontinuities in θ_0 at $\phi = -2a$ and $\phi = 2a$. Elsewhere θ_0 is continuous and the right-hand side of (13) remains finite. Thus (34)–(36) can be written

$$0 = 2a(\theta_b - \theta_a) + \int_{-2a}^{2a} \frac{\phi\,d\phi}{R_0 q_0}, \tag{37}$$

$$\Omega_0(\phi^*) = -\frac{1}{\pi}\int_{-2a}^{2a} \frac{1}{R_0 q_0}\ln\left|\sinh\frac{\pi}{2h}(\phi-\phi^*)\right| d\phi$$

$$-\frac{\theta_a}{\pi}\ln\left|\sinh\frac{\pi}{2h}(2a+\phi^*)\right| -\frac{\theta_b}{\pi}\ln\left|\sinh\frac{\pi}{2h}(2a-\phi^*)\right|, \tag{38}$$

and
$$0 = \theta_b + \theta_a + \int_{-2a}^{2a} \frac{d\phi}{R_0 q_0}, \tag{39}$$

where the range of the (Riemann) integrals here is the *open* interval $-2a < \phi < 2a$.

Perhaps the easiest method of solving the integral equation implicit in (38) is to replace the integral by an approximating sum. Let the range $-2a \leqslant \phi \leqslant 2a$ be divided into n parts

$$\phi_{n+\frac{1}{2}}\,(= 2a), \quad \phi_{n-\frac{1}{2}}, \quad \phi_{n-1\frac{1}{2}}, \ldots, \phi_{1\frac{1}{2}}, \quad \phi_{\frac{1}{2}}\,(= -2a),$$

such that $1/R_0 q_0$ can be taken as constant in the ith interval $(\phi_{i+\frac{1}{2}}, \phi_{i-\frac{1}{2}})$, $(i = 1, 2, \ldots, n)$, with negligible error. With this approximation the set of n constants defined by

$$B_i = -\frac{h}{R_0 q_0} \quad (\phi_{i-\frac{1}{2}} \leqslant \phi \leqslant \phi_{i+\frac{1}{2}}), \quad (i = 1, 2, \ldots, n), \tag{40}$$

can be introduced.

Instead of regarding (40) as being simply an approximation permitting the reduction of the integral in (38), we shall take (40) to be the *exact* specification of a cylinder, which is itself an approximation to the original cylinder. As this approximating cylinder must exactly satisfy the closure conditions (37) and (39), it is clear that the angles θ_a and θ_b must be modified from the values they have on the original cylinder.

Let $-B_0\,(= \theta_a)$, $-B_{n+1}\,(= \theta_b)$ be these modified values of θ_a, θ_b, then from (37), (39) and (40) B_0 and B_{n+1} will be given by the pair of equations

$$\left.\begin{aligned}
B_{n+1} - B_0 + \frac{1}{4ah} \sum_{i=1}^{n} B_i [\phi_{i+\frac{1}{2}}^2 - \phi_{i-\frac{1}{2}}^2] = 0, \\[2mm]
B_{n+1} + B_0 + \frac{1}{h} \sum_{i=1}^{n} B_i [\phi_{i+\frac{1}{2}} - \phi_{i-\frac{1}{2}}] = 0.
\end{aligned}\right\} \tag{41}$$

Equation (40) permits (38) to be written in the form

$$\Omega_0(\phi_j) = \sum_{i=0}^{n+1} B_i A_{ij}, \tag{42}$$

where
$$\left.\begin{aligned}
A_{ij} &= \frac{2}{\pi^2}\left[F\left\{\frac{\pi}{2h}(\phi_{i+\frac{1}{2}} - \phi_j)\right\} - F\left\{\frac{\pi}{2h}(\phi_{i-\frac{1}{2}} - \phi_j)\right\} \right], \\[2mm]
F(x) &= \int_0^x \ln|\sinh x|\, dx, \\[2mm]
A_{0j} &= \frac{1}{\pi}\ln\left|\sinh\frac{\pi}{2h}(2a + \phi_j)\right|, \\[2mm]
A_{n+1\,j} &= \frac{1}{\pi}\ln\left|\sinh\frac{\pi}{2h}(2a - \phi_j)\right|,
\end{aligned}\right\} \tag{43}$$

and θ_a, θ_b have been replaced by $-B_0$, $-B_{n+1}$.

Near the stagnation points q is small, and in many cases R also becomes relatively small, e.g. when the cylinder is an aerofoil. This means that if (40) is to be an adequate representation of the original cylinder in these neighbourhoods the intervals $\phi_{i+\frac{1}{2}} - \phi_{i-\frac{1}{2}}$ must be considerably reduced.

The perimeter distance $s(\phi_j)$ from $\phi = -2a$ to $\phi = \phi_j$ is given by

$$Us(\phi_j) = \int_{-2a}^{\phi_j} \frac{U}{q}\,d\phi = \int_{-2a}^{\phi_1} \frac{U}{q}\,d\phi + \int_{\phi_1}^{\phi_j} \frac{U}{q}\,d\phi. \tag{44}$$

Provided $j \neq n + \frac{1}{2}$, i.e. we are not integrating from one stagnation point to the other, the second term in (44) has no singularities in its integrand and can be evaluated from a knowledge of U/q at ϕ_j, $j = 1, 2, \ldots$, by numerical integration. If the intervals $\phi_{i+\frac{1}{2}} - \phi_{i-\frac{1}{2}}$ are all equal the simplest method having sufficient accuracy for this quadrature is to use the Simpson rule (see Comrie, 1949) to find $s(\phi_j)$ $(j = 1, 3, 5, \ldots)$. One use of the three-eighths rule plus further use of the Simpson rule then enables $s(\phi_j)$ $(j = 2, 4, \ldots)$ to be found.

The first integral in (44) requires special attention as $U/q = \infty$ at $\phi = -2a$. From (6.3) q is small where Ω is large and positive. Hence from (6.3) and (6.4)

$$\frac{q}{U} \approx \frac{2\sinh\Omega^*}{e^{\Omega^*}}\,e^{-\Omega} = \frac{2\beta_a}{1+\beta_a}\,e^{-\Omega},$$

near stagnation points. Near $\phi^* = -2a$ equation (38) can be written

$$\Omega_0(\phi^*) = -\frac{\theta_a}{\pi}\ln|2a + \phi^*| + \text{terms finite at } \phi^* = -2a;$$

therefore
$$\frac{q}{U} \simeq K\,|2a + \phi|^{\theta_a/\pi},$$

where K is a constant. Substituting this in the first term of (44) we have

$$Us(\phi_1) = K\int_{-2a}^{\phi_1} (2a + \phi)^{-\theta_a/\pi}\,d\phi = \frac{2a + \phi_1}{1 - \theta_a/\pi}\,K(2a + \phi_1)^{-\theta_a/\pi}$$

so
$$Us(\phi_1) = \left(\frac{U}{q}\right)_1 \frac{2a + \phi_1}{1 - \theta_a/\pi}, \tag{45}$$

where $(U/q)_1$ is the value of U/q at $\phi = \phi_1$, and the appropriate value to use for θ_a/π is $-B_0/\pi$. A similar equation holds for $Us(\phi_{n+\frac{1}{2}}) - Us(\phi_n)$.

The total perimeter distance from the front to the rear stagnation point, S, is given by

$$US = \frac{2a + \phi_1}{1 - \theta_a/\pi}\left(\frac{U}{q}\right)_1 + \int_{\phi_1}^{\phi_n} \frac{U}{q}\,d\phi + \frac{2a - \phi_n}{1 - \theta_b/\pi}\left(\frac{U}{q}\right)_n, \tag{46}$$

the integral in which can be calculated numerically.

Equations (41)–(46) are used as follows (cf. the discussion at the end of §7.2):

(1) Assume a mode of subdivision $(2a, \phi_{n-\frac{1}{2}}, ..., \phi_{1\frac{1}{2}}, -2a)$, and a value for a/h. (For aerofoil shapes $a \approx c/4$, where c is the linear distance between the stagnation points, the *chord* length. Hence from (33) $a/h \approx c/(4\beta_a HU)$ for aerofoils.)

(2) Compute A_{ij} from (43).

(3) Assume a set of values for $q^{(1)}(\phi_j)/U$ (unity will serve if no better estimate is available).

(4) Deduce $s(\phi_j)$ and $S(a)$ from (44) and (46).

(5) Find $R(\phi_j)$ from the known $R = R(s)$ relation.

(6) Calculate B_i from (40), then B_0, B_{n+1} from (41).

(7) Equation (42), together with

$$\frac{q}{U} = \frac{\beta_a}{M_a} \operatorname{cosech}\left(\Omega + \ln\frac{1+\beta_a}{M_a}\right),$$

which follows from (6.3) and (6.4), now enables a new set of values $(q^{(2)}(\phi_j)/U)$ to be deduced.

(8) Steps 4–7 are repeated until the difference

$$\frac{q^{(r+1)}}{U}(\phi_j) - \frac{q^{(r)}}{U}(\phi_j)$$

between successive solutions is negligible.

In the method just described $a/h = a/HU\beta_a$ remains constant throughout the iterations, and this means that until the calculation is complete, and the final $S = f(a/U)$, or $a/U = f^{-1}(S)$ relation is found from (46), the ratio $f^{-1}(S)/H$, i.e. the ratio σ of a typical cylinder dimension to the tunnel width, is unknown. When a particular value of σ is required, one can either modify a/h after each iteration to make σ correct at that stage, or repeat the complete calculation for a number of fixed values of a/h, find the σ, a/h relation, and then interpolate for the desired σ. The second method is to be preferred for the first requires a recalculation of A_{ij} after each iteration, and it does not yield as much information as the second.

When convergence is slow—this is the case with narrow channels and broad obstacles such as circular cylinders—one can often estimate a much improved $q^{(r+1)}(\phi_j)$ over that obtained by an iteration starting from $q^{(r)}(\phi_j)$, by an extrapolation from $q^{(r)}(\phi_j)$, $q^{(r-1)}(\phi_j)$, $q^{(r-2)}(\phi_j)$, ... The successive iterates tend to oscillate about the final value, and after a certain stage the crude process of averaging $q^{(r-1)}$, and $q^{(r)}$—giving $q^{(r)}$ more weight than $q^{(r-1)}$—speeds up the convergence remarkably. This is unnecessary with aerofoil shapes for if $q^{(1)} = U$, $q^{(3)}$ and $q^{(4)}$ will be almost indistinguishable.

7.7   A numerical example: the circular cylinder

To illustrate the method we shall calculate the flow about a circular cylinder centrally placed in a channel with a width $(2H)$ about three times the cylinder diameter $(2B)$. The Mach number of the flow at infinity, M_∞, will be 0·4.

Although the example we have selected is somewhat unpractical (Dirichlet flow is not a good model of the real flow about a circular cylinder, see § 11.13) it will serve to test thoroughly the iterative method described above, for the larger the thickness:length ratio of the cylinder cross-section the more slowly does the method converge. Furthermore the convergence rate is found to be reduced by the presence of a high local Mach number on the cylinder. This will certainly be the case in the example we have selected, for sonic speed is just achieved on a circular cylinder in a stream of an *unlimited* width at $M_a \simeq 0·4$ (Sears, 1954, p. 355), and the presence of channel walls will considerably increase the velocity near the cylinder.

Taking the tangency conditions to be those in the undisturbed stream at infinity, we have $M_a = 0·4$, and $\beta_a = 0·9165$. We shall divide $(-2a, 2a)$ into 16 equal intervals so that in accordance with the notation of § 7.6, $\phi_\frac{1}{2} = -2a$, $\phi_1 = -1\frac{7}{8}a$, ..., $\phi_{15} = 1\frac{5}{8}a$, $\phi_{16} = 1\frac{7}{8}a$, $\phi_{16\frac{1}{2}} = 2a$.

A value of a/h which gives $H/B \simeq 3$ can be selected as follows. For the incompressible flow of an unlimited stream about a circular cylinder of radius B it is a standard result that

$$w = U\left(z + \frac{B^2}{z}\right) \tag{47}$$

(see Milne-Thomson, 1949, p. 171). The stagnation points are at $z = \pm B$, hence $4a = w(z = B) - w(z = -B) = 4UB$, i.e. $a = UB$. If the flow is compressible and the stream width limited, the average velocity over the cylinder will be increased, so for our example we can write $a = kUB$, where k is a correction factor greater than unity. Experience with a similar example suggests that in the present application k may be about 1.25. Hence by (33)

$$\frac{a}{h} = \frac{kUB}{\beta_a UH} = \frac{k}{3\beta_a} \simeq 0·455.$$

In the construction of A_{ij} we need

$$F\left\{\frac{\pi}{2h}(\phi_{j+\frac{1}{2}} - \phi_j)\right\} = F\left\{\frac{\pi a}{16h}\right\} \simeq F(0·089),$$

and for convenience in using a published table (Woods, 1955b) of the function $F(x)$ this suggests we take $\pi a/16h = 0·1$, i.e.

$$\frac{a}{h} = \frac{1·6}{\pi}. \tag{48}$$

The array A_{ij} can now be obtained from the first differences of $(2/\pi^2) F(0\cdot1)$, $(2/\pi^2) F(0\cdot3)$, $(2/\pi^2) F(0\cdot5)$, ... (see (43)).

A suitable initial distribution of $q(\phi_j)/U$ is

$$\frac{q^{(1)}}{U} = 2\left\{1 - \left(\frac{\phi_j}{2a}\right)^2\right\}^{-\frac{1}{2}}, \tag{49}$$

which is easily verified from (47) to be incompressible infinite stream distribution. The radius of curvature R_0 is constant and equal to $-B$ (convex surfaces have a negative R_0 in accordance with the definition of R_0 contained in (13)), so (33) and (40) give

$$B_j = \frac{\beta_a H}{B}\left(\frac{U}{q}\right)_{\phi=\phi_j},$$

i.e.
$$B_j = \frac{\beta_a H}{B}\frac{1}{2}\left\{1 - \left(\frac{\phi_j}{2a}\right)^2\right\}^{-\frac{1}{2}}, \tag{50}$$

by (49).

The distribution in (49) gives $a = UB$, so $H\beta_a/B = \pi/1\cdot6 \simeq 1\cdot964$ by (33) and (48); this is the value to use in (50).

Table 7.7. Successive iterates

$\dfrac{\phi_j}{a}$ (1)	$\dfrac{U^{(1)}}{q}$ (2)	$\dfrac{U^{(2)}}{q}$ (3)	$\dfrac{U^{(3)}}{q}$ (4)	$\dfrac{U^{(4)}}{q}$ (5)	$\dfrac{U^{(5)}}{q}$ (6)	$\dfrac{U^{(6)}}{q}$ (7)	$\dfrac{U^{(\infty)}}{q}$ (8)	$\dfrac{q^{(\infty)}}{U}$ (9)	Angular displacement in degrees (10)
$\frac{1}{8}$	0·502	0·320	0·362	0·323	0·352	0·338	0·340	2·941	2·78
$\frac{3}{8}$	0·509	0·330	0·370	0·332	0·359	0·346	0·348	2·874	10·04
$\frac{5}{8}$	0·527	0·351	0·385	0·369	0·373	0·371	0·366	2·732	16·49
$\frac{7}{8}$	0·556	0·386	0·412	0·385	0·404	0·395	0·396	2·525	24·47
$1\frac{1}{8}$	0·605	0·434	0·452	0·431	0·446	0·439	0·440	2·273	32·10
$1\frac{3}{8}$	0·689	0·531	0·536	0·526	0·532	0·529	0·529	1·890	42·00
$1\frac{5}{8}$	0·858	0·621	0·600	0·617	0·606	0·612	0·610	1·639	52·78
$1\frac{7}{8}$	1·437	1·271	1·135	1·186	1·153	1·170	1·166	0·858	69·00
—	3·142	2·332	2·319	2·292	—	2·300	2·297	—	—
—	1·964	2·645	2·660	2·691	—	2·682	2·685	—	—

Equations (42) and (6.3) now give the second approximation $U^{(2)}/q$, shown in column 3 of Table 7.7. ($(U/q)(\phi_j)$ is a symmetrical function of ϕ_j and so only the values for $\phi_j > 0$ are shown.) At the foot of the column the values of $SU/a = \pi BU/a$ and $H\beta_a/B$ are shown. These are computed from (46), (33) and (48).

Columns 4–6 of the table show the successive iterates $U^{(3)}/q$, $U^{(4)}/q$ and $U^{(5)}/q$. Comparing these with the final value $U^{(\infty)}/q$ shown in column 8, we see that they oscillate about the final solution and converge slowly towards it. The average of two successive iterates (excepting $U^{(1)}/q$ and $U^{(2)}/q$) is usually closer to the true value than either iterate. Column 7,

$U^{(6)}/q$, is the average of $U^{(4)}/q$ and $U^{(5)}/q$, and it yields a $U^{(7)}/q$ differing by at most one unit from $U^{(\infty)}/q$. Clearly an application of this averaging process at an earlier stage would have been profitable. Column 9 gives $q^{(\infty)}/U$, and column 10 the angular displacement of ϕ_j measured from the rear stagnation point. This last column is computed by using (44) and (45).

The final value of H/B follows from the last entry in column 7. We have $H/B = 2\cdot685/\beta_a = 2\cdot930$, which is reasonably close to the specified value of $3\cdot0$. To increase H/B to a value very close to 3 we must reduce a/h from $1\cdot6/\pi$ to $1\cdot6/\pi \times 3/2\cdot93$, recalculate A_{ij}, then repeat the iterative process, starting of course with the $U^{(\infty)}/q$ of column 7. We shall omit this further calculation, as no new principles are involved.

7.8   Wind-tunnel blockage

The free-stream characteristics of full-scale aerofoils must often be obtained from experiments on models between wind-tunnel walls. Two basic corrections must be applied to these wind-tunnel results, namely Reynolds number corrections (outside the scope of this text), and corrections arising from the constraining effect of the tunnel walls on the flow. This constraint modifies both the lift and drag of the aerofoil, but we shall defer consideration of the lift correction (see ch. 13) at this stage, and suppose that the aerofoil is symmetrical and at zero incidence on the tunnel centre-line (Fig. 7.8). The tunnel walls will be assumed to be straight and solid (other cases are considered in §§ 7.21, 7.22 and 7.28), and hence they will cause a speeding-up of the flow over the aerofoil surface. It is the magnitude of this velocity increment—the so-called *blockage factor*—which we calculate in this section.

Let values appropriate to the flow of an infinite stream past the aerofoil be distinguished from corresponding values in the tunnel by an asterisk, and let values at $x = \pm\infty$ be indicated as usual by the subscripts $\pm\infty$. Then the blockage factor ϵ is defined by

$$\epsilon(x, y) \equiv \frac{q(x, y)}{q_{-\infty}} - \frac{q^*(x, y)}{q^*_{-\infty}}, \tag{51}$$

i.e. ϵ is the velocity increment per unit velocity due to the presence of the walls. The importance of ϵ is enhanced by the fact that on the aerofoil surface its dominant term is constant; this is proved below. In this section it is convenient to put $q_{-\infty}$ and $q^*_{-\infty}$ both equal to the reference velocity U—there is no essential loss in generality in this—and write (51)

$$\epsilon = (q - q^*)/U. \tag{52}$$

Two distinct components of blockage occur. These are (i) the *solid* blockage, ϵ_s, due to the thickness of the aerofoil, and (ii) the *wake* blockage,

ϵ_w, caused by the displacement thickness δ of the aerofoil wake (see § 1.16). The total blockage is the sum of these components.

Let H be the tunnel width (see Fig. 7.8), then the symmetry about the tunnel axis $y = 0$ permits us to regard the flow as being through two adjacent channels of width $\frac{1}{2}H$. Thus we have to find the flow through a channel such that (in the notation of § 7.2)

$$\theta_h = 0 \; (-\infty < x < \infty), \qquad \theta_0 = \begin{cases} \tan^{-1}\left(\dfrac{dy}{dx}\right)_{\text{aerofoil}} & (-\tfrac{1}{2}c \leqslant x \leqslant \tfrac{1}{2}c) \\ 0 & (\text{elsewhere}), \end{cases}$$

$$H_{-\infty} = \tfrac{1}{2}H, \qquad H_{\infty} = \tfrac{1}{2}(H - \delta), \tag{53}$$

where the origin of the z-plane is taken to be at the aerofoil's mid-chord point.

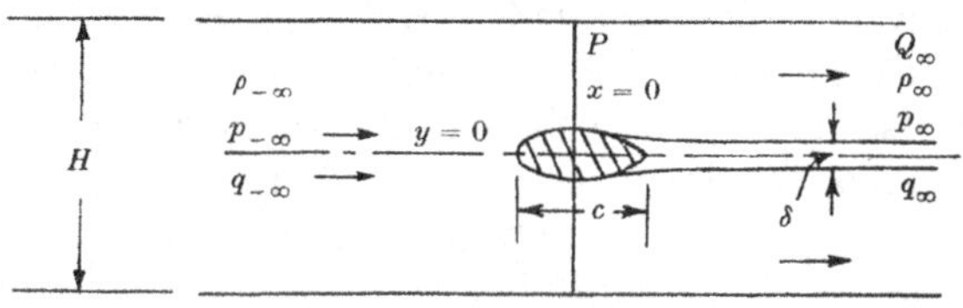

Fig. 7.8

It will be assumed that the aerofoil is slender enough to permit the use of linear perturbation theory, i.e. to permit us to use the approximations (see § 6.1)

$$\left. \begin{aligned} \Omega &= \tfrac{1}{2}\beta_a C_p = \beta_a\left(1 - \frac{q}{U}\right), \\ \theta &= \left(\frac{dy}{dx}\right), \\ UZ &= w \quad (Ux = \phi, \quad \beta_a U y = \psi), \end{aligned} \right\} \tag{54}$$

to be used in (8) and (9). Further, as $q_{-\infty} = U$ it follows from (53) and (4) that $\Omega_{-\infty} = 0$ and $h = \tfrac{1}{2}\beta_a H U$. Also $\beta_a = \beta_{-\infty}$, but we shall use the more convenient subscript 'a' here.

With these results (8) and (9) can be written

$$\beta_a\left(1 - \frac{q}{U}\right) + i\theta = \frac{1}{\beta_a H} \int_{-\frac{1}{2}c}^{\frac{1}{2}c} \left(\frac{dy}{dx}\right)_0 \coth \frac{\pi}{\beta_a H}(x - Z)\,dx + \tfrac{1}{2}\Omega_{\infty}, \tag{55}$$

$$\Omega_{\infty} = -\frac{2}{\beta_a H} \int_{-\frac{1}{2}c}^{\frac{1}{2}c} \left(\frac{dy}{dx}\right)_0 dx$$

A glance at Fig. 7.8 shows that $y = 0$ at $x = -\tfrac{1}{2}c$, and $y = \tfrac{1}{2}\delta$ at $x = \tfrac{1}{2}c$, hence the last equation reduces to

$$\Omega_\infty = -\frac{\delta}{\beta_a H}. \tag{56}$$

The displacement thickness δ is related to the drag coefficient C_D by (1.53), viz.

$$\delta = \tfrac{1}{2}cC_D. \tag{57}$$

Therefore combining (54), (56) and (57) we find that the pressure coefficient downstream at infinity is given by

$$C_{p\infty} = -\frac{cC_D}{\beta_a^2 H}. \tag{58}$$

In a real flow this pressure is larger in magnitude owing to the non-isentropic flow in the wake—flow in which there is an additional pressure drop due to the irreversible transfer of pressure energy into heat energy. In view of this Allen & Vincenti (1944) replaced (57) by $\delta = \tfrac{1}{2}\beta_a^2 H(-C_p)$, taking C_p to be the larger (non-isentropic) coefficient. However, the wake diffuses slowly (see (1.51)), and the energy transfer will not be complete for many chord lengths down the tunnel, whereas the pressure given by (58) is achieved only two or three chord lengths downstream owing to the exponential terms in the integrand of (55). It therefore seems reasonable to assume that the pressure gradient and increased velocity at the aerofoil due to the wake is correctly given by using (57).

In an infinite stream $h = \infty$, then (8) and (9) yield

$$\tau^*(w) = \frac{1}{\pi}\int_{-\infty}^{\infty} \frac{\theta_0(\phi)\,d\phi}{\phi - w} + \Omega_\infty^*, \tag{59}$$

the asterisk having the same meaning as in (51). In a free stream $q_\infty^* = q_{-\infty}^*$, and as we have set $q_{-\infty}^* = U$ it follows that $\Omega_\infty^* = 0$. Hence by (54)

$$\beta_a\left(1 - \frac{q^*}{U}\right) + i\theta^* = \frac{1}{\pi}\int_{-\frac{1}{2}c}^{\frac{1}{2}c} \left(\frac{dy}{dx}\right)_0 \frac{dx}{x - Z}. \tag{60}$$

It only remains to subtract (55) from (60) and to eliminate q^*, q and Ω_∞ by (52) and (56) to obtain an equation for $\epsilon(x, y)$. The value of ϵ on the aerofoil is all we require; this is found to be

$$\epsilon(x^*) = \frac{1}{\beta_a^2 H}\int_{-\frac{1}{2}c}^{\frac{1}{2}c} \left(\frac{dy}{dx}\right)_0 \left\{\frac{1}{(\pi/\beta_a H)(x - x^*)} - \coth\frac{\pi}{\beta_a H}(x - x^*)\right\} dx + \frac{\delta}{2\beta_a^2 H}. \tag{61}$$

An approximate formula for $\epsilon(x^*)$ is found by making use of the expansion

$$\frac{1}{\alpha} - \coth\alpha = -\frac{\alpha}{3}\{1 + O(\alpha^2)\},$$

and the fact that $|x - x^*| \leqslant c$ on the aerofoil surface. We find

$$\epsilon(x^*) = \frac{-\pi}{3\beta_a^3 H^2} \int_{-\frac{1}{2}c}^{\frac{1}{2}c} \left(\frac{dy}{dx}\right)_0 (x-x^*) \left\{1 + O\left(\frac{c}{\beta_a H}\right)^2\right\} dx + \frac{\delta}{2\beta_a^2 H}$$

$$= \frac{-\pi}{3\beta_a^3 H^2} \left\{\frac{\delta}{2}(\tfrac{1}{2}c - x^*) - \int_{-\frac{1}{2}c}^{\frac{1}{2}c} y_0 dx + O\left(\frac{c}{\beta_a H}\right)^2\right\} + \frac{\delta}{2\beta_a^2 H},$$

by an integration by parts. The integral in this last expression is $\frac{1}{2}A$, where A is the cross-sectional area of the profile. Rearranging this equation, and using (57) to eliminate δ, we get the approximate formula

$$\epsilon(x) = \frac{\pi A}{6\beta_a^3 H^2} + \frac{cC_D}{4\beta_a^2 H}\left\{1 + \frac{\pi}{3\beta_a H}(x - \tfrac{1}{2}c)\right\}. \tag{62}$$

The dominant term here is constant as stated earlier. The mean value of $\epsilon(x)$ over the aerofoil is clearly

$$\bar{\epsilon} = \epsilon(0) = \frac{\pi A}{6\beta_a^3 H^2} + \frac{cC_D}{4\beta_a^2 H}\left(1 - \frac{\pi c}{6\beta_a H}\right). \tag{63}$$

The pressure gradient, l say, due to blockage can be calculated as follows. From (6.8) $p = p_a + \frac{1}{2}\rho_a U^2 C_p$, hence from (51) and (54)

$$l \equiv \frac{\partial}{\partial x}(p - p^*) = -\rho_a U \frac{\partial}{\partial x}(q - q^*) \simeq -\rho_a U q^* \frac{\partial\epsilon}{\partial x} \simeq -\rho_a U^2 \frac{\partial\epsilon}{\partial x},$$

on using the linear approximation $q^* = U$ in the coefficient of this small derivative. Equation (62) now gives

$$l = -\frac{\rho_a U^2 c C_D \pi}{12\beta_a^3 H^2} = -\frac{\pi D}{6\beta_a^3 H^2}, \tag{64}$$

D being the drag on the aerofoil.

An experimental method of obtaining $\bar{\epsilon}$ is based on the following theory. Let the pressure coefficient at $Z = i\beta_a H/2$, i.e. on the wall opposite the aerofoil (point P in Fig. 7.8), be denoted by C_{p_0}, then (54)–(56) give

$$\tfrac{1}{2}\beta_a C_{p_0} = \frac{1}{\beta_a H}\int_{-\frac{1}{2}c}^{\frac{1}{2}c}\left(\frac{dy}{dx}\right)_0 \tanh\frac{\pi x}{\beta_a H} dx - \frac{\delta}{2\beta_a H}$$

$$= \frac{\pi}{\beta_a^2 H^2}\int_{-\frac{1}{2}c}^{\frac{1}{2}c}\left(\frac{dy}{dx}\right)_0 x\left\{1 + O\left(\frac{c}{\beta_a H}\right)^2\right\} dx - \frac{\delta}{2\beta_a H}$$

$$= \frac{\pi}{\beta_a^2 H^2}\{\tfrac{1}{4}\delta c - \tfrac{1}{2}A\} - \frac{\delta}{2\beta_a H}$$

$$= -\frac{\pi A}{2\beta_a^2 H^2} - \frac{cC_D}{4\beta_a H} + \frac{\pi c^2 C_D}{8\beta_a^2 H^2},$$

by (57). Equations (58) and (63) now give the very simple result

$$\bar{\epsilon} = -\tfrac{1}{6}(C_{p_0} + C_{p_\infty}),\tag{65}$$

showing that it is necessary only to make pressure measurements on the tunnel wall opposite the aerofoil and well downstream to obtain an experimental estimate of $\bar{\epsilon}$.

7.9 The increase in drag due to blockage

To illustrate the application of the theory of the last section we shall now calculate the correction term to be applied to wind-tunnel drag measurements to reduce these to free stream values.

The drag on an aerofoil in a wind tunnel is larger than that in an infinite stream for two reasons: (i) because of the increase in the average velocity of the aerofoil caused by the tunnel walls, and (ii) because of the existence of a pressure gradient at the aerofoil due to the presence of the wake.

The increase in the average velocity at the aerofoil in the tunnel is equivalent in its effect to an increase in the velocity at infinity in the infinite stream to $U^* = U(1+\bar{\epsilon})$. If this velocity change is made, the drag in the tunnel, D, will differ from the drag in the free stream, D^*, only by ΔD, the increment to the drag due to the change in the pressure gradient caused by the wake blockage. Therefore $D - D^* = \Delta D$, i.e.

$$\Delta D = \tfrac{1}{2}c\rho_a U^2 C_D - \tfrac{1}{2}c\rho_\infty^* U^2(1+\bar{\epsilon})^2 C_D^*.\tag{66}$$

The pressure gradient due to the wake is given by (64), and is constant. Hence $p(x) = p^*(x) + lx$, and

$$D = \int_C p(x)\,dy = \int_C p^*(x)\,dy + l\int_C x\,dy,$$

where C is the aerofoil contour. The last integral is $-A$, where A is the profile cross-sectional area. Therefore

$$D - D^* = \Delta D = -lA = \frac{\pi A D}{6\beta_a^3 H^2},\tag{67}$$

by (64).

It remains to find the relation between ρ_a, i.e. $\rho_{-\infty}$ and ρ_∞^*; these differ because they correspond to velocities U and $U(1+\bar{\epsilon})$ respectively. Equation (6.27) gives

$$\rho_\infty^* = \rho_a\left\{1 + \frac{M_a^2}{\beta_a}\Omega_\infty^*\right\} = \rho_a\left\{1 - M_a^2\left(\frac{U(1+\bar{\epsilon})}{U} - 1\right)\right\},$$

where the second form comes from an application of the first of (54). Thus

$$\rho_\infty^* = \rho_a\{1 - \bar{\epsilon}M_a^2\}.\tag{68}$$

Equations (66)–(68) give

$$C_D^* = C_D\left\{1 - (2 - M_a^2)\,\bar{\epsilon} - \frac{\pi A}{6\beta_a^3 H^2}\right\},\tag{69}$$

on ignoring terms $O(\bar{\epsilon}^2)$. Eliminating $\bar{\epsilon}$ by (63) we get the required correction formula for the drag coefficient

$$C_D^* = C_D\left\{1 - \frac{(3 - M_a^2)\,\pi A}{6\beta_a^3 H^2} - \frac{(2 - M_a^2)}{4\beta_a^2}\left(\frac{c}{H}\right)C_D\left(1 - \frac{\pi c}{6\beta_a H}\right)\right\}.\tag{70}$$

CHANNEL DESIGN

7.10   General theory

We now turn our attention to the conjugate problem, i.e. to the problem of *designing* a channel to have a prescribed pressure distribution over its walls. The problem considered above—that of finding the flow through a *given* channel—is sometimes termed the *direct* problem, while the design of channels is termed the *indirect* problem. These names might suggest that designing channels is more difficult than finding the flow through given channels, but the opposite is true, as will be seen below.

The conjugate equations to (8) and (9) are

$$\tau(w) = \frac{-i}{2h}\int_{-\infty}^{\infty}\left\{\Omega_0(\phi)\coth\frac{\pi}{2h}(\phi - w) - \Omega_h(\phi)\tanh\frac{\pi}{2h}(\phi - w)\right\}d\phi$$

$$+ \tfrac{1}{2}i(\theta_\infty + \theta_{-\infty}),\tag{71}$$

and

$$\theta_\infty - \theta_{-\infty} = \frac{1}{h}\int_{-\infty}^{\infty}\{\Omega_0(\phi) - \Omega_h(\phi)\}\,d\phi\tag{72}$$

(cf. (3.103) and (3.112)). On $\psi = 0$ and $\psi = h$ (71) gives

$$\theta_0(\phi^*) = -\frac{1}{2h}\int_{-\infty}^{\infty}\left\{\Omega_0(\phi)\coth\frac{\pi}{2h}(\phi - \phi^*)\right.$$

$$\left. - \Omega_h(\phi)\tanh\frac{\pi}{2h}(\phi - \phi^*)\right\}d\phi + \tfrac{1}{2}(\theta_\infty + \theta_{-\infty}),\tag{73}$$

and

$$\theta_h(\phi^*) = -\frac{1}{2h}\int_{-\infty}^{\infty}\left\{\Omega_0(\phi)\tanh\frac{\pi}{2h}(\phi - \phi^*)\right.$$

$$\left. - \Omega_h(\phi)\coth\frac{\pi}{2h}(\phi - \phi^*)\right\}d\phi + \tfrac{1}{2}(\theta_\infty + \theta_{-\infty}),\tag{74}$$

respectively.

In design problems we would normally be given the wall pressures

p_0, p_h as functions of the distances s_0 and s_h along the channel walls. From these and (6.8), viz.

$$p = p_a + \rho_a U^2 \sinh \Omega \cosh \Omega^* \operatorname{cosech}(\Omega + \Omega^*), \qquad (75)$$

$\Omega_0(s_0)$, $\Omega_h(s_h)$ can be deduced. Bernoulli's equation or alternatively (6.3) gives the velocities $q_0(s_0)$, $q_h(s_h)$, then inversion of

$$\phi(s_0) = \int^{s_0} q_0 \, ds, \quad \phi(s_h) = \int^{s_h} q_h \, ds, \qquad (76)$$

yields $s_0(\phi)$, $s_h(\phi)$. These enable $\Omega_0(\phi)$, $\Omega_h(\phi)$ to be found, and then $\theta_0(\phi)$, $\theta_h(\phi)$ follow from (73) and (74). Finally, as $dx_0 = \cos\theta_0 \, ds_0$, $dy_0 = \sin\theta_0 \, ds_0$,

$$x_0 = \int^\phi \frac{\cos\theta_0}{q_0} d\phi_0, \quad y_0 = \int^\phi \frac{\sin\theta_0}{q_0} d\phi_0, \qquad (77)$$

and we have a parametric equation for the coordinates of the wall $\psi = 0$; similar equations give $x_h(\phi)$ and $y_h(\phi)$.

Equation (72) enables the overall deflexion angle of the channel to be calculated directly from $\Omega_0(\phi)$ and $\Omega_h(\phi)$. If it is desired to specify this angle initially, the pressure distributions can be defined only to within some arbitrary parameter, which must be selected in order to satisfy (72).

The indirect problem therefore is essentially only a matter of three integrations—integrations which, in most cases, must be performed numerically.

Some examples of incompressible flow channel designs will be found in a paper by Lighthill (1945 c).

7.11 Sectionally constant wall pressures

Suppose that the wall pressures are prescribed as step-functions, such that Ω_0 has jumps α_j at $\phi = \phi_j$ $(j = 1, 2, ..., m)$ and Ω_h has jumps α_i at $\phi = \phi_i$ $(i = 1, 2, ..., n)$, then (71) gives

$$\tau(w) = \frac{i}{\pi}\left\{\sum_j \alpha_j \ln \sinh \frac{\pi}{2h}(\phi_j - w) - \sum_i \alpha_i \ln \cosh \frac{\pi}{2h}(\phi_i - w)\right\}$$

$$+ \tfrac{1}{2}i(\theta_\infty + \theta_{-\infty}), \quad (78)$$

where from (72) the channel is turned through an angle

$$\theta_\infty - \theta_{-\infty} = \frac{1}{h}\{\sum_i \alpha_i \phi_i - \sum_j \alpha_j \phi_j\}. \qquad (79)$$

As a simple example suppose Ω_0 jumps from 0 to σ at $\phi = -a$, and from σ to 0 at $\phi = a$, and that $\Omega_h = 0$ for all ϕ; then, from (78) and (79),

$$\tau = \frac{i\sigma}{\pi} \ln \left\{ \frac{\sinh \dfrac{\pi}{2h}(w+a)}{\sinh \dfrac{\pi}{2h}(w-a)} \right\} + \tfrac{1}{2}i(\theta_\infty + \theta_{-\infty}), \tag{80}$$

and
$$\theta_\infty - \theta_{-\infty} = \frac{2a\sigma}{h}. \tag{81}$$

7.12   Design of a bend in a channel

Equations (80) and (81) can be used to design a bend in a channel. Let $\Omega_0 = \sigma$ in $-a \leqslant \phi \leqslant a$ correspond to a velocity V ($V > U$) over a section AB on the channel wall, then from (6.3) and (6.4)

$$\frac{V}{U} = \frac{\beta_a}{M_a} \operatorname{cosech}\left(\sigma + \ln\left(\frac{1+\beta_a}{M_a}\right)\right), \tag{82}$$

where the subscript a refers to values on the channel walls outside AB. Equation (33) applies here; also if S is the total length of AB, $2a = VS$. Hence (81) can be written

$$\sigma = \left(\frac{HU}{SV}\right)\beta_a\alpha, \tag{83}$$

where α is the angle through which the channel is deflected by the bend. On eliminating V/U from (82) and (83) we obtain a complicated relation for σ in terms of H/S, β_a and α, which can be solved graphically.

On AB, $\phi = Vs$, where s is distance measured from $\phi = 0$, and so (80) yields the intrinsic equation

$$\theta = \tfrac{1}{2}\alpha + \frac{\sigma}{\pi}\ln\left|\frac{\sinh\{\pi V(s+\tfrac{1}{2}S)/H\beta_a U\}}{\sinh\{\pi V(s-\tfrac{1}{2}S)/H\beta_a U\}}\right|, \tag{84}$$

from which the shape of AB can be deduced. Similar intrinsic relations can be written down for the remaining sections of the wall.

Lighthill (1945b) has used this method to design a right-angled bend for incompressible flow; Fig. 7.12 gives some idea of the type of bend achieved. The infinite adverse pressure gradient which occurs at B is counteracted in its tendency to cause the boundary layer to separate by applying suction at B (see § 1.19).

Notice that according to (84) θ becomes infinite at $s = \pm \tfrac{1}{2}S$, i.e. at the points A and B, and so (84) is useless as a practical design equation in the immediate neighbourhoods of these singularities. These logarithmic infinities in θ—corresponding to the discontinuities in the conjugate function Ω (see § 3.10)—are effective only very close to the singular points, and can therefore be blurred over in an actual design without much resultant change in the pressure distribution. Of course in the neighbourhood of the slot at B (84) is, in any case, not appropriate.

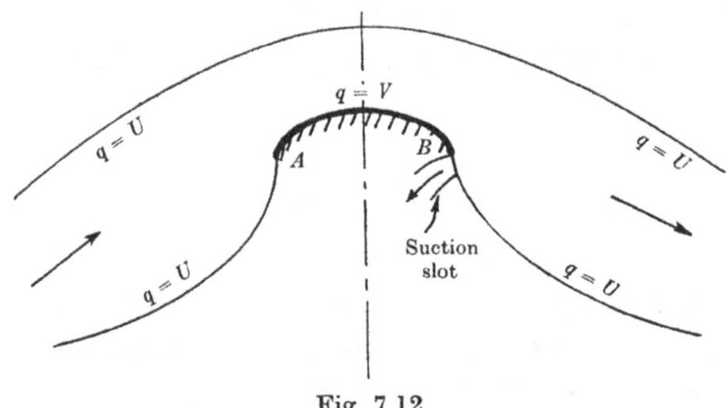

Fig. 7.12

SIMPLY-MIXED BOUNDARY CONDITIONS

7.13 General theory

The simplest type of mixed boundary condition occurs when θ is prescribed over one connected section of the channel walls, and Ω is prescribed over the remaining section. In §4.1 we have termed these 'simply-mixed' boundary conditions. In this definition the points $\phi = -\infty$ and $\phi = \infty$ are to be regarded as connecting $\psi = 0$ to $\psi = h$. Figs. 7.13 (b), (c), (d), (e) and (f) show the five distinct ways in which the boundaries of the infinite strip $-\infty < \phi < \infty$, $0 \leqslant \psi \leqslant h$ in the w-plane can be divided into two parts. In these diagrams we have employed a solid line to indicate the boundary condition $\theta = \theta(s)$, and a dotted line to indicate $\Omega = \Omega(s)$. Figs. 7.13 (a) and (g) apply to channel design, and flow through given channels respectively; these could be regarded as limiting cases of the present theory. Fig. 7.13 (h) is an example of a more complicated type of mixed boundary condition outside the scope of the equations to be given in this and subsequent sections. It will be considered in §7.24.

The next step (cf. §6.14) is to select a suitable auxiliary ζ-plane for which $\tau(\zeta)$ can be readily calculated. Fig. 7.13 (i) shows such a plane, the infinite strip $-\infty < \gamma < \infty$, $0 \leqslant \eta \leqslant \tfrac{1}{2}\pi$—a plane identical in appearance with that of Fig. 7.13 (d). A new notation to distinguish boundary values is now needed, as the subscripts 0 and h used hitherto might cause confusion if applied to the ζ-plane. We shall therefore write

$$\lim_{\epsilon \to +0} \tau(\gamma, \epsilon) = \tau_1(\gamma) = \Omega_1(\gamma) + i\theta_1(\gamma), \ \lim_{\epsilon \to +0} \tau\left(\gamma, \frac{\pi}{2} - \epsilon\right) = \tau_2(\gamma)$$

$$= \Omega_2(\gamma) + i\theta_2(\gamma). \quad (85)$$

The solution $\tau(\zeta)$ appropriate to prescribed values of $\theta_1(\gamma)$ and $\Omega_2(\gamma)$ is

$$\tau(\zeta) = \frac{1}{\pi} \int_{-\infty}^{\infty} \{\theta_1(\gamma) \operatorname{cosech}(\gamma - \zeta) + \Omega_2(\gamma) \operatorname{sech}(\gamma - \zeta)\}\, d\gamma, \qquad (86)$$

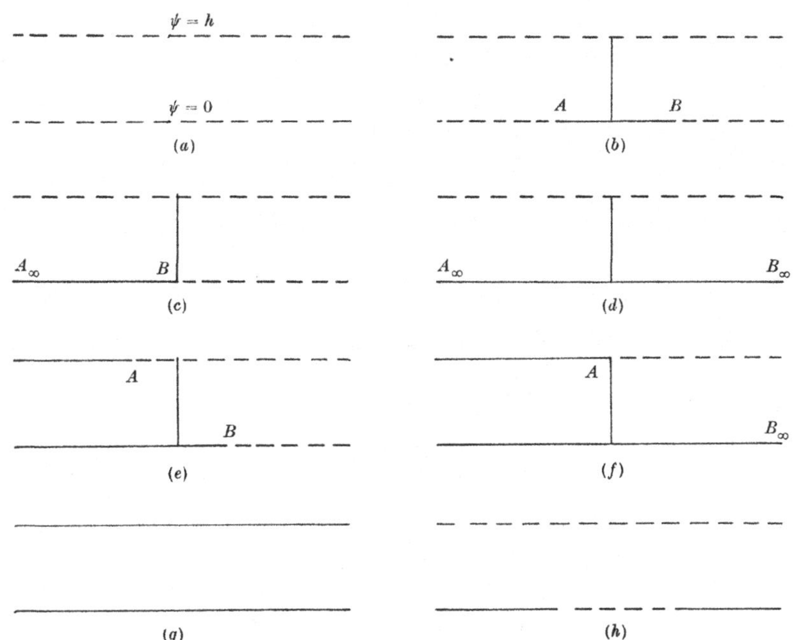

Figs. 7.13a to h

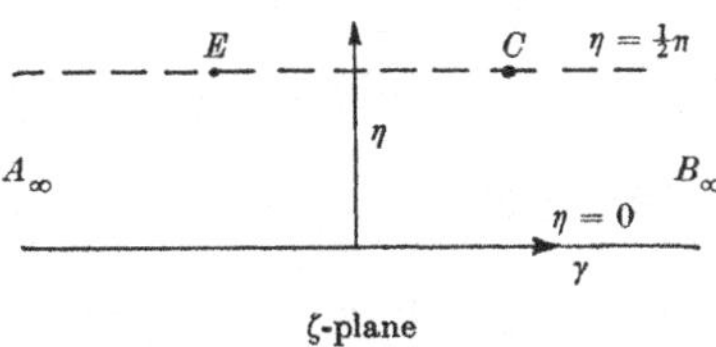

Fig. 7.13i

which follows from (4.9) on replacing h, x and z by the appropriate symbols for the present application, namely $\tfrac{1}{2}\pi$, γ and ζ. By integrating (86) by parts we obtain the alternative and sometimes more useful form

$$\tau(\zeta) = \tau(\zeta = \infty) + \frac{2}{\pi} \int_{-\infty}^{\infty} [\tanh^{-1}\{\exp(\gamma - \zeta)\}\, d\theta_1(\gamma) - \tan^{-1}$$
$$\times \{\exp(\gamma - \zeta)\}\, d\Omega_2(\gamma)]. \qquad (87)$$

To complete the general theory we need the $\zeta(w)$ relations between the five w-planes (b), (c), (d), (e) and (f) and the ζ-plane (i).

Consider the first of these w-planes, which we have shown in more detail in Fig. 7.13 (j). It has already been established in §5.12 that this plane is mapped into the ζ-plane by

$$\tanh \zeta = \coth \frac{\pi \phi_0}{2h} \tanh \frac{\pi w}{2h}, \tag{88}$$

where $2\phi_0$ is the potential difference between A and B. A more useful form of (88) is

$$\exp \zeta = \left\{ \frac{\sinh \left[\pi(\phi_0 + w)/2h \right]}{\sinh \left[\pi(\phi_0 - w)/2h \right]} \right\}^{\frac{1}{2}}. \tag{89}$$

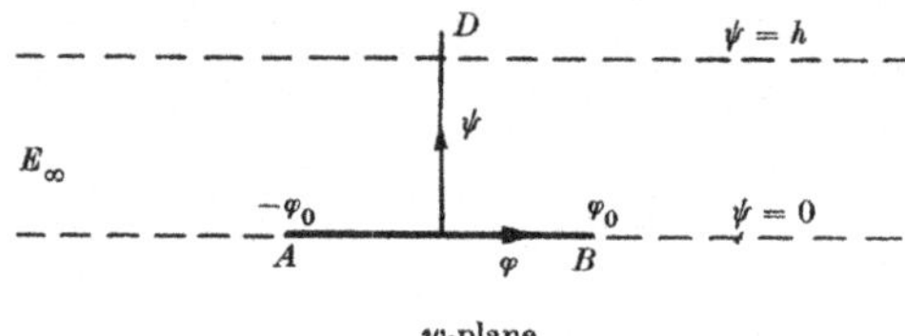

Fig. 7.13j

Similar calculations to that of §5.12 give

$$\left. \begin{aligned} \exp \zeta &= \{ e^{-\pi w/h} - 1 \}^{-\frac{1}{2}}, & (c) \\[1ex] \zeta &= \frac{\pi w}{2h}, & (d) \\[1ex] \exp \zeta &= \left\{ \frac{e^{-\pi(w+\phi_0)/h} + 1}{e^{-\pi(w-\phi_0)/h} - 1} \right\}^{\frac{1}{2}}, & (e) \\[1ex] \exp \zeta &= \{ e^{\pi w/h} + 1 \}^{\frac{1}{2}}, & (f) \end{aligned} \right\} \tag{90}$$

as the appropriate mapping equations for the w-planes (c), (d), (e) and (f). In the third of these $2\phi_0$ is the potential difference between points A and B.

Dozens of useful potential flows can be solved by using (87) and the appropriate $\zeta(w)$ relation. In the following sections we shall solve a few of these.

7.14　Flow of a jet, deflected around a given curved surface

Consider the flow problem illustrated in Fig. 7.14. The jet meets the surface tangentially at A, and leaves it tangentially at B, after being deflected through an angle. The w-plane is of the type shown in Fig. 7.13 (b). We find from (87) and (89) that

$$\tau(w) = \tau_B$$

$$+\frac{2}{\pi}\int_{-\phi_0}^{\phi_0} \tanh^{-1}\left\{\frac{\sinh\{\pi(\phi_0+\phi^*)/2h\}\sinh\{\pi(\phi_0-w)/2h\}}{\sinh\{\pi(\phi_0-\phi^*)/2h\}\sinh\{\pi(\phi_0+w)/2h\}}\right\}^{\frac{1}{2}} d\theta_0(\phi^*)$$

$$+\frac{2}{\pi}\left\{\int_{-\infty}^{-\phi_0} + \int_{\phi_0}^{\infty}\right\} \tan^{-1}\left\{\frac{\sinh\{\pi(\phi^*+\phi_0)/2h\}\sinh\{\pi(\phi_0-w)/2h\}}{\sinh\{\pi(\phi^*-\phi_0)/2h\}\sinh\{\pi(\phi_0+w)/2h\}}\right\}^{\frac{1}{2}} d\Omega_0(\phi^*)$$

$$-\frac{2}{\pi}\int_{-\infty}^{\infty} \tan^{-1}\left\{\frac{\cosh\{\pi(\phi_0+\phi^*)/2h\}\sinh\{\pi(\phi_0-w)/2h\}}{\cosh\{\pi(\phi_0-\phi^*)/2h\}\sinh\{\pi(\phi_0+w)/2h\}}\right\}^{\frac{1}{2}} d\Omega_h(\phi^*), \qquad (91)$$

where we have separated Ω_2 into its components Ω_0 on $\psi = 0$, and Ω_h on $\psi = h$, and $\tau_B = \tau\,(\zeta = \infty)$, the value of τ at point B in Fig. 7.14.

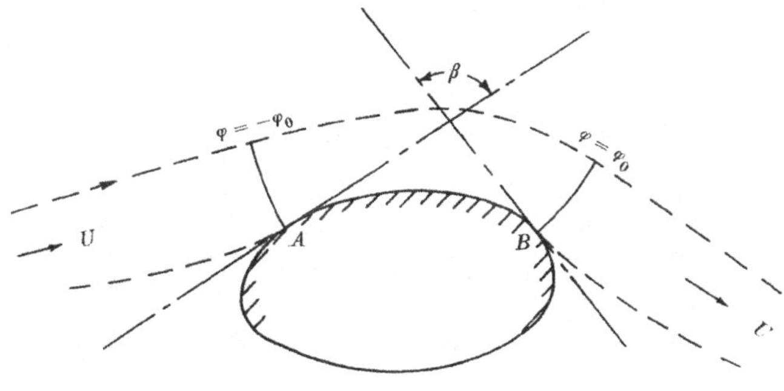

Fig. 7.14

When the jet boundaries are at the same pressure we can set $q = U$ on $\psi = h$, and on $\psi = 0$ outside AB, i.e. set $\Omega_0 = \Omega_h = 0$ in (91). Then using (13), we find that at the point $w = \phi$ where $-\phi_0 \leqslant \phi \leqslant \phi_0$,

$$\Omega_0(\phi) = \frac{2}{\pi}\int_{-\phi_0}^{\phi_0} \frac{1}{R_0 q_0} \tanh^{-1}\left\{\frac{\sinh\{\pi(\phi_0+\phi^*)/2h\}\sinh\{\pi(\phi_0-\phi)/2h\}}{\sinh\{\pi(\phi_0-\phi^*)/2h\}\sinh\{\pi(\phi_0+\phi)/2h\}}\right\}^{\frac{1}{2}} d\phi^*. \tag{92}$$

In general (92) is an integral equation which must be solved numerically.

A similar method to that used in the solution of (38) can be employed here. Using the same mode of subdivision for $(-\phi_0, \phi_0)$ as employed for $(-2a, 2a)$ in §7.6 we obtain

$$\Omega_0(\phi_j) = \sum_{i=1}^{n} B_i A_{ij}, \tag{93}$$

where

$$A_{ij} \equiv \int_{\phi_{i-\frac{1}{2}}/\phi_0}^{\phi_{i+\frac{1}{2}}/\phi_0} \tanh^{-1}\left\{\frac{\sinh\{\pi\phi_0(1+t)/2h\}\sinh\{\pi\phi_0(1-t_j)/2h\}}{\sinh\{\pi\phi_0(1-t)/2h\}\sinh\{\pi\phi_0(1+t_j)/2h\}}\right\}^{\frac{1}{2}} dt, \quad (94)$$

$$B_i \equiv \frac{2}{\pi}\left(\frac{\phi_0}{R_0 q_0}\right) \quad (\phi_{i-\frac{1}{2}} \leqslant \phi \leqslant \phi_{i+\frac{1}{2}}), \quad (95)$$

and $t_j = \phi_j/\phi_0$.

In place of (36) we have

$$\int_{-\phi_0}^{\phi_0} d\theta(\phi) = \frac{\pi}{2\phi_0}\int_{-\phi_0}^{\phi_0} B_i\, d\phi = \beta,$$

where β is the angle between the tangents at A and B (Fig. 7.14). Hence

$$\frac{\pi}{2\phi_0}\sum_{i=1}^{n} B_i(\phi_{i+\frac{1}{2}} - \phi_{i-\frac{1}{2}}) = \beta. \quad (96)$$

The perimeter distance s is calculated as in §7.6 except that there are no stagnation point difficulties to consider.

Table 7.14. *Velocity distributions*

ϕ/ϕ_0	$q^{(1)}/U$	$q^{(2)}/U$	$q^{(3)}/U$	$q^{(4)}/U$	$q^{(5)}/U$	ω	q/U
	(1)	(2)	(3)	(4)	(5)	(6)	(7)
0·1	1·000	1·570	1·534	1·539	1·538	2·76°	1·570
0·3	1·000	1·542	1·513	1·516	1·516	8·24°	1·547
0·5	1·000	1·482	1·466	1·468	1·468	14·04°	1·494
0·7	1·000	1·384	1·384	1·384	1·384	20·00°	1·404
0·9	1·000	1·220	1·231	1·230	1·230	26·54°	1·237
1·0	1·000	1·000	1·000	1·000	1·000	30·00°	1·000

It is necessary to assign an initial value to the ratio ϕ_0/h to commence the iterative solution of (93), and this means that the ratio of the length of AB to the jet width H at infinity will not be known exactly until the solution has been found. If a fixed value of AB/H is required then a method similar to that described in the penultimate paragraph of §7.6 must be used.

Table 7.14 (taken from Woods, 1954a) sets out the successive iterates $q^{(1)}/U$, $q^{(2)}/U$, ..., $q^{(5)}/U$ of a calculation based on (93) and (96) for the particular case of incompressible flow past a circular cylinder with $\beta = -\frac{1}{3}\pi$, the jet width such that $\pi\phi_0/2h = 1$, and $(-\phi_0, \phi_0)$ divided into ten equal intervals. The convergence is clearly quite rapid. Column 6 gives the angular coordinate ω measured from the mid-point of the arc AB. Column 7 will be explained shortly. The ratio of the depth of the jet at infinity, H, to the radius B of the circular cylinder is found to be 1·166.

An approximate analytical solution of the problem just considered can be obtained as follows. Let q under the integral sign in (92) be approximated to by $\bar{q}$, the mean value of q in $(-\phi_0, \phi_0)$, then in the ζ-plane this equation becomes (see (87))

$$\Omega(\gamma) = \frac{-2}{\pi B \bar{q}} \int_{-\infty}^{\infty} \tanh^{-1}\{\exp(\gamma^* - \gamma)\} \frac{d\phi^*}{d\gamma^*} d\gamma^*,$$

as the radius of curvature R for the cylinder is $-B$ ($R = ds/d\theta$). Using (88) to evaluate $d\phi^*/d\gamma^*$, we find we can integrate the resulting integral to obtain

$$\Omega(\gamma) = -\frac{2h}{\pi B \bar{q}} \tan^{-1}\left\{\frac{\sinh(\pi\phi_0/2h)}{\cosh\gamma}\right\}. \tag{97}$$

Let s_0 be the surface distance on the cylinder from $\phi = 0$ to $\phi = \phi_0$, then $\phi_0 \simeq \bar{q}s_0 = \bar{q}|\beta|\frac{1}{2}B$, which with (33) and (88) enables us to write (97) as

$$\Omega(\phi) = \frac{\beta}{2\sigma} \tan^{-1}\left\{\sinh^2\sigma - \cosh^2\sigma \tanh^2\left(\frac{\sigma\phi}{\phi_0}\right)\right\}^{\frac{1}{2}}, \tag{98}$$

where

$$\sigma = \frac{\pi\phi_0}{2H\beta_a U}. \tag{99}$$

Assuming $\phi/\phi_0 \simeq s/s_0$, and putting $\sigma = 1$, $\beta_a = 1$, we find from $q = e^{-\Omega}$ (98) and (99) that

$$\frac{q}{U} = \exp\left\{-\tfrac{1}{2}\beta \tan^{-1}\left(\sinh^2(1) - \cosh^2(1)\tanh^2\left(\frac{s}{s_0}\right)^{\frac{1}{2}}\right)\right\},$$

and

$$\frac{H}{B} = \frac{\pi\bar{q}}{4U}|\beta|.$$

These are the equations appropriate to the problem solved numerically earlier in this section. When $\beta = -\frac{1}{2}\pi$ these equations yield the values given in columns 6 and 7 of Table 7.14. A comparison of columns 5 and 7 shows that the maximum error of (98) for this example does not exceed 6 % of the velocity increment.

The phenomena of jet adherence to curved bodies has been investigated experimentally by Squire (1945), but there is no evidence, to the author's knowledge, which would help in the assessment of the relevance of the inviscid flow model studied above. It may well be wide of the mark.

7.15   An example of a channel design

Consider the channel shown in Fig. 7.15. To avoid boundary-layer separation, the inner wall of the bend, AB, is to be designed to be a constant pressure wall, with boundary-layer suction applied at B (see § 1.19).

The w-plane is similar to the one shown in Fig. 7.13 (j); it is only necessary to replace ϕ_0 in (89) by $\phi_0 + ih$ to obtain the required $\zeta(w)$ relation. The general solution follows from (87) and this modified $\zeta(w)$ equation.

Alternatively, the conjugate nature of the w-plane of Fig. 7.13 (j) and that implicit in Fig. 7.15 suggests that the required general solution can be obtained by replacing w by $ih-w$ in the harmonic conjugate of (91), and interchanging the subscripts 0 and h.

From Fig. 7.15 we observe that $d\Omega_h = 0$ on AB (q being constant), $d\theta_h = 0$ outside AB, and θ_0 is a step function with a jump of σ at $\phi = 0$. Hence from (91), modified as just described, we find that

$$\tau(w) = \Omega_B + i\sigma + \frac{2\sigma}{\pi}\tanh^{-1}\left\{\frac{\cosh\{\pi(\phi_0-w)/2h\}}{\cosh\{\pi(\phi_0+w)/2h\}}\right\}^{\frac{1}{2}}, \tag{100}$$

as $\theta_B = \sigma$. Now Ω_B corresponds to a velocity $q = V$, and can be calculated from (6.3).

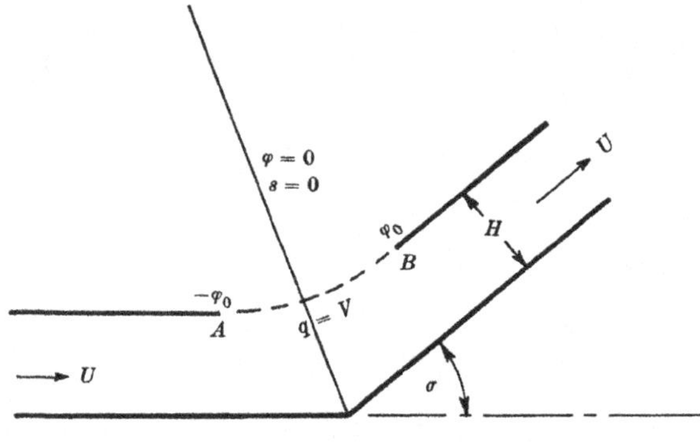

Fig. 7.15

As $\tau = 0$ at $\phi = -\infty$ it follows from (100) that

$$\tanh\frac{\pi\phi_0}{2h} = \operatorname{sech}\frac{\pi\Omega_B}{\sigma}, \tag{101}$$

which means that if H, U and σ are given, then V and the length of AB cannot be assigned independently.

The shape of the 'free' surface AB follows from (100) by putting $w = \phi + ih$, which results in

$$\theta(\phi) = \sigma + \frac{2\sigma}{\pi}\tan^{-1}\left\{\frac{\sinh\{\pi(\phi_0-\phi)/2h\}}{\sinh\{\pi(\phi_0+\phi)/2h\}}\right\}^{\frac{1}{2}},$$

or
$$\tanh(\pi\phi/2h) = -\tanh(\pi\phi_0/2h)\cos(\pi\theta/\sigma) \tag{102}$$

on the free surface $\phi = Vs$. Hence by (33) and (101), (102) yields the intrinsic equation

$$\tanh(\pi Vs/2HU\beta_a) = -\operatorname{sech}(\pi\Omega_A/\sigma)\cos(\pi\theta/\sigma) \tag{103}$$

for AB. As $s = s_0$, the value of s at $\phi = \phi_0$, when $\theta = \sigma$, it follows from (102) that the total length of AB is given by

$$2s_0 = (4HU\beta_a/\pi V)\tanh^{-1}\{\operatorname{sech}(\pi\Omega_A/\sigma)\}. \tag{104}$$

The required shape can now be deduced from (103) and (104).

7.16   The flow of a stream up a step, with boundary-layer separation

Let a stream flowing along a straight wall meet a step EC (Fig. 7.16), then the boundary layer on the wall in front of E will be separated from the wall at some point B due to the adverse pressure gradient caused by the step. In our inviscid model of this flow we shall assume that the separated streamline BC is one of constant pressure up to C, where re-attachment occurs. This is followed by immediate separation at the lower pressure applying at $\phi = \pm\infty$. In terms of velocity we have $q = V$ on AC, $q = U$ on CF_∞ and $q = U$ on the free surface, $\psi = h$, of the stream.

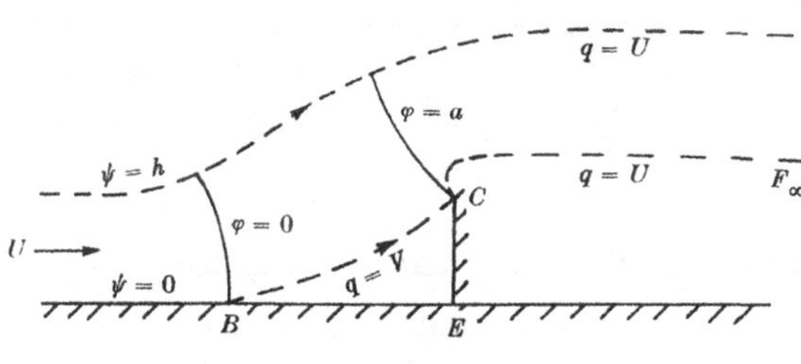

Fig. 7.16

Let $\phi = 0$ at B and $\phi = a$ at C, then the w-plane is that shown in Fig. 7.13 (c). Combining (87) and (90c) we find the general solution

$$\begin{aligned}
\tau(w) = \Omega_B + i\theta_B + \frac{2}{\pi}\int_{-\infty}^{0} &\tanh^{-1}\left\{\frac{1-\exp(-\pi w/h)}{1-\exp(-\pi\phi^*/h)}\right\}^{\frac{1}{2}} d\theta_0(\phi^*) \\
+\frac{2}{\pi}\int_{0}^{\infty} &\tan^{-1}\left\{\frac{\exp(-\pi w/h)-1}{1-\exp(-\pi\phi^*/h)}\right\}^{\frac{1}{2}} d\Omega_0(\phi^*) \\
+\frac{2}{\pi}\int_{-\infty}^{\infty} &\tan^{-1}\left\{\frac{\exp(-\pi w/h)-1}{\exp(-\pi\phi^*/h)+1}\right\}^{\frac{1}{2}} d\Omega_h(\phi^*).
\end{aligned} \tag{105}$$

In the present application $\theta_0 = \theta_B = 0$, $\Omega_h = 0$, and there is a jump in $\Omega_0(\phi^*)$ at $\phi^* = a$. Let $\Omega = \sigma$ when $q = V$, then this jump is of magnitude $-\sigma$, and $\Omega_B = \sigma$. Hence

$$\tau(w) = \sigma\left[1 - \frac{2}{\pi}\tan^{-1}\left\{\frac{\exp(-\pi w/h)-1}{1-\exp(-\pi a/h)}\right\}^{\frac{1}{2}}\right]. \tag{106}$$

If s_0 is the length of the curve BC, then $s_0 = a/V$. The lengths of BE and EC are given by the real and imaginary parts of

$$I = \int_0^a \exp(i\theta)\, ds\,(\phi) = \frac{1}{V}\int_0^a \exp(i\theta)\, d\phi,$$

respectively. On $0 \leqslant \phi \leqslant a$, (106) gives

$$\theta = \frac{2\sigma}{\pi}\tanh^{-1}\left\{\frac{1-\exp(-\pi\phi/h)}{1-\exp(-\pi a/h)}\right\}^{\frac{1}{2}},$$

therefore

$$I = \frac{1-e^{-\pi a/h}}{V\sigma}\int_0^\infty \frac{e^{i\theta}\tanh(\pi\theta/2\sigma)\, d\theta}{\cosh^2(\pi\theta/2\sigma) - (1-e^{-\pi a/h})\sinh^2(\pi\theta/2\sigma)}.$$

The imaginary part of this integral can be obtained in an algebraic form. If $\frac{1}{2}l$ is the length of EC, it is found that

$$l = \frac{2h}{V}\operatorname{cosech}\sigma\cosh\left\{\frac{2\sigma}{\pi}\tanh^{-1}[1-\exp(-\pi a/h)]^{\frac{1}{2}}\right\}.$$

Lighthill (1953) has employed this model of separating flow (in an infinite stream, $h = \infty$) in order to study the upstream influence of a step on boundary-layer separation. Boundary-layer theory is used to obtain an estimate of V/U.

7.17   Helmholtz flow of a jet impinging normally on a flat plate

If the boundary layer of the flow considered in §7.16 separates only at C, then the flow shown in Fig. 7.17 results. As EC can be regarded as being one half of a flat plate placed midway in a jet impinging normally on it, we have here a special case of the Helmholtz model of flow past bluff bodies.

The solution can be written down directly from (105); there is a single jump in θ_0 of $\frac{1}{2}\pi$ at $\phi = -a$ (see Fig. 7.17), and therefore

$$\tau(w) = \frac{i}{2}\pi + \tanh^{-1}\left\{\frac{1-\exp(-\pi w/h)}{1-\exp(\pi a/h)}\right\}^{\frac{1}{2}}. \tag{107}$$

From (26) $H = h/\beta_a U$, and it remains to calculate the length of EC.

On EC (107) gives

$$\Omega(\phi) = \tanh^{-1}[\{\exp(-\pi\phi/h)-1\}/p]^{\frac{1}{2}}, \tag{108}$$

where

$$p \equiv \exp(\pi a/h) - 1.$$

If $\frac{1}{2}l$ is the length of EC

$$l = 2\int_{-a}^0 ds(\phi) = \frac{2}{U}\int_{-a}^0 \frac{U}{q}\, d\phi = \frac{2}{U\sinh\Omega^*}\int_{-a}^0 \sinh(\Omega+\Omega^*)\, d\phi, \tag{109}$$

where Ω is given by (108). This can be integrated analytically in the case of incompressible flow ($U/q = e^{\Omega}$, see §6.1). The result is

$$l = \frac{2h}{U}\left[\frac{1}{\pi}\left(\frac{p}{p+1}\right)^{\frac{1}{2}}\ln\left\{\frac{1+\{p/(p+1)\}^{\frac{1}{2}}}{1-\{p/(p+1)\}^{\frac{1}{2}}}\right\}+1-(1+p)^{-\frac{1}{2}}\right]. \qquad (110)$$

As the pressure behind the plate is $p_a = p_{\infty}$, the drag force acting on it is

$$D = 2\int_{-a}^{0}(p-p_{\infty})\,ds\,(\phi) = \frac{2}{U}\int_{-a}^{0}(p-p_{\infty})\frac{U}{q}\,d\phi.$$

By (6.3), (6.4) and (6.8) this can be expressed

$$D = \frac{2\rho_{\infty}U}{\beta_{\infty}}\int_{-a}^{0}\sinh\Omega\,d\phi, \qquad (111)$$

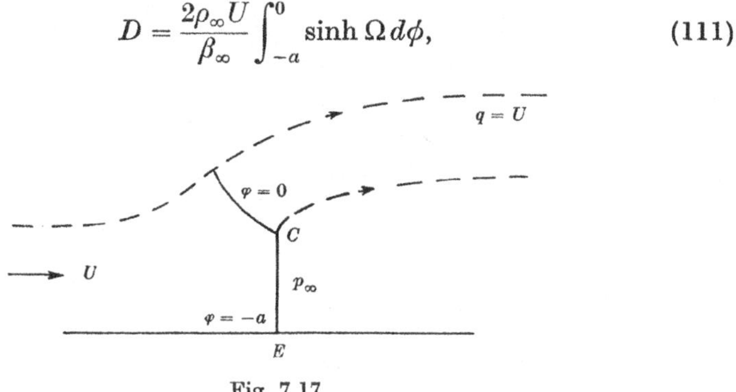

Fig. 7.17

where Ω is given by (108). This can be integrated to give

$$D = \frac{2H\rho_{\infty}}{\beta_{\infty}^{2}}\{1-\exp\left(-\pi a/2H\beta_{\infty}U\right)\}, \qquad (112)$$

on replacing h by $H\beta_{\infty}U$.

For small values of the ratio l/H in incompressible flow ($\beta_{\infty} = 1$), (110) and (112) yield

$$D = \frac{\rho_{\infty}U^{2}\pi l}{4+\pi}\left\{1-\frac{\pi l}{3(4+\pi)^{2}H}\right\}. \qquad (113)$$

7.18   A generalization of Borda's mouthpiece

Borda's mouthpiece is a long, straight, sharp-edged tube, inserted well into a large vessel containing fluid. This fluid flows out of the vessel, separated at the sharp edges of the tube to form a free jet. A generalization of the mouthpiece (in two dimensions), which has a curved surface and sloping sides, is shown in Fig. 7.18a. The interior of the vessel lies to the right, and the fluid flows out of the gap, separating at points A and B to form a free jet ABE_{∞}. The wetted walls of the mouthpiece (i.e. the walls over which the fluid flows) are curved near the exit AB, but are straight, and inclined at angles $2\pi - \epsilon$ and ϵ to the jet axis elsewhere. The mouthpiece is assumed to be symmetrical about its axis, i.e.

$$\theta_h = 2\pi - \theta_0. \tag{114}$$

The walls of the vessel are supposed to be far enough away from the exit to have a negligible effect on the mouthpiece flow.

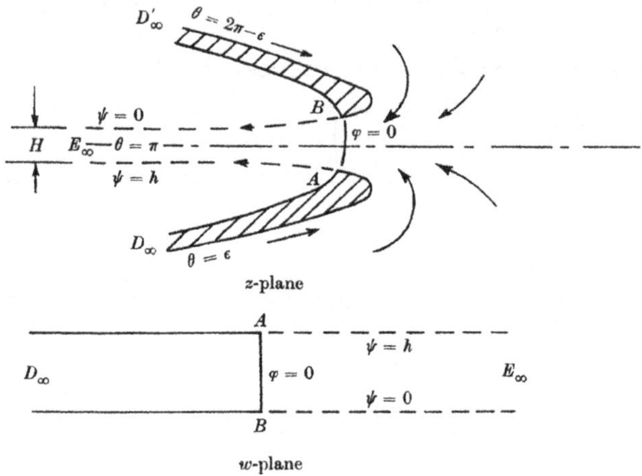

Fig. 7.18a

The w-plane is a special case of that shown in Fig. 7.13e, so the $\zeta(w)$ relation is obtained by putting $\phi_0 = 0$ in (90e). This yields

$$\exp \zeta = \left[\coth \left(-\frac{\pi w}{2h} \right) \right]^{\frac{1}{2}}. \tag{115}$$

The bounding streamlines of the jet will be at constant pressure; let this be the reference pressure p_a so that $q = U$ on these streamlines. In the notation employed in (87) we thus have $\Omega_2 = 0$, and therefore

$$\tau(\zeta) = \tau_B + \frac{2}{\pi} \int_{-\infty}^{\infty} \tanh^{-1} \{ \exp (\gamma - \zeta) \} \, d\theta_1(\gamma). \tag{116}$$

The integral in this equation can be transformed to the w-plane by means of (115), but before doing this the discontinuity in $\theta_1(\gamma)$ at $\gamma = 0$ must be removed. Referring to Fig. 7.18a we see that continuity of the streamline direction θ means that if $\theta = \epsilon$ near $\phi = -\infty$ on the lower surface of the mouthpiece, and it increases to π on the axis, it must further increase to $2\pi - \epsilon$ near $\phi = -\infty$ on the upper surface of the mouthpiece. As γ increases in the direction $AD_\infty D'_\infty B$, we conclude that θ_1 has a jump of $-(2\pi-2\epsilon)$ at $\gamma=0$—by (115) the point corresponding to $\phi=-\infty$. By (116) this jump contributes a term

$$-\frac{2}{\pi}(2\pi-2\epsilon)\tanh^{-1}\{\exp(-\zeta)\} = -\frac{4}{\pi}(\pi-\epsilon)\tanh^{-1}\left\{\tanh\left(\frac{-\pi w}{2h}\right)\right\}^{\frac{1}{2}}$$

$$= -\frac{4i}{\pi}(\pi-\epsilon)\tan^{-1}\left\{\tanh\left(\frac{\pi w}{2h}\right)\right\}^{\frac{1}{2}}$$

to $\tau(w)$. Having separated this term from the integral in (116) we can transform the rest of the integral in the usual way.† Making use of the fact that $\Omega_B=0$ ($q=U$ at B), and (114) we arrive at

$$\tau(w) = i\theta_B - \frac{4i}{\pi}(\pi-\epsilon)\tan^{-1}\left\{\tanh\frac{\pi w}{2h}\right\}^{\frac{1}{2}}$$
$$-\frac{2}{\pi}\int_{-\infty}^{0}\left[\tanh^{-1}\left\{\frac{\tanh(\pi w/2h)}{\coth(\pi\phi^*/2h)}\right\}^{\frac{1}{2}}\right.$$
$$\left.+\tanh^{-1}\left\{\frac{\tanh(\pi w/2h)}{\tanh(\pi\phi^*/2h)}\right\}^{\frac{1}{2}}\right]d\theta_h(\phi^*). \quad (117)$$

The theory for a curved mouthpiece is now developed from (117) by a numerical method similar to that already fully illustrated in §§ 7.7 and 7.14.

We shall restrict our attention to the straight-sided, sharp-edged mouthpiece, for which the last term of (117) is zero, and $\theta_B=2\pi-\epsilon$ (cf. Fig. 7.18a). In this case

$$\tau(w) = \pi i + i(\pi-\epsilon)\left\{1-\frac{4}{\pi}\tan^{-1}\left(\tanh\frac{\pi w}{2h}\right)^{\frac{1}{2}}\right\}. \quad (118)$$

On the free streamline AE (see Fig. 7.18a) $\psi=h$ and $\phi=Us$, s being distance measured from A, and therefore (118) yields the intrinsic equation

$$\theta = 2\pi-\epsilon-\frac{4}{\pi}(\pi-\epsilon)\tan^{-1}\left\{\coth\frac{\pi Us}{2h}\right\}^{\frac{1}{2}}. \quad (119)$$

Let the width of the jet at E_∞ be H, then by (33) $h=H\beta_a U$ ($\beta_a=\beta_{-\infty}$ in the present case). Thus inverting (119)

$$s = \frac{2H\beta_a}{\pi}\coth^{-1}\left\{\frac{1+\sin\left(\dfrac{\pi}{2}\dfrac{\pi-\theta}{\pi-\epsilon}\right)}{1-\sin\left(\dfrac{\pi}{2}\dfrac{\pi-\theta}{\pi-\epsilon}\right)}\right\}$$

---

† The point here is that as $\int_{-\infty} f(\phi)\,dg(\phi)$ is defined to be $\lim_{R\to\infty}\int_{-R} f(\phi)\,dg(\phi)$, a discontinuity in g at the point, $\phi=-\infty$ (i.e. at $\gamma=0$) contributes nothing to the integral in this form; it is for this reason that the discontinuity must first be removed.

so

$$ds = \frac{H\beta_a}{2(\pi-\epsilon)}\cot\left\{\frac{\pi}{2}\left(\frac{\pi-\theta}{\pi-\epsilon}\right)\right\}d\theta.$$

Let the origin of the z-plane be at A where $\theta = \epsilon$, then as

$$dx = \cos\theta\,(ds/d\theta)\,d\theta, \quad dy = \sin\theta\,(ds/d\theta)\,d\theta,$$

the coordinates of the free streamline are given in parametric form by

$$\left.\begin{aligned}
x(\theta) &= \frac{H\beta_a}{2(\pi-\epsilon)}\int_\epsilon^\theta \cos\theta\cot\left\{\frac{\pi}{2}\left(\frac{\pi-\theta}{\pi-\epsilon}\right)\right\}d\theta,\\
y(\theta) &= \frac{H\beta_a}{2(\pi-\epsilon)}\int_\epsilon^\theta \sin\theta\cot\left\{\frac{\pi}{2}\left(\frac{\pi-\theta}{\pi-\epsilon}\right)\right\}d\theta.
\end{aligned}\right\} \tag{120}$$

These integrals are easily evaluated for certain values of ϵ/π, such as 0, $\frac{1}{2}$ and 1; in other cases numerical methods are necessary.

The contraction ratio of the jet, $C_C = H/A$ where A is the width AB of the mouthpiece, is of some interest. On the free streamline $\psi = h$, $y = \frac{1}{2}(A-H)$ and $\theta = \pi$ at $\phi = \infty$. Hence from the second of (120)

$$\tfrac{1}{2}(A-H) = \frac{H\beta_a}{2(\pi-\epsilon)}\int_\epsilon^\pi \sin\theta\cot\left\{\frac{2}{\pi}\left(\frac{\pi-\theta}{\pi-\epsilon}\right)\right\}d\theta. \tag{121}$$

Let

$$\begin{aligned}
I &\equiv \frac{1}{\pi-\epsilon}\int_\epsilon^\pi \sin\theta\cot\left\{\frac{\pi}{2}\left(\frac{\pi-\theta}{\pi-\epsilon}\right)\right\}d\theta\\
&= \frac{1}{\pi}\int_0^\pi \sin\left\{p\left(1-\frac{\epsilon}{\pi}\right)\right\}\cot\tfrac{1}{2}p\,dp\\
&= -\frac{i}{2\pi}\int_{-\pi}^\pi e^{ip(1-(\epsilon/\pi))}\cot\tfrac{1}{2}p\,dp,
\end{aligned}$$

by the properties of odd and even integrands. Put $z = e^{ip}$ then

$$2\pi iI = \int_C z^{-\epsilon/\pi}\frac{z+1}{z-1}\,dz,$$

where C is unit circle. The integrand here is analytic within the indented contour shown in Fig. 7.18b. Hence by Cauchy's theorem, and the extension of the residue theorem given in §3.11,

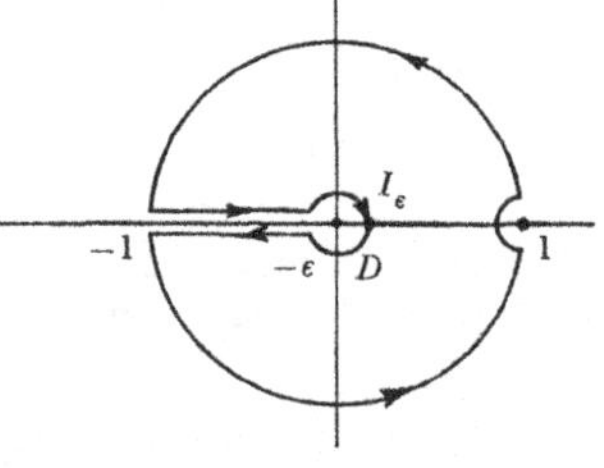

Fig. 7.18b

$$2\pi iI + \int_{-1}^{-\epsilon} z^{-\epsilon/\pi}\frac{z+1}{z-1}\,dz + \int_{-\epsilon}^{-1} z^{-\epsilon/\pi}\frac{z+1}{z-1}\,dz + I_\epsilon - \pi iR = 0,$$

where I_ϵ is the contribution from the small circle of radius ϵ about the origin, and R is the residue at $z = 1$, namely 2. It is easily verified that

$\lim_{\epsilon \to 0} I_\epsilon = 0$. We put $z = e^{\pi i} t$, $e^{-\pi i} t$ on the upper and lower surface of the negative real axis respectively, and find

$$2\pi i I + (e^{-i\epsilon} - e^{i\epsilon}) \int_0^1 t^{-\epsilon/\pi} \frac{t-1}{t+1} \, dt - 2\pi i = 0,$$

or
$$I = 1 - \frac{\sin \epsilon}{\pi} \left\{ \int_0^1 \frac{t^{-\epsilon/\pi}}{t+1} \, dt - \int_0^1 \frac{t^{1-\epsilon/\pi}}{t+1} \, dt \right\}$$

$$= 1 - \frac{\sin \epsilon}{2\pi} \left\{ 2\Psi \left(1 - \frac{\epsilon}{2\pi} \right) - \Psi \left(\frac{3}{2} - \frac{\epsilon}{2\pi} \right) - \Psi \left(\frac{1}{2} - \frac{\epsilon}{2\pi} \right) \right\},$$

where $\Psi(x)$ is the logarithmic derivative of the gamma function, $\Psi(x) \equiv \frac{d}{dx} \{ \ln \Gamma(x) \}$, and we have used the formula

$$\int_0^1 \frac{t^{x-1}}{t+1} \, dt = \tfrac{1}{2}\Psi \left(\frac{x+1}{2} \right) - \tfrac{1}{2}\Psi \left(\frac{x}{2} \right).$$

Now $\Gamma(x) = (x-1) \Gamma(x-1)$, and therefore $\Psi(x) = (x-1)^{-1} + \Psi(x-1)$. Hence

$$I = 1 + \frac{\sin \epsilon}{\pi - \epsilon} - \frac{\sin \epsilon}{\pi} \left\{ \Psi \left(1 - \frac{\epsilon}{2\pi} \right) - \Psi \left(\frac{1}{2} - \frac{\epsilon}{2\pi} \right) \right\}.$$

Equation (121) can now be written in the form

$$\frac{A}{H} \equiv \frac{1}{C_C} = 1 + \beta_a + \frac{\beta_a \sin \epsilon}{\pi - \epsilon} - \frac{\beta_a}{\pi} \sin \epsilon \left\{ \Psi \left(1 - \frac{\epsilon}{2\pi} \right) - \Psi \left(\tfrac{1}{2} - \frac{\epsilon}{2\pi} \right) \right\}, \quad (122)$$

and C_C is determined.

Particular cases of (122) of some interest are

(i) $\epsilon = 0$ (the classical Borda mouthpiece), $C_C = 1/(1 + \beta_a)$ and

(ii) $\epsilon = \tfrac{1}{2}\pi$ (efflux from a slit), $C_C = \pi/(\pi + 2\beta_a)$.

7.19 Flow from a nozzle

Fig. 7.19a shows fluid flowing from a symmetrical nozzle to form a free jet, the boundaries of which are at constant pressure p_∞. The w- and ζ-planes shown are for the half of the flow above the axis of symmetry. On this axis, say $\psi = 0$, $\theta_0 = 0$, while on the other boundary of the w-plane, $\psi = h$, $\theta_h(\phi)$ is prescribed in $-\infty \leqslant \phi \leqslant 0$, and $\Omega_h(\phi) = \Omega_\infty$ in $0 \leqslant \phi \leqslant \infty$.

The w-plane is of the type shown in Fig. 7.13 (f), and so by 90 (f) it is related to the ζ-plane by
$$\exp \zeta = \{ e^{\pi w/h} + 1 \}^{\frac{1}{2}}. \quad (123)$$

At $w = \phi^* + ih$ (123) yields
$$\exp \gamma = | 1 - e^{\pi \phi^*/h} |^{\frac{1}{2}}. \quad (124)$$

Transforming (87) into the w-plane with the aid of (123) and (124) we get

$$\tau(w) = \Omega_\infty - \frac{2}{\pi} \int_{-\infty}^{0} \tanh^{-1} \left\{ \frac{1 - e^{\pi\phi^*/h}}{e^{\pi w/h} + 1} \right\}^{\frac{1}{2}} d\theta_h(\phi^*)$$

$$- \frac{2}{\pi} \int_{0}^{\infty} \tan^{-1} \left\{ \frac{e^{\pi\phi^*/h} - 1}{e^{\pi w/h} + 1} \right\}^{\frac{1}{2}} d\Omega_h(\phi^*), \tag{125}$$

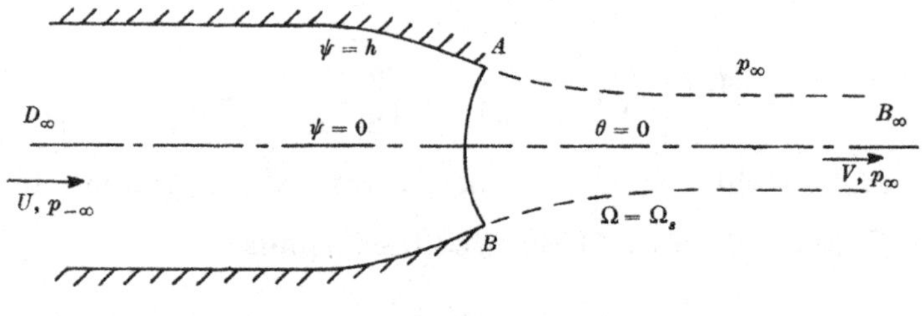

Fig. 7.19a

as $\theta_0(\phi) = 0$, and $\tau(\zeta = \infty) = \tau(w = \infty) = \Omega_\infty$. In the case under consideration

$$d\Omega_h(\phi^*) = d\Omega_\infty = 0, \tag{126}$$

and so

$$\tau(w) = \Omega_\infty - \frac{2}{\pi} \int_{-\infty}^{0} \tanh^{-1} \left\{ \frac{1 - e^{\pi\phi^*/h}}{e^{\pi w/h} + 1} \right\}^{\frac{1}{2}} d\theta_h(\phi^*). \tag{127}$$

If $q = U$ at $\phi = -\infty$, $\Omega_{-\infty} = 0$, which from (127) requires that $\theta_h(\phi^*)$ satisfies

$$\Omega_\infty = \frac{2}{\pi} \int_{-\infty}^{0} \tanh^{-1} \left\{ 1 - e^{\pi\phi^*/h} \right\}^{\frac{1}{2}} d\theta_h(\phi^*). \tag{128}$$

This means that if p_∞ and $p_{-\infty}$ are given, the curved surface of the nozzle cannot be chosen completely arbitrarily. For example, in the simple case shown in Fig. 7.19b it follows from (128) that for given values of p_∞, $p_{-\infty}$ and θ_1, a must equal $(2h/\pi)\ln\{\cosh(\pi\Omega_\infty/2\theta_1)\}$.

At first sight this limitation is rather surprising, and it raises the question 'suppose p_∞, $p_{-\infty}$ (i.e. Ω_∞) and the shape of the nozzle *are* prescribed arbitrarily, does there exist a corresponding inviscid flow?' If we adhere to (127) the answer must be 'no'. However, if the possibility of a sudden jump in the value of $\Omega_h(\phi^*)$ at $\phi^* = +0$ is admitted, an inviscid flow can be found as follows.

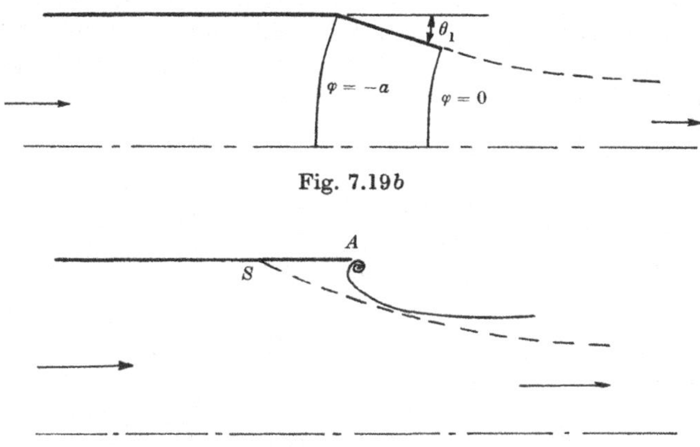

Fig. 7.19b

Fig. 7.19c

Let $d\Omega_h(\phi^*) = 0$ except at $\phi^* = \epsilon$ where Ω_h has a jump of σ, then from the last term of (125) this jump contributes

$$-(2\sigma/\pi)\tan^{-1}\{(e^{\pi\epsilon/h}-1)/(e^{\pi w/h}+1)\}^{\frac{1}{2}} \quad \text{to} \quad \tau(w).$$

Now let $\epsilon \to 0$ and $\sigma \to \infty$ such that

$$\frac{2\sigma}{\pi}\sqrt{\frac{\pi\epsilon}{h}} \to a, \tag{129}$$

where a is non-zero and finite, then the contribution due to the jump becomes $-a(e^{\pi w/h}+1)^{-\frac{1}{2}}$. Thus in place of (127) and (128) we have

$$\tau(w) = \Omega_\infty - \frac{a}{(e^{\pi w/h}+1)^{\frac{1}{2}}} - \frac{2}{\pi}\int_{-\infty}^{0}\tanh^{-1}\left\{\frac{1-e^{\pi\phi^*/h}}{e^{\pi w/h}+1}\right\}^{\frac{1}{2}} d\theta_h(\phi^*), \tag{130}$$

and

$$a = \Omega_\infty - \frac{2}{\pi}\int_{-\infty}^{0}\tanh^{-1}\{1-e^{\pi\phi^*/h}\}^{\frac{1}{2}} d\theta_h(\phi^*). \tag{131}$$

To examine the significance of this solution consider the simplest case, i.e. $d\theta_h = 0$ (see Fig. 7.19c). Equations (130) and (131) give

$$\tau(w) = \Omega_\infty\{1-(e^{\pi w/h}+1)^{-\frac{1}{2}}\}.$$

On $\psi = h$, $\phi > 0$, this gives $\Omega = \Omega_\infty$, $\theta = \Omega_\infty \, (e^{\pi\phi/h} - 1)^{-\frac{1}{2}}$, and therefore $\theta \simeq \Omega_\infty/\sqrt{(\pi\phi/h)}$ near the separation point $\phi = 0$. Similarly we find that just before separation ($\phi < 0$), $\Omega \simeq -\Omega_\infty/\sqrt{(-\pi\phi/h)}$, and the separation point is either a stagnation point (if $\Omega_\infty < 0$, i.e. if $p_\infty < p_{-\infty}$), or a point of infinite velocity.

In the practical case $p_\infty < p_{-\infty}$, and so according to our inviscid flow theory the separation point A (see Fig. 7.19c) will be a stagnation point, unless the nozzle is properly designed. A real fluid will therefore tend to become separated from the nozzle wall at a point S upstream of A owing to the adverse pressure gradient. Unsteadiness and turbulence would accompany this separation, and it is for this reason that nozzles need to be properly designed.

SYMMETRICAL JET FLOWS

7.20   A symmetrical cylinder in a jet

In the last six sections we have given examples of flows which have w-planes of the types (b), (c), (e) and (f) (Fig. 7.13), but we have yet to consider the simplest type of w-plane, namely type (d) for which the transformation is $\zeta = \pi w/2h$. The most important problem involving this type of w-plane is the determination of the flow of a jet about a symmetrical cylinder placed on the jet axis (figure 7.20). Let $\psi = h$ be the upper boundary of the jet, and $\psi = 0$ be the jet axis and cylinder boundary, and let $q = U$ on $\psi = h$. Then the boundary conditions are (i) $\Omega_h = 0$, (ii) $\theta_0 = 0$ outside $-2a \leqslant \phi \leqslant 2a$, and (iii) $\theta_0(\phi)$ given on the cylinder, which we assume to lie in $-2a \leqslant \phi \leqslant 2a$. Transforming (86) and (87) to the w-plane and using these boundary conditions we get

$$\tau(w) = \frac{1}{2h} \int_{-2a}^{2a} \theta_0(\phi^*) \operatorname{cosech}\left\{\frac{\pi}{2h}(\phi^* - w)\right\} d\phi^* \tag{132}$$

and
$$\tau(w) = \frac{2}{\pi} \int_{-2a}^{2a} \tanh^{-1}\left\{\exp\frac{\pi}{2h}(\phi^* - w)\right\} d\theta_0(\phi^*). \tag{133}$$

As
$$\tanh^{-1}\{\exp(-x)\} = \tfrac{1}{2}\ln\frac{1+e^{-x}}{1-e^{-x}} = -\tfrac{1}{2}\ln\tanh\tfrac{1}{2}x,$$

there is the alternative form to (133), viz.

$$\tau(w) = -\frac{1}{\pi} \int_{-2a}^{2a} \ln\left\{\tanh\frac{\pi}{4h}(w - \phi^*)\right\} d\theta_0(\phi^*). \tag{134}$$

On the cylinder this gives

$$\Omega_0(\phi^*) = -\frac{1}{\pi} \int_{-2a}^{2a} \ln\left|\tanh\frac{\pi}{4h}(\phi^* - \phi)\right| d\theta_0(\phi^*), \tag{135}$$

which should be compared with (35) for a cylinder in a straight channel. Equation (36), viz.

$$0 = \int_{-2a}^{2a} d\theta_0(\phi),\tag{136}$$

also applies in the present case, but not (34). Instead of (34) we have the somewhat more complicated equation

$$\int_{-2a}^{2a} dy\,(\phi_0) = \int_{-2a}^{2a} \frac{dy}{ds}\frac{ds}{d\phi}\,d\phi = \int_{-2a}^{2a} \frac{\sin\theta_0}{q_0}\,d\phi = 0.\tag{137}$$

In linear perturbation theory θ_0 is small, $q_0 \simeq U$, and the closure condition (137) reduces to (34).

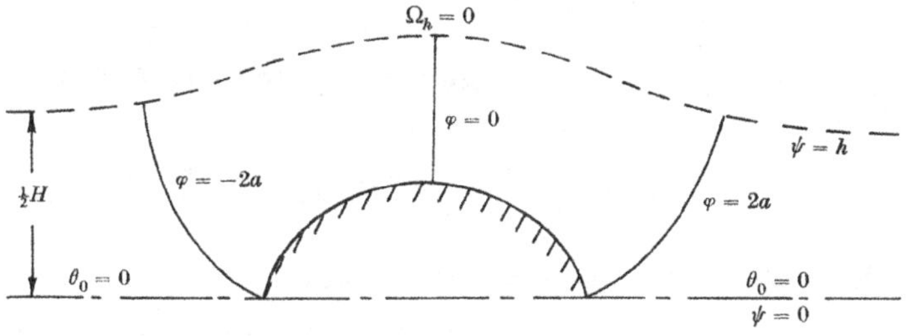

Fig. 7.20

Equations (135)–(137) can now be solved by a method exactly parallel to that used for (34)–(36). The only difference is that the function $F(x)$ in (43) is replaced by

$$F(x) = 2\int_0^x \ln|\tanh\tfrac{1}{2}x|\,dx,$$

and $\sinh\{\pi(2a \pm \phi_j)/2h\}$ in (43) is replaced by $\tanh\{\pi(2a \pm \phi_j)/4h\}$. (For a tabulation of $F(x)$ see Woods, 1955b.)

By putting $w = ih + \phi$ in (133) we find that the slope of the free surface is

$$\theta_h(\phi) = -\frac{2}{\pi}\int_{-2a}^{2a} \tan^{-1}\left\{\exp\frac{\pi}{2h}(\phi^* - \phi)\right\}d\theta_0(\phi^*).\tag{138}$$

As the Gudermannian function $gd(x)$ is

$$gd(x) \equiv 2\tan^{-1}\tanh\tfrac{1}{2}x = 2\tan^{-1}e^x - \tfrac{1}{2}\pi = \sinh^{-1}(\tanh x),\tag{139}$$

(138) has the alternative form

$$\theta_h(\phi) = -\frac{1}{\pi}\int_{-2a}^{2a} gd\left\{\frac{\pi}{2h}(\phi^* - \phi)\right\}d\theta_0(\phi^*),\tag{140}$$

on taking advantage of (136). Once (135) has been solved by iteration, and $\theta_0(\phi^*)$ found, $\theta_h(\phi)$ follows directly from (140). The coordinates of the constant pressure surface then follow from

$$Ux = \int^\phi \cos\theta_h \, d\phi, \quad Uy = \int^\phi \sin\theta_h \, d\phi. \tag{141}$$

If either h is large or the cylinder is slender, θ_h will be small, and the second of (141) can be replaced by

$$y = \tfrac{1}{2}H + \frac{1}{U}\int_{-\infty}^{\phi} \theta_h \, d\phi,$$

where y is measured from the jet axis, and H is the total width of the jet. Integrating (140) by parts $\{(d/dx)(g\,dx) = \operatorname{sech} x\}$, and substituting the result in the equation for y we get

$$y = \tfrac{1}{2}H + \frac{1}{2hU}\int_{-\infty}^{\phi}\left\{\int_{-2a}^{2a} \theta_0(\phi^*)\operatorname{sech}\frac{\pi}{2h}(\phi^*-\phi)\,d\phi^*\right\}d\phi,$$

i.e.
$$y = \tfrac{1}{2}H + \frac{1}{\pi U}\int_{-2a}^{2a} \theta_0(\phi^*)\left\{\frac{\pi}{2} + gd\,\frac{\pi}{2h}(\phi-\phi^*)\right\}d\phi^*. \tag{142}$$

7.21 Blockage in an open 'wind-tunnel'

As free jets are sometimes used in place of closed tunnels for testing aerofoils, it is important to know the blockage (see §7.8) in a free jet. We shall use the notation of §7.8.

In linear perturbation theory (132) becomes

$$\beta_a\left(1-\frac{q}{U}\right) + i\theta = \frac{1}{\beta_a H}\int_{-\frac{1}{2}c}^{\frac{1}{2}c}\left(\frac{dy}{dx}\right)_0 \operatorname{cosech}\frac{\pi}{\beta_a H}(x-Z)\,dx, \tag{143}$$

which corresponds to (55) and (60) for the closed tunnel and infinite stream respectively. Subtracting (143) from (60), and eliminating $q-q^*$ by (52), we find that at $Z = x^*$, a point on the aerofoil surface, the blockage factor is given by

$$\epsilon(x^*) = \frac{1}{\beta_a^2 H}\int_{-\frac{1}{2}c}^{\frac{1}{2}c}\left(\frac{dy}{dx}\right)_0\left\{\frac{1}{(\pi/\beta_a H)(x-x^*)} - \operatorname{cosech}\frac{\pi}{\beta_a H}(x-x^*)\right\}dx.$$

From this and the expansion $\alpha^{-1} - \operatorname{cosech}\alpha = \tfrac{1}{6}\alpha\{1 + O(\alpha^2)\}$ we find corresponding to (62) the approximate formula

$$\epsilon(x) \simeq -\frac{\pi A}{12\beta_a^3 H^2} + \frac{\pi c C_D}{24\beta_a^3 H^2}(\tfrac{1}{2}c - x). \tag{144}$$

Comparison of (63) and (144) shows that the average blockage factors ϵ_T, ϵ_J for the tunnel and jet are related by

$$\epsilon_T = -2\epsilon_J + \frac{cC_D}{4\beta_a^2 H}. \tag{145}$$

The theory for the correction to the drag coefficient is the same as in § 7.9—we only need to change ϵ_T to ϵ_J. The result of this is

$$C_D^* = C_D\left\{1 + \frac{(3 - M_a^2)\pi A}{12\beta_a^3 H^2} - \frac{\pi(2 - M_a^2)}{48\beta_a^3}\left(\frac{C}{H}\right)^2 C_D\right\},\qquad(146)$$

which corresponds to (70).

7.22　The setting of 'streamline' walls

Equation (142) has an important application to the theory of flexible-wall wind tunnels. In such tunnels the wall shape can be changed as required, and the principle adopted is to set them, as nearly as possible, to lie along the streamlines corresponding to an infinitely deep stream. This eliminates automatically all wind-tunnel corrections, and, more important, largely avoids that defect of straight-walled tunnels at transonic Mach numbers known as 'choking' (see Howarth, 1953, pp. 534–5).

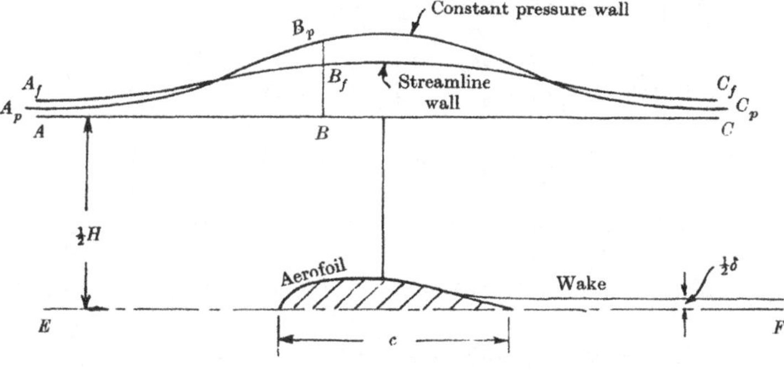

Fig. 7.22

The method adopted of setting these walls is illustrated in Fig. 7.22. In this figure ABC is the constant pressure wall in the absence of the aerofoil, $A_p B_p C_p$ is the constant pressure wall in the presence of the aerofoil (assumed to be at zero incidence on the tunnel centre-line), while $A_f B_f C_f$ is the corresponding free streamline about the aerofoil. ABC will be almost parallel to the tunnel centre-line, diverging very slowly to allow for the slow thickening of the tunnel-wall boundary layer. These curves have a common point at an infinite distance (in practice several aerofoil chord lengths) upstream, but downstream the existence of the aerofoil's wake causes a displacement of the constant pressure and free streamline walls from the wall position of the empty tunnel.

The constant pressure wall is obtained experimentally, so the displacement BB_p at any point will be known. The required streamline wall

$A_f B_f C_f$ can therefore be determined from $BB_f = \sigma_1 BB_p + \sigma_2$ if σ_1 and σ_2 are known. We shall now calculate the theoretical values of σ_1 and σ_2. The method is due to Woods (1955b).

By (54) the linear perturbation form of (142) is

$$y_p = \tfrac{1}{2}H + \frac{1}{\pi}\int_{-\frac{1}{2}c}^{\frac{1}{2}c} \left(\frac{dy}{dx^*}\right)_0 \left\{\tfrac{1}{2}\pi + gd\,\frac{\pi}{H\beta_a}(x - x^*)\right\} dx^*, \qquad (147)$$

where $y_p - \tfrac{1}{2}H$ is the displacement BB_p.

The displacement $BB_f\;(= y_f - \tfrac{1}{2}H)$ can be deduced from the approximations

$$y_f = \tfrac{1}{2}H + \int_{-\infty}^{s} \sin\theta_f\,ds \simeq \tfrac{1}{2}H + \int_{-\infty}^{x} \theta_f\,dx,$$

and

$$\theta_f \simeq \frac{1}{\pi}\int_{-\frac{1}{2}c}^{\frac{1}{2}c}\left(\frac{dy}{dx^*}\right)_0 \mathscr{I}\left(\frac{1}{x^* - x - \tfrac{1}{2}i\beta_a H}\right) dx^*,$$

which are both consistent with the derivation of (147). The equation for θ_f is obtained by putting $Z = x + \tfrac{1}{2}i\beta_a H$ in (60). Thus

$$y_f = \tfrac{1}{2}H + \frac{1}{\pi}\int_{-\frac{1}{2}c}^{\frac{1}{2}c}\left(\frac{dy}{dx^*}\right)_0 \left\{\int_{-\infty}^{x}\frac{\tfrac{1}{2}\beta_a H\,dx}{(x-x^*)^2 + \tfrac{1}{4}\beta_a^2 H^2}\right\} dx^*,$$

i.e.

$$y_f = \tfrac{1}{2}H + \frac{1}{\pi}\int_{-\frac{1}{2}c}^{\frac{1}{2}c}\left(\frac{dy}{dx^*}\right)_0 \left\{\tfrac{1}{2}\pi + \tan^{-1}\frac{2(x-x^*)}{\beta_a H}\right\} dx^*. \qquad (148)$$

Now on the aerofoil surface $y_0 = 0$ at $x = -\tfrac{1}{2}c$ (the leading edge), and $y_0 = \tfrac{1}{2}\delta = \tfrac{1}{4}cC_D$ (see (57)) at $x = \tfrac{1}{2}c$ (the trailing edge). For large z

$$gd(z) = \tfrac{1}{2}\pi - \operatorname{sech} z - \tfrac{1}{6}\operatorname{sech}^3 z + \ldots, \quad \tan^{-1} z = \tfrac{1}{2}\pi - \frac{1}{z} + \frac{1}{6z^3} - \ldots,$$

and it therefore follows from (147) and (148) that well downstream near $x = \infty$ both y_f and y_p tend to $\tfrac{1}{2}H + \tfrac{1}{4}cC_D$, i.e. BB_f and BB_p tend to $\tfrac{1}{4}cC_D$. Hence $\lim_{x\to\infty}\sigma_1 = 1$, $\lim_{x\to\infty}\sigma_2 = 0$. Similarly, well upstream σ_1 tends to infinity like $(\cosh x)/x$.

The values of σ_1 and σ_2 are next calculated in the neighbourhood of the aerofoil. From (147) and (148)

$$(y_f - \tfrac{1}{2}H) = \frac{2}{\pi}(y_p - \tfrac{1}{2}H) + \frac{\pi - 2}{4\pi}cC_D$$

$$+ \frac{1}{\pi}\int_{-\frac{1}{2}c}^{\frac{1}{2}c}\left\{\tan^{-1}\frac{2(x-x^*)}{\beta_a H} - gd\,\frac{\pi}{\beta_a H}(x - x^*)\right\} dy_0(x^*). \qquad (149)$$

For small z,

$$gd(z) = \tfrac{1}{2}z(1 - \tfrac{1}{6}z^2 + \ldots), \quad \tan^{-1} z = z(1 - \tfrac{1}{3}z^2 + \ldots),$$

so if x/H is small (149) gives

$$BB_f = \frac{2}{\pi} BB_p + \frac{\pi - 2}{4\pi} cC_D + O\left(\frac{c}{H}\right)^3.$$

Hence
$$\sigma_1 = \frac{2}{\pi}, \quad \sigma_2 = \frac{\pi - 2}{4\pi} cC_D, \tag{150}$$

which can be used in the experimental setting of streamline walls.

It is of some interest to determine the value of x at which $BB_f = BB_p$. From (147) and (148) this occurs when

$$\int_{-\frac{1}{2}c}^{\frac{1}{2}c} F\left\{\frac{2(x - x^*)}{\beta_a H}\right\} dy_0(x^*) = 0, \tag{151}$$

where $F(y) = gd(\frac{1}{2}\pi y) - \tan^{-1} y$. This equation can be solved numerically for a given aerofoil if C_D is known. A good estimate for x can be found as follows.

For the rectangle $y_0 = \frac{1}{2}t$, $-\frac{1}{2}c \leqslant x \leqslant \frac{1}{2}c$, (151) gives

$$F\left(\frac{2x + c}{\beta_a H}\right) = \left(1 - \frac{cC_D}{4t}\right) F\left(\frac{2x - c}{\beta_a H}\right), \tag{152}$$

which is easily solved graphically. When $cC_D/4t$ can be ignored, corresponding values of $c/(\beta_a H)$ and $x/(\beta_a H)$ satisfying (152) are:

$c/(\beta_a H)$	0	0·1	0·2	0·3	0·4	0·5	0·6
$x/(\beta_a H)$	$\pm 0·71$	$\pm 0·72$	$\pm 0·72$	$\pm 0·73$	$\pm 0·74$	$\pm 0·76$	$\pm 0·77$

This shows that $(x/\beta_a H)$ is almost independent of $(c/\beta_a H)$, and as an aerofoil can be constructed from a number of rectangular strips of varying chord, it is reasonable to apply this table to any aerofoil shape. In fact even for the extreme case of the circular cylinder discussed in the next section, for which $(c/\beta_a H) = 0·528$, the table yields $x/(\beta_a H) = 0·76$, whereas the true value of this ratio (see Fig. 7.23) is $0·766$.

When C_D is not zero the values of $(x/\beta_a H)$ are reduced downstream and increased upstream, but this effect appears to be small for typical values of the ratios $(c/\beta_a H)$ and $cC_D/(4t)$. Thus it would probably be a satisfactory rule in experimental work to take σ_1 as unity outside the range $-\frac{3}{4}\beta_a H \leqslant x \leqslant \frac{3}{4}\beta_a H$, say, as the adjusting of the constant pressure wall to the 'streamline' position has to stop somewhere upstream and downstream, and the points where $\sigma_1 = 1$ are most convenient for this purpose.

7.23   An example: the circular cylinder

For the reasons given in §7.7 the theory of §§7.20–7.22 will be illustrated by considering the case of a circular cylinder in a jet. The integral equation to be solved is (135), and the numerical method of doing this is outlined in the paragraph following (137). As the numerical details are very similar to those presented in §7.7 we shall omit them, and simply give the final solution.

The fluid was assumed to be incompressible; and corresponding to (48) it was assumed that

$$\frac{a}{h} = \frac{1}{2},\tag{153}$$

where $h = \frac{1}{2}HU$. In an infinite stream it is shown in the fourth paragraph of §7.7 that $a = BU$, where B is the radius of the cylinder, so (153) ought to yield a ratio of H to B of about 4.

The initial velocity distribution was taken to be that for an infinite stream, i.e. that given in (49). The final values of q/U, obtained after five iterations, are given in column 2 of Table 7.23a (the notation is the same as in §7.7). Column 3 gives the angular displacement α (defined in Fig. 7.23), while column 4 gives the corresponding value of q/U for a free stream, q_f/U. The solid blockage factor, $\epsilon = (q - q_f)/U$ is given in column 5. The value of H/B was found by the method of §7.7 to be 3.79.

Table 7.23a. *Blockage factor*

$\dfrac{\phi_f}{a}$	$\dfrac{q^{(\infty)}}{U}$	α (degrees)	$\dfrac{q_f}{U}$	ϵ
(1)	(2)	(3)	(4)	(5)
$\frac{1}{8}$	1·851	3·58	1·996	−0·145
$\frac{3}{8}$	1·824	11·11	1·963	−0·139
$\frac{5}{8}$	1·768	18·51	1·897	−0·129
$\frac{7}{8}$	1·679	26·52	1·790	−0·111
$1\frac{1}{8}$	1·550	34·76	1·643	−0·093
$1\frac{3}{8}$	1·367	44·20	1·434	−0·077
$1\frac{5}{8}$	1·102	55·02	1·147	−0·045
$1\frac{7}{8}$	0·653	70·85	0·656	−0·003

Averaging column 5 we find $\bar{\epsilon} = -0.093$. It is interesting to compare this with the value given by the first term of (144). For the present application

$$\bar{\epsilon} = -\pi A/(12H^2) = -(\pi B)^2/(12H^2) \simeq 0.057,$$

but as (144) is based on linear perturbation theory the error here is not unexpected.

Table 7.23b gives in columns 2 and 3 the coordinates of the free boundary $\psi = h$ of the jet. These were computed by using (140)–(142). The function $\theta_0(\phi)$ to be used in these equations follows from $s_0 = \int d\phi/q^{(\infty)}$ ($q^{(\infty)}$ is given in column 2 of Table 7.23a) and the known $\theta_0(s_0)$ relation. Column 4 of Table 7.23b gives the y-coordinate of the streamline $\psi = \frac{1}{2}HU$ in an infinite stream. This was obtained as follows. In an infinite stream (47) gives

$$\phi + i\psi = U\left\{x + iy + \frac{B^2(x - iy)}{x^2 + y^2}\right\},$$

and so on $\psi = \frac{1}{2}HU$

$$\frac{H}{y_f} = 2\left\{1 - \frac{(B/H)^2}{(x/H)^2 + (y_f/H)^2}\right\},$$

which enables column 4 to be calculated from column 3.

Table 7.23b. *Streamline-wall factor*

$\dfrac{\phi_j}{a}$	$\dfrac{y_p}{H}-\frac{1}{2}$	$\dfrac{x}{H}$	$\dfrac{y_f}{H}-\frac{1}{2}$	σ_1	$\dfrac{\phi_j}{a}$	$\dfrac{y_p}{H}-\frac{1}{2}$	$\dfrac{x}{H}$	$\dfrac{y_f}{H}-\frac{1}{2}$	σ_1
(1)	(2)	(3)	(4)	(5)	(1)	(2)	(3)	(4)	(5)
$\frac{1}{8}$	0·162	0·031	0·113	0·698	$2\frac{5}{8}$	0·061	0·647	0·053	0·882
$\frac{3}{8}$	0·159	0·094	0·111	0·699	$3\frac{1}{8}$	0·042	0·772	0·043	1·005
$\frac{5}{8}$	0·153	0·156	0·108	0·710	$3\frac{5}{8}$	0·029	0·895	0·034	1·177
$\frac{7}{8}$	0·145	0·218	0·102	0·706	$4\frac{1}{8}$	0·020	1·020	0·027	1·41
$1\frac{1}{8}$	0·134	0·279	0·096	0·713	$4\frac{5}{8}$	0·009	1·269	0·019	2·11
$1\frac{3}{8}$	0·122	0·341	0·089	0·724	$5\frac{1}{8}$	0·004	1·519	0·014	3·35
$1\frac{5}{8}$	0·109	0·402	0·081	0·741	$5\frac{5}{8}$	0·002	1·769	0·010	5·5
$1\frac{7}{8}$	0·096	0·463	0·074	0·764	$6\frac{1}{8}$	0·002	2·019	0·008	9·7
$2\frac{1}{8}$	0·083	0·524	0·067	0·799	$6\frac{5}{8}$	0·001	2·269	0·007	19·0

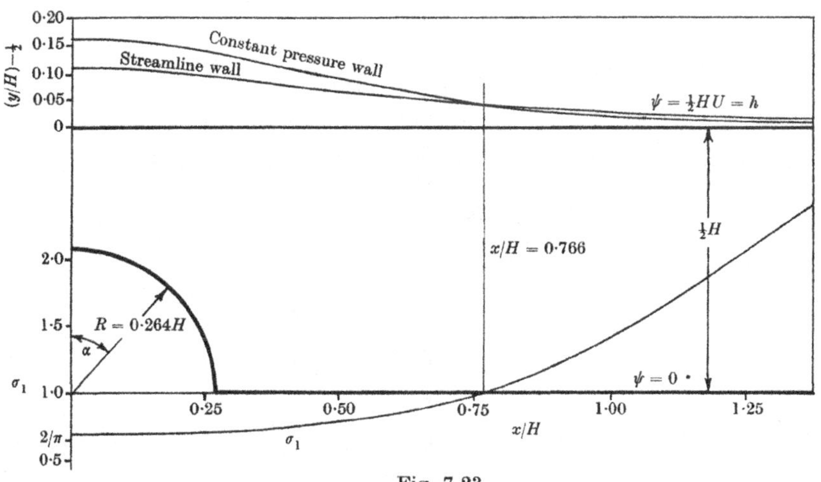

Fig. 7.23

In order to test the theory of §7.22 on the setting of streamline walls we have given the value of $\sigma_1 = (y_f - \frac{1}{2}H)/(y_p - \frac{1}{2}H)$ in column 5. In §7.22 we used linear perturbation theory to establish that $\sigma_1 = 2/\pi \simeq 0.637$ on the wall opposite slender obstacles. Table 7.23b shows that this result is only 10–15 % in error for a circular cylinder, which is certainly the opposite of 'slender'. The wide range of applicability implied by this result is one of the advantages of the flexible-wall tunnel.

DOUBLY-MIXED BOUNDARY CONDITIONS

7.24　General theory

The next stage of complexity with mixed boundary conditions occurs when θ is prescribed on *two* distinct sections of the boundary of a simply-connected domain, and Ω is prescribed on the remaining two sections.

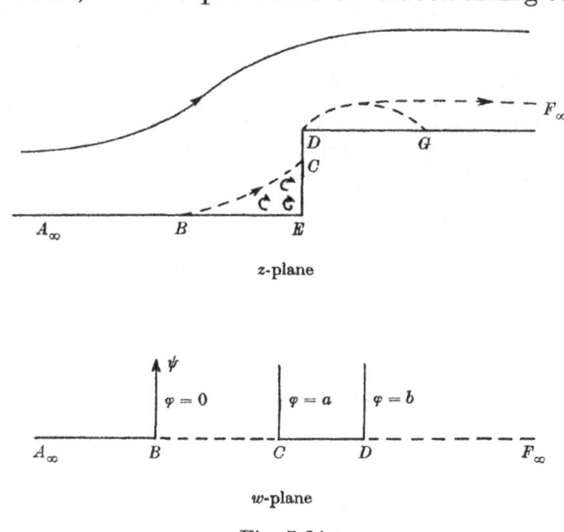

Fig. 7.24a

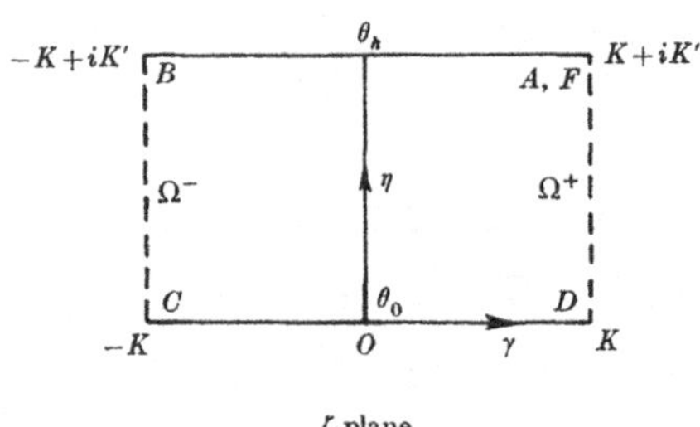

Fig. 7.24b

Fig. 7.24 shows a simple example—the flow past a step with two boundary-layer separations and one re-attachment. The flow separates at B, re-attaches at C, a point on the step ED, and forms a 'dead-air pocket' BEC. The flow separates again at the corner D to form a free streamline DF_∞. The case when C and D coincide has been considered in § 7.16. The pressure (and hence the value of Ω) on BC must be estimated from boundary-layer theory, while that on DF_∞ must be estimated from wake theory. These difficult tasks are outside the scope of this text; we shall be content to assume pressure distributions on these free streamlines. Thus we have the boundary value problem: θ given on $A_\infty B$ and CD, Ω given on BC and DF_∞. In an alternative model of the same flow we assume re-attachment of the flow at a point G on DF_∞, then θ is prescribed on the three intervals $A_\infty B$, CD and GF_∞. However, the intervals $A_\infty B$, CF_∞ should be regarded as being just one interval $GF_\infty A_\infty B$, as the points A_∞ and F_∞ are the *same* point—the point at infinity.

The general method of dealing with flows having 'doubly-mixed' boundary conditions like that just described (another example is to be found in §5.16) is to map the w-plane into the rectangle $-K \leqslant \gamma \leqslant K$, $0 \leqslant \eta \leqslant K'$ in the ζ-$(= \gamma + i\eta)$ plane in such a way that the intervals over which θ is prescribed map into opposite sides of the rectangle. Let these be the sides $\eta = 0$, $\eta = K'$, then the ζ-plane is that shown in Fig. 7.24*b*. Comparison of this figure with Fig. 4.13*b* leads to the conclusion that the solution to our doubly-mixed boundary value problem is

$$\tau(\zeta) = \frac{1}{\pi}\operatorname{cn}\zeta\,\operatorname{dn}\zeta\left[\int_{-K}^{K}\left\{\frac{\theta_0(\gamma)}{\operatorname{sn}\gamma - \operatorname{sn}\zeta} - \frac{k\theta_n(\gamma)}{\operatorname{ns}\gamma - k\operatorname{sn}\zeta}\right\}d\gamma \right.$$
$$\left. + \int_0^{K'}\left\{\frac{\Omega^+(\eta)}{\operatorname{cd}(i\eta) - \operatorname{sn}\zeta} + \frac{\Omega^-(\eta)}{\operatorname{cd}(i\eta) + \operatorname{sn}\zeta}\right\}d\eta\right], \quad (154)$$

which follows from (4.138) on replacing z, x, y, l and h by ζ, γ, η, K and K'. (K and K' are the real and imaginary quarter-periods of the elliptic functions in (154).)

On $\zeta = \gamma$, $\zeta = \gamma + iK'$, (154) yields a pair of integral equations for Ω_0 and Ω_h, which must be treated in a manner similar to that described in §7.2; while on $\zeta = K + i\eta$, $\zeta = -K + i\eta$, (154) gives two 'design' equations for θ^+ and θ^-, which may be treated by the method of §7.10. We shall not give examples of the theory requiring these complicated calculations, but shall restrict our attention to cases in which θ_0, θ_h, Ω^- and Ω^+ are constants.

7.25   Flow over a step

We shall complete the problem illustrated in Fig. 7.24*a*. The w-plane shown in this figure is first mapped into the t-plane of Fig. 7.25. There is an 'angle' of 2π at A_∞, F_∞ in the w-plane, and as this maps into the point $t = 1/k$, it follows from the Schwarz–Christoffel theorem that $dw/dt = A(t - 1/k)^{-2}$, where A is a constant. Integrating this and using the fact that the points $\phi = 0$, a, b map into $t = -1/k$, -1, 1, we find

$$w = \sqrt{(ab)}\,\frac{1+kt}{1-kt}, \quad (155)$$

where
$$k = \frac{1 - \sqrt{(a/b)}}{1 + \sqrt{(a/b)}}.\tag{156}$$

The t-plane is now mapped into the ζ-plane of Fig. 7.24b by (cf. § 5.16) $t = \mathrm{sn}\,(\zeta, k)$, hence
$$w = \sqrt{(ab)}\,\frac{1 + k\,\mathrm{sn}\,\zeta}{1 - k\,\mathrm{sn}\,\zeta}\tag{157}$$

is the required mapping formula. The general solution in the ζ-plane is given by (154).

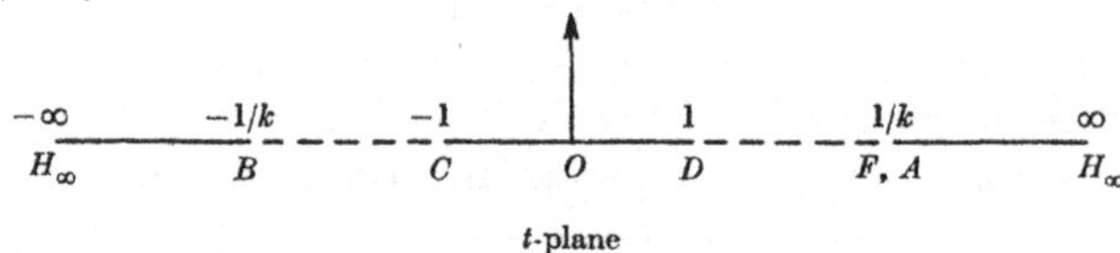

Fig. 7.25

We shall assume that the free streamlines BC and DF_∞ are at constant pressure—this gives a rough approximation to a real flow—so that we can write $q = V$ on BC and $q = U$ on DF_∞. Now $\Omega = 0$ when $q = U$, and let $\Omega = \sigma$ when $q = V$, so that (see (6.3))
$$\frac{V}{U} = \frac{\sinh \Omega^*}{\sinh(\Omega^* + \sigma)}.$$

Comparing the z-plane of figure 7.24a and the ζ-plane of Fig. 7.24b we conclude that
$$\Omega^+ = 0, \quad \Omega^- = \sigma, \quad \theta_0 = \tfrac{1}{2}\pi, \quad \theta_h = 0.\tag{158}$$

It follows from (154) that
$$\tau(\zeta) = \tfrac{1}{2}\,\mathrm{cn}\,\zeta\,\mathrm{dn}\,\zeta \int_{-K}^{K} \frac{d\gamma}{\mathrm{sn}\,\gamma - \mathrm{sn}\,\zeta} + \frac{\sigma}{\pi}\,\mathrm{cn}\,\zeta\,\mathrm{dn}\,\zeta \int_{0}^{K'} \frac{d\eta}{\mathrm{cd}\,i\eta + \mathrm{sn}\,\zeta}.$$

Now it is easily verified by differentiation (see §§ 4.6, 4.8) that
$$\mathrm{cn}\,\zeta\,\mathrm{dn}\,\zeta \int \frac{d\gamma}{\mathrm{sn}\,\gamma - \mathrm{sn}\,\zeta} = \tfrac{1}{2}\ln\frac{\mathrm{cd}\,\gamma - \mathrm{cd}\,\zeta}{\mathrm{cd}\,\gamma + \mathrm{cd}\,\zeta} + \Pi(\gamma + iK', \zeta).\tag{159}$$

Therefore
$$\mathrm{cn}\,\zeta\,\mathrm{dn}\,\zeta \int_{-K}^{K} \frac{d\gamma}{\mathrm{sn}\,\gamma - \mathrm{sn}\,\zeta} = \tfrac{1}{2}\ln(-1) - \tfrac{1}{2}\ln\frac{1}{(-1)}$$
$$+ \Pi(K + iK', \zeta) - \Pi(-K + iK', \zeta) = i\pi + 2KZ_4(\zeta),$$

by equation (4.80). Again, by making the substitution $i\eta = -K + x$, we find that
$$\mathrm{cn}\,\zeta\,\mathrm{dn}\,\zeta \int_{0}^{K'} \frac{d\eta}{\mathrm{cd}\,(i\eta) + \mathrm{sn}\,\zeta} = -i\,\mathrm{cn}\,\zeta\,\mathrm{dn}\,\zeta \int_{K}^{K+iK'} \frac{dx}{\mathrm{sn}\,x - \mathrm{sn}\,\zeta}$$
$$= \tfrac{1}{2}\pi - K'Z_4(\zeta) - \frac{\pi\zeta}{2K},$$

using (159). Therefore

$$\tau(\zeta) = i\frac{\pi}{2} + \left(K - \frac{\sigma K'}{\pi}\right) Z_4(\zeta) + \frac{\sigma}{2}\left(1 - \frac{\zeta}{K}\right), \tag{160}$$

and it remains only to eliminate ζ from this equation to obtain the desired $\tau(w)$ relation. The problem is now completed by deducing $z(w, \overline{w})$ from (6.13), but clearly numerical methods will be necessary for this last step.

The above theory is a generalization of that given in §7.16; the limit $b \to a$ reduces (160) to the case $h = \infty$ of (106).

7.26   Design of a channel bend

Next consider the problem described in §5.16, i.e. the problem of designing a bend in a channel so that large adverse pressure gradients are avoided on both walls. Referring to Fig. 5.16 we shall let $q = V_1$ on AB, and $q = V_2$ on CD, where V_1 and V_2 are constants. Let σ_1 and σ_2 be the corresponding values of Ω. Our problem is that of determining the shape of AB and CD.

It is shown in §5.16 that the w-plane is mapped into the ζ-plane by

$$\tanh\frac{\pi w}{2h} = \tanh\frac{\pi a}{2h}\,\mathrm{sn}\left(\zeta, \tanh\frac{\pi a}{2h}\tanh\frac{\pi b}{2h}\right), \tag{161}$$

where a, b and h are the 'distances' in the w-plane as defined in Fig. 5.16. If S_1, S_2 are the lengths of AB, CD then clearly

$$2b = V_1 S_1, \quad 2a = V_2 S_2. \tag{162}$$

Comparison of the ζ-planes of Figs. 5.16 and 7.24b shows that the harmonic conjugate of (154), i.e.

$$\tau(\zeta) = \frac{1}{\pi i}\,\mathrm{cn}\,\zeta\,\mathrm{dn}\,\zeta\left[\int_{-K}^{K}\left\{\frac{\Omega_0(\gamma)}{\mathrm{sn}\,\gamma - \mathrm{sn}\,\zeta} - \frac{k\Omega_h(\gamma)}{\mathrm{ns}\,\gamma - k\,\mathrm{sn}\,\zeta}\right\}d\gamma\right.$$
$$\left. - \int_0^{K'}\left\{\frac{\theta^+(\eta)}{\mathrm{cd}\,(i\eta) - \mathrm{sn}\,\zeta} + \frac{\theta^-(\eta)}{\mathrm{cd}\,(i\eta) + \mathrm{sn}\,\zeta}\right\}d\eta\right] \tag{163}$$

is the equation appropriate in the present application.

Let the channel be deflected through an angle α then it is apparent from Fig. 5.16 that the boundary values are $\Omega_0 = \sigma_2$, $\Omega_h = \sigma_1$, $\theta^- = 0$, $\theta^+ = \alpha$. Substituting these values in (163) and making use of (159) we arrive at

$$\tau(\zeta) = \sigma_2 + \frac{i}{\pi}\{2K(\sigma_1 - \sigma_2) + \alpha K'\}Z_4(\zeta) + \tfrac{1}{2}i\alpha\left(1 + \frac{\zeta}{K}\right). \tag{164}$$

On CD this yields

$$\theta(\gamma) = \frac{1}{\pi}\{2K(\sigma_1 - \sigma_2) + \alpha K'\}Z_4(\gamma) + \tfrac{1}{2}\alpha\left(1 + \frac{\gamma}{K}\right), \tag{165}$$

where from (161)

$$\gamma = \mathrm{sn}^{-1}\left(\tanh\frac{\pi\phi}{2h}\coth\frac{\pi a}{2h}, \quad \tanh\frac{\pi a}{2h}\tanh\frac{\pi b}{2h}\right),$$

and
$$\phi = V_2 s,$$

s being the distance measured along CD from $\phi = 0$. The required intrinsic equation of CD now follows on eliminating γ and ϕ from (165).

A similar equation can be deduced for AB, and the design is complete.

POROUS AND PERFORATED WALLS

7.27　General theory

If the channel has a porous wall, then because of the removal of fluid from the main stream through the wall, this wall is no longer a streamline. Fig. 7.27a shows the w-plane for a channel with solid walls everywhere, except in an interval CD. (This particular w-plane corresponds to the

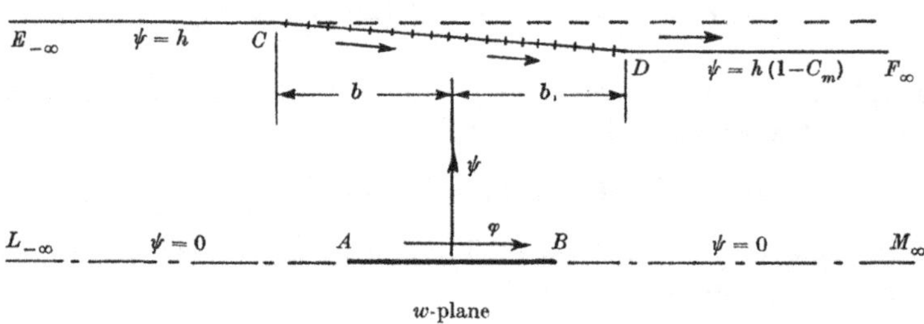

Fig. 7.27a

z-plane shown in Fig. 7.28a.) On the solid wall upstream of the porous section $\psi = h$, while downstream of this section $\psi = h(1-C_m)$ where C_m is a factor proportional to the mass Q sucked through the wall. If the theory of porous walls summarized in §6.3 is adopted, then the outflow mass Q must be kept small compared with the total mass flowing through the channel; this is equivalent to restricting C_m to be a small (compared with unity) first-order quantity. Second-order terms in C_m shall be neglected.

With this restriction on C_m, the boundary $\psi = h(1-C_m)$ will be very close to the extension of the boundary $\psi = h$, so that if we make the approximation of applying the boundary conditions on $\psi = h$ over the *whole* range $-\infty < \phi < \infty$, we can expect to incur only a small second-order error term. (For the example under consideration this point can be verified *a posteriori* by replacing H by $H(1-C_m)$ in (189) for the blockage factor ϵ, and noting that the change in ϵ is of second order only.)

We shall accept the approximation just described in relation to the particular problem illustrated in Fig. 7.28a, as applying to the general case of a channel with porous (or perforated) walls. Then, if these walls have the linear characteristic described in § 6.3, the boundary conditions on $\psi = 0$ and $\psi = h$ can be written

$$\left.\begin{aligned}\psi &= 0, \quad \theta_0 + \lambda_0\,\Omega_0 = \theta_{s0} + \lambda_0\,\Omega_{s0}, \\ \psi &= h, \quad \theta_h - \lambda_h\,\Omega_h = \theta_{sh} - \lambda_h\,\Omega_{sh},\end{aligned}\right\} \tag{166}$$

where θ_{s0}, θ_{sh}, Ω_{s0}, Ω_{sh} are known functions,

$$\lambda_0 = \frac{\lambda_{s0}}{\beta_a}\,\frac{\sinh\Omega^*}{\sinh(\Omega^*+\Omega_{s0})}, \quad \lambda_h = \frac{\lambda_{sh}}{\beta_a}\,\frac{\sinh\Omega^*}{\sinh(\Omega^*+\Omega_{sh})}, \tag{167}$$

and λ_{s0}, λ_{sh} are the 'porosities' of the walls as defined in § 1.20. The angles θ_h, θ_{sh} and pressures p_h, p_{sh} are defined in Fig. 7.27b; Ω_h, Ω_{sh} are the

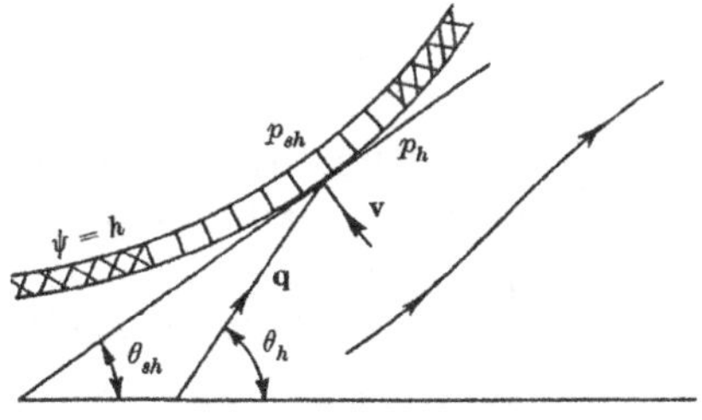

Fig. 7.27b

values of the speed parameter Ω corresponding to p_h, p_{sh}. A similar notation applies on $\psi = 0$. The change in sign of λ in the first of (166) means that λ_0 and λ_h are both positive quantities—see discussion on the sign of λ in § 6.3.

The boundary conditions given above have been studied in §§ 4.15–4.20, and the general solution for $\tau = \Omega + i\theta$ in an infinite strip of width h in the w-plane can be written down immediately from (4.153) and (4.157). Let

$$\epsilon_0 = \frac{1}{\pi}\tan^{-1}\lambda_0, \quad \epsilon_h = \frac{1}{\pi}\tan^{-1}\lambda_h, \tag{168}$$

then the function $\epsilon(t)$ occurring in (4.153) will be $\epsilon_h(\phi)$ on $\psi = h$ and $-\epsilon_0(\phi)$ on $\psi = 0$. The contour C is now $\psi = 0$, $-\infty < \phi < \infty$ plus $\psi = h$,

$-\infty < \phi < \infty$ traversed in the negative direction. These facts and (4.157) enable us to write (4.153) in the form

$$\tau(w) = \exp\left\{-\frac{\pi}{2h}\int_{-\infty}^{\infty}\left[\epsilon_0(\phi)\coth\frac{\pi}{2h}(\phi-w)+\epsilon_h(\phi)\tanh\frac{\pi}{2h}(\phi-w)\right]d\phi\right\}$$

$$\times\left[K+\frac{1}{2h}\int_{-\infty}^{\infty}(\theta_{s0}\cos\pi\epsilon_0+\Omega_{s0}\sin\pi\epsilon_0)\exp\left\{\frac{\pi}{2h}\right.\right.$$

$$\times\,\mathscr{R}\int_{-\infty}^{\infty}\left[\epsilon_0(\phi^*)\coth\frac{\pi}{2h}(\phi^*-\phi)+\epsilon_h(\phi^*)\tanh\frac{\pi}{2h}(\phi^*-\phi)\right]d\phi^*\right\}$$

$$\times\coth\frac{\pi}{2h}(\phi-w)\,d\phi-\frac{1}{2h}\int_{-\infty}^{\infty}(\theta_{sh}\cos\pi\epsilon_h-\Omega_{sh}\sin\pi\epsilon_h)$$

$$\times\exp\left\{\frac{\pi}{2h}\mathscr{R}\int_{-\infty}^{\infty}\left[\epsilon_0(\phi^*)\tanh\frac{\pi}{2h}(\phi^*-\phi)\right.\right.$$

$$\left.\left.+\epsilon_h(\phi^*)\coth\frac{\pi}{2h}(\phi^*-\phi)\right]d\phi^*\right\}\tanh\frac{\pi}{2h}(\phi-w)\,d\phi\right], \tag{169}$$

where from (4.161)

$$K = \tfrac{1}{2}(\Omega\cos\pi\epsilon+\theta\sin\pi\epsilon)_{\infty}\exp\left\{-\frac{\pi}{2h}\int_{-\infty}^{\infty}[\epsilon_0(\phi)+\epsilon_h(\phi)]\,d\phi\right\}$$

$$+(\Omega\cos\pi\epsilon+\theta\sin\pi\epsilon)_{-\infty}\exp\left\{\frac{\pi}{2h}\int_{-\infty}^{\infty}[\epsilon_0(\phi)+\epsilon_h(\phi)]\,d\phi\right\}, \tag{170}$$

the subscripts ∞ and $-\infty$ indicating values at $\phi=\infty$ and $\phi=-\infty$. In addition to these equations there is the restriction

$$(\Omega\cos\pi\epsilon+\theta\sin\pi\epsilon)_{\infty}\exp\left\{-\frac{\pi}{2h}\int_{-\infty}^{\infty}[\epsilon_0(\phi)+\epsilon_h(\phi)]\,d\phi\right\}$$

$$= (\Omega\cos\pi\epsilon+\theta\sin\pi\epsilon)_{-\infty}\exp\left\{\frac{\pi}{2h}\int_{-\infty}^{\infty}[\epsilon_0(\phi)+\epsilon_h(\phi)]\,d\phi\right\}$$

$$-\frac{1}{h}\int_{-\infty}^{\infty}(\theta_{s0}\cos\pi\epsilon_0+\Omega_{s0}\sin\pi\epsilon_0)\exp\left\{\frac{\pi}{2h}\right.$$

$$\times\,\mathscr{R}\int_{-\infty}^{\infty}\left[\epsilon_0(\phi^*)\coth\frac{\pi}{2h}(\phi^*-\phi)+\epsilon_h(\phi^*)\tanh\frac{\pi}{2h}(\phi^*-\phi)\right]d\phi^*\right\}d\phi$$

$$+\frac{1}{h}\int_{-\infty}^{\infty}(\theta_{sh}\cos\pi\epsilon_h-\Omega_{sh}\sin\pi\epsilon_h)\exp\left\{\frac{\pi}{2h}\right.$$

$$\times\,\mathscr{R}\int_{-\infty}^{\infty}\left[\epsilon_0(\phi^*)\tanh\frac{\pi}{2h}(\phi^*-\phi)+\epsilon_h(\phi^*)\coth\frac{\pi}{2h}(\phi^*-\phi)\right]d\phi^*\right\}d\phi, \tag{171}$$

which follows from (162).

This completes the general theory; in §§ 7.28 and 7.29 we shall study two important special cases.

7.28   Blockage in a porous-walled tunnel

In §§ 7.8 and 7.21 it was shown that the average value of the blockage factor in a solid-walled and open-jet wind 'tunnel' is given by

$$\epsilon_T = \frac{\pi A}{6\beta_a^3 H^2} + \frac{cC_D}{4\beta_a^2 H}\left(1 - \frac{\pi c}{6\beta_a H}\right),\tag{172}$$

and

$$\epsilon_J = -\frac{\pi A}{12\beta_a^3 H^2} + \frac{\pi c C_D}{48\beta_a^3 H^2},\tag{173}$$

respectively. These equations suggest the possibility of designing a wind tunnel which combines the characteristics of the solid-walled and open-jet tunnel in such a way that the resulting blockage factor vanishes. This leads to the 'ventilated' wind tunnel in which the walls are either 'slotted'

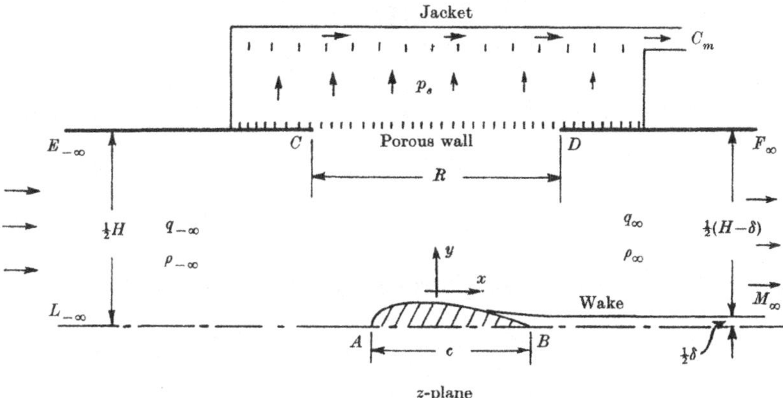

Fig. 7.28 a

by gaps running the length of the tunnel (see § 7.30), or are porous in some or all sections. However, it should be noted that the real advantage of a correctly designed ventilated wind-tunnel lies not so much in the fact that the blockage is zero, but in that it avoids that serious defect of conventional tunnels occurring in transonic flow known as 'choking'. (An excellent introduction to the subject as a whole is to be found in a group of papers in the *Journal of the Royal Aeronautical Society*, January 1958.)

We shall follow Woods (1955 h) and consider a tunnel of the construction shown in Fig. 7.28 a. In this tunnel there is a porous section of (adjustable) length R covered by a jacket assumed to be at constant pressure p_s. Let the porous section be of constant porosity, then we can write

$$\epsilon_0(\phi) = 0, \quad \epsilon_h(\phi) = \begin{cases} k & (|\phi| < b), \\ 0 & (|\phi| > b), \end{cases}\tag{174}$$

where b is defined in Fig. 7.27b, and where k is a constant. Our problem is to find the functional relation $\epsilon_p = f(R/H, k, M_a)$, where ϵ_p is the mean blockage factor of our porous tunnel.

As the tunnel is straight we can put

$$\theta_{-\infty} = \theta_{\infty} = 0, \tag{175}$$

and we can choose the reference velocity U to be equal to the velocity upstream at infinity, i.e. $U = q_{-\infty}$. Then

$$\Omega_{-\infty} = 0. \tag{176}$$

The value of Ω_{∞} can be calculated by putting $H_{-\infty} = H$ and $H_{\infty} = H - \delta$ in (2). Retaining only first-order terms in δ and C_m, the result is

$$\Omega_{\infty} = \frac{1}{\beta_a}\left(C_m - \frac{\delta}{H}\right), \tag{177}$$

where C_m is defined by (3), and δ by (57).

If the jacket pressure p_s is assumed not to differ greatly from the pressure upstream at infinity, $p_a = p_{-\infty}$, then

$$\Omega_{sh} \simeq \tfrac{1}{2}\beta_a C_p = \beta_a \frac{p_s - p_a}{\rho_a U^2} \equiv \Omega_J, \quad \text{say} \tag{178}$$

(see (6.23)). Let the aerofoil slope be θ_a, then the boundary values are

$$\left.\begin{aligned}
&\theta_{sh} = 0; \quad \theta_{s0} = \theta_a\,(\phi\ \text{in}\ AB), \quad \theta_{s0} = 0\,(\phi\ \text{outside}\ AB),\\
&\Omega_{s0} = 0; \quad \Omega_{sh} = \Omega_J\,(|\phi| < b).
\end{aligned}\right\} \tag{179}$$

On substituting (174) to (176) and (179) into (169), (170) and (171) we find

$$
\begin{aligned}
\tau(w) = \frac{1}{2h}&\left\{\frac{\cosh\dfrac{\pi}{2h}(b+w)}{\cosh\dfrac{\pi}{2h}(b-w)}\right\}^k\left[\,h\Omega_{\infty}\,e^{-b\pi k/h}\right.\\[2ex]
&+\int_{AB}\theta_a\left\{\frac{\cosh\dfrac{\pi}{2h}(b-\phi)}{\cosh\dfrac{\pi}{2h}(b+\phi)}\right\}^k\coth\frac{\pi}{2h}(\phi-w)\,d\phi\\[2ex]
&\left.+\sin\pi k\int_{-b}^{b}\Omega_J\left\{\frac{\sinh\dfrac{\pi}{2h}(b-\phi)}{\sinh\dfrac{\pi}{2h}(b+\phi)}\right\}^k\tanh\frac{\pi}{2h}(\phi-w)\,d\phi\right], \tag{180}
\end{aligned}
$$

and

$$h\Omega_\infty e^{-b\pi k/h} = -\int_{AB} \theta_a \left\{ \frac{\cosh\dfrac{\pi}{2h}(b-\phi)}{\cosh\dfrac{\pi}{2h}(b+\phi)} \right\}^k d\phi$$

$$-\sin\pi k \int_{-b}^{b} \Omega_J \left\{ \frac{\sinh\dfrac{\pi}{2h}(b-\phi)}{\sinh\dfrac{\pi}{2h}(b+\phi)} \right\}^k d\phi. \qquad (181)$$

We have not yet assumed that the jacket pressure p_s is constant, so that (180) and (181) are applicable if, for instance, we divide the jacket into a number of compartments with different pressures in each. If p_s is constant, so is Ω_J. In this case the integrals involving this factor in (180) and (181) can be evaluated by the substitution

$$x = \frac{\sinh\dfrac{\pi}{2h}(b-\phi)}{\sinh\dfrac{\pi}{2h}(b+\phi)},$$

and then using the well-known result

$$\int_0^\infty \frac{x^k\,dx}{1+x} = -\frac{\pi}{\sin\pi k}.$$

It will be found that (180) and (181) reduce to

$$\tau(w) = \Omega_J + \frac{1}{2h}\left\{ \frac{\cosh\dfrac{\pi}{2h}(b+w)}{\cosh\dfrac{\pi}{2h}(b-w)} \right\}^k \left[\int_{AB} \theta_a \left\{ \frac{\cosh\dfrac{\pi}{2h}(b-\phi)}{\cosh\dfrac{\pi}{2h}(b+\phi)} \right\}^k \right.$$

$$\left. \times \coth\frac{\pi}{2h}(\phi-w)\,d\phi + h(\Omega_\infty - \Omega_J)e^{-b\pi k/h} - h\Omega_J e^{b\pi k/h} \right], \qquad (182)$$

and

$$h\Omega_\infty e^{-b\pi k/h} = -\int_{AB} \theta_a \left\{ \frac{\cosh\dfrac{\pi}{2h}(b-\phi)}{\cosh\dfrac{\pi}{2h}(b+\phi)} \right\}^k d\phi - 2h\Omega_J \sinh\frac{k\pi b}{h}. \qquad (183)$$

In calculating the blockage factor we shall follow the method of §7.8. The first step is to linearize (182) by replacing h, b, ϕ, w, Ω and θ by $\beta_a HU/2$, $RU/2$, Ux, UZ, $1-q/U$ and dy/dx respectively. (The length R

is defined in Fig. 7.28 a.) We then subtract the form the resulting equation takes from (60) to find that the blockage factor on the aerofoil is given by

$$
\epsilon(x^*) = \frac{1}{\beta_a^2 H} \int_{-\frac{1}{2}c}^{\frac{1}{2}c} \left(\frac{dy}{dx}\right)_0 \left[\frac{1}{\dfrac{\pi}{\beta_a H}(x-x^*)} \right.
$$

$$
\left. - \left\{ \frac{\cosh \dfrac{\pi}{\beta_a H}(\tfrac{1}{2}R+x^*)\cosh \dfrac{\pi}{\beta_a H}(\tfrac{1}{2}R-x)}{\cosh \dfrac{\pi}{\beta_a H}(\tfrac{1}{2}R-x^*)\cosh \dfrac{\pi}{\beta_a H}(\tfrac{1}{2}R+x)} \right\}^{k} \coth \frac{\pi}{\beta_a H}(x-x^*) \right] dx
$$

$$
- \frac{\Omega_J}{\beta_a} \left[1 - \cosh \frac{R\pi k}{\beta_a H} \left\{ \frac{\cosh \dfrac{\pi}{\beta_a H}(\tfrac{1}{2}R+x^*)}{\cosh \dfrac{\pi}{\beta_a H}(\tfrac{1}{2}R-x^*)} \right\}^{k} \right]
$$

$$
- \frac{1}{2\beta_a^2} e^{-R\pi k/\beta_a H} \left(C_m - \frac{\delta}{H}\right) \left\{ \frac{\cosh \dfrac{\pi}{\beta_a H}(\tfrac{1}{2}R+x^*)}{\cosh \dfrac{\pi}{\beta_a H}(\tfrac{1}{2}R-x^*)} \right\}^{k}, \tag{184}
$$

where we have used (177) to eliminate Ω_∞. From (177) and (183) it is found that the jacket-pressure parameter is given by

$$
\Omega_J = -\operatorname{cosech} \frac{R\pi k}{\beta_a H} \left[\frac{1}{\beta_a H} \int_{-\frac{1}{2}c}^{\frac{1}{2}c} \left(\frac{dy}{dx}\right)_0 \left\{ \frac{\cosh \dfrac{\pi}{\beta_a H}(\tfrac{1}{2}R-x)}{\cosh \dfrac{\pi}{\beta_a H}(\tfrac{1}{2}R+x)} \right\}^{k} dx \right]
$$

$$
- \frac{1}{2\beta_a} e^{-R\pi k/\beta_a H} \left(C_m - \frac{\delta}{H}\right). \tag{185}
$$

In writing the limits of the integrals in these equations as $\frac{1}{2}c$ and $-\frac{1}{2}c$, we have tacitly assumed that the mid-chord point of the aerofoil coincides with the origin of x, which means that the porous wall will extend equal distances up and downstream of the aerofoil. The more general case is too algebraically complicated to present here; the reader should have no difficulty in developing a more general theory if he needs it.

Equation (184) can be simplified by expanding the integrand and ignoring terms $O(c/H)^3$, just as in the derivation of (62) from (61). Let

$$
\sigma \equiv \tanh \frac{\pi R}{2\beta_a H}, \tag{186}
$$

then

$$\left\{\frac{\cosh\dfrac{\pi}{\beta_a H}(\tfrac{1}{2}R-x)}{\cosh\dfrac{\pi}{\beta_a H}(\tfrac{1}{2}R+x)}\right\}^k = 1 - 2\frac{k\sigma\pi}{\beta_a H}x + 2\left(\frac{k\sigma\pi}{\beta_a H}\right)^2 x^2 + O\left(\frac{x}{H}\right)^3.$$

With the aid of this expansion we find that (184) and (185) can be written

$$\epsilon(x^*) = \frac{2k\sigma}{\beta_a^2 H}\int_{-\frac{1}{2}c}^{\frac{1}{2}c}\left(\frac{dy}{dx}\right)_0 dx - \frac{\pi}{3\beta_a^3 H^2}(1+6k^2\sigma^2)\int_{-\frac{1}{2}c}^{\frac{1}{2}c}\left(\frac{dy}{dx}\right)_0(x-x^*)\,dx$$

$$- \frac{\Omega_J}{\beta_a}\left(1 - \cosh\frac{R\pi k}{\beta_a H} - 2\frac{k\sigma\pi}{\beta_a H}x^*\cosh\frac{R\pi k}{\beta_a H}\right)$$

$$- \frac{1}{2\beta_a^2}\left(C_m - \frac{\delta}{H}\right)e^{-R\pi k/\beta_a H}\left(1 + 2\frac{k\sigma\pi}{\beta_a H}x^*\right), \tag{187}$$

and

$$\Omega_J = -\left\{\frac{1}{\beta_a H}\int_{-\frac{1}{2}c}^{\frac{1}{2}c}\left(\frac{dy}{dx}\right)_0 dx - 2\frac{k\sigma\pi^2}{\beta_a^2 H^2}\int_{-\frac{1}{2}c}^{\frac{1}{2}c}\left(\frac{dy}{dx}\right)_0 x\,dx\right\}\operatorname{cosech}\frac{R\pi k}{\beta_a H}$$

$$- \frac{1}{2\beta_a}e^{-R\pi k/\beta_a H}\left(C_m - \frac{\delta}{H}\right)\operatorname{cosech}\frac{R\pi k}{\beta_a H}, \tag{188}$$

ignoring the higher-order terms. As $y = 0$ at $x = -\tfrac{1}{2}c$ and $y = \tfrac{1}{2}\delta$ at $x = \tfrac{1}{2}c$ (see Fig. 7.28a), then

$$\int_{-\frac{1}{2}c}^{\frac{1}{2}c}\left(\frac{dy}{dx}\right)_0 dx = \tfrac{1}{2}\delta$$

and

$$\int_{-\frac{1}{2}c}^{\frac{1}{2}c}\left(\frac{dy}{dx}\right)_0 x\,dx = \int_{-\frac{1}{2}c}^{\frac{1}{2}c}x\,dy = \frac{c\delta}{4} - \int_{-\frac{1}{2}c}^{\frac{1}{2}c}y\,dx = \frac{c\delta}{4} - \tfrac{1}{2}A,$$

where A is the cross-sectional area of the profile. Substituting these values in (187) and (188), and using (57) to eliminate δ, we arrive at the relations

$$\epsilon(x) = \frac{\pi A}{6\beta_a^3 H^2}\left(1 + 6k^2\sigma^2 - 6k\sigma\tanh\frac{R\pi k}{2\beta_a H}\right)$$

$$+ \frac{cC_D}{4\beta_a^2 H}\left\{1 + 2k\sigma - 2\tanh\frac{R\pi k}{2\beta_a H}\right.$$

$$+ \frac{\pi(x-\tfrac{1}{2}c)}{3\beta_a H}\left(1 + 6k^2\sigma^2 - 6k\sigma\tanh\frac{R\pi k}{2\beta_a H}\right)\right\}$$

$$- \frac{1}{2\beta_a^2}C_m\left(1 - \tanh\frac{R\pi k}{2\beta_a H} + \frac{2\pi k\sigma}{\beta_a H}x\operatorname{cosech}\frac{R\pi k}{\beta_a H}\right), \tag{189}$$

and

$$p_s = p_a - \frac{\rho_a U^2}{2\beta_a^2}\left\{\frac{cC_D}{2H} + \frac{2k\sigma\pi}{\beta_a H^2}(A - \tfrac{1}{4}c^2 C_D)\right.$$

$$\left. + \left(C_m - \frac{cC_D}{2H}\right)e^{-R\pi k/\beta_a H}\right\}\operatorname{cosech}\frac{R\pi k}{\beta_a H}, \tag{190}$$

where (178) has been used to eliminate Ω_J. This last result shows the dependence of the jacket pressure on the mass flow through the jacket, and the drag acting on the aerofoil. The value of the mean blockage factor, ϵ_p is simply

$$\epsilon_p = \epsilon(0). \tag{191}$$

Three special cases are worth mentioning. First if $k = 0$ (i.e. $\epsilon_h = 0$), we will have the case of a conventional solid-wall tunnel. In this case it follows from (190) that $C_m = 0$, and (191) correctly reduces to (172).

Secondly, if $k = \frac{1}{2}$ it follows from (174) and §6.4 that the porous section is replaced by an open-jet section of the same length. If in this case there are no facilities for removing jacket air, i.e. if $C_m = 0$, then the first term of (189)—the solid-blockage factor—vanishes when

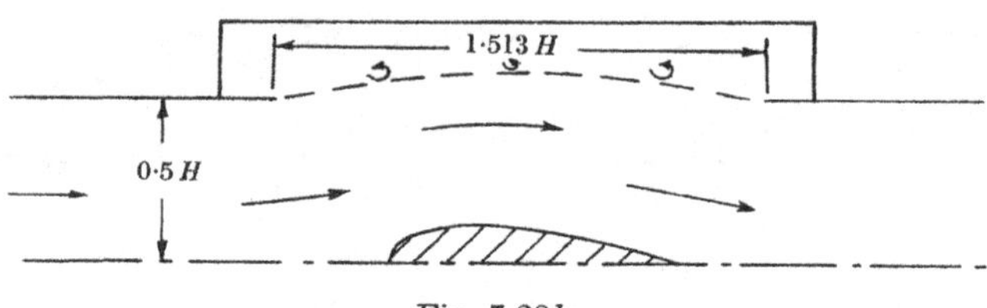

Fig. 7.28*b*

$R \simeq 1\cdot513\beta_a H$. This special zero-blockage tunnel is illustrated in Fig. 7.28*b* for the case $\beta_a = 1$. Thirdly there is the case when the length of the porous section is effectively infinite, i.e. $R = \infty$. In this limit $\sigma = 1$ by (186), and (189) reduces to

$$\epsilon(x) = \frac{\pi A}{6\beta_a^3 H^2}(1 + 6k^2 - 6k) + \frac{cC_D}{4\beta_a^2 H}\left\{2k - 1 + \frac{\pi(x - \tfrac{1}{2}c)}{3\beta_a H}(1 + 6k^2 - 6k)\right\}. \tag{192}$$

The solid-blockage factor is found to vanish if $\lambda_h \simeq 0\cdot781$ (see (168) and (174)); from (6.33) and (167) this requires that the porosity of the wall be given by $\lambda_{sh} \simeq 0\cdot781\beta_a$. Many authors have considered this case; Maeder & Wood (1956) have recently reviewed the topic of ventilated wind-tunnels and provide many references.

The correction to the drag coefficient can be calculated just as in §7.9.

7.29 Perforated wind-tunnels

In §6.4 it was postulated that the effective value of ϵ for a closely *perforated* wall is given by

$$\bar{\epsilon} = \frac{\frac{1}{2}S_{\frac{1}{2}}}{S_0 + S_{\frac{1}{2}}}, \tag{193}$$

where $S_{\frac{1}{2}}$, S_0 are the areas of the perforations and solid surface respectively. In this section we shall verify this directly for the case of an aerofoil in a wind tunnel.

Fig. 7.29 shows a symmetrical aerofoil on the centre-line of a tunnel, the boundaries of which are 'open' in equal gaps, occupying the intervals $(2r - \sigma)\,a \leqslant x \leqslant (2r + \sigma)\,a$, $r = 0, \pm 1, \pm 2, \ldots$. The number σ is the ratio of the length of a gap to the distance from the centre of one gap to the next. Thus $0 \leqslant \sigma \leqslant 1$, the limits $\sigma = 0$, $\sigma = 1$ corresponding to the completely solid and completely open boundaries respectively. It will be assumed that the pressure p_s in the gaps is equal to p_a, the pressure in the undisturbed flow at infinity.

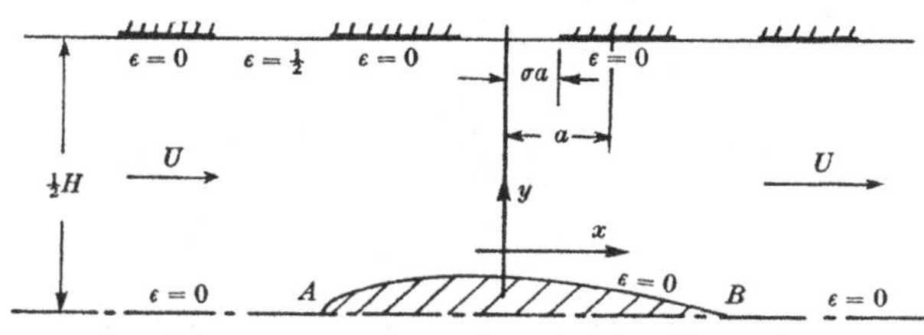

Fig. 7.29

The boundary conditions are

(i) $\epsilon_h = \tfrac{1}{2}$ in $(2r - \sigma)\,a \leqslant x \leqslant (2r + \sigma)\,a$ $(r = 0, \pm 1, \pm 2, \ldots)$,

(ii) $\epsilon_h = 0$ elsewhere,

(iii) $\epsilon_0 = 0$,

(iv) $\theta_s = 0$, except on the aerofoil surface, $y = 0$, $A \leqslant x \leqslant B$,

(v) $\Omega_s = 0$ in the gaps on $y = \tfrac{1}{2}H$.

To apply the general solution (169) to this problem it is first necessary to evaluate

$$I \equiv \lim_{n \to \infty} \exp\left\{ -\frac{\pi}{\beta_a H} \int_{-(2n+\sigma)a}^{(2n+\sigma)a} \epsilon_h(x) \tanh \frac{\pi}{\beta_a H} (x - Z)\, dx \right\},$$

which is the linearized form of the first factor of (169) (see paragraph following (183)). From condition (i)

$$I = \lim_{n \to \infty} \exp\left\{ -\frac{\pi}{2\beta_a H} \sum_{r=-n}^{n} \int_{(2r-\sigma)a}^{(2r+\sigma)a} \tanh \frac{\pi}{\beta_a H} (x - Z)\, dx \right\}$$

$$= \lim_{n \to \infty} \prod_{r=-n}^{n} \left[\frac{\cosh \dfrac{\pi}{\beta_a H}\{(2r - \sigma)\,a - Z\}}{\cosh \dfrac{\pi}{\beta_a H}\{(2r + \sigma)\,a - Z\}} \right]^{\frac{1}{2}}$$

$$= \left[\frac{\cos \dfrac{i\pi}{\beta_a H}(\sigma a + Z) \displaystyle\prod_{r=1}^{\infty} \left\{ 1 + q^{2r} \cos \dfrac{2i\pi}{\beta_a H}(\sigma a + Z) + q^{4r} \right\}}{\cos \dfrac{i\pi}{\beta_a H}(\sigma a - Z) \displaystyle\prod_{r=1}^{\infty} \left\{ 1 + q^{2r} \cos \dfrac{2i\pi}{\beta_a H}(\sigma a - Z) + q^{4r} \right\}} \right]^{\frac{1}{2}},$$

where $q = \exp\left(-2\pi a/\beta_a H\right)$. Therefore (see p. 470 of Whittaker & Watson, 1952)

$$I = \left[\frac{\vartheta_2\left\{\dfrac{i\pi}{\beta_a H}(\sigma a + Z)\middle| i\dfrac{2a}{\beta_a H}\right\}}{\vartheta_2\left\{\dfrac{i\pi}{\beta_a H}(\sigma a - Z)\middle| i\dfrac{2a}{\beta_a H}\right\}}\right]^{\frac12},$$

where ϑ is the theta function defined in §4.4. Jacobi's imaginary transformation—(4.29)—enables us to rewrite this result in the form

$$I = \exp\left(\pi\sigma Z/\beta_a H\right)\left[\frac{\vartheta_4\left\{\dfrac{\pi}{2a}(\sigma a + Z)\middle| i\dfrac{\beta_a H}{2a}\right\}}{\vartheta_4\left\{\dfrac{\pi}{2a}(\sigma a - Z)\middle| i\dfrac{\beta_a H}{2a}\right\}}\right]^{\frac12}. \tag{194}$$

An expression for the constant K in (169) can be found by eliminating the last term of (170) by (171), and then evaluating the integrals as above. Using this expression for K and (194) in (169) we find that

$$\tau(Z) = \frac{1}{2a}\left[\frac{\vartheta_4\left\{\dfrac{\pi}{2a}(\sigma a + Z)\right\}}{\vartheta_4\left\{\dfrac{\pi}{2a}(\sigma a - Z)\right\}}\right]^{\frac12}\int_{AB}\theta_s\, e^{\pi\sigma(Z-x)/\beta_a H}$$

$$\times\left|\frac{\vartheta_4\left\{\dfrac{\pi}{2a}(\sigma a - x)\right\}}{\vartheta_4\left\{\dfrac{\pi}{2a}(\sigma a + x)\right\}}\right|^{\frac12}\left\{\coth\frac{\pi}{\beta_a H}(x - Z) - 1\right\}dx, \tag{195}$$

on using boundary conditions (iv) and (v).

If the perforations are relatively small, $H/a \simeq \infty$, and the theta functions are nearly zero. In the limit $H/a \to \infty$ (195) reduces to

$$\tau(Z) = \frac{1}{\beta_a H}\int_{AB}\theta_s\, e^{\pi\sigma(Z-x)/\beta_a H}\left\{\coth\frac{\pi}{\beta_a H}(x - Z) - 1\right\}dx. \tag{196}$$

The wall we now have behaves just like a porous wall for which $\epsilon_h = \tfrac12\sigma$; this can be shown as follows. Set $\Omega_J = 0$ in (182), and use (183) to eliminate Ω_∞. Then the limit $b \to 0$ and the linearization of the resulting equation yields (196), except that σ is replaced by $2k = 2\epsilon_h$. Thus the effective value of ϵ for our tunnel wall is $\tfrac12\sigma$, and (193) is verified for this case. The results given in this section are due to Woods (1957a).

SLOTTED WALLS

7.30　Mathematical formulation

If the gaps in the wall of the tunnel shown in Fig. 7.29 are turned through a right-angle, they will then lie parallel to the tunnel axis. In this case the walls are described as being 'slotted' rather than perforated, and a quite different mathematical approach is required. This is because the gaps or slots now render the flow three-dimensional by imposing a periodic circulating 'cross-flow' on the basic two-dimensional flow. This is illustrated in Fig. 7.30. Efforts have been made to find exact solutions of this three-dimensional flow (Tomlinson, 1954), but even for the simple case of a symmetrical aerofoil at zero lift on the tunnel axis the result is a very complicated set of equations, which require numerical methods for their complete solution.

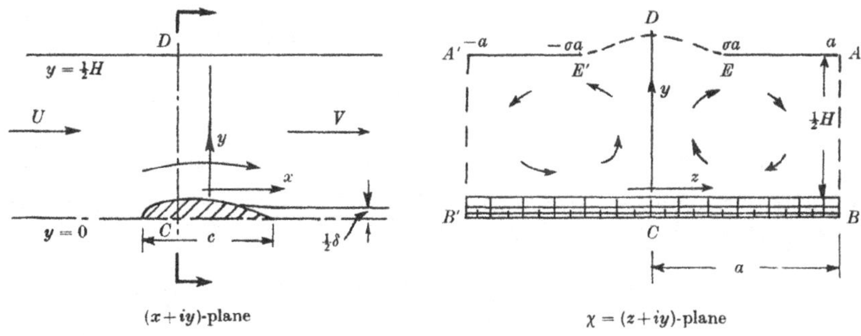

Fig. 7.30

A much simpler treatment of the problem can be obtained if it is assumed that the flow is quasi-plane (§ 2.23), i.e. that the cross-flow produced by the slots is only a small perturbation of the basic two-dimensional flow. These flows are assumed to be independent, except for a linking through the boundary conditions at the wall. The method thus has much in common with slender body theory (see pp. 96–100, Sears, 1954). When the cross-flow problem has been solved an average boundary condition for the mean or basic two-dimensional flow can be deduced. Several authors have given various derivations of this averaged boundary condition (see Maeder & Wood, 1956, who give references to the work of Guderly, Goethert, Davis and Moore on this topic; also see Baldwin, Turner & Knechtel, 1954) but these derivations are not very satisfactory for the following reason. The averaged boundary condition is calculated on the assumption that $2a/H \simeq 0$, where $2a$ is the distance from the centre of one slot to the next, and H is the tunnel height (this assumption is wrongly accepted by some authors as being essential to the validity of using averaged boundary conditions), but later this boundary condition is used to derive a relation involving $2a/H$ and σ, the ratio of open to total boundary (see Fig. 7.30). We shall give a new derivation, which overcomes this defect.

Let the axes Ox, Oy, Oz be chosen as shown in Fig. 7.30, and let (u, v, w) be the velocity components in these directions. The velocity upstream at infinity will be U, and that downstream at infinity, V. We shall adopt linear perturbation theory and ignore second-order terms in $u - U$, v and w. In steady irrotational three-dimensional flow it follows from (1.11), (1.16), (1.18) and (1.22) that

$$\frac{\partial u}{\partial x} + \frac{\partial v}{\partial y} + \frac{\partial w}{\partial z} + \frac{1}{\rho}\left(u\frac{\partial \rho}{\partial x} + v\frac{\partial \rho}{\partial y} + w\frac{\partial \rho}{\partial z}\right) = 0, \tag{197}$$

$$\frac{\partial v}{\partial x} = \frac{\partial u}{\partial y}, \quad \frac{\partial w}{\partial x} = \frac{\partial u}{\partial z}, \quad \frac{\partial w}{\partial y} = \frac{\partial v}{\partial z}, \tag{198}$$

$$u = \frac{\partial \phi}{\partial x}, \quad v = \frac{\partial \phi}{\partial y}, \quad w = \frac{\partial \phi}{\partial z}, \tag{199}$$

and

$$\frac{dp}{\rho} = \frac{a^2}{\rho}d\rho = -q\,dq. \tag{200}$$

In linear perturbation theory the last two terms of (197) can be ignored, while by (200) the first can be written

$$\frac{u}{\rho}\frac{\partial \rho}{\partial x} = -\frac{u}{2a^2}\frac{\partial q^2}{\partial x} \simeq -\frac{u}{2a^2}\frac{\partial u^2}{\partial x} \simeq -\frac{U^2}{a_a^2}\frac{\partial u}{\partial x} = -M_a^2\frac{\partial u}{\partial x},$$

M_a being the Mach number upstream at infinity. Therefore, as $\beta_a^2 = 1 - M_a^2$, (197) becomes

$$\beta_a^2\frac{\partial u}{\partial x} + \frac{\partial v}{\partial y} + \frac{\partial w}{\partial z} = 0. \tag{201}$$

As we are assuming the slots to be all of equal width, and to be uniformly distributed across the tunnel wall, the flow pattern must be periodic in the Oz-direction. Let $(\bar{u}, \bar{v}, \bar{w})$ denote average values over one period $-a \leqslant z \leqslant a$, i.e.

$$\bar{u} = \frac{1}{2a}\int_{-a}^{a} u\,dz, \quad \bar{v} = \frac{1}{2a}\int_{-a}^{a} v\,dz, \quad \overline{w} = \frac{1}{2a}\int_{-a}^{a} w\,dz, \tag{202}$$

then these quantities will depend only on x and y. It is clear from Fig. 7.30 that if we chose the Oz-axis to lie on a line of symmetry passing through the centre of a slot, then w will be an odd function of z. Thus

$$\overline{w} = 0. \tag{203}$$

If the first of (198) and (201) are now averaged over one period the result is

$$\frac{\partial \overline{v}}{\partial x} - \frac{\partial \overline{u}}{\partial y} = 0, \quad \beta_a^2 \frac{\partial \overline{u}}{\partial x} + \frac{\partial \overline{v}}{\partial y} = 0. \tag{204}$$

Let

$$\tau \equiv \Omega + i\theta = \beta_a \left(1 - \frac{\overline{u}}{U}\right) + i\frac{\overline{v}}{U}, \tag{205}$$

then as $\overline{u} \simeq \overline{q}$, Ω and θ are the speed parameter and the flow direction of the mean two-dimensional flow, and it follows from (204) and (205) that for this flow

$$\tau = \tau(Z), \quad (Z = x + i\beta_a y) \tag{206}$$

(cf. (6.22) and (6.23)). This equation plus the corresponding averaged boundary conditions determine the mean flow. However before these boundary conditions can be obtained it is necessary to calculate the cross-flow.

To find a suitable equation for the cross-flow, let

$$u = \overline{u} + \tilde{u}, \quad v = \overline{v} + \tilde{v}, \quad w = \tilde{w}, \quad \phi = \overline{\phi} + \tilde{\phi}. \tag{207}$$

Substituting these into (201), and taking (204) into account we get

$$\beta_a^2 \frac{\partial \tilde{u}}{\partial x} + \frac{\partial \tilde{v}}{\partial y} + \frac{\partial \tilde{w}}{\partial z} = 0. \tag{208}$$

We now introduce the approximation that $\partial \tilde{u}/\partial x$ is small compared with $\partial \tilde{v}/\partial y$ and $\partial \tilde{w}/\partial z$. This is the usual approximation of slender body theory, and is valid if a typical downstream distance is much greater than a typical cross-stream distance, e.g. if $2a/H$ is small. In the present case, however, this restriction is not really essential; this can be shown as follows. The velocity u has two distinct components, viz. u_a due to the aerofoil *alone*, i.e. with the walls at infinity, and u_w the increment to u caused by the walls. This increment u_w can be written $u_w = \overline{u}_w + \tilde{u}$ as u_a is obviously independent of z. With walls that are either completely solid or completely open $\partial u_w/\partial x$ is a very small quantity over the aerofoil surface. In fact from (62) and (144) we see that this quantity is approximately $k(\pi c U C_D/\beta_a^3 H^2)$, where $k = \frac{1}{12}$ with solid walls and $\frac{1}{48}$ with open 'walls'. Thus with slotted walls the variations in $\partial u_w/\partial x$ with z will certainly not exceed $(\pi c U C_D/16\beta_a^3 H^2)$, and will in fact be a good deal less. This means that $\partial \tilde{u}/\partial x$ is quite small, regardless of the magnitude of the ratio $2a/H$, and as the variations we have described are greater at the aerofoil than elsewhere in the tunnel, this remark applies generally.

We are now able to replace (208) by

$$\frac{\partial \tilde{v}}{\partial y} + \frac{\partial \tilde{w}}{\partial z} = 0,$$

which by (199) and the last of (207) is equivalent to

$$\frac{\partial^2 \tilde{\phi}}{\partial y^2} + \frac{\partial^2 \tilde{\phi}}{\partial z^2} = 0. \tag{209}$$

This is the equation governing the cross-flow.

The boundary conditions for the cross-flow are:

(i) On the solid sections of the walls,

$$0 = v = \partial \phi / \partial y = \partial(\tilde{\phi} + \bar{\phi}) / \partial y,$$

or
$$\frac{\partial \tilde{\phi}}{\partial y} = -\frac{\partial \bar{\phi}}{\partial y}. \tag{210}$$

(ii) In the gaps the pressure is constant, hence $q = U$ or, ignoring second-order terms, $u = \partial(\bar{\phi} + \tilde{\phi}) / \partial x = U$. Let the origins of ϕ and x coincide, then this boundary condition can be written

$$\tilde{\phi} = -\bar{\phi} + Ux. \tag{211}$$

(iii) On $z = \pm a$ the periodicity of the flow (see Fig. 7.30), and resulting symmetry, requires that $w = \tilde{w} = 0$, so

$$\frac{\partial \tilde{\phi}}{\partial z} = 0. \tag{212}$$

(iv) On the aerofoil surface $v = U\theta_s$, where θ_s is the slope of the surface. But as θ_s is independent of z, $\bar{v} = U\theta_s$, whence $\tilde{v} = 0$, or

$$\frac{\partial \tilde{\phi}}{\partial y} = 0. \tag{213}$$

This completes the mathematical formulation of the problem.

7.31   The mean boundary conditions

The first step in determining the mean boundary conditions is to calculate the cross-flow. Fig. 7.31 shows the χ- ($= z + iy$) plane, and the boundary conditions in this plane. It will be noticed that the normal derivative is specified everywhere, except on the gap $E'E$.

The χ-plane is mapped on to the upper half of the t-plane shown by

$$t = \text{sn} \, (\lambda \chi, k) \tag{214}$$

(cf. (5.109)), where the modulus k and parameter λ are determined by

$$\lambda = \frac{K}{a} = \frac{2K'}{H}. \tag{215}$$

At E, $t = t_0$ and $\chi = \sigma a + \dfrac{i}{2} H = \sigma a + iK'/\lambda$, so

$$t_0 = \operatorname{sn}(\lambda \sigma a + iK') = \frac{1}{k}\operatorname{ns}(\lambda \sigma a) = \frac{1}{k}\operatorname{ns}(K\sigma), \qquad (216)$$

by (4.48).

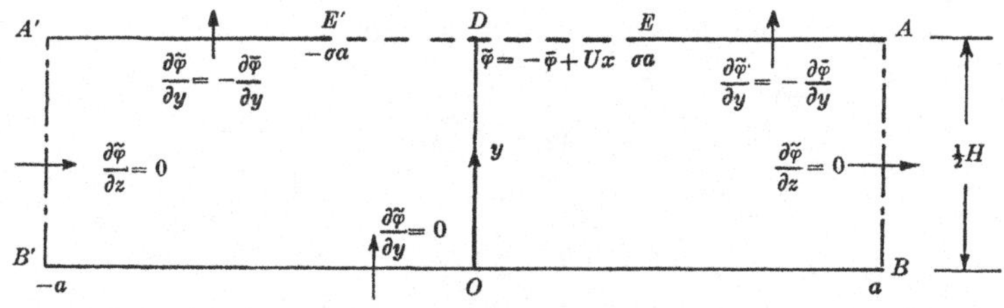

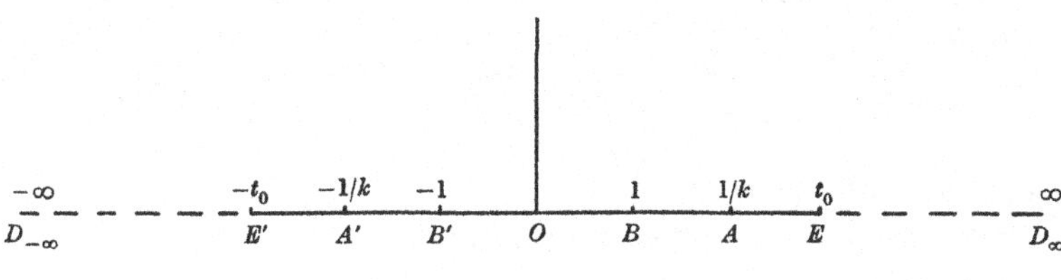

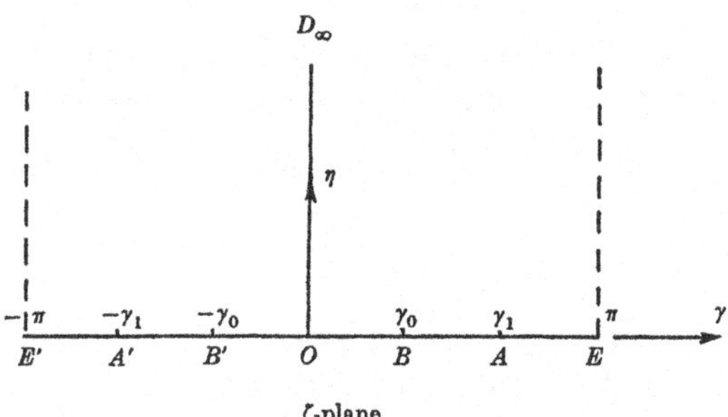

Fig. 7.31

The t-plane is mapped into the ζ-plane shown in Fig. 7.31 by

$$t = t_0 \sin \tfrac{1}{2}\zeta. \qquad (217)$$

as is readily verified from the Schwarz–Christoffel mapping theorem
(§ 5.3).

Combining (214) to (217) we now have

$$\sin \tfrac{1}{2}\zeta = k \operatorname{sn}(K\sigma) \operatorname{sn}(K\chi/a). \qquad (218)$$

The points B and A at $\chi = a$ and $\chi = a + iK'a/K$ map on to $\zeta = \gamma_0$ and $\zeta = \gamma_1$, and hence by results given in § 4.6

$$\sin \tfrac{1}{2}\gamma_0 = k \operatorname{sn}(K\sigma), \tag{219}$$

and

$$\sin \tfrac{1}{2}\gamma_1 = \operatorname{sn}(K\sigma). \tag{220}$$

Let $\bar{\psi}$ be the harmonic conjugate of $\bar{\phi}$, then

$$\bar{\phi} + i\bar{\psi} = \frac{\cos \tfrac{1}{2}\zeta}{2\pi} \left\{ \int_{-\pi}^{\pi} \frac{\bar{\psi}(\gamma)\, d\gamma}{\sin \tfrac{1}{2}\gamma - \sin \tfrac{1}{2}\zeta} \right.$$
$$\left. + \int_0^\infty \left(\frac{\bar{\phi}^+(\eta)}{\cosh \tfrac{1}{2}\eta - \sin \tfrac{1}{2}\zeta} + \frac{\bar{\phi}^-(\eta)}{\cosh \tfrac{1}{2}\eta + \sin \tfrac{1}{2}\zeta} \right) d\eta \right\}, \tag{221}$$

a result readily deduced from (4.138) (see derivation of (11.8)). Now on $\gamma = \pm\pi$ both $\phi^+(\eta)$ and $\phi^-(\eta)$ have the value $-\bar{\phi} + Ux$, which is independent of η. Substituting this value in (221), and integrating the first term by parts, we get

$$\bar{\phi}(\gamma^*) = -\frac{1}{\pi} \int_{-\pi}^{\pi} \left(\frac{\partial \bar{\psi}}{\partial \gamma} \right) \ln \left| \frac{\sin \tfrac{1}{4}(\gamma - \gamma^*)}{\cos \tfrac{1}{4}(\gamma + \gamma^*)} \right| d\gamma - \bar{\phi} + Ux, \tag{222}$$

at a point $\gamma = \gamma^*$ on $\eta = 0$.

By the Cauchy–Riemann equation, on $\eta = 0$, we have

$$\frac{\partial \bar{\psi}}{\partial \gamma} = -\frac{\partial \bar{\phi}}{\partial \eta} = -\frac{\partial \bar{\phi}}{\partial n} \left(\frac{\partial n}{\partial \eta} \right),$$

where n is measured along the normal to the boundary in the χ-plane. It is apparent from Fig. 7.31 that $\partial \bar{\psi}/\partial \gamma$ vanishes except in $-\pi < \gamma < -\gamma_1$, $\gamma_1 < \gamma < \pi$, where

$$\frac{\partial \bar{\psi}}{\partial \gamma} = \frac{\partial \bar{\phi}}{\partial y} \frac{\partial y}{\partial \eta}.$$

Differentiation of (218) gives

$$\frac{d\chi}{d\zeta} = \frac{a \operatorname{ns}(K\sigma) \cos \tfrac{1}{2}\zeta}{2Kk \operatorname{cn}(K\chi/a) \operatorname{dn}(K\chi/a)}, \tag{223}$$

by a result given in § 4.6. On $\zeta = \gamma$, $\chi = x + \dfrac{i}{2}H$, this gives

$$\frac{\partial y}{\partial \eta} = -\frac{a \operatorname{ns}(K\sigma) \cos \tfrac{1}{2}\gamma}{2K \operatorname{ds}(Kx/a) \operatorname{cs}(Kx/a)},$$

where by (218) and (220) $\operatorname{ns}(Kx/a) = \sin \tfrac{1}{2}\gamma / \sin \tfrac{1}{2}\gamma_1$. Hence

$$\frac{\partial \bar{\psi}}{\partial \gamma} = -\frac{\partial \bar{\phi}}{\partial y} \frac{a \operatorname{ns}(K\sigma)}{2K} \frac{\cos \tfrac{1}{2}\gamma}{\left\{ \left(\dfrac{\sin \tfrac{1}{2}\gamma}{\sin \tfrac{1}{2}\gamma_1} \right)^2 - k^2 \right\}^{\frac{1}{2}} \left\{ \left(\dfrac{\sin \tfrac{1}{2}\gamma}{\sin \tfrac{1}{2}\gamma_1} \right)^2 - 1 \right\}^{\frac{1}{2}}}.$$

Substituting this value in (222) we obtain

$$\check{\phi}(\gamma^*) = -\overline{\phi} + Ux + A(\gamma^*)\frac{\partial\overline{\phi}}{\partial y}, \qquad (224)$$

where

$$A(\gamma^*) \equiv \frac{a\,\mathrm{ns}\,(K\sigma)}{2\pi K}\int_{\gamma_1}^{\pi}\frac{\cos\tfrac{1}{2}\gamma\ln\left|\dfrac{\cos\tfrac{1}{2}\gamma - \cos\tfrac{1}{2}\gamma^*}{\cos\tfrac{1}{2}\gamma + \cos\tfrac{1}{2}\gamma^*}\right|d\gamma}{\left\{\left(\dfrac{\sin\tfrac{1}{2}\gamma}{\sin\tfrac{1}{2}\gamma_1}\right)^2 - k^2\right\}^{\frac{1}{2}}\left\{\left(\dfrac{\sin\tfrac{1}{2}\gamma}{\sin\tfrac{1}{2}\gamma_1}\right)^2 - 1\right\}^{\frac{1}{2}}}. \qquad (225)$$

From the definition of $\check{\phi}$ its average value over $-a < z < a$ is zero, so taking the average of (224) with respect to z we get

$$\overline{A}\frac{\partial\overline{\phi}}{\partial y} - \overline{\phi} + Ux = 0, \qquad (226)$$

where

$$\overline{A} = \frac{1}{2a}\int_{-\gamma_0}^{\gamma_0} A(\gamma^*)\frac{dz}{d\gamma^*}d\gamma^*$$

$$= \frac{\mathrm{ns}\,(K\sigma)}{4Kk}\int_{-\gamma_0}^{\gamma_0}\frac{A(\gamma)\cos\tfrac{1}{2}\gamma\,d\gamma}{\left\{1 - \left(\dfrac{\sin\tfrac{1}{2}\gamma}{\sin\tfrac{1}{2}\gamma_0}\right)^2\right\}^{\frac{1}{2}}\left\{1 - k^2\left(\dfrac{\sin\tfrac{1}{2}\gamma}{\sin\tfrac{1}{2}\gamma_0}\right)^2\right\}^{\frac{1}{2}}}, \qquad (227)$$

by (219), (220) and (223).

The required averaged boundary conditions now follow by differentiating (226) with respect to x:

$$\overline{A}\frac{\partial\overline{v}}{\partial x} + U - \overline{u} = 0,$$

or by (205)

$$\lambda\frac{\partial\theta}{\partial x} - \Omega = 0, \qquad (228)$$

where

$$\lambda = -\beta_a\overline{A}. \qquad (229)$$

The published work cited in the second paragraph of § 7.30 applies only to the limiting case $2a/H = 0$. By (215) this is the limit $K/K' = 0$, and from (4.65), (4.66) and (4.57) this means that $K = \tfrac{1}{2}\pi$, $k = 0$ and $\mathrm{sn}\,(K\sigma) = \sin(\tfrac{1}{2}\pi\sigma)$. It now follows from (219), (220), (225) and (227) that in this limit

$$\lambda = -\beta_a\overline{A} = -\beta_a A(0)$$

$$= \beta_a\frac{a\sin\tfrac{1}{2}\gamma_1}{\pi^2}\int_{\gamma_1}^{\pi}\cot\tfrac{1}{2}\gamma\,\frac{\ln\{(1+\cos\tfrac{1}{2}\gamma)/(1-\cos\tfrac{1}{2}\gamma)\}}{\{\sin^2\tfrac{1}{2}\gamma - \sin^2\tfrac{1}{2}\gamma_1\}^{\frac{1}{2}}}d\gamma,$$

whence

$$\lambda - \frac{2a\beta_a}{\pi}\left|\ln\sin\frac{\pi\sigma}{2}\right|. \qquad (230)$$

A more useful special case is obtained by assuming σ to be small, for it is shown in the next section that the values of σ which eliminate tunnel blockage are always less than $0\cdot1$ regardless of the value of $2a/H$. If σ is small it follows from (219) and (220) that γ_0 and γ_1 are small. Then by (225) and (227) $\overline{A} = A(0)$, which reduces to the same integral as obtained in the derivation of (230), except that $K \neq \frac{1}{2}\pi$ and $\frac{1}{2}\gamma_1 \simeq K\sigma$. Thus in this case

$$\lambda = \frac{a\beta_a}{K}\,|\ln K\sigma|. \tag{231}$$

7.32   Blockage in a slotted tunnel

We have now reduced the problem of determining the flow past an aerofoil in a slotted tunnel to that of finding $\tau(Z)$ in the strip $-\infty < x < \infty$, $0 < y < \frac{1}{2}H$, subject to the boundary conditions

$$\theta = \theta_0(x) \quad \text{on} \quad y = 0, \quad \lambda\left(\frac{\partial\theta}{\partial x}\right)_h - \Omega_h = 0 \quad \text{on} \quad y = \tfrac{1}{2}H, \tag{232}$$

where the subscripts 0 and h have their usual significance (see §3.16). This problem is readily solved by the transform method described in §4.23.

We shall put $q = U$ at $x = -\infty$, so that $\Omega_{-\infty} = 0$, then as $H_{-\infty} = \frac{1}{2}H$ (see Fig. 7.2) it follows from (4) that

$$h = \tfrac{1}{2}\beta_a HU. \tag{233}$$

Now substitute $\zeta = \pi w/2h \simeq \pi Z/\beta_a H$ in (86), then from (54) and the theory of §7.13 it follows that the resulting equation will be the linear-theory solution to the problem 'find $\tau(Z)$ given $\theta_0(x)$ and $\Omega_h(x)$'. On $y = 0$ and $y = \frac{1}{2}H$ this solution gives

$$\Omega_0(x^*) = \frac{1}{2b}\int_{-\infty}^{\infty}\left\{\theta_0(x)\operatorname{cosech}\frac{\pi}{2b}(x-x^*) + \Omega_h(x)\operatorname{sech}\frac{\pi}{2b}(x-x^*)\right\}dx,$$

and

$$\theta_h(x^*) = \frac{1}{2b}\int_{-\infty}^{\infty}\left\{\theta_0(x)\operatorname{sech}\frac{\pi}{2b}(x-x^*) + \Omega_h(x)\operatorname{cosech}\frac{\pi}{2b}(x-x^*)\right\}dx,$$

$$\tag{234}$$

respectively, where $\qquad\qquad b \equiv \tfrac{1}{2}\beta_a H.$ $\qquad\qquad$ (235)

From transforms given in §4.23 we have

$$\mathscr{L}\left\{\operatorname{sech}\frac{\pi x}{2b};p\right\} = 2p\sec pb,$$

and $\quad \mathscr{L}\left\{\operatorname{cosech}\frac{\pi x}{2b};p\right\} = p\mathscr{L}\left\{\int_{-\infty}^{x}\operatorname{cosech}\frac{\pi x}{2b}\,dx;p\right\}$

$$= \frac{2pb}{\pi}\mathscr{L}\left\{\ln\left|\tanh\frac{\pi x}{4b}\right|;p\right\} = -2pb\tan pb.$$

With these and with the aid of (4·199) we can transform (234) into

$$\mathscr{L}\{\Omega_0(x);\, p\} = \tan pb\, \mathscr{L}\{\theta_0(x);\, p\} + \sec pb\, \mathscr{L}\{\Omega_h(x);\, p\},$$

$$\mathscr{L}\{\theta_h(x);\, p\} = \sec pb\, \mathscr{L}\{\theta_0(x);\, p\} + \tan pb\, \mathscr{L}\{\Omega_h(x);\, p\}.$$

The second boundary condition in (232) transforms by (4.198) into

$$p\lambda\, \mathscr{L}\{\theta_h(x);\, p\} - \mathscr{L}\{\Omega_h(x);\, p\} = 0.$$

On eliminating Ω_h and θ_h we get

$$\mathscr{L}\{\Omega_0(x);\, p\} = \mathscr{L}\{\theta_0(x);\, p\}\left\{\frac{p\lambda + \tan pb}{1 - p\lambda \tan pb}\right\}$$

$$= -\frac{|p|}{p}\,\mathscr{L}\{\theta_0(x);\, p\} + \frac{1}{p}\,\mathscr{L}\{\theta_0(x);\, p\}\,\mathscr{L}\{F(x);\, p\}, \quad (236)$$

where from (4.194)

$$F(x) = \frac{1}{2\pi i}\int_{c-i\infty}^{c+i\infty} e^{px}\left\{\frac{|p|}{p} + \frac{p\lambda + \tan pb}{1 - p\lambda \tan pb}\right\} dp. \quad (237)$$

Van der Pol & Bremmer (1950) give $\mathscr{L}\left\{\dfrac{1}{x};\, p\right\} = \pi\,|p|$, consequently using (4·199) to invert (236) we obtain

$$\Omega_0(x) = \frac{1}{\pi}\int_{-\infty}^{\infty} \frac{\theta_0(x^*)\, dx^*}{x^* - x} + \int_{-\infty}^{\infty} \theta_0(x^*)\, F(x - x^*)\, dx^*. \quad (238)$$

As the first term on the right-hand side is the result for an aerofoil in an infinite stream (cf. (60)), the second term must represent the increment to Ω_0 due to the slotted wall.

Equation (237) for $F(x)$ can be expressed in real form as follows. The real numbers c and $\mathscr{R}p$ must be chosen to secure the convergence of the integral, otherwise they are arbitrary. Let $c = 0$, and $p = -i\omega/b$, where ω is real, then (237) can be transformed into

$$F(x) = \frac{1}{\pi b}\int_0^{\infty} \frac{(b - \omega\lambda)\, e^{-w}\sin(\omega x/b)}{b\cosh\omega + \omega\lambda\sinh\omega}\, d\omega. \quad (239)$$

The blockage factor defined in (52) can now be written down directly from (238). Comparing (238) with (55), (60) and (61) we conclude that

$$\epsilon(x) = -\frac{1}{\beta_a}\int_{-\frac{1}{2}c}^{\frac{1}{2}c}\left(\frac{dy}{dx^*}\right)_0 F(x - x^*)\, dx^*,$$

on replacing θ_0 by $(dy/dx^*)_0$. Integration by parts gives

$$\epsilon(x) = \frac{\delta}{2\beta_a} F\left(\frac{c}{2} - x\right) - \frac{1}{\beta_a}\int_{-\frac{1}{2}c}^{\frac{1}{2}c} y_0(x^*)\, F'(x^* - x)\, dx^*, \quad (240)$$

as $y_0 = 0$ at $x = -\frac{1}{2}c$ and $y_0 = \frac{1}{2}\delta$ at $x = \frac{1}{2}c$, where δ is given by (57) (see Fig. 7.30).

A useful approximation for $\epsilon(x)$ can be found by following the method given in §7.8 of expanding the integrand in (240) in a series in $(x - x^*)/H$ and ignoring terms $O(H^{-3})$. The result is

$$\epsilon(x) = \frac{GA}{\beta_a^3 H^2} + \frac{\delta}{2\beta_a} F\left(\frac{c}{2} - x\right), \tag{241}$$

where from (235) and (239) G is a function of λ/b given by

$$G = -\frac{\beta_a^2 H^2}{2} F'(0) = \frac{2}{\pi} \int_0^\infty \frac{\omega(\omega\lambda/b - 1)\, e^{-\omega}\, d\omega}{\cosh\omega + \omega\,(\lambda/b)\sinh\omega}. \tag{242}$$

The ratio λ/b is $2(-\bar{A})/H$ by (229) and (235), and is therefore independent of the compressibility factor β_a. With small slots it follows from (231) that

$$\frac{\lambda}{b} = \frac{2a}{H}\frac{|\ln K\sigma|}{K}. \tag{243}$$

With completely solid walls $\sigma = 0$, so by (219), (220) and (227) $\gamma_0 = \gamma_1 = 0$, and $-\beta_a\bar{A} = \lambda = \infty$. Hence

$$G = \frac{2}{\pi} \int_0^\infty \frac{\omega\, e^{-\omega}}{\sinh\omega}\, d\omega = \tfrac{1}{6}\pi,$$

while with completely open 'walls', $\sigma = 1$, so (219), (220) and (225) give $\gamma_0 = \gamma_1 = \pi$, and $-\beta_a\bar{A} = \lambda = 0$. Thus

$$G = -\frac{2}{\pi} \int_0^\infty \frac{\omega\, e^{-\omega}}{\cosh\omega}\, d\omega = -\tfrac{1}{12}\pi.$$

These results show (241) to be in agreement with results given in §§7.8 and 7.21.

Zero average blockage can be obtained by choosing λ/b so as to make $\epsilon(0)$ vanish. Suppose C_D is negligible, then the wake blockage term in (241) can be ignored, and the condition for zero blockage reduces to $G = 0$. This equation can be solved numerically; let its solution be $\lambda/b = r$. If σ is small, this value can be substituted in (243) to give

$$\sigma = \frac{1}{K} e^{-rK(H/2a)} = \frac{1}{K} e^{-rK'}, \tag{244}$$

by (215). From a graph given by Baldwin *et al.* (1954) it appears that $r \simeq 1.2$. This value, (215) and a table of elliptic functions enables us to draw up the following table:

$2a/H$	4	2	1	$\frac{1}{2}$	$\frac{1}{4}$	$\frac{1}{8}$
σ	0·024	0·047	0·084	0·0143	0·0003	0·0000

There are two points to note here: first that a remarkably small value of σ is sufficient to eliminate solid blockage, and secondly that σ has a maximum value at about $2a/H = 1$.

BRANCHED CHANNELS

7.33　Transformation of the w-plane

Consider the flow through the branched channel of given shape shown in Fig. 7.33. For the flow to be uniquely defined we prove below that it is necessary to define (1) the directions of flow in all branches, (2) the positions of all stagnation points, S, and (3) the velocity of *one* point in the channel (this is usually the velocity at infinity in one of the upstream branches). Of course if we are able to prescribe velocities at more points, e.g. at the three points at infinity in the downstream branches, then the stagnation points must be free to assume their appropriate positions, otherwise the flow will be overdetermined. However, as the stagnation points will tend to move to the points of maximum curvature on the channel walls under the action of viscosity (see discussion of Joukowski's hypothesis in § 1.17), they must normally be regarded as being fixed points. If there are sharp corners appropriately placed on the boundaries, we can assume that the stagnation points coincide with them. In this discussion we are of course assuming that the flow does not separate from the channel wall at any point.

In Fig. 7.33 there are shown two upstream channels and three downstream channels. We shall generalize this and suppose there are m upstream and n downstream channels, the upstream channels being separated by $\psi = h'_r$, $-\infty \leqslant \phi \leqslant a'_r$ ($r = 1, 2, ..., m-1$), and the downstream channels by $\psi = h_r$, $a_r \leqslant \phi \leqslant \infty$ ($r = 1, 2, ..., n-1$). The w-plane of this flow is mapped into the upper half of the ζ-plane by

$$\frac{dw}{d\zeta} = K \frac{\prod\limits_{i=1}^{n-1}(\zeta-\delta_i)\prod\limits_{i=1}^{m-1}(\zeta-\delta'_i)}{\prod\limits_{i=1}^{n}(\zeta-\gamma_i)\prod\limits_{i=1}^{m}(\zeta-\gamma'_i)},$$

as is readily verified from the Schwarz–Christoffel mapping formula (5.25). This can be factorized and then integrated to give

$$w = \frac{1}{\pi}\sum_{r=1}^{n}h_r\ln(1-\zeta/\gamma_r) - \frac{1}{\pi}\sum_{r=1}^{m}h'_r\ln(1-\zeta/\gamma'_r), \tag{245}$$

where

$$h_r = \pi K \frac{\prod\limits_{i=1}^{n-1} (\gamma_r - \delta_i) \prod\limits_{i=1}^{m-1} (\gamma_r - \delta_i')}{\prod\limits_{\substack{i=1 \\ i \neq r}}^{n} (\gamma_r - \gamma_i) \prod\limits_{i=1}^{m} (\gamma_r - \gamma_i')} \quad (r = 1, 2, \ldots, n),$$

and

$$h_r' = -\pi K \frac{\prod\limits_{i=1}^{n-1} (\gamma_r' - \delta_i) \prod\limits_{i=1}^{m-1} (\gamma_r' - \delta_i')}{\prod\limits_{i=1}^{n} (\gamma_r' - \gamma_i) \prod\limits_{\substack{i=1 \\ i \neq r}}^{m} (\gamma_r' - \gamma_i')} \quad (r = 1, 2, \ldots, m).$$

$$(246)$$

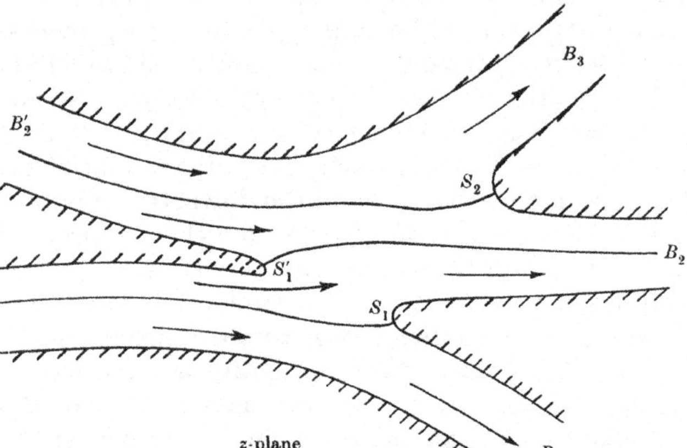

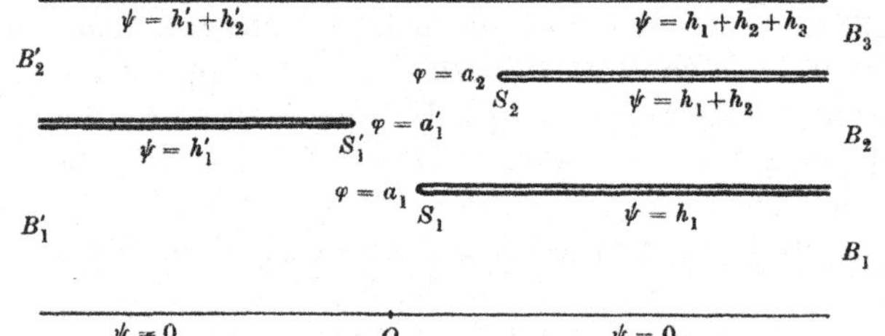

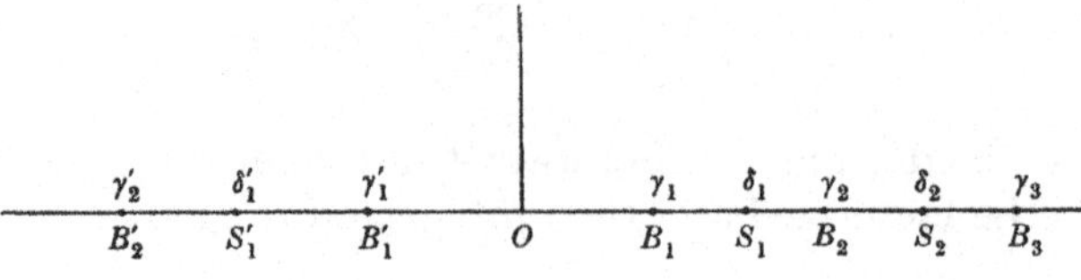

Fig. 7.33

As $\mathscr{I}\ln(1-\gamma/\gamma_r)$ increases by π when γ increases through γ_r, it follows that $\mathscr{I}w = \psi$ increases by h_r, and the interpretation of h_r shown in Fig. 7.33 is verified. In addition to (245) we have the continuity equation

$$\sum_{r=1}^{m} h'_r = \sum_{r=1}^{n} h_r,\qquad(247)$$

and

$$\left.\begin{aligned}
a_r &= \frac{1}{\pi}\sum_{s=1}^{n} h_s \ln\left|1-\delta_r/\gamma_s\right| - \frac{1}{\pi}\sum_{s=1}^{m} h'_s \ln\left|1-\delta_r/\gamma'_s\right| \quad (r = 1, 2, \ldots, n-1), \\
a'_r &= \frac{1}{\pi}\sum_{s=1}^{n} h_s \ln\left|1-\delta'_r/\gamma_s\right| - \frac{1}{\pi}\sum_{s=1}^{m} h'_s \ln\left|1-\delta'_r/\gamma'_s\right| \quad (r = 1, 2, \ldots, m-1).
\end{aligned}\right\}$$
$$(248)$$

If a_r, h_r, a'_r and h'_r are given, (246)–(248) fix the values of K, γ_r, δ_r, γ'_r and δ'_r.

In the next section it will be shown how h_r and h'_r can be calculated. Relations involving a_r, a'_r can be obtained by integration of $ds = d\phi/q$ along the boundaries up to the stagnation points, but if these boundaries are curved, an integral equation must first be solved (cf. theory given in §5.4).

7.34   Solution in the ζ-plane

If the channel shape is known then from §3.18 the appropriate solution for $\tau(\zeta)$ is

$$\tau(\zeta) = \frac{1}{\pi}\int_{-\infty}^{\infty} \frac{\theta(\gamma)\,d\gamma}{\gamma-\zeta} + A,\qquad(249)$$

where A is a real constant. To evaluate this constant we need to know the velocity at some point in the flow. Suppose q equals the reference velocity U at infinity in the sth upstream branch, i.e. at $\gamma = \gamma'_s$, then Ω vanishes at this point, whence

$$A = -\frac{1}{\pi}\int_{-\infty}^{\infty} \frac{\theta(\gamma)\,d\gamma}{\gamma-\gamma'_s}.\qquad(250)$$

Let Ω_r, Ω'_r denote the values of Ω at the points at infinity, $\gamma = \gamma_r$ and $\gamma = \gamma'_r$, then the $(m+n)$ equations

$$\Omega_r = \frac{1}{\pi}\int_{-\infty}^{\infty} \frac{\theta(\gamma)\,d\gamma}{\gamma-\gamma_r} + A, \quad \Omega'_r = \frac{1}{\pi}\int_{-\infty}^{\infty} \frac{\theta(\gamma)\,d\gamma}{\gamma-\gamma'_r} + A,\qquad(251)$$

enable us to determine the velocities q_r, q'_r in the branches at infinity. Let H_r, H'_r denote the physical widths of the branches at infinity, then by (4)

$$h_r = \beta_a H_r U \frac{\cosh\Omega^*}{\cosh(\Omega^* + \Omega_r)} \quad (r = 1, 2, \ldots, n).\qquad(252)$$

Similar equations hold for h'_r. We can now eliminate h_r, h'_r from (246) in favour of H_r, H'_r, and the resulting complicated relations will serve to define γ_r, γ'_r provided $\theta(\gamma)$ is known completely, i.e. provided the positions of the stagnation points are known.

This completes our outline of the theory of the direct problem of branched-channel flow. With the indirect problem, or with mixed boundary conditions we must replace (249) by the appropriate formula, which can be deduced for any particular type of boundary condition from (4.164).

7.35 Flow into a cascade of channels

The case of an infinite stream flowing into a grid of identical channels can be treated by the method described in §§ 12.9 and 12.10.

CHAPTER 8

AEROFOIL THEORY

8.1 Introduction

In this chapter we study the steady inviscid flow of an infinite stream past a closed aerofoil. The aerofoil will be assumed to be of 'moderate' thickness : chord ratio (e.g. less than 15 %), and to be at incidences less than the stalling angle, i.e. less than the angle at which the flow becomes separated from the aerofoil upper surface and forms a thick turbulent wake. The first approximation in this case is the closed Dirichlet flow about the aerofoil in which the boundary layer enters only through the Joukowski condition (§ 1.17) to fix the circulation. Helmholtz or well-separated flows past thick cylinders and aerofoils at incidences above the stalling angle will be considered in a later chapter.

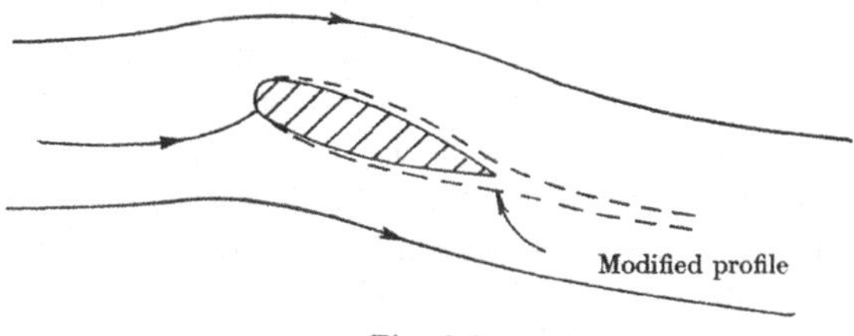

Fig. 8.1

A second approximation to the inviscid flow can be obtained by allowing for the (second-order) effects of the boundary layer and wake displacement thickness. This is done by using the first approximation to compute the boundary layer and wake displacement thickness (§ 1.16), and then superimposing these thicknesses on the original profile in the manner indicated in Fig. 8.1. This yields a semi-infinite profile, and the next step is to calculate the inviscid flow about this modified shape. In the third approximation we recompute the boundary layer and wake thickness, using the second approximation to do so, and so on. The calculation iterates between the inviscid flow and the boundary layer until it converges—or diverges. Divergence would probably indicate that the initial assumption of a Dirichlet flow was at fault, and that some flow separation occurs, but this is not likely to happen at small incidences. For the greater part of this chapter we shall be concerned with the first approximation, i.e. the Dirichlet flow about the aerofoil.

Quite a number of distinct methods have been devised to compute the inviscid low-speed flow about an aerofoil of given general shape. Notable contributors include Glauert (1924), Theodorsen (1932), Goldstein (1952), Lighthill (1951), Preston (1953) and Timman (1948): there are many others. The effects of compressibility on aerofoil characteristics have also been exhaustively investigated—references to some of the principal contributors have been given in ch. 2. On the design of aerofoils with given pressure distributions important contributions have been made by Goldstein (1952), Allen (1945), Lighthill (1945c) and Thwaites (1945).

A number of texts have devoted many chapters to aerofoil theory. The classic in this field is undoubtedly Glauert's (first edition 1926) famous *Elements of Aerofoil and Airscrew Theory*. More recent publications include *Theoretical Aerodynamics* (Milne-Thomson, 1948), *Theory of Wing Sections* (Abbott & von Doenhoff, 1949), *Wing Theory* (Robinson & Laurmann, 1956) and *Incompressible Aerodynamics* (Thwaites (ed.), 1960). Of these the text by Abbott & von Doenhoff is by far the most detailed in its treatment of the subject; experimental and theoretical results for dozens of aerofoil profiles are given.

The treatment of aerofoil theory given below—based on the function $\tau = \Omega + i\theta$ defined in ch. 2—differs somewhat from those just mentioned, although it is related to the methods of Timman (1948), Lighthill (1945, 1951), Kármán (1941) and Tsien (1939). From it can be derived the principal equations of most other methods, and further it has a clear advantage over those methods originally devised for incompressible flow and based on the relation $z = z(w)$ (cf. (2.84) and (2.86)). This advantage is its generality and adaptability, for essentially the one method deals with incompressible flow, compressible flow, with the 'direct' problem of finding the flow about a given aerofoil, and with the 'indirect' problem of aerofoil design. And as is shown below the method can be used to deal with aerofoils with mixed boundary conditions, having porous surfaces, or with sources and sinks on their surfaces. In fact the principal techniques of the method—the use of the function $\tau(w)$, conformal transformation of the w-plane, and general solutions derived from the application of Cauchy integrals—can be used consistently throughout the whole range of plane subsonic flow problems, as is evidenced by the topics covered in this book.

GENERAL AEROFOIL THEORY

8.2 The conformal transformations

It will be assumed that there is a circulation of $-\Gamma$ about the aerofoil, so that the z- and w-planes will appear as in Fig. 8.2 (cf. Fig. 5.14). A contour C closed in the z-plane will fail to close in the w-plane by an interval $\phi = \Gamma$ on $\psi = 0$. As shown in §5.14 the w-plane is mapped on to the ζ-plane shown in Fig. 8.2 by

$$w = -2a\cos(\zeta + \alpha_0) + (\zeta + \alpha_0 - \tfrac{1}{2}\pi)\, 2a\sin\alpha_0, \qquad (1)$$

where a is a constant closely related to the aerofoil chord length c, α_0 is a constant related to the circulation $-\Gamma$ by

$$\Gamma = 4\pi a \sin\alpha_0, \qquad (2)$$

and the origin of the w-plane is at $\zeta = \tfrac{1}{2}\pi - \alpha_0$.

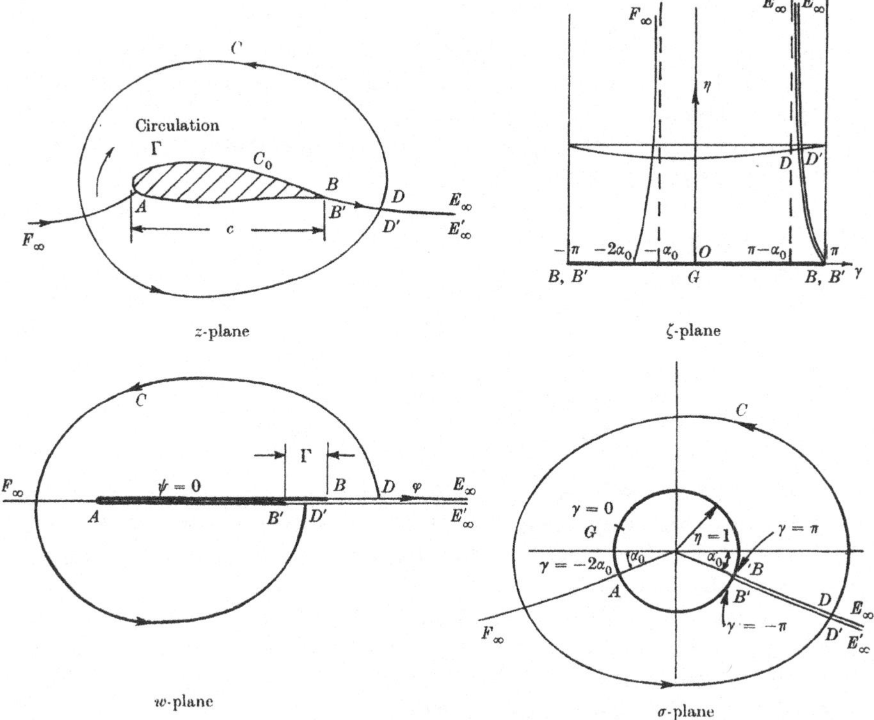

Fig. 8.2

Notice that the aerofoil is mapped on to $\eta = 0$ ($-\pi \leqslant \gamma \leqslant \pi$), and that the front stagnation point (which lies at the minimum value of ϕ on $\eta = 0$) lies at $\gamma = -2\alpha_0$. When $\psi = 0$ (1) gives that either $\eta = 0$, or

$$\frac{\eta}{\sinh\eta} = \frac{\sin(\gamma + \alpha_0)}{\sin\alpha_0}, \qquad (3)$$

which is therefore the equation of the leading and trailing edge streamlines in the ζ-plane. In the limit $\eta \to \infty$ it follows from (3) that $\gamma \to \pm(\pi - \alpha_0)$

on the trailing edge streamline, and $\gamma \to -\alpha_0$ on the leading edge streamline.

If w changes by an amount Γ exactly along a curved path in the z-plane it is obvious that this path must be a contour enclosing the aerofoil. But from (1) and (2) $w(\pi, \eta) - w(-\pi, \eta) = \Gamma$, from which we conclude that any curve in the ζ-plane joining $(-\pi, \eta)$ to (π, η) corresponds to some closed contour about the aerofoil. Thus a single-valued function $f(x, y)$ in the z-plane must satisfy $f(-\pi, \eta) = f(\pi, \eta)$ in the ζ-plane.

The σ-plane in Fig. 8.2 is defined by

$$\sigma = -e^{-i(\zeta + \alpha_0)}, \tag{4}$$

a transformation already considered in §5.7. The aerofoil maps on to $|\sigma| = 1$, and as $\sigma(-\pi, \eta) = \sigma(\pi, \eta)$ it follows that closed contours in the σ-plane correspond to closed contours in the z-plane. This last property makes the plane useful in calculating the lift and moment, as it permits the use of the residue theorem.

8.3   The solution for a given aerofoil

Suppose the aerofoil shape is given in the form $y = y(x)$ then

$$\theta = \theta_s(s) \tag{5}$$

(the subscript s denoting surface values), can be deduced from

$$\theta_s = \tan^{-1}\left(\frac{dy}{dx}\right), \quad s = \int \sqrt{\left\{1 + \left(\frac{dy}{dx}\right)^2\right\}}\, dx.$$

We now have the difficulty met in §7.2 with channels, and we shall deal with it in the same way as in §7.2. Thus we assume *pro tempore* that $\theta_s(\phi)$ is known, and subsequently employ (5) to derive an integral equation for $\theta_s'(\phi)$ from the theory (cf. (5.33)).

The function $\tau(w)$ defined in §6.1 must be an analytic function of the auxiliary variables ζ and σ. In the ζ-plane τ satisfies the following boundary conditions.

(1) On $\eta = 0$ it follows from (1) that

$$\phi = -2a \cos(\gamma + \alpha_0) + (\gamma + \alpha_0 - \tfrac{1}{2}\pi)\, 2a \sin \alpha_0, \tag{6}$$

hence from an assumed $\theta_s(\phi)$ we can deduce the boundary distribution

$$\theta = \theta_s(\gamma) \quad (-\pi \leqslant \gamma \leqslant \pi). \tag{7}$$

(2) As $\tau(x, y)$ is single valued in the z-plane it follows that

$$\tau(-\pi, \eta) = \tau(\pi, \eta). \tag{8}$$

(3) Let q have the (average) value U at $z = \infty$, then as $z = \infty$ maps into $w = \infty$, which is in turn mapped into $\eta = \infty$ by (1), it follows from (6.3)

that $\lim\limits_{\eta\to\infty} \Omega \equiv \Omega_\infty = 0$. Let the Ox-axis be selected so that it is parallel to the stream direction at infinity, then $\lim\limits_{\eta\to\infty} \theta \equiv \theta_\infty = 0$. Thus the boundary condition at infinity can be written

$$\lim_{\eta\to\infty} \tau(\gamma,\eta) \equiv \Omega_\infty + i\theta_\infty = 0. \tag{9}$$

It is established in §4.3 ((4.20) and (4.21)) that with these boundary conditions $\tau(\zeta)$ is given by

$$\tau(\zeta) = \frac{1}{2\pi} \int_{-\pi}^{\pi} \theta_s(\gamma) \cot \tfrac{1}{2}(\gamma - \zeta)\, d\gamma, \tag{10}$$

with

$$0 = \frac{1}{2\pi} \int_{-\pi}^{\pi} \theta_s(\gamma)\, d\gamma. \tag{11}$$

On $\eta = 0$ (10) yields (see footnote on p. 104)

$$\Omega_s(\gamma^*) = \frac{1}{2\pi} \int_{-\pi}^{\pi} \theta_s(\gamma) \cot \tfrac{1}{2}(\gamma - \gamma^*)\, d\gamma = -\frac{1}{\pi} \int_{-\pi}^{\pi} \ln \left| \sin \tfrac{1}{2}(\gamma - \gamma^*) \right| d\theta_s(\gamma), \tag{12}$$

on integrating by parts. Let R be the radius of curvature, then by (6)

$$d\theta_s(\gamma) = \frac{d\theta_s}{ds} \frac{ds}{d\phi} \frac{d\phi}{d\gamma} d\gamma = T\left(\frac{c}{R}\right) \frac{U}{q_s} \sin \tfrac{1}{2}(\gamma + 2\alpha_0) \cos \tfrac{1}{2}\gamma\, d\gamma, \tag{13}$$

where c is the chord length,

$$T \equiv \frac{4a}{Uc}, \tag{14}$$

and the right-hand side of (13) is to be given the value

$$\alpha_i \delta(\gamma - \gamma_i)\, d\gamma = \alpha_i\, dU(\gamma - \gamma_i)$$

at discontinuities of magnitude α_i in θ_s at $\gamma = \gamma_i$ (see §7.2). The number T defined in (14) is a non-dimensional quantity we shall term the 'thickness factor', the reason for which will be explained in §8.13.

From (6.3)

$$\Omega = \sinh^{-1}\left(\frac{U}{q} \sinh \Omega^*\right) - \Omega^*, \tag{15}$$

and substituting (13) and (15) into (12) we arrive at the integral equation

$$\sinh^{-1}\left(\frac{U}{q_s} \sinh \Omega^*\right)$$
$$= \Omega^* - \frac{T}{\pi} \int_{-\pi}^{\pi} \left(\frac{c}{R}\right) \left(\frac{U}{q_s}\right) \sin \tfrac{1}{2}(\gamma + 2\alpha_0) \cos \tfrac{1}{2}\gamma \ln \left| \sin \tfrac{1}{2}(\gamma - \gamma^*) \right| d\gamma. \tag{16}$$

In this $R(\gamma)$ can be regarded as being a complicated function of q and γ defined by the relation $R = R(s)$—which can be deduced from the profile shape—and the relation

$$s(\gamma) = \int^{\gamma} \frac{1}{q_s} \frac{d\phi}{d\gamma} d\gamma = cT \int^{\gamma} \frac{U}{q_s} \sin \tfrac{1}{2}(\gamma + 2\alpha_0) \cos \tfrac{1}{2}\gamma\, d\gamma, \tag{17}$$

which follows from $q = d\phi/ds$ and (6). In general the integral equation is much too complicated to yield to analytical methods, so numerical methods must be employed.

The solution to (16) would give the velocity distribution $q_s(s)$ over a given profile; the pressure distribution follows from Bernoulli's theorem, and then integrations yield the force and moment acting on the aerofoil. However, rather simple formulae can be obtained for the lift and moment, which appear to avoid the need to solve (16). These are deduced in §§ 8.10 and 8.11.

8.4 The closure conditions

Let C be a contour enclosing the aerofoil, then the closure condition is $\int_C dz = 0$. Let C also denote the corresponding contour in the σ-plane (see Fig. 8.2), then it follows from (6.42) that the closure condition is equivalent to

$$(1+\beta_a)\int_C e^{\tau(\sigma)}\left(\frac{dw}{d\sigma}\right) d\sigma = (1-\beta_a)\int_C \overline{e^{-\tau(\sigma)}\left(\frac{dw}{d\sigma}\right)} \, d\sigma, \tag{18}$$

the bar denoting, as usual, the conjugate number. Each side of (18) can be evaluated from the values of the residues of the integrands at $\sigma = \infty$ (§ 3.3). For this we shall need expansions near the point at infinity.

From (1) and (4),

$$\frac{dw}{d\sigma} = a\left\{1 + \frac{2i\sin\alpha_0}{\sigma} - \frac{1}{\sigma^2}\right\}, \tag{19}$$

exactly. Now

$$\cot\tfrac{1}{2}(\gamma - \zeta) = i\left\{1 + 2\sum_{n=1}^{\infty} e^{ni(\zeta-\gamma)}\right\}$$

$$= i\left\{1 + 2\sum_{n=1}^{\infty} (-1)^n e^{-ni(\gamma+\alpha_0)}/\sigma^n\right\}, \tag{20}$$

and it therefore follows from (10) and (11) that

$$\tau(\sigma) = i\sum_{n=1}^{\infty} (-1)^n \frac{a_n}{\sigma^n}, \tag{21}$$

where

$$a_n \equiv \frac{1}{\pi}\int_{-\pi}^{\pi} \theta_s e^{-ni(\gamma+\alpha_0)} \, d\gamma. \tag{22}$$

Equations (19) and (21), and their conjugates, are now combined to yield the expansions

$$e^{\tau(\sigma)} dw = a\left\{1 + \frac{A_1}{\sigma} + \frac{B_1}{\sigma^2} + O(\sigma^{-3})\right\} d\sigma, \tag{23}$$

and

$$e^{-\overline{\tau(\sigma)}} d\overline{w} = a\left\{1 + \frac{\overline{A}_2}{\overline{\sigma}} - \frac{\overline{B}_2}{\overline{\sigma}^2} + O(\overline{\sigma}^{-3})\right\} d\overline{\sigma}, \tag{24}$$

where $\quad A_1 \equiv 2i \sin \alpha_0 - ia_1, \quad B_1 \equiv ia_2 - \tfrac{1}{2}a_1^2 - 1 + 2a_1 \sin \alpha_0,$

$$A_2 \equiv 2i \sin \alpha_0 + ia_1, \quad B_2 \equiv ia_2 + \tfrac{1}{2}a_1^2 + 1 + 2a_1 \sin \alpha_0. \tag{25}$$

From (18), (24) and the residue theorem we see that the closure condition is equivalent to

$$(1 + \beta_a) A_1 + (1 - \beta_a) \bar{A}_2 = 0, \tag{26}$$

or $\quad \dfrac{1}{\pi} \displaystyle\int_{-\pi}^{\pi} \theta_s \{\cos(\gamma + \alpha_0) - i \sin(\gamma + \alpha_0)\} \, d\gamma = a_1 = 2\beta_a \sin \alpha_0, \tag{27}$

from (22) and (25). These two restrictions can be written in the alternative forms

$$2\beta_a \sin^2 \alpha_0 = -\frac{1}{\pi} \int_{-\pi}^{\pi} \theta_s(\gamma) \sin \gamma \, d\gamma, \quad \beta_a \sin 2\alpha_0 = \frac{1}{\pi} \int_{-\pi}^{\pi} \theta_s(\gamma) \cos \gamma \, d\gamma, \tag{28}$$

which are often more convenient than (27).

We now have three restrictions on the function $\theta_s(\gamma)$, namely (11) and (28), but one of (28) can be regarded as defining the value of α_0, i.e. as defining the position of the front stagnation point (see Fig. 8.2). The two remaining restrictions on $\theta_s(\gamma)$ are naturally satisfied with *given* closed profiles. However, when approximate methods are adopted to solve the integral (16), these methods are usually equivalent to finding an *exact* solution of (16) for an approximating profile, and then it is important to ensure that the restrictions are satisfied exactly by the new profile. The point here is that the approximate profile is not deduced from the original profile in the z-plane, but is obtained simply by modifying the distributions $\theta_s(\gamma)$; it will therefore not usually close unless the modifications to $\theta_s(\gamma)$ are selected so that the resulting distribution satisfies (28). Errors arising from the non-closure of the approximating profile can be quite appreciable. A simple method of incorporating (28) into the numerical treatment of (16) will be described later (§ 8.23).

8.5   The profile centre and the circulation centroid

The expansions given above can be used to establish an important relation which connects the points z_c with z_a in the physical plane, where

$$z_c \equiv \int_{C_0} z \, d\gamma \Big/ \int_{C_0} d\gamma = \frac{1}{2\pi} \int_{-\pi}^{\pi} z(\gamma, 0) \, d\gamma, \tag{29}$$

and $\quad z_a \equiv \displaystyle\int_{C_0} z \, d\phi \Big/ \int_{C_0} d\phi = \frac{1}{\Gamma} \int_{-\pi}^{\pi} z(\gamma, 0) \frac{d\phi}{d\gamma} \, d\gamma, \tag{30}$

C_0 being the aerofoil surface contour (see Fig. 8.2). The points z_c and z_a are termed the 'profile centre' and the 'centroid of the circulation' respectively. Their importance in aerofoil theory will emerge later, when we deal with the lift and moment acting on the aerofoil.

Notice that as $d\phi = q\,ds = d\Gamma$,

$$z_a = x_a + iy_a = \int_{C_0} x\,d\Gamma \Big/ \iint_{C_0} d\Gamma + i\int_{C_0} y\,d\Gamma \Big/ \iint_{C_0} d\Gamma,$$

so z_a is appropriately named. The centre of the profile, z_c, derives its name from the fact that in incompressible flow, when z is an analytic function of w, z_c maps on to $\sigma = 0$, the centre of the circle on to which C_0 is (conformally) mapped. While this property does not hold in compressible flow, it is convenient to retain the name, for it is not really important which point in the σ-plane corresponds to $z = z_c$.

Let $\sigma = \sigma_0$ at a point $z = z_0$ on C_0, then (6.13) yields

$$z = z_0 + \frac{1+\beta_a}{2\beta_a U}\int_{\sigma_0}^{\sigma} e^{\tau}\frac{dw}{d\sigma}d\sigma - \frac{1-\beta_a}{2\beta_a U}\overline{\int_{\sigma_0}^{\sigma} e^{-\tau}\frac{dw}{d\sigma}d\sigma}. \tag{31}$$

On C_0, $d\gamma = d\zeta = d\bar\zeta$, hence we can integrate (26) along C_0 and obtain

$$(1+\beta_a)\int_{\sigma_0}^{\sigma} A_1\frac{d\zeta}{d\sigma}d\sigma + (1-\beta_a)\overline{\int_{\sigma_0}^{\sigma} A_2\frac{d\zeta}{d\sigma}d\zeta} = 0,$$

or as (4) gives $$\frac{d\zeta}{d\sigma} = \frac{i}{\sigma}, \tag{32}$$

$$(1+\beta_a)\int_{\sigma_0}^{\sigma} A_1\frac{d\sigma}{\sigma} - (1-\beta_a)\overline{\int_{\sigma_0}^{\sigma} A_2\frac{d\sigma}{\sigma}} = 0. \tag{33}$$

Let z in (31) be a point on C_0, then (31) and (33) can be combined to give

$$z_{\eta=0} = z_0 + \frac{1+\beta_a}{2\beta_a U}\int_{\sigma_0}^{\sigma}\left(e^{\tau}\frac{dw}{d\sigma} - \frac{aA_1}{\sigma}\right)d\sigma - \frac{1-\beta_a}{2\beta_a U}\overline{\int_{\sigma_0}^{\sigma}\left(e^{-\tau}\frac{dw}{d\sigma} - \frac{aA_2}{\sigma}\right)d\sigma}. \tag{34}$$

Integrating this over $-\pi \leqslant \gamma \leqslant \pi$, and using (29) and the fact that on C_0 $d\gamma = d\zeta = d\bar\zeta$, we obtain

$$2\pi z_c = 2\pi z_0 - \frac{1+\beta_a}{2\beta_a U}\int_{C_0}\left\{\int_{\sigma_0}^{\sigma}\left(e^{\tau}\frac{dw}{d\sigma} - \frac{aA_1}{\sigma}\right)d\sigma\right\}\frac{d\zeta}{d\sigma}d\sigma$$

$$+ \frac{1-\beta_a}{2\beta_a U}\overline{\int_{C_0}\left\{\int_{\sigma_0}^{\sigma}\left(e^{-\tau}\frac{dw}{d\sigma} - \frac{aA_2}{\sigma}\right)d\sigma\right\}\frac{d\zeta}{d\sigma}d\sigma}, \tag{35}$$

where the change in sign is because C_0 is traversed in the (usual) anticlockwise direction—the direction of γ decreasing. Each integrand in (35) is analytic so C_0 can be replaced by any contour C enclosing the aerofoil. If C is near the point at infinity we can use (23), (24) and (32) to find

$$z_c = z_0 - \frac{1+\beta_a}{2\beta_a U}\, a \int_C \left\{ \sigma + K_1 - \frac{B_1}{\sigma} + O(\sigma^{-2}) \right\} \frac{i}{\sigma}\, d\sigma$$

$$+ \frac{1-\beta_a}{2\beta_a U}\, a \overline{\int_C \left\{ \sigma + K_2 + \frac{B_2}{\sigma} + O(\sigma^{-2}) \right\} \frac{i}{\sigma}\, d\sigma}, \quad (36)$$

where K_1 and K_2 are constants of integration. The residue theorem now gives

$$z_c = z_0 + \frac{a}{2\beta_a U}\{(1+\beta_a)\, K_1 - (1-\beta_a)\, \bar{K}_2\}, \qquad (37)$$

and subtracting this from (34) we get

$$z_{\eta=0} = z_c + \frac{1+\beta_a}{2\beta_a U}\left\{ \int_{\sigma_0}^{\sigma} \left(e^{\tau}\frac{dw}{d\sigma} - \frac{aA_1}{\sigma} \right) d\sigma - aK_1 \right\}$$

$$- \frac{1-\beta_a}{2\beta_a U}\left\{ \int_{\sigma_0}^{\sigma} \left(e^{-\tau}\frac{dw}{d\sigma} - \frac{aA_2}{\sigma} \right) d\sigma - aK_2 \right\}.$$

To obtain the relation between z_c and z_a we integrate this around C_0 with respect to ϕ. It follows fom (30) that

$$\Gamma z_a = \Gamma z_c - \frac{1+\beta_a}{2\beta_a U}\int_{C_0} \left\{ \int_{\sigma_0}^{\sigma} \left(e^{\tau}\frac{dw}{d\sigma} - \frac{aA_1}{\sigma} \right) d\sigma - aK_1 \right\} \frac{dw}{d\sigma}\, d\sigma$$

$$+ \frac{1-\beta_a}{2\beta_a U}\overline{\int_{C_0} \left\{ \int_{\sigma_0}^{\sigma} \left(e^{-\tau}\frac{dw}{d\sigma} - \frac{aA_2}{\sigma} \right) d\sigma - aK_2 \right\} \frac{dw}{d\sigma}\, d\sigma}, \quad (38)$$

as $d\phi = dw = d\bar{w}$ on C_0. We again have analytic integrands, and can therefore evaluate the integrals from the residues at infinity. From (19), (23) and the definitions of K_1, K_2 implicit in (36) we have

$$\left\{ \int_{\sigma_0}^{\sigma} \left(e^{\tau}\frac{dw}{d\sigma} - \frac{aA_1}{\sigma} \right) d\sigma - aK_1 \right\} \frac{dw}{d\sigma} = a^2\left\{ \sigma - \frac{B_1}{\sigma} + O(\sigma^{-2}) \right\}\left\{ 1 + \frac{2i\sin\alpha_0}{\sigma} - \frac{1}{\sigma^2} \right\},$$

$$(39)$$

so the residue is $-a^2(1+B_1)$. Similarly,

$$\left\{ \int_{\sigma_0}^{\sigma} \left(e^{-\tau}\frac{dw}{d\sigma} - \frac{aA_2}{\sigma} \right) d\sigma - aK_2 \right\} \frac{dw}{d\sigma}$$

$$= a^2\left\{ \sigma + \frac{B_2}{\sigma} + O(\sigma^{-2}) \right\}\left\{ 1 + \frac{2i\sin\alpha_0}{\sigma} - \frac{1}{\sigma^2} \right\}, \quad (40)$$

and the corresponding residue is $a^2(B_2-1)$. Thus (38) reduces to

$$\Gamma z_a = \Gamma z_c + \frac{\pi a^2}{\beta_a U} i\{(1+\beta_a)(1+B_1) - (1-\beta_a)(\bar{B}_2-1)\},$$

which by (2), (14), (25) and (27) is equivalent to

$$z_a = z_c - \frac{cT}{8\beta_a \sin \alpha_0} \{\mathscr{R}a_2 + i\beta_a \mathscr{I}a_2\} + (\tfrac{1}{4}i)Tc\beta_a \sin \alpha_0. \tag{41}$$

This is the required relation between z_a and z_c.

8.6 Conjugate equations for aerofoil design

On interchanging the roles of Ω and $i\theta$ in the theory of §§ 8.3–8.5, we obtain the harmonic conjugate equations, which are the basis of the method of designing aerofoils. Corresponding to (10), (11), (12), (27) and (28) we find

$$\tau(\zeta) = -\frac{i}{2\pi} \int_{-\pi}^{\pi} \Omega_s(\gamma) \cot \tfrac{1}{2}(\gamma - \zeta)\, d\gamma, \tag{42}$$

$$0 = \frac{1}{2\pi} \int_{-\pi}^{\pi} \Omega_s(\gamma)\, d\gamma, \tag{43}$$

$$\theta_s(\gamma^*) = -\frac{1}{2\pi} \int_{-\pi}^{\pi} \Omega_s(\gamma) \cot \tfrac{1}{2}(\gamma - \gamma^*)\, d\gamma, \tag{44}$$

$$2\beta_a \sin \alpha_0 = -\frac{1}{\pi} \int_{-\pi}^{\pi} \Omega_s(\gamma) \sin(\gamma + \alpha_0)\, d\gamma, \quad 0 = \frac{1}{\pi} \int_{-\pi}^{\pi} \Omega_s(\gamma) \cos(\gamma + \alpha_0)\, d\gamma, \tag{45}$$

and

$$2\beta_a \sin^2 \alpha_0 = -\frac{1}{\pi} \int_{-\pi}^{\pi} \Omega_s(\gamma) \cos \gamma\, d\gamma, \quad \beta_a \sin 2\alpha_0 = -\frac{1}{\pi} \int_{-\pi}^{\pi} \Omega_s(\gamma) \sin \gamma\, d\gamma. \tag{46}$$

These equations are used in the design problem as follows. Initially we are usually given the pressure p_s on the aerofoil surface as a function of the perimeter distance s. From $p_s(s)$ and Bernoulli's equation we deduce $q_s(s)$, and then use

$$\phi = -\tfrac{1}{2} TUc\{\cos(\gamma + \alpha_0) - \gamma \sin \alpha_0\} = \int^s q_s\, ds, \tag{47}$$

—which follows from (6), (14) and the equation $d\phi = q\, ds$—to calculate the relation between γ and s. The next step is to transform $q_s(s)$ into $q_s(\gamma)$, and then to deduce $\Omega_s(\gamma)$ with the aid of (15). This function is now introduced under the integral sign in (44) and the conjugate function $\theta_s(\gamma)$ obtained.

As
$$dx = \cos \theta_s\, ds, \quad dy = \sin \theta_s\, ds,$$

it follows from (17) that

$$x + iy = cT \int^\gamma e^{i\theta_s(\gamma)} \frac{U}{q_s} \sin \tfrac{1}{2}(\gamma + 2\alpha_0) \cos \tfrac{1}{2}\gamma\, d\gamma, \tag{48}$$

from which the coordinates of the aerofoil surface can next be calculated.

The initial pressure distribution cannot be chosen arbitrarily, but must be selected so that the corresponding distribution $\Omega_s(\gamma)$ satisfies (43) and the two closure conditions in (46). There are several ways of doing this, the most obvious of which is to choose a distribution having three free parameters, and then to select these so that the three conditions are satisfied. For example, (43) can be satisfied quite simply by taking one of the constants to be an additive constant, and this will not affect (46); the second of (46) can serve to define the value of α_0, and then the first of (46) can be satisfied by adding any acceptable (from the design standpoint) *even* function of γ to $\Omega_s(\gamma)$, as such a function will have no effect on the other two equations.

One difficulty with the method just outlined is that an estimate of α_0 is required in (47) before $\Omega(\gamma)$ has been deduced. This can be overcome either by using an iterative process in which the second of (46) is used to recalculate α_0 and (47) to recalculate $\Omega_s(\gamma)$ alternatively, or by finding another method of satisfying the second of (46), keeping α_0 fixed in value.

8.7　The effect of a change of incidence

The notation we shall adopt for aerofoil incidence is shown in Fig. 8.7; α is the *absolute* or *effective* incidence, i.e. the angle swept out by the chord DB when the aerofoil moves from the attitude at which $\Gamma = 0$ to that at which $\Gamma = 4\pi \sin \alpha_0$ (see (2)). The *physical* incidence α^* is the angle between DB and the Ox-axis, and $-\alpha'$ is the *no-lift* angle, i.e. the value of α^* at $\Gamma = 0$ (the relation between lift and Γ is established later).

Thus
$$\alpha = \alpha^* + \alpha'. \tag{49}$$

If it is assumed that the Joukowski condition applies, then the rear stagnation point B will remain fixed at $\gamma = \pm \pi$ for all incidences. The front stagnation point A is at $\gamma = -2\alpha_0$ when $\Gamma = 4\pi a \sin \alpha_0$; let A be displaced to $\gamma = \lambda$ when $\alpha = 0$. Then the slope of the aerofoil surface θ_s at incidence α is related to the slope θ_0 at zero incidence by

$$\theta_s(\gamma) = \theta_0(\gamma) - \alpha + \pi\{\mathrm{U}(\gamma + 2\alpha_0) - \mathrm{U}(\gamma - \lambda)\}, \tag{50}$$

where U is the unit function. The graph of $\theta_s - \theta_0$ is shown in Fig. 8.7.

On substitution of (50) into (10) there results

$$\tau(\zeta) = \tau_0(\zeta) - i\alpha - \ln \frac{\sin \frac{1}{2}(\zeta + 2\alpha_0)}{\sin \frac{1}{2}(\zeta - \lambda)}, \tag{51}$$

where τ_0 is due to the aerofoil thickness alone.

As the incidence of the aerofoil can have no effect on the flow speed or direction at infinity, it follows from (9) that

$$\lim_{\eta \to \infty} \tau(\gamma, \eta) = \lim_{\eta \to \infty} \tau_0(\gamma, \eta) = 0.$$

Therefore in the limit $\eta \to \infty$ (51) yields

$$\lambda = 2(\alpha - \alpha_0). \tag{52}$$

Substituting (50) into (11), and taking (52) into account we get

$$0 = \int_{-\pi}^{\pi} \theta_0(\gamma)\, d\gamma. \tag{53}$$

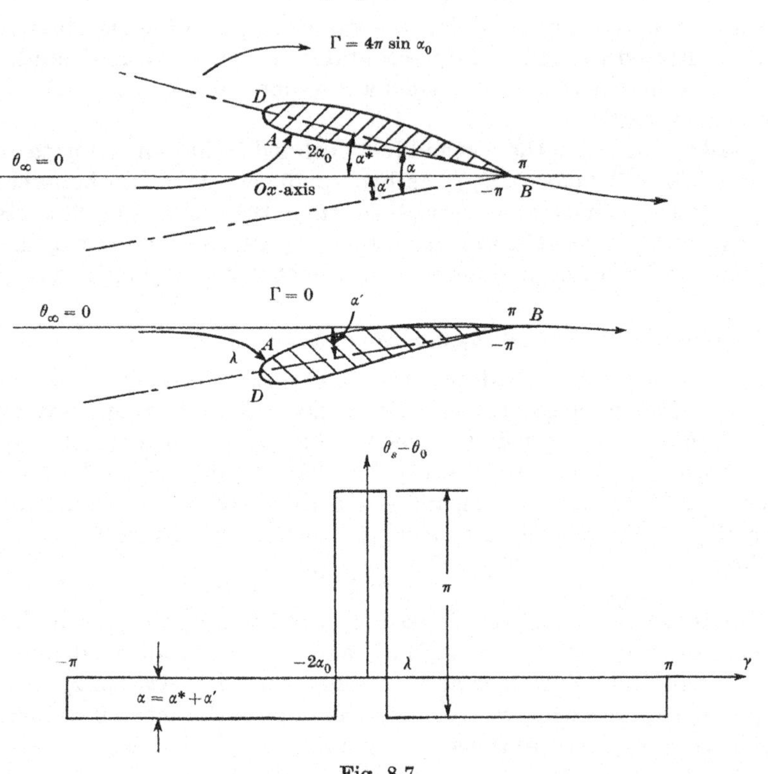

Fig. 8.7

Further relations between α and α_0 follow on substituting (50) into (28). These are

$$2\beta_a \sin^2 \alpha_0 + \cos 2\alpha_0 = \cos 2(\alpha - \alpha_0) - \frac{1}{\pi}\int_{-\pi}^{\pi} \theta_0(\gamma)\sin\gamma\, d\gamma, \tag{54}$$

and

$$-(1-\beta_a)\sin 2\alpha_0 = \sin 2(\alpha - \alpha_0) + \frac{1}{\pi}\int_{-\pi}^{\pi} \theta_0(\gamma)\cos\gamma\, d\gamma, \tag{55}$$

where (52) has been used to eliminate λ.

In §2.16 it was shown that in incompressible flow τ could be written as the sum of two independent terms, namely one due to the aerofoil thickness and the other due to the aerofoil's incidence. A first glance at (51) suggests that we have already written τ as the sum of two independent terms without requiring that the flow be incompressible. But this is really not so, for the 'independent' variable ζ is shown in (1) to depend on α_0. The function $\theta_0(\gamma)$ implicit in (47) cannot be determined until γ is known as a function of s, and for a given aerofoil this function will depend on both α_0 and α. Hence the first term on the right-hand side of (51) is clearly not independent of the second.

Let the dependence of γ on s, α and α_0 for a given aerofoil be indicated by the usual functional notation, then in *incompressible* flow

$$\gamma(s, \alpha, \alpha_0) = \gamma(s, 0, 0) \equiv \gamma_0(s). \tag{56}$$

This important result, which is established in the next section, means that in incompressible flow the terms on the right-hand side of (51) are really independent.

Returning to compressible flow, we note that in the notation just defined (54) and (55) reduce to

$$\int_{-\pi}^{\pi} \theta_0(\gamma_0)\, d\gamma_0 = \int_{-\pi}^{\pi} \theta_0(\gamma_0) \sin \gamma_0\, d\gamma_0 = \int_{-\pi}^{\pi} \theta_0(\gamma_0) \cos \gamma_0\, d\gamma_0 = 0,$$

where $\Gamma = \alpha = \alpha_0 = 0$. These equations and (53) enable (55) to be expressed

$$-(1 - \beta_a) \sin 2\alpha_0 = \sin 2(\alpha - \alpha_0) - \frac{1}{\pi} \int_{-\pi}^{\pi} [\theta_0(\gamma) - \theta_0\{\gamma_0(\gamma)\}] (1 - \cos \gamma)\, d\gamma.$$
$$\tag{57}$$

Now at the trailing edge $\gamma = \pm\pi$ at all incidences, and at the leading edge $\gamma = -2\alpha_0 \simeq 0$ for small incidences. From this we conclude that γ will not depend strongly on α and α_0 at small incidences, and that the approximation $\gamma = \gamma_0$ is reasonable, except where small changes in the γ, s relation could cause large changes in θ. With conventional aerofoils this could occur only near the leading edge, i.e. near $\gamma = 0$. Hence $\theta_0(\gamma) - \theta_0\{\gamma_0(\gamma)\}$ is small except near $\gamma = 0$, and the integrand in (57) is small everywhere. The integral can thus be neglected, and if terms $O(\alpha^3)$ and $O(\alpha_0^3)$ are also neglected, (57) reduces to

$$\alpha_0 = \alpha/\beta_a. \tag{58}$$

Let $\Delta\phi$ be the potential difference between the positions occupied by the front stagnation point at incidence α and at zero incidence, then from (6) and (52) (see Fig. 8.7) $\Delta\phi = \phi(\lambda) - \phi(-2\alpha_0) \simeq 4a\alpha^2$. Consequently, if (58) is substituted into (54), and only the highest order terms are retained, then

$$\frac{1}{\pi} \int_{-\pi}^{\pi} \theta_0(\gamma) \sin \gamma\, d\gamma = 2\alpha^2 \left(\frac{1}{\beta_a} - 1 \right) = \frac{\Delta\phi}{2a} \left(\frac{1}{\beta_a} - 1 \right),$$

or
$$\frac{1}{U}\int_{-\pi}^{\pi}\theta_0(\gamma)\,d\phi(\gamma) = \frac{\Delta\phi}{U}\pi\left(\frac{1}{\beta_a}-1\right),\tag{59}$$

by (6), (53) and (55). Equation (59) will be found to have some importance in the theory of jet-flapped aerofoils (see § 10.16).

Equation (58) is very important in the linear perturbation theory of compressible flow, which will be studied in some detail in §§ 8.14–8.21. In incompressible flow the approximations just introduced are unnecessary, as shown below.

8.8 Incompressible flow

In incompressible flow $\beta_a = 1$ and (54) and (55) reduce to

$$1 = \cos 2(\alpha-\alpha_0) - \frac{1}{\pi}\int_{-\pi}^{\pi}\theta_0(\gamma)\sin\gamma\,d\gamma,$$

$$0 = \sin 2(\alpha-\alpha_0) + \frac{1}{\pi}\int_{-\pi}^{\pi}\theta_0(\gamma)\cos\gamma\,d\gamma.$$

So we have a pair of equations for the single number $(\alpha-\alpha_0)$. The only possible solution is

$$\alpha = \alpha_0,\tag{60}$$

and
$$\int_{-\pi}^{\pi}\theta_0(\gamma)\sin\gamma\,d\gamma = \int_{-\pi}^{\pi}\theta_0(\gamma)\cos\gamma\,d\gamma = 0.\tag{61}$$

From (6.17)
$$\frac{U}{q} = e^{\Omega}.\tag{62}$$

On the aerofoil, $\eta = 0$, (51), (52) and (60) give

$$\Omega_s(\gamma) = \Omega_0(\gamma) - \ln\left|\frac{\sin\tfrac{1}{2}(\gamma+2\alpha_0)}{\sin\tfrac{1}{2}\gamma}\right|,$$

consequently
$$q_s = q_0\left|\frac{\sin\tfrac{1}{2}(\gamma+2\alpha_0)}{\sin\tfrac{1}{2}\gamma}\right|,$$

q_0 being the velocity distribution at $\alpha_0 = 0$.

From $d\phi = q_s\,ds$, (6) and (14) it follows that

$$UcT\sin\tfrac{1}{2}(\gamma+2\alpha_0)\cos\tfrac{1}{2}\gamma\,d\gamma = q_s\,ds.$$

Eliminating α_0 and integrating:

$$\cos\gamma = -\frac{2}{T}\int^s\frac{q_0}{U}\,d\left(\frac{s}{c}\right).\tag{63}$$

Let $s = s_1$ at the trailing edge, where $\gamma = \pi$, then

$$T = 2\int^{s_1}\frac{q_0}{U}\,d\left(\frac{s}{c}\right).\tag{64}$$

The right-hand side of (64) is independent of α_0, and so the same must be true of (63). Hence γ is independent of α_0, and (56) is established.

This result means that the ζ-plane can be derived from *any* w-plane, regardless of the circulation, the γ, s relation being unaffected. Also the thickness factor T is a constant for a given aerofoil. The discussion preceding equation (58) shows that these conclusions cannot apply in general to compressible flow.

8.9 The displacement function

In §6.1 it is shown that the function

$$\delta(w) \equiv \frac{1}{\beta_a} \int (e^\tau - 1)\, dw = Ux - \phi + \frac{i}{\beta_a^2}(\beta_a Uy - \psi), \tag{65}$$

applies to linear perturbations of compressible flows, and to all incompressible flows. We can write

$$\delta(\sigma) = \delta_0 + \frac{1}{\beta_a} \int_{\sigma_0}^{\sigma} (e^{\tau(\sigma)} - 1) \frac{dw}{d\sigma}\, d\sigma, \tag{66}$$

where $\delta_0 = \delta(\sigma_0)$, and then use (19), (23), (25) and (27) to find the expansion

$$\delta(\sigma) = \delta_0 + \frac{a}{\beta_a} \int_{\sigma_0}^{\sigma} \left\{ \frac{-2i\beta_a \sin \alpha_0}{\sigma} + \frac{(B_1 + 1)}{\sigma^2} + O(\sigma^{-3}) \right\} d\sigma,$$

near infinity. It appears from this that $\lim_{\sigma \to \infty} \delta(\sigma) \equiv \delta_\infty$ is infinite unless α_0 vanishes, i.e. the circulation is zero. We must therefore confine the use of the displacement function to this case. Then (25) and (27) give $(B_1 + 1) = ia_2$, and the expansion becomes

$$\delta(\sigma) = \delta_\infty - \frac{ia\,a_2}{\beta_a}\frac{1}{\sigma} + \frac{ia\,a_3}{2\beta_a}\frac{1}{\sigma^2} + O(\sigma^{-3}), \tag{67}$$

on using (19)–(22) to take the expansion to one more term.

As α_0 is zero (6) gives $\phi = -2a\cos\gamma = -\tfrac{1}{2}cUT\cos\gamma$ by (14); hence on the aerofoil surface ($\psi = 0$) it follows from (65) that

$$\mathscr{R}\delta_s = U(x_s + \tfrac{1}{2}cT\cos\gamma), \quad \mathscr{I}\delta_s = Uy_s/\beta_a, \tag{68}$$

the subscript 's' denoting surface values. As there is no circulation, $\delta(x, y)$ is single valued in the z-plane, and therefore $\delta(-\pi, \eta) = \delta(\pi, \eta)$. It now follows from (68), (29) and the theory given in §4.3 that the boundary values of $\delta(\zeta)$ satisfy

$$\mathscr{R}\delta_\infty - \frac{1}{2\pi}\int_{-\pi}^{\pi} \mathscr{R}\delta_s(\gamma)\, d\gamma = \frac{U}{2\pi}\int_{-\pi}^{\pi} x_s(\gamma)\, d\gamma = Ux_c, \left.\begin{array}{c} \\ \\ \\ \end{array}\right\}$$

$$\tag{69}$$

$$\text{and} \qquad \mathscr{I}\delta_\infty = \frac{1}{2\pi}\int_{-\pi}^{\pi} \mathscr{I}\delta_s(\gamma)\, d\gamma = \frac{U}{2\pi\beta_a}\int_{-\pi}^{\pi} y_s(\gamma)\, d\gamma = \frac{Uy_c}{\beta_a}.$$

It now follows from (68), (69) and (4.14) applied to $\delta(\zeta)$ that

$$\delta(\zeta) = Ux_c + \frac{U}{2\pi\beta_a} \int_{-\pi}^{\pi} y_s(\gamma) \cot \tfrac{1}{2}(\gamma-\zeta)\, d\gamma. \tag{70}$$

The harmonic conjugate to (70) is

$$\delta(\zeta) = i\frac{Uy_c}{\beta_a} - \frac{i}{2\pi} \int_{-\pi}^{\pi} (Ux-\phi) \cot \tfrac{1}{2}(\gamma-\zeta)\, d\gamma, \tag{71}$$

which is sometimes useful in aerofoil design.

From (20) and (69) we find that (70) has the expansion

$$\delta(\sigma) = \delta_{\infty} + \frac{Ui}{\beta_a} \sum_{n=1}^{\infty} \frac{(-1)^n}{\sigma^n} b_n, \tag{72}$$

where
$$b_n \equiv \frac{1}{\pi} \int_{-\pi}^{\pi} y_s(\gamma)\, e^{-ni\gamma}\, d\gamma, \tag{73}$$

as $\alpha_0 = 0$ here. Comparison of (67) and (72) yields the relations

$$b_1 = a\,a_2/U, \quad b_2 = a\,a_3/2U, \tag{74}$$

which from (14), (22) and (73) can be written in the form

$$\int_{-\pi}^{\pi} \frac{y_s}{c}(\gamma)\,(\cos\gamma - i\sin\gamma)\,d\gamma = \tfrac{1}{4}T \int_{-\pi}^{\pi} \theta_s(\gamma)\,(\cos 2\gamma - i\sin 2\gamma)\,d\gamma, \tag{75}$$

$$\int_{-\pi}^{\pi} \frac{y_s}{c}(\gamma)\,(\cos 2\gamma - i\sin 2\gamma)\,d\gamma = \tfrac{1}{8}T \int_{-\pi}^{\pi} \theta_s(\gamma)\,(\cos 3\gamma - i\sin 3\gamma)\,d\gamma. \tag{76}$$

Similarly, relations between b_n and a_n for $n > 3$ can be found, but these are not much use, being rather complicated and non-linear. It should be noted that (75) and (76) are exact in incompressible flows.

Several more relations can be derived from (75) and (76) by replacing y_s by its harmonic conjugate (see (68)) or by replacing θ_s by its harmonic conjugate, $-i\Omega_s$.

FORCES ACTING ON AEROFOILS

8.10   The lift

The lift force L acting on an aerofoil is that component of force which is at right angles to the stream direction at infinity, i.e. the force component along Oy. The drag force D acts along Ox. We can therefore put $L = Y$ and $D = X$ in (6.14) to find

$$L + iD = -\frac{\rho_a U}{1+\beta_a} \int_C e^{-\tau(\sigma)} \frac{dw}{d\sigma}\, d\sigma, \tag{77}$$

where C is any contour enclosing the aerofoil. From (25), (27) and the conjugate of (24) we see that the residue of $e^{-\tau}dw/d\sigma$ is

$$aA_2 = 2ia(1+\beta_a)\sin\alpha_0.$$

Hence

$$L+iD = -\frac{\rho_a U}{1+\beta_a}.2\pi i.2ia(1+\beta_a)\sin\alpha_0,$$

giving
$$D = 0, \tag{78}$$

and
$$L = \pi T c\rho_a U^2 \sin\alpha_0, \tag{79}$$

on using (14). Of course (78) follows directly from the theory of §1.15, where it is established (as D'Alembert's paradox) under somewhat more general conditions.

We see from (79) that the lift coefficient,

$$C_L = \frac{L}{\tfrac{1}{2}c\rho_a U^2}, \tag{80}$$

is given by
$$C_L = 2\pi T \sin\alpha_0, \tag{81}$$

or by (27)
$$C_L = \frac{1}{\beta_a}T\int_{-\pi}^{\pi}\theta_s(\gamma)\cos(\gamma+\alpha_0)\,d\gamma. \tag{82}$$

Equations (2) and (79) yield the famous Kutta–Joukowski theorem, namely
$$L = \rho_a U\Gamma. \tag{83}$$

In incompressible flow (60) and (81) give the well-known result

$$C_L = 2\pi T \sin\alpha, \tag{84}$$

whereas in compressible flow the somewhat complicated relation between α and α_0, implicit in the simultaneous equations (54) and (55), prevents a similar reduction of (81), except in the case of linear perturbations (see §8.15).

8.11   The moment

Equation (6.15) is not the most convenient form for the calculation of the moment, the difficulty being the occurrence of terms involving $\ln\sigma$ in the expansions of the integrands near infinity (cf. (23) and (24)). This can be overcome by a method similar to that used to deal with (31).

In the present case (2.103) gives

$$M = \frac{\rho_a U}{1+\beta_a}\mathscr{R}\int_{C_0} z\,e^{-\tau(\sigma)}\frac{dw}{d\sigma}\,d\sigma,$$

where C_0 is the aerofoil surface. This can be written

$$M = \frac{\rho_a U}{1+\beta_a} \mathscr{R} \int_{C_0} (z-z_c) \left(e^{-\tau}\frac{dw}{d\sigma} + ia\, A_2\frac{d\zeta}{d\sigma} \right) d\sigma$$

$$+ \frac{\rho_a \dot{U}}{1+\beta_a} \mathscr{R} z_c \int_{C_0} \left(e^{-\tau}\frac{dw}{d\sigma} + ia\, A_2\frac{d\zeta}{d\sigma} \right) d\sigma + \frac{\rho_a U}{1+\beta_a} \mathscr{R} ia\, A_2 \int_{C_0} z\, d\zeta. \quad (85)$$

The second integrand is analytic, and therefore the integral can be evaluated from the residue at infinity. It follows from (32) and the conjugate of (24) that this residue is zero, and hence so is the integral in question. By (14), (25), (27), (29) and (79) the third integral in (85) is

$$\frac{\rho_a U}{1+\beta_a} \mathscr{R} ia\, \{2i\sin\alpha_0(1+\beta_a)\}(-1)\int_{-\pi}^{\pi} z\, d\gamma = -4\pi a\rho_a U \sin\alpha_0\, \mathscr{R} z_c$$

$$= -Lx_c.$$

Furthermore, on C_0 we can eliminate $z-z_c$ from (85) by using the equation following (37). On using (32) we find that (85) now reduces to

$$M = \frac{\rho_a}{2\beta_a} \mathscr{R} \int_{C_0} \left\{ \int_{\sigma_0}^{\sigma} \left(e^{\tau}\frac{dw}{d\sigma} - \frac{aA_1}{\sigma} \right) d\sigma - aK_1 \right\} \left(e^{-\tau}\frac{dw}{d\sigma} - \frac{aA_2}{\sigma} \right) d\sigma$$

$$- \frac{\rho_a}{2\beta_a}\frac{1-\beta_a}{1+\beta_a} \mathscr{R} \int_{C_0} \bar{G}\, dG - Lx_c, \quad (86)$$

where $$G \equiv \int_{\sigma_0}^{\sigma} \left(e^{-\tau}\frac{dw}{d\sigma} - \frac{aA_2}{\sigma} \right) d\sigma - aK_2.$$

Now $$\mathscr{R}\int_{C_0} \bar{G}\, dG = \tfrac{1}{2}\int_{C_0} (\bar{G}\, dG + G\, d\bar{G}) = \tfrac{1}{2}[G]_{C_0}\overline{[G]}_{C_0},$$

and from the definition of G and (24) it follows that this integral vanishes. The first integral in (86) also has an analytic integrand enabling C_0 to be replaced by C. From (23), the conjugate of (24) and the definition of K_1 implicit in (36) we find that this integrand has the expansion

$$a^2\left(\sigma - \frac{B_1}{\sigma} + \dots \right)\left(1 - \frac{B_2}{\sigma^2} + \dots \right);$$

consequently its residue at infinity is $-a^2(B_1+B_2)$. Hence from (14), (27) and (86) we obtain

$$M = -Lx_c + \frac{\pi T^2}{8\beta_a} c^2\rho_a U^2 \mathscr{R}\, a_2. \quad (87)$$

The moment coefficient,

$$C_M \equiv \frac{M}{\tfrac{1}{2}c^2\rho_a U^2}, \quad (88)$$

is found from (22), (80) and (87) to be given by

$$C_M = -\frac{x_c}{c}C_L + \frac{T^2}{4\beta_a}\int_{-\pi}^{\pi}\theta_s(\gamma)\sin 2(\gamma+\alpha_0)\,d\gamma, \tag{89}$$

which should be compared with (82), the corresponding equation for C_L.

The corresponding equation for use in aerofoil design is

$$C_M = -\frac{x_c}{c}C_L - \frac{T^2}{4\beta_a}\int_{-\pi}^{\pi}\Omega_s(\gamma)\sin 2(\gamma+\alpha_0)\,d\gamma, \tag{90}$$

as is readily verified from the harmonic conjugate of (22) and (87). In this case C_L is given by

$$C_L = -\frac{1}{\beta_a}T\int_{-\pi}^{\pi}\Omega_s(\gamma)\sin(\gamma+\alpha_0)\,d\gamma, \tag{91}$$

from (45) and (81).

Equations (89) and (90) give the moment coefficient about the origin in the z-plane. Let C_M' be the moment about the point (x',y') then clearly (see Fig. 8.11)

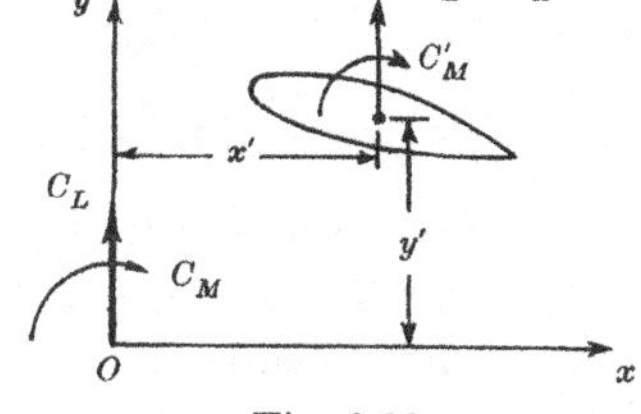

Fig. 8.11

$$C_M' = \frac{x'}{c}C_L + C_M, \tag{92}$$

and (89) gives

$$C_M' = \frac{x'-x_c}{c}C_L + \frac{T^2}{4\beta_a}\int_{-\pi}^{\pi}\theta_s(\gamma)\cos 2(\gamma+\alpha_0)\,d\gamma. \tag{93}$$

8.12   The centre of pressure and the aerodynamic centre

The centre of pressure is defined to be that point on the aerofoil chord about which the moment vanishes. From (81) and (93) it is apparent that this lies on the straight line

$$\frac{x}{c} = \frac{x_c}{c} - \frac{T}{8\pi\beta_a\sin\alpha_0}\int_{-\pi}^{\pi}\theta_s(\gamma)\cos 2(\gamma+\alpha_0)\,d\gamma = \frac{x_a}{c}, \tag{94}$$

the last form of which comes from (22) and (41). Hence the centre of pressure lies on the intersection of the chord and the straight line parallel to the lift force passing through the centroid of the circulation.

A much more important centre is the point about which the *rate of change* of the moment with lift is zero. This point is called the *aerodynamic centre*. It follows from (93) that this centre is at $x = x_e$, where

$$\frac{x_e}{c} = \frac{x_c}{c} - \frac{1}{4\beta_a}\frac{\partial}{\partial C_L}\left\{T^2\int_{-\pi}^{\pi}\theta_s(\gamma)\cos 2(\gamma+\alpha_0)\,d\gamma\right\}. \tag{95}$$

To progress further we shall restrict the analysis to the cases when γ is independent of α and α_0, i.e. (see § 8.7) either to incompressible flow, or to flow at small absolute incidences. In these cases T is independent of C_L, and

$$\alpha = \alpha_0 \beta_a, \tag{96}$$

from (58) and (60).

Differentiating (50) with respect to α_0, and using (52), (96) and (3.30) we get

$$\frac{\partial \theta_s}{\partial \alpha_0} = -\beta_a + 2\pi\{\delta(\gamma + 2\alpha_0) - \delta(\gamma - 2\alpha + 2\alpha_0) + \beta_a \delta(\gamma - 2\alpha + 2\alpha_0)\},$$

as $\theta_0(\gamma)$ is independent of α_0. Further from (81)

$$\frac{\partial}{\partial C_L} = \frac{1}{2\pi T \cos \alpha_0} \frac{\partial}{\partial \alpha_0}.$$

Using these in (95) we arrive at

$$\frac{x_e}{c} = \frac{x_c}{c} + \frac{T}{4\pi \beta_a \cos \alpha_0} \int_{-\pi}^{\pi} \theta_s(\gamma) \sin 2(\gamma + \alpha_0)\, d\gamma - \frac{T \cos 2\alpha_0}{4 \cos \alpha_0}, \tag{97}$$

where we have used $(1/\beta_a)\{\cos 2\alpha_0 + (\beta_a - 1)\cos 2(\alpha - \alpha_0)\} = \cos 2\alpha_0$ to reduce the last term, a result obviously valid in each of the two types of flow under consideration.

It is interesting to note that the integral in (97) can be eliminated by using the imaginary part of (41); the result is

$$\frac{x_e}{c} = \frac{x_c}{c} + \frac{y_a - y_c}{\beta_a c} \tan \alpha_0 - \tfrac{1}{4} T \cos \alpha_0, \tag{98}$$

from which a few interesting geometrical relations follow (cf. pp. 115–22 of Milne-Thomson, 1948).

The value of x_e at zero lift is particularly important. From (97)

$$\left(\frac{x_e}{c}\right)_{C_L=0} \equiv \frac{h}{c} = \frac{x_c}{c} - \frac{T}{4} + \frac{T}{4\pi \beta_a} \int_{-\pi}^{\pi} \theta_s(\gamma) \sin 2\gamma\, d\gamma, \tag{99}$$

an alternative and more useful form of which is

$$\frac{h}{c} = \frac{x_c}{c} - \frac{T}{4} + \frac{1}{\pi \beta_a} \int_{-\pi}^{\pi} \frac{y_s}{c} \sin \gamma\, d\gamma, \tag{100}$$

which follows from (75). From (89) and (95)

$$\frac{h}{c} = -\left(\frac{\partial C_M}{\partial C_L}\right)_{C_L=0}, \tag{101}$$

and therefore to first order

$$C_M = -h C_L. \tag{102}$$

8.13   Estimates for the thickness factor

The accurate method of calculating T is based on (17), which gives

$$T = \left(\frac{P}{c}\right)\left\{\left(\int_{-2\alpha_0}^{\pi} + \int_{-2\alpha_0}^{-\pi}\right)\frac{U}{q_s}\sin\tfrac{1}{2}(\gamma + 2\alpha_0)\cos\tfrac{1}{2}\gamma\,d\gamma\right\}^{-1}, \qquad (103)$$

where P is the total perimeter length of the profile. In the calculation of the flow about a given aerofoil from the integral equation (16), (103) is used to estimate T afresh in each iteration, and this value is then introduced into (17) to complete the calculation of the estimate for $s(\gamma)$.

Another formula for T, which is more convenient in some situations, can be deduced from (70), which the reader will recall applies to both linear perturbation flows (slender bodies) and to incompressible flows (bodies of any shape), at zero circulation.

We make a slight approximation by identifying the distance between the stagnation points, resolved in the Ox-direction, as the chord length c, i.e.

$$x(\zeta = \pi) - x(\zeta = 0) = c. \qquad (104)$$

This is certainly true with symmetrical bodies—the circulation being zero—and in other cases only a trifling error will be introduced. (Of course (104) could be taken as the definition of c.)

At $\zeta = 0, \pi$ it follows from (1) that $w = -2a, 2a$. Substituting these values in (70), subtracting the results, and using (104) we find

$$Uc - 4a = -\frac{U}{2\pi\beta_a}\int_{-\pi}^{\pi} y_s(\gamma)\,(\tan\tfrac{1}{2}\gamma + \cot\tfrac{1}{2}\gamma)\,d\gamma,$$

which can be rearranged to read

$$T = 1 + \frac{1}{2\pi\beta_a}\int_{-\pi}^{\pi}\frac{y_s}{c}(\gamma)\frac{d\gamma}{\sin\gamma},$$

or

$$T = 1 + \frac{1}{\pi\beta_a}\int_{0}^{\pi}\frac{t(\gamma)\,d\gamma}{\sin\gamma}, \qquad (105)$$

where

$$t(\gamma) = \frac{y_s}{c}(\gamma) - \frac{y_s}{c}(-\gamma)$$

is the aerofoil 'thickness function'. With symmetrical aerofoils $t(\gamma)$ is the *local* thickness:chord ratio. The maximum value of t, say t_m, is then the so-called thickness:chord ratio of the aerofoil, an important parameter in aerofoil theory. With a flat plate $t = 0$, and hence $T = 1$; in other cases T exceeds unity by an amount which is roughly proportional to t_m.

We shall now calculate T for two simple profiles in order to derive a rough estimate for use in thin aerofoil theory. In the present case $(\alpha_0 = 0)$ (1) can be written

$$w = -\tfrac{1}{2}Uc\,T\cos\zeta. \qquad (106)$$

Now the approximation of linear perturbation (see § 2.17) permits us to replace the *independent* variable w by UZ (assuming the origins of the two planes to coincide) so (106) becomes $Z = -\tfrac{1}{2}cT\cos\zeta$. Combining this with (104) we conclude that to be consistent we must put $T = 1$ in this approximation. Hence

$$Z = -\tfrac{1}{2}c\cos\zeta, \tag{107}$$

and in particular on $\eta = 0$

$$x = -\tfrac{1}{2}c\cos\gamma. \tag{108}$$

The origin of the Z-plane is thus at the mid-chord point, where $\gamma = \pm\tfrac{1}{2}\pi$.

For the first example consider the slender *elliptic* profile

$$y/c = \tfrac{1}{2}t_m\{1-(2x/c)^2\}^{\frac{1}{2}} \quad (t_m \ll 1).$$

From (108) $t = 2y/c = t_m\sin\gamma$, and substituting this in (105) we get

$$T = 1+\frac{t_m}{\beta_a}. \tag{109}$$

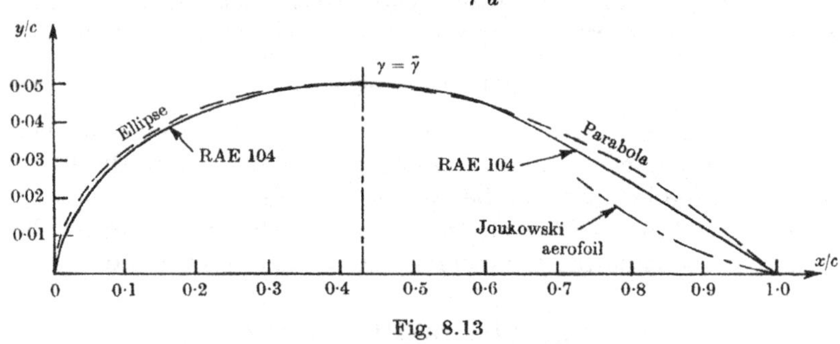

Fig. 8.13

The second example is the parabolic profile $y/c = \tfrac{1}{2}t_m\{1-(2x/c)^2\}$, for which (105) yields

$$T = 1+\frac{2}{\pi}\frac{t_m}{\beta_a}. \tag{110}$$

Many conventional symmetrical aerofoils can be fairly closely approximated to by an ellipse over the front section up to the point of maximum thickness, and by a parabola over the rear section. This is illustrated in Fig. 8.13, in which the full line is the profile (vertical scale magnified 3 times) of an aerofoil known as RAE 104 (Pankhurst & Squire, 1950). This aerofoil is a reasonably typical modern symmetrical aerofoil. The broken curves in the figure are an ellipse in $0 \leqslant x/c \leqslant 0.42$, and a parabola in $0.42 \leqslant x/c \leqslant 1.00$ (origin at the leading edge here). Excepting that the parabola overestimates y towards the trailing edge, the approximation is quite good for our present purposes. Let $\bar\gamma$ be the position of maximum thickness t_m, then from (105)

$$T = 1+\frac{t_m}{\beta_a}\left[\frac{1}{\pi}(1+\bar\gamma+\cos\bar\gamma)\right], \tag{111}$$

which contains (109) and (110) as the special cases $\gamma = 0$ and $\gamma = \pi$. The factor in square brackets in (111) varies only from 0·804 to 0·818 as the point of maximum thickness moves from $x = 0·2c$ to $x = 0·5c$. In view of this and the fact that (111) slightly overestimates T, it seems reasonable to take

$$T \simeq 1 + 0·8\frac{t_m}{\beta_a}, \tag{112}$$

as a fair approximation for aerofoils of moderate thickness. Furthermore, it would have little effect on (112) if the aerofoil had a small degree of camber; we shall therefore use this equation for all aerofoils of conventional shape when a more accurate estimate of T is lacking.

A final check on (112) is provided by the well-known exact result for a Joukowski aerofoil (see Robinson & Laurmann, 1956, p. 98), viz. $T = 1 + 0·77\, t_m/\beta_a$, for small camber and thickness. The Joukowski aerofoil has a cusped trailing edge (see Fig. 8.13), which must clearly result in smaller values of T than those of normal aerofoils with finite trailing edge angles. Hence it is a fair conclusion that the coefficient 0·8 in (112) is well supported by Joukowski aerofoil theory.

THIN AEROFOIL THEORY

8.14 The iterative solution of the basic integral equation

A convenient approach to *thin* aerofoil theory is to regard it as being the solution achieved after only the first step in an iterative method of solving the 'exact' integral equation (16). The iterative method we have in mind is similar to that already described in several earlier sections of this book (e.g. § 7.6). It is as follows:

Let $q_s^{(n)}(\gamma)$ denote the solution after $(n-1)$ steps, so that $q_s^{(1)}$ is the starting distribution, then $q_s^{(n+1)}(\gamma)$ is found from $q_s^{(n)}(\gamma)$ by:

(1) Calculating $s^{(n)}(\gamma)$ and $T^{(n)}$ by substituting $q_s^{(n)}(\gamma)$ in (17) and (103).

(2) Finding $R^{(n)}(\gamma)$ from $s^{(n)}(\gamma)$ and the known $R(s)$.

(3) Substituting $R^{(n)}$, $T^{(n)}$ and $q_s^{(n)}$ in the right-hand side of (16). The left-hand side of (16) then yields $q_s^{(n+1)}(\gamma)$.

The process only requires some starting assumption for $q_{(s)}^{(1)}$. Thin aerofoil theory is based on

$$q_s^{(1)} = U, \tag{113}$$

and it stops at $q_s^{(2)}$. The virtue of this approach is that it immediately suggests how thin aerofoil theory can be improved. Second ($q_s^{(3)}$) and even third approximations ($q_s^{(4)}$) might be worth seeking if the aerofoil is thicker than usual.

The assumption $q_s^{(1)} = U$, i.e. $\Omega_s^{(1)} = 0$, is equivalent to putting $w = UZ$ (see §6.1) so it is clear that thin aerofoil theory can also be regarded as being the result of applying linear perturbation theory to aerofoils.

8.15   The principal equations of thin aerofoil theory

From the remark just made it follows that the principal equations of the theory are as follows:

From (107) and (108),

$$Z = -\tfrac{1}{2}c\cos\zeta \quad (Z = x + i\beta_a y), \tag{114}$$

and
$$x = -\tfrac{1}{2}c\cos\gamma \quad (\eta = 0). \tag{115}$$

From (6·23) we have that for linear perturbation theory

$$\tau = \tfrac{1}{2}\beta_a C_p + i\frac{dy}{dx}. \tag{116}$$

It has already been shown (see (58)) that

$$\alpha_0 = \alpha/\beta_a \tag{117}$$

is also applicable in linear perturbation theory.

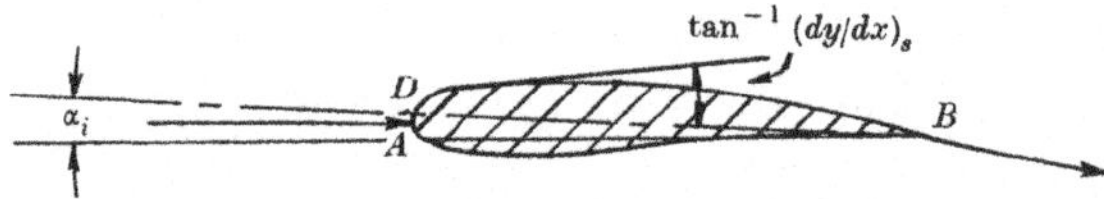

Fig. 8.15

Let α_i be the physical incidence—i.e. the value of α^* (see Fig. 8.7)—at which the front stagnation point A coincides with the leading edge D, and $\theta_i(\gamma)$ be the corresponding surface slope. If this incidence is taken as the reference attitude in place of the no-lift attitude employed in §8.7, we find that (50), (52) and (53) are replaced by

$$\theta_s(\gamma) = \theta_i(\gamma) - (\alpha^* - \alpha_i) + \pi\{U(\gamma + 2\alpha_0) - U(\gamma - \lambda_i)\}, \tag{118}$$

$$\lambda_i = 2(\alpha^* - \alpha_i - \alpha_0), \tag{119}$$

$$0 = \int_{-\pi}^{\pi} \theta_i(\gamma)\,d\gamma, \tag{120}$$

where λ_i is the value of γ at D, and $(\alpha^* - \alpha_i)$ is the incidence change required to move A from $\gamma = -2\alpha_0$ to $\gamma = \lambda_i$ (cf. Figs. 8.7 and 8.15). Let $(dy/dx)_s$ denote the slope of the aerofoil surface measured from the aerofoil chord direction DB, then the approximation

$$\theta_i \simeq \left(\frac{dy}{dx}\right)_s - \alpha_i \tag{121}$$

is consistent with linear perturbation theory, and is adequate for thin aerofoils except in the immediate neighbourhood of a rounded nose.

Equations (120) and (121) give

$$\alpha_i = \frac{1}{2\pi} \int_{-\pi}^{\pi} \left(\frac{dy}{dx}\right)_s d\gamma, \tag{122}$$

an incidence known as the 'ideal' or 'design' incidence (Theodorsen, 1931), for this is clearly the incidence at which the flow past the nose is as 'smooth' as possible.

Eliminating θ_i from (118) and (121) and substituting the result in the second of (28) we get

$$\beta_a \sin 2\alpha_0 = \frac{1}{\pi} \int_{-\pi}^{\pi} \left(\frac{dy}{dx}\right)_s \cos \gamma \, d\gamma + \sin \lambda_i + \sin 2\alpha_0.$$

To first order in incidence this reduces to

$$2\beta_a \alpha_0 = 2\alpha = 2(\alpha^* + \alpha') = \frac{1}{\pi} \int_{-\pi}^{\pi} \left(\frac{dy}{dx}\right)_s \cos \gamma \, d\gamma + 2(\alpha^* - \alpha_i),$$

by (49), (117) and (119). Hence on eliminating α_i by (122) we obtain

$$\alpha' = \frac{1}{2\pi} \int_{-\pi}^{\pi} \left(\frac{dy}{dx}\right)_s^{1} (\cos \gamma - 1) \, d\gamma. \tag{123}$$

To the same accuracy it follows from (49), (81) and (117) that the lift coefficient is given by

$$C_L = \frac{2\pi T}{\beta_a} (\alpha^* + \alpha'), \tag{124}$$

where T is given by (105).

The moment coefficient is given by

$$C_M = -\frac{x_c}{c} C_L + \frac{T^2}{4\beta_a} \int_{-\pi}^{\pi} \left(\frac{dy}{dx}\right)_s \cos 2\gamma \, d\gamma$$

$$- \frac{T^2(\alpha^* + \alpha')}{2\beta_a^2} \int_{-\pi}^{\pi} \left(\frac{dy}{dx}\right)_s \sin 2\gamma \, d\gamma + \frac{\pi T^2}{2\beta_a} (\alpha^* - \alpha_i), \tag{125}$$

which is obtained by substituting (118) and (121) in (89) and retaining only first-order terms. By (115)

$$\int_{-\pi}^{\pi} \left(\frac{dy}{dx}\right)_s \sin 2\gamma \, d\gamma = -\frac{8}{c^2} \int_{-\pi}^{\pi} \left(\frac{dy}{dx}\right)_s x \, dx = \frac{8A}{c^2},$$

where A is the cross-sectional area of the profile. The second term on the right-hand side of (125) is thus a second-order term—A/c^2 being small for a thin aerofoil—and can therefore be neglected. Similarly,

$$\frac{1}{2\pi}\int_{-\pi}^{\pi}\left(\frac{dy}{dx}\right)_s \cos 2\gamma\, d\gamma = \alpha_i - \frac{2}{\pi}\int_{-\pi}^{\pi}\sin\gamma\, d\left(\frac{y_s}{c}\right)$$

$$= \alpha_i + \frac{2}{\pi}\int_{-\pi}^{\pi}\frac{y_s}{c}\cos\gamma\, d\gamma,$$

on making use of (122) and an integration by parts. Hence (125) becomes

$$C_M = -\frac{x_c}{c}C_L + \frac{\pi\alpha^*}{2\beta_a} + \frac{1}{\beta_a}\int_{-\pi}^{\pi}\frac{y_s}{c}\cos\gamma\, d\gamma,$$

where the thickness factor T is put equal to unity (cf. (112)). From (69) and (115) we have $x_c = 0$. Therefore

$$C_M = \tfrac{1}{4}C_L - \frac{\pi\alpha'}{2\beta_a} + \frac{1}{\beta_a}\int_{-\pi}^{\pi}\frac{y_s}{c}\cos\gamma\, d\gamma, \tag{126}$$

on using (124) to eliminate α^*. Comparing (101) and (126) we conclude that the aerodynamic centre is at the 'quarter-chord' point, $x = -\tfrac{1}{4}c$.

It will be noticed that both C_M and C_L depend on α' and not α_i. This means that as far as these forces are concerned the error in the approximation (121) near $\gamma = 0$, i.e. near the nose of the aerofoil, is somewhat mitigated by the presence of the factor $(\cos\gamma - 1)$ in (123). For even if $(dy/dx)_s$ is $O(1/\gamma)$ over a small range $-\epsilon \leqslant \gamma < \epsilon$, the resulting contribution to α' is only $O(\epsilon^2)$. However, the approximation (121) does cause a serious error in the velocity distribution near $\gamma = 0$ as we shall now show.

From (118), (119), (121), (122) and (10) it is found that

$$\tau(\zeta) = -(\alpha^* - \alpha_i)\cot\tfrac{1}{2}\zeta - i(\alpha^* - \alpha_i) + \frac{1}{2\pi}\int_{-\pi}^{\pi}\left(\frac{dy}{dx}\right)_s \{\cot\tfrac{1}{2}(\gamma - \zeta) - i\}\, d\gamma, \tag{127}$$

on ignoring second-order terms in incidence. Hence on using the expansion given in (20) we get

$$\tau(\zeta) = -(\alpha^* - \alpha_i)\cot\tfrac{1}{2}\zeta - i(\alpha^* - \alpha_i) + \frac{i}{\pi}\sum_{n=1}^{\infty}e^{ni\zeta}\int_{-\pi}^{\pi}\left(\frac{dy}{dx}\right)_s e^{-ni\gamma}\, d\gamma. \tag{128}$$

Equation (128) involves the assumption that $(dy/dx)_s$ does possess a Fourier expansion. This is not true with an aerofoil having a rounded nose, and the theory will be slightly modified to allow for this case in § 8.18.

On $\eta = 0$ at $\zeta = \gamma^*$ the real part of (128) yields

$$C_p = -\frac{2}{\beta_a}(\alpha^* - \alpha_i)\cot\tfrac{1}{2}\gamma^* + \frac{2}{\pi\beta_a}\left\{\sum_{n=2}^{\infty}\cos n\gamma^* \int_{-\pi}^{\pi}\left(\frac{dy}{dx}\right)_s \sin n\gamma\, d\gamma\right.$$

$$\left. - \sum_{n=1}^{\infty}\sin n\gamma^* \int_{-\pi}^{\pi}\left(\frac{dy}{dx}\right)_s \cos n\gamma\, d\gamma\right\}, \quad (129)$$

where we have made use of the closure condition $\int_{-\pi}^{\pi} dy_s = 0$ to eliminate the first term of the first infinite sum. This equation gives that $C_p \to -\infty$ as $\gamma^* \to 0$, except when $\alpha^* = \alpha_i$, i.e. except when the aerofoil is at the ideal incidence. This failure of linear perturbation theory at stagnation points is the most serious of thin aerofoil theory. A simple method of modifying the theory to render it 'uniformly valid' at $\gamma = 0$ (see discussion in § 2.12) will be given in § 8.16.

8.16   Round-nosed aerofoils

The following method of adjusting thin aerofoil theory to allow for rounded noses is due in essence to Riegels (1949). Other methods giving much the same result have been given by Goldstein (1952) and Lighthill (1951). The method given here is based on a comparison of the exact theory for the flow past an elliptic cylinder (having the same nose radius as the aerofoil) with the linear perturbation theory for this cylinder. The flow is assumed to be incompressible, but some remarks about compressible flow are made later.

Assuming there is no circulation about the cylinder we have

$$w = -2a\cos\zeta, \quad (130)$$

from (1). The relation between z and ζ is easily verified to be

$$z = -\frac{c}{2\cosh t}\cos(\zeta - \lambda_0 + it), \quad (131)$$

where t and λ_0 are real constants, for on $\eta = 0$ (131) gives the ellipse

$$\frac{x^2}{(\tfrac{1}{2}c)^2} + \frac{y^2}{(\tfrac{1}{2}c)^2\tanh^2 t} = 1.$$

The thickness ratio and nose radius are given in turn by

$$t_m = \tanh t, \quad \rho_n = \tfrac{1}{2}c\tanh^2 t. \quad (132)$$

From (6.16), (130) and (131),

$$e^\tau = \frac{U\,dz}{dw} = \frac{U}{q}e^{i\theta} = \left(\frac{cU}{4a}\right)\operatorname{sech} t\,\frac{\sin(\zeta - \lambda_0 + it)}{\sin\zeta}. \quad (133)$$

From this and the limit $\lim\limits_{\eta\to\infty} q = U$ it follows that $\lim\limits_{\eta\to\infty} \theta = \lambda_0$, and

$$\frac{4a}{Uc} = e^t \operatorname{sech} t. \tag{134}$$

We shall rotate the z-plane through an angle $-\lambda_0$ in order to make $\lim\limits_{\eta\to\infty} \tau = 0$ as in (9). Thus modified (133) becomes

$$e^\tau = e^{-i\lambda_0} e^{-t} \cosh t \frac{\sin(\zeta-\lambda_0)}{\sin\zeta} \{1 + i\tanh t \cot(\zeta-\lambda_0)\}. \tag{135}$$

On $\eta = 0$ it follows from this equation that

$$\tan(\theta+\lambda_0) = \tanh t \cot(\gamma-\lambda_0),$$

$$\frac{U}{q} = \left(\frac{Uc}{4a}\right)\frac{\sin(\gamma-\lambda_0)}{\sin\gamma}\{1 + \tanh^2 t \cot^2(\gamma-\lambda_0)\}^{\frac{1}{2}},$$

whence
$$\frac{q}{U} = \left(\frac{4a}{Uc}\right)\frac{\sin\gamma}{\sin(\gamma-\lambda_0)}\cos(\theta+\lambda_0). \tag{136}$$

The front stagnation point is at $\gamma = 0$, and the nose of the ellipse at $\gamma = \lambda_0$.

A linear perturbation theory solution of this problem would yield only first-order terms in the thickness t_m and the displacement λ_0 of the front stagnation point from the leading edge. To this order (132) and (135) give

$$\tau = -i\lambda_0 - t - \lambda_0\cot\zeta + it\cot\zeta. \tag{137}$$

Let $(q/U)_0$ denote the linear perturbation theory solution then it follows from (137) that
$$(q/U)_0 = 1 + t + \lambda_0\cot\gamma,$$
on $\eta = 0$. By (136)

$$\frac{q}{U} = \left(\frac{4a}{Uc}\right)\frac{\sin\gamma}{\sin(\gamma-\lambda_0)}\frac{\cos(\theta+\lambda_0)}{(1+t+\lambda_0\cot\gamma)}\left(\frac{q}{U}\right)_0.$$

Ignoring second-order terms in λ_0 and t (see (134)) we get the simple rule

$$\frac{q}{U} = \cos\theta\left(\frac{q}{U}\right)_0, \tag{138}$$

where $\hat\theta = \theta + \lambda_0 = \tan^{-1}(dy/dx)_s$ is the angle between the surface tangent and the chord line. When applied to aerofoils (138) becomes Riegels's rule for modifying linear perturbation theory.

Lighthill's rule, namely

$$\frac{q}{U} = \left(\frac{x^*}{x^*+\frac{1}{2}\rho_n}\right)^{\frac{1}{2}}\left(\frac{q}{U}\right)_0, \tag{139}$$

where x^* is the chord distance measured from the leading edge, can be deduced from (138) by eliminating γ and t from (132), $x^* = \tfrac{1}{2}c(1 - \cos\gamma)$, and $\cos\theta = \{1 + \tanh^2 t \cot^2(\gamma - \lambda_0)\}^{-\frac{1}{2}}$, and then retaining only first-order terms in ρ_n/x^*. Riegels's rule is the more accurate of the two.

It remains to make some comments about compressible flow. It is rather difficult to extend the above method to tangent gas theory, and in any case such an extension would scarcely be justifiable, for near stagnation points tangent gas theory is less accurate than incompressible flow theory. Van Dyke (1954) has extended the rules given in (138) and (139) to subsonic compressible flow by using the Janzen–Rayleigh theory (expansion in a power series in M^2; see Robinson & Laurmann, 1956, pp. 345–8), which is quite accurate near stagnation points, but the theory is too complicated to give here. Furthermore, the effect of compressibility in the rules is not large (compared with the effects of other possible sources of error), e.g. at $x^*/\rho_n = 0{\cdot}25$ $(q/U)_0$ is multiplied by $0{\cdot}58$ at $M = 0$, $0{\cdot}52$ at $M = 0{\cdot}6$ and $0{\cdot}49$ at $M = 0{\cdot}8$ (this is the worst case in a table given by Van Dyke). We conclude that (138) will be a satisfactory rule for compressible subsonic flow.

8.17   Aerofoils with finite trailing edge angles

Thin aerofoil theory also breaks down at sharp corners on aerofoils where q either becomes infinite (projecting corners) or zero (re-entrant corners). Tangent gas theory fails in the first case and is inaccurate in the second, so we shall again restrict the calculations to incompressible flows, but by the same arguments as given in the last paragraph of § 8.16 we shall also use the theory for all subsonic (compressible) flows.

The most important 'corner' on an aerofoil is that occurring at a finite trailing edge, but it is often not considered worth while adjusting thin aerofoil theory to allow for this, because in real flows boundary-layer thickening or separation near the trailing edge masks the effect of the corner and eliminates the stagnation point. Nevertheless, it is still important to calculate inviscid flows accurately near trailing edges, if only to provide a datum solution from which the effects of the boundary layer can be measured. Furthermore, the trailing edge angle does have a general influence on the (real) flow at points outside its neighbourhood.

From (10) it follows that if $\Delta\theta$ is the jump in θ at a corner on the aerofoil at $\gamma = \gamma^*$, then we can write

$$\tau(\zeta) = -\frac{\Delta\theta}{\pi}\ln\sin\tfrac{1}{2}(\gamma^* - \zeta) + F(\zeta), \tag{140}$$

where $F(\zeta)$ is regular at $\zeta = \gamma^*$. At a re-entrant corner (a stagnation point) $\Delta\theta$ is positive, and at a sharp protruding corner $\Delta\theta$ is negative. Equation (6·17) now gives that

$$\frac{q}{U} = \left|\sin \tfrac{1}{2}(\gamma^* - \gamma)\right|^{\Delta\theta/\pi} \exp\left(-\mathscr{R}F(\gamma)\right), \tag{141}$$

in an exact theory, and

$$\left(\frac{q}{U}\right)_0 = 1 + \frac{\Delta\theta}{\pi} \ln\left|\sin \tfrac{1}{2}(\gamma^* - \gamma)\right| - \mathscr{R}F(\gamma),$$

in a linear perturbation theory ($\Delta\theta$ small). Consequently the rule for adjusting thin aerofoil theory will read

$$\frac{q}{U} = \left|\sin \tfrac{1}{2}(\gamma^* - \gamma)\right|^{\Delta\theta/\pi} \left\{1 - \frac{\Delta\theta}{\pi} \ln\left|\sin \tfrac{1}{2}(\gamma^* - \gamma)\right|\right\} \left(\frac{q}{U}\right)_0. \tag{142}$$

Let θ_t be the trailing edge angle, then $\Delta\theta = \theta_t$ at $\gamma^* = \pi$, and (141) becomes

$$\frac{q}{U} = \left|\cos \tfrac{1}{2}\gamma\right|^{\theta_t/\pi} \left\{1 + O(\pi - \gamma)\right\}.$$

Retaining only first-order terms (e.g. ignoring circulation) we can write this in the form

$$\frac{q}{U} = \left|\frac{1}{4a}(2a - \phi)\right|^{\theta_t/2\pi}, \tag{143}$$

by (1). Substituting this in $s = \displaystyle\int_{2a}^{\phi} \frac{d\phi}{q}$, where s is distance measured from the trailing edge along the aerofoil surface, we find that $s \propto (2a - \phi)^{1 - \theta_t/2\pi}$, and therefore

$$\frac{q}{U} = \left(\frac{q}{U}\right)_1 \left(\frac{s}{s_1}\right)^{\theta_t/(2\pi - \theta_t)}, \tag{144}$$

where $q = q_1$ at $s = s_1$. When θ_t is small this can be replaced by the approximation

$$\frac{q}{U} = \left(\frac{q}{U}\right)_1 \left(\frac{x}{x_1}\right)^{\theta_t/2\pi}, \tag{145}$$

where x is distance measured forwards from the trailing edge along the chord line.

If a rule corresponding to (142) is derived from (145) it will read

$$\frac{q}{U} = x^{\theta_t/2\pi} \left\{1 - \frac{\theta_t}{2\pi} \ln|x|\right\} \left(\frac{q}{U}\right)_0. \tag{146}$$

The rules contained in (142) and (146) can be applied to compressible subsonic flow without serious error. Remember that the main effects of compressibility are already accounted for in $(q/U)_0$.

The effect of finite curvature near the trailing edge can be obtained by studying the symmetrical profile comprised of two circular arcs shown in Fig. 8.17. (This is a special example of a Kármán–Trefftz profile; see Milne-Thomson, 1948, p. 125). It is easily verified that this profile is mapped into the ζ-plane by

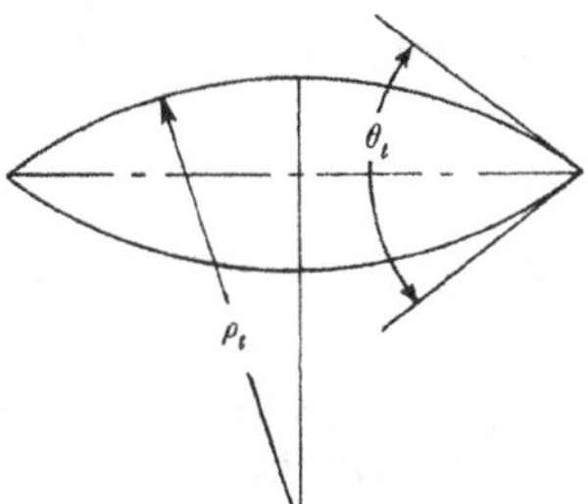

$$z = -\frac{c}{2}\left\{\frac{1 - e^{i\theta_t/2}(\tan\tfrac12\zeta)^{2-\theta_t/\pi}}{1 + e^{i\theta_t/2}(\tan\tfrac12\zeta)^{2-\theta_t/\pi}}\right\}^2, \quad (147)$$

where θ_t is the trailing (and leading) edge angle. On $\eta = 0$ this equation yields

Fig. 8.17

$$x^2 + (y \pm \cot\tfrac12\theta_t)^2 = (\tfrac12 c\,\mathrm{cosec}\,\tfrac12\theta_t)^2, \tag{148}$$

where the plus and minus signs refer to the upper and lower surfaces respectively. Thus the radius of curvature of the arcs is

$$\rho_t = \tfrac12 c\,\mathrm{cosec}\,\tfrac12\theta_t. \tag{149}$$

We find from (130) and (147) that

$$\frac{dw}{dz} = U\frac{4\sin^2\zeta}{\left(1 - \dfrac{\theta_t}{2\pi}\right)^2}\left\{e^{-i\theta_t/4}\left(\cot\frac{\zeta}{2}\right)^{1-\theta_t/2\pi} + e^{i\theta_t/4}\left(\tan\frac{\zeta}{2}\right)^{1-\theta_t/2\pi}\right\}^2,$$

where we have used $\displaystyle\lim_{\eta\to\infty}\frac{dw}{dz} = U$ to find

$$\frac{4a}{Uc} = \frac{1}{\left(1 - \dfrac{\theta_t}{2\pi}\right)}, \tag{150}$$

and then eliminated this ratio. Ignoring second-order terms in θ_t we find

$$\tau \equiv \ln\frac{U\,dz}{dw} = -\frac{\theta_t}{\pi} + \frac{i\theta_t}{2}\cos\zeta + \frac{\theta_t}{\pi}\cos\zeta\ln\left(\cot\frac{\zeta}{2}\right). \tag{151}$$

This equation can be written

$$\tau = \left\{-\frac{\theta_t}{2\pi} + \frac{i\theta_t}{2}\cos^2\frac{\zeta}{2} + \frac{\theta_t}{\pi}\cos^2\frac{\zeta}{2}\ln\cot\frac{\zeta}{2}\right\}$$

$$+ \left\{-\frac{\theta_t}{2\pi} - \frac{i\theta_t}{2}\sin^2\frac{\zeta}{2} + \frac{\theta_t}{\pi}\sin^2\frac{\zeta}{2}\ln\tan\frac{\zeta}{2}\right\},$$

which separates the contributions to τ from the leading and trailing edges of the profile. We conclude from this equation that near the trailing edge of an aerofoil we can write

$$\tau = -\frac{\theta_t}{\pi}\sin^2\frac{\zeta}{2}\left\{\frac{i\pi}{2} - \ln\tan\frac{\zeta}{2}\right\} + F(\zeta), \tag{152}$$

where $F(\zeta)$ is regular at $\zeta = \pi$. We shall use this result in the next section.

8.18   The effect of round noses and trailing edge angles on thin aerofoil theory

The thin aerofoil theory for subsonic flow, given in §8.15, will now be modified to allow for a rounded nose and a finite trailing edge angle. First consider the rounded nose.

Equation (137) can be written

$$\tau = \left\{-i\tfrac{1}{2}\lambda_0 - \tfrac{1}{2}t - \tfrac{1}{2}\lambda_0\cot\tfrac{1}{2}\zeta + i\tfrac{1}{2}t\cot\tfrac{1}{2}\zeta\right\} + \left\{-i\tfrac{1}{2}\lambda_0 - \tfrac{1}{2}t \right.$$
$$\left. + \tfrac{1}{2}\lambda_0\tan\tfrac{1}{2}\zeta - i\tfrac{1}{2}t\tan\tfrac{1}{2}\zeta\right\},$$

from which it follows that near the nose of the ellipse we can write

$$\tau = -\frac{\lambda_0}{2}\cot\frac{\zeta}{2} + i\frac{t}{2}\cot\frac{\zeta}{2} + F(\zeta), \tag{153}$$

where $F(\zeta)$ is regular at $\zeta = 0$.

Now consider the flow about an aerofoil with the same nose radius. In this case the displacement λ_0 will equal $2(\alpha^* - \alpha_i)$ (cf. (118)), for we should regard zero incidence for the ellipse as corresponding to the ideal incidence for the aerofoil.

Combining (152) and (153) we conclude that with an aerofoil having a rounded nose and a finite trailing edge angle we can write

$$\tau = -(\alpha^* - \alpha_i)\cot\frac{\zeta}{2} + i\frac{t}{2}\cot\frac{\zeta}{2} - \frac{\theta_t}{\pi}\sin^2\frac{\zeta}{2}\left\{\frac{i\pi}{2} - \ln\tan\frac{\zeta}{2}\right\} + F(\zeta), \tag{154}$$

where $F(\zeta)$ is regular everywhere in the ζ-plane, and by (132) t is related to the nose radius by

$$t = \sqrt{\left(\frac{2\rho_n}{c}\right)}. \tag{155}$$

It may be argued that we have not allowed for compressibility in (154), but it is clear from (116) that (154) does allow for compressibility to the usual extent of linear perturbation theory, and from the comments made at the end of §8.16 further allowance is not really justified in this case.

As $F(\zeta)$ is regular it can be expanded in the complex Fourier series

$$F(\zeta) = \sum_{n=0}^{\infty}(a_n + ib_n)e^{ni\zeta}, \tag{156}$$

where a_n and b_n are real constants. On the aerofoil surface (154) and (156) give

$$\Omega_s = -(\alpha^* - \alpha_i)\cot \tfrac{1}{2}\gamma + \pi t\boldsymbol{\delta}(\gamma) + \frac{\theta_t}{\pi}\sin^2\frac{\gamma}{2}\ln|\tan\tfrac{1}{2}\gamma|$$

$$+ \sum_{n=0}^{\infty}(a_n\cos n\gamma - b_n\sin n\gamma), \quad (157)$$

$$\theta_s = 2\pi(\alpha^* - \alpha_i)\,\boldsymbol{\delta}(\gamma) + \frac{t}{2}\cot\tfrac{1}{2}\gamma + \theta_t\sin^2\frac{\gamma}{2}\{\mathbf{U}(\gamma - \pi) - \tfrac{1}{2}\}$$

$$+ \sum_{n=0}^{\infty}(b_n\cos n\gamma + a_n\sin n\gamma), \quad (158)$$

where the delta and unit function here arise because

$$\ln(\tan\tfrac{1}{2}\gamma) = \ln|\tan\tfrac{1}{2}\gamma| + i\pi\mathbf{U}(\gamma - \pi),$$

$$\ln(\sin\tfrac{1}{2}\gamma) = \ln|\sin\tfrac{1}{2}\gamma| + i\pi\mathbf{U}(-\gamma),$$

and the delta function is the derivative of the unit function. (Most accounts of thin aerofoil theory neglect to include the delta function terms, with the confusing result that the equations given for Ω_s and θ_s are not genuine conjugate harmonics.)

We note from (118), (119) and (121) that

$$\theta_s = \left(\frac{dy}{dx}\right)_s - \alpha^* + 2\pi(\alpha^* - \alpha_i)\,\boldsymbol{\delta}(\gamma), \quad (159)$$

as $\quad \mathbf{U}(\gamma + 2\alpha_0) - \mathbf{U}(\gamma - \lambda_i) = 2(\alpha^* - \alpha_i)\,\mathbf{U}'(\gamma + 2\alpha_0) = 2(\alpha^* - \alpha_i)\,\boldsymbol{\delta}(\gamma),$

neglecting second-order terms.

From (11), (28), (49) and (117) $\theta_s(\gamma)$ must satisfy

$$0 = \int_{-\pi}^{\pi}\theta_s\,d\gamma, \quad 0 = \int_{-\pi}^{\pi}\theta_s\sin\gamma\,d\gamma, \left.\begin{array}{r}\\ \\ \\ \\ \end{array}\right\} \quad (160)$$

and $\quad 2\pi\alpha_0\beta_a = 2\pi(\alpha^* + \alpha') = \int_{-\pi}^{\pi}\theta_s\cos\gamma\,d\gamma,$

to first order in incidence. Substituting (158) in these we find that

$$b_0 = -(\alpha^* - \alpha_i), \quad a_1 = \frac{\theta_t}{\pi} - t, \quad b_1 = 2(\alpha' + \alpha_i). \quad (161)$$

The equations conjugate to (160) are from (43) and (46)

$$0 = \int_{-\pi}^{\pi}\Omega_s\,d\gamma, \quad 0 = \int_{-\pi}^{\pi}\Omega_s\cos\gamma\,d\gamma, \left.\begin{array}{r}\\ \\ \\ \\ \end{array}\right\} \quad (162)$$

and $\quad 2\pi(\alpha^* + \alpha') = -\int_{-\pi}^{\pi}\Omega_s\sin\gamma\,d\gamma.$

Substituting (157) in these we obtain the second two of (161) together with

$$a_0 = -\frac{\theta_t}{2\pi} - \frac{t}{2}. \tag{163}$$

Equations (158) and (159) give

$$\left(\frac{dy}{dx}\right)_s = \alpha_i + \frac{t}{2}\cot\tfrac{1}{2}\gamma + \theta_t \sin^2\frac{\gamma}{2}\{\mathrm{U}(\gamma-\pi) - \tfrac{1}{2}\} + 2(\alpha' + \alpha_i)\cos\gamma$$

$$+ \left(\frac{\theta_t}{\pi} - t\right)\sin\gamma + \sum_{n=2}^{\infty}(b_n\cos n\gamma + a_n\sin n\gamma), \tag{164}$$

with the help of (161) and (163). On multiplying this through by $\cos n\gamma$ and integrating over $(-\pi, \pi)$ we recover (122), (123) and also

$$b_n = \frac{1}{\pi}\int_{-\pi}^{\pi}\left(\frac{dy}{dx}\right)_s \cos\gamma\, d\gamma \quad (n \geqslant 2). \tag{165}$$

Similarly, $\quad a_n = \dfrac{1}{\pi}\displaystyle\int_{-\pi}^{\pi}\left(\frac{dy}{dx}\right)_s \sin n\gamma\, d\gamma - t$

$$-\frac{\theta_t}{\pi}\frac{1}{n(n^2-1)}\begin{cases} 1 + (-1)^{(n+1)/2}\,n & (n\,\text{odd}), \\ n^2 & (n\,\text{even}). \end{cases} \tag{166}$$

We have now evaluated all the constants in (157) (except α^* which is of course an independent variable), and it remains only to multiply Ω_s by $2/\beta_a$ to find the pressure distribution over a thin aerofoil of given shape. Equations (124) and (126) for C_L and C_M are not affected by the extension of the theory given above.

8.19   Thin aerofoil design

At the ideal or 'design' incidence (157), (161) and (163) give

$$\Omega_s = \frac{\beta_a C_p}{2} = \pi t\,\delta(\gamma) + \frac{\theta_t}{\pi}\sin^2\frac{\gamma}{2}\ln|\tan\tfrac{1}{2}\gamma| - \frac{1}{2}\left(\frac{\theta_t}{\pi} + t\right) + \left(\frac{\theta_t}{\pi} - t\right)\cos\gamma$$

$$- 2(\alpha' + \alpha_i)\sin\gamma + \sum_{n=2}^{\infty}(a_n\cos n\gamma - b_n\sin n\gamma), \tag{167}$$

from which it follows that

$$b_n = -\frac{\beta_a}{2\pi}\int_{-\pi}^{\pi}C_p(\gamma)\sin n\gamma\, d\gamma \quad (n \geqslant 2),$$

$$a_n = \frac{1}{2\pi}\int_{-\pi}^{\pi}\left(\beta_a C_p - \frac{2\theta_t}{\pi}\sin^2\frac{\gamma}{2}\ln|\tan\tfrac{1}{2}\gamma|\right)\cos n\gamma\, d\gamma - t \quad (n \geqslant 2).$$

When these values are substituted in (164), an integration yields $y_s(\gamma)$ and hence the profile shape corresponding to the given pressure distribution.

As written (167) satisfies the closure conditions (162), but in a *true* design problem θ_t, t and $(\alpha' + \alpha_i)$ will not be given. Also the given distribution of C_p will not (usually) contain the two singular terms on the right-hand side of (167). We can proceed as follows. The Fourier coefficients a_0, a_1 and b_1 of the given distribution are calculated and from these are deduced θ_t, t and $(\alpha' + \alpha_i)$ (see (161) and (162)). The given distribution is now modified by adding the first two terms on the right-hand side of (167). This modified distribution now satisfies the closure conditions, and deviates appreciably from the given distribution only in the immediate neighbourhood of the trailing edge (the delta function has zero range of influence).

In a *mixed* design problem, e.g. in which one or more of θ_t, ρ_n, α' and α_i are prescribed in addition to C_p, C_p must of course have sufficient degrees of freedom to satisfy the closure conditions (162).

8.20   Aerofoils with hinged flaps

There are numerous methods of altering the basic shape of an aerofoil so as to increase the lift it will give at a particular attitude. The simplest of these is the *plain* flap, in which a section of the aerofoil, called the flap, is hinged to the rest as shown in Fig. 8.20a. Two flaps are shown in the figure—a 'nose' flap and a trailing edge flap. Many other types and combinations of flaps are possible (see ch. 8 of Abbott & van Doenhoff, 1949), however we shall confine our attention here to the system illustrated.

Downwards deflexion of the trailing edge and nose flaps increases the camber of the aerofoil, and as for a given attitude this will plainly increase the lift on the aerofoil, the net effect is a reduction in the no-lift angle, $-\alpha'$ (i.e. an increase in α'). This conclusion also follows directly from (123), for the deflexion of the flaps increases $(dy/dx)_s$ near the nose, $\gamma = 0$, and reduces $(dy/dx)_s$ near the trailing edge, $\gamma = \pi$. From (122) it follows that the ideal incidence α_i is reduced by the trailing edge flap, and increased by the nose flap. A sensible method of operating this pair of flaps will be to select the deflexion angles δ_n and δ_t—defined in the figure—so that α_i is unchanged as α' increases, for if the aerofoil remains at the ideal incidence sharp velocity peaks are avoided near the nose, and the possibility of flow separation is reduced. Of course there is a chance of flow separation at the hinges if the deflexion angles δ_n and δ_t are too large, or if the hinges are designed badly, so that the change in direction is too rapid on the upper surface. The theory of hinged flaps that produce separation at their hinges will be given in § 11.10.

In Fig. 8.20 there is shown the increment $\Delta\theta_s$ to θ_s resulting from the deflexion of the flaps at a fixed incidence α^*. In drawing this figure we have assumed that the hinges produce sudden changes in the surface slope, a convenient idealization of the problem. A more careful analysis would allow for finite curvatures at γ_t and γ_n. It is also assumed that the discontinuities caused in $\Delta\theta_s$ by each flap occur opposite each other at the same values of x. It follows from (159) and the fact that α^* is not changed

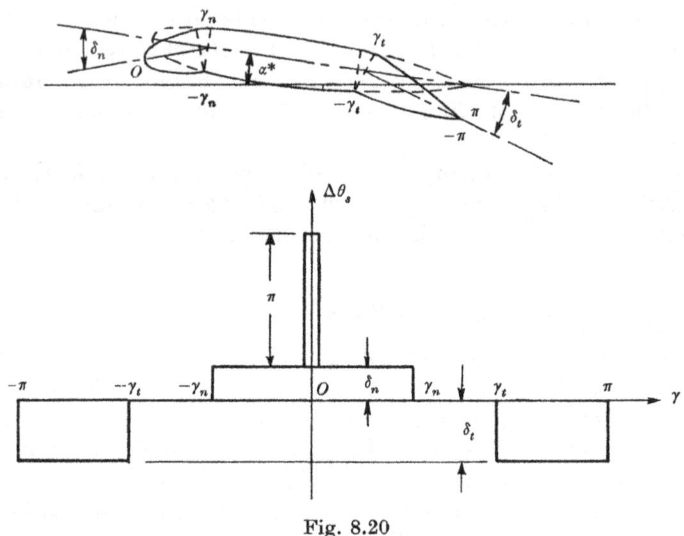

Fig. 8.20

that $\Delta\,(dy/dx)_s$ equals $\Delta\theta_s$ excepting that it does not include the front stagnation point term, i.e. that corresponding to the region of length π shown in Fig. 8.20. Therefore

$$\Delta\left(\frac{dy}{dx}\right)_s = -\delta_t\{\mathbf{U}(\gamma_t) - \mathbf{U}(\pi) + \mathbf{U}(-\pi) - \mathbf{U}(-\gamma_t)\} + \delta_n\{\mathbf{U}(-\gamma_n) - \mathbf{U}(\gamma_n)\},$$

$$(168)$$

and substituting this in (122) and (123) we arrive at

$$\Delta\alpha_i = \delta_n\frac{\gamma_n}{\pi} - \delta_t\left(1 - \frac{\gamma_t}{\pi}\right), \tag{169}$$

and
$$\Delta\alpha' = \delta_n\frac{1}{\pi}(\sin\gamma_n - \gamma_n) + \delta_t\frac{1}{\pi}(\pi - \gamma_t + \sin\gamma_t). \tag{170}$$

Equations (124), (126) and the approximation $T = 1$ give

$$\Delta C_L = (2\pi/\beta_a)\,\Delta\alpha',$$

and
$$\Delta C_M = \frac{1}{\beta_a}\int_{-\pi}^{\pi}\Delta\left(\frac{y_s}{c}\right)\cos\gamma\,d\gamma = -\frac{1}{4\beta_a}\int_{-\pi}^{\pi}\Delta\left(\frac{dy}{dx}\right)_s(1 - \cos 2\gamma)\,d\gamma,$$

by (115) and an integration by parts. Thus

$$\Delta C_L = \frac{2}{\beta_a}\{\delta_n(\sin\gamma_n - \gamma_n) + \delta_t(\pi - \gamma_t + \sin\gamma_t)\}, \tag{171}$$

and
$$\Delta C_M = \frac{1}{4\beta_a}\{\delta_n(\sin 2\gamma_n - 2\gamma_n) + \delta_t(2\pi - 2\gamma_t + \sin 2\gamma_t)\} \tag{172}$$

are the increments to C_L and C_M caused by deflecting the flaps.

On $\eta = 0$ (116), (127) and (168) give

$$\Delta C_p = \frac{2}{\beta_a}\left\{\Delta\alpha_i\cot\tfrac{1}{2}\gamma + \frac{\delta_n}{\pi}\ln\left|\frac{\sin\tfrac{1}{2}(\gamma_n - \gamma)}{\sin\tfrac{1}{2}(\gamma_n + \gamma)}\right|\right.$$
$$\left. + \frac{\delta_t}{\pi}\ln\left|\frac{\sin\tfrac{1}{2}(\gamma_t - \gamma)}{\sin\tfrac{1}{2}(\gamma_t + \gamma)}\right|\right\}, \tag{173}$$

as the increment to the pressure coefficient; and with the help of this equation it is a simple matter to calculate the force on each flap, and the flap-hinge moments.

In ch. 10 we shall discuss another type of 'flap' — the so-called *jet-flap*— which is in essence a high-speed jet of air ejected from the trailing edge at an angle to the main stream. The study of this high-lift device is deferred because the mathematical theory involved is more akin to unsteady than to steady aerofoil theory.

8.21   Motion in a curved path

Suppose that an aerofoil is moving in a curved path having an instantaneous radius of curvature R at the mid-chord point O. Take the origin of the z-plane to be at O, with the plane oriented so that the centre

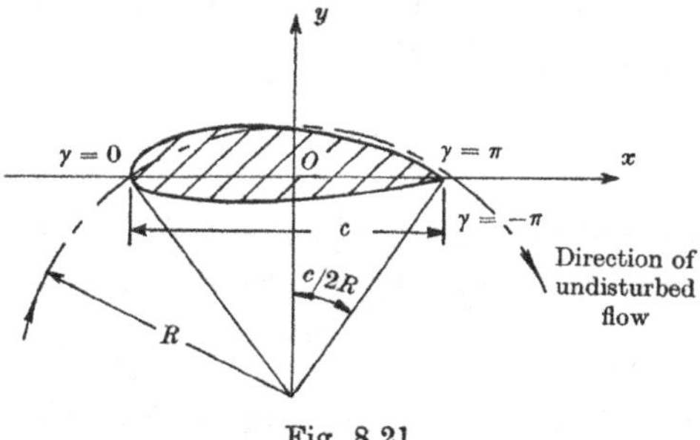

Fig. 8.21

of curvature of the path is at $y = -R$. It will be assumed that R is large compared with the chord c. Then a glance at Fig. 8.21 shows that the aerofoil would produce no disturbance in the flow only if it were the circular arc $y = (c^2/8R)\{1 - (2x/c)^2\}^{\frac{1}{2}}$. From this and $x = -\tfrac{1}{2}\cos\gamma$ (see (115)) it follows that $(dy/dx) = (c/2R)\cos\gamma$.

The conclusion we draw from this result is that the effect on the flow, of an aerofoil moving in a curved path, is (to first order) equivalent to the effect of a distorted aerofoil moving rectilinearly where the distorted aerofoil has a surface slope $(dy/dx)_m$ given by

$$\left(\frac{dy}{dx}\right)_m = \left(\frac{dy}{dx}\right)_s - \frac{c}{2R}\cos\gamma.$$

Replacing $(dy/dx)_s$ in (123) by $(dy/dx)_m$ we find that the effect is equivalent to an increase in the no-lift angle $-\alpha'$ by an amount $c/4R$. This is also apparent from (164). Equations (157) and (161) show that Ω_s is increased by $(c/2R)\sin\gamma$, while (124) and (126) give that C_L is reduced by $(\pi c/2\beta_a R)$ and C_M is not altered.

THICK AEROFOIL THEORY

8.22   Theodorsen's theory for incompressible flow

There are several methods of calculating the incompressible flow about a thick aerofoil, and the best known of these is one due to Theodorsen (1932). Theodorsen's method has been developed further by Theodorsen & Garrick (1933), Pinkerton (1936), Goldstein (1952) and many others. We shall just outline the theory, for it cannot be extended to compressible flow. A full account of the theory appears in ch. 8 of Pope (1951).

The first step is to select the orientation and origin of the z-plane so that Ox lies on the line joining the point of minimum radius of curvature at the nose (ρ_n) to the trailing edge, B, and O lies at distance $\frac{1}{2}b = \frac{1}{2}c - \frac{1}{4}\rho_n$ from B. (This is just for computational convenience, and is not essential.)

Next we apply the transformation

$$z = -\frac{b}{2}\cos\tilde{\zeta} \quad (\tilde{\zeta} = \tilde{\gamma} + i\tilde{\eta}), \tag{174}$$

which transforms the straight line $-\frac{1}{2}b \leqslant x \leqslant \frac{1}{2}b, y = 0$ into $-\pi < \tilde{\gamma} < \pi$, $\tilde{\eta} = 0$. In the $\tilde{\zeta}$-plane the aerofoil will become a known curve

$$\tilde{\eta} = g(\tilde{\gamma}), \tag{175}$$

with the nose at $\tilde{\gamma} = 0$, and the trailing edge at $\tilde{\gamma} = \pm\pi$. As B is at $x = \frac{1}{2}b$ it follows that

$$g(\pi) = g(-\pi) = 0. \tag{176}$$

The choice of the scale and origin of the elliptic transformation (174) confers the computational advantage that $g(\tilde{\gamma})$ is a slowly varying function near $\tilde{\gamma} = 0$, instead of the rapidly varying function it would otherwise be. This can be seen as follows. If k is small, the curve $\tilde{\eta} = k$ is an ellipse in the z-plane of nose radius $\rho = \frac{1}{2}bk^2$ (cf. (132)). Furthermore, the real part of (174) gives $x = -\frac{1}{2}b\cosh k \simeq -\frac{1}{2}b - \frac{1}{2}\rho$ at the nose of this ellipse. Thus if $k = \sqrt{(2\rho_n/b)}$, the ellipse in question will pass through the leading edge of the aerofoil, and have the same radius of curvature. It will not necessarily 'osculate' the aerofoil, for the axis of the osculating ellipse will in general be inclined to DB. However, this inclination is usually quite small, and so we can conclude that $g(\tilde{\gamma}) \simeq g(0) \simeq \sqrt{(2\rho_n/c)}$ near the nose.

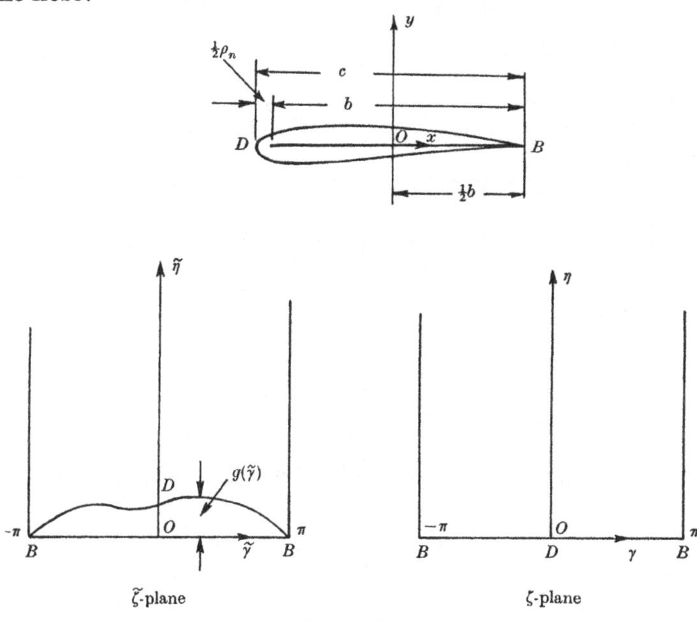

Fig. 8.22a

The ζ-plane shown in Fig. 8.22a is related to the w-plane by (1), viz.

$$w = -2a\cos(\zeta + \alpha_0) + (\zeta + \alpha_0 - \tfrac{1}{2}\pi)\,2a\sin\alpha_0, \tag{177}$$

and it now remains to establish the relation between the $\tilde{\zeta}$- and ζ-planes.

As the flow is incompressible, $w = w(z)$, and therefore $\tilde{\zeta}$ is an analytic function of ζ. Consider the function

$$\delta(\zeta) = \tilde{\zeta} - \zeta = \epsilon + i(\tilde{\eta} - \eta) \quad (\epsilon = \tilde{\gamma} - \gamma). \tag{178}$$

On $\eta = 0$ $\mathscr{I}\delta = \tilde{\eta} = g(\tilde{\gamma})$, and as $\delta(\zeta) = \delta(\zeta + 2\pi)$ (clearly both $\tilde{\zeta}$ and ζ increase by 2π on each circuit of the aerofoil), it follows from the theory of § 4.3 that

$$\delta = \epsilon_\infty + \frac{1}{2\pi}\int_{-\pi}^{\pi} g(\tilde{\gamma})\cot\tfrac{1}{2}(\gamma - \zeta)\,d\gamma, \tag{179}$$

where $\epsilon_\infty = \lim\limits_{\eta \to \infty} \mathscr{R}\delta$. On the aerofoil at $\zeta = \gamma^*$ this gives

$$\epsilon(\gamma^*) = \epsilon_\infty + \frac{1}{2\pi} \int_{-\pi}^{\pi} g(\tilde{\gamma}) \cot \tfrac{1}{2}(\gamma - \gamma^*)\, d\gamma. \tag{180}$$

If $\tilde{\gamma}$ is replaced by $\gamma + \epsilon$ we obtain an integral equation for ϵ. This can be solved (cf. § 5.8) by taking the limit $n \to \infty$ in the set of equations

$$\epsilon^{(n+1)} = \epsilon_\infty + \frac{1}{2\pi} \int_{-\pi}^{\pi} g(\gamma + \epsilon^{(n)}) \cot \tfrac{1}{2}(\gamma - \gamma^*)\, d\gamma, \quad \epsilon^{(0)} = 0, \tag{181}$$

but in practical problems it is usually unnecessary to proceed beyond $\epsilon^{(3)}$.

Comparing (65) and (178) we note that Theodorsen's method is essentially the displacement function method applied to the ζ-plane instead of the w-plane.

From (174) and (177) it follows that

$$\frac{dw}{dz} = q\, e^{-i\theta} = U \left(\frac{4a}{bU}\right) \frac{(\sin(\zeta + \alpha_0) + \sin \alpha_0)}{\sin \tilde{\zeta}} \left(\frac{d\zeta}{d\tilde{\zeta}}\right). \tag{182}$$

Now $\lim\limits_{\eta \to \infty} \dfrac{q}{U} e^{-i\theta} = e^{-i\alpha^*}$, where α^* is the incidence of the chord DB, therefore

$$e^{-i\alpha^*} = \left(\frac{4a}{bU}\right) e^{-i\alpha_0} \lim_{\eta \to \infty} \frac{\sin \zeta\, d\zeta}{\sin \tilde{\zeta}\, d\tilde{\zeta}} = \left(\frac{4a}{bU}\right) e^{-i\alpha_0} \lim_{\eta \to \infty} e^{i\delta},$$

on making use of (178). Thus from the limit of (179)

$$e^{-i\alpha^*} = \left(\frac{4a}{bU}\right) e^{i(\epsilon_\infty - \alpha_0)} e^{-g_0},$$

where
$$g_0 = \frac{1}{2\pi} \int_{-\pi}^{\pi} g(\tilde{\gamma})\, d\gamma. \tag{183}$$

It follows that
$$\frac{4a}{bU} = e^{g_0}, \quad \alpha_0 = \alpha^* + \epsilon_\infty. \tag{184}$$

From (49), (60) and the second of (184) we observe that $-\epsilon_\infty = -\alpha'$, the no-lift angle.

On $\eta = 0$ we can write

$$\frac{d\zeta}{d\tilde{\zeta}} = \left(\frac{d\tilde{\zeta}}{d\gamma}\right)^{-1} = \left(\frac{d\gamma}{d\tilde{\gamma}}\right)\left(\frac{d\tilde{\zeta}}{d\tilde{\gamma}}\right)^{-1} = \frac{1 - \epsilon'(\tilde{\gamma})}{1 + ig'(\tilde{\gamma})},$$

by (175) and (178). Also it follows from the imaginary part of (174) that

$$|\sin \tilde{\zeta}| = \sqrt{\{\sin^2 \tilde{\gamma} \cosh^2 \tilde{\eta} + \cos^2 \tilde{\gamma} \sinh^2 \tilde{\eta}\}} = \sqrt{\{\sin^2 \tilde{\gamma} + \sinh^2 \tilde{\eta}\}}$$

$$= \sqrt{\left\{\sin^2 \tilde{\gamma} + \left(\frac{2y}{c \sin \tilde{\gamma}}\right)^2\right\}},$$

on $\eta = 0$. Therefore the velocity on the aerofoil can be expressed

$$q = U\, e^{g_0} \frac{1 - \epsilon'(\tilde{\gamma})}{\sqrt{[1 + \{g'(\tilde{\gamma})\}^2]}} \frac{\{\sin(\tilde{\gamma} - \epsilon + \alpha^* + \alpha') + \sin(\alpha^* + \alpha')\}}{\sqrt{\left\{\sin^2\tilde{\gamma} + \left(\dfrac{2y}{c\sin\tilde{\gamma}}\right)^2\right\}}}, \quad (185)$$

by (182), (178), (184) and the result $\epsilon_\infty = \alpha'$.

In drawing Fig. 8.22a it has been tacitly assumed that the rear stagnation point $\gamma = \pm\pi$ coincides with the trailing edge $\tilde{\gamma} = \pm\pi$ for all incidences, i.e. that Joukowski's condition holds. Unfortunately this assumption results in a circulation about the aerofoil which may be 10–20 % in excess of the experimental value. This failure is due to the action of the boundary layer near the trailing edge, which renders the Joukowski condition an unattainable ideal. A more general (though physically

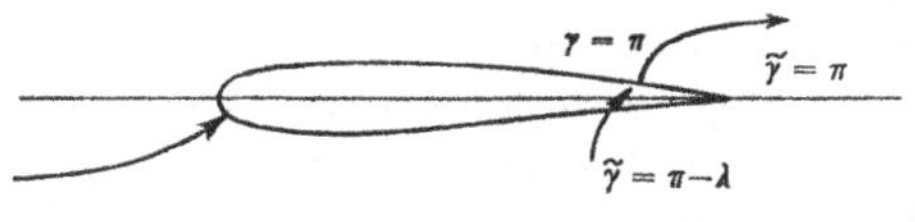

Fig. 8.22b

unrealistic) situation is that shown in Fig. 8.22b, where $\tilde{\gamma} = \pi - \lambda$ at $\gamma = \pi$. In this case (178) and (180) give

$$\epsilon_\infty = \alpha' = -\lambda + \frac{1}{2\pi}\int_{-\pi}^{\pi} g(\tilde{\gamma})\tan\tfrac{1}{2}\gamma\,d\gamma,$$

which leaves α' indeterminant. This means that $\alpha_0 = \alpha^* + \alpha'$ can be given its experimental value without altering α^* for the purpose. This semi-empirical theory gives a distribution of q in good agreement with experiment, except near the trailing edge. Pinkerton (1936) has given a better semi-empirical theory, based on Theodorsen's method, which avoids the infinity in q that must arise in the flow shown in Fig. 8.22b. Pinkerton's method is to modify $g(\tilde{\gamma})$ near the trailing edge so as to change slightly the profile shape in this region. This change turns the trailing edge upwards, and reduces the circulation; its scale is selected so that the experimental (C_L, α^*) relation is achieved.

Goldstein (1952) has developed a series of approximations based on Theodorsen's exact theory, and suitable for aerofoils of moderate thickness. We shall omit details of this work, for it involves no new principles, and is well presented in an earlier volume of the Cambridge Aeronautical Series (Robinson & Laurmann, 1956). We shall just note that Goldstein's third and most accurate approximation consists in replacing ϵ and g_0 in

(185) by $\epsilon^{(1)}$ and $\left\{\int_{-\pi}^{\pi} g(\gamma)\,d\gamma\right\}\Big/ 2\pi$ (cf. (181) and (183)).

8.23   Woods's method for subsonic flow

Another method of dealing with thick aerofoils has already been outlined in §§ 8.3 and 8.14. It is based on the variable τ, which reduces to $\ln(U\,dz/dw)$ in incompressible flow (cf. Theodorsen's function $\tilde{\zeta} = \cos^{-1}(-2z/b)$). Woods's (1950, 1952$a$) method[†] has the advantage over Theodorsen's method of being applicable to subsonic compressible flows. It is also rather more convenient for the general development of the theory, and is indispensible in mixed boundary value problems. Thin aerofoil theory follows from either theory, but it does emerge more naturally from the theory based on τ.

However the method is probably not as convenient for calculating the steady incompressible flow about a given solid aerofoil as is Theodorsen's method, for it does not take advantage of the near-elliptical nose shape of most aerofoil profiles in the way indicated in Fig. 8.22a. The radius of curvature function under the integral sign in (16) will vary rapidly in the neighbourhood of the aerofoil nose, and so will add to the difficulty of the iterative solution of (16) outlined in § 8.14.

The best method of applying the theory to the problem of finding the velocity distribution over an aerofoil at a given incidence α^* is probably as follows.

(1) The thin aerofoil theory of § 8.18 is applied, and the $\Omega_s(\gamma)$ distribution obtained from (157) is used in (15) to give q_0.

(2) Equations (138) and (146) are now used to modify q_0, to allow for the rounded nose and the trailing edge angle. The resulting distribution, which we shall label $q_s^{(1)}$, almost corresponds in accuracy to Goldstein's third approximation.

(3) Next $\alpha_0^{(1)}$ is deduced from $\alpha_0 = (\alpha^* + \alpha')/\beta_a$ (see (49), (58)) and (123).

(4) Equations (17), (103) and $\alpha_0^{(1)}$, $q_s^{(1)}$ now yield $s^{(1)}(\gamma)$ and $T^{(1)}$.

(5) $R^{(1)}(\gamma)$ follows from $s^{(1)}(\gamma)$ and the known $R(s)$, which can be deduced from the profile shape.

(6) We now substitute $R^{(1)}$, $T^{(1)}$ and $q_s^{(1)}$ in the right-hand side of (16) and deduce $q_s^{(2)}$ from the left-hand side.

(7) $\alpha_0^{(2)}$ is calculated from

$$\beta_a \sin 2\alpha_0 = -\frac{T}{\pi}\int_{-\pi}^{\pi}\left(\frac{c}{R}\right)\left(\frac{U}{q_s}\right)\sin\gamma\,\sin\tfrac{1}{2}(\gamma+2\alpha_0)\cos\tfrac{1}{2}\gamma\,d\gamma, \quad (186)$$

which follows when the second of equations (28) is integrated by parts and (13) is used to eliminate $d\theta_s(\gamma)$.

(8) Steps 4 to 7 are repeated until $q_s^{(n+1)} - q_s^{(n)}$ is negligible.

[†] The *incompressible* flow form of τ has of course been used extensively in problems involving free streamlines for a very long time. It was first (?) applied to (symmetrical)

Most of the work in this iterative scheme must be carried out numerically. Equation (16) can be handled by a numerical technique similar to that described in § 7.6 (see (7.34) to (7.43)). If this method is adopted then the closure conditions (11) and the first of (28)—when expressed in a form similar to (186)—are best employed to determine the magnitudes of the discontinuities θ_n, θ_t in θ at the nose and trailing edge of the approximating aerofoil. The terms corresponding to these discontinuities must be made explicit in (16) (see comment following (14)), (186) and the two closure conditions just mentioned. But they cannot be given their exact physical values (i.e. $\theta_n = \pi$, $\theta_t = $ T.E. angle) otherwise the approximating aerofoil will not close.

The above theory has been extended (Woods, 1952b) to deal with hinged flaps on thick aerofoils.

8.24   Other methods

In incompressible flow the displacement function theory of § 8.9 is exact, and can be applied to thick aerofoils in the no-lift attitude. In this case (65) and (70) yield

$$Uz - w = Ux_c + \frac{U}{2\pi} \int_{-\pi}^{\pi} y_s(\gamma) \cot \tfrac{1}{2}(\gamma - \zeta)\, d\gamma, \tag{187}$$

where
$$w = -2a \cos \zeta. \tag{188}$$

Integrating (187) by parts, and then differentiating it with respect to w we find

$$U\frac{dz}{dw} = \frac{U \cos \theta}{q} + i\frac{U \sin \theta}{q}$$

$$= 1 + \frac{U}{2\pi} \int_{-\pi}^{\pi} \frac{\sin \gamma}{\sin \zeta} \cot \tfrac{1}{2}(\gamma - \zeta) \left(\frac{dy}{dx}\frac{\cos \theta}{q}\right)_s d\gamma, \tag{189}$$

on using
$$dy_s(\gamma) = \left(\frac{dy}{dx}\right)_s \frac{dx}{d\phi}\frac{d\phi}{d\gamma} d\gamma = 2a \sin \gamma \left(\frac{dy}{dx}\frac{\cos \theta}{q}\right)_s.$$

On $\eta = 0$ (189) reduces to an integral equation for $(\{U/q\} \cos \theta)$, which can be solved by a method similar to that employed for (16) (Woods 1950).

The effect of incidence on (189) is shown in the equation

$$\frac{U\, dz}{dw} = e^{-i\alpha} \frac{\sin \tfrac{1}{2}\zeta}{\sin \tfrac{1}{2}(\zeta + 2\alpha^* + 2\alpha')}$$

$$\times \left\{ 1 + \frac{U}{2\pi} \int_{-\pi}^{\pi} \frac{\sin \gamma}{\sin \zeta} \cot \tfrac{1}{2}(\gamma - \zeta) \left(\frac{\sin \theta}{q}\right)_s d\gamma \right\}, \tag{190}$$

aerofoil problems by Thom (1948) in his numerical technique called 'squaring', and later Lighthill made it the basis of his method of aerofoil design (see § 8.25). Timman (1948) also used this function for the direct and inverse problem of aerofoil theory.

which follows from (51), (60), (189) and (6.16). (We have used $dy/dx = \tan\theta$ to modify the integrand slightly.)

The conjugate equation to (190), which is obtained by replacing $(\sin\theta)/q$ by $-i(\cos\theta)/q$, can be used as a basis for aerofoil design, but the method to be described in § 8.25 is preferable.

The displacement function method is due in essence to Thom & Klanfer (1947), who developed it for the case of a symmetrical aerofoil in a channel (see § 7.3), although they used numerical rather than analytic methods to obtain final results. Woods (1950) has used both (187) and (190) to find the flow about thick aerofoils, but the methods prove to be inferior in some respects to those of §§ 8.22 and 8.23. However, the equations are often useful in theoretical investigations.

Another method of dealing with thick aerofoils in incompressible flow is based on

$$\frac{dw}{dz} = \frac{q}{U}e^{-i\theta} = u - iv = U + \frac{1}{2\pi}\int_{-\pi}^{\pi} v_s(\gamma)\cot\tfrac{1}{2}(\gamma-\zeta)\,d\gamma, \qquad (191)$$

where u, v are the Ox, Oy velocity components. This equation is an obvious consequence of applying the theory of § 4.3 to the analytical function $u - iv$. It is also inferior to the methods described in earlier sections. It can be easily verified from (6.23) that, to the order of accuracy of linear perturbation theory, (191) is indistinguishable from the incompressible flow form of (10).

Several other methods of dealing with thick aerofoils in subsonic flow have been developed, but we shall be content here just to give the reader some references, for these methods (i) are usually restricted to incompressible flow, or (ii) do not differ in principle from the methods already described, or (iii) are mathematically extremely complicated and not in keeping with the approach of this text.

In the first category we have the well-known Joukowski and Kármán–Trefftz aerofoils (see Milne-Thomson, 1948) based on analytic (in the algebraic sense) mapping equations. While of great theoretical interest these analytical methods are useless either to find the flow about a given aerofoil, or to design aerofoils.

In the second category is the work of Weber (1953) and of Spence & Routledge (1955), which are based respectively on the symmetrical-aerofoil forms of (191) and of (187) (put $y_s(\gamma) = y_s(-\gamma)$ and $v_s(\gamma) = v_s(-\gamma)$). Interesting numerical procedures have been introduced in the application of these methods, and the reader is referred to the original papers for these.

In the third category there is the exact hodograph method to which reference has been made in § 2.7.

8.25 Thick aerofoil design

Several distinct methods exist for the design of thick aerofoils to yield prescribed pressure distributions in incompressible flow. First the thin aerofoil theory given in § 8.19 can be modified to give an exact theory, secondly it is possible to use the theory of § 8.22 as the basis for aerofoil design (Theodorsen, 1944), thirdly the conjugate of (190) can be employed, and finally there is the incompressible flow form of the theory of § 8.6. Many variants of these methods have been developed, for example Goldstein (1952) has adapted Theodorsen's theory to produce a design method for aerofoils of moderate thickness, and Thwaites (1945) has improved Goldstein's method; Allen (1945) has also given an approximate method, based on the conjugate of (190), and there are others.

In compressible flow there are three distinct design methods, namely: (i) the linear perturbation method of modifying an aerofoil, previously designed for incompressible flow, into one giving approximately the same pressure distribution in a compressible flow; (ii) the more accurate theory given in § 8.6; and (iii) an exact theory based on the hodograph transformation (§ 2.7).

The first method is very simple, and can be understood on referring to (164) and (167). If θ_t, t, α_i and α' are written in the form $\theta_t = \beta_a \theta_{t0}$, $t = \beta_a t_0$, $\alpha_i = \beta_a \alpha_{i0}$, $\alpha' = \beta_a \alpha_0'$, where the subscript 0 refers to an aerofoil designed for incompressible flow, it is apparent from the equations following (167) and from (164) that $(dy/dx)_s = \beta_a (dy/dx)_{s0}$, whence

$$y_s = \beta_a y_{s0}. \tag{192}$$

It also follows from the last of (162) and $\Omega_s \simeq \tfrac{1}{2}\beta_a C_p$ that $\alpha^* = \beta_a \alpha_0^*$. Thus an aerofoil designed for incompressible flow must have its ordinates and incidence reduced by a factor β_a if it is to have the same pressure distribution in a compressible flow. This theory is of course applicable to thick aerofoils only at relatively low Mach numbers.

As mentioned earlier the hodograph method lies outside the scope of this text.

The second method listed above, i.e. the (tangent gas) method outlined in § 8.6, is actually a generalization of Lighthill's (1945c) exact design method for incompressible flow, and most of the extensions of Lighthill's work (e.g. see Glauert (1947) and Williams (1950)) are equally applicable to the tangent gas.

MIXED BOUNDARY CONDITIONS

8.26   Generalized aerofoil theory

The theory given in §§ 8.2–8.12 is capable of a remarkable generalization (Woods, 1956a), which we shall now give. In this generalization the aerofoil surface is replaced by a porous surface, which is assumed to have the linear pressure velocity characteristic described in § 6.3. The boundary condition on the aerofoil is therefore

$$\theta - \Omega \tan \pi\epsilon = \theta_s - \Omega_s \tan \pi\epsilon, \tag{193}$$

where θ_s, Ω_s are given functions and ϵ is related to the aerofoil porosity λ_s by (6.32) and (6.34). This boundary condition involves the assumption that the mass of air sucked into or emitted from the aerofoil through the porous surface is relatively small. The remarkable feature of the theory is that, despite the relative complexity of (193) compared with the boundary condition given in (5), it has almost exactly the same form as the theory based on (5).

The main difference between the theory of § 8.2 *et seq.* and the present theory is that in the w-plane the aerofoil is no longer the slit on $\psi = 0$ shown in Fig. 8.2. Instead we have the situation shown in Fig. 8.26, where (see (6.43)) there is a jump of $-Q/\rho_s$ in ψ across the trailing edge streamline owing to the withdrawal of air from the main stream through the porous surface. The actual shape of the curve $B'AB$ in the w-plane will be initially unknown, since the mass flow across the porous surface at any point depends on the pressure difference across the surface at that point. We shall prescribe the pressure distribution on the *inside* of the porous surface, but the pressure just outside the surface is part of the final solution of the problem. However, it will become evident below that it is only necessary to know the *total* mass Q sucked in by the aerofoil per second, in order to calculate the forces and moment accurate to first order in the mass coefficient,

$$C_Q \equiv \frac{Q}{cU\rho_a}, \tag{194}$$

of the sucked air.

The w_s-plane shown in Fig. 8.26 is a conformal mapping of the w-plane, and has the same circulation as the w-plane, that is the potential difference between B and B' is unchanged. The w_s- and w-planes coincide if $C_Q = 0$. In preserving the circulation in the transformation from w to w_s it is not, in general, possible to transform the front stagnation point A on to the point labelled A' in the w_s-plane. The actual position of A in the w_s-plane will be quite close to A', but its precise position could be calculated only if the shape $B'AB$ in the w-plane were known. For a theory correct to first order in incidence it proves unnecessary to know the position of A in the w_s-plane, but this restriction in the incidence is not necessary with a solid aerofoil (for which $C_Q = 0$ and A coincides with A') and so it will not be made until it is essential to do so.

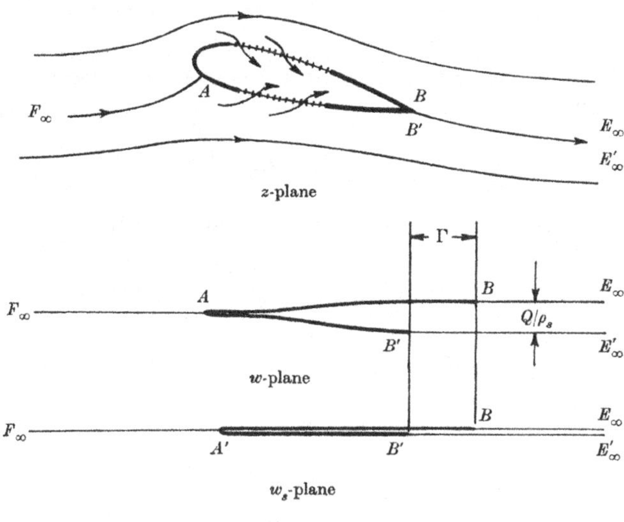

Fig. 8.26

A comparison of Figs. 8.26 and 8.2 shows that w_s now replaces w in (1), i.e.

$$w_s = -2a\cos(\zeta + \alpha_0) + (\zeta + \alpha_0 - \pi/2)\,2a\sin\alpha_0, \tag{195}$$

where the circulation is still $\Gamma = 4\pi a\sin\alpha_0$. The definition of σ given in (4), viz.

$$\sigma = -e^{-i(\zeta + \alpha_0)} \tag{196}$$

is unchanged, but $dw/d\sigma$ will not now satisfy (19) near infinity. Instead we shall have

$$\frac{dw}{d\sigma} = a\left(1 + \frac{2i\sin\alpha_0 - 2\beta_a C_Q/\pi T}{\sigma} + \frac{D}{\sigma^2}\right), \tag{197}$$

because from (6.44), (2), (14), (194) and $\rho_s = \rho_a/\beta_a$ it follows that

$$\int_C \left(\frac{dw}{d\sigma}\right) d\sigma = -\Gamma - iQ/\rho_s = 2\pi i\,.\,a(2i\sin\alpha_0 - 2\beta_a C_Q/\pi T),$$

which verifies the coefficient of $1/\sigma$ in (197). The constant D in (197) will depend on C_Q, but we shall not need its actual value in the following theory.

It has been shown in §4.20 that the analytic function $\tau = \Omega + i\theta$ that satisfies (193) on $\eta = 0$, and which is periodic (period 2π) and continuous in the γ-direction, is given by

$$\tau = \frac{1}{2\pi}\exp\left\{\frac{1}{2}\int_{-\pi}^{\pi}\epsilon\cot\tfrac{1}{2}(\gamma-\zeta)\,d\gamma\right\}\left[A + \int_{-\pi}^{\pi}(\theta_s\cos\pi\epsilon - \Omega_s\sin\pi\epsilon)\right.$$

$$\left. \times\exp\left\{-\tfrac{1}{2}\mathscr{R}\int_{-\pi}^{\pi}\epsilon\cot\tfrac{1}{2}(\gamma^*-\gamma)\,d\gamma^*\right\}\cot\tfrac{1}{2}(\gamma-\zeta)\,d\gamma\right], \quad (198)$$

where A is a real constant. In the limit $\eta \to \infty$ we find from (9) and (198) that $A = 0$ and

$$\int_{-\pi}^{\pi}(\theta_s\cos\pi\epsilon - \Omega_s\sin\pi\epsilon)\exp\left\{-\tfrac{1}{2}\mathscr{R}\int_{-\pi}^{\pi}\epsilon\cot\tfrac{1}{2}(\gamma^*-\gamma)\,d\gamma^*\right\}d\gamma = 0. \quad (199)$$

These results and (20) enable us to expand (198) thus:

$$\tau = i\,e^{ib_0}\exp\left\{i\sum_{n=1}^{\infty}(-1)^n\frac{b_n}{\sigma^n}\right\}\sum_{n=1}^{\infty}(-1)^n\frac{a_n}{\sigma^n}, \quad (200)$$

where $\qquad b_0 \equiv \dfrac{1}{2}\displaystyle\int_{-\pi}^{\pi}\epsilon\,d\gamma, \quad b_n \equiv \int_{-\pi}^{\pi}\epsilon\,e^{-ni(\gamma+\alpha_0)}\,d\gamma \quad (n \geq 1), \quad (201)$

and

$$a_n \equiv \frac{1}{\pi}\int_{-\pi}^{\pi}(\theta_s\cos\pi\epsilon - \Omega_s\sin\pi\epsilon)$$

$$\times\exp\left\{-\tfrac{1}{2}\mathscr{R}\int_{-\pi}^{\pi}\epsilon\cot\tfrac{1}{2}(\gamma^*-\gamma)\,d\gamma^*\right\}e^{-ni(\gamma+\alpha_0)}\,d\gamma. \quad (202)$$

Equations (200)–(202) give the required generalization of (21) and (22). Combining (197) and (200) we arrive at (23) and (24), except that A_1, A_2, B_1 and B_2 are now defined by

$$\left.\begin{aligned}
A_1 &\equiv 2i\sin\alpha_0 - \frac{2\beta_a C_Q}{\pi T} - ia_1 e^{ib_0},\\[2mm]
A_2 &\equiv 2i\sin\alpha_0 - \frac{2\beta_a C_Q}{\pi T} + ia_1 e^{ib_0},
\end{aligned}\right\} \quad (203)$$

$$\left.\begin{aligned}
B_1 &\equiv i(a_2 + ia_1 b_1)\,e^{ib_0} - \tfrac{1}{2}a_1^2 e^{2ib_0} + D + a_1 e^{ib_0}\left(2\sin\alpha_0 + \frac{2i\beta_a C_Q}{\pi T}\right),\\[2mm]
B_2 &\equiv i(a_2 + ia_1 b_1)\,e^{ib_0} + \tfrac{1}{2}a_1^2 e^{2ib_0} - D + a_1 e^{ib_0}\left(2\sin\alpha_0 + \frac{2i\beta_a C_Q}{\pi T}\right),
\end{aligned}\right\} \quad (204)$$

with a_n defined by (202).

The closure condition is still given by (26), but with A_1, A_2 as just defined. In place of (27) we find

$$C_Q = \frac{\pi}{2}T\,\mathscr{I}\{a_1 e^{ib_0}\}, \quad (205)$$

and $\qquad\qquad\qquad 2\beta_a\sin\alpha_0 = \mathscr{R}\{a_1 e^{ib_0}\}, \quad (206)$

which can be rewritten in the form

$$\frac{2C_Q}{\pi T}\cos\alpha_0 + 2\beta_a\sin^2\alpha_0 = \frac{1}{\pi}\int_{-\pi}^{\pi}(\theta_s\cos\pi\epsilon - \Omega_s\sin\pi\epsilon)$$

$$\times\exp\left\{-\tfrac{1}{2}\mathscr{R}\int_{-\pi}^{\pi}\epsilon\cot\tfrac{1}{2}(\gamma^*-\gamma)\,d\gamma^*\right\}\sin(b_0-\gamma)\,d\gamma, \quad (207)$$

$$-\frac{2C_Q}{\pi T}\sin\alpha_0 + \beta_a\sin 2\alpha_0 = \frac{1}{\pi}\int_{-\pi}^{\pi}(\theta_s\cos\pi\epsilon - \Omega_s\sin\pi\epsilon)$$

$$\times\exp\left\{-\tfrac{1}{2}\mathscr{R}\int_{-\pi}^{\pi}\epsilon\cot\tfrac{1}{2}(\gamma^*-\gamma)\,d\gamma^*\right\}\cos(b_0-\gamma)\,d\gamma. \quad (208)$$

The theory given in §§ 8.5, 8.10 and 8.11 depends on the use of the relation $dw = d\overline{w} = d\phi$ on the aerofoil surface. This relation is only approximately true in the case of flow past a porous aerofoil, for the boundary is no longer the streamline $\psi = 0$. However, if the boundary conditions are applied on $\psi = 0$ instead of on the actual boundary—which deviates from $\psi = 0$ by only a small amount owing to the restriction on the magnitude of the mass flow into the aerofoil—the error produced will be only second order in C_Q. On adopting this approximation we can immediately apply the results of §§ 8.5, 8.10 and 8.11 to the present case.

In place of the equation preceding (41) we find

$$\Gamma z_a = \Gamma z_c + \frac{\pi a^2}{\beta_a U}i\{(1+\beta_a)(B_1-D) - (1-\beta_a)(\bar{B}_2+\bar{D})\},$$

which by (2), (204)–(206) and the neglect of a term of order C_Q^2 is equivalent to

$$z_a = z_c - \frac{cT}{8\beta_a\sin\alpha_0}[\mathscr{R}\{(a_2+ia_1b_1)e^{ib_0}\} + i\beta_a\mathscr{I}\{(a_2+ia_1b_1)e^{ib_0}\}]$$

$$+\frac{i}{4}Tc\left(\beta_a\sin\alpha_0 + \frac{2iC_Q}{\pi T\beta_a}\right). \quad (209)$$

Equation (81), i.e. $C_L = 2\pi T\sin\alpha_0$, is unchanged in form, but $\sin\alpha_0$ is now calculated by eliminating C_Q from (207) and (208), which are to be regarded as simultaneous equations fixing the values of α_0 and C_Q.

Equation (89) is replaced by

$$C_M = -\frac{x_c}{c}C_L + \frac{1+\beta_a^2}{\beta_a}TC_Q\sin\alpha_0 + \frac{\pi T^2}{4\beta_a}\mathscr{R}\{(a_2+ia_1b_1)e^{ib_0}\}, \quad (210)$$

which could be expressed entirely in terms of the boundary values θ_s, Ω_s and ϵ by using (81), (201), (202), (207) and (208). From (92), (209) and (210) we find that the centre of pressure is now at

$$x' = x_a - \frac{\beta_a cC_Q}{2\pi}. \quad (211)$$

On $\eta = 0$ the real and imaginary parts of (198) give

$$\left.\begin{matrix}\Omega(\gamma^*)\\ \theta(\gamma^*)\end{matrix}\right\} = \begin{matrix}\mathscr{R}\\ \mathscr{I}\end{matrix}\, \frac{1}{2\pi}\exp\left\{\frac{1}{2}\int_{-\pi}^{\pi}\epsilon\cot\tfrac{1}{2}(\gamma - \gamma^*)\,d\gamma\right\}\int_{-\pi}^{\pi}(\theta_s\cos\pi\epsilon - \Omega_s\sin\pi\epsilon)$$

$$\times\exp\left\{-\tfrac{1}{2}\mathscr{R}\int_{-\pi}^{\pi}\epsilon\cot\tfrac{1}{2}(\gamma^* - \gamma)\,d\gamma^*\right\}\cot\tfrac{1}{2}(\gamma - \gamma^*)\,d\gamma, \quad (212)$$

from which the velocity and direction of flow at the aerofoil surface can be deduced.

There are two points to note in regard to this generalized aerofoil theory, namely (i) that it is exact—within the limits of tangent gas theory—if there is no flow through the aerofoil surface, i.e. if C_Q is zero and ϵ takes only the limit values 0 or $\pm\tfrac{1}{2}$ on each section of the aerofoil surface, and (ii) that otherwise it is valid only to first order in C_Q.

In the next four sections we shall give some important special cases of the theory.

8.27   An aerofoil of uniform porosity

From (6.32) and (6.34)

$$\epsilon = \frac{1}{\pi}\tan^{-1}\left\{\frac{\lambda_s}{\beta_a}\,\frac{\sinh\Omega^*}{\sinh(\Omega^* + \Omega_s)}\right\}, \quad (213)$$

whence ϵ is constant in each of the following cases:

(i) λ_s is equal to $\pm\infty$, i.e. $\epsilon = \pm\tfrac{1}{2}$,

(ii) λ_s is constant and the internal pressure p_s (and hence Ω_s) is constant,

(iii) λ_s is constant and $p_s - p_a$ is small so that Ω_s is small and negligible in (213).

Case (iii) is applicable in linear perturbation theory for if $p_s - p_a$ is small so is $p - p_a$, as $p - p_s$ was assumed small in deducing the boundary condition (193).

On the *upper* surface of the aerofoil an increase in $(p - p_s)$, i.e. an increase in $\Omega - \Omega_s$, will reduce $\theta - \theta_s$, and consequently ϵ will be *negative* (cf. (193)), while on the lower surface the same argument shows ϵ to be positive. More precisely

$$-\tfrac{1}{2}\leqslant\epsilon\leqslant 0 \quad\text{if}\quad -2\alpha_0 < \gamma < \pi, \qquad 0\leqslant\epsilon\leqslant\tfrac{1}{2} \quad\text{if}\quad -\pi < \gamma < -2\alpha_0,$$
$$(214)$$

for it is really the positions of the stagnation points that distinguish the two cases, as the reader can easily verify.

Now let $\epsilon = 0$ in $-\pi\leqslant\gamma < \gamma_0$, $\gamma_1 < \gamma\leqslant\pi$, and $\epsilon = $ constant in

$\gamma_0 < \gamma < \gamma_1$, then we will have a solid aerofoil, excepting for one section into which a porous material has been inserted. If conditions (i), (ii) or (iii) given above hold, then this material will have a uniform porosity. We shall restrict attention to the case $\gamma_0 > -2\alpha_0$, i.e. the porous section is to lie entirely on the upper surface of the aerofoil, for this is the case of greatest practical interest. Aerofoils designed for distributed suction experiments are often of this type.

In this case

$$\exp\left\{\frac{1}{2}\int_{-\pi}^{\pi}\epsilon\cot\tfrac{1}{2}(\gamma^*-\gamma)\,d\gamma^*\right\} = \left\{\frac{\sin\tfrac{1}{2}(\gamma_1-\gamma)}{\sin\tfrac{1}{2}(\gamma_0-\gamma)}\right\}^{\epsilon} \quad (-\tfrac{1}{2} \leqslant \epsilon \leqslant 0),$$

and from (201), $b_0 = \lambda_0\epsilon$, where

$$\lambda_0 = \tfrac{1}{2}(\gamma_1-\gamma_0). \tag{215}$$

We shall also write

$$\lambda_1 = \tfrac{1}{2}(\gamma_1+\gamma_0). \tag{216}$$

These results enable us to write (199), (207), (208), (210) and (212) as

$$\frac{2C_Q}{T}\left\{\begin{array}{c} 0 \\ \cos\alpha_0 \\ -\sin\alpha_0 \end{array}\right\} + \pi\beta_a\left\{\begin{array}{c} 0 \\ 2\sin^2\alpha_0 \\ \sin 2\alpha_0 \end{array}\right\}$$

$$= \left(\int_{-\pi}^{\gamma_0}+\int_{\gamma_1}^{\pi}\right)\theta_s\left[\frac{\sin\tfrac{1}{2}(\gamma_0-\gamma)}{\sin\tfrac{1}{2}(\gamma_1-\gamma)}\right]^{\epsilon}\left\{\begin{array}{c} 1 \\ \sin(\lambda_0\epsilon-\gamma) \\ \cos(\lambda_0\epsilon-\gamma) \end{array}\right\}d\gamma$$

$$+\int_{\gamma_0}^{\gamma_1}(\theta_s\cos\pi\epsilon-\Omega_s\sin\pi\epsilon)\left[\frac{\sin\tfrac{1}{2}(\gamma-\gamma_0)}{\sin\tfrac{1}{2}(\gamma_1-\gamma)}\right]^{\epsilon}\left\{\begin{array}{c} 1 \\ \sin(\lambda_0\epsilon-\gamma) \\ \cos(\lambda_0\epsilon-\gamma) \end{array}\right\}d\gamma,$$

$$\tag{217}$$

$$C_M = -\frac{x_c}{c}C_L + \frac{1+\beta_a^2}{\beta_a}TC_Q\sin\alpha_0 - \frac{\epsilon}{\beta_a}TC_Q\sin\lambda_0\cos(\lambda_1+\alpha_0)$$

$$+\epsilon T^2\pi\sin\alpha_0\sin\lambda_0\sin(\lambda_1+\alpha_0)$$

$$+\frac{T^2}{4\beta_a}\left(\int_{-\pi}^{\gamma_0}+\int_{\gamma_1}^{\pi}\right)\theta_s\left[\frac{\sin\tfrac{1}{2}(\gamma_0-\gamma)}{\sin\tfrac{1}{2}(\gamma_1-\gamma)}\right]^{\epsilon}\cos 2(\gamma+\alpha_0-\lambda_0\epsilon)\,d\gamma$$

$$+\frac{T^2}{4\beta_a}\int_{\gamma_0}^{\gamma_1}(\theta_s\cos\pi\epsilon-\Omega_s\sin\pi\epsilon)\left[\frac{\sin\tfrac{1}{2}(\gamma-\gamma_0)}{\sin\tfrac{1}{2}(\gamma_1-\gamma)}\right]^{\epsilon}$$

$$\times\cos 2(\gamma+\alpha_0-\lambda_0\epsilon)\,d\gamma, \tag{218}$$

and

$$\left.\begin{aligned}\Omega(\gamma^*)\\ \theta(\gamma^*)\end{aligned}\right\} = \begin{matrix}\mathscr{R}\\ \mathscr{I}\end{matrix} \frac{1}{2\pi}\left\{\frac{\sin\frac{1}{2}(\gamma_1-\gamma^*)}{\sin\frac{1}{2}(\gamma_0-\gamma^*)}\right\}^\epsilon$$

$$\times\left\{\left(\int_{-\pi}^{\gamma_0}+\int_{\gamma_1}^{\pi}\right)\theta_s\left[\frac{\sin\frac{1}{2}(\gamma_0-\gamma)}{\sin\frac{1}{2}(\gamma_1-\gamma)}\right]^\epsilon\cot\tfrac{1}{2}(\gamma-\gamma^*)\,d\gamma\right.$$

$$\left.+\int_{\gamma_0}^{\gamma_1}(\theta_s\cos\pi\epsilon-\Omega_s\sin\pi\epsilon)\left[\frac{\sin\frac{1}{2}(\gamma-\gamma_0)}{\sin\frac{1}{2}(\gamma_1-\gamma)}\right]^\epsilon\cot\tfrac{1}{2}(\gamma-\gamma^*)\,d\gamma\right\},$$

$$(219)$$

the first three of which have been compressed into a single (vector) equation to save space.

8.28   A thin aerofoil with uniform suction pressure

The above theory will now be applied to a thin aerofoil; in this case

$$\theta_s = \left(\frac{dy}{dx}\right)_s - \alpha^* + 2\pi(\alpha^*-\alpha_i)\,\delta(\gamma), \qquad (220)$$

from (159). It will also be assumed that the 'internal' pressure p_s, and hence Ω_s, is constant.

When these values of θ_s and Ω_s are substituted into (217)–(219) it will be found that it is necessary to evaluate the following integrals:

$$\left.\begin{aligned}I_r\\ J_r\\ K\end{aligned}\right\} \equiv \int_{\gamma_0}^{\gamma_1}\left[\frac{\sin\frac{1}{2}(\gamma-\gamma_0)}{\sin\frac{1}{2}(\gamma_1-\gamma)}\right]^\epsilon\left\{\begin{aligned}\cos r\gamma\\ \sin r\gamma\\ \cot\tfrac{1}{2}(\gamma-\gamma^*)\end{aligned}\right\}d\gamma \quad (r=0,1,2),$$

I_r^*, J_r^* and K^*, where the latter are defined like I_r, J_r and K, except that $\int_{\gamma_0}^{\gamma_1}$ is replaced by $\left(\int_{-\pi}^{\gamma_0}+\int_{\gamma_1}^{\pi}\right)$ and $(\gamma-\gamma_0)$ by $(\gamma_0-\gamma)$. These integrals are all reducible to the form $I\equiv\int_0^\infty x^\epsilon Q(x)\,dx$, where $Q(x)$ is a rational function of x having no poles on the positive part of the real axis, and such that $x^{1+\epsilon}Q(x)$ tends to zero in each of the limits $x\to 0$, $x\to\infty$. The substitution required for this is just $x = |\sin\frac{1}{2}(\gamma-\gamma_0)/\sin\frac{1}{2}(\gamma_1-\gamma)|$. It is easily shown by contour integration that $I = -\pi\operatorname{cosec}\pi\epsilon\,\Sigma R$, where ΣR is the sum of the residues of $(-z)^\epsilon Q(z)$ at all its poles (see Whittaker & Watson, 1952, p. 117). For example the substitution reduces I_0 to

$$I_0 = 2\sin\lambda_0\int_0^\infty\frac{x^\epsilon\,dx}{1+2x\cos\lambda_0+x^2},$$

the integrand of which has poles at $z = e^{\pm i\lambda_0}$, and the result just quoted yields

$$I_0 = -\pi\operatorname{cosec}\pi\epsilon\,(-2\sin\lambda_0\epsilon).$$

Similarly, $\quad I_1 = 2\pi\epsilon \operatorname{cosec} \pi\epsilon \sin \lambda_0 \cos (\lambda_1 + \epsilon\lambda_0),$

$$I_2 = 2\pi\epsilon \operatorname{cosec} \pi\epsilon \sin \lambda_0 \{\cos \lambda_0 \cos (\lambda_0\epsilon + 2\lambda_1) - \epsilon \sin \lambda_0 \sin (\lambda_0\epsilon + \lambda_1)\},$$

$$J_1 = 2\pi\epsilon \operatorname{cosec} \pi\epsilon \sin \lambda_0 \sin (\lambda_1 + \epsilon\lambda_0),$$

and

$$J_2 = 2\pi\epsilon \operatorname{cosec} \pi\epsilon \sin \lambda_0 \{\cos \lambda_0 \sin(\lambda_0\epsilon + 2\lambda_1) + \epsilon \sin \lambda_0 \cos (\lambda_0\epsilon + 2\lambda_1)\}.$$

The values of I_r^* and J_r^* can now be written down directly with the aid of the easily established relations $I_r^*(\lambda_0, \lambda_1) = I_r(\pi - \lambda_0, \pi - \lambda_1)$, and $J_r^*(\lambda_0, \lambda_1) = -J_r(\pi - \lambda_0, \pi - \lambda_1)$. Finally, similar calculations give

$$K = \frac{2\pi \cos \lambda_0 \epsilon}{\sin \pi\epsilon} \begin{cases} -\dfrac{2\pi}{\sin \pi\epsilon} \left\{\dfrac{\sin \frac{1}{2}(\gamma_0 - \gamma^*)}{\sin \frac{1}{2}(\gamma_1 - \gamma^*)}\right\}^\epsilon & \text{(if } \gamma^* \text{ lies outside } (\gamma_0, \gamma_1)\text{),} \\[4mm] -2\pi \cot \pi\epsilon \left\{\dfrac{\sin \frac{1}{2}(\gamma^* - \gamma_0)}{\sin \frac{1}{2}(\gamma_1 - \gamma^*)}\right\}^\epsilon & \text{(if } \gamma^* \text{ lies inside } (\gamma_0, \gamma_1)\text{),} \end{cases}$$

and $K^*(\lambda_0, \lambda_1, \gamma^*) = -K(\pi - \lambda_0, \pi - \lambda_1, -\gamma^*)$.

Retaining only first-order terms in C_Q and α_0, and making use of the results just given, we can write (217) as

$$\frac{C_Q}{\pi T} \begin{Bmatrix} 0 \\ 1 \\ 0 \end{Bmatrix} + \beta_a \alpha_0 \begin{Bmatrix} 0 \\ 0 \\ 1 \end{Bmatrix} = \alpha^* \begin{Bmatrix} -\cos \lambda_0 \epsilon \\ \epsilon \sin \lambda_0 \cos \lambda_1 \\ \epsilon \sin \lambda_0 \sin \lambda_1 \end{Bmatrix} + \Omega_s \begin{Bmatrix} -\sin \lambda_0 \epsilon \\ \epsilon \sin \lambda_0 \sin \lambda_1 \\ -\epsilon \sin \lambda_0 \cos \lambda_1 \end{Bmatrix}$$

$$+ (\alpha^* - \alpha_1) \left[\frac{\sin \frac{1}{2}\gamma_0}{\sin \frac{1}{2}\gamma_1}\right]^\epsilon \begin{Bmatrix} 1 \\ \sin \lambda_0 \epsilon \\ \cos \lambda_0 \epsilon \end{Bmatrix}$$

$$- \left(\int_{-\pi}^{\gamma_0} + \int_{\gamma_1}^{\pi}\right) \left(\frac{dy}{dx}\right)_s \left[\frac{\sin \frac{1}{2}(\gamma_0 - \gamma)}{\sin \frac{1}{2}(\gamma_1 - \gamma)}\right]^\epsilon \begin{Bmatrix} 1 \\ \sin (\lambda_0\epsilon - \gamma) \\ \cos (\lambda_0\epsilon - \gamma) \end{Bmatrix} d\gamma$$

$$- \cos \pi\epsilon \int_{\gamma_0}^{\gamma_1} \left(\frac{dy}{dx}\right)_s \left[\frac{\sin \frac{1}{2}(\gamma - \gamma_0)}{\sin \frac{1}{2}(\gamma_1 - \gamma)}\right]^\epsilon \begin{Bmatrix} 1 \\ \sin (\lambda_0\epsilon - \gamma) \\ \cos (\lambda_0\epsilon - \gamma) \end{Bmatrix} d\gamma.$$

$$(221)$$

The terms involving α_i and $(dy/dx)_s$ in (221) can be eliminated as follows. The first of (221), i.e. the equation with zero on the left-hand side, is used to eliminate α_i from the other two. Now let $C_Q = C_Q'$ and $\Omega_s = \Omega_s'$ at $\alpha_0 = 0$, i.e. at $\alpha^* = -\alpha'$ (the no-lift angle, see §8.7), then we can substitute these values into (221), and eliminate α_i as before, to obtain another pair of equations. Subtraction of these from the first pair now eliminates the $(dy/dx)_s$ term, and leaves

$$\frac{1}{\pi T}(C_Q - C'_Q) = (\alpha^* + \alpha')\{\sin\lambda_0\,\epsilon\cos\lambda_0\,\epsilon + \epsilon\sin\lambda_0\cos\lambda_1\}$$
$$+ (\Omega_s - \Omega'_s)\{\sin^2\lambda_0\,\epsilon + \epsilon\sin\lambda_0\sin\lambda_1), \quad (222)$$

and
$$\beta_a\alpha_0 = (\alpha^* + \alpha')\{\cos^2\lambda_0\,\epsilon + \epsilon\sin\lambda_0\sin\lambda_1\}$$
$$+ (\Omega_s - \Omega'_s)\{\sin\lambda_0\,\epsilon\cos\lambda_0\,\epsilon - \epsilon\sin\lambda_0\cos\lambda_1\}. \quad (223)$$

From (222) we can calculate the change in total mass sucked into the aerofoil due to changes in the incidence and the internal pressure, while (223) and $C_L \simeq 2\pi\alpha_0$ yield the lift coefficient. (In studying these equations it should be remembered that ϵ is a negative number.)

A similar treatment of (218) gives

$$C_M - C'_M = -\frac{x_c}{c}C_L$$
$$+ \frac{\pi}{2\beta_a}T^2[(\alpha^* + \alpha')\{\cos\lambda_0\,\epsilon\cos 2\lambda_0\,\epsilon + \epsilon\sin\lambda_0\,(\cos\lambda_0\sin 2\lambda_1$$
$$+ 2\cos\lambda_0\,\epsilon\sin(\lambda_1 - \lambda_0\,\epsilon)) - \epsilon^2\sin^2\lambda_0\cos 2\lambda_1\}$$
$$+ (\Omega_s - \Omega'_s)\{\sin\lambda_0\,\epsilon\cos 2\lambda_0\,\epsilon - \epsilon\sin\lambda_0\,(\cos\lambda_0\cos 2\lambda_1$$
$$- 2\sin\lambda_0\,\epsilon\sin(\lambda_1 - \lambda_0\,\epsilon)) - \epsilon^2\sin^2\lambda_0\sin 2\lambda_1\}], \quad (224)$$

on making use of (222) and (223) to eliminate $C_Q - C'_Q$ and α_0; C'_M is the moment coefficient at zero lift.

Similarly, from (219)

$$\Omega - \Omega' = \Omega_s - \Omega'_s - \left\{\frac{\sin\frac{1}{2}(\gamma_1 - \gamma^*)}{\sin\frac{1}{2}(\gamma_0 - \gamma^*)}\right\}^\epsilon \left\{(\alpha^* + \alpha')\frac{\cos(\lambda_0\epsilon + \frac{1}{2}\gamma^*)}{\sin\frac{1}{2}\gamma^*}\right.$$
$$\left. + (\Omega_s - \Omega'_s)\frac{\sin(\lambda_0\epsilon + \frac{1}{2}\gamma^*)}{\sin\frac{1}{2}\gamma^*}\right\} \quad (\text{provided } \gamma^* \text{ is outside } (\gamma_0, \gamma_1)),$$

$$\Omega - \Omega' = \Omega_s - \Omega'_s - \cos\pi\epsilon\left\{\frac{\sin\frac{1}{2}(\gamma_1 - \gamma^*)}{\sin\frac{1}{2}(\gamma^* - \gamma_0)}\right\}^\epsilon \left\{(\alpha^* + \alpha')\frac{\cos(\lambda_0\epsilon + \frac{1}{2}\gamma^*)}{\sin\frac{1}{2}\gamma^*}\right.$$
$$\left. + (\Omega_s - \Omega'_s)\frac{\sin(\lambda_0\epsilon + \frac{1}{2}\gamma^*)}{\sin\frac{1}{2}\gamma^*}\right\},$$

$$\theta - \theta' = -(\alpha^* + \alpha') - \sin\pi\epsilon\left\{\frac{\sin\frac{1}{2}(\gamma_1 - \gamma^*)}{\sin\frac{1}{2}(\gamma^* - \gamma_0)}\right\}^\epsilon \left\{(\alpha^* + \alpha')\frac{\cos(\lambda_0\epsilon + \frac{1}{2}\gamma^*)}{\sin\frac{1}{2}\gamma^*}\right.$$
$$\left. + (\Omega_s - \Omega'_s)\frac{\sin(\lambda_0\epsilon + \frac{1}{2}\gamma^*)}{\sin\frac{1}{2}\gamma^*}\right\} \quad (\text{provided } \gamma^* \text{ is inside } (\gamma_0, \gamma_1)),$$

$$(225)$$

where $\Omega = \Omega'$ at $\alpha_0 = 0$.

If the values of α_i, α', C'_Q and Ω'_s are required, equations relating these to $(dy/dx)_s$ follow from (221) on setting $C_Q = C'_Q$, $\Omega_s = \Omega'_s$, $\alpha_0 = 0$ and $\alpha^* = -\alpha'$.

Notice from (225) that $\Omega - \Omega'$ is finite at $\gamma^* = \gamma_1$ only if

$$\Omega_s - \Omega'_s = -(\alpha^* + \alpha')\cot(\lambda_0 \epsilon + \tfrac{1}{2}\gamma_1), \tag{226}$$

and this can be regarded as being a restriction on $\Omega_s - \Omega'_s$—although it is permissible to ignore it, for it is only a 'local' condition, and not an essential 'over-all' condition like (222) or (223).

The relations between γ_1, γ_0, λ_0, λ_1 and distances measured from the mid-chord point are found from $x = -\tfrac{1}{2}c\cos\gamma$, which is appropriate to thin aerofoil theory (see § 8.15). Thus the width of the porous section is

$$b = -\tfrac{1}{2}c\,(\cos\gamma_1 - \cos\gamma_0) = c\sin\lambda_0\sin\lambda_1, \tag{227}$$

and the mid-point of the porous section is at

$$x = -\tfrac{1}{2}c\,(\cos\gamma_1 + \cos\gamma_0) = -\tfrac{1}{2}c\cos\lambda_0\cos\lambda_1. \tag{228}$$

8.29   Sources and sinks on aerofoils

Let the porous strip be narrow, then $\tfrac{1}{2}(\gamma_1 - \gamma_0) \equiv \lambda_0$ will be small, and (227) and (228) can be written

$$b = c\lambda_0\sin\lambda_1, \quad x = -\frac{c}{2}\cos\lambda_1, \tag{229}$$

on ignoring terms $O(\lambda_0^2)$. With this approximation (222) and (223) become

$$\frac{1}{\pi T}(C_Q - C'_Q) = \lambda_0\epsilon\{(\alpha^* + \alpha')(1 + \cos\lambda_1) + (\Omega_s - \Omega'_s)\sin\lambda_1\}, \tag{230}$$

and $\quad \beta_a\alpha_0 = \alpha^* + \alpha' + \lambda_0\epsilon\{(\alpha^* + \alpha')\sin\lambda_1 + (\Omega_s - \Omega'_s)(1 - \cos\lambda_1)\}. \tag{231}$

Notice from (229) and (230) that the change ΔC_Q in C_Q due to a change in internal pressure alone is

$$\Delta C_Q = \pi T\frac{b}{c}\epsilon\Delta\Omega_s = -\pi T\frac{b}{c}(-\epsilon)\frac{\beta_a}{2}\Delta C_{p_s}, \tag{232}$$

as $\Omega \approx \tfrac{1}{2}\beta_a C_p$. This relation provides a means of determining the number ϵ by experiment.

From (81) $C_L \simeq 2\pi T\alpha_0$, therefore on eliminating $\lambda_0\epsilon$ from (230) and (231) we find

$$C_L = \frac{2\pi}{\beta_a}T(\alpha^* + \alpha') + \frac{2}{\beta_a}(C_Q - C'_Q)\tan\tfrac{1}{2}\lambda_1. \tag{233}$$

Similarly, (224) reduces to

$$C_M - C'_M = -\frac{x_c}{c}C_L + \frac{\pi}{2\beta_a}T^2(\alpha^* + \alpha') + \frac{1}{\beta_a}T(C_Q - C'_Q)\sin\lambda_1. \quad (234)$$

As the width of the strip does not appear in (233) and (234) they must also be valid for a concentrated sink ($b \to 0$). Thus these equations give the effect of a (line) sink on aerofoil surface on the force and moment acting on the aerofoil (e.g. see Ehlers, 1946, where a different derivation is given).

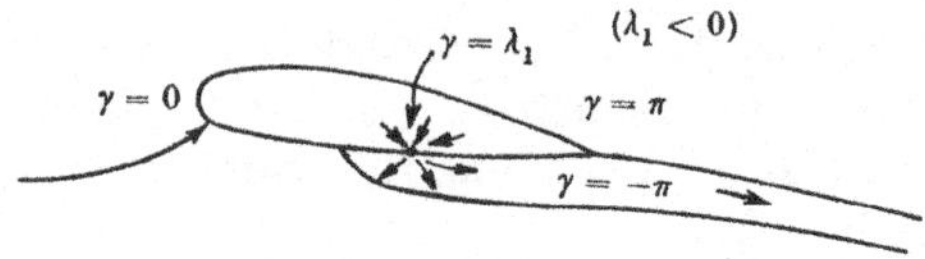

Fig. 8.29a

The change in pressure coefficient, $C_p - C'_p$, due to circulation and the presence of a sink at $\gamma = \lambda_1$ is found from the first of (225), (230) and $C_p \approx 2\Omega/\beta_a$ to be given by

$$C_p - C'_p = -\frac{2}{\beta_a}(\alpha^* + \alpha')\cot\tfrac{1}{2}\gamma - \frac{(C_Q - C'_Q)}{2\pi T \cos\tfrac{1}{2}\lambda_1 \sin\tfrac{1}{2}\gamma \sin\tfrac{1}{2}(\lambda_1 - \gamma)}. \quad (235)$$

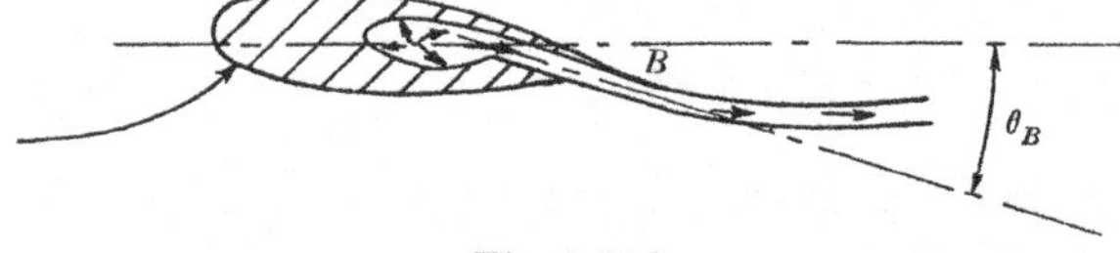

Fig. 8.29b

If there is a *source* at $\gamma = \lambda_1$, the equations have exactly the same form except that the sign of C_Q will be changed. Therefore if C_Q is now the (positive) mass coefficient of the source, (233) will read

$$C_L = \frac{2\pi}{\beta_a}T(\alpha^* + \alpha') - \frac{2}{\beta_a}(C_Q - C'_Q)\tan\tfrac{1}{2}\lambda_1, \quad (236)$$

and the increment to C_L will be positive only if the source is on the lower surface of the aerofoil (see Fig. 8.29a), a result which is physically obvious.

It will be noticed from (233) and (236) that the theory fails if $\lambda_1 = \pi$, i.e. if the source or sink is at the trailing edge. The effect of a source at or near the trailing edge depends critically on how the fluid is assumed to emerge from the aerofoil, and no realistic conclusions can be drawn from the simple source model. The shape of the duct leading the fluid from the interior of the aerofoil to the trailing edge must be taken into account. Suppose the source fluid emerges at an angle θ_B to the chord after flowing along a parallel duct as shown in Fig. 8.29b, then it has been shown that (Woods, 1956b)

$$C_L = 2\pi\,(\alpha^* + \alpha') + 4\theta_B \sqrt{\frac{2C_Q}{\pi}}\,,$$

in incompressible flow. This result is of some interest in the theory of jet-flapped aerofoils (see ch. 10).

8.30   Bubbles on thin aerofoils and aerofoils with spoilers

It has long been known that thin aerofoils tend to stall from the leading edge rather than the trailing edge. At a certain incidence the flow separates from a point close to the leading edge, and then re-attaches a little farther downstream, forming a short closed bubble of almost stagnant air. As the incidence is increased, the bubble length grows continuously, until eventually it fails to re-attach. There may exist a closed wake for a further short incidence range (see Fig. 8.32), but in any case the final result of a further incidence increase is an infinite open wake, and a completely stalled aerofoil. For an experimental investigation of the phenomena see McCullogh & Gault (1951).

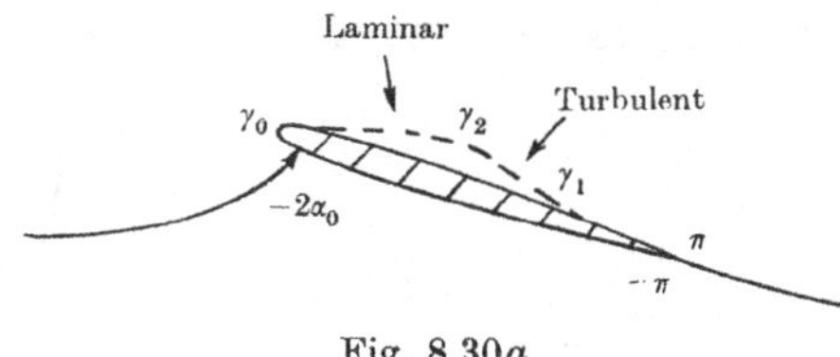

Fig. 8.30a

There are two distinct types of bubble, namely (i) the short leading edge bubble (having a length less than about $0\cdot1c$), and (ii) the long bubble, an example of which is shown in Fig. 8.30a. The short bubble is usually separated from the main stream by a laminar 'boundary' layer, whereas with the long bubble this separating layer is laminar only over the first half or so of the bubble length; transition then occurs, and a turbulent layer closes the bubble back on to the aerofoil surface. As a consequence the short bubble is at almost constant pressure, while in the long bubble the pressure is approximately constant only up to the transition point, and then steadily increases up to the re-attachment point. The 'dead' air inside a long bubble therefore cannot be regarded as being a constant pressure region; energy is fed into the bubble through the turbulent part of the separating layer, producing a region of slowly eddying fluid in the rear portion of the bubble, which is thus rendered capable of maintaining a pressure gradient.

A similar type of flow sometimes occurs behind a spoiler on an aerofoil surface (Pearcey, Pankhurst & Lee, 1952). The spoiler is a short bluff projection causing flow separation, and if it is set far enough forward on the aerofoil, at certain incidences the flow is able to re-attach in the manner shown in Fig. 8.30b. If there is an upwards deflected flap behind the aerofoil (see Fig. 8.30c) re-attachment is very likely. The bubble in this case will have the characteristics of a long rather than a short bubble.

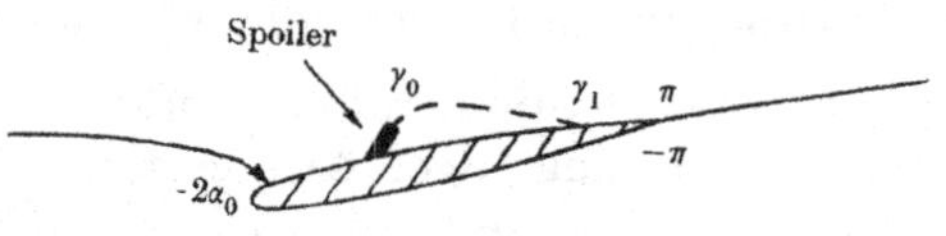

Fig. 8.30b

(i) *The short leading edge bubble*

The theory of the short bubble is easily deduced from equations given in § 8.28 by setting

$$\epsilon = -\tfrac{1}{2}, \quad C_Q - C'_Q = 0, \tag{237}$$

which changes the porous surface into an impermeable free streamline. The leading edge of the aerofoil is at $\gamma = 0$, and (215) and (216) give $\lambda_0 = \lambda_1 = \tfrac{1}{2}\gamma_1$. With these values (222) and (226) become

$$\Omega_s - \Omega'_s = -(\alpha^* + \alpha')\tan\tfrac{1}{2}\gamma_1 \frac{1 + \cos\tfrac{1}{2}\gamma_1}{1 - \cos\tfrac{1}{2}\gamma_1}, \tag{238}$$

and

$$\Omega_s - \Omega'_s = -(\alpha^* + \alpha')\cot\tfrac{1}{4}\gamma_1, \tag{239}$$

which are easily verified to be incompatible for all values of γ_1. The first of these equations is the condition that the bubble is closed, while the second is the condition that $\Omega_s - \Omega'_s$ remains finite at the re-attachment point.

As $x = -\tfrac{1}{2}c\cos\gamma$ the length l of the bubble is

$$l = \tfrac{1}{2}c(1 - \cos\gamma_1) = c\sin^2\tfrac{1}{2}\gamma_1. \tag{240}$$

As $\Omega_s = \tfrac{1}{2}\beta_a C_p$, for small bubbles each of (238) and (239) can be expressed

$$\frac{l}{c} = \frac{k^2}{\beta_a^2}\left(\frac{\Delta C_p}{\Delta\alpha}\right)^{-2}, \tag{241}$$

where k has the values 8 and 4 respectively. The experimental results at present available (to the author) are not sufficient to distinguish between these values.

Similarly, (223) reduces to

$$\beta_a \alpha_0 = \tfrac{1}{2}(\alpha^* + \alpha') \cos \tfrac{1}{2}\gamma_1 (1 + \cos \tfrac{1}{2}\gamma_1) - \tfrac{1}{2}(\Omega_s - \Omega_s') \sin \tfrac{1}{2}\gamma_1 (1 - \cos \tfrac{1}{2}\gamma_1).$$

Eliminating $\Omega_s - \Omega_s'$ by using the combined form of (238) and (239) given in (241), and recalling that $C_L \simeq 2\pi\alpha_0$, we find

$$C_L \simeq \frac{2\pi}{\beta_a}(\alpha^* + \alpha')\left\{1 + \frac{l}{8c}(k-6)\right\}, \tag{242}$$

for short bubbles. The effect of the bubble on the lift therefore depends significantly on which of the two bubble conditions dominates. A similar result for C_M can be deduced from (224), and an equation for the shape of the bubble can be found from the third of (225).

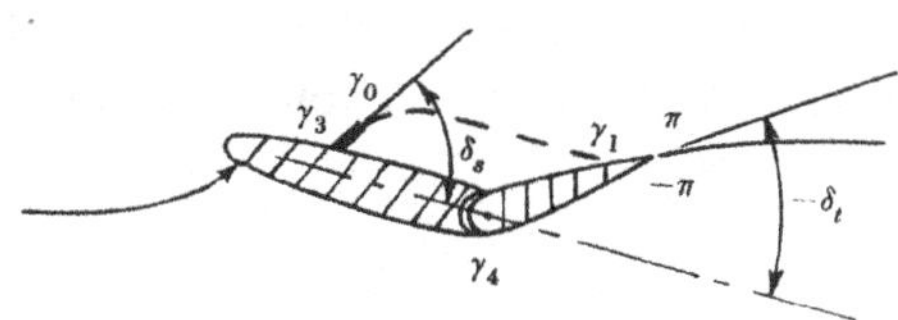

Fig. 8.30c

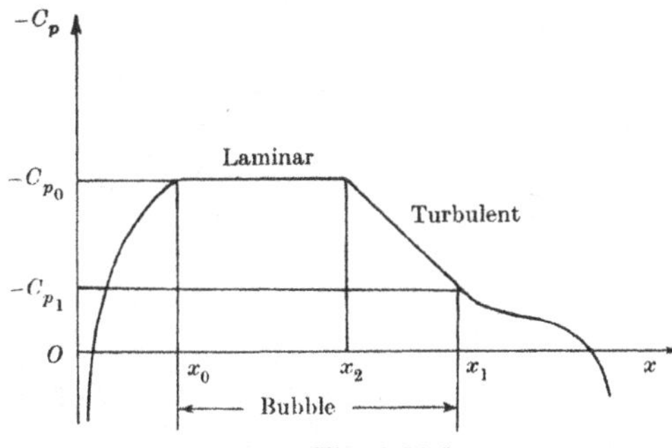

Fig. 8.30d

(ii) *The long bubble*

The simple constant pressure model used for the short bubble fails completely with long bubbles, e.g. (238) shows that $\Omega_s - \Omega_s' \to \infty$ as $l \to c$, and it is necessary to vary the pressure so that both the bubble conditions can be satisfied. (The theory given in §14 of Woods (1956a) is wrong, being based on the limit $\epsilon \to \tfrac{1}{2}$ instead of $\epsilon \to -\tfrac{1}{2}$.)

Let the flow separate at $x = x_0$ and re-attach at $x = x_1$, then the experimental evidence supports a 'roof-top' pressure distribution (see Fig. 8.30d):

$$C_p = \begin{cases} C_{p_0} & (x_0 \leqslant x \leqslant x_2), \\ C_{p_0} - \dfrac{x - x_1}{x_2 - x_1}(C_{p_0} - C_{p_1}) & (x_2 \leqslant x \leqslant x_1), \end{cases} \tag{243}$$

where C_{p_0} and C_{p_1} are constants.

As $C_p \propto \Omega_s$, and $x \propto -\cos\gamma$ in thin aerofoil theory, we can write (243) as

$$\Omega_s = \begin{cases} \Omega_0 & (\gamma_0 \leqslant \gamma \leqslant \gamma_2), \\ \Omega_0 - \dfrac{\cos\gamma_1 - \cos\gamma}{\cos\gamma_1 - \cos\gamma_2}(\Omega_1 - \Omega_0) & (\gamma_2 \leqslant \gamma \leqslant \gamma_1). \end{cases} \quad (244)$$

From (219) we observe that Ω will be finite at $\gamma^* = \gamma_1$ provided that

$$\left(\int_{-\pi}^{\gamma_0} + \int_{\gamma_1}^{\pi}\right)\theta_s \left[\frac{\sin\frac{1}{2}(\gamma_1 - \gamma)}{\sin\frac{1}{2}(\gamma_0 - \gamma)}\right]^{\frac{1}{2}} \cot\frac{1}{2}(\gamma - \gamma_1)\,d\gamma$$

$$+ \int_{\gamma_0}^{\gamma_1} \Omega_s \left[\frac{\sin\frac{1}{2}(\gamma_1 - \gamma)}{\sin\frac{1}{2}(\gamma - \gamma_0)}\right]^{\frac{1}{2}} \cot\frac{1}{2}(\gamma - \gamma_1)\,d\gamma = 0, \quad (245)$$

as $\epsilon = -\frac{1}{2}$ in the present application. Substitution of (244) and (237) into this result and into (217) yields the four conditions

$$2\pi\alpha_0\beta_a \begin{Bmatrix} 0 \\ 0 \\ 1 \\ 0 \end{Bmatrix} = \left(\int_{-\pi}^{\gamma_0} + \int_{\gamma_1}^{\pi}\right)\theta_s \left[\frac{\sin\frac{1}{2}(\gamma_1 - \gamma)}{\sin\frac{1}{2}(\gamma_0 - \gamma)}\right]^{\frac{1}{2}} \begin{Bmatrix} 1 \\ -\sin(\gamma + \frac{1}{2}\lambda_0) \\ \cos(\gamma + \frac{1}{2}\lambda_0) \\ \cot\frac{1}{2}(\gamma - \gamma_1) \end{Bmatrix} d\gamma$$

$$+ \Omega_0 \int_{\gamma_0}^{\gamma_2} \left[\frac{\sin\frac{1}{2}(\gamma_1 - \gamma)}{\sin\frac{1}{2}(\gamma - \gamma_0)}\right]^{\frac{1}{2}} \begin{Bmatrix} 1 \\ -\sin(\gamma + \frac{1}{2}\lambda_0) \\ \cos(\gamma + \frac{1}{2}\lambda_0) \\ \cot\frac{1}{2}(\gamma - \gamma_1) \end{Bmatrix} d\gamma$$

$$+ \left(\Omega_0 - \frac{\cos\gamma_1(\Omega_1 - \Omega_0)}{\cos\gamma_1 - \cos\gamma_2}\right) \int_{\gamma_2}^{\gamma_1} \left[\frac{\sin\frac{1}{2}(\gamma_1 - \gamma)}{\sin\frac{1}{2}(\gamma - \gamma_0)}\right]^{\frac{1}{2}} \begin{Bmatrix} 1 \\ -\sin(\gamma + \frac{1}{2}\lambda_0) \\ \cos(\gamma + \frac{1}{2}\lambda_0) \\ \cot\frac{1}{2}(\gamma - \gamma_1) \end{Bmatrix} d\gamma$$

$$+ \left(\frac{\Omega_1 - \Omega_0}{\cos\gamma_1 - \cos\gamma_2}\right) \int_{\gamma_2}^{\gamma_1} \left[\frac{\sin\frac{1}{2}(\gamma_1 - \gamma)}{\sin\frac{1}{2}(\gamma - \gamma_0)}\right]^{\frac{1}{2}} \cos\gamma \begin{Bmatrix} 1 \\ -\sin(\gamma + \frac{1}{2}\lambda_0) \\ \cos(\gamma + \frac{1}{2}\lambda_0) \\ \cot\frac{1}{2}(\gamma - \gamma_1) \end{Bmatrix} d\gamma, \quad (246)$$

the last three integrals of which are easily evaluated by the substitution used in § 8.28.

With an aerofoil of given shape and incidence, and known to have a separation bubble on its upper surface, six quantities are initially unknown, viz. (1) the position of the front stagnation point $(-2\alpha_0)$, (2) the separation point (γ_0), (3) the transition point (γ_2), (4) the re-attachment point (γ_1), (5) the pressure at separation (C_{p_0}), and (6) the pressure at separation (C_{p_1}) (or equivalently, the pressure gradient over the turbulent part of the bubble). Equations (246) fix four of these, and so only two conditions need to be provided by boundary-layer theory (or more precisely *separated* 'boundary' layer theory). For example the separation point γ_0 and the transition point γ_2 might be the most convenient quantities to calculate from boundary-layer theory, but at the moment of writing an adequate theory of separated boundary layers has yet to be developed.

Once the six quantities just listed have been fixed, the bubble shape and pressure distribution follow from (219). Substituting (244) and $\epsilon = -\frac{1}{2}$ in (219) we find

$$
\begin{aligned}
\frac{\Omega(\gamma^*)}{\theta(\gamma^*)} &= \frac{\mathscr{R}}{\mathscr{I}} \frac{1}{2\pi} \left[\frac{\sin\frac{1}{2}(\gamma_0 - \gamma^*)}{\sin\frac{1}{2}(\gamma_1 - \gamma^*)} \right]^{\frac{1}{2}} \\
&\times \left\{ \left(\int_{-\pi}^{\gamma_0} + \int_{\gamma_1}^{\pi} \right) \theta_s \left[\frac{\sin\frac{1}{2}(\gamma_1 - \gamma)}{\sin\frac{1}{2}(\gamma_0 - \gamma)} \right]^{\frac{1}{2}} \cot\tfrac{1}{2}(\gamma - \gamma^*) \, d\gamma \right. \\
&+ \left(\Omega_0 \int_{\gamma_0}^{\gamma_2} + \left[\Omega_0 - \frac{\cos\gamma_1 (\Omega_1 - \Omega_0)}{\cos\gamma_1 - \cos\gamma_2} \right] \int_{\gamma_2}^{\gamma_1} \right) \left[\frac{\sin\frac{1}{2}(\gamma_1 - \gamma)}{\sin\frac{1}{2}(\gamma - \gamma_0)} \right]^{\frac{1}{2}} \cot\tfrac{1}{2}(\gamma - \gamma^*) \, d\gamma \\
&+ \left. \left(\frac{\Omega_1 - \Omega_0}{\cos\gamma_1 - \cos\gamma_2} \right) \int_{\gamma_2}^{\gamma_1} \left[\frac{\sin\frac{1}{2}(\gamma_1 - \gamma)}{\sin\frac{1}{2}(\gamma - \gamma_0)} \right]^{\frac{1}{2}} \cos\gamma \cot\tfrac{1}{2}(\gamma - \gamma^*) \, d\gamma \right\}.
\end{aligned}
\tag{247}
$$

Similarly, equations for the lift and moment can be found from (81) and (218).

The theory given above can also be used to study the flow depicted in Fig. 8.30c; the only change required is to replace θ_s by (cf. (168))

$$
\theta_s + \delta_s\{U(\gamma_3) - U(\gamma_0)\} - \delta_l\{U(\gamma_1) - U(\pi) + U(-\pi) - U(\gamma_4)\},
$$

where θ_s refers to the aerofoil without spoiler $(\delta_s = 0)$ or flap $(\delta_l = 0)$, and the remaining symbols are defined in the figure.

8.31   Aerofoil design

The theory of long bubbles suggests a method of aerofoil design in which only a portion of the aerofoil surface is designed to give a prescribed pressure distribution. This may in certain circumstances have advantages over the 'overall' method of design described in §8.25. For example in many cases of aerofoil design it would be convenient to prescribe the pressure distribution over the front half or so of the profile— to avoid adverse pressure gradients—and the shape of the rear section of the profile—where such gradients are, in any case, unavoidable. The design equation for the free surface follows on putting $\epsilon = -\frac{1}{2}$ in the imaginary part of (219). (In Woods's (1956a) earlier theory the wrong limit for ϵ has been used.) As the restrictions contained in (217) and (245) must be satisfied for a closed aerofoil having a continuous velocity distribution to result, there must be at least four disposal parameters in the theory. A related theory for the 'sectional' design of symmetrical aerofoils has been given by Woods (1955a).

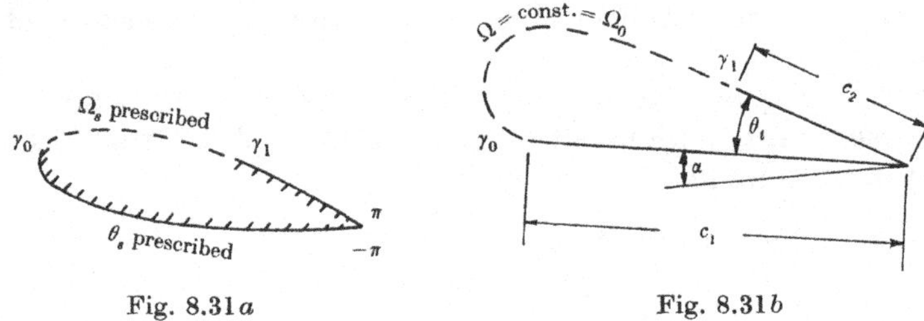

Fig. 8.31a Fig. 8.31b

The simplest example of this theory is the aerofoil shown in Fig. 8.31b —an aerofoil comprised of two straight unequal arms inclined at an angle θ_t (the trailing edge angle) and containing a constant pressure bubble between them. Notice that it is not possible to prescribe α, θ_t and c_1/c_2, for this leaves only *three* free parameters γ_0, γ_1 and Ω_0. In a practical application one would probably choose values for α and c_1/c_2, and let θ_t be fixed by the closure conditions. By a method based on conformal mapping of the hodograph plane, Hurley (1957) has studied this aerofoil in an incompressible flow with a view to designing thick aerofoils for high lift. Boundary-layer control, in the form of blowing over the upper straight surface of the aerofoil, is suggested as a means of overcoming the tendency to flow separation from this surface, which will exist when θ_t is large. The design equation for Hurley's aerofoil follows from (247) on putting $\Omega_0 = \Omega_1$ and $\theta_s = -\alpha - \theta_t\{\mathbf{U}(\gamma_1) - \mathbf{U}(\pi)\}$, where θ_t, γ_0, γ_1 and Ω_1 have been chosen to satisfy the closure conditions.

8.32   Finite cavities behind hydrofoils

Riabouchinsky's (1921) well-known model of finite cavities (see §11.23) is another special case of generalized aerofoil theory. In this model (actually a generalization of Riabouchinsky's original conception) a constant-pressure cavity extending behind a body moving through a liquid is terminated by another body downstream of the first. The second body is not an essential part of the problem, but is introduced simply to obviate mathematical difficulties at the rear end of the cavity (see discussion in §11.4). Fig. 8.32 shows an example of this model, which has an obvious application to hydrofoils moving at speeds high enough to produce cavitation.

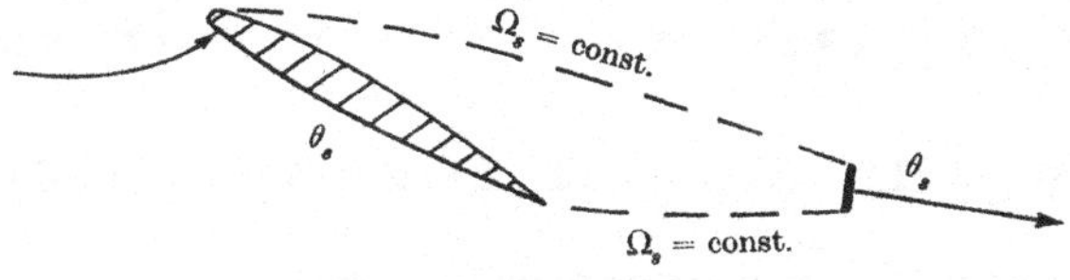

Fig. 8.32

The equations appropriate to this type of flow are easily derived from the equations given in § 8.26 by putting $\epsilon = -\frac{1}{2}$ and $\epsilon = \frac{1}{2}$ on the upper and lower surfaces of the cavitation bubble respectively, with $\Omega_s = $ constant on these surfaces.

CHAPTER 9

UNSTEADY AEROFOIL THEORY

MATHEMATICAL FORMULATION

9.1 Basic concepts and approximations

In the following pages we shall study the unsteady motion of a closed aerofoil†, where the unsteadiness consists of small perturbations about a mean motion, which is itself not necessarily steady (see §6.8). As stated in §2.20 the boundary conditions in this case can be applied at the mean boundary position without introducing errors of significant order, except possibly at a few exceptional points. The unsteady motion of the aerofoil will be assumed at first to be quite general in character—within the restriction of small displacements just mentioned—and the possibility of non-rigid movements will also be admitted. Towards the end of the chapter we will study a few special types of unsteady motion of practical importance.

It is shown in §2.21 that provided the flow is either (i) incompressible, or (ii) compressible, but varying with time slowly enough for *second*-order time derivatives to be ignored, the differential equation satisfied by $\tau = \Omega + i\theta$ is

$$\nabla^2 \tau_1 = \frac{2}{(a_a \beta_a)^2} \frac{\partial^2 \tau}{\partial \phi_0 \partial t}, \tag{1}$$

where
$$\nabla^2 \tau = 0, \tag{2}$$

and τ_1 satisfies the same boundary conditions as τ. It is assumed in the derivation of (1) that a mean *steady* flow exists; the independent variables involved in ∇^2 are ϕ_0, ψ_0, which are the potential and stream functions for this mean steady flow. Let $w_0 = \phi_0 + i\psi_0$, then it follows from (2) that τ is an analytic function of w_0. We shall denote the solutions of (2) and (1) by $\tau(w_0)$ and $\tau_1(\tilde{w}_0)$ $(= \tau_1(w_0, \overline{w}_0)$, see §3.19) respectively.

The first step in finding the solution of (1) is to calculate $\tau(w_0)$; then $\tau_1(\tilde{w}_0)$ can be calculated from $\tau(w_0)$ quite easily with the help of a formula we shall give in §9.13. If the flow is incompressible or varying slowly enough for the right-hand side of (1) to be negligible compared with the left-hand side, then $\tau_1 = \tau$, and the theory of §9.13 is unnecessary.

With aerofoils that are not excessively thick or at large incidences at any instant in their motion, both the mean flow and unsteady perturba-

---

† The unsteady motion of 'open aerofoils', or rather aerofoils behind which extend infinite wakes of finite and non-zero width will be studied in §11.14 *et seq.*

tions can be treated by linear perturbation theory; this is equivalent to replacing the independent variable w_0 by UZ. As we shall adopt the semi-infinite strip $-\pi < \gamma < \pi$, $0 < \eta < \infty$, used in steady aerofoil theory, to be the plane of the independent variables, then (see § 8.15)

$$Z = -\tfrac{1}{2}c \cos \zeta \quad (Z = x + i\beta_a y, \zeta = \gamma + i\eta), \tag{3}$$

and in particular, $$x = -\tfrac{1}{2}c \cos \gamma, \tag{4}$$

on $\eta = 0$. This linearization is by no means essential to mathematical progress, and we make it only to avoid some rather involved algebra (see Woods, 1954b for an account of the unsteady motion of thick aerofoils).

Some of the boundary conditions of our problem have been stated in § 6.8. On the aerofoil we have from (6.46) that

$$\theta = \theta_s + \frac{n}{U} - \alpha + \pi\lambda\delta(\gamma), \tag{5}$$

where the adoption of linear perturbation theory has enabled us to replace (i) $|\mathbf{q}|$ by U, the velocity at infinity (which is not necessarily steady), and (ii) $\mathbf{n.v}$ by n, the velocity of the aerofoil in the Oy-direction. In writing (5) we have also assumed that the front stagnation point lies in the immediate neighbourhood of the leading edge $\gamma = 0$. The number λ in (5) is initially unknown; its value is fixed by the condition that the unsteady motion of the aerofoil can have no influence on the flow direction at infinity, as will become apparent below.

Another important boundary condition on the aerofoil is that Joukowski's condition holds, i.e. that the rear stagnation point is fixed in position at the sharp trailing edge. The reasons for adopting this condition are the same as for its adoption in steady flow (see § 1.17). A consequence of Joukowski's condition is that the circulation Γ_1 about a fixed circuit C_1 enclosing the aerofoil and passing through a point just downstream of the trailing edge (see Fig. 9.1) will be time dependent, for otherwise it would be impossible to change the circulation about the aerofoil by any change in aerofoil attitude or speed. Kelvin's theorem, namely $d\Gamma/dt = 0$, therefore does not apply to C_1. Of course this theorem is strictly applicable only to a circuit C which *moves* with the fluid (§ 1.8), but as C can be deformed in any way provided it crosses no singularities in the fluid, it is clear that the theorem should also be applicable to a fixed circuit like C_1. The escape from this contradiction is obviously that C cannot be deformed into C_1 (see Fig. 9.1) without crossing some singularity in the flow. As it is a *circulation* balance we seek, this singularity must be an isolated vortex or, more generally, a vortex sheet. This sheet must spring from the trailing edge for several reasons, the simplest of which is just that the selection of any other point for the start of the sheet could not be unique.

A more cogent reason is that in a motion which commences from rest the vorticity must remain zero on all streamlines, except possibly those springing from solid boundaries (see § 1.7). We therefore have the flow shown in Fig. 9.1, with the varying circulations about each of the aerofoil and sheet adding to give a total circulation about C which is independent of t.

The position adopted by the sheet is initially unknown, but as we shall use linear perturbation theory and shall therefore apply the aerofoil boundary conditions on $y = 0$, $-\tfrac{1}{2}c \leqslant x \leqslant \tfrac{1}{2}c$, it is consistent to assume that the sheet lies on $y = 0$, $\tfrac{1}{2}c \leqslant x \leqslant \infty$. It has been shown by Greidanus & Van Heemert (1948) that the errors arising from this approximation in the position of the sheet are negligible.

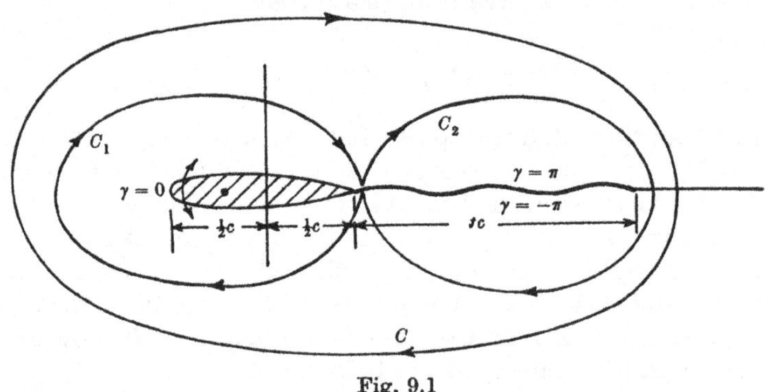

Fig. 9.1

The vorticity in the sheet is convected downstream with a speed $\bar{q}$ equal to the average of the speeds on each side of the sheet, and except very close to the trailing edge this speed is almost the same as the velocity at infinity, U. We shall therefore assume that $\bar{q} = U$ over the whole of the sheet, which means that while $\bar{q}$ may still vary with time, it will be independent of distance x along the sheet. A theory which allows for the slower initial motion of the vorticity from the trailing edge of thick aerofoils has been given by Woods (1954b).

These approximations enable us to write (6.49) and (6.50) in the forms

$$\frac{1}{c}\frac{\partial \chi}{\partial t} + \frac{\partial \chi}{\partial x} = 0, \tag{6}$$

$$\chi(x,t) = f(x - ct), \tag{7}$$

where

$$\chi = \varpi U, \tag{8}$$

ϖ is the jump in Ω across the sheet, and t is the reduced time,

$$t = \frac{1}{c}\int_0^t U\, dt. \tag{9}$$

Equation (7) can be expressed in the more convenient form

$$\chi(x,t) = \chi(\tfrac{1}{2}c, t + \tfrac{1}{2} - \{x/c\}), \tag{10}$$

which relates the value of χ at any point and time to its value at the trailing edge, $x = \tfrac{1}{2}c$, at an earlier time.

Notice from (9) that the physical significance of t is that it is the total length of the vortex sheet in units of chord length for an unsteady motion that commenced at $t = 0$. In the following pages we shall assume that this is the case, i.e. that the flow is steady for $t < 0$, and unsteady for $t > 0$.

9.2　Solution of the boundary value problem

The transformation (3) maps the aerofoil on to $\eta = 0$, and the vortex sheet on to $\gamma = \pm\pi$. Our first boundary value problem is therefore to find $\tau = \tau(\zeta)$, given θ on $\eta = 0$, and the discontinuity ϖ in Ω across $\gamma = \pm\pi$. The solution to this problem has been given in § 4.3. On transforming (4.20) and (4.21) into the notation of the present section ($l = \pi$, $z = \zeta$) we find

$$\tau(\zeta,t) = \frac{1}{2\pi}\int_{-\pi}^{\pi} \theta(\gamma,t)\cot\tfrac{1}{2}(\gamma-\zeta)\,d\gamma + \frac{\sin\zeta}{2\pi}\int_0^{\infty}\frac{\varpi(\eta,t)\,d\eta}{\cosh\eta + \cos\zeta}, \tag{11}$$

and

$$0 = \int_{-\pi}^{\pi}\theta(\gamma,t)\,d\gamma + \int_0^{\infty}\varpi(\eta,t)\,d\eta, \tag{12}$$

as $q_\infty = U$ means that $\Omega_\infty = 0$, and we have put $\theta_\infty = 0$, i.e. taken the Ox-axis to be parallel to the undisturbed stream direction at infinity.

It is easily established from the expansion in (8.20) that

$$\frac{\sin\zeta}{\cosh\eta + \cos\zeta} = i\left\{1 + 2\sum_{n=1}^{\infty}\frac{\cosh n\eta}{\sigma^n}\right\}\quad (\sigma = -e^{-i\zeta}),$$

which together with (8.20) and (12) enables us to expand (11) as

$$\tau = \frac{i}{\pi}\sum_{n=1}^{\infty}\frac{1}{\sigma^n}\left[(-1)^n\int_{-\pi}^{\pi}\theta\,e^{-ni\gamma}\,d\gamma + \int_0^{\infty}\varpi\cosh n\eta\,d\eta\right]. \tag{13}$$

The total circulation about the aerofoil plus wake is easily verified to be

$$\Gamma = \mathscr{R}\int_C q\,e^{-i\theta}\,dz,$$

where C is a contour large enough to enclose both aerofoil and wake. In linear perturbation theory it follows from (6.23) that

$$\mathscr{R}q\,e^{-i\theta}\,dz = \mathscr{R}U\left\{1 - \frac{\Omega}{\beta_a} - i\theta\right\}\{dx + i\,dy\} = U\,dx - \frac{1}{\beta_a}\mathscr{R}\tau\,dZ,$$

as $dZ = dx + i\beta_a dy$. Therefore

$$\Gamma = -\frac{1}{\beta_a}\mathscr{R}\int_C \tau\, dZ = \mathscr{R}\frac{c}{4\beta_a}\int_C \tau(\sigma)\left(1 - \frac{1}{\sigma^2}\right)d\sigma, \tag{14}$$

where the last form of writing follows from (3) and $\sigma = -e^{-i\zeta}$.

We shall assume that before the unsteady motion commences Γ is zero; Kelvin's circulation theorem then gives that $\Gamma = 0$ throughout the unsteady motion. Consequently it follows from (13), (14) and the residue theorem that the real part of the coefficient of $1/\sigma$ in the expansion for τ must vanish, i.e.

$$\int_{-\pi}^{\pi}\theta\cos\gamma\,d\gamma - \int_0^{\infty}\varpi\cosh\eta\,d\eta = 0. \tag{15}$$

Another important restriction is that the unsteady motion of the aerofoil cannot affect the aerofoil closure condition, $\int_C dy = 0$. Hence as $dy/dx \simeq \theta$ in linear perturbation theory, $\int_C \theta\, dx = 0$, or

$$\int_{-\pi}^{\pi}\theta\sin\gamma\,d\gamma = 0. \tag{16}$$

from (4).

Notice that (15) and (16) enable us to change $\sum_{n=1}^{\infty}$ in (13) to $\sum_{n=2}^{\infty}$.

A final basic result we shall require is the relation between Ω and the pressure on the aerofoil surface. This can be derived from (1.62) which can be written

$$\frac{1}{\rho}\frac{\partial p}{\partial x} = -\frac{1}{2}\frac{\partial q^2}{\partial x} - \frac{\partial q}{\partial t},$$

as $\partial/\partial s \simeq \partial/\partial x$ on the surface, and the external force potential χ is negligible in the present application. In linear perturbation theory

$$q \simeq U(1 - \Omega/\beta_a), \quad \rho = \rho_a(1 + O(\Omega)),$$

hence ignoring second-order terms,

$$\frac{\partial p}{\partial x} = \frac{\rho_a U^2}{\beta_a}\frac{\partial\Omega}{\partial x} - \frac{\rho_a U}{c}\frac{dU}{dt} + \frac{\rho_a U}{c\beta_a}\frac{\partial}{\partial t}(U\Omega),$$

using (9) and the fact that U is independent of x. Integration gives

$$p = G(t) + \frac{\rho_a U^2}{\beta_a}\Omega - \frac{\rho_a U x}{c}\frac{dU}{dt} + \frac{\rho_a U}{c\beta_a}\int_0^x \frac{\partial(U\Omega)}{\partial t}\,dx, \tag{17}$$

where G is a function of t only. Equation (4) enables us to write this in the form

$$p(\gamma) = G(t) + \frac{\rho_a U^2}{\beta_a}\Omega(\gamma) + \tfrac{1}{2}\rho_a U\frac{dU}{dt}\cos\gamma + \frac{\rho_a U}{2\beta_a}\int_0^{\gamma}\frac{\partial(U\Omega)}{\partial t}\sin\gamma\,d\gamma. \tag{18}$$

NON-UNIFORM PERTURBATIONS

9.3 The strength of the vortex sheet

In steady flow there is no vortex sheet, i.e. $\varpi = 0$, and (5) reduces to $\theta = \theta_s$. Hence (12), (15) and (16) simplify to

$$\int_{-\pi}^{\pi} \theta_s(\gamma)\,d\gamma = \int_{-\pi}^{\pi} \theta_s(\gamma)\cos\gamma\,d\gamma = \int_{-\pi}^{\pi} \theta_s(\gamma)\sin\gamma\,d\gamma = 0. \tag{19}$$

Consequently if we substitute (5) into the unsteady forms of (12), (15) and (16) and make use of (8) to eliminate ϖ the resulting equations can be written

$$\lambda U = 2\alpha U - \frac{1}{\pi}\int_{-\pi}^{\pi} n(\gamma,t)\,d\gamma - \frac{1}{\pi}\int_{0}^{\infty} \chi(\eta,t)\,d\eta, \tag{20}$$

$$0 = 2\alpha U - \frac{1}{\pi}\int_{-\pi}^{\pi} n(\gamma,t)(1-\cos\gamma)\,d\gamma - \frac{1}{\pi}\int_{0}^{\infty} \chi(\eta,t)(1+\cosh\eta)\,d\eta, \tag{21}$$

and

$$0 = \int_{-\pi}^{\pi} n(\gamma,t)\sin\gamma\,d\gamma. \tag{22}$$

The first of these fixes the value of λ, the second is an integral equation for the vortex sheet strength χ, while the third is a (physical) restriction on the unsteady perturbations.

In addition to (21) χ also satisfies the functional relation given in (10), which may be written

$$\chi(\xi,t) = \chi(1,t+1-\xi), \tag{23}$$

where ξ is related to the distance x measured along the sheet, and to the variable η by

$$\xi = \frac{1}{2} + \frac{x}{c} = \tfrac{1}{2}(1+\cosh\eta), \tag{24}$$

as is easily verified by putting $\zeta = \pi + i\eta$ (see Fig. 9.1) in (3). Equations (23) and (24) permit us to write (20) and (21) as

$$\lambda U = 2\alpha U - 2a_0 - \frac{1}{\pi}\int_{1}^{\infty} \chi(1,t+1-\xi)\frac{d\xi}{\{\xi(\xi-1)\}^{\frac{1}{2}}}, \tag{25}$$

$$0 = 2\alpha U - 2a_0 + 2a_1 - \frac{1}{\pi}\int_{1}^{\infty} \chi(1,t+1-\xi)\frac{2\xi\,d\xi}{\{\xi(\xi-1)\}^{\frac{1}{2}}}, \tag{26}$$

where

$$a_r = \frac{1}{2\pi}\int_{-\pi}^{\pi} n(\gamma,t)\cos r\gamma\,d\gamma \quad (r = 0, 1, 2, \ldots). \tag{27}$$

The end of the wake is at $x = \tfrac{1}{2}c + ct$ (see Fig. 9.1), i.e. at $\xi = 1+t$, and consequently $\chi = 0$ if $1+t \leqslant \xi < \infty$. This fact and the further change of variable

$$\sigma = t+1-\xi \tag{28}$$

enables us to write (25) and (26) as

$$\lambda U = 2\alpha U - 2a_0 - \frac{1}{\pi}\int_0^{\jmath} \frac{\chi(1,\sigma)\,d\sigma}{\{(\jmath+1-\sigma)(\jmath-\sigma)\}^{\frac{1}{2}}}, \tag{29}$$

$$0 = \alpha U - a_0 + a_1 - \frac{1}{\pi}\int_0^t \chi(1,\sigma)\left\{\frac{\jmath+1-\sigma}{\jmath-\sigma}\right\}^{\frac{1}{2}}d\sigma. \tag{30}$$

Equation (30) is an integral equation for the strength of the vortex sheet at the aerofoil's trailing edge, and as it has a difference kernel we can use the Laplace transform method of solution (see § 4.23).

By making use of the unit function $U(x)$ we are able to write (30) in the form

$$0 = A(\jmath) - \frac{1}{\pi}\int_{-\infty}^{\infty} [U(\sigma)\,\chi(1,\sigma)]\left[U(\jmath-\sigma)\left\{\frac{\jmath-\sigma+1}{\jmath-\sigma}\right\}^{\frac{1}{2}}\right]d\sigma, \tag{31}$$

where

$$A(\jmath) \equiv \begin{cases} \alpha(\jmath)U(\jmath) - a_0(\jmath) + a_1(\jmath) & (\jmath>0), \\ 0 & (\jmath<0). \end{cases} \tag{32}$$

By the composition product theorem, (4.199), (31) can be transformed into

$$\mathscr{L}\{A(\jmath);p\} = \frac{1}{\pi p}\mathscr{L}\{U(\jmath)\,\chi(1,\jmath);p\}\,\mathscr{L}\left\{U(\jmath)\left(\frac{\jmath+1}{\jmath}\right)^{\frac{1}{2}};p\right\}. \tag{33}$$

Van der Pol & Bremmer (1950) give the transform

$$\mathscr{L}\{U(t)\,(t^2-1)^{\nu-\frac{1}{2}};p\} = \frac{2^{\nu}\Gamma(\nu+\frac{1}{2})}{\sqrt{\pi}}\frac{K_{\nu}(p)}{p^{\nu-1}} \quad (0<\mathscr{R}p<\infty,\ \mathscr{R}\nu>-\tfrac{1}{2}),$$
$$\tag{34}$$

where $\Gamma(\nu)$ is the gamma function $(\Gamma(\nu+1)=\nu\Gamma(\nu),\ \Gamma(\tfrac{1}{2})=\sqrt{\pi})$ and $K_{\nu}(p)$ is the cylinder function defined by

$$K_{\nu}(p) = \int_0^{\infty} e^{-p\cosh x}\cosh\nu x\,dx \quad (\mathscr{R}p>0).$$

Rules (4.195) and (4.196) enable us to transform (34) into

$$\mathscr{L}\,[U(t)\,\{t(t+1)\}^{\nu-\frac{1}{2}};p] = \frac{\Gamma(\nu+\frac{1}{2})}{\sqrt{\pi}}\,e^{\frac{1}{2}p}\frac{K_{\nu}(\frac{1}{2}p)}{p^{\nu-1}},$$

of which we shall need the special cases

$$\mathscr{L}\,[U(t)\,\{t(t+1)\}^{\frac{1}{2}};p] = \tfrac{1}{2}e^{\frac{1}{2}p}K_1(\tfrac{1}{2}p),$$

$$\mathscr{L}\,[U(t)\,\{t(t+1)\}^{-\frac{1}{2}};p] = p\,e^{\frac{1}{2}p}K_0(\tfrac{1}{2}p).$$

Now

$$\left(\frac{t+1}{t}\right)^{\frac{1}{2}} = \tfrac{1}{2}\{t(t+1)\}^{-\frac{1}{2}} + \frac{d}{dt}\{t(t+1)\}^{\frac{1}{2}}$$

and therefore

$$\mathscr{L}\left\{U(t)\left(\frac{t+1}{t}\right)^{\frac{1}{2}}; p\right\} = \tfrac{1}{2}p\,e^{\frac{1}{2}p}\{K_0(\tfrac{1}{2}p) + K_1(\tfrac{1}{2}p)\}, \tag{35}$$

by an application of (4.198). All these transforms are subject to the restriction $\mathscr{R}p > 0$.

Applying this result to (33) we get

$$\mathscr{L}(A) = \frac{1}{2\pi}\mathscr{L}(\chi)\{K_0(\tfrac{1}{2}p) + K_1(\tfrac{1}{2}p)\}e^{\frac{1}{2}p},$$

using an abbreviated transform notation. Thus

$$\mathscr{L}(\chi) = \frac{2\pi\mathscr{L}(A)\,e^{-\frac{1}{2}p}}{K_0(\tfrac{1}{2}p) + K_1(\tfrac{1}{2}p)} \quad (\mathscr{R}p > 0). \tag{36}$$

Similarly the transform of (29) yields

$$\mathscr{L}(\lambda U) = 2\mathscr{L}(A) - 2\mathscr{L}(a_1) - \frac{1}{\pi}\mathscr{L}(\chi)\,K_0(\tfrac{1}{2}p)\,e^{\frac{1}{2}p},$$

or, by (36),

$$\mathscr{L}(\lambda U) = \frac{2\mathscr{L}(A)\,K_1(\tfrac{1}{2}p)}{K_0(\tfrac{1}{2}p) + K_1(\tfrac{1}{2}p)} - 2\mathscr{L}(a_1) \quad (\mathscr{R}p > 0). \tag{37}$$

This result suggests that a function of some importance in the theory will be that defined by

$$C(z) \equiv \frac{K_1(iz)}{K_1(iz) + K_0(iz)} = \frac{H_1^{(2)}(z)}{H_1^{(2)}(z) + iH_0^{(2)}(z)}, \tag{38}$$

the second form of which follows from the relation (Watson, 1952, p. 78)

$$K_n(z) = -\tfrac{1}{2}i\pi\,e^{-\frac{1}{2}ni\pi}\,H_n^{(2)}(-iz), \tag{39}$$

between the K-functions and the Hankel functions. The function $C(z)$ is known as Theodorsen's function† (Theodorsen, 1935). A brief tabulation of $C(\omega) \equiv A(\omega) - iB(\omega)$ for real argument ω is given in Table 9.3a.

Table 9.3a. $C(\omega) = A - iB$

ω	0·00	0·01	0·02	0·05	0·1	0·2	0·3
A	1·0000	0·9824	0·9637	0·9090	0·8319	0·7276	0·6650
B	∞	2·7826	1·8802	1·3064	0·8615	0·4716	0·3989

ω	0·5	1	2	3	5	∞
A	0·5979	0·5394	0·5130	0·5063	0·5024	0·5000
B	0·1507	0·0501	0·0144	0·0067	0·0025	0·0000

We are now able to write (37) in the form

$$\mathscr{L}\{\lambda(t)\,U(t); p\} = \mathscr{L}\{A(t); p\}\,\mathscr{L}\{k(t); p\} - 2\mathscr{L}\{a_1(t); p\}, \tag{40}$$

† The theory can be equally well developed in terms of Küssner's function, $T(z)$ (Küssner, 1940), which is related to Theodorsen's function by $T = 2C - 1$.

where
$$\mathscr{L}\{k(\mathit{t}); p\} = 2C(-\tfrac{1}{2}ip) \quad (\mathscr{R}p > 0), \tag{41}$$

i.e. from (4.194)
$$k(\mathit{t}) = \mathbf{U}(\mathit{t})\frac{1}{\pi i}\int_{c_0-i\infty}^{c_0+i\infty} C(-\tfrac{1}{2}ip)\, e^{p\mathit{t}}\frac{dp}{p}, \tag{42}$$

replacing c by c_0 to avoid confusion with the chord length c, and making k vanish in the steady motion, $\mathit{t} < 0$. With the aid of the rules given in (4.198) and (4.199) we can write (40) as

$$\mathscr{L}\{\lambda(\mathit{t})\, U(\mathit{t}); p\} = \frac{1}{p}\mathscr{L}\left\{\frac{d}{d\mathit{t}}[A(\mathit{t})]; p\right\}\mathscr{L}\{k(\mathit{t}); p\} - 2\mathscr{L}\{a_1(\mathit{t}); p\}$$

$$= \mathscr{L}\left\{\int_{-\infty}^{\infty} \dot{A}(\mathit{t}^*)\, k(\mathit{t}-\mathit{t}^*)\, d\mathit{t}^*; p\right\} - 2\mathscr{L}\{a_1(\mathit{t}); p\},$$

where the dot denotes a derivative with respect to t. Hence

$$\lambda(\mathit{t})\, U(\mathit{t}) = \int_0^{\mathit{t}} \dot{A}(\mathit{t}^*)\, k(\mathit{t}-\mathit{t}^*)\, d\mathit{t}^* - 2a_1(\mathit{t}), \tag{43}$$

on taking advantage of the fact that both A and k vanish for negative values of their arguments. It should be noted that the integral in (43) must include any discontinuity in A at $\mathit{t} = 0$, e.g. if $A(\mathit{t}^*) = b\mathbf{U}(\mathit{t}^*)$, $\dot{A}(\mathit{t}^*) = b\delta(\mathit{t}^*)$, and (43) gives $\lambda(\mathit{t})\, U(\mathit{t}) = bk(\mathit{t}) - 2a_1(\mathit{t})$.

The function $k(\mathit{t})$ is known as the *Wagner*, or '*growth of lift*' function. It has been tabulated by Söhngen (1940), Küssner (1940) and others. Küssner gives the expansion

$$k(\mathit{t}) = 1 + \frac{\mathit{t}}{(2\mathit{t}+2)} + \frac{\mathit{t}^2}{(2\mathit{t}+2)^2} + O(\mathit{t}^3). \tag{44}$$

The following brief table will give the reader some idea of this important function.

Table 9.3b. $k(\mathit{t})$

t	0	1	2	3	4	5	∞
$k(\mathit{t})$	1·000	1·339	1·516	1·625	1·698	1·750	2·000

On applying a calculation to (36) similar to that given for (40), we find that the strength of the vortex sheet is

$$\chi(1,\mathit{t}) = \int_0^{\mathit{t}} \dot{A}(\mathit{t}^*)\, k_0(\mathit{t}-\mathit{t}^*)\, d\mathit{t}^*, \tag{45}$$

where
$$k_0(\mathit{t}) \equiv -i\mathbf{U}(\mathit{t})\int_{c_0-i\infty}^{c_0+i\infty} \frac{e^{p(\mathit{t}-\frac{1}{2})}}{K_0(\tfrac{1}{2}p) + K_1(\tfrac{1}{2}p)}\frac{dp}{p}. \tag{46}$$

Further progress in the calculation of the strength of the vortex sheet requires the specification of the type of unsteady motion; we shall return to this in §§ 9.6 and 9.8.

9.4　The pressure distribution on the aerofoil

Equations (5), (8) and the real part of (11) yield that

$$U\Omega(\gamma,t) = U\Omega_s(\gamma) - \tfrac{1}{2}\lambda U \cos\tfrac{1}{2}\gamma + \frac{1}{2\pi}\int_{-\pi}^{\pi} n(\gamma^*,t)\cot\tfrac{1}{2}(\gamma^*-\gamma)\,d\gamma^*$$

$$+ \frac{\sin\gamma}{2\pi}\int_0^\infty \frac{\chi(\eta,t)\,d\eta}{\cosh\eta+\cos\gamma}, \qquad (47)$$

where
$$\Omega_s \equiv \frac{1}{2\pi}\int_{-\pi}^{\pi}\theta_s\cot\tfrac{1}{2}(\gamma^*-\gamma)\,d\gamma^*$$

is the value Ω had in the steady motion, $t < 0$. Equation (18) can be written

$$p(\gamma,t) = p_s(\gamma) + G(t) + \frac{1}{\beta_a}\rho_a U^2(\Omega-\Omega_s) + \tfrac{1}{2}\rho_a U\frac{dU}{dt}\cos\gamma$$

$$+ \frac{\rho_a U}{2\beta_a}\int_0^\gamma \frac{\partial}{\partial t}(U\Omega - U\Omega_s)\sin\gamma\,d\gamma, \qquad (48)$$

on ignoring a small term $\Omega_s(dU/dt)$. The required equation for the pressure on the aerofoil now follows on using (47) to eliminate $(\Omega-\Omega_s)$ from (48). We shall now deduce a remarkably simple form of this pressure equation in which the vortex sheet strength χ does not appear.

As
$$\frac{\sin\gamma}{\cosh\eta+\cos\gamma} = \frac{\cosh\eta-\cos\gamma}{\sin\gamma} - \frac{\sinh^2\eta}{\sin\gamma(\cosh\eta+\cos\gamma)}$$

the last term of (47) can be written

$$\frac{\sin\gamma}{2\pi}\int_0^\infty \frac{\chi\,d\eta}{\cosh\eta+\cos\gamma} = (\tfrac{1}{2}\lambda U + a_1)\operatorname{cosec}\gamma + (\tfrac{1}{2}\lambda U + a_0 - \alpha U)\cot\gamma$$

$$- \frac{1}{2\pi\sin\gamma}\int_0^\infty \frac{\chi\sinh^2\eta\,d\eta}{\cosh\eta+\cos\gamma},$$

on taking advantage of equations (20), (21) and (27). Consequently

$$\sin\gamma\frac{\partial}{\partial t}(U\Omega - U\Omega_s) = \dot a_1 + \{\dot a_0 - (\dot{\alpha U})\cos\gamma\} + \frac{\sin\gamma}{2\pi}\int_{-\pi}^{\pi}\dot n\cot\tfrac{1}{2}(\gamma^*-\gamma)\,d\gamma^*$$

$$- \frac{1}{2\pi}\int_0^\infty \frac{\dot\chi\sinh^2\eta\,d\eta}{\cosh\eta+\cos\gamma}. \qquad (49)$$

From (6) and the fact that $x = \tfrac{1}{2}c\cosh\eta$ on the vortex sheet,

$$\dot\chi = -c\frac{\partial\chi}{\partial x} = -2\operatorname{cosech}\eta\frac{\partial\chi}{\partial\eta}, \qquad (50)$$

and so the last term of (49) can be written

$$\frac{1}{\pi}\int_0^\infty \frac{\partial\chi}{\partial\eta}\frac{\sinh\eta\,d\eta}{\cosh\eta+\cos\gamma} = -\frac{1}{\pi}\int_0^\infty \chi\frac{\partial}{\partial\eta}\left(\frac{\sinh\eta}{\cosh\eta+\cos\gamma}\right)d\eta$$

$$= -\frac{1}{\pi}\int_0^\infty \chi\frac{\partial}{\partial\gamma}\left(\frac{\sin\gamma}{\cosh\eta+\cos\gamma}\right)d\eta,$$

the second form of which follows by an integration by parts and $\chi(\infty)=0$, and the third from the easily established identity of the derivatives in the integrands.

Now
$$\sin\gamma\cot\tfrac{1}{2}(\gamma^*-\gamma) = \sin\gamma^*\cot\tfrac{1}{2}(\gamma^*-\gamma)-\cos\gamma^*-\cos\gamma,$$

thus the first integral in (49) can be written

$$\frac{1}{2\pi}\int_{-\pi}^\pi \dot{n}(\gamma^*,t)\sin\gamma^*\cot\tfrac{1}{2}(\gamma^*-\gamma)\,d\gamma^* - (\dot{a}_1+\dot{a}_0\cos\gamma),$$

by (27). Let
$$N(\gamma,t) \equiv \frac{1}{2}\int_0^\gamma n(\gamma,t)\sin\gamma\,d\gamma, \tag{51}$$

then from (22)
$$N(\pi,t) = N(-\pi,t).$$

Therefore an integration by parts gives

$$\frac{1}{2\pi}\int_{-\pi}^\pi \dot{n}(\gamma^*,t)\sin\gamma^*\cot\tfrac{1}{2}(\gamma^*-\gamma)\,d\gamma^*$$

$$= -\frac{1}{\pi}\int_{-\pi}^\pi \dot{N}(\gamma^*,t)\frac{\partial}{\partial\gamma^*}\{\cot\tfrac{1}{2}(\gamma^*-\gamma)\}\,d\gamma^*$$

$$= \frac{1}{\pi}\int_{-\pi}^\pi \dot{N}(\gamma^*,t)\frac{\partial}{\partial\gamma}\{\cot\tfrac{1}{2}(\gamma^*-\gamma)\}\,d\gamma^*.$$

We are now able to integrate (49) and simplify the result to find

$$\int_0^\gamma \sin\gamma\frac{\partial}{\partial t}(U\Omega-U\Omega_s)\,d\gamma = -(\dot{\alpha}U)\sin\gamma+B(t)$$

$$+\frac{1}{\pi}\int_{-\pi}^\pi \dot{N}\cot\tfrac{1}{2}(\gamma^*-\gamma)\,d\gamma^* - \frac{\sin\gamma}{\pi}\int_0^\infty \frac{\chi\,d\eta}{\cosh\eta+\cos\gamma}, \tag{52}$$

where $B(t)$ is independent of γ. The final step is to substitute this result and (47) into (48) to find

$$p(\gamma,t) = p_s(\gamma)+C(t)+\frac{\rho_a U}{2\beta_a}\{\beta_a \dot{U}\cos\gamma-(\dot{\alpha}U)\sin\gamma-\lambda U\cot\tfrac{1}{2}\gamma\}$$

$$+\frac{\rho_a U}{2\beta_a\pi}\int_{-\pi}^\pi \{n(\gamma^*,t)+\dot{N}(\gamma^*,t)\}\cot\tfrac{1}{2}(\gamma^*-\gamma)\,d\gamma^*,$$

where λU is given by (43). This is the simple equation for the pressure mentioned above.

9.5   The lift and moment

The lift, L, and moment about the mid-chord point, M, are given by

$$L = L_s - \int_{-\pi}^{\pi} (p - p_s)\, dx(\gamma) = L_s - \tfrac{1}{2} c \int_{-\pi}^{\pi} (p - p_s) \sin \gamma \, d\gamma, \qquad (53)$$

and

$$M = M_s + \int_{-\pi}^{\pi} (p - p_s)\, x \, dx(\gamma) = M_s - \frac{c^2}{8} \int_{-\pi}^{\pi} (p - p_s) \sin 2\gamma \, d\gamma, \qquad (54)$$

where L_s, M_s are their values in the steady flow. Substituting (52) in (53) and (54) we obtain

$$L - L_s = \frac{\pi \rho_a c U}{4\beta_a} \{(\alpha \dot{U}) + 2\lambda U + 4a_1\} + \frac{\rho_a c U}{2\beta_a} \int_{-\pi}^{\pi} \dot{N}(\gamma^*, t) \cos \gamma^* \, d\gamma^*, \qquad (55)$$

$$M - M_s = \frac{\pi \rho_a c^2 U}{16\beta_a} \{2\lambda U + 4a_2\} + \frac{\rho_a c^2 U}{8\beta_a} \int_{-\pi}^{\pi} \dot{N}(\gamma^*, t) \cos 2\gamma^* \, d\gamma^*, \qquad (56)$$

where we have made use of (27).

These results, together with (27), (31), (43) and (51), enable us to express L and M completely in terms of the known functions $U(t)$ and $n(\gamma, t)$. Further progress requires that U and n be specified, i.e. that the details of the unsteady motion be defined.

HARMONIC OSCILLATIONS

9.6   A rigid aerofoil

We shall now study some special cases of the general theory developed above, and the first of these will be the harmonic oscillations of a rigid aerofoil about a mean motion which is steady. In this case the flapping and pitching displacements can be expressed

$$y = y_0 e^{i\nu t}, \quad \alpha = \alpha_0 e^{i\nu t}, \qquad (57)$$

where $\nu/2\pi$ is the frequency of the oscillations, and y_0, α_0 are their amplitudes. Now U is constant and from (9) the reduced time is $t = Ut/c$, and hence if (57) are replaced by

$$y = y_0 e^{2i\omega t}, \quad \alpha = \alpha_0 e^{2i\omega t}, \qquad (58)$$

then

$$\omega \equiv \frac{c\nu}{2U} \qquad (59)$$

will be termed the 'reduced' frequency of the oscillation.

As any combined flapping and pitching motion can be analysed into a transverse motion of the mid-chord point plus a pitching moment

about this point, there is no loss in generality in writing the 'upwash'
velocity n as (cf. Fig. 9.6)

$$n = v - x\frac{d\alpha}{dt},$$

where
$$v = \frac{dy}{dt} = iv y_0\, e^{ivt} = i\frac{2U}{c}\,\omega y_0\, e^{2i\omega t}, \tag{60}$$

and
$$\frac{d\alpha}{dt} = iv\alpha_0\, e^{ivt} = i\frac{2U}{c}\,\omega\alpha_0\, e^{2i\omega t}. \tag{61}$$

Therefore
$$n = i\frac{2U}{c}\,\omega(y_0 + \tfrac{1}{2}c\alpha_0\cos\gamma)\,e^{2i\omega t}, \tag{62}$$

and
$$\dot{n} = -\frac{4U}{c}\,\omega^2(y_0 + \tfrac{1}{2}c\alpha_0\cos\gamma)\,e^{2i\omega t} \tag{63}$$

(the dots denoted derivatives with respect to $\mathfrak{t}$ and *not* t).

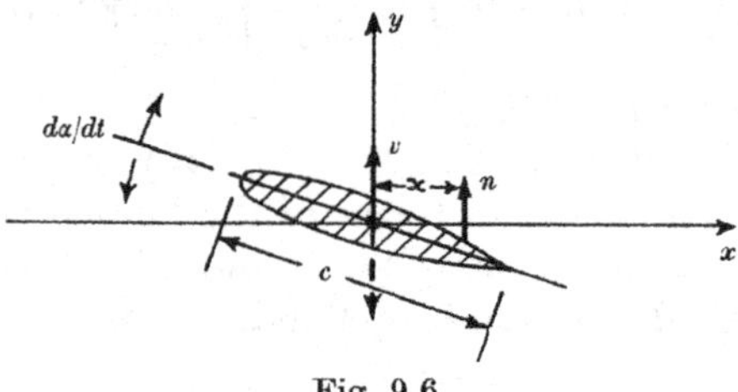

Fig. 9.6

Equations (58), (62) and the fact that in the present application $\dot{U} = 0$,
enables us to evaluate the functions defined in (27), (32) and (51). The
results are

$$a_0 = 2iU\omega\frac{y_0}{c}\,e^{2i\omega t}, \quad a_1 = \frac{i}{2}U\omega\alpha_0\, e^{2i\omega t}, \quad a_r = 0 \quad (r \geqslant 2),$$

$$A(\mathfrak{t}) = U\left\{\alpha_0(1 + \tfrac{1}{2}i\omega) - 2i\omega\frac{y_0}{c}\right\}e^{2i\omega t},$$

and
$$N(\gamma,\mathfrak{t}) = iU\omega\left[\tfrac{1}{8}\alpha_0(1 - \cos 2\gamma) + \frac{y_0}{c}(1 - \cos\gamma)\right]e^{2i\omega t}.$$

Substituting these values in (43) we obtain

$$\lambda = 2i\omega\left\{\alpha_0(1 + \tfrac{1}{2}i\omega) - 2i\omega\frac{y_0}{c}\right\}\int_0^t e^{2i\omega \mathfrak{t}^*}k(\mathfrak{t} - \mathfrak{t}^*)\,d\mathfrak{t}^* - i\omega\alpha_0\, e^{2i\omega t},$$

which can be written

$$\lambda = \left[2\left\{\alpha_0(1 + \tfrac{1}{2}i\omega) - 2i\omega\frac{y_0}{c}\right\}C(\omega,\mathfrak{t}) - i\omega\alpha_0\right]e^{2i\omega t},$$

where
$$C(\omega,\mathfrak{t}) \equiv i\omega\int_0^{\mathfrak{t}} e^{-2i\omega s}k(s)\,ds. \tag{64}$$

Equations (55) and (56) yield

$$L - L_s = \frac{\pi}{\beta_a}\rho_a c U^2 e^{i\nu t}\left[\left\{(1+\tfrac{1}{2}i\omega)\,C(\omega,f)+\tfrac{1}{2}i\omega\right\}\alpha_0\right.$$
$$\left. +\left\{\omega^2 - 2i\omega C(\omega,f)\right\}\frac{y_0}{c}\right], \quad (65)$$

$$M - M_s = \frac{\pi}{4\beta_a}\rho_a c^2 U^2 e^{i\nu t}\left[\left\{(1+\tfrac{1}{2}i\omega)\,C(\omega,f)-\tfrac{1}{2}i\omega+\tfrac{1}{8}\omega^2\right\}\alpha_0\right.$$
$$\left. +\left\{-2i\omega C(\omega,f)\right\}\frac{y_0}{c}\right], \quad (66)$$

as $\nu t = 2\omega f$.

The 'air-load coefficients' l_{12}, m_{12}, l_{34}, m_{34} are defined by

$$\left.\begin{aligned}
L - L_s &= \frac{\pi}{\beta_a}\rho_a c U^2 e^{i\nu t}\left\{l_{12}\frac{y_0}{c}+l_{34}\alpha_0\right\}, \\[2mm]
M - M_s &= \frac{\pi}{4\beta_a}\rho_a c^2 U^2 e^{i\nu t}\left\{m_{12}\frac{y_0}{c}+m_{34}\alpha_0\right\},
\end{aligned}\right\} \quad (67)$$

and hence

$$\left.\begin{aligned}
l_{12} &= (1+\tfrac{1}{2}i\omega)\,C(\omega,f)+\tfrac{1}{2}i\omega, & l_{34} &= \omega^2 - 2i\omega C(\omega,f), \\[2mm]
m_{12} &= (1+\tfrac{1}{2}i\omega)\,C(\omega,f)-\tfrac{1}{2}i\omega+\tfrac{1}{8}\omega^2, & m_{34} &= -2i\omega C(\omega,f).
\end{aligned}\right\} \quad (68)$$

It should be remembered that these coefficients apply to motion about an axis of rotation passing through the aerofoil's mid-chord point. By transforming the moments and displacements to an axis at $x = \tilde{x}$, it is found that these coefficients transform according to

$$\left.\begin{aligned}
\tilde{l}_{12} &= l_{12}, & \tilde{l}_{34} &= l_{34}+\frac{\tilde{x}}{c}l_{12}, \\[2mm]
\tilde{m}_{12} &= m_{12}+\frac{\tilde{x}}{c}l_{12}, & \tilde{m}_{34} &= m_{34}+\frac{\tilde{x}}{c}(l_{34}+m_{12})+\left(\frac{\tilde{x}}{c}\right)^2 l_{12}.
\end{aligned}\right\} \quad (69)$$

If ω is replaced by $-i\mu$ the oscillations are changed into the divergent motion

$$y = y_0 e^{2\mu f}, \quad \alpha = \alpha_0 e^{2\mu f},$$

but the theory is unchanged in form. More generally we can replace ω by $\omega - i\mu$ and obtain the divergent oscillation

$$y = y_0 e^{2i\omega f}e^{2\mu f}, \quad \alpha = \alpha_0 e^{2i\omega f}e^{2\mu f} \quad (\mu > 0), \quad (70)$$

from which more general displacements can be obtained by superposition.

9.7   Oscillations of long duration

Let the oscillation be slightly divergent then (64) becomes

$$2C(\omega - i\mu, t) = 2(\mu + i\omega)\int_0^t e^{-2(\mu+i\omega)s}\, k(s)\, ds,$$

where μ is small. If the oscillation is of very long duration we can set $t = \infty$ in this equation, and then the right side becomes the Laplace transform of $k(s)$. Hence from (41)

$$C(\omega - i\mu, \infty) = C(\omega - i\mu), \tag{71}$$

the condition $\mathcal{R}p > 0$ being met by having $\mu > 0$.

We now take the limit $\mu \to 0$ and get

$$\lim_{\mu \to 0} C(\omega - i\mu, \infty) = C(\omega) = \frac{H_1^{(2)}(\omega)}{H_1^{(2)}(\omega) + iH_0^{(2)}(\omega)}, \tag{72}$$

from (38). This limit is finite in spite of the restriction on the argument of the K-functions.

Thus the theory for oscillations of long duration is obtained by replacing the 'incomplete Theodorsen function' $C(\omega, t)$ by the 'complete' function $C(\omega)$.

9.8   Oscillations of hinged flaps

The theory for the oscillation of a hinged flap of length Ec is obtained by putting (see Fig. 9.8)

$$n = -\{x - (\tfrac{1}{2} - E)\,c\}\,\{i\nu\delta_T\, e^{i\nu t}\}\,\mathrm{U}(x - \tfrac{1}{2}c + Ec),$$

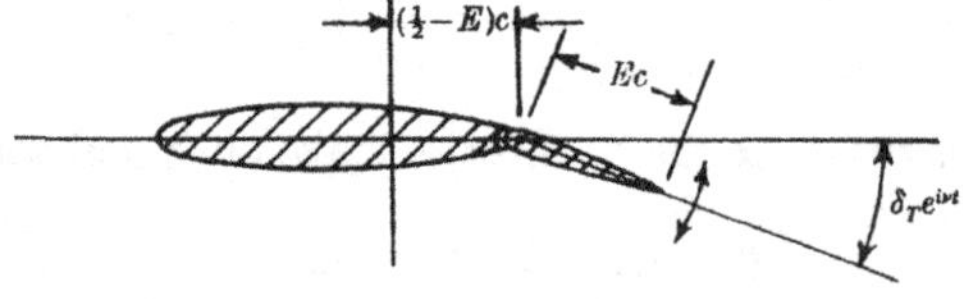

Fig. 9.8

in the appropriate equations of §§ 9.3 − 9.5. We shall not pursue the algebra further—it is straightforward but rather complicated—the reader may consult a paper by Minhinnick (1950), where further references are given.

GUSTS

9.9 Sharp-edged upgusts

It is of some practical importance to be able to calculate the forces
acting on an aerofoil which enters a so-called sharp-edged upgust. This is
an idealized flow in which the fluid upstream of a given surface S possesses
a uniform upwards velocity, v say, with respect to the fluid downstream
of S; S moves downstream with the main-stream velocity U, so the aerofoil passes smoothly across S until it is completely in the gust. The problem is to calculate the transient forces in this period.

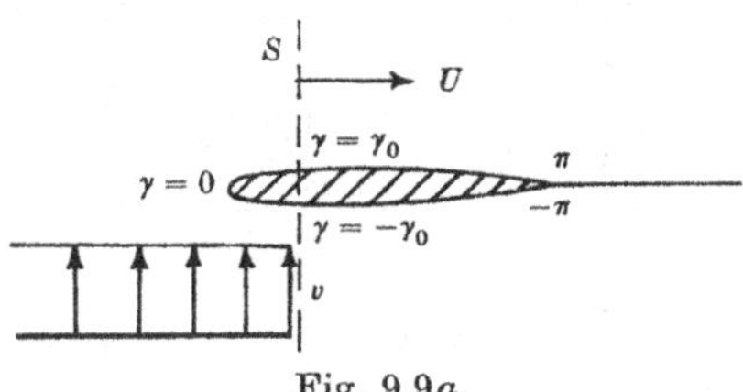

Fig. 9.9a

We shall assume $v \ll U$, so that the small perturbation theory developed above is applicable; it is also assumed that any vorticity the gust
may possess has a negligible effect on the flow. Let the aerofoil nose
($\gamma = 0$) enter the gust at $t = 0$, then at a time t a length $Ut = ct$ will have
entered the gust, and this length of the aerofoil behaves as if it has
a downwards velocity of $n = -v$ relative to a flow along the x-axis.
Consequently we can write

$$n = -v \ (-\gamma_0 < \gamma < \gamma_0), \quad n = 0 \ (-\pi < \gamma < -\gamma_0, \gamma_0 < \gamma < \pi), \quad (73)$$

where $Ut = \tfrac{1}{2}c(1 - \cos \gamma_0)$, or

$$\gamma_0 = \cos^{-1}(1 - 2t) \ (t < 1), \qquad \gamma_0 = \pi \ (t > 1). \tag{74}$$

On substituting (73) in (27), (32) and (51) we obtain

$$a_0 = -\frac{v}{\pi}\gamma_0, \quad a_1 = -\frac{v}{\pi}\sin \gamma_0, \quad a_2 = -\frac{v}{2\pi}\sin 2\gamma_0, \quad A = \frac{v}{\pi}(\gamma_0 - \sin \gamma_0),$$

$$\int_{-\pi}^{\pi} N \cos \gamma \, d\gamma = \tfrac{1}{2}v(\gamma_0 - \tfrac{1}{2}\sin 2\gamma_0), \qquad \int_{-\pi}^{\pi} N \cos 2\gamma \, d\gamma = \tfrac{1}{3}v \sin^3 \gamma_0.$$

By (74)

$$\dot{A} = \frac{v}{\pi}(1 - \cos \gamma_0)\dot{\gamma}_0 = \frac{2v}{\pi}\left(\frac{t}{1-t}\right)^{\tfrac{1}{2}} \ (t < 1), \qquad \dot{A} = 0 \ (t > 1).$$

Similarly

$$\int_{-\pi}^{\pi} \dot{N} \cos\gamma\, d\gamma = \begin{cases} 4v\{f(1-f)\}^{\frac{1}{2}} & (f < 1), \\ 0 & (f > 1), \end{cases}$$

$$\int_{-\pi}^{\pi} \dot{N} \cos 2\gamma\, d\gamma = \begin{cases} 4v\{f(1-f)\}^{\frac{1}{2}}(1-2f) & (f < 1), \\ 0 & (f > 1). \end{cases}$$

All these values vanish for $f < 0$, i.e. before the aerofoil enters the gust.

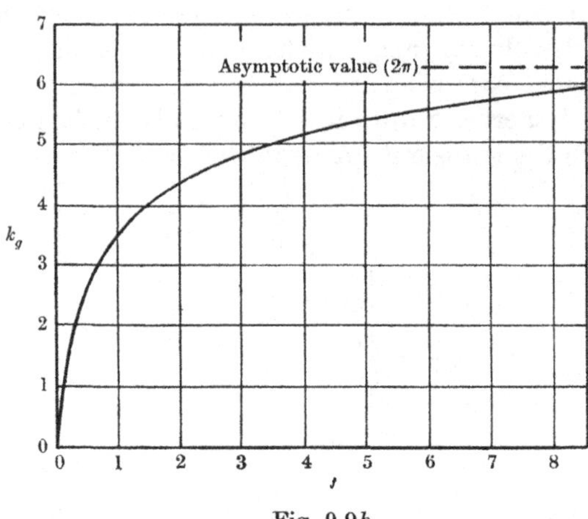

Fig. 9.9b

Substituting these values in (55) and (56) we arrive at

$$L - L_s = \frac{1}{2\beta_a}\rho_a c U v\, k_g(f), \tag{75}$$

$$M - M_s = \frac{1}{8\beta_a}\rho_a c^2 U v\, k_g(f), \tag{76}$$

where

$$k_g(f) \equiv \begin{cases} 0 & (f < 0), \\ 2\int_0^f \left(\frac{f^*}{1-f^*}\right)^{\frac{1}{2}} k(f-f^*)\, df^* + 4\{f(1-f)\}^{\frac{1}{2}} & (0 < f < 1), \\ 2\int_0^1 \left(\frac{f^*}{1-f^*}\right)^{\frac{1}{2}} k(f-f^*)\, df^* & (f > 1). \end{cases} \tag{77}$$

When

$$f \gg 1, \quad k_g(f) = k(f) \times 2\int_0^1 \left(\frac{f^*}{1-f^*}\right)^{\frac{1}{2}} df^* = \pi k(f),$$

and therefore $\lim_{f\to\infty} k_g(f) = 2\pi$ (see Table 9.3b). A graph of $k_g(f)$ is shown in Fig. 9.9.b.

It will be noticed that in the limit $f \to \infty$, (75) and (76) yield changes of lift and moment appropriate to an increase of incidence of v/U, as would be expected from steady aerofoil theory.

9.10 Graded gusts

The theory for sharp-edged gusts is easily extended to gusts of a general type by an application of Duhamel's theorem, but we shall derive the general theory from a simple argument based on (75).

Let $dv(\mathcal{I}^*)$ be an infinitesimal gust, incident on the leading edge at a reduced time $\mathcal{I}^*$, then from (75) the increment to the lift will be

$$dL = \frac{1}{2\beta_a}\rho_a cU\, k_g(\mathcal{I}-\mathcal{I}^*)\, dv(\mathcal{I}^*).$$

Summing over all the elemental gusts into which any smoothly graded gust may be resolved, we find

$$L - L_s = \frac{1}{2\beta_a}\rho_a cU \int_0^{\mathcal{I}} k_g(\mathcal{I}-\mathcal{I}^*)\, dv(\mathcal{I}^*), \tag{78}$$

where we have assumed the edge of the gust to reach the aerofoil at $\mathcal{I} = 0$, and have taken into account that k_g vanishes for negative values of its argument.

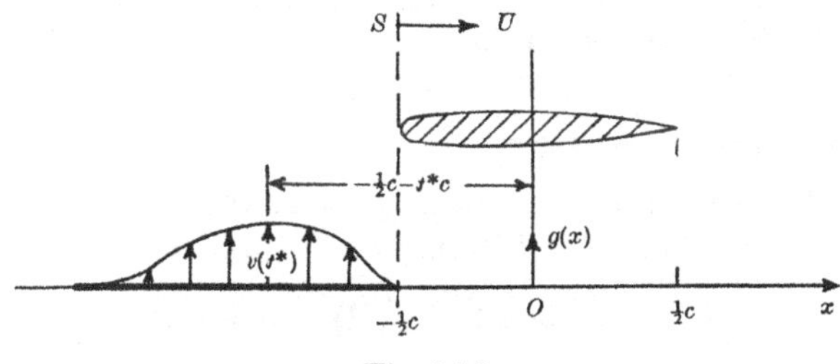

Fig. 9.10

Suppose the gust profile is $g(x)$ at the instant $\mathcal{I} = 0$, then clearly (see Fig. 9.10) $v(\mathcal{I}^*) = g(-\tfrac{1}{2}c-\mathcal{I}^*c)$. Substituting this value into (78), and changing the variable of integration we obtain

$$L - L_s = -\frac{1}{2\beta_a}\rho_a cU \int_{-\frac{1}{2}c-Ut}^{-\frac{1}{2}c} k_g\left(\frac{Ut}{c}+\frac{1}{2}+\frac{\xi}{c}\right) g'(\xi)\, d\xi. \tag{79}$$

The corresponding result for $M - M_s$ is $\tfrac{1}{4}c$ times the right-hand side of this equation.

ACCELERATED MOTION

9.11 General theory

Consider a rigid aerofoil placed at a fixed incidence α, and accelerating in an arbitrary fashion. In this case the normal velocity n vanishes; hence the a_r defined in (27) vanish, and (32) reduces to $A(t) = \alpha U(t)$. Equations (43), (55) and (56) then become

$$\lambda(t)\,U(t) = \alpha \int_0^t \dot{U}(t^*)\,k(t - t^*)\,dt^*, \tag{80}$$

$$L - L_s = \frac{\pi\rho_a cU}{4\beta_a}(\alpha\dot{U} + 2\lambda U), \tag{81}$$

and

$$M - M_s = \frac{\pi\rho_a c^2 U}{8\beta_a}\lambda U. \tag{82}$$

Suppose the aerofoil is suddenly given a velocity U_0 starting from rest at $t = 0$, then $U(t) = U_0 \mathsf{U}(t)$, $\dot{U}(t) = U_0\,\delta(t)$, and (80) and (81) yield

$$C_L = \frac{L}{\frac{1}{2}\rho_a cU_0^2} = \frac{\pi}{\beta_a}\alpha k(t). \tag{83}$$

A similar result holds for C_M. Notice from the table following (44) that at $t = +0$, $C_L = \pi\alpha/\beta_a$, which is half of the final value obtained at $t = \infty$. This lift deficiency just after the start of the motion is known as the Wagner effect (Wagner, 1925).

9.12 Uniform acceleration

The special case of constant acceleration is important. In this case we can write $U = \mu t$, where μ is a constant. Equation (9) gives $t = \mu t^2/2c$, whence

$$U(t) = (2c\mu t)^{\frac{1}{2}}, \quad \dot{U}(t) = (c\mu/2t)^{\frac{1}{2}}.$$

Equation (80) now gives

$$\lambda U = \alpha\left(\frac{c\mu}{2}\right)^{\frac{1}{2}}\int_0^t (t^*)^{-\frac{1}{2}}k(t - t^*)\,dt^*. \tag{84}$$

Substituting these values in (81) we arrive at

$$L = \frac{\pi\rho_a c^2\mu\alpha}{4\beta_a}\left\{1 + 2t^{\frac{1}{2}}\int_0^t (t^*)^{-\frac{1}{2}}k(t - t^*)\,dt^*\right\}. \tag{85}$$

An alternative form for the integral

$$I(t) \equiv \int_0^t (t^*)^{-\frac{1}{2}}k(t - t^*)\,dt^*, \tag{86}$$

can be found as follows. Its transform is $\mathscr{L}(I) = \dfrac{1}{p}\mathscr{L}(t^{-\frac{1}{2}})\mathscr{L}(k)$ by (4.199). Hence by the transform

$$\mathscr{L}\{t^\nu \mathsf{U}(t); p\} = \frac{\Gamma(\nu + 1)}{p}, \tag{87}$$

and by (41) we have $\mathscr{L}(I) = 2\pi^{\frac{1}{2}}p^{-\frac{1}{2}}C(-\frac{1}{2}ip)$, i.e. by (4.194),

$$I = \frac{1}{i\sqrt{\pi}} \int_{c_0-i\infty}^{c_0+i\infty} C(-\tfrac{1}{2}ip)\frac{e^{pJ}}{p^{\frac{3}{2}}}\,dp. \tag{88}$$

On page 124 of their text Van der Pol & Bremmer (1950) give the Abel theorem

$$\lim_{t\to\infty}\left(\frac{h(t)}{t^{\nu}}\right) = \lim_{p\to 0}\frac{p^{\nu}f(p)}{\Gamma(\nu+1)} \quad (\nu > -1), \tag{89}$$

where

$$\mathscr{L}\{h(t)\,\mathrm{U}(t);\,p\} \equiv f(p).$$

Now (88) is equivalent to $\mathscr{L}\{I(J)\,\mathrm{U}(J);\,p\} = 2\sqrt{(\pi)}\,p^{-\frac{1}{2}}C(-\frac{1}{2}ip)$, consequently an application of (89) yields

$$\lim_{J\to\infty}\frac{I(J)}{J^{\frac{1}{2}}} = \frac{2\sqrt{\pi}}{\Gamma(\frac{3}{2})}\lim_{p\to 0}C(-\tfrac{1}{2}ip) = 4C(0).$$

From Table 9.3a we note that $\mathscr{R}C(0) = 1$, therefore $\mathscr{R}I(J) \sim 4J^{\frac{1}{2}}$ for large J. Substituting this result in (85), and ignoring unity compared with J, we get

$$L = \frac{\pi\rho_a c}{\beta_a}(2c\mu J) = \frac{\pi\rho_a}{\beta_a}cU^2, \tag{90}$$

which is the value that would apply if the aerofoil were moving with constant velocity U.

COMPRESSIBLE FLOW

9.13   Generalization of the theory

The theory present so far is valid in compressible flow only if the time rate of change of the flow is small, e.g. in oscillating flow only if the reduced frequency ω is small. It will be noticed that we have allowed for compressibility correctly in the pressure equation (18); the wake equations (6) and (10) are also correct in compressible flow. The restriction to small rates of change in compressible flow arises because we have taken the variable τ to satisfy (2) instead of the accurate equation (2.116). An accurate theory based on the latter equation leads to rather difficult analysis, and lies outside the range of the complex variable methods of this book. For a short survey of the exact theory of Timman (1946) and van de Vooren on this topic the reader should consult van Spiegel & van de Vooren (1953). Another approach to the problem, with references to the main contributors to the subject, is to be found in a paper by Jones (1956).

An intermediate step, which allows partially for the effect of compressibility plus high time rates of change on the differential equation satisfied by τ, is to use (1) to define a more accurate value of τ, a value which we have distinguished by the subscript 1. Complex variable theory can be used to determine the relation between τ—the solution of (2)—and τ_1, the more accurate value defined by (1). First linearize (1) by putting $w_0 = UZ$ (see 2.17). The result is

$$\nabla^2 \tau_1 = \frac{8\kappa}{U} \frac{\partial}{\partial x}\left(\frac{\partial \tau}{\partial t}\right),$$

where
$$\kappa \equiv \left(\frac{U}{2a_a\beta_a}\right)^2 = \left(\frac{M_a}{2\beta_a}\right)^2. \tag{91}$$

By (3.116), (3.117) and (9) the equation for τ_1 can be written

$$\frac{\partial^2 \tau_1}{\partial Z \partial \bar{Z}} = \frac{2\kappa}{c}\left(\frac{\partial \dot{\tau}}{\partial Z} + \frac{\partial \dot{\tau}}{\partial \bar{Z}}\right) = \frac{2\kappa}{c}\frac{\partial \dot{\tau}}{\partial Z}, \tag{92}$$

where the dot denotes a derivative with respect to t, and we have taken advantage of the fact that τ depends only on Z, being an analytic function. Integrating (92) with respect to Z we find

$$\frac{\partial \tau_1}{\partial \bar{Z}} = \kappa\left\{\frac{2}{c}\dot{\tau}(Z) + f'(\bar{Z})\right\},$$

where f' is the derivative of an arbitrary function of $\bar{Z}$. Further integration gives

$$\tau_1 = h(Z) + \kappa\left\{\frac{2\bar{Z}}{c}\dot{\tau}(Z) + f(\bar{Z})\right\}, \tag{93}$$

where $h(Z)$ is another arbitrary function. Let $h(Z)$ be independent of κ, then as the flow becomes incompressible, and $\tau_1 \to \tau$, in the limit $\kappa \to 0$, we have $h(Z) = \tau(Z)$. Transforming (93) into the ζ-plane by (3) we obtain

$$\tau_1(\bar{\zeta},t) = \tau(\zeta,t) - \kappa\{\dot{\tau}(\zeta,t)\cos\bar{\zeta} - f(\bar{\zeta},t)\}, \tag{94}$$

on making the dependence of τ_1, τ and f on t explicit.

On $\eta = 0$ the imaginary part of (94) is

$$\theta_1 = \theta - \kappa(\dot{\theta}\cos\gamma - \mathscr{I}f). \tag{95}$$

Now the only possible difference between θ_1 and θ on the aerofoil surface is that in the compressible flow the front stagnation point may not occupy the same point as in incompressible flow. Thus (see (5))

$$\theta_1 = \theta_s + \frac{n}{U} - \alpha + \pi\lambda_1\delta(\gamma), \tag{96}$$

whence
$$\theta_1 - \theta = \pi(\lambda_1 - \lambda)\delta(\gamma).$$

Substitution of this value in (95) gives

$$\mathscr{I}f = \frac{\pi}{\kappa}(\lambda_1 - \lambda)\delta(\gamma) + \dot{\theta}\cos\gamma. \tag{97}$$

Next we need an expression for the discontinuity $[\mathscr{R}f]$, in the real part of f across the wake $\gamma = \pm\pi$. Let ϖ_1, ϖ be the discontinuities in $\mathscr{R}\tau_1 = \Omega_1$. and $\mathscr{R}\tau = \Omega$, then from (94)

$$[\mathscr{R}f] = \frac{1}{\kappa}(\varpi_1 - \varpi) - \dot{\varpi}\cosh\eta. \tag{98}$$

If we now apply (11) to the function $f(\zeta, \jmath)$, we find

$$f(\zeta, \jmath) = -\frac{1}{2\kappa}(\lambda_1 - \lambda)\cot\tfrac{1}{2}\zeta + \frac{\sin\zeta}{2\pi}\int_0^\infty \frac{(\varpi_1 - \varpi)\,d\eta}{\cosh\eta + \cos\zeta}$$

$$+\frac{1}{2\pi}\int_{-\pi}^{\pi}\dot\theta(\gamma)\cos\gamma\cot\tfrac{1}{2}(\gamma - \zeta)\,d\gamma$$

$$-\frac{\sin\zeta}{2\pi}\int_0^\infty \frac{\dot\varpi\cosh\eta\,d\eta}{\cosh\eta + \cos\zeta}, \tag{99}$$

with the aid of (97) and (98). Now differentiate (11) with respect to $\jmath$, replace ζ in (99) by $\bar\zeta$ and substitute the results into (94) to find

$$\tau_1(\bar\zeta, \jmath) = \tau(\zeta, \jmath) - \tfrac{1}{2}(\lambda_1 - \lambda)\cot\tfrac{1}{2}\bar\zeta + \frac{\sin\bar\zeta}{2\pi}\int_0^\infty \frac{(\varpi_1 - \varpi)\,d\eta}{\cosh\eta + \cos\bar\zeta}$$

$$-\frac{\kappa}{2\pi}\left[\int_{-\pi}^{\pi}\dot\theta\{\cos\bar\zeta\cot\tfrac{1}{2}(\gamma - \zeta) - \cos\gamma\cot\tfrac{1}{2}(\gamma - \bar\zeta)\}\,d\gamma\right.$$

$$\left.+\int_0^\infty \dot\varpi\left\{\frac{\sin\zeta\cos\bar\zeta}{\cosh\eta + \cos\zeta} + \frac{\sin\bar\zeta\cosh\eta}{\cosh\eta + \cos\bar\zeta}\right\}d\eta\right]. \tag{100}$$

This is the required relation between τ_1 and τ. The functions $\varpi(\eta, \jmath)$ and $\lambda(\jmath)$ and their derivatives can be calculated from results given in §9.3. Formulae for the new functions $\varpi_1(\eta, \jmath)$ and $\lambda_1(\jmath)$ are given in the next section.

9.14   The strength of the vortex sheet

By using the expansions employed in (13), and also the result $\cos\bar\zeta = -\tfrac{1}{2}(\bar\sigma + 1/\bar\sigma)$, we can expand (100) near infinity in the form

$$\tau_1 = \frac{i}{\pi}\sum_{n=2}^{\infty}\frac{1}{\sigma^n}\left\{(-1)^n\int_{-\pi}^{\pi}\theta\,e^{-ni\gamma}\,d\gamma + \int_0^\infty \varpi\cosh n\eta\,d\eta\right\}$$

$$-\tfrac{1}{2}i(\lambda_1 - \lambda) - \frac{i}{2\pi}\int_0^\infty (\varpi_1 - \varpi)\,d\eta$$

$$-\frac{i}{\pi}\sum_{n=1}^{\infty}\frac{1}{\bar\sigma^n}\left\{(-1)^n\pi(\lambda_1 - \lambda) + \int_0^\infty (\varpi_1 - \varpi)\cosh n\eta\,d\eta\right\}$$

$$+\frac{i\kappa}{\pi}\left[\frac{1}{2}\left(\bar\sigma + \frac{1}{\bar\sigma}\right)\sum_{n=2}^{\infty}\frac{1}{\sigma^n}\left\{(-1)^n\int_{-\pi}^{\pi}\dot\theta\,e^{-ni\gamma}\,d\gamma + \int_0^\infty \dot\varpi\cosh n\eta\,d\eta\right\}\right.$$

$$\left.-\sum_{n=1}^{\infty}\frac{1}{\bar\sigma^n}\left\{(-1)^n\int_{-\pi}^{\pi}\dot\theta\cos\gamma\,e^{ni\gamma}\,d\gamma - \int_0^\omega \dot\varpi\cosh\eta\cosh n\eta\,d\eta\right\}\right],$$

$$\tag{101}$$

on taking advantage of (12) and (15).

Now τ_1 vanishes at infinity, and so

$$\pi(\lambda_1 - \lambda) + \int_0^\infty (\varpi_1 - \varpi)\, d\eta = 0. \tag{102}$$

The analytic character of τ was not used in the derivation of (14) and (15), and so the same argument applies to τ_1; hence

$$\mathscr{R}\int_C \tau_1\left(1 - \frac{1}{\sigma^2}\right) d\sigma = 0.$$

Substituting (101) into this equation, and taking C to be the large circle $|\sigma| = R$ we find

$$-\pi(\lambda_1 - \lambda) + \int_0^\infty (\varpi_1 - \varpi)\cosh\eta\, d\eta = 2\pi\kappa\dot{b}, \tag{103}$$

where

$$b \equiv \frac{1}{2\pi}\int_{-\pi}^{\pi} \theta\cos^2\gamma\, d\gamma + \frac{1}{2\pi}\int_0^\infty \varpi\cosh^2\eta\, d\eta. \tag{104}$$

Let $\chi_1 = \varpi_1 U$, then by (8), (20), (21) and (27) we can transform (102) and (103) into

$$\lambda_1 U = 2\alpha U - 2a_0 - \frac{1}{\pi}\int_0^\infty \chi_1(\eta, \jmath)\, d\eta, \tag{105}$$

and

$$0 = 2\alpha U - 2a_0 + 2a_1 + 2\kappa\dot{b}U - \frac{1}{\pi}\int_0^\infty \chi_1(\eta, \jmath)\,(1 + \cosh\eta)\, d\eta. \tag{106}$$

These are the extensions of (20) and (21) to compressible flow, the only difference is the introduction of the term $2\kappa\dot{b}U$ on the right-hand side of the second of these equations.

The strength of the vortex sheet is now χ_1 instead of χ, and so the theory which gave (23) applies equally well when χ is replaced by χ_1. Thus $\lambda_1 U$ and χ_1 satisfy equations of precisely the same form as do λU and χ, and therefore they equal the right-hand side of (43) and (45) respectively, except that $A(\jmath)$ must be replaced by

$$A_1(\jmath) = \begin{cases} \alpha(\jmath)\, U(\jmath) - a_0(\jmath) + a_1(\jmath) + \kappa\dot{b}(\jmath)\, U(\jmath) & (\jmath > 0), \\ 0 & (\jmath < 0). \end{cases} \tag{107}$$

The number $\dot{b}$ can be evaluated from the solution to the incompressible flow. Consider, for example, the case of unaccelerated motion; i.e. motion in which U is constant. Then by (8), (12) and (104)

$$2\dot{b}U = -\frac{1}{\pi}\int_{-\pi}^{\pi} U\theta\sin^2\gamma\, d\gamma + \frac{1}{\pi}\int_0^\infty \dot{\chi}\sinh^2\eta\, d\eta$$

$$= \dot{a}_2 - \dot{a}_0 + \dot{\alpha}U - \frac{2}{\pi}\int_0^\infty \sinh\eta\left(\frac{\partial\chi}{\partial\eta}\right) d\eta,$$

on making use of (5), (27) and (50). Integration by parts and (20) and (21) now give

$$2\dot{b}U = \dot{a}_2 - \dot{a}_0 + \dot{\alpha}U + 4\dot{a}_1 + 2\dot{\lambda}U. \tag{108}$$

9.15 The pressure distribution on the aerofoil

At a point $\zeta = \bar{\zeta} = \gamma^*$ on the aerofoil surface (100) reduces to

$$\Omega_1(\gamma^*,t) = \Omega(\gamma^*,t) - \tfrac{1}{2}(\lambda_1 - \lambda)\cot\tfrac{1}{2}\gamma^* + \frac{\sin\gamma^*}{2\pi}\int_0^\infty \frac{(\varpi_1 - \varpi)\,d\eta}{\cosh\eta + \cos\gamma^*},$$

where we have used (12) and (16) to eliminate the term involving κ. Comparison of this with (47) shows that Ω_1 is given by an equation of the same form as (47), with λ and replaced by λ_1 and χ_1. We conclude that the whole of the theory of §§ 9.4 and 9.5 can be immediately extended to compressible flow—it is only necessary to replace λ, χ and Ω by λ_1, χ_1 and Ω_1.

CHAPTER 10

FLOW PAST SEMI-INFINITE PROFILES

10.1 Introduction

In this chapter we shall study the flow of an infinite stream about a certain type of semi-infinite profile, namely one which extends an infinite distance downstream as illustrated in Fig. 10.1a. Such a profile will, in general, have a stagnation point A near its leading edge. We shall take the origin of the w-plane to be at this point, and $\psi = +0$, $\psi = -0$ will indicate the upper and lower surfaces of the profile respectively. Let $\theta_s(\phi)$ denote the slope of the profile surface, and let the velocity vector (q, θ) have the value $(U, 0)$ upstream at infinity, i.e.

$$\lim_{\phi \to -\infty} q(\phi, \psi) = U, \quad \lim_{\phi \to -\infty} \theta(\phi, \psi) = 0. \tag{1}$$

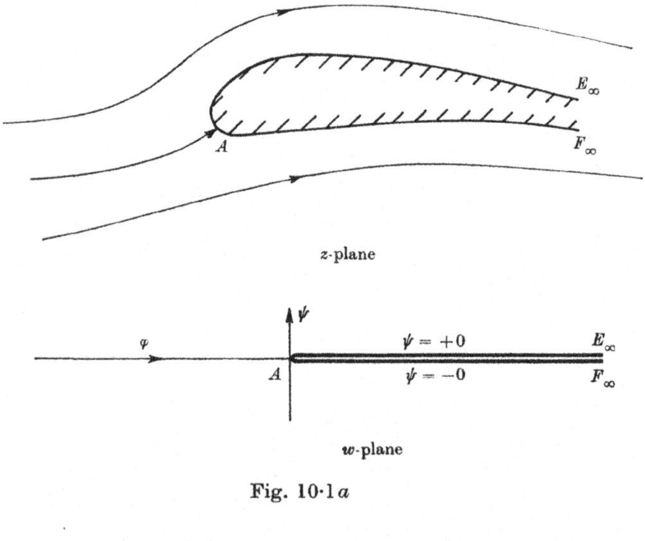

Fig. 10·1a

Semi-infinite profiles can be divided into those that satisfy the restrictions

$$\lim_{\phi \to \infty} \theta_s(\phi, +0) = 0, \quad \lim_{\phi \to \infty} \theta_s(\phi, -0) = 0, \tag{2}$$

and those which do not. However, divergent profiles, i.e. those not satisfying (2), can be obtained as limiting cases of profiles satisfying (2) by initially setting $\theta_s = 0$ in $R < \phi < \infty$, and allowing R to tend to infinity at a later stage in the calculation.

Doing this for the simple profile shown in Fig. 10.1b it becomes immediately apparent that as $q(R, +0)$ and $q(R, -0)$ are infinite (assuming $\alpha_1, \alpha_2 > 0$), so will be the limits $\lim_{R \to \infty} q(R, \pm 0)$, i.e. $U(\infty, \pm 0)$. It therefore follows that (2) is a necessary condition that U be finite. If α_2 were negative the same reasoning gives $U(\infty, +0) = \infty$, $U(\infty, -0) = 0$, and the singularity at infinity is rather complicated. Equation (2) secures the continuity of θ at infinity, and hence ensures that the corresponding velocity is finite and non-zero. From these results we conclude that (2) is the necessary and sufficient condition that U be finite and non-zero.

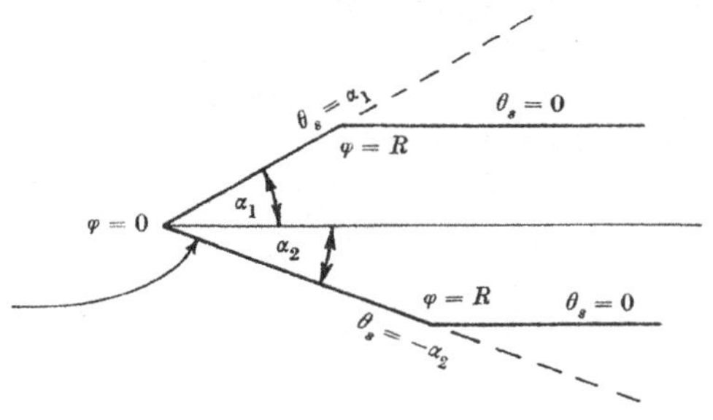

Fig. 10.1b

Most of the profiles of interest to us have a finite and constant width well downstream, and therefore satisfy the restriction

$$y(s, +0) - y(s, -0) = b \quad (S \leqslant s \leqslant \infty), \tag{3}$$

where b is a positive finite number and s is the distance measured along the profile. This restriction implies that

$$\theta_s(\phi^+, +0) = \theta_s(\phi^-, -0) \quad (R \leqslant \phi^+, \phi^- \leqslant \infty), \tag{4}$$

where ϕ^+, ϕ^- are values of ϕ on the upper and lower surfaces at geometrically opposite points.

Several flow problems of practical interest involve semi-infinite profiles. Perhaps the most interesting—and complex—of these occurs in the jet-flap problem. The jet flap (Davidson, 1956) is a high-speed jet of air ejected from the trailing edge of an aerofoil at some angle to the main stream (see § 1.14 and Fig. 1.14). The conventional methods of obtaining aerofoil lift are (a) by increasing the incidence, or (b) by increasing the camber—which is most easily effected by the downwards deflexion of a nose or trailing edge hinged flap. The jet flap is essentially another (relatively new) method of obtaining aerofoil lift, which ideally at least has several advantages over the conventional methods, but there are some practical difficulties not yet solved (see discussion at end of Davidson's paper). An outline of the theory of the jet flap will be given in § 10.14 *et seq.*

A similar problem arises when we allow for the existence of an infinite wake extending behind an aerofoil (see Fig. 8.1). As described in § 1.16 this wake can be represented by an infinitely long solid sting or 'tail' of width $b = \frac{1}{2}cC_D$. However, the effect of such a wake on the lift of an aerofoil arises because of its asymmetry at the trailing edge and not because of its thickness.

The symmetrical flow over a step or cliff is another example of the type of flow considered in this chapter. Such flows are frequently found to be of theoretical and even practical value (e.g. see Thwaites, 1954; Newmark, 1955). The mathematical theory of symmetrical flows is very simple; we shall deal with these before studying the more general case of asymmetric flow.

SYMMETRICAL FLOWS

10.2 General theory

If the flow is symmetrical about $y = 0$ we need consider only the upper half of the flow pattern, the w-plane of which is just the half plane $\psi \geqslant 0$. As $(q, \theta) = (U, 0)$ at infinity then

$$\lim_{w \to \infty} \tau = \Omega_\infty + i\theta_\infty = 0. \tag{5}$$

The solution to our boundary value problem has been given in § 3.18. It is

$$\tau(w) = \frac{1}{\pi} \int_{-\infty}^{\infty} \frac{\theta_s(\phi)\, d\phi}{\phi - w}. \tag{6}$$

In writing down (5) we have of course tacitly assumed (1) and (2) to hold, but divergent profiles can be obtained by the limiting process described in § 10.1, so the following theory, based on (6), is not as restricted as it at first appears.

By (6.12) the z-plane can be calculated from

$$z = \frac{1}{2\beta_a U}\left\{(1+\beta_a)\int e^\tau\, dw - (1-\beta_a)\overline{\int e^{-\tau}\, dw}\right\}. \tag{7}$$

In many of the cases of interest to us there will be a value a of ϕ such that $\theta_s = 0$ in $a < \phi < \infty$ (a can always be as large as we please). In this case we can expand (6)

$$\tau(w) = \frac{1}{\pi} \int_0^a \frac{\theta_s(\phi)\, d\phi}{\phi - w} = -\sum_{n=0}^{\infty} \frac{A_n}{w^{n+1}} \quad (|w| > a), \tag{8}$$

where
$$A_n \equiv \frac{1}{\pi} \int_0^a \theta_s(\phi)\,\phi^n\,d\phi, \tag{9}$$

as θ_s vanishes in $-\infty < \phi < 0$. Substituting this expansion in (7) we have

$$z = \frac{1+\beta_a}{2\beta_a U} \int \left\{1 - \frac{A_0}{w} + O(w^{-2})\right\} dw - \frac{1-\beta_a}{2\beta_a U} \int \overline{\left\{1 + \frac{A_0}{w} + O(w^{-2})\right\} dw}$$

provided $|w| > a$. An integration about a closed contour C lying outside $|w| = a$ yields $[z]_C = -ib$, where b is the profile width downstream near infinity; there is no net change in x about the contour because of the symmetry of the flow. The negative sign arises because C starts at $y = \frac{1}{2}b$ and ends at $y = -\frac{1}{2}b$ (see Fig. 10.2). The residue theorem now gives

$$b = \frac{2\pi}{U} A_0 = \frac{2}{U} \int_0^a \theta_s(\phi)\,d\phi, \tag{10}$$

a result which is immediately obvious for thin profiles, as $\theta_s \simeq dy/dx$, and $d\phi \simeq U dx$. This equation will be proved under more general conditions in §10.5.

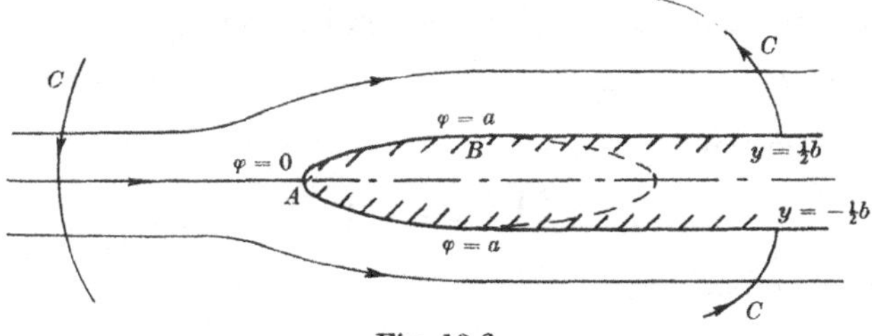

Fig. 10.2

It has been shown in §1.15 that the drag force on a semi-infinite body is

$$D = bp_a, \tag{11}$$

where p_a is the pressure at infinity. This result also applies to asymmetric profiles.

10.3   Example. Incompressible flow past a wedge

For the wedge shown in Fig. 10.3 $\theta_s = \alpha$ in $0 < \phi < a$, therefore (6) and (10) give

$$\tau = \frac{\alpha}{\pi} \ln\left(\frac{w-a}{w}\right),$$

where $a = bU/2\alpha$. In incompressible flow $\tau = \ln(U/q) + i\theta$, and so

$$q e^{-i\theta} = \frac{U}{a^{\alpha/\pi}} \left(\frac{w}{1-w/a}\right)^{\alpha/\pi} e^{-i\alpha}.$$

We can deduce the corresponding result for the infinitely diverging wedge by taking the limit $a \to \infty$. This limit will also give the flow in a corner. In this limit we find

$$q\,e^{-i\theta} = K(w)^{\alpha/\pi}\,e^{-i\alpha},$$

where
$$K = \lim_{a \to \infty}(Ua^{-\alpha/\pi}).$$

If K is to be finite and non-zero we require that $U \to \infty$ for $\alpha > 0$, and $U \to 0$ for $\alpha < 0$, which verifies some remarks made in the second paragraph of § 10.1.

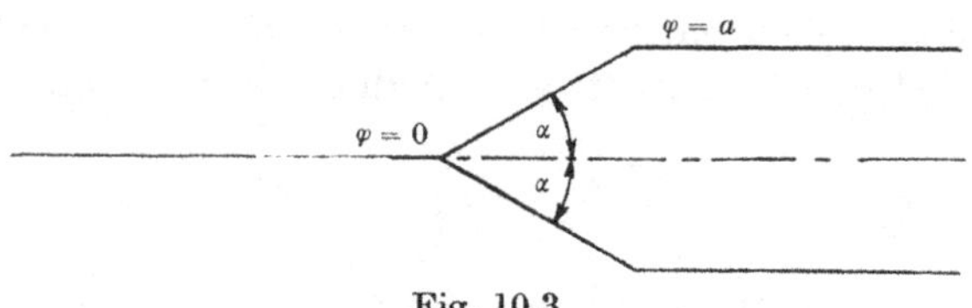

Fig. 10.3

10.4   Linear perturbation theory

Linear perturbation theory is obtained by setting

$$w = UZ,\ (Z = x + i\beta_a y),\ a = Uc \quad \text{and} \quad \tau = \beta_a(1 - \{q/U\}) + i\theta.$$

Equation (6) becomes

$$\beta_a\left(1 - \frac{q}{U}\right) + i\theta = \frac{1}{\pi}\int_0^c \frac{\theta_s(x)\,dx}{x - Z}, \tag{12}$$

and hence on the profile,

$$\frac{q}{U} = 1 - \frac{1}{\pi\beta_a}\int_0^c \frac{\theta_s(x^*)\,dx^*}{x^* - x}. \tag{13}$$

This can be used to establish a simple but interesting result due to Thwaites (1954) and Newmark (1955) concerning the change in velocity at the thickest point on a closed doubly-symmetric profile caused by replacing the rear half by a semi-infinite parallel body (see Fig. 10.2). Equation (13) applies to both closed profiles and semi-infinite profiles. Let the doubly-symmetric body be of length c, and take the origin of the Z-plane to be at its centre, then from (13) and the symmetry $\{\theta_s(x^*) = -\theta_s(-x^*)\}$,

$$\frac{q}{U} - 1 = -\frac{1}{\pi\beta_a}\int_{-\frac{1}{2}c}^{\frac{1}{2}c} \frac{\theta_s(x^*)\,dx^*}{x^* - x} = -\frac{2}{\pi\beta_a}\int_{-\frac{1}{2}c}^{0} \frac{\theta_s(x^*)\,x^*\,dx^*}{x^{*2} - x^2},$$

whereas the modified (infinite) profile gives

$$\frac{q}{U} - 1 = -\frac{1}{\pi\beta_a}\int_{-\frac{1}{2}c}^{0} \frac{\theta_s(x^*)\,dx^*}{x^* - x}.$$

The thickest point is at $x = 0$, and so we conclude that the velocity increment at this point is halved by changing the rear half of the closed profile into a semi-infinite parallel body. It should be remembered that this result is true only for slender profiles.

ASYMMETRIC PROFILES

10.5   The basic equations

Next consider the flow about an asymmetric aerofoil like that shown in Fig. 10.1a. To obtain the basic equations for this profile we shall start with the equations given in §§ 8.2–8.4 for (finite) aerofoil theory and then stretch the aerofoil indefinitely in the downstream direction. This approach has the virtue of showing quite clearly the fate of the aerofoil closure conditions in the limit, and yields results very difficult to obtain by the alternative method of starting directly with the 'infinite' aerofoil.

To carry out the stretching process we shall replace the variable ζ defined in (8.1) by χ, where

$$\zeta = \gamma + i\eta = \frac{\pi}{T}\chi = \frac{\pi}{T}(\xi + i\delta), \tag{14}$$

and then take the limit $T \to \infty$. The front stagnation point for the finite aerofoil is at $\zeta = -2\alpha_0$ (see § 8.2), and so if the w-plane origin is moved to this point, (8.1) is transformed into

$$w = 2a\cos\alpha_0(1 - \cos\zeta) + 2a\sin\alpha_0(\sin\zeta + \zeta + 2\alpha_0). \tag{15}$$

The number $2a$ here is proportional to the length of the aerofoil, and will tend to infinity in the stretching. Let

$$\alpha_0 = \frac{\pi\xi_0}{T}, \tag{16}$$

then (14) and (15) give

$$w = \frac{a\pi^2}{T^2}(\chi + \xi_0)^2 + O\left(\frac{a}{T^4}\right).$$

Therefore if a tends to infinity with T^2, in the limit

$$w = K(\chi + \xi_0)^2, \tag{17}$$

where

$$K = \lim_{T\to\infty} \frac{a\pi^2}{T^2}. \tag{18}$$

Equation (17) maps the w-plane on to the upper half of the χ-plane.

Equations (8.10) and (14) give

$$\left| \tau(\chi) - \frac{1}{2T} \int_{-R}^{R} \theta_s(\xi) \cot \frac{\pi}{2T}(\xi - \chi)\, d\xi \right|$$

$$= \left| \frac{1}{2T} \left(\int_{R}^{T} + \int_{-T}^{-R} \right) \theta_s(\xi) \cot \frac{\pi}{2T}(\xi - \chi)\, d\xi \right|,$$

where $0 < R < T$. Now it is easily verified that an absolute constant C can be found such that the right-hand side of this equation is less than

$$C \left(\int_{R}^{\infty} + \int_{-\infty}^{-R} \right) \theta_s(\xi) \frac{d\xi}{\xi},$$

and therefore provided $\theta_s = O(\xi^{-\mu})\,(\mu > 0)$ as $\xi \to \infty$, we can find a value of R which makes this side of the equation less than any given small positive number ϵ. This ϵ is independent of T, and therefore in the limit $T \to \infty$ we get

$$\left| \tau(\chi) - \frac{1}{\pi} \int_{-R}^{R} \theta_s(\xi) \frac{d\xi}{\xi - \chi} \right| < \epsilon,$$

whence

$$\tau(\chi) = \frac{1}{\pi} \int_{-\infty}^{\infty} \frac{\theta_s(\xi)\, d\xi}{\xi - \chi}. \tag{19}$$

Similarly, it follows from (8.42) that the harmonic conjugate of (19) is

$$\tau(\chi) = -\frac{i}{\pi} \int_{-\infty}^{\infty} \frac{\Omega_s(\xi)\, d\xi}{\xi - \chi}. \tag{20}$$

10.6   The auxiliary conditions

It is next necessary to study the behaviour of the aerofoil closure conditions in the limit $T \to \infty$.

First (8.11) and (14) yield

$$\int_{-T}^{T} \theta_s(\xi)\, d\xi = 0. \tag{21}$$

Thus in the limit $T \to \infty$ we obtain

$$\int_{-\infty}^{\infty} \theta_s(\xi)\, d\xi = 0, \tag{22}$$

which is to be interpreted as a Cauchy principal value at infinity (see §3.12). Convergence of the integral in (22) is secured if $\theta_s = O(\xi^{-(1+\mu)})$, $(\mu > 0)$, but as it is a Cauchy principal value it is sufficient for this order restriction to be satisfied by only the symmetrical part of θ_s, i.e. the integral converges if

$$\theta_s(\xi) + \theta_s(-\xi) = O(\xi^{-(1+\mu)}) \quad (\mu > 0), \tag{23}$$

near $\xi = \infty$. The distribution $\theta_s(\xi)$ depends not only on the profile slope, but also on the position of the front stagnation point $\chi = -2\xi_0$, relative to other points on the profile. Thus (22) fixes ξ_0, and renders θ_s unique.

The aerofoil closure conditions (8.27) are too restrictive a starting-point from which to develop the equivalent equations for the semi-infinite profile, for such a profile need not close at all. This limitation can be removed as follows. Suppose a source of strength Q is placed at the aerofoil trailing edge, then the closure condition is replaced by

$$\frac{1}{\pi} \int_{-\pi}^{\pi} \theta_s(\gamma) \, e^{-i(\gamma+\alpha_0)} \, d\gamma = 2\beta_a \sin \alpha_0 - \frac{iQ}{2\pi a \rho_a}, \tag{24}$$

as may be verified from (8.194), (8.205) and (8.206). The change in sign in Q is necessary because in (8.194) Q is the mass sucked *into* the aerofoil per second. Well downstream of the aerofoil Q equals $b\rho_a U$, where b is the width of the source stream; hence for the semi-infinite profile comprised of the aerofoil plus source displacement, b is the profile thickness near infinity—the number defined in (3).

Equations (14), (16) and the real part of (24) give

$$2\beta_a T \sin \frac{\pi \xi_0}{T} = \int_{-T}^{T} \theta_s(\xi) \cos \frac{\pi}{T}(\xi + \xi_0) \, d\xi,$$

and retaining only the highest order terms (in $1/T^2$) and using (21) we can write this

$$2\pi \beta_a \xi_0 = \int_{-T}^{T} \theta_s \left(\cos \frac{\pi}{T} \xi - 1 \right) d\xi = -\frac{\pi^2}{2} \int_{-T}^{T} \theta_s \left(\frac{\xi}{T} \right)^2 d\xi.$$

From this it follows that if

$$\theta_s(\xi) + \theta_s(-\xi) = O(\xi^{-(1+\mu)}) \quad \text{near} \quad \xi = \infty,$$

then $\qquad\qquad\qquad T\xi_0 = O(T^{1-\mu}) \quad \text{near} \quad T = \infty. \tag{25}$

Therefore, provided $\mu > 0$, which is required in any case for the convergence of the integral in (22),

$$\lim_{T \to \infty} \xi_0 = 0, \tag{26}$$

and (17) can be replaced by

$$w = \phi + i\psi = K\chi^2 = K(\xi + i\delta)^2. \tag{27}$$

The imaginary part of (24) yields

$$T^2 \frac{Q}{2\pi a \rho_a} = T \int_{-T}^{T} \theta_s \sin \frac{\pi}{T}(\xi + \xi_0) \, d\xi,$$

by (14) and (16). Hence by using (18), (26), (27) and $\lim_{T\to\infty}(Q/\rho_a) = bU$ (see remarks following (24)) we find

$$b = \frac{2K}{U}\int_{-\infty}^{\infty}\theta_s(\xi)\,\xi\,d\xi = \frac{1}{U}\int_C \theta_s(\phi)\,d\phi, \tag{28}$$

where C is the profile contour. This is obviously a generalization of (10). Convergence is secured if

$$\theta_s(\xi) - \theta_s(-\xi) = O(\xi^{-(2+\mu)}) \quad (\mu > 0). \tag{29}$$

The conjugate harmonic equations are also very useful. These are obtained by replacing θ_s by $-i\Omega_s$ in the above theory. Then in place of (22) there is

$$\int_{-\infty}^{\infty}\Omega_s(\xi)\,d\xi = 0. \tag{30}$$

The conjugate of (24) is

$$\frac{1}{\pi}\int_{-\pi}^{\pi}\Omega_s(\gamma)\,\{\sin(\gamma+\alpha_0) + i\cos(\gamma+\alpha_0)\}\,d\gamma = -2\beta_a\sin\alpha_0 + \frac{iQ}{2\pi a\rho_a}. \tag{31}$$

From the real part of this equation and an argument similar to that used in deriving (28) we find

$$\lim_{T\to\infty}(4K\xi_0 T) = \lim_{T\to\infty}(4\pi^2 a\xi_0/T) = -\frac{2K}{\beta_a}\int_{-\infty}^{\infty}\Omega_s(\xi)\,\xi\,d\xi, \tag{32}$$

which converges provided

$$\Omega_s(\xi) - \Omega_s(-\xi) = O(\xi^{-(2+\mu)}) \quad (\mu > 0). \tag{33}$$

Similarly, corresponding to (25), the imaginary part of (31) yields that if

$$\Omega_s(\xi) + \Omega_s(-\xi) = O(\xi^{-(1+\mu)}) \quad \text{near} \quad \infty,$$

then

$$b = O(T^{1-\mu}) \quad \text{near} \quad T = \infty. \tag{34}$$

10.7   The drag and lift forces

As the formula given in (11) is quite general, it follows from (28) that

$$D = p_a b = p_a\frac{2K}{U}\int_{-\infty}^{\infty}\theta_s(\xi)\,\xi\,d\xi = \frac{p_a}{U}\int_C \theta_s(\phi)\,d\phi. \tag{35}$$

A corresponding expression for the total lift force, L_∞, acting in the Oy-direction on the semi-infinite profile, can be deduced by taking the limit $T \to \infty$ of

$$L = \rho_a U\Gamma = \rho_a U 4\pi a\sin\alpha_0 = \rho_a U 4\pi a\sin\frac{\pi\xi_0}{T}, \tag{36}$$

which follows from (8.2), (8.83) and (16). (A source right at the trailing edge of an aerofoil cannot affect the lift or circulation Γ.) Equations (18), (32) and (36) now yield

$$L_\infty \equiv \lim_{T\to\infty} L = \rho_a U \lim_{T\to\infty} \Gamma = \rho_a U \lim_{T\to\infty} (4K\xi_0 T) = \rho_a U \Gamma_\infty, \qquad (37)$$

hence $\qquad L_\infty = -\dfrac{\rho_a U^2}{\beta_a}\dfrac{2K}{U}\displaystyle\int_{-\infty}^{\infty} \Omega_s(\xi)\,\xi\,d\xi = -\dfrac{\rho_a U}{\beta_a}\displaystyle\int_C \Omega_s(\phi)\,d\phi, \qquad (38)$

by (27).

It is interesting to note from (35) and (38) the conjugate nature of (D/ρ_a) and $(\beta_a L_\infty/\rho_a U^2)$. This and the fact that it is physically possible for b to be zero, finite or infinite, suggests that L_∞ can be zero, finite or infinite depending on the profile shape. In fact from (25) and (37) we observe that $L = O(T^{1-\mu})$, so that $L_\infty = 0, 0 < L_\infty < \infty, L_\infty = \infty$ as $\mu > 1$, $\mu = 1, \mu < 1$. (Strictly $L = O(1)$ does not exclude $L_\infty = 0$, but as shown below L_∞ cannot vanish if $\mu = 1$.)

If $\mu = 1$ it follows from (25) and (27) that the average slope, $\theta_{sa} \equiv \frac{1}{2}(\theta_s(\xi) + \theta_s(-\xi))$ of the profile satisfies $\theta_{sa} = O(\phi^{-1})$ near $\phi = \infty$. Near infinity $\theta_{sa} \simeq (dy/dx)_{sa}$, and $\phi \simeq Ux$, therefore L_∞ is finite and non-zero only if the average slope satisfies $(dy/dx)_{sa} = O(x^{-1})$. More generally we have:

$$\text{if } \left(\frac{dy}{dx}\right)_{sa} = O(x^{-\nu}) \text{ near } x = \infty, \text{ then } L_\infty = \rho_a U \Gamma_\infty = \begin{cases} \infty\,(\nu < 1), \\ k_1\,(\nu = 1), \\ 0\,(\nu > 1), \end{cases} \qquad (39)$$

where k_1 is a finite non-zero number. Similarly, it follows from $\Omega_s \simeq \frac{1}{2}\beta_a C_p$ (see (6.23)) and (34) that

$$\text{if } C_{p_{sa}} = O(x^{-\nu}) \text{ near } x = \infty, \text{ then } D = p_a b = \begin{cases} \infty\,(\nu < 1), \\ k_2\,(\nu = 1), \\ 0\,(\nu > 1), \end{cases} \qquad (40)$$

where k_2 is another finite non-zero number.

Notice from (39) that if the force acting on a semi-infinite profile is finite and non-zero then

$$y_{sa} \sim A \ln|x|, \quad (0 < |A| < \infty), \quad \text{near} \quad x = \infty, \qquad (41)$$

i.e. the centre-line of the profile diverges infinitely far from the x-axis with increasing downstream distance. If the rate of increase is greater than that given in (41), then L_∞ is infinite, whereas if it is less L_∞ is zero.

We shall consider the case of finite drag and lift in more detail. From (5) and (27) τ is regular at infinity in the χ-plane and so can be expanded (see §3.3)

$$\tau = \Omega + i\theta = \sum_{n=1}^{\infty} \frac{a_n + ib_n}{\chi^n}$$

$(a_n, b_n$ real). On the profile this gives that Ω_s and θ_s are $O(\xi^{-1})$, and therefore that D and L_∞ are infinite unless $a_1 = b_1 = 0$. Hence for the case of interest

$$\tau = \sum_{n=2}^{\infty} \frac{a_n + ib_n}{\chi^n}. \tag{42}$$

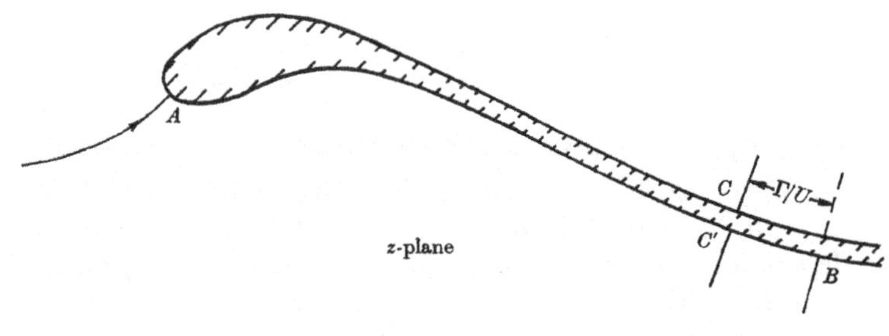

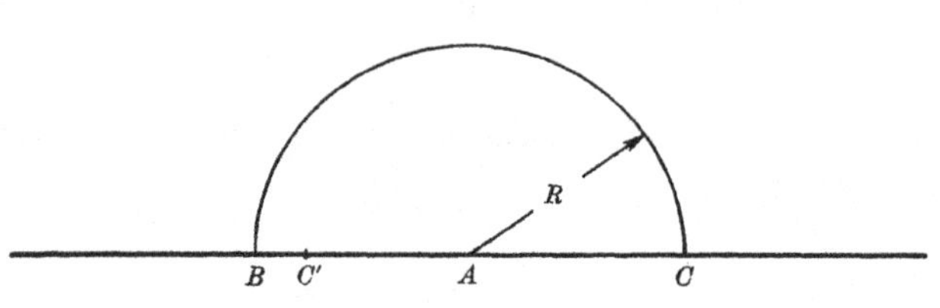

Fig. 10.7

The lift and drag can be calculated directly from (6.14). Applying this to the contour CAB shown in Fig. 10.7, where C and B are at the same value of ϕ, we have

$$X - iY = i\frac{\rho_a U}{1+\beta_a}\int_{CB} e^{-\tau}\frac{dw}{d\chi}d\chi - i(\bar{z}_B - \bar{z}_C)\left\{p_a + \frac{\rho_a U^2}{1+\beta_a}\right\}. \tag{43}$$

Replace CB by the semicircle $|\chi| = Re^{i\epsilon}$, $(0 \leqslant \epsilon \leqslant \pi)$—which is permissible owing to the analytic integrand—then by (27) and (42)

$$\int_{CB} e^{-\tau}\frac{dw}{d\chi}d\chi = 2K\int_0^\pi \left\{1 - \frac{a_2 + ib_2}{R^2 e^{2i\epsilon}} + O(R^{-3})\right\} iR^2 e^{2i\epsilon}\,d\epsilon.$$

Therefore
$$\lim_{R \to \infty} \int_{CB} e^{-\tau} \frac{dw}{d\chi} d\chi = -2K\pi i(a_2 + ib_2).$$
(44)

Also as the jump in ϕ across the profile at a given value of x tends to Γ_∞, and $\phi \simeq Ux$, the jump in x at a given value of ϕ tends to Γ_∞/U. Hence (cf. Fig. 10.7)
$$\lim_{R \to \infty} (\bar{z}_B - \bar{z}_C) = \frac{\Gamma_\infty}{U} + ib.$$
(45)

The force X in (43) tends to the total drag D on the profile, but Y does not tend to L_∞, the total lift. The reason for this is clear from Fig. 10.7, which shows that $\lim_{R \to \infty} Y$ will exceed L_∞ by the limit of the additional force acting on $C'B$, i.e. by $p_a \Gamma_\infty/U$. Thus
$$\lim_{R \to \infty} (X - iY) = D - i(L_\infty + p_a \Gamma_\infty/U).$$
(46)

On substituting (44) to (46) in (43), and also eliminating b and Γ_∞ in favour of D and L_∞ (see (35) and (37)) we arrive at
$$a_2 = -\frac{DU}{p_a 2K\pi} = -\frac{bU}{2K\pi} = -\frac{1}{\pi} \int_{-\infty}^{\infty} \theta_s(\xi)\,\xi\,d\xi,$$
(47)

and
$$b_2 = -\frac{\beta_a L_\infty}{\rho_a U 2K\pi} = -\frac{\beta_a \Gamma_\infty}{2K\pi} = \frac{1}{\pi} \int_{-\infty}^{\infty} \Omega_s(\xi)\,\xi\,d\xi.$$
(48)

Near infinity $\phi \simeq Ux = K\xi^2$, and $\theta_s = b_2/\xi^2 + \ldots = Kb_2/Ux + \ldots$ by (42). Hence the constant A in (41) equals Kb_2/U, and therefore
$$y_{sa} \sim -\frac{\beta_a L_\infty}{2\pi \rho_a U^2} \ln|x|.$$
(49)

10.8   The moment about the leading edge

An expression for the moment about the leading edge of the pressures acting on a semi-infinite profile can be deduced from (8.14) and (8.88) to (8.91). Let $x_c = \frac{1}{2}cT$, then the origin of the z-plane will be near the leading edge, and the equations referred to will yield

$$M = 2a^2 \frac{\rho_a}{\beta_a} \int_{-\pi}^{\pi} \Omega_s(\gamma) \{2\sin(\gamma + \alpha_0) - \sin 2(\gamma + \alpha_0)\}\,d\gamma.$$

If we now make the transformations introduced in § 10.5, and take the limit $T \to \infty$, the result is

$$M_\infty \equiv \lim_{T \to \infty} M = 2K^2 \frac{\rho_a}{\beta_a} \int_{-\infty}^{\infty} \Omega_s(\xi)\,\xi^3\,d\xi,$$
(50)

where use has been made of (18) and (26). It is obvious from the form of the integrand in (50) that M_∞ is the moment about the front stagnation point, $\xi = 0$. It is also apparent from (42) that M_∞ will be infinite in the general case, for near $\xi = \infty$, $\Omega_s = a_2 \xi^{-2} + a_3 \xi^{-3} + \ldots$, and the integral in (50) is finite only if $a_2 = a_3 = 0$.

10.9   An example: a small nose-flap

As a simple example of the theory consider the profile comprised of two straight portions shown in Fig. 10.9. Let the 'nose-flap' DBD' be at an incidence α to the main stream, and let the points D, B, D' be at $1, \xi_1, -1+\lambda$ respectively. By (26) the front stagnation point is at $\xi = 0$. It is obvious from the figure that

$$\theta_s(\xi) = \begin{cases} -\alpha & (-1+\lambda < \xi < 0, \ \xi_1 < \xi < 1), \\ \pi - \alpha & (0 < \xi < \xi_1). \end{cases} \tag{51}$$

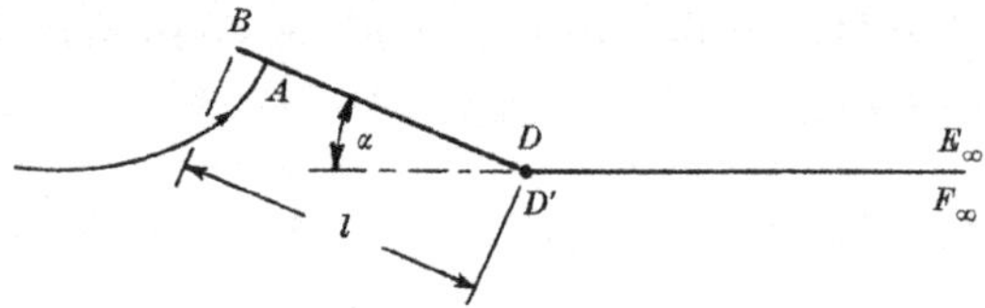

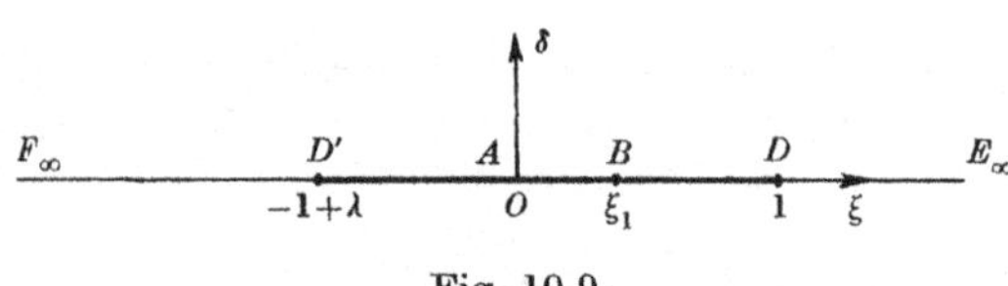

Fig. 10.9

The downstream width b of the profile is zero. Therefore from (22), (28) and (43),

$$\xi_1 = \frac{\alpha}{\pi}(2-\lambda), \quad \xi_1^2 = \frac{\alpha\lambda}{\pi}(2-\lambda),$$

whence
$$\xi_1 = \lambda = \frac{2\alpha}{\pi+\alpha}.$$

From this equation, (19) and (51)

$$\tau(\chi) = \ln\left(1 - \frac{\lambda}{\chi}\right) - \frac{\alpha}{\pi}\ln\left(\frac{\chi-1}{\chi+1-\lambda}\right). \tag{52}$$

Equations (7) and (27) give

$$z = \frac{K}{\beta_a U}\left\{(1+\beta_a)\int e^\tau \chi \, d\chi - (1-\beta_a)\overline{\int e^{-\tau} \chi \, d\chi}\right\},$$

and substituting (52) into this we get the required solution.

If second-order terms in α are negligible, $\lambda = 2\alpha$, and

$$\tau(\chi) = -\frac{\alpha}{\pi}\left\{\frac{2}{\chi} + \ln\left(\frac{\chi-1}{\chi+1}\right)\right\}. \tag{53}$$

Near $\chi = \pm\infty$ this gives

$$\tau(\chi) = \frac{2\alpha}{3\pi}\frac{1}{\chi^3} + \frac{2\alpha}{5\pi}\frac{1}{\chi^5} + \cdots,$$

and it follows from (42), (47) and (48) that D and L_∞ vanish.

The lift force on the flap alone is

$$L \simeq \int_{DBD'} (p - p_a)\,dx = -\tfrac{1}{2}\rho_a U \int_{D'BD} C_{p_s}\,d(Ux) = -\frac{\rho_a U}{\beta_a}\int_{D'BD} \Omega_s\,d\phi,$$

by the approximations $C_{p_s} = 2\Omega_s/\beta_a$ and $\phi = Ux$ of linear perturbation theory. This approximation also shows that if l is the length of the flap, $l \simeq K[1^2 - \lambda^2]/U \simeq K/U$, as λ^2 is of second order. Hence

$$d\phi = 2K\xi\,d\xi = 2Ul\xi\,d\xi,$$

and

$$L = -\frac{2}{\beta_a} l\rho_a U^2 \int_{-1+\lambda}^{1} \Omega_s \xi\,d\xi$$

$$\simeq \frac{2}{\pi\beta_a} l\rho_a U^2 \alpha \int_{-1}^{1} \xi\left\{\frac{2}{\xi} + \ln\left(\frac{1-\xi}{1+\xi}\right)\right\}d\xi, \tag{54}$$

from the real part of (53) on the profile surface. Thus the lift and lift coefficient, $C_L \equiv L/(\tfrac{1}{2}\rho_a lU^2)$, are

$$L = \frac{4\alpha}{\pi\beta_a}\rho_a lU^2, \quad C_L = \frac{8\alpha}{\pi\beta_a}. \tag{55}$$

The total lift, L_∞, on the profile is zero, and so there must be a downwards force of $4\alpha\rho_a lU^2/(\pi\beta_a)$ acting on the horizontal portion of the profile shown in Fig. 10.9. This may be verified directly from (53).

A similar calculation shows that the (nose-up) moment about the hinge CD of the flap is

$$M = \int_{DBD'} (p - p_a)\,x\,dx = \frac{2}{\beta_a}\rho_a l^2 U^2 \int_{-1}^{1} \Omega_s(\xi^2 - 1)\,\xi\,d\xi = \frac{8\alpha}{\pi\beta_2}\rho_a l^2 U^2,$$

and therefore the moment coefficient, $C_M \equiv M/(\tfrac{1}{2}\rho_a l^2 U^2)$, is

$$C_M = \frac{16\alpha}{\pi\beta_a}. \tag{56}$$

Equations (55) and (56) show that the centre of pressure is a distance l upstream of the leading edge of the flap.

10.10   Profiles of constant downstream width

An important class of semi-infinite profiles have a constant width b downstream of some given section DD', say (see Fig. 10.10). Such profiles satisfy (4), which can be written

$$\theta_s(\phi, +0) = \theta_s(\phi - \Gamma, -0) \quad (K \leqslant \phi < \infty), \tag{57}$$

where Γ is the potential difference between two geometrically opposite points, and we have taken $\phi = K$ on $\psi = +0$ as the point D at the

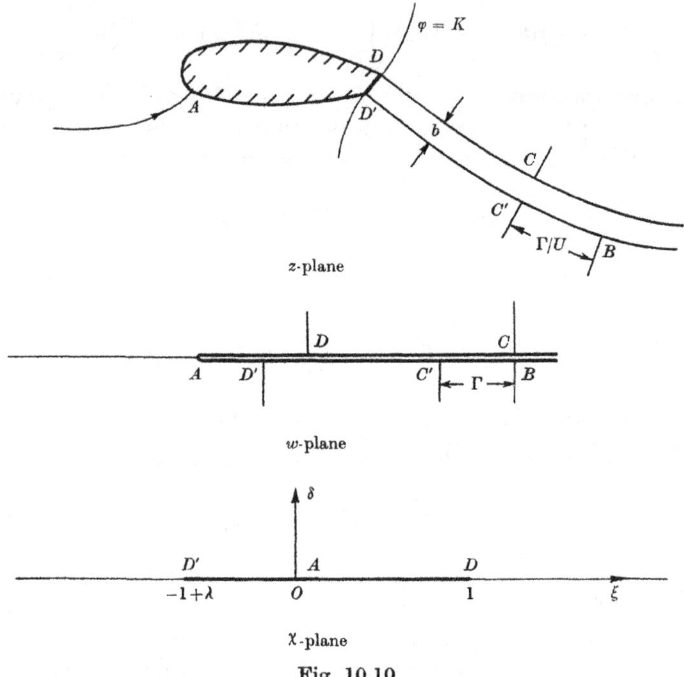

Fig. 10.10

beginning of the parallel section of the profile. Thus from (27) $\xi = 1$ at D; let $\xi = -1+\lambda$ at the geometrically opposite point D', then at DD'

$$\Gamma = \Gamma_0 \equiv K\{1 - (1-\lambda)^2\} = K\lambda(2-\lambda). \tag{58}$$

Generally Γ will vary along the profile from Γ_0 to the value at infinity given in (48), viz.

$$\Gamma_\infty = \frac{L_\infty}{\rho_a U} = -\frac{2K\pi b_2}{\beta_a}$$

$$= -\frac{2K}{\beta_a} \int_{-1}^{1} \Omega_s(\xi)\,\xi\,d\xi - \frac{2K}{\beta_a} \int_{1}^{\infty} \varpi(\xi)\,\xi\,d\xi, \tag{59}$$

where
$$\varpi(\xi) \equiv \Omega_s(\xi) - \Omega_s(-\xi). \tag{60}$$

A linear approximation for Γ is

$$\Gamma(\xi) = \Gamma_0 - \frac{2K}{\beta_a} \int_1^\infty \varpi(\xi)\,\xi\,d\xi, \tag{61}$$

which can be derived as follows. Substituting the linear approximation $q = U(1 - \Omega/\beta_a)$ (see 6.23) into

$$s = \int d\phi/q = 2K \int \xi\,d\xi/q$$

we get
$$sU = \phi + \frac{1}{\beta_a} \int \Omega_s(\xi)\,\xi\,d\xi, \tag{62}$$

for the distance s measured along the profile surface. Applying this to the surfaces DC and $D'B$ shown in Fig. 10.10 we have

$$s_{DC}U = \phi_C - K + \frac{1}{\beta_a} \int_1^\xi \Omega_s(\xi)\,\xi\,d\xi, \tag{63}$$

$$s_{D'B}U = \phi_B - K(1-\lambda)^2 + \frac{1}{\beta_a} \int_{-1}^{-\xi} \Omega_s(\xi)\,\xi\,d\xi. \tag{64}$$

Now $U(s_{D'B} - s_{DC}) \simeq \phi_B - \phi_{C'} = \Gamma$, and $\phi_C = \phi_B$. Hence if the sign of ξ is changed in (64) and the result subtracted from (63), (61) is obtained.

We shall refer to the parallel section of the profile as being its 'sting', then (57) is the sting boundary condition. This is rather difficult to apply so we shall replace it by a more convenient, approximate relation. Provided the sting has a large and slowly varying curvature (57) can be replaced by

$$\theta_s(\phi, +0) = \theta_s(\phi, -0) - \Gamma\frac{d}{d\phi}\{\theta_s(\phi, -0)\} = \theta_s(\phi, -0) - \Gamma\frac{d}{d\phi}\{\theta_{sa}(\phi)\}, \tag{65}$$

where
$$\theta_{sa}(\phi) = \tfrac{1}{2}\{\theta_s(\phi, +0) + \theta_s(\phi, -0)\}, \tag{66}$$

and second-order derivatives have been neglected. By (27) we can write (65)

$$\theta_s(\xi) - \theta_s(-\xi) = -\frac{\Gamma}{2K\xi}\theta'_{sa}(\xi) \quad (|\xi| > 1). \tag{67}$$

Equation (19) can be written

$$\tau(\chi) = \frac{1}{\pi}\int_{-1}^1 \frac{\theta_s(\xi)\,d\xi}{\xi - \chi} + \frac{1}{\pi}\int_1^\infty \left\{\frac{\theta_s(\xi)}{\xi - \chi} - \frac{\theta_s(-\xi)}{\xi + \chi}\right\} d\chi,$$

and substituting (67) in this we get

$$\tau(\chi) = \frac{1}{\pi}\int_{-1}^1 \frac{\theta_s(\xi)\,d\xi}{\xi - \chi} + \frac{2\chi}{\pi}\int_1^\infty \frac{\theta_{sa}(\xi)\,d\xi}{\xi^2 - \chi^2} - \frac{1}{2\pi K}\int_1^\infty \frac{\Gamma\theta'_{sa}(\xi)\,d\xi}{\xi^2 - \chi^2}. \tag{68}$$

Similarly, (22) and (28) can be written

$$0 = \int_{-1}^{1} \theta_s(\xi)\, d\xi + 2\int_{1}^{\infty} \theta_{sa}(\xi)\, d\xi, \tag{69}$$

$$b = \frac{2K}{U} \int_{-1}^{1} \theta_s(\xi)\, \xi\, d\xi - \frac{1}{U}\int_{1}^{\infty} \Gamma\theta_{sa}'(\xi)\, d\xi. \tag{70}$$

Let χ be a point on the profile sting, then it follows from (68) and (60) that

$$\varpi(\chi) = \frac{2\chi}{\pi}\int_{-1}^{1} \frac{\theta_s(\xi)\, d\xi}{\xi^2 - \chi^2} + \frac{4\chi}{\pi}\int_{1}^{\infty} \frac{\theta_{sa}(\xi)\, d\xi}{\xi^2 - \chi^2}. \tag{71}$$

If this value is substituted in (61) and the resulting value of $\Gamma(\xi)$ substituted in (68) there will be obtained the required relation between τ and the boundary values of θ. Of course this is still implicitly an integral equation as $\theta_s(\xi)$ is initially unknown, and so an iterative method of solution similar to that described in §8.3 is necessary. The first step of this iteration yields the linear perturbation theory described below.

Note that the use of (61) implies that as far as the sting is concerned the problem is already partially linearized. This approximation could be eliminated, but there is little point in doing this in view of the use of the approximate sting boundary condition (67). Another point to notice is that the range $-1 < \xi < 1$ of the leading integrals in (68) to (71) includes a small portion $(-1 < \xi < -1+\lambda)$ of the sting itself—an asymmetry in the presentation of the theory which seems unavoidable with the method adopted.

10.11 Linear theory

If θ_s, θ_{sa}, θ_{sa}' are small quantities of first order, ϵ say, (71), (59) and (61) show that this is also true of Γ, Γ_0 and Γ_∞. Hence the last integrals in (68) and (70) are $O(\epsilon^2)$ and can be neglected in a linear theory. Furthermore, as λ is $O(\epsilon)$ by (58), integrals of the form $\int_{-1}^{-1+\lambda} \theta_s \ldots$ are $O(\epsilon^2)$, which means we can assume the 'head' of the semi-infinite profile lies in $-1 \leqslant \xi \leqslant 1$, and that its sting lies in $1 \leqslant \xi < \infty$.

It is also consistent to replace the independent variable w by UZ (see §6.1), whence (27) becomes

$$Z = x + i\beta_a y = c\chi^2 = c(\xi + i\delta)^2, \tag{72}$$

where

$$c \equiv K/U, \tag{73}$$

is the length of the head.

As this step of the linearization is equivalent to the assumption that the boundary condition on the surface $\delta = 0$ can be applied on $y = 0$, it is important to check the validity of this for the sting, which we have seen diverges infinitely from the axis $y = 0$ (see (41)). Such a possibility has been envisaged in (2.51), which can be written

$$E \equiv \Delta y\left(\frac{d\theta}{dy}\right) \simeq \Delta y\left(\frac{d\Omega}{dx}\right) = O(\epsilon^2), \tag{74}$$

by the Cauchy–Riemann relation. From §10.7 $\Delta y = O(\ln x)$, Ω_s, $\theta_s = O(1/x)$, hence if $\epsilon = 1/x$, $E = O(\epsilon^2 \ln \epsilon)$, and (74) is not quite satisfied. However, the important point (see §2.11) is that the error term E be of higher order than the boundary value θ_s, and this is just achieved in the present case. We conclude that linear perturbation theory is valid, but that its accuracy may be low. If the sting is of zero width the position is improved for (42), (47) and (72) show that $\Omega_s = O(x^{-\frac{3}{2}})$, whence $E = O(\epsilon^{2\frac{1}{2}} \ln \epsilon)$, and (74) is satisfied.

MIXED BOUNDARY CONDITIONS

10.12   The general equations for a sting of given pressure distribution

Instead of being given the shape of the sting, in some problems we know or can deduce the pressure distribution over the sting. From this distribution and (6.8) we can deduce Ω_s. Referring to Fig. 10.10 we see that our problem now is to find $\tau(\chi)$ given θ_s in $-1+\lambda < \xi < 1$, and Ω_s in $-\infty < \xi < -1+\lambda, 1 < \xi < \infty$. The asymmetry in these ranges caused by λ could be eliminated by a trivial transformation of χ (see §11.5), but in order to have a theory complementary to that given in §10.10 we shall adopt a different procedure. This is to leave χ as the independent variable, but to assume θ_s given in $|\xi| < 1$, and Ω_s given in $|\xi| > 1$. This amounts to assuming that the initial slope of the sting is given.

The solution to our problem can be deduced from (4.166), or more precisely from its harmonic conjugate (cf. Figs. 4.18 and 10.10). It is necessary only to replace Ω_∞, θ_∞, Ω, θ, a, b, t and z in (4.166) by the appropriate variables for the present problem, viz. $0, 0, i\theta, -i\Omega, -1, 1, \xi$ and χ respectively, to obtain the required solution:

$$\tau(\chi) = \frac{i}{\pi}\left(\frac{\chi+1}{\chi-1}\right)^{\frac{1}{2}}\left\{\int_{-1}^{1} \theta_s\left(\frac{1-\xi}{1+\xi}\right)^{\frac{1}{2}}\frac{d\xi}{\xi-\chi} - \left(\int_{1}^{\infty}+\int_{-\infty}^{-1}\right)\Omega_s\left(\frac{\xi-1}{\xi+1}\right)^{\frac{1}{2}}\frac{d\xi}{\xi-\chi}\right\},$$

as $\tau_\infty = 0$ by (5). This can be written

$$\tau(\chi) = -\frac{i}{\pi}\left(\frac{\chi+1}{\chi-1}\right)^{\frac{1}{2}}\left\{\int_{-1}^{1}\frac{\theta_s(\xi)}{(1-\xi^2)^{\frac{1}{2}}}\frac{\xi-1}{\xi-\chi}d\xi \right.$$

$$\left. + \int_{1}^{\infty}\frac{\varpi(\xi)}{(\xi^2-1)^{\frac{1}{2}}}\frac{\xi^2-\chi}{\xi^2-\chi^2}d\xi + 2\int_{1}^{\infty}\frac{\Omega_{sa}(\xi)}{(\xi^2-1)^{\frac{1}{2}}}\frac{\xi(\chi-1)}{\xi^2-\chi^2}d\xi\right\}, \tag{75}$$

where
$$\Omega_{sa} \equiv \tfrac{1}{2}\{\Omega_s(\xi) + \Omega_s(-\xi)\} \tag{76}$$

and $\varpi(\xi)$ is defined in (60).

From (42) we observe that τ is $O(\chi^{-2})$ near infinity, and therefore from (75) we must have

$$\int_{-1}^{1} \frac{\theta_s(\xi)}{(1-\xi^2)^{\frac{1}{2}}}(1-\xi)\,d\xi + \int_{1}^{\infty} \frac{\varpi(\xi)\,d\xi}{(\xi^2-1)^{\frac{1}{2}}} - 2\int_{1}^{\infty} \frac{\Omega_{sa}(\xi)\,\xi\,d\xi}{(\xi^2-1)^{\frac{1}{2}}} = 0. \tag{77}$$

Let χ be a point on the sting, then χ is real and the imaginary part of (75) is $\theta_s(\chi)$. From this we find that

$$\tfrac{1}{2}\{\theta_s(\xi)+\theta_s(-\xi)\} = \frac{\xi}{\pi(\xi^2-1)^{\frac{1}{2}}}\left\{\int_{-1}^{1} \frac{\theta_s(\xi^*)\,(1-\xi^{*2})^{\frac{1}{2}}}{\xi^{*2}-\xi^2}\,d\xi^* \right.$$
$$\left. -\int_{1}^{\infty} \frac{\varpi(\xi^*)\,(\xi^{*2}-1)^{\frac{1}{2}}}{\xi^{*2}-\xi^2}\,d\xi^*\right\}, \tag{78}$$

$$\theta_s(\xi)-\theta_s(-\xi) = \frac{2}{\pi(\xi^2-1)^{\frac{1}{2}}}\left\{\int_{-1}^{1} \frac{\theta_s(\xi^*)\,\xi^*(1-\xi^{*2})^{\frac{1}{2}}}{\xi^{*2}-\xi^2}\,d\xi^* \right.$$
$$\left. -2\int_{1}^{\infty} \frac{\Omega_{sa}(\xi^*)\,\xi^*(\xi^{*2}-1)^{\frac{1}{2}}}{\xi^{*2}-\xi^2}\,d\xi^*\right\}, \tag{79}$$

on making some use of (77).

To this point the theory is quite general. We shall now apply it to the profile shown in Fig. 10.10. First notice from (78) that the average $\theta_{sa}(\xi) \equiv \tfrac{1}{2}(\theta_s(\xi)+\theta_s(-\xi))$ is finite at the beginning of the sting, $\xi = 1$, only if

$$\int_{-1}^{1} \frac{\theta_s(\xi)\,d\xi}{(1-\xi^2)^{\frac{1}{2}}} + \int_{1}^{\infty} \frac{\varpi(\xi)\,d\xi}{(\xi^2-1)^{\frac{1}{2}}} = 0. \tag{80}$$

In this case it follows from (77) that

$$\int_{-1}^{1} \frac{\theta_s(\xi)\,\xi\,d\xi}{(1-\xi^2)^{\frac{1}{2}}} + 2\int_{1}^{\infty} \frac{\Omega_{sa}(\xi)\,\xi\,d\xi}{(\xi^2-1)^{\frac{1}{2}}} = 0, \tag{81}$$

and a glance at (67) and (79) shows this to be the condition that $\theta'_{sa}(\xi)$ be finite at $\xi = 1$.

10.13 Expressions for the pressure, lift and moment

Let ξ be a point on the head of the profile, i.e. $-1 \leqslant \xi \leqslant 1$, then the real part of (75) yields

$$\Omega_s(\xi) = -\frac{1}{\pi}\left(\frac{1+\xi}{1-\xi}\right)^{\frac{1}{2}}\left\{\int_{-1}^{1} \frac{\theta_s(\xi^*)}{(1-\xi^{*2})^{\frac{1}{2}}}\frac{\xi^*-1}{\xi^*-\xi}\,d\xi^* \right.$$
$$\left. +\int_{1}^{\infty} \frac{\varpi(\xi^*)}{(\xi^{*2}-1)^{\frac{1}{2}}}\frac{\xi^{*2}-\xi}{\xi^{*2}-\xi^2}\,d\xi^* + 2\int_{1}^{\infty} \frac{\Omega_{sa}(\xi^*)}{(\xi^{*2}-1)^{\frac{1}{2}}}\frac{\xi^*(\xi-1)}{\xi^{*2}-\xi^2}\,d\xi^*\right\}, \tag{82}$$

and the corresponding pressure can be deduced from this and the p_s, Ω_s relation given in (6.8). If linear perturbation theory is valid we can put $C_{p_s} \simeq \Omega_s/\beta_a$.

By (59)
$$L_\infty = -\frac{2K\rho_a U}{\beta_a}\left\{\int_{-1}^{1}\Omega_s(\xi)\,\xi\,d\xi + \int_{1}^{\infty}\varpi(\xi)\,\xi\,d\xi\right\}, \tag{83}$$

and the first integral in this expression can be evaluated directly from (82). It is found that

$$\int_{-1}^{1}\Omega_s(\xi)\,\xi\,d\xi = -\int_{-1}^{1}\theta_s(\xi)\,(1-\xi^2)^{\frac{1}{2}}\,d\xi$$
$$+ \int_{1}^{\infty}\varpi(\xi)\,\{(\xi^2-1)^{\frac{1}{2}} - \xi\}\,d\xi,$$

the validity of the interchange of the order of integration involved here being easily established. Hence

$$L_\infty = \frac{2K\rho_a U}{\beta_a}\left\{\int_{-1}^{1}\theta_s(\xi)\,(1-\xi^2)^{\frac{1}{2}}\,d\xi - \int_{1}^{\infty}\varpi(\xi)\,(\xi^2-1)^{\frac{1}{2}}\,d\xi\right\}. \tag{84}$$

Turning now to the moment, we can express (50) in the form

$$M_\infty = \frac{2K^2\rho_a}{\beta_a}\left\{\int_{-1}^{1}\Omega_s(\xi)\,\xi^3\,d\xi + \int_{1}^{\infty}\varpi(\xi)\,\xi^3\,d\xi\right\}, \tag{85}$$

and then find from (82) that

$$\int_{-1}^{1}\Omega_s(\xi)\,\xi^3\,d\xi = -\int_{-1}^{1}\theta_s(\xi)\,(1-\xi^2)^{\frac{1}{2}}\,(\xi^2+\tfrac{1}{2})\,d\xi$$
$$+ \int_{1}^{\infty}\varpi(\xi)\,\{(\xi^2-1)^{\frac{1}{2}}\,(\xi^2+\tfrac{1}{2}) - \xi^3\}\,d\xi. \tag{86}$$

Therefore

$$M_\infty = -\frac{2K^2\rho_a}{\beta_a}\left\{\int_{-1}^{1}\theta_s(\xi)\,(1-\xi^2)^{\frac{1}{2}}\,\xi^2\,d\xi - \int_{1}^{\infty}\varpi(\xi)\,(\xi^2-1)^{\frac{1}{2}}\,\xi^2\,d\xi\right\} - \frac{K}{2U}L_\infty. \tag{87}$$

As noted in §10.8, in the general case M_∞ is infinite. However in the application we shall make of the theory below it is only the moment over the *head* of the profile that interests us. This moment is proportional to the right-hand side of (86), which is finite, the last integrand being $O(\xi^{-2})$ near $\xi = \infty$.

THE JET FLAP

10.14   Description

The jet flap is an aerofoil lifting device having some similarity to the solid flap described in §8.20. In place of the solid flap there is a thin sheet of high-velocity air ejected at an angle to the main stream from a slot at the trailing edge of the aerofoil. This jet of air produces lift and thrust on the aerofoil, which depend principally on the momentum $\mathcal{M}$ of the jet, its angle $\mathcal{T}$ to the aerofoil chord, and α the aerofoil incidence. This lifting device was first proposed by Hagedorn & Ruden (1938), but the advent of the jet engine was required to make the scheme really feasible. Davidson (1956) and Stratford (1956) have recently stimulated fresh interest in it, and since then much experimental and theoretical work has been done.

An important idealization of the problem is obtained on assuming, as in § 1.14, that there is no mixing of the jet stream and main-stream fluids. In this case the flow will appear as shown in Fig. 10.14—also see Fig. 1.14. This model of the flow, which ignores the effects of turbulence and viscous

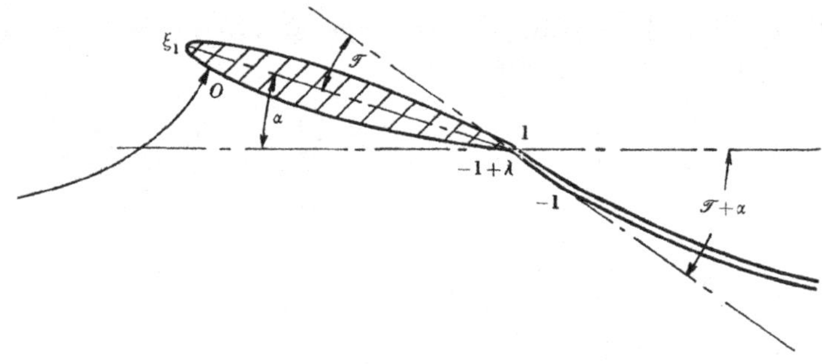

Fig. 10.14

mixing of the jet and main stream, is the one adopted below. The lift and thrust derived from it can be regarded as the 'ideal' values, and intuitively we would expect these values to exceed the real values.

Let the jet width downstream at infinity be b, then if ρ_0, q_0 are the corresponding jet density and velocity, the ideal thrust T on the aerofoil has been shown in § 1.14 to be

$$T = b\rho_0 q_0^2 = \mathcal{M}. \tag{88}$$

Hence if momentum and thrust coefficients are defined by

$$C_J \equiv \frac{b\rho_0 q_0^2}{\frac{1}{2}c\rho_a U^2}, \quad C_T \equiv \frac{T}{\frac{1}{2}c\rho_a U^2}, \tag{89}$$

where c is the aerofoil chord, and the subscript 'a' denotes conditions at infinity as usual ($U = q_a$), then

$$C_T = C_J. \tag{90}$$

This is the jet thrust paradox (see last paragraph of § 1.14), first noted by Stratford (1956), and proved generally by Woods (1956c).

Comparison of Figs. 10.10 and 10.14 shows that the aerofoil plus jet combination can be regarded as being a semi-infinite profile, and so all the theory of earlier sections is applicable, but first we need the jet boundary condition.

10.15   The jet boundary condition

The jet is assumed for our purposes to be a narrow irrotational stream of high-speed fluid, not mixing with the main stream, but separated from it by two vortex sheets as shown in Fig. 10.15. As the jet stream is irrotational it follows from (2.21) that in the jet $\partial q/\partial n = q/R$, where R is the radius of curvature of the jet streamlines, and n is measured in a direction normal to these streamlines. As the jet is narrow R can be assumed

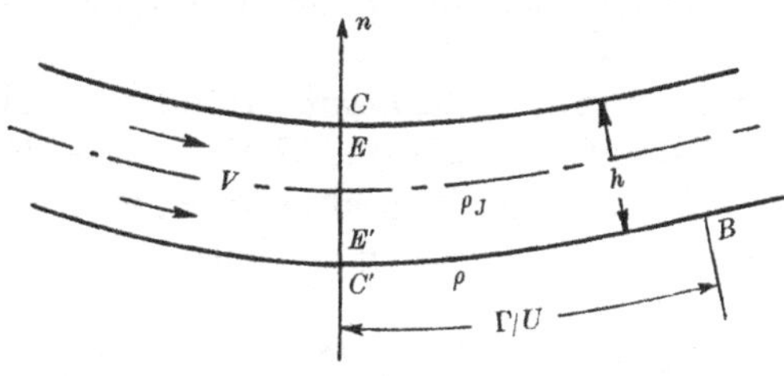

Fig. 10.15

constant across any section EE', and consequently integration from the centre line of the jet, where $q = V$ say, to E and E', points just inside the jet boundaries (see figure), gives

$$q_E = V\, e^{h/2R} \simeq V\left(1+\frac{h}{2R}\right), \quad q_{E'} = V\, e^{-h/2R} \simeq V\left(1-\frac{h}{2R}\right), \quad (91)$$

where h is the local jet width, which is assumed small compared with R.

From (1.3) and (1.23) we find a form of Bernoulli's equation, which gives for the jet

$$\rho_J q_E^2 + \frac{2n}{n+1}p_E = \rho_J q_{E'}^2 + \frac{2n}{n+1}p_{E'},$$

where ρ_J is the jet fluid density and we have neglected density variations across the jet. Similarly, for two points C, C' in the main stream just outside E and E',

$$\rho_a q_C^2 + \frac{2n}{n+1}p_C = \rho_a q_{C'}^2 + \frac{2n}{n+1}p_{C'}.$$

The continuity of pressure across the vortex sheets enables us to subtract these equations and find

$$q_C^2 - q_{C'}^2 = \frac{\rho_J}{\rho_a}(q_E^2 - q_{E'}^2) \simeq \frac{2\rho_J V^2 h}{\rho_a R},$$

the last form of which follows from (91) and neglect of terms $O(h/R)^3$. At this point we adopt the approximation $q = U(1 - \Omega/\beta_a)$ of linear perturbation theory (see § 2.17), and find that

$$\Omega_C - \Omega_{C'} = -\beta_a \frac{\rho_J V^2 h}{\rho_a U^2 R} = -\beta_a \tfrac{1}{2} c C_J \frac{1}{R}, \tag{92}$$

where C_J is here the *local* momentum coefficient.

Let the pressure on the jet centre-line equal p_J then except possibly in the immediate neighbourhood of the jet exit, p_J will vary only slowly along the jet. We shall simplify the theory by assuming that $p_J = p_a$, which is equivalent to assuming V is constant and equal to its value at infinity, q_0. Conservation of mass now gives that the jet width h is constant and equal to b, and it therefore follows that C_J is constant along the jet, and equal to the value defined in (89).

An important consequence of the assumption $p_J = p_a$ is that $\Omega_C + \Omega_{C'}$ is of order $(h/R)^2$, and can be neglected. This can be shown as follows. Adding the Bernoulli equations for points E, E', and also for the points C, C', then subtracting the resulting equations we get

$$\rho_a(q_C^2 + q_{C'}^2) - \rho_J(q_E^2 + q_{E'}^2) = 2\rho_a U^2 - 2\rho_J V^2 + \frac{4n}{n+1}(p_a - p_J).$$

Equations (91) and $q = U(1 - \Omega/\beta_a)$ enable us to reduce this to

$$\frac{\rho U^2}{\beta_a}(\Omega_C + \Omega_{C'}) = \frac{2n}{n+1}(p_J - p_a) + O\left(\frac{h}{R}\right)^2 = 0. \tag{93}$$

Employing the linear approximations $\phi = K\xi^2 \simeq sU$, $K \simeq cU$ (see (27) and (73)) we have

$$\frac{c}{R} = c\frac{d\theta_s}{ds} \simeq cU\frac{d\theta_s}{d\phi} = \frac{cU}{2K\xi}\theta_s'(\xi) \simeq \frac{1}{2\xi}\theta_{sa}'(\xi), \tag{94}$$

on ignoring second-order derivatives of θ (cf. (65)). Similarly (cf. (57) and (65))

$$\Omega_C - \Omega_{C'} = \Omega_s(\phi, +0) - \Omega_s(\phi - \Gamma, -0) \simeq \Omega_s(\phi, +0)$$
$$- \Omega_s(\phi, -0) + \Gamma\Omega_{sa}'(\phi),$$

where Ω_{sa} is defined in (76), and

$$\Omega_C + \Omega_{C'} = \Omega_s(\phi, +0) + \Omega_s(\phi, -0) - \Gamma\Omega_{sa}'(\phi) = 2\Omega_{sa}(\phi) - \Gamma\Omega_{sa}'(\phi).$$

This last result and (93) shows $\Gamma\Omega_{sa}'$ to be of order $\{\Gamma(\Gamma\Omega_{sa}')'\}$ and so it can be consistently ignored as a second-order derivative. Hence

$$\Omega_C - \Omega_{C'} = \Omega_s(\phi, +0) - \Omega_s(\phi, -0) = \varpi(\phi), \tag{95}$$

by (60).

Substituting (94) and (95) into (92) we arrive at the required jet boundary condition, namely

$$\varpi(\xi) = -\tfrac{1}{4}\beta_a C_J \frac{1}{\xi}\theta'_{sa}(\xi). \tag{96}$$

At this point it is pertinent to inquire if (96) is uniformly valid over the length of the jet. Clearly it cannot hold right at the trailing edge, for in an ideal flow the velocity in the fluid just outside the jet at the trailing edge will be infinite on one side and zero on the other (see Fig. 10.14), whereas (96) is based on the assumption that the difference in these velocities is small. This failure of (96) is akin to the failure of classical linear perturbation theory at solid flap hinges, and in a first-order theory can be ignored.

10.16   The leading edge potential

Let ϕ_1 be the velocity potential at the 'leading edge',[†] and let ξ_1 be the corresponding value of ξ, then by (73)

$$\phi_1 = K\xi_1^2 = cU\xi_1^2. \tag{97}$$

We shall now establish the important result

$$2\pi\frac{\phi_1}{cU} = 2\pi\xi_1^2 = \beta_a\{\alpha C_{L\infty} + \tfrac{1}{2}C_J(\mathscr{T}^2 - \alpha^2)\}, \tag{98}$$

where $\mathscr{T}$ and α are defined in Fig. 10.14, and $C_{L\infty}$ is the lift coefficient corresponding to the total lift force L_∞.

With thin aerofoils we observe from Fig. 10.14 that

$$\theta_s = \begin{cases} \left(\dfrac{dy}{dx}\right)_s -\alpha + \pi & (0 < \xi < \xi_1), \\[2ex] \left(\dfrac{dy}{dx}\right)_s -\alpha & (-1+\lambda < [\xi < 0, \quad \xi_1 < \xi < 1), \\[2ex] -\mathscr{T} -\alpha & (-1 < \xi < -1+\lambda), \end{cases} \tag{99}$$

where $(dy/dx)_s$ is the surface slope measured from the chord line, and we have taken into account that the front stagnation point is at $\xi = 0$, and that the interval $-1 < \xi < -1+\lambda$ lies on the jet (see § 10.10). Hence by (58) and (73)

$$\frac{2K}{U}\int_{-1}^{1}\theta_s(\xi)\,\xi\,d\xi = \frac{\mathscr{T}K}{U}\{1^2 - (1-\lambda)^2\} + I + \frac{\pi K}{U}\xi_1^2$$

$$= \frac{\mathscr{T}\Gamma_0}{U} + I + \pi c\xi_1^2, \tag{100}$$

[†] By 'leading edge' here we mean the point occupied by the front stagnation point when both the absolute incidence α and jet deflexion angle $\mathscr{T}$ vanish.

where
$$I \equiv \frac{1}{U}\int_{-1+\lambda}^{1}\left(\frac{dy}{dx}\right)_s d\phi(\xi),$$

is an integral taken over the aerofoil surface. If the linear approximation $d\phi \simeq U\,ds \simeq dx$ is used here, the result is

$$I = \int_{-1+\lambda}^{1} dy_s(\xi) = b, \tag{101}$$

because the discontinuity in y_s at the trailing edge equals the (constant) jet width b. However, this value of I is not correct in compressible flow, as a glance at (8.59) will show. This equation is

$$I = \frac{\Delta\phi}{U}\pi\left(\frac{1}{\beta_a}-1\right), \tag{102}$$

but applies to the case of a closed aerofoil ($b = 0$). Here $\Delta\phi$ is the difference in potential between the points occupied by the stagnation points at zero lift (i.e. at $\alpha = 0$ as α is the absolute incidence) and at incidence α. In the present case this equals the leading edge potential, ϕ_1, hence $\Delta\phi/U = K\xi_1^2/U \simeq c\xi_1^2$ by (73). Combining (101) and (102) we conclude that

$$I = b + \pi c\xi_1^2\left(\frac{1}{\beta_a}-1\right). \tag{103}$$

Substituting (96) in (61),

$$\Gamma(\xi) = \Gamma_0 + \tfrac{1}{2}cUC_J\{\theta_{sa}(\xi) - \theta_{sa}(1)\}$$

$$= \Gamma_0 + \tfrac{1}{2}cUC_J(\mathscr{T}+\alpha) + \tfrac{1}{2}cUC_J\theta_{sa}(\xi),$$

as
$$\theta_{sa}(1) = -(\mathscr{T}+\alpha) \tag{104}$$

(see Fig. 10.14). Therefore

$$\Gamma_\infty = \Gamma_0 + \tfrac{1}{2}cUC_J(\mathscr{T}+\alpha), \tag{105}$$

and
$$\Gamma(\xi) = \Gamma_\infty + \tfrac{1}{2}cUC_J\theta_{sa}(\xi). \tag{106}$$

It follows that

$$\frac{1}{U}\int_{1}^{\infty}\Gamma(\xi)\,\theta'_{sa}(\xi)\,d\xi = \frac{1}{U}\Gamma_\infty(\mathscr{T}+\alpha) + \tfrac{1}{2}cC_J\int_{1}^{\infty}\theta_{sa}\theta'_{sa}\,d\xi$$

$$= \frac{1}{U}\Gamma_\infty(\mathscr{T}+\alpha) - \tfrac{1}{4}cC_J(\mathscr{T}+\alpha)^2.$$

On substituting this result, (100) and (105) in (70) we obtain

$$2\pi\xi_1^2 = \beta_a\left\{\left(\frac{2\Gamma_\infty}{cU}\right)\alpha + \tfrac{1}{2}C_J(\mathscr{T}^2-\alpha^2)\right\},$$

and it remains only to observe that

$$C_{L\infty} \equiv \frac{2L_\infty}{\rho_a c U^2} = \frac{2\Gamma_\infty}{cU},\tag{107}$$

to complete the proof of (98). In $\xi\,10.18$ we shall establish that the lift force L on the aerofoil is equal to the total lift L_∞ acting on the semi-infinite profile of aerofoil plus jet. Anticipating this result we can drop the subscript ∞ in (98), and write

$$2\pi\xi_1^2 = \beta_a\{\alpha C_L + \tfrac{1}{2}C_J(\mathscr{T}^2 - \alpha^2)\}.\tag{108}$$

10.17   A relation between the derivatives of the lift coefficient

A most remarkable and important conclusion can be deduced from the equation for ξ_1.

For a given value of C_J both ξ_1 and C_L are functions of α and $\mathscr{T}$, which vanish when α and $\mathscr{T}$ are zero. We shall assume α and $\mathscr{T}$ to be small first-order quantities, $O(\epsilon)$ say, then we can write

$$C_L = a_\alpha \alpha + a_\mathscr{T} \mathscr{T} + O(\epsilon^2),\tag{109}$$

where
$$a_\alpha \equiv \frac{\partial C_L}{\partial\alpha}, \quad a_\mathscr{T} \equiv \frac{\partial C_L}{\partial\mathscr{T}} \quad \text{at} \quad \alpha = \mathscr{T} = 0.\tag{110}$$

A similar expansion holds for ξ_1. Substitution in (108) gives

$$\left\{\left(\frac{\partial\xi_1}{\partial\alpha}\right)\alpha + \left(\frac{\partial\xi_1}{\partial\mathscr{T}}\right)\mathscr{T}\right\}^2 = \frac{\beta_a}{2\pi}\{(a_\alpha - \tfrac{1}{2}C_J)\alpha^2 + a_\mathscr{T}\alpha\mathscr{T} + \tfrac{1}{2}C_J\mathscr{T}^2\} + O(\epsilon^3),\tag{111}$$

the derivatives of ξ_1 being those at $\alpha = \mathscr{T} = 0$. This equation must hold identically for all (small) values of α and $\mathscr{T}$. Hence

$$\frac{\partial\xi_1}{\partial\alpha} = \left\{\frac{\beta_a}{2\pi}(a_\alpha - \tfrac{1}{2}C_J)\right\}^{\frac{1}{2}}, \quad \frac{\partial\xi_1}{\partial\mathscr{T}} = \left(\frac{\beta_a C_J}{4\pi}\right)^{\frac{1}{2}},\tag{112}$$

and
$$a_\mathscr{T}^2 = 2C_J(a_\alpha - \tfrac{1}{2}C_J).\tag{113}$$

This last equation is just the condition that the quadratic form on the right-hand side of (111) be an exact square. It can be written

$$\frac{\partial C_L}{\partial\alpha} = \frac{1}{2C_J}\left(\frac{\partial C_L}{\partial\mathscr{T}}\right)^2 + \tfrac{1}{2}C_J.\tag{114}$$

This surprising result is due to Spence (1956), who proved it by a rather different argument based on a study of the suction force at the nose of the aerofoil. The above proof is shorter and more general than Spence's, whose contributions to jet-flap theory will be described in § 10.19.

10.18   The lift and moment coefficients

In the calculation of the forces acting on the aerofoil allowance must be made for the pressure acting on the walls of the (interior) duct conveying the jet to the trailing edge. Fortunately this does not require a detailed knowledge of the actual duct shape. For consider the flow shown in Fig. 10.18. The plane surface ef shown in the duct is at right-angles to the duct at a point where the duct is parallel to the x-axis, i.e. ef is parallel to the lift force. We shall assume that the source of the jet fluid, lying upstream of ef, ejects fluid symmetrically into the duct, so that no contribution to the lift or moment comes from the pressures acting on the portion of the duct upstream of ef. This means that the lift and moment can be derived from integrals involving the pressure and momentum flux taken over the surfaces S_1 and S_2, where S_1 is the external surface hag, and S_2 is the interior surface $hefg$.

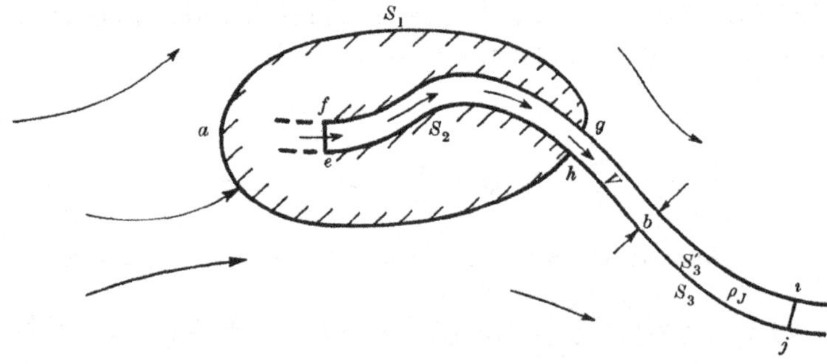

Fig. 10.18

The theorems given in § 1.13 allow us to replace the surface $S_1 + S_2$ in these integrals by the surface $S_1 + S_3'$, where S_3' is the interior surface $gijh$. In turn S_3' can be replaced by S_3, the external surface $gijh$, as the pressure is continuous across S_3. The surface ij can be taken anywhere along the jet, but for the purposes of calculation the most convenient position is either at the jet exit gh or downstream at infinity, for elsewhere the direction of the jet flow is initially unknown. We shall take the section ij to be at infinity in our calculations, then S_3 becomes the total (external) surface of the jet.

First consider the lift force. From the above remarks the total lift L on the aerofoil equals the lift force on $S_1 + S_3$, because the horizontal momentum flux through ij (at infinity) makes no contribution to L. The lift on $S_1 + S_3$ is the total lift L_∞ on the semi-infinite profile of aerofoil plus jet. Consequently we have the simple results

$$L = L_\infty, \quad C_L = C_{L\infty}, \tag{115}$$

and it follows from (84) and (96) that

$$L = \frac{2K\rho_a U}{\beta_a}\left\{\int_{-1}^{1}\theta_s(\xi)\,\sqrt{(1-\xi^2)}\,d\xi + \frac{\beta_a C_J}{4}\int_{1}^{\infty}\theta'_{sa}(\xi)\,\frac{\sqrt{(\xi^2-1)}}{\xi}\,d\xi\right\}.$$

Integrating the last term by parts, and putting $K = Uc$ (see (73)) we arrive at

$$C_L = \frac{L}{\tfrac{1}{2}\rho_a c U^2} = \frac{4}{\beta_a}\int_{-1}^{1}\theta_s(\xi)\,\sqrt{(1-\xi^2)}\,d\xi - C_J\int_{1}^{\infty}\frac{\theta_{sa}(\xi)\,d\xi}{\xi^2\sqrt{(\xi^2-1)}}. \tag{116}$$

The moment calculation is a little more difficult because the horizontal momentum flux through ij is by (49) a logarithmically infinite distance below the x-axis, and so exerts an infinite nose-up moment about the aerofoil's leading edge. This is offset by a logarithmic infinity in M_∞ (see last paragraph of § 10.13), and our first task is to evaluate the combined effect of these two infinities.

Instead of having the section ij at infinity let it be at $\xi = R$, where R is large, then ignoring terms of second order in $\theta_{sa}(R)$, we have that the momentum flux through ij has Ox, Oy components $\rho_J b V^2$, $-\rho_J b V^2\theta_{sa}(R)$ and consequently produces a nose-up moment about the leading edge of

$$M_J(R) = \rho_J b V^2\{y_{sa}(R) - x(R)\,\theta_{sa}(R)\},$$

assuming the origin of the xy-plane to be at the leading edge. Now consider the function

$$I(R) \equiv M_J(R) + \tfrac{1}{2}\rho_a c^2 U^2\frac{4}{\beta_a}\int_{1}^{R}\varpi(\xi)\,\xi^3\,d\xi. \tag{117}$$

By (96) and the definition of C_J we can write this in the form

$$I(R) = -\tfrac{1}{2}\rho_a c U^2 C_J\left\{c\int_{1}^{R}\theta'_{sa}(\xi)\,\xi^2\,d\xi + y_{sa}(R) - x(R)\,\theta_{sa}(R)\right\}.$$

The linearizations $\theta_{sa} = (dy/dx)_{sa}$ and

$$x = \frac{\phi}{U} = \frac{K}{U}\xi^2 = c\xi^2, \tag{118}$$

plus integration by parts enable us to reduce $I(R)$ to

$$I(R) = -\tfrac{1}{2}\rho_a c U^2 C_J[y_{sa}(1) - c\theta_{sa}(1)]$$
$$= -\tfrac{1}{2}\rho_a c^2 U^2 C_J\mathscr{T},$$

as $y_{sa}(1) = -\alpha c$ and $\qquad \theta_{sa}(1) = -(\mathscr{T}+\alpha),$ (119)

(see Fig. 10.14). Therefore

$$M_J(R) = -\tfrac{1}{2}\rho_a c^2 U^2\left\{C_J\mathscr{T} + \frac{4}{\beta_a}\int_{1}^{R}\varpi(\xi)\,\xi^3\,d\xi\right\}. \tag{120}$$

Now the moment on the aerofoil equals the moment acting on the surface $S_1 + S_3$ of the semi-infinite profile plus the moment just calculated due to the momentum flux through ij. From (85) and (86) the moment on the profile up to $\xi = R$ is

$$M_R = -\frac{2c^2 U^2 \rho_a}{\beta_a}\left\{\int_{-1}^{1} \theta_s(\xi)\,\xi^2\,\sqrt{(1-\xi^2)}\,d\xi\right.$$
$$\left. -\int_{1}^{\infty} \varpi(\xi)\{\xi^2\,\sqrt{(\xi^2-1)}-\xi^3\}\,d\xi+\int_{1}^{R} \varpi(\xi)\,\xi^3\,d\xi\right\} -\tfrac{1}{2}cL, \quad (121)$$

making use of (73), (84) and (115). In the limit $R \to \infty$, $M_R \to M_\infty$, which is infinite owing to the last integral of (121).

It now remains only to add (120) and (121) to obtain the moment on the aerofoil, M. The moment coefficient is

$$C_M \equiv \frac{M}{\tfrac{1}{2}\rho_a c^2 U^2} = -\frac{4}{\beta_a}\left\{\int_{-1}^{1}\theta_s(\xi)\,\xi^2\,\sqrt{(1-\xi^2)}\,d\xi\right.$$
$$\left. -\int_{1}^{\infty}\varpi(\xi)\,[\xi^2\,\sqrt{(\xi^2-1)}-\xi^3]\,d\xi\right\} -C_J\mathscr{T}-\tfrac{1}{2}C_L.$$

Substituting (96) in this equation and then integrating by parts we find

$$C_M = -\frac{4}{\beta_a}\int_{-1}^{1}\theta_s(\xi)\,\xi^2\,\sqrt{(1-\xi^2)}\,d\xi$$
$$+C_J\int_{1}^{\infty}\theta_{sa}(\xi)\left\{\frac{2\xi^2-1}{\sqrt{(\xi^2-1)}}-2\xi\right\}d\xi+C_J\alpha-\tfrac{1}{2}C_L. \quad (122)$$

Equations (116) and (122) are the most convenient forms we can find for the coefficients C_L and C_M, short of calculating the jet slope θ_{sa} and evaluating the integrals containing it. We shall show how to calculate θ_{sa} in the next section.

The aerofoil slope θ_s is defined completely by (99) provided ξ_1 and λ are known. Now ξ_1 is given by (108), and from (58), (73), (105), (107) and (115)

$$(2-\lambda)\lambda = \frac{\Gamma_0}{K} = \tfrac{1}{2}C_L-\tfrac{1}{2}C_J(\mathscr{T}+\alpha). \quad (123)$$

Thus ξ_1 and λ are functions of C_L, and (116) is not really an explicit equation for C_L. However once θ_{sa} is determined (108), (116) and (123) do define C_L uniquely in terms of C_J, $\mathscr{T}$ and α. With C_L evaluated, (122) is an explicit equation for C_M.

10.19   Integro-differential equations for the jet slope

There are two integro-differential equations that can be written down immediately for the jet slope θ_{sa}. The first of these is

$$-\tfrac{1}{4}\beta_a C_J \theta'_{sa}(\xi) = \frac{2\xi^2}{\pi}\int_{-1}^{1}\frac{\theta_s(\xi^*)\,d\xi^*}{\xi^{*2}-\xi^2}+\frac{4\xi^2}{\pi}\int_{1}^{\infty}\frac{\theta_{sa}(\xi^*)\,d\xi^*}{\xi^{*2}-\xi^2}, \quad (124)$$

which follows from (71) and (96). Integrating the last term by parts, and using (119), we can convert (124) into the quasi-regular Fredholm integral equation (see §4.21) for θ'_{sa}

$$-\tfrac{1}{4}\beta_a C_J \theta'_{sa}(\xi) = \frac{2\xi^2}{\pi}\int_{-1}^{1}\frac{\theta_s(\xi^*)\,d\xi^*}{\xi^{*2}-\xi^2} + \frac{2}{\pi}(\mathscr{T}+\alpha)\,\xi\ln\left|\frac{1-\xi}{1+\xi}\right|$$

$$-\frac{2\xi}{\pi}\int_{1}^{\infty}\theta'_{sa}(\xi^*)\ln\left|\frac{\xi^*-\xi}{\xi^*+\xi}\right|d\xi^*. \quad (125)$$

The second integro-differential equation follows from (78) and (96). It is

$$\theta_{sa}(\xi) = \frac{\xi}{\pi(\xi^2-1)^{\frac{1}{2}}}\left\{\int_{-1}^{1}\frac{\theta_s(\xi^*)(1-\xi^{*2})^{\frac{1}{2}}}{\xi^{*2}-\xi^2}\,d\xi^*\right.$$

$$\left.+\tfrac{1}{4}\beta_a C_J\int_{1}^{\infty}\frac{\theta'_{sa}(\xi^*)(\xi^{*2}-1)^{\frac{1}{2}}}{\xi^*(\xi^{*2}-\xi^2)}\,d\xi^*\right\}, \quad (126)$$

and this could be changed into an integral equation for θ_{sa} by integrating the last term by parts. Unfortunately this step cannot be taken as it yields a divergent integral, as noted for the general case in the last paragraph of §4.21.

Both (125) and (126) are due to Woods (1954c, 1956d), who was unable to find analytical solutions for them. He found that iterative methods of solution could not be carried far owing to the emergence of divergent integrals in the process. The first step of the Liouville–Neumann iterative method (see Whittaker & Watson, 1952, p. 221), applied to (126) for small C_J, was found to yield a jet slope θ_{sa}, which on being substituted in (116) gave the approximate result (for incompressible flow)

$$C_L \simeq 2\pi\alpha + \frac{4\mathscr{T}}{\sqrt{\pi}}C_J^{\frac{1}{2}}\left\{1+0\cdot76\left(1-\frac{2C_Q}{C_J}\right)\right\}\quad (2C_Q < C_J < 1), \quad (127)$$

where C_Q is the jet mass-coefficient

$$C_Q \equiv \frac{\rho_J bV}{\rho_a cU}. \quad (128)$$

Similarly, it was deduced from (125) that for large C_J,

$$C_L \simeq 2\pi\alpha + \frac{\pi^2}{4}C_J(\mathscr{T}+\alpha)\quad (1 < C_J). \quad (129)$$

If $V = U$, the discontinuity in velocity across the jet boundary vanishes, and we have a source flow of the type described in the last paragraph of §8.29. It follows from (89) and (128) that in this case $C_J = 2C_Q$, and so (127) correctly reduces to the exact result quoted at the end of §8.29.

In Fig. 10.19 we have compared (127) and (129) at $\mathscr{T} = 31\frac{1}{2}°$, $\alpha = 0°$ with some experimental values obtained by Dimmock (1957). The agreement is fair, especially if a constant is added to (129) to make it agree with (127) at $C_J = 1$.

A different approach was adopted by Spence (1956), who used a numerical method of solving (126). Briefly, and slightly simplified, his method is first to eliminate the logarithmic singularity at the trailing

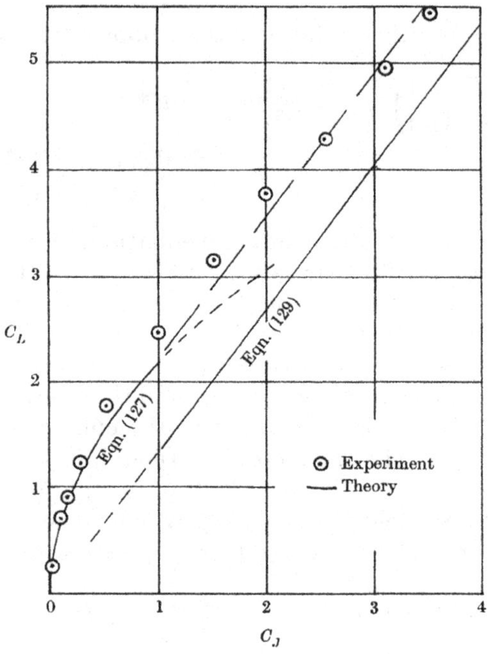

Fig. 10.19

edge (cf. the second right-hand term of (125)) by transforming from $\theta_{sa}(\xi)$ to a new well-behaved function $F(\gamma)$, with $\gamma = 2\sec^{-1}\xi$ $(0 < \gamma < \pi)$. A Fourier expansion is now assumed for $F(\gamma)$, and this permits the first and last terms in (126) to be replaced by sums involving the unknown Fourier coefficients a_n of F. If the Fourier series is terminated at the Nth term, the modified form of (126) will now involve only the first N of the coefficients a_n. This equation is next satisfied exactly at the N pivotal points $\gamma_m = m\pi/N$, $m = 0, 1, 2, ..., N-1$, which provides sufficient relations to determine the N coefficients a_n. When these coefficients have been evaluated, and $F(\gamma)$ found, an expression for $\theta_{sa}(\xi)$ can be deduced and substituted into (116) and (122) to give the values of C_L and C_M. For further details the reader should consult Spence's paper. The method, with 9 pivotal points, yields the following table of derivatives of C_L and C_M at $\mathscr{T} = \alpha = 0$.

Table 10.19.   *Derivatives at $\mathscr{T} = \alpha = 0$*

C_J	$\dfrac{\partial C_L}{\partial \alpha}$	$\dfrac{\partial C_L}{\partial \mathscr{T}}$	$\dfrac{\partial C_M}{\partial \alpha}$	$\dfrac{\partial C_M}{\partial \mathscr{T}}$
0	6·283	0	-1·571	0
0·05	6·460	0·803	-1·626	-0·408
0·1	6·614	1·137	-1·668	-0·588
0·2	6·885	1·641	-1·739	-0·863
0·4	7·365	2·391	-1·850	-1·295
0·5	7·588	2·707	-1·898	-1·485
1·0	8·605	4·026	-2·097	-2·315
1·5	9·532	5·133	-2·258	-3·054
2·0	10·405	6·135	-2·396	-3·745
3·0	12·050	7·958	-2·633	-5·055
4·0	13·609	9·638	-2·836	-6·307
5·0	15·109	11·231	-3·016	-7·525
10·0	22·116	18·504	-3·727	-13·341

Spence finds that the derivatives of C_L are given fairly closely by

$$\frac{\partial C_L}{\partial \mathscr{T}} = 3\cdot54C_J^{\frac{1}{2}} + 0\cdot325C_J + 0\cdot156C_J^{\frac{3}{2}}, \tag{130}$$

and
$$\frac{\partial C_L}{\partial \alpha} = 2\pi + 1\cdot152C_J^{\frac{1}{2}} + 1\cdot106C_J + 0\cdot051C_J^{\frac{3}{2}}. \tag{131}$$

The comparison of these results with experiment will be deferred until another approach to the problem of determining the jet slope has been presented.

10.20   An approximation for the jet slope

While the numerical method just described successfully solves the jet-flap problem, it has two defects from a theoretical viewpoint. These are (i) it throws no light on the role of the obviously important restrictions (49), (69) and (114) in determining the jet slope, and (ii) it cannot be easily extended to deal with jet-flapped aerofoils near a solid surface or in a wind tunnel. We shall therefore give an alternative, approximate treatment, which, with only a small loss in accuracy, avoids the considerable numerical work involved in Spence's approach, and which overcomes the defects just mentioned. This is not meant to detract from the importance and value of Spence's solution, which we shall take as being the exact datum solution with which to compare our approximate result.

In the derivation of (127) the relatively crude assumption was made that the jet position is the same as that adopted by the (homogeneous) flow from a 'directed' source at the aerofoil's trailing edge (Fig. 8.29b). This fact and the success of (127) exhibited in Fig. 10.19, strongly suggests that as far as the forces acting on the aerofoil are concerned, a detailed and accurate knowledge of the jet position might not be essential.

This belief is given support by the surprising success of Stratford's (1956) jet-flap analogy, in which the effect of the jet on the aerofoil is assumed to be equivalent to that of a 'corresponding' mechanical flap. If this flap has such a length and deflexion angle θ_F that it carries the same proportion of the total lift acting on the aerofoil-flap system as does the jet in the aerofoil-jet system, then with the help of the theory of § 8.20 it is readily shown that the lift coefficient at $\alpha = 0$ is given by

$$C_L \simeq 5 \left(\frac{\theta_F \sin \mathscr{T}}{\mathscr{T}^2} \right)^{\frac{1}{2}} \mathscr{T} C_J^{\frac{1}{2}} + O(C_J^{\frac{3}{2}}). \tag{132}$$

At $\mathscr{T} = 31\frac{1}{2}°$ Davidson & Stratford (1954) found experimentally that the ratio $(\theta_F/\mathscr{T})^{\frac{1}{2}}$ was fairly constant and equal to about 0·9 over a wide range of C_J. Thus with $\mathscr{T} = 31\frac{1}{2}°$ the coefficient of $\mathscr{T} C_J^{\frac{1}{2}}$ in (132) reduces to about 4.3 compared with 4·0 in (127) (assuming $C_J \gg 2C_Q$) and 3.54 in (130). Even if it were assumed that $\theta_F = \mathscr{T}$ (132) would not be far out in spite of the complete neglect of all but the gross over-all effect of the jet.

A similar point arises in ordinary aerofoil theory, in which the lift depends almost entirely on the boundary conditions at infinity plus the boundary condition fixing the position of one stagnation point, and not— or very little—on the details of the aerofoil shape.

The foregoing discussion implies that an approximate equation for the jet slope is worth seeking, and that such an equation should at least satisfy the jet boundary conditions:

$$\theta_{sa}(1) = -(\mathscr{T} + \alpha), \tag{133}$$

and
$$\theta_{sa}(\xi) = -\frac{\beta_a C_L}{4\pi} \frac{1}{\xi^2} + O(\xi^{-4}) \tag{134}$$

—which follow from (119) and the theory of § 10.7 respectively—together with the 'over-all' restriction (69), viz.

$$0 = \int_{-1}^{1} \theta_s(\xi)\,d\xi + 2\int_{1}^{\infty} \theta_{sa}(\xi)\,d\xi. \tag{135}$$

In addition the jet slope should be such that (114)—

$$\frac{\partial C_L}{\partial \alpha} = \frac{1}{2C_J} \left(\frac{\partial C_L}{\partial \mathscr{T}} \right)^2 + \tfrac{1}{2} C_J \tag{136}$$

—is satisfied. We could combine (80) and (96) to derive a further restriction on θ_{sa}, but this would not be essentially different from (135), for, as is easily verified by multiplying (71) by $(\xi^{*2}-1)^{-\frac{1}{2}}$ and integrating over $(1,\infty)$, (80) and (135) are equivalent relations.

Equations (133)–(136) still leave a wide variety of functions open to $\theta_{sa}(\xi)$. To help narrow these down it is profitable to have a look at the classical theory of an aerofoil *without* a jet flap, i.e. the special case $\mathscr{T} = C_J = 0$. In this case it follows from (80) and (126) that the trailing edge streamline has a slope given by

$$\theta_{sa}(\xi) = \frac{\xi}{\pi(\xi^2-1)^{\frac{1}{2}}} \int_{-1}^{1} \theta_s(\xi^*) \left\{ \frac{(1-\xi^{*2})^{\frac{1}{2}}}{\xi^{*2}-\xi^2} + \frac{1}{(1-\xi^{*2})^{\frac{1}{2}}} \right\} d\xi^*$$

$$= -\frac{\xi(\xi^2-1)^{\frac{1}{2}}}{\pi} \int_{-1}^{1} \frac{\theta_s(\xi^*)\,d\xi^*}{(\xi^{*2}-\xi^2)(1-\xi^{*2})^{\frac{1}{2}}},$$

or
$$\theta_{sa}(\xi) = \frac{1}{\pi} \int_{-1}^{1} \theta_s(\xi^*)\,d\left[\tan^{-1}\frac{\xi^*}{\xi} \left(\frac{\xi^2-1}{1-\xi^{*2}} \right)^{\frac{1}{2}} \right]. \tag{137}$$

From $\mathscr{T} = 0$, (99) and (137) it follows that

$$\theta_{sa}(\xi) = -\alpha + \tan^{-1}\left[\frac{\xi_1}{\xi} \left(\frac{\xi^2-1}{1-\xi_1^2} \right)^{\frac{1}{2}} \right] + g(\xi),$$

where $g(\xi)$ is the contribution due to the term involving $(dy/dx)_s$. If the incidence is small, and the aerofoil is thin, then

$$\theta_{sa}(\xi) \simeq \alpha \left\{ \frac{(\xi^2-1)^{\frac{1}{2}}}{\xi} - 1 \right\}, \tag{138}$$

where we have used $\theta_{sa}(\infty) = 0$ to establish that $\xi_1 = \alpha$. Notice from (138) that $\theta_{sa}(\xi) = -\frac{1}{2}\alpha\xi^{-2} + O(\xi^{-4})$ and comparison of this expansion with (134) yields the classical result, $C_L = 2\pi\alpha/\beta_a$. We also observe that $\theta_{sa}(1) = -\alpha$.

The above theory suggests that in the general case a suitable expansion for the jet slope will have the form

$$\theta_{sa} = (\mathscr{T}+\alpha)\left\{ \frac{(\xi^2-1)^{\frac{1}{2}}}{\xi} - 1 \right\} + \sum_{r=2}^{\infty} a_r \frac{(\xi^2-1)^{\frac{1}{2}}}{\xi^{2r-1}}, \tag{139}$$

where
$$a_2 = \frac{1}{2}(\mathscr{T}+\alpha) - \frac{\beta_a C_L}{4\pi}. \tag{140}$$

This expansion satisfies (133) and (134), and furthermore it has the advantage that the coefficients a_r, $r > 2$ depend principally on the jet parameters $\mathscr{T}$ and C_J, for it follows from (138) that these coefficients tend to zero with $\mathscr{T}$ and C_J.

Equation (139) does, however, possess the disadvantage of not giving the correct form of singularity in $\theta_{sa}'(\xi)$ at $\xi = 1$. (By substituting (99) into (125), and ignoring second-order terms, it is found that

$\theta'_{sa}(\xi) \sim \{(8\mathscr{T})/(\pi\beta_a C_J)\}\ln|\xi-1|$ near $\xi = 1$.) However, as noted in the last paragraph of § 10.15, the jet boundary condition fails at $\xi = 1$ in any case, and the logarithmic singularity in θ'_{sa} is fictitious. There is therefore little point in introducing it in our approximate theory.

In practical computations it will be necessary to put $a_r = 0$ for $r > N$, where N is sufficiently large to ensure that θ_{sa} is adequately described. Also two of the coefficients a_r, $2 < r \leqslant N$ must be chosen so that (135) and (136) are satisfied. Equations for the remaining $N-4$ coefficients could be derived by substituting (139) into (125), and then satisfying the resulting equation at $N-4$ pivotal points. This is similar to the technique used by Spence (see p. 418), but has the advantage that full use is made of the preliminary knowledge available on the jet slope.

The minimum value of N possible in this theory is 4. In this case (133) to (136) are satisfied, but (125) is ignored completely. It is found that (135) is satisfied if

$$a_3 = \frac{\beta_a C_L}{\pi} + \frac{2}{\pi}(3\pi - 8)(\mathscr{T}+\alpha) - \frac{8}{\pi}\int_{-1}^{1}\theta_s(\xi)\,d\xi - \tfrac{1}{2}a_4, \qquad (141)$$

$$a_r = 0 \quad (r > 4). \qquad (142)$$

This leaves one parameter a_4 to be selected later in order to satisfy (136).

10.21 Explicit equations for the lift and moment coefficients and their derivatives

Substituting (139)–(142) into (116) we find

$$C_L = \frac{4}{\beta_a}\int_{-1}^{1}\theta_s(\xi)(1-\xi^2)^{\frac{1}{2}}\,d\xi + \frac{4C_J}{3\pi}\int_{-1}^{1}\theta_s(\xi)\,d\xi$$
$$+ \left(\frac{8}{3\pi} - \frac{5}{8}\right)C_J(\mathscr{T}+\alpha) - \frac{5}{48\pi}\beta_a C_J C_L - \frac{C_J}{24}a_4. \quad (143)$$

Let α and $\mathscr{T}$ be small quantities of $O(\epsilon)$ as assumed in § 10.7, then ξ_1 and λ are both $O(\epsilon)$, and this is also true of C_L (cf. (144) below). Terms $O(\epsilon^2)$ and higher will be ignored in the following theory, although no restriction is imposed on C_J.

From (99) and (143)

$$C_L = \frac{4\pi}{\beta_a}(\xi_1 - \tfrac{1}{2}\alpha) + C_J\left\{\tfrac{4}{3}\xi_1 - \tfrac{5}{8}\alpha + \left(\frac{8}{3\pi} - \frac{5}{8}\right)\mathscr{T}\right\} - \frac{5}{48\pi}\beta_a C_J C_L - \frac{C_J}{24}a_4, \quad (144)$$

where the terms involving $(dy/dx)_s$ have vanished on account of the fact that $C_L = 0$ at $\alpha = \mathscr{T} = 0$.

The only unknown quantity here is a_4, which must be chosen so that (136) is satisfied. There is some degree of freedom in the choice of a value of a_4 which will do this. However, this does not matter very much for it is found that a_4 adds only a small correction term to the final result at large values of C_J. If we put $a_4 = 0$ and ignore (136) completely the error committed is negligible in $0 < C_J < 2$.

The simplest method is to put

$$a_4 = k\alpha, \tag{145}$$

where k is a constant to be evaluated.

If we now differentiate (144) with respect to τ and make use of (112), we obtain

$$a_{\mathscr{T}} = \frac{\partial C_L}{\partial \mathscr{T}} = \frac{1}{\beta_a} \left\{ \frac{2\sqrt{\pi}(\beta_a C_J)^{\frac{1}{2}} + \left(\dfrac{8}{3\pi} - \dfrac{5}{8}\right)(\beta_a C_J) + \dfrac{2}{3\sqrt{\pi}}(\beta_a C_J)^{\frac{3}{2}}}{1 + \dfrac{5}{48\pi}(\beta_a C_J)} \right\}. \tag{146}$$

Ignoring terms of $O(C_J^{2\frac{1}{2}})$ and higher, we can write this as

$$\frac{\partial C_L}{\partial \mathscr{T}} = \frac{1}{\beta_a}\{3\cdot545(\beta_a C_J)^{\frac{1}{2}} + 0\cdot224(\beta_a C_J) + 0\cdot259(\beta_a C_J)^{\frac{3}{2}} - 0\cdot074(\beta_a C_J)^2\}. \tag{147}$$

Next differentiate (144) with respect to α; (110), (112) and (145) enable us to write the result in the form

$$a_\alpha\left(1 + \frac{5}{48\pi}\beta_a C_J\right) + \frac{2\pi}{\beta_a}\left(1 + \frac{15+k}{48\pi}\beta_a C_J\right) = 2\left(\frac{2\pi}{\beta_a}\right)^{\frac{1}{2}}(a_\alpha - \tfrac{1}{2}C_J)^{\frac{1}{2}}\left(1 + \frac{\beta_a C_J}{3\pi}\right).$$

We can use (113) to replace a_α by $a_{\mathscr{T}}$ in this equation, and then eliminate $a_{\mathscr{T}}$ by (146). Solving the resulting equation for k we get

$$k = -0\cdot602 - 1\cdot273\,\beta_a C_J, \tag{148}$$

on ignoring higher order terms in C_J. This value of k ensures that $(\partial C_L/\partial\alpha)$ is given by (114). From (114) and (147) it follows that

$$\frac{\partial C_L}{\partial\alpha} = \frac{1}{\beta_a}\{2\pi + 0\cdot793(\beta_a C_J)^{\frac{1}{2}} + 1\cdot442(\beta_a C_J)$$
$$+ 0\cdot031(\beta_a C_J)^{\frac{3}{2}} + 0\cdot001(\beta_a C_J)^2\}. \tag{149}$$

Turning now to the moment coefficient, we find from (122) and (139) to (142) that

$$C_M = -\frac{4}{\beta_a}\int_{-1}^{1}\theta_s(\xi)\,\xi^2(1-\xi^2)^{\frac{1}{2}}\,d\xi - \frac{2C_J}{3\pi}\int_{-1}^{1}\theta_s(\xi)\,d\xi - \tfrac{1}{2}C_L$$
$$+ \left(\frac{5}{4} - \frac{4}{3\pi}\right)C_J\alpha - \left(\frac{4}{3\pi} - \frac{1}{4}\right)C_J\mathscr{T} + \frac{1}{2\pi}(\ln 2 - \tfrac{7}{12})\beta_a C_J C_L, \tag{150}$$

the term in a_4 making no contribution here. Substituting (99) into (150) we find corresponding to (144) that

$$C_M = C_{M_0} - \tfrac{1}{2}C_L + \frac{\pi\alpha}{2\beta_a} - C_J\left\{\tfrac{2}{3}\xi_1 - \tfrac{5}{4}\alpha + \left(\frac{4}{3\pi} - \frac{1}{4}\right)\mathscr{T}\right\} + \frac{1}{2\pi}(\ln 2 - \tfrac{7}{12})\beta_a C_J C_L, \tag{151}$$

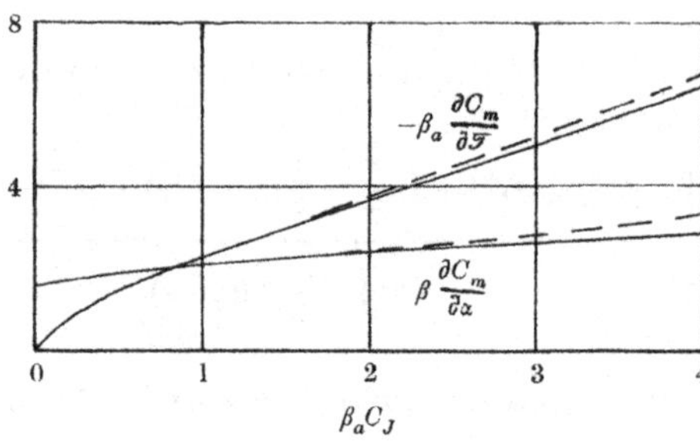

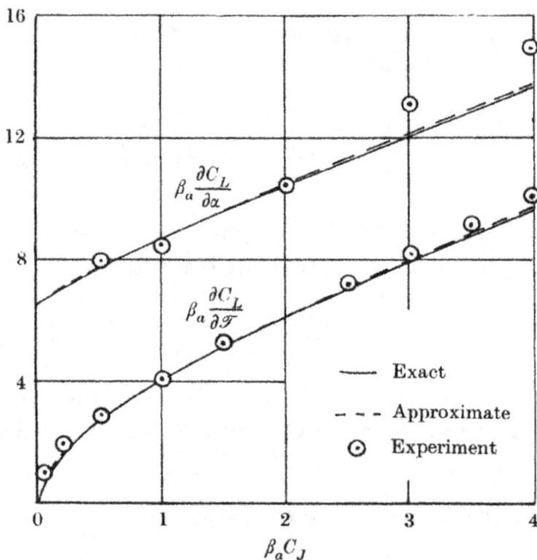

Fig. 10.21 a

where C_{M_0} is the moment coefficient at $\mathscr{T} = \alpha = 0$. Differentiating this with respect to $\mathscr{T}$, and using (112) and (147) we obtain

$$\frac{\partial C_M}{\partial \mathscr{T}} = -\frac{1}{2}\frac{\partial C_L}{\partial \mathscr{T}} - \frac{1}{\beta_a}\{0{\cdot}174(\beta_a C_J) + 0{\cdot}126(\beta_a C_J)^{\frac{3}{2}} - 0{\cdot}004(\beta_a C_J)^2\}, \tag{152}$$

on ignoring terms of order $C_J^{2\frac{1}{2}}$ and higher. Similarly, it follows that

$$\frac{\partial C_M}{\partial \alpha} = -\frac{1}{2}\frac{\partial C_L}{\partial \alpha} + \frac{\pi}{2\beta_a} + \frac{1}{\beta_a}\{0\cdot693\beta_a C_J - 0\cdot047(\beta_a C_J)^2\}. \tag{153}$$

Equations (147), (149), (152) and (153) have been used to draw the broken lines shown in Fig. 10.21a. The full lines refer to Spence's theory.

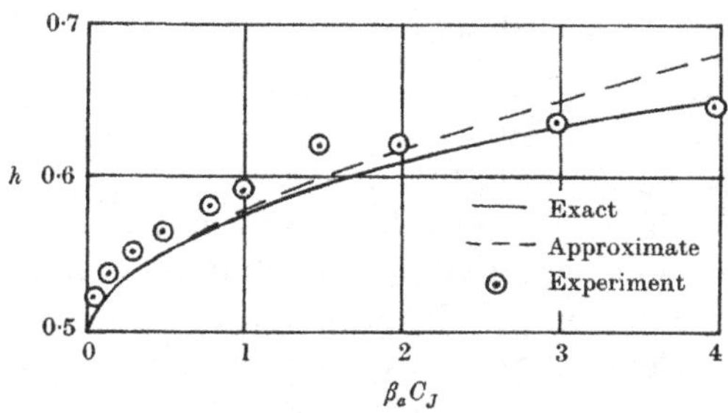

Fig. 10.21b

The agreement between our approximate theory and Spence's accurate theory is excellent. The experimental points shown are from the same source as those given in Fig. 10.19. In Fig. 10.21b we have plotted $h = -(\partial C_M/\partial \mathscr{T})/(\partial C_L/\partial \mathscr{T})$ against C_J. This is the aerodynamic centre at zero incidence (cf. (8.102)).

CHAPTER 11

WAKES AND CAVITIES

GENERAL DESCRIPTION

11.1 Cavities, bubbles and wakes

In this chapter we shall give a mathematical theory for the potential flow past a body which is sufficiently bluff or angular to cause the separation of the flow from both the upper and lower surfaces, in circumstances when re-attachment is unlikely. The principal result of this is the existence of a large wake of slowly eddying fluid extending downstream to infinity (see discussion in § 1.16), although in some special cases the wake may be finite in length. Separated flows which re-attach have much in common with the fully separated flows described below—e.g. in their behaviour near the separation points—but we shall exclude them from this chapter, as they have received a brief treatment in §§ 7.25 and 8.30. Inasmuch as the following theory is applicable to re-attaching flows, it can be regarded as completing this earlier work.

The flow separation just described is a boundary-layer phenomena, and the wake is filled with eddying main-stream fluid in which pressure variations in both the downstream and transverse directions exist. In the Helmholtz model (Helmholtz, 1868) of this flow the real wake is replaced by a region of stationary fluid, bounded by two 'free' streamlines, and in which the pressure is constant. We shall adopt a modified version of this model (Woods, 1955c), which allows wake pressure variations in the downstream direction. As described in § 1.16 this pressure variation should be such that the normal distance b between the free streamlines tends to the wake displacement thickness δ, where

$$\delta = \tfrac{1}{2}cC_D. \tag{1}$$

There is another, quite distinct, type of flow separation, which yields a flow pattern similar in general appearance to that just described. This is caused by the vaporization at the point of minimum pressure in the high-speed flow of a *liquid* past a body. In this case a *cavity* forms behind the body, and the phenomena is termed 'cavitation'. The pressure in the vapour-filled cavity is constant, being almost equal to the vapour pressure of the liquid. These cavities have finite lengths which are functions of the cavitation number $Q = (p_\infty - p_c)/\tfrac{1}{2}\rho_\infty U^2$, where p_c is the pressure in the cavity and the other symbols refer to conditions at infinity.

It is convenient to restrict the words 'bubble' and 'cavity' to the cases of boundary-layer separation and cavitation respectively. Bubbles and cavities which extend an 'infinite' distance downstream are usually termed 'wakes', without distinction. (This term is also used in the literature to refer to the vortex sheets which may extend behind wings and aerofoils.)

Flows with bubbles or cavities present have in common (considering only the flow external to the bubble or cavity) that they are flows with mixed boundary conditions. Consider the separating or cavitating flow past a circular cylinder as shown in Fig. 11.1 (cf. plate 31 of Goldstein, 1938). The surface CBE is termed the 'wetted surface', while the streamlines CD and ED are the 'separation streamlines' or the 'free boundaries'. The pressures on the free boundaries are either given, or must be deduced from some hypothesis; in the case of a cavity they are constant, but this is not usually so for a bubble. The (mixed) boundary conditions then are

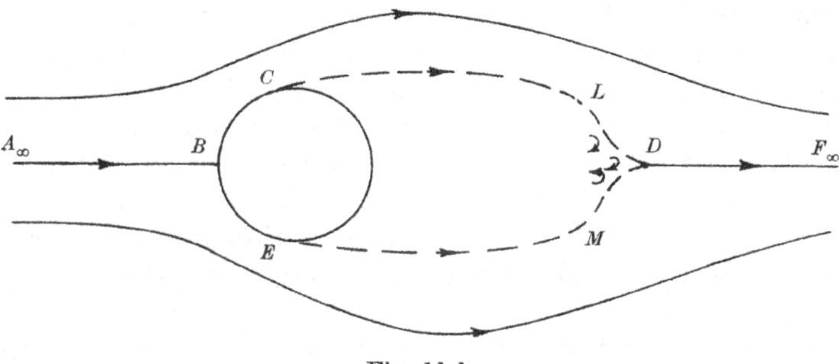

Fig. 11.1

that the shape of the wetted surface is known, and the pressures on the separation streamlines are known. The usual problem is to find the pressure distribution on the wetted surface, but occasionally the shape of the bubble or cavity is of some interest.

From the theoretical viewpoint the main difference between bubbles and cavities is that significant pressure gradients can occur in bubbles but not in cavities. A bubble is filled with eddying fluid in which pressure gradients are clearly possible, whereas a cavity is filled with comparatively low-density stuff, necessarily at constant pressure. Another important difference between bubble and cavity flows is that the minimum pressure occurring in the flow field is in the cavity itself, while for bubbles this minimum pressure occurs on the wetted surface—the boundary layer withstanding some adverse pressure gradient before separating. Because of the constant pressure in the cavity, cavity flows are easier to treat theoretically, in fact until recently almost all theoretical work on free boundary problems has been concerned with cavity flow rather than bubble flow. In aerodynamics it is the latter type of flow that is of greater interest. The reader requiring an extensive account of cavity flows—especially of the mathematical theory of these flows—can study a recent text by Birkhoff & Zarantonello (1957), which also deals with the related jet flows (see § 7.14 *et. seq.*). However, apart from a chapter in Thwaites (1960) there is little available on the theory of bubble flows, the real difficulty being, as already noted in § 1.24, that such a theory requires to be supplemented by a theory of separated boundary layers, and this is not yet available.

11.2   The points of flow separation

First consider the case of bubble flow. A knowledge of the pressure distribution just outside the boundary layer on the wetted surface enables us to use boundary-layer theory to calculate the points of flow separation. On the other hand (assuming here that a method of determining the free streamline pressure distribution is available), knowledge of the points of flow separation and inviscid flow theory (given in §§ 11.5–11.9) enable us to determine the pressure distribution on the wetted surface. Hence the following iterative process suggests itself. Likely separation points are selected on the obstacle surface, and the inviscid theory is applied to give the pressure distribution on the wetted surface. From this pressure distribution and boundary-layer theory new points of flow separation are deduced, and the inviscid flow is then recalculated with these new separation points, and so on. The calculation alternates between the boundary-layer calculation and the external flow calculation until the process converges to a definite final solution. Of course this process may not converge in all cases, but convergence seems likely for those obstacles of interest in aerodynamics. Although this iterative process has been mentioned occasionally in the literature (e.g. see Stewartson, 1953), apparently it has not yet been applied to a curved obstacle. Lighthill (1953) has considered the flow over a spoiler or projection on a flat plate by this method.

If a sharp projecting corner occurs on the obstacle, then boundary-layer calculations are unnecessary as the flow will separate from the surface at the corner (assuming that it has not already separated upstream of the corner). Examples of this type of separation important in aerodynamics are (i) the separation deliberately produced by the use of a spoiler or split flap on an aerofoil surface (see § 11.11), and (ii) the separation that occurs at the nose of a thin aerofoil at moderate incidences (see § 8.30).

While for the general case inviscid theory does not alone determine the separation points, it does impose an important restriction on their positions. It will be proved below (§ 11.6) that as a separation point S is approached along the free streamline, the curvature of the streamline either tends to infinity or remains finite and equal to the curvature of the wetted surface at S. Suppose the free streamline curvature is infinite at S, then the streamline can only be *concave* when viewed from outside the bubble, as a convex streamline of infinite curvature would cut into the obstacle, and so be physically impossible. Clearly this restriction does not apply when S coincides with a sharp corner. It appears that at least for convex obstacles (see Birkhoff & Zarantonello, 1957, p. 167) there exists a point S_0 on the surface such that only those flows separating at or downstream of S_0 satisfy this curvature restriction. If the flow separates at the limiting point S_0 the streamline curvature is finite and continuous there. It has not yet been established that boundary-layer theory always yields separation points at or downstream of these limiting points S_0, but this is probably the case.

An important conclusion about cavity flow can be deduced from the curvature restriction. As the pressure is a minimum in the cavity, the pressure gradient outside is negative along normals to the streamlines, the normals being directed in towards the cavity. Thus from Euler's equation (1.12) the acceleration along these normals of the fluid particles is positive so the cavity must be convex viewed from outside. This convexity requirement can be reconciled with the earlier curvature restriction only if separation occurs at the limiting point S_0 (assuming it exists for the obstacle in question). Hence for the case of cavitating flow inviscid theory *does* fix the positions of the separation points. Such a fluid motion has been termed 'proper cavitation' (see Milne-Thomson, 1949), but 'smooth separation' (Armstrong, 1953) is perhaps a more suitable term.

There is a special case of bubble flow in which the separation points can be determined without a boundary-layer calculation, although the result depends on boundary-layer theory. This is the case when the boundary layer remains laminar right up to the separation point S. It is shown in § 11.6 that if S is downstream (upstream) of the limiting point S_0 the pressure gradient at S is infinite and positive (negative), while if S and S_0 coincide the pressure gradient at S is finite and continuous. We have already seen that S cannot be *upstream* of S_0, for then our solution would be physically impossible. Also a laminar boundary layer cannot adhere to the obstacle surface right up to S if at S there is an infinite and adverse (positive) pressure gradient. Hence it follows that in this case S and S_0 must coincide. This result, which applies quite generally, was deduced by Squire (1934) for the special case of a constant pressure wake behind a circular cylinder.

11.3 Conditions in the wake

First consider the simplest case, namely cavitating flows for which the cavity pressure, and hence the pressure on the free streamlines bounding the cavity, is constant. It is an experimental fact that the cavitation number,

$$Q \equiv \frac{p_\infty - p_c}{\frac{1}{2}\rho_\infty U^2}, \tag{2}$$

is always positive, the vapour pressure p_c always being less than p_∞. It can be shown (see § 11.23) that the maximum cavity width is of order $1/Q$, and its length is of order $1/Q^2$. Hence as the speed U increases, Q decreases, and the cavity width and length increase. In fact for a given obstacle the cavity length and shape is determined uniquely by Q. As Q tends to zero the cavity length and maximum width both tend to infinity, and we then have the constant pressure wake of Helmholtz and Kirchhoff. For bluff obstacles Helmholtz flow is a great improvement over Dirichlet (i.e. non-separating) flow, the most significant advance being the introduction of a drag force, which, by D'Alembert's Paradox, is not present in the older Dirichlet flow.

The simplicity of the cavitation wake model leads us to ask if it has any value as a representation of the infinite 'bubbles' occurring in aerodynamics, hereafter termed 'aerodynamic wakes'. However, from this point of view the model has two serious defects. First there is the infinite cavity width instead of the finite value given by (1). When such quantities as the wake blockage in tunnels are determined, using the Helmholtz model of a wake, this defect is responsible for quite absurd values of the tunnel-wall corrections (see Birkhoff, 1950). A simple source placed at the beginning of the wake (see § 1.16) provides a much better representation, the significant point being that such a system (source in a stream) yields a wake displacement thickness that is essentially constant well downstream (see § 6.11). Secondly, the constant pressure cavity makes no allowance for the so-called 'wake under-pressure' $(p_c - p_\infty < 0)$ immediately behind the obstacle. Nor can it be adjusted to do so, for the wake pressure, if constant, must necessarily be equal to the pressure outside the wake at infinity. The suction behind the obstacle can be quite large. For example, the pressure coefficient defined by $C_p = (p - p_\infty)/\frac{1}{2}\rho_\infty U^2$ is about $-1\cdot4$ behind a perpendicular flat plate (Fage & Johansen, 1927), whereas in Helmholtz's model it is zero. This accounts for the large discrepancy in the corresponding values of the drag coefficient—the Helmholtz model yields $C_D \simeq 0\cdot88$ (see (63)), while experiment gives $C_D \simeq 1\cdot95$.

The theory given in § 11.5 permits a varying pressure distribution over the free streamlines, and so enables us to overcome both these defects of the classical theory. The distribution can be chosen in many ways; we shall take the simplest possible that makes the free streamlines parallel and a finite distance apart at infinity. One other important restriction can be imposed on the streamline pressure distribution; this is that the pressures at the two separation points are equal. This result follows from the well-known fact (see Milne-Thomson, 1949, p. 526) that in steady flow the total vorticity shed from a body is zero. Hence the vorticity shed from the separation points in two-dimensional flows must be opposite in sign, and equal in magnitude (Fage & Johansen, 1927, 1928 have verified this experimentally). This implies that in our model of aerodynamic wake flow we should make the velocities (and hence pressures) equal at the separation points.

Unfortunately, in spite of the two restrictions just described, one arbitrary constant is unavoidable; we shall determine this empirically from the experimental value of C_p immediately behind the obstacle. It may be possible to eliminate this empirical element from the theory by some kind of extension of boundary-layer theory to the flow neighbouring the free streamline downstream of separation. Such an extension has been suggested by Stewartson (1953), but his method is incomplete and does not yield the value of the wake under-pressure. As noted in § 8.30 a similar difficulty arises with the finite bubbles that occur on thin aerofoils at incidences just below the stalling incidence.

11.4 Physical reality

The important restriction imposed by physical reality on the positions of the separation points, S, has already been mentioned in § 11.2; this will be given a mathematical form in § 11.7. Similarly, if the shape of the obstacle downstream of S is such that the free streamline would pass into the obstacle at some point (see Fig. 1.25), the solution is physically unreal and must be discarded. In general, physical unreality of this latter type can be discovered only by calculating the positions of the free streamlines in the neighbourhood of the obstacle. This can be done from the inviscid theory, but it is a laborious calculation, and in many cases will be clearly unnecessary. Some general existence theorems have been established (see Weinstein, 1949, also Birkhoff & Zarantonello, 1957) specifying a few types of obstacle for which this type of physical unreality does not occur.

Another type of physical unreality occurs if the free streamlines cross each other behind the obstacle and so produce a 'doubly-covered' region of flow. Both types of unreality are shown in Fig. 11.4. Except for appealing to the existence theorems already mentioned, which cover very few cases of importance in aerodynamics, our only method of discovering this type of unreality is, as before, to calculate the free streamline shape from the theory. A necessary condition that the streamlines do not cross is given in §11.7.

With bubbles and aerodynamic wakes these physical unrealities could be eliminated by a suitable variation of the pressure distribution on the free streamlines in the neighbourhood of the body, whereas with cavities and cavitation wakes the only possibility, apart from altering the whole character of the flow, is to alter the positions of the points of flow separation.

There is another type of physical unreality which occurs with finite cavities. It was noted by Brillouin (1911) that since a cavity with a positive value of the cavitation number Q is convex, it must, if it does not reattach, have a stagnation point at its rear end. But at a stagnation point the pressure is a maximum, which contradicts the fact that the

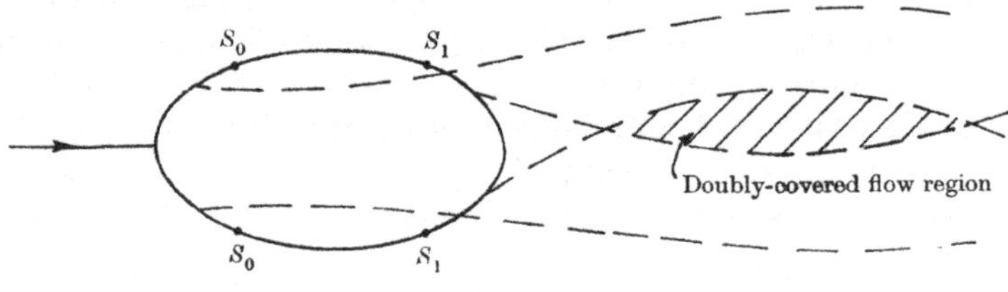

Fig. 11.4

minimum pressure occurs in the cavity. This difficulty can be avoided only by making artificial alterations to the rear of the cavity, and assuming that the cavity length is great enough for these changes to have negligible effect on the flow near the obstacle. Three methods of doing this are as follows. First a second body can be introduced at the rear of the cavity to close it, so that the cavity now extends between two obstacles as shown in Fig. 11.23. This device, due to Riabouchinsky (1921), has already been described in §8.32. Secondly, a re-entrant jet, as shown in Fig. 11.25, can be used. This model (Gilbarg & Rock, 1945; see also Gilbarg & Serrin, 1950) is physically unrealistic as the jet passes through the obstacle from the rear producing a doubly-covered region of flow in front of it. Nevertheless, it appears that this flow closely resembles real cavitating flows. Re-entrant jets have been observed in experiments, but with a horizontal cavity of reasonable length the action of gravity prevents them reaching the back of the obstacle. Thirdly, a pressure gradient can be introduced over the rear portion of the cavity so as to change the sign of the curvature, and produce a cusped end with no stagnation point, as shown in Fig. 11.1. The rear section of the cavity is thus made to behave like a bubble. Cavities whose rear ends are filled with a turbulent mixture of vapour and water (foam) have been observed, and give some support to this model.

All the models of cavitating flows described above have positive values of Q in agreement with experiment, however models with $Q < 0$ have been devised (Lighthill, 1949a; Southwell & Vaisey, 1946). These flows avoid the difficulty at the rear end of the cavity by having cusped ends, but they are only of academic interest as they have little resemblance to real cavities.

THEORY OF STEADY WAKE FLOW

11.5   The integral equation

Fig. 11.5 shows the w-plane for the flow past a curved obstacle behind which there is a wake extending to infinity. The surface S_2AS_1 is the 'wetted' surface of the obstacle, and the dotted lines are the edges of the wake. We assume that the front stagnation point A and the separation points S_1, S_2 are at $\phi = 0$, ϕ_1 and ϕ_2 respectively. It is readily verified from the Schwartz–Christoffel mapping theorem (see §5.3) that the w-plane is mapped into the semi-infinite strip shown in the ζ-plane by

$$w = 4a(\sin \tfrac{1}{2}\zeta - \sin \tfrac{1}{2}\lambda)^2, \tag{3}$$

where

$$4a = \tfrac{1}{4}(\sqrt{\phi_2} + \sqrt{\phi_1})^2 \tag{4}$$

and

$$\sin \tfrac{1}{2}\lambda = \frac{\sqrt{\phi_2} - \sqrt{\phi_1}}{\sqrt{\phi_2} + \sqrt{\phi_1}}. \tag{5}$$

The wetted surface now lies on $-\pi \leqslant \gamma \leqslant \pi$, $\eta = 0$ with the front stagnation point at $\gamma = \lambda$, while the free streamlines are on $\gamma = \pm\pi$. On the wetted surface $\psi = 0$ and $\eta = 0$, and (3) becomes

$$\phi = 4a(\sin \tfrac{1}{2}\gamma - \sin \tfrac{1}{2}\lambda)^2. \tag{6}$$

The boundary conditions in the ζ-plane are given on the wetted surface $-\pi \leqslant \gamma \leqslant \pi$, and the pressures, or equivalently $\Omega_s(\eta)$, given on the free streamlines $\gamma = \pi$, $\gamma = -\pi$. As usual we shall put $q_\infty = U$ and $\theta_\infty = 0$, so that $\tau = \Omega + i\theta$ vanishes at infinity. From (3) this can be expressed

$$\lim_{\eta \to \infty} \tau(\eta + i\gamma) = 0. \tag{7}$$

These boundary conditions are 'simply-mixed', so the value of $\tau(\zeta)$ can be calculated from the theory given in §4.2; the relation between the z-plane used in §4.2 and the ζ-plane defined above is easily shown to be $\tanh \dfrac{\pi z}{2h} = \sin \tfrac{1}{2}\zeta$, and this can be used to transform equation (4.9) into the ζ-plane. Alternatively the equation for $\tau(\zeta)$ can be deduced as the special case $l = \pi$, $h = \infty$ of (4.138). In this limit the basic rectangle shown in Fig. 4.14 becomes the semi-infinite strip of Fig. 11.5. From

(4.65), (4.66) and (4.92) we see that if $h = \infty$, then $K = \pi/2$ and $k = 0$, whence from (4.57) $\operatorname{sn}(Kx/l) = \sin\frac{1}{2}x$, $\operatorname{cd}(Kiy/l) = \cosh\frac{1}{2}y$, etc. With these limit values (4.138) becomes

$$\tau(\zeta) = \frac{\cos\frac{1}{2}\zeta}{2\pi}\left\{\int_{-\pi}^{\pi}\frac{\theta_s(\gamma)\,d\gamma}{\sin\frac{1}{2}\gamma - \sin\frac{1}{2}\zeta}\right.$$

$$\left. + \int_0^{\infty}\left(\frac{\Omega^+(\eta)}{\cosh\frac{1}{2}\eta - \sin\frac{1}{2}\zeta} + \frac{\Omega^-(\eta)}{\cosh\frac{1}{2}\eta + \sin\frac{1}{2}\zeta}\right)d\eta\right\}, \quad (8)$$

on replacing z, x, y by the corresponding variables ζ, γ, η of the present application. Here Ω^+ and Ω^- denote the values of Ω_s on $\psi = +0$ and $\psi = -0$ respectively (see Fig. 11.5).

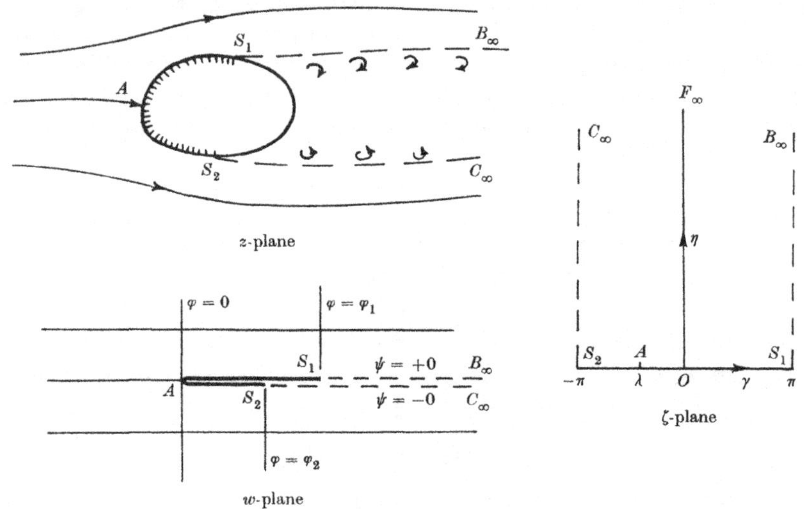

Fig. 11.5

If s is the distance measured along the wetted surface then, from (6),

$$\frac{s}{c} = \frac{1}{c}\int^{\phi}\frac{d\phi}{q} = \left(\frac{4a}{Uc}\right)\int^{\gamma}\frac{U}{q}(\sin\tfrac{1}{2}\gamma - \sin\tfrac{1}{2}\lambda)\cos\tfrac{1}{2}\gamma\,d\gamma, \quad (9)$$

where c is a typical length associated with the obstacle.

On $\eta = 0$ (8) plus an integration by parts yields

$$\Omega_s(\gamma^*) = -\frac{1}{\pi}\int_{-\pi}^{\pi}\ln\left|\frac{\sin\frac{1}{4}(\gamma^* - \gamma)}{\cos\frac{1}{4}(\gamma^* + \gamma)}\right|d\theta_s(\gamma)$$

$$+ \frac{\cos\frac{1}{2}\gamma^*}{2\pi}\int_0^{\infty}\left(\frac{\Omega^+(\eta)}{\cosh\frac{1}{2}\eta - \sin\frac{1}{2}\gamma^*} + \frac{\Omega^-(\eta)}{\cosh\frac{1}{2}\eta + \sin\frac{1}{2}\gamma^*}\right)d\eta. \quad (10)$$

Now
$$d\theta_s(\gamma) = \frac{d\theta_s}{ds}\frac{ds}{d\phi}\frac{d\phi}{d\gamma}d\gamma = \frac{4a}{Rq_s}(\sin\tfrac{1}{2}\gamma - \sin\tfrac{1}{2}\lambda)\cos\tfrac{1}{2}\gamma\,d\gamma, \qquad (11)$$

where R is the radius of curvature of the wetted surface. If (11) is substituted in (10) and Ω_s expressed in terms of q_s (see (8.15)) the result will be a complicated integral equation for q_s, very similar to that obtained in §8.3 for a closed aerofoil. This can be solved by an iterative method like that described in §8.23 for the aerofoil case. Further details will be given in §11.12.

In Helmholtz or cavitation wake flow $p = p_\infty$ in the wake, consequently $q = U$ on the free streamlines and $\Omega^+ = \Omega^- = 0$. This gives a much simpler theory.

11.6   Flow in the neighbourhood of a separation point

We shall now examine the nature of the flow in the neighbourhood of the separation point S_1 where $\zeta = \pi$ (see Fig. 11.5). The flow near S_2 will be similar.

Integration by parts of both integrals in (8) gives

$$\tau(\zeta) = \tau(-\pi) - \frac{1}{\pi}\int_{-\pi}^{\pi}\ln\frac{\sin\tfrac{1}{4}(\gamma-\zeta)}{\cos\tfrac{1}{4}(\gamma+\zeta)}d\theta_s(\gamma)$$

$$+\frac{i}{\pi}\int_0^\infty\ln\frac{\sin\tfrac{1}{4}(\pi+i\eta-\zeta)}{\cos\tfrac{1}{4}(\pi+i\eta+\zeta)}d\Omega^+(\eta) + \frac{i}{\pi}\int_0^\infty\ln\frac{\sin\tfrac{1}{4}(\pi+i\eta+\zeta)}{\cos\tfrac{1}{4}(\pi+i\eta-\zeta)}d\Omega^-(\eta).$$

If we set $\zeta = \pi + \zeta'$, and expand about $\zeta' = 0$ we obtain

$$\tau(\zeta') = \tau(0) - \frac{A}{2\pi}\zeta' + O(\zeta'^2), \qquad (12)$$

where

$$A \equiv \int_{-\pi}^{\pi}\tan\tfrac{1}{4}(\gamma+\pi)\,d\theta_s(\gamma) + \int_0^\infty\{\coth\tfrac{1}{4}\eta\,d\Omega^+(\eta) - \tanh\tfrac{1}{4}\eta\,d\Omega^-(\eta)\}. \quad (13)$$

On $\psi = +0$ (3) gives $\phi' = -a\zeta'^2(1-\sin\tfrac{1}{2}\lambda) + O(\zeta'^4)$, where ϕ' is measured from S_1. Hence

$$\zeta' \simeq \begin{cases} i\sqrt{\phi'}/\sqrt{\{a(1-\sin\tfrac{1}{2}\lambda)\}} & (\phi' > 0), \\[2mm] -\sqrt{(-\phi')}/\sqrt{\{a(1-\sin\tfrac{1}{2}\lambda)\}} & (\phi' < 0), \end{cases}$$

as the imaginary part of ζ' is positive on the free streamline, and its real part is negative on the wetted surface (cf. Fig. 11.5).

It therefore follows from (12) that on the free streamline ($\phi' > 0$),

$$\theta(\phi') = \theta(0) - \frac{A}{2\pi}\frac{\sqrt{\phi'}}{\sqrt{\{a(1-\sin\tfrac{1}{2}\lambda)\}}} + O(\phi')^{\frac{3}{2}}, \qquad (14)$$

while on the wetted surface ($\phi' < 0$),

$$\Omega(\phi') = \Omega(0) + \frac{A}{2\pi} \frac{\sqrt{(-\phi')}}{\sqrt{\{a(1-\sin\frac{1}{2}\lambda)\}}} + O(-\phi')^{\frac{3}{2}}. \tag{15}$$

The curvature of the free surface is given by

$$\frac{1}{R} \equiv \frac{d\theta}{ds} = \frac{d\theta}{d\phi'}\frac{d\phi'}{ds} = -\frac{A}{4\pi} \frac{q_s}{\sqrt{\{a(1-\sin\frac{1}{2}\lambda)\,\phi'\}}} + O(\phi')^{\frac{1}{2}},$$

where q_s is the velocity at S_1. From this we see that at S_1 ($\phi' = 0$) the curvature is minus infinity, finite or plus infinity, as A is respectively positive, zero or negative.

If the body has a finite curvature at S_1 it is apparent that a separating streamline of curvature minus infinity (concave downwards) will be physically impossible as it will cut into the obstacle. Thus in this case the condition of physical reality at the separation point is $A \leqslant 0$. When $A = 0$ the curvature of the free streamline remains finite at S_1, and we have the case of 'smooth separation'. Furthermore, in this case, the curvature is usually continuous at S_1, which can be shown as follows. The velocity is continuous at S_1 and

$$1/(qR) = (d\theta/ds)\,(ds/d\phi') = (d\theta/d\phi) = (d\theta/dw) = \mathscr{I}(d\tau/dw)$$

on $\psi = 0$. Thus a simple discontinuity in $1/R$ is reproduced in the imaginary part of $(d\tau/dw)$, and so implies a logarithmic singularity in the conjugate function $(d\Omega/dw) = (d\Omega/d\phi')$ (see §3.10), and this is only possible if the distribution of Ω on the free streamline is deliberately chosen to have this character at S_1. With cavity flows $(d\Omega/d\phi') = 0$ on the free streamline, so $A = 0$ always implies that the curvature is continuous at S_1.

The pressure gradient on the wetted surface in the neighbourhood of S_1 can be calculated by (2.63), which can be written

$$\frac{\partial p}{\partial s} = \frac{\rho_a q^3}{\beta_a} \frac{\partial \Omega}{\partial \phi'}, \tag{16}$$

so from (15),

$$\frac{\partial p}{\partial s} = -\frac{A\rho_a q_s^3}{4\pi\beta_a} \frac{1}{\sqrt{\{a(1-\sin\frac{1}{2}\lambda)\,(-\phi')\}}} + O(-\phi')^{\frac{1}{2}}.$$

Hence there is an infinite and adverse (positive) pressure gradient on the wetted surface at S_1 unless $A \geqslant 0$.

As mentioned in the last paragraph of §11.2 the physical reality condition ($A \leqslant 0$) and the result just given for the pressure gradient to be favourable ($A \geqslant 0$) means that for laminar separation we must have $A = 0$. In this case the pressure gradient is continuous at S_1 by an argument similar to that given above for the curvature at S_1. This result has an interesting corollary. It is that if the pressure distribution on the free streamline is selected so that the pressure increases continuously along the wake from the value p_s at S_1 to p_∞ at infinity, and such that it has a finite non-zero gradient at S_1, then for smooth separation the minimum pressure in the flow field must occur on the wetted surface upstream of S_1. This is of course just what happens in real *bubble* flows.

Some idea of the significance of the condition $A \leqslant 0$ can be obtained by considering the case of flow about a symmetrical obstacle, convex viewed from the flow field and with a continuous tangent. At the front stagnation point, $\gamma = \lambda = 0$ owing to the symmetry, and θ_s has a jump in value of π, which from (13) contributes $\pi \tan \frac{1}{4}\pi = \pi$ to A. The gradient $d\theta_s/d\gamma$ is always negative in the present case and hence reduces A by an amount depending directly on the distance between the front stagnation point and the separation points. This certainly suggests the existence of a point S_0 corresponding to $A = 0$, such that only those flows separating at or downstream of S_0 satisfy $A \leqslant 0$, and hence are physically possible. Now consider the effect of an increase in the wake under-pressure on the position of S_0, assuming it exists. The resulting increase in the positive pressure gradient down the wake leads to a positive value of $d\Omega^+/d\eta$, and hence from (13), to an *increase* in A. Hence a corresponding *downstream* movement in S_0 is necessary in order to compensate for this change.

Finally it should be noticed that if the body has a reduced curvature downstream of S physical unreality may still occur when $A \leqslant 0$. In the general case this possibility could be checked only by an actual calculation of the streamline position from (8). It has been proved (Leray, 1935) that difficulties of this kind do not occur for symmetrical convex obstacles with a curvature which does not decrease with increasing distance from the stagnation point.

11.7   Flow in the neighbourhood of infinity

The second kind of physical unreality, namely the crossing of the free streamlines behind the obstacles shown in Fig. 11.4, can be investigated directly by calculating the streamline shapes from (8). However a *necessary* condition that the streamlines do not cross can be deduced by expanding (8) near infinity.

First note from (7) and (8) that in all cases we must have

$$\int_{-\pi}^{\pi} \theta_s \, d\gamma + \int_{0}^{\infty} (\Omega^+ - \Omega^-) \, d\eta = 0; \tag{17}$$

this fixes the position of the front stagnation point.

Now $z = \infty$, $w = \infty$ and $\eta = \infty$ are corresponding points, and in this neighbourhood (3) gives

$$\frac{w}{a} = -e^{-i\zeta}\{1 + 4i\,e^{\frac{1}{2}i\zeta}\sin \tfrac{1}{2}\lambda + O(e^{i\zeta})\}. \tag{18}$$

To expand (8) similarly it is necessary to assume that there exists a value of η, say η_0, such that $\Omega^+ = \Omega^- = 0$ for $\eta > \eta_0$. (The general case will be treated later.) With this restriction (8) can be expanded

$$\tau = B\,e^{\frac{1}{2}i\zeta} + iC\,e^{i\zeta} + O(e^{\frac{3}{2}i\zeta}), \tag{19}$$

where
$$B \equiv \frac{1}{\pi}\int_{-\pi}^{\pi}\theta_s\sin\tfrac{1}{2}\gamma\,d\gamma + \frac{1}{\pi}\int_0^{\infty}(\Omega^+ + \Omega^-)\cosh\tfrac{1}{2}\eta\,d\eta, \tag{20}$$

and
$$C \equiv \frac{1}{\pi}\int_{-\pi}^{\pi}\theta_s\cos\gamma\,d\gamma - \frac{1}{\pi}\int_0^{\infty}(\Omega^+ - \Omega^-)\cosh\eta\,d\eta. \tag{21}$$

Equations (18) and (19) can be combined to give

$$\tau(w) = iB\left(\frac{a}{w}\right)^{\frac{1}{2}} - i(C + 2B\sin\tfrac{1}{2}\lambda)\left(\frac{a}{w}\right) + O\left(\frac{a}{w}\right)^{\frac{3}{2}}. \tag{22}$$

Thus, provided ϕ is large enough, it follows that the free streamline slope is

$$\theta = B\left(\frac{a}{\phi}\right)^{\frac{1}{2}} - (C + 2B\sin\tfrac{1}{2}\lambda)\left(\frac{a}{\phi}\right) + O\left(\frac{a}{\phi}\right)^{\frac{3}{2}}. \tag{23}$$

Well downstream $\theta \simeq (dy/dx)$, and $\phi \simeq Ux$ hence integration gives the distance between the free streamlines to be

$$y^+ - y^- = 2B\left(\frac{a}{U}\right)^{\frac{1}{2}} x^{\frac{1}{2}} + b + O(x^{-\frac{1}{2}}), \tag{24}$$

where b is the constant of integration. Near the obstacle $y^+ - y^- > 0$, and so if $B < 0$ the streamlines must cross according to (24), and the solution will be unreal. Therefore it can be concluded from (20) that a necessary condition for physical reality is

$$\int_{-\pi}^{\pi}\theta_s\sin\tfrac{1}{2}\gamma\,d\gamma + \int_0^{\infty}(\Omega^+ + \Omega^-)\cosh\tfrac{1}{2}\eta\,d\eta \geqslant 0. \tag{25}$$

By an argument similar to that applied to the condition $A \leqslant 0$ in §11.6 this result implies the existence of limiting points S_0^* on the obstacle surface such that only those flows separating at or *upstream* of S_0^* are physically possible. The effect of an increase in the wake underpressure on S_0^* is to move it upstream.

If this result is combined with the similar one for S_0, we can conclude that (i) for physically possible wakes the separation point S must lie in the interval (S_0, S_0^*), and (ii) an increase in the wake under pressure reduces (S_0, S_0^*).

We shall now show that these conclusions hold with more general distributions of Ω^+ and Ω^-, i.e. with distributions for which the simple expansion method we used to obtain (19) is not valid. In §10.7 it was shown that

$$\tau = \sum_{n=1}^{\infty} \frac{a_n + ib_n}{\chi^n} \quad (\mathscr{R}\chi = \xi), \tag{26}$$

where

$$-\frac{bU}{2K\pi} = a_2 = -\frac{1}{\pi}\int_{-\infty}^{\infty} \theta_s \xi \, d\xi, \tag{27}$$

$$-\frac{\beta_a L_\infty}{\rho_a U 2K\pi} = b_2 = \frac{1}{\pi}\int_{-\infty}^{\infty} \Omega_s \xi \, d\xi, \tag{28}$$

and

$$\chi = \left(\frac{w}{K}\right)^{\frac{1}{2}} = \left(\frac{4a}{K}\right)^{\frac{1}{2}} (\sin \tfrac{1}{2}\zeta - \sin \tfrac{1}{2}\lambda), \tag{29}$$

by (3). On $\gamma = \pi$ (29) gives

$$\xi = \left(\frac{4a}{K}\right)^{\frac{1}{2}} (\cosh \tfrac{1}{2}\eta - \sin \tfrac{1}{2}\lambda). \tag{30}$$

This, and similar equations for $\gamma = -\pi$ and $\eta = 0$ enable us to write (27) as

$$\frac{bU}{4a} = -\frac{\pi K a_2}{2a} = \int_{-\pi}^{\pi} \theta_s(\sin \tfrac{1}{2}\gamma - \sin \tfrac{1}{2}\lambda)\cos \tfrac{1}{2}\gamma \, d\gamma$$

$$+ \int_0^{\infty} \theta^+(\cosh \tfrac{1}{2}\eta - \sin \tfrac{1}{2}\lambda)\sinh \tfrac{1}{2}\eta \, d\eta$$

$$- \int_0^{\infty} \theta^-(\cosh \tfrac{1}{2}\eta + \sin \tfrac{1}{2}\lambda)\sinh \tfrac{1}{2}\eta \, d\eta, \tag{31}$$

where θ^+, θ^- are the slopes of the upper and lower free streamlines. It follows from (8) that these slopes are given by

$$\theta^{\pm} = \mp \frac{\sinh \tfrac{1}{2}\eta}{2\pi}\left\{\int_{-\pi}^{\pi} \frac{\theta_s \, d\gamma}{\sin \tfrac{1}{2}\gamma \mp \cosh \tfrac{1}{2}\eta}\right.$$

$$\left. + \int_0^{\infty} \left(\frac{\Omega^+}{\cosh \tfrac{1}{2}\eta^* \mp \cosh \tfrac{1}{2}\eta} + \frac{\Omega^-}{\cosh \tfrac{1}{2}\eta^* \pm \cosh \tfrac{1}{2}\eta}\right) d\eta^*\right\}. \tag{32}$$

The integral in (27) is to be interpreted as a Cauchy Principal Value, i.e. as $\lim\limits_{\xi_0 \to \infty} \int_{-\xi_0}^{\xi_0} \dots$, and it is readily verified from (29) that this implies that the corresponding limits in (31) are coupled, i.e. are to be taken as $\lim\limits_{\eta_0 \to \infty} \left(\int_0^{\eta_0} - \int_0^{\eta_0}\right)$. This fact enables us to combine the last two terms of (31) under one integral sign. Now (17) and (32) give

$$\int_0^{\infty} (\theta^+ - \theta^-)\cosh \tfrac{1}{2}\eta \sinh \tfrac{1}{2}\eta \, d\eta - \frac{1}{2\pi}\int_0^{\infty} \sinh^2 \tfrac{1}{2}\eta \cosh \tfrac{1}{2}\eta \, d\eta$$

$$\times \left(2\int_{-\pi}^{\pi} \frac{\theta_s \sin \tfrac{1}{2}\gamma \, d\gamma}{\sin^2 \tfrac{1}{2}\gamma - \cosh^2 \tfrac{1}{2}\eta} + \int_0^{\infty} \frac{(\Omega^+ + \Omega^-)\cosh \tfrac{1}{2}\eta^* \, d\eta^*}{\cosh^2 \tfrac{1}{2}\eta^* - \cosh^2 \tfrac{1}{2}\eta}\right),$$

which shows that unless the equality in (25) holds, i.e. unless $B = 0$, the right-hand side of (31) is infinite. In fact it is clear that $b = \infty$ if $B > 0$, and $b = -\infty$ if $B < 0$, and we have a more general proof of the results deduced from (24).

When $B = 0$ we can write

$$\int_0^\infty (\theta^* - \theta^-) \cosh \tfrac{1}{2}\eta \sinh \tfrac{1}{2}\eta \, d\eta$$

$$= \pi^2 \phi(\infty, \infty) - \frac{1}{\pi} \int_{-\pi}^{\pi} \theta_s \sin^3 \tfrac{1}{2}\gamma \, d\gamma \int_0^\infty \frac{\sinh^2 \tfrac{1}{2}\eta \, d\eta}{\cosh \tfrac{1}{2}\eta (\sin^2 \tfrac{1}{2}\gamma - \cosh^2 \tfrac{1}{2}\eta)}$$

$$- \frac{1}{2\pi} \int_0^\infty (\Omega^+ + \Omega^-) \cosh^3 \tfrac{1}{2}\eta^* \, d\eta^* \int_0^\infty \frac{\sin^2 \tfrac{1}{2}\eta \, d\eta}{\cosh \tfrac{1}{2}\eta (\cosh^2 \tfrac{1}{2}\eta^* - \cosh^2 \tfrac{1}{2}\eta)},$$

$$(33)$$

where $\pi^2 \phi(\infty, \infty)$ is the additional term arising from the reversal of the order of integration of the repeated Cauchy integral (see § 3.9). The substitutions $t = \cosh \tfrac{1}{2}\eta$, $t_1 = \cosh \tfrac{1}{2}\eta^*$ reduce the integral in question (before reversal) to

$$\int_1^\infty \frac{dt}{t} \int_1^\infty \frac{\phi(t, t_1)}{t_1 - t} \, dt_1,$$

where

$$\phi(t, t_1) = -\frac{2}{\pi} \left(\frac{t^2 - 1}{t_1^2 - 1} \right)^{\tfrac{1}{2}} \frac{t_1^3 (\Omega^+ + \Omega^-)}{t_1 + t}.$$

Hence

$$\pi^2 \phi(\infty, \infty) = -\pi \lim_{t \to \infty} t^2 (\Omega^+ + \Omega^-) = -\frac{\pi K}{4a} \lim_{\xi \to \infty} \xi \{\Omega(\xi) + \Omega(-\xi)\},$$

and from (26) this term equals $-\pi K a_2/2a$. Equation (33) now reduces to

$$\int_0^\infty (\theta^+ - \theta^-) \sinh \tfrac{1}{2}\eta \cosh \tfrac{1}{2}\eta \, d\eta = -\int_{-\pi}^{\pi} \theta_s \sin \tfrac{1}{2}\gamma \cos \tfrac{1}{2}\gamma \, d\gamma - \frac{\pi K a_2}{2a}, \quad (34)$$

where we have taken advantage of $B = 0$ to eliminate some terms.

A similar type of calculation yields

$$-\int_0^\infty (\theta^+ - \theta^-) \sinh \tfrac{1}{2}\eta \, d\eta = \int_{-\pi}^{\pi} \theta_s \cos \tfrac{1}{2}\gamma \, d\gamma + 2\pi \left(\frac{K}{4a} \right)^{\tfrac{1}{2}} a_1; \quad (35)$$

and it remains only to substitute (34) and (35) into (31) to obtain $a_1 = 0$. We conclude that b is finite only if both B and a_1 vanish.

The right-hand side of (28) can be evaluated by a similar method to that just employed for (27); the calculation is more straightforward, and results in

$$\frac{L_\infty}{2\rho_a \pi a U} = -\frac{K b_2}{a} = (C + 2B \sin \tfrac{1}{2}\lambda). \quad (36)$$

11.8　The lift and drag forces

We are now in a position to write down expressions for the lift and drag acting on the body. There are two distinct cases to consider, viz. (*a*) aerodynamic wakes, and (*b*) cavitation wakes.

(*a*) *Aerodynamic wakes.* As described in §§ 11.1 and 11.3 aerodynamic wakes have their free streamlines a distance $b = \frac{1}{2}cC_D$ apart near infinity. Consequently the drag coefficient is $C_D = 2b/c = -4\pi Ka_2/cU$, and so it depends on the velocity distribution chosen for the free streamlines. From (26) and (30) we see that this distribution must have an asymptotic form such that

$$\Omega^+, \Omega^- \sim -\frac{cU}{16\pi a}\frac{C_D}{\cosh^2\frac{1}{2}\eta}.\tag{37}$$

It appears that our theory cannot be used to determine the drag due to an aerodynamic wake.

As b is finite, $B = 0$, and so Ω^+, Ω^- also satisfy the restriction

$$\int_{-\pi}^{\pi}\theta_s\sin\tfrac{1}{2}\gamma\,d\gamma+\int_0^{\infty}(\Omega^++\Omega^-)\cosh\tfrac{1}{2}\eta\,d\eta = 0.\tag{38}$$

An expression for the lift coefficient can be obtained by regarding an aerodynamic wake as a kind of 'negative jet'. In this case the theory of § 10.18 shows that there is no net lift on the wake, and so $C_L = C_{L\infty}$. Therefore by (21), (36) and (38)

$$C_L = \frac{1}{\pi}\left(\frac{4a}{Uc}\right)\int_{-\pi}^{\pi}\theta_s\cos\gamma\,d\gamma.\tag{39}$$

(*b*) *Cavitation wakes.* With cavitation wakes the pressure in the wake is constant and equal to its value at infinity, p_a, and $\Omega^+ = \Omega^- = 0$. Thus the pressure coefficient, $C_p \equiv (p-p_a)/(\frac{1}{2}\rho_a U^2)$, vanishes on the free streamlines, so that

$$C_L - iC_D = \frac{1}{c}\int_{\Sigma}C_p(\cos\theta_s+i\sin\theta_s)\,ds = \frac{1}{c}\int_{\Sigma}C_p\,dz,$$

where Σ is any open arc, like FS_1AS_2E shown in Fig. 11.8, which lies on the wetted surface and extends some distance along both the free streamlines. Hence by (2.100)

$$C_L - iC_D = \frac{1}{cU\beta_a}\int_{\Sigma}e^{\tau}\,dw - \frac{1}{cU\beta_a}\int_{\Sigma}e^{-\bar{\tau}}\,d\bar{w}.\tag{40}$$

In the χ-plane ($w = K\chi^2$, see (29)) the flow is confined to the upper half-plane, and the wetted surface and free streamlines lie on the real axis. On the free streamlines $\Omega = 0$, so $\tau = -\bar{\tau}$, and hence $\tau(\chi)$ can be continued analytically to be the lower half-plane by $\tau(\bar{\chi}) = -\overline{\tau(\chi)}$ (see § 3.15). The second integral in (40) can now be taken as being the integral of $e^{\tau(\chi)}\,dw/d\chi$ along the *lower* surface of the real axis in the χ-plane, and the two integrals can be combined to give

$$C_L - iC_D = \frac{1}{cU\beta_a}\int_{\Sigma'} e^{\tau(\chi)}\frac{dw}{d\chi}\,d\chi, \tag{41}$$

where, since the integrand is analytic, Σ' is any contour enclosing the strip representing the wetted surface (see Fig. 11.8).

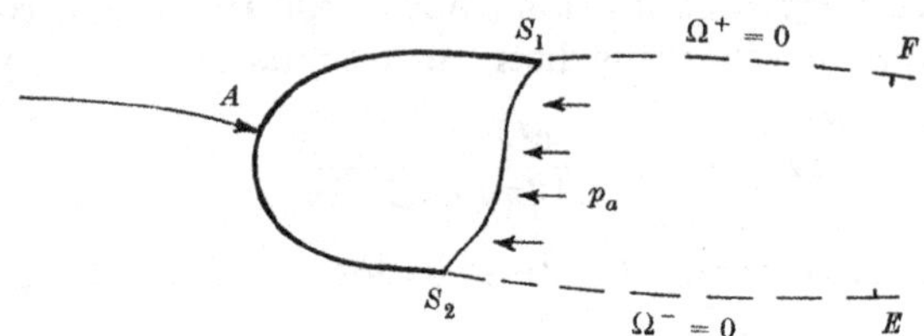

z-plane

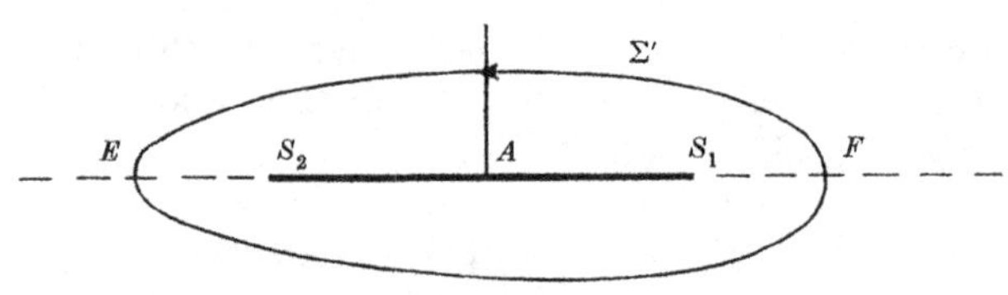

χ-plane

Fig. 11.8

From $w = K\chi^2$ and (22)

$$e^{\tau(\chi)}\frac{dw}{d\chi} = 2K\chi\left\{1 + iB\left(\frac{a}{K}\right)^{\frac{1}{2}}\frac{1}{\chi} - i\left(C + 2B\sin\tfrac{1}{2}\lambda - \frac{i}{2}B^2\right)\left(\frac{a}{K}\right)\frac{1}{\chi^2} + O(\chi^{-3})\right\},$$

and therefore by (41) and the residue theorem,

$$C_L = \pi\left(\frac{4a}{Uc}\right)(C + 2B\sin\tfrac{1}{2}\lambda) = \left(\frac{4a}{Uc}\right)\int_{-\pi}^{\pi}\theta_s(\cos\gamma + 2\sin\tfrac{1}{2}\lambda\sin\tfrac{1}{2}\gamma)\,d\gamma, \tag{42}$$

$$C_D = \frac{\pi}{2}\left(\frac{4a}{Uc}\right)B^2 = \frac{1}{2\pi}\left(\frac{4a}{Uc}\right)\left\{\int_{-\pi}^{\pi}\theta_s\sin\tfrac{1}{2}\gamma\,d\gamma\right\}^2, \tag{43}$$

the wake terms in B and C vanishing in Helmholtz flow.

Levi-Cività (1907) has given similar, but less explicit, results for these coefficients.

A similar expression can be found for the moment coefficient in incompressible Helmholtz flow (see Brodetsky, 1923), but this does not appear capable of being generalized to either compressible flow or aerodynamic wakes, and will not be given here.

11.9   Pressure distribution in aerodynamic wakes

The pressure distribution in an aerodynamic wake must be such that (at least) three conditions are satisfied. The first is that the pressures at the separation points are equal (see §11.3); this is equivalent to

$$\Omega^+(\eta = 0) = \Omega^-(\eta = 0) = \Omega_w, \tag{44}$$

where Ω_w is the (constant) value of Ω immediately behind the body. The second and third conditions have already been given a mathematical form in (37) and (38). As we have no further information on the nature of the wake pressure distribution, we must be content to select a simple mathematical form which satisfies these three conditions.

A suitable form is

$$\Omega^- = \Omega^+ = (\Omega_w - k_1 - k_2)\operatorname{sech}^2 \tfrac{1}{2}\eta + k_1\operatorname{sech}^4 \tfrac{1}{2}\eta + k_2\operatorname{sech}^6 \tfrac{1}{2}\eta, \tag{45}$$

where from (37) and (38) Ω_w, k_1 and k_2 are related by

$$\left.\begin{aligned}
\Omega_w - k_1 - k_2 &= -\frac{1}{4\pi}\left(\frac{cU}{4a}\right)C_D, \\
\Omega_w - \tfrac{1}{2}k_1 - \tfrac{5}{8}k_2 &= -\frac{1}{2\pi}\int_{-\pi}^{\pi}\theta_s\sin\tfrac{1}{2}\gamma\,d\gamma.
\end{aligned}\right\} \tag{46}$$

and

Other equally suitable expressions for Ω^+ and Ω^- can be found.

If our main interest lies in calculating the pressure distribution over the wetted surface, it is important to satisfy (44), but (37)—which gives the precise form of Ω^+ and Ω^- near infinity—can be ignored. Woods (1955c) adopted the distribution

$$\Omega^+ = \Omega^- = \frac{\Omega_w}{1 + k^2\sinh^2 \tfrac{1}{2}\eta}, \tag{47}$$

where

$$\frac{\Omega_w}{k} = -\frac{1}{2\pi}\int_{-\pi}^{\pi}\theta_s\sin\tfrac{1}{2}\gamma\,d\gamma, \tag{48}$$

which satisfies (44) and (38).

A rather different approach has been used by Roshko (1954), who adopted the simple distribution

$$\Omega^+ = \Omega_w \quad (0 < \eta < \eta_0), \qquad \theta^+ = 0 \quad (\eta_0 < \eta < \infty),$$

where η_0 is selected so that $\theta^+(\eta)$ is a continuous function; a similar distribution holds on the lower free streamline. An interesting recent study of the problem of choosing free streamline pressures has been made by Mimura (1958).

The theory unfortunately depends at the least on knowing the wake 'under-pressure' immediately behind the body. Further development of viscous wake theory, and of the theory of detached 'boundary' layers, might lead to the elimination of this empirical element.

11.10   Two simple examples of Helmholtz flow

The first example is the flow illustrated in Fig. 11.10a, in which the obstacle is meant to be an approximate representation of an aerofoil with a hinged flap. Four distinct cases of some practical interest are (i) $E_1 = E_2$ —a normal split flap, (ii) $E_1 = 0$—an ordinary flap, with flow separation at the hinge, (iii) $E_1 = 0$, $\xi = 0$—a crude model of the trailing edge stall phenomenon, and (iv) E_2 small, $\xi \simeq \tfrac{1}{2}\pi$—an aerofoil spoiler.

The value of γ at various points on the wetted surface is marked on the figure. It is clear that

$$\theta_s = -\alpha + \pi\{\mathbf{U}(\gamma - \lambda) - \mathbf{U}(\gamma - \lambda - \epsilon)\} - \xi\{\mathbf{U}(\gamma + \pi) - \mathbf{U}(\gamma - \gamma_0)\}.$$

Also $\Omega^+ = \Omega^- = 0$ in the present case (Helmholtz flow). Equations (8) and (17) give

$$\Omega(\gamma) = \ln\left|\frac{\sin\tfrac{1}{4}(\lambda + \epsilon - \gamma)\cos\tfrac{1}{4}(\lambda + \gamma)}{\sin\tfrac{1}{4}(\lambda - \gamma)\cos\tfrac{1}{4}(\lambda + \epsilon + \gamma)}\right| - \frac{\xi}{\pi}\ln\left|\frac{\sin\tfrac{1}{4}(\gamma_0 - \gamma)}{\cos\tfrac{1}{4}(\gamma_0 + \gamma)}\right|, \quad (49)$$

and

$$2\alpha - \epsilon + \xi\left(1 + \frac{\gamma_0}{\pi}\right) = 0. \qquad (50)$$

Let α be small, and suppose that either ξ is small, or that E_2 is small (i.e. $\gamma_0 \simeq -\pi$), then (50) shows that ϵ is also small. Linear perturbation theory is applicable, and we can write $\phi = Ux$. In particular from a comparison of Figs. 11.5 and 11.10a it follows that

$$\phi_1 = (E + E_1)\,Uc, \quad \phi_2 = (E + E_2)\,Uc, \qquad (51)$$

whence

$$\frac{4a}{Uc} = \tfrac{1}{4}\{\sqrt{(E + E_1)} + \sqrt{(E + E_2)}\}^2, \qquad (52)$$

and

$$\sin\tfrac{1}{2}\lambda = \frac{\sqrt{(E + E_2)} - \sqrt{(E + E_1)}}{\sqrt{(E + E_2)} + \sqrt{(E + E_1)}}, \qquad (53)$$

by (4) and (5). Also as $\phi = EUc$ at $\gamma = \gamma_0$, (6), (52) and (53) give

$$\sin\tfrac{1}{2}\gamma_0 = \frac{-2\sqrt{E} + \sqrt{(E + E_2)} - \sqrt{(E + E_1)}}{\sqrt{(E + E_2)} + \sqrt{(E + E_1)}}, \qquad (54)$$

the negative sign being selected as γ_0 is on the lower surface. The last equation is not valid with a spoiler for the velocity varies rapidly from

$q = 0$ at the root of the spoiler ($\gamma = \gamma_0$) to $q = U$ at the top of the spoiler where the flow separates ($\gamma = -\pi$). Spoilers will be treated separately in the next section. The x, γ relation is

$$\frac{x}{c} = \left(\frac{4a}{Uc}\right)(\sin \tfrac{1}{2}\gamma - \sin \tfrac{1}{2}\lambda)^2, \tag{55}$$

by (6).

If E, E_1 and E_2 are given, (49), (50), (52)–(55) enable us to find the pressure distribution as a function of α, ξ and x over the aerofoil surface.

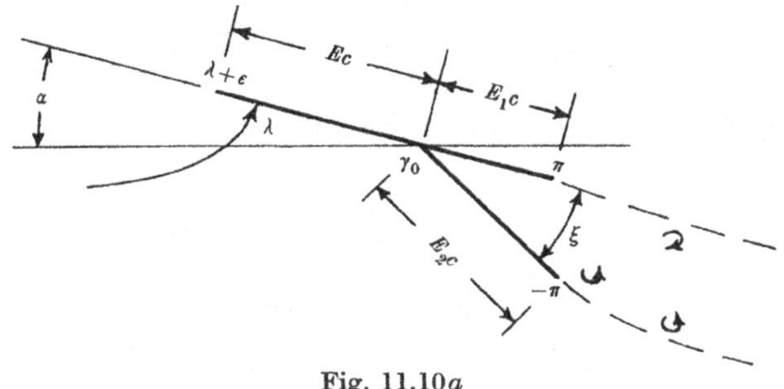

Fig. 11.10a

The lift and drag coefficients are found from (42) and (43) to be

$$C_L = \pi\left(\frac{4a}{Uc}\right)\left\{\sin(\lambda+\epsilon) + \sin\lambda - 4\sin\tfrac{1}{2}\lambda\cos\tfrac{1}{2}(\lambda+\epsilon)\right.$$
$$\left. - \frac{\xi}{\pi}[\sin\gamma_0 - 4\sin\tfrac{1}{2}\lambda\cos\tfrac{1}{2}\gamma_0]\right\}, \tag{56}$$

$$C_D = 2\pi\left(\frac{4a}{Uc}\right)\left\{\cos\tfrac{1}{2}(\lambda+\epsilon) - \cos\tfrac{1}{2}\lambda - \frac{\xi}{\pi}\cos\tfrac{1}{2}\gamma_0\right\}^2. \tag{57}$$

Fig. 11.10b

To make these results a little more significant consider the particular case of the trailing edge stall (Fig. 11.10b). In this case $\xi = 0$, $E_1 = 0$ and $E_2 = 1 - E$. Substituting these values in (50), (52), (53), (56) and (57) and retaining only the highest order terms in α we find

$$C_L = 2\pi\left(\frac{1+\sqrt{E}}{2}\right)^2\alpha, \quad C_D = 2\pi\left(\frac{1-\sqrt{E}}{2}\right)^2\alpha^2. \tag{58}$$

To complete this model of the trailing edge stall it remains to find the relationship between E and α. To do this boundary-layer theory must be employed, and the relation will depend very much on the aerofoil shape.

A more elaborate theory would take into account the fact that the values of C_p on the free streamline near the separation point is usually appreciably less than zero. Because of this (58) will underestimate C_L.

For a second example consider the incompressible flow past a double wedge 'aerofoil', at zero incidence (see Fig. 11.10c). This is a classical problem, which was solved about eighty years ago by Bobyleff (see comments on p. 104 of Lamb, 1932).

We have

$$\theta_s = \begin{cases} \epsilon & (0 < \gamma < \pi), \\ -\epsilon & (-\pi < \gamma < 0), \end{cases}$$

and so from (8)
$$\tau(\zeta) = i\epsilon - \frac{2\epsilon}{\pi}\ln(\tan\tfrac{1}{4}\zeta). \tag{59}$$

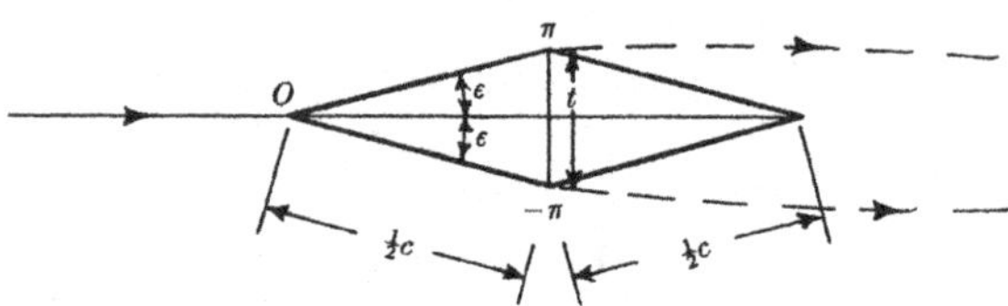

Fig. 11.10c

From the symmetry of the flow $\lambda = 0$, and (3) gives

$$w = 2a(1 - \cos\zeta). \tag{60}$$

Equations (59) and (60) can be combined to give

$$\frac{z}{c} = \frac{1}{2}\left(\frac{4a}{Uc}\right)e^{i\epsilon}\int_0^\xi (\tan\tfrac{1}{4}\zeta)^{-2\epsilon/\pi}\sin\zeta\,d\zeta,$$

the origin of the z-plane being at the leading edge of the wedge.

Notice from Fig. 11.10c that $z = \tfrac{1}{2}c\,e^{i\epsilon}$ at the separation point $\gamma = \pi$, so

$$\frac{cU}{4a} = \int_0^\pi (\tan\tfrac{1}{2}\gamma)^{-2\epsilon/\pi}\sin\gamma\,d\gamma = 4\int_0^1 x^{-\epsilon/\pi}\left\{\frac{2}{(1+x)^3} - \frac{1}{(1+x)^2}\right\}dx.$$

This integral can be evaluated in terms of the logarithmic derivative of the gamma function, $\Psi(z)$, defined by

$$\Psi(z) + \gamma = \int_0^1 \frac{1 - u^{z-1}}{1 - u}\,du,$$

where γ is Euler's constant. It is found that

$$\frac{cU}{4a} = 2 + \frac{4\epsilon}{\pi} + \frac{4\epsilon^2}{\pi^2}\left\{\Psi\left(1 - \frac{\epsilon}{2\pi}\right) - \Psi\left(\frac{1}{2} - \frac{\epsilon}{2\pi}\right)\right\}. \tag{61}$$

From (43) the drag coefficient is

$$C_D = \frac{8}{\pi}\left(\frac{4a}{cU}\right)\epsilon^2.$$ (62)

Two special cases of this result are of interest in aerodynamics. First the classical result for a flat plate normal to the stream is obtained by putting $\epsilon = \frac{1}{2}\pi$. Equations (61) and (62) yield

$$C_D = \frac{2\pi}{4+\pi} \simeq 0\cdot88,$$ (63)

on using the relation $\Psi'(1-z)-\Psi'(z) = \pi \cot \pi z$. As mentioned in §11.3 this is less than half the experimental value in air, due almost entirely to the neglect of the wake under-pressure in the Helmholtz model. Secondly, consider the case when ϵ is a small first-order quantity. To first order (61) gives $(4a/Uc) = \frac{1}{2}$, so

$$C_D = \frac{4}{\pi}\left(\frac{t}{c}\right)^2,$$ (64)

where t is the wedge thickness. This result leads us to expect that the form drag of a double-wedge aerofoil, which has theoretical advantages at supersonic speeds, will increase as the square of the thickness : chord ratio. It is interesting to note that this law of variation of form drag also holds for conventional aerofoil shapes (Squire & Young, 1937).

11.11 Theory of aerofoil spoilers

Fig. 11.11 shows a thin aerofoil fitted with a spoiler; the relevant theory is a special case of that given in §11.10 for split flaps. Comparison of Figs. 11.10a and 11.11 shows that for the present case $E + E_1 = 1$ and $E_2 = h/c$, where h is the spoiler height. As h is assumed to be small relative to c then $\gamma_0 + \pi = \epsilon_0$ say, must also be small. We shall adopt linear perturbation theory and write $C_p = 2\Omega/\beta_a$. Then neglecting second-order terms in ϵ and ϵ_0, we find from (49) and (50) that

$$C_p = \frac{1}{\beta_a}\left(2\alpha + \frac{\xi\epsilon_0}{\pi}\right)\frac{\cos\frac{1}{2}\gamma}{\sin\frac{1}{2}\lambda - \sin\frac{1}{2}\gamma} + \frac{\xi\epsilon_0}{\beta_a\pi}\frac{\cos\frac{1}{2}\gamma}{1+\sin\frac{1}{2}\gamma}.$$ (65)

The relation between γ and x is given by (55), where

$$\frac{4a}{Uc} = \tfrac{1}{4}(1+\sqrt{E})^2, \quad \sin\tfrac{1}{2}\lambda = -\frac{1-\sqrt{E}}{1+\sqrt{E}},$$ (66)

neglecting some small terms in h/c.

This theory is obviously not valid on the spoiler itself, i.e. in the range $-\pi < \gamma < -\pi + \epsilon_0$. In this range the last term of (49) is dominant, and so

$$\Omega(\gamma) \simeq -\frac{\xi}{\pi}\ln\frac{\sin\frac{1}{4}(\epsilon_0-\pi-\gamma)}{\sin\frac{1}{4}(\epsilon_0+\pi+\gamma)}$$

$$\simeq -\frac{\xi}{\pi}\ln\left(\frac{\epsilon_0-\pi-\gamma}{\epsilon_0+\pi+\gamma}\right) \quad (-\pi < \gamma < -\pi+\epsilon_0),$$

as ϵ_0 is small. The velocity varies from $q = 0$ at $\gamma = -\pi + \epsilon_0$ to $q = U$ at $\gamma = -\pi$, whence $\beta = \sqrt{(1-M^2)}$ varies from 1 to β_a. We shall give β its average value in the definition of Ω (see (2.46)), and write $\Omega = \frac{1}{2}(\beta_a+1)\ln(U/q)$. Then on the spoiler

$$\frac{U}{q} = \left(\frac{\epsilon_0+\pi+\gamma}{\epsilon_0-\pi-\gamma}\right)^\sigma, \quad \sigma \equiv \frac{2\xi}{\pi(1+\beta_a)}. \tag{67}$$

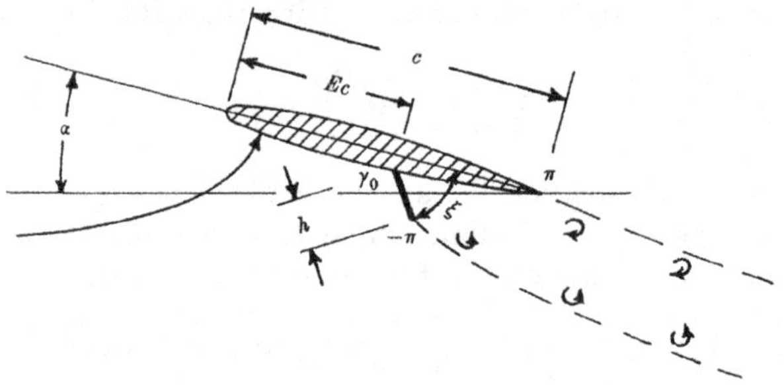

Fig. 11.11

The spoiler height can now be calculated from (9), which can be written

$$\frac{h}{c} = \frac{\epsilon_0^2}{2}\left(\frac{4a}{Uc}\right)(1+\sin\tfrac{1}{2}\lambda)\int_0^1\left(\frac{1+t}{1-t}\right)^\sigma t\,dt,$$

on putting $t = (\pi+\gamma)/\epsilon_0$ and using the fact that ϵ_0 is small. This can be written

$$\epsilon_0 = F\left\{\frac{2h/c}{E+\sqrt{E}}\right\}^{\frac{1}{2}}, \tag{68}$$

where

$$F \equiv \left\{\frac{1}{2}\int_0^1\left(\frac{1+t}{1-t}\right)^\sigma t\,dt\right\}^{-\frac{1}{2}}. \tag{69}$$

The function $F(\sigma)$ behaves as indicated by the following table:

σ	0	0·1	0·2	0·3	0·4	0·5	0·6	0·7	0·8	0·9	1·0
F	2·00	1·81	1·61	1·42	1·24	1·06	0·88	0·71	0·53	0·35	0

With a spoiler at right angles to the aerofoil surface in incompressible flow, $\sigma = \frac{1}{2}$ and $F = \sqrt{\{8/(4+\pi)\}}$. Equations (65) and (68) give the required relation between C_p, α, ξ, h and γ.

From (50) and (56) it is found that

$$C_L = 2\pi\left(\frac{4a}{Uc}\right)\left\{\alpha + \frac{\xi\epsilon_0}{\pi}(1+\sin\tfrac{1}{2}\lambda)\right\},\tag{70}$$

on neglecting second-order terms. Similarly, (57) yields

$$C_D = 2\pi\left(\frac{4a}{Uc}\right)\left\{\alpha\sin\tfrac{1}{2}\lambda + \frac{\xi\epsilon_0}{2\pi}(1+\sin\tfrac{1}{2}\lambda)\right\}^2.\tag{71}$$

The moment coefficient is easily calculated from (55) and (65). It is found that the nose-up moment about the leading edge has the coefficient

$$C_M = -\frac{\pi}{2\beta_a}\left(\frac{4a}{Uc}\right)^2\left\{(1+4\sin^2\tfrac{1}{2}\lambda)\,\alpha\right.$$
$$\left.+\frac{2\xi\epsilon_0}{\pi}(1+\sin\tfrac{1}{2}\lambda)(1+2\sin\tfrac{1}{2}\lambda+2\sin^2\tfrac{1}{2}\lambda)\right\}.\tag{72}$$

The main defect of the above theory is that if the spoiler is forward of the trailing edge, i.e. if $E < 1$, the pressure distribution in the wake immediately behind the spoiler makes a significant contribution to the force and moment acting on the aerofoil. This can be allowed for by a semi-empirical method similar to that described in §11.9. A further difficulty is that the boundary layer upstream of the spoiler tends to mask the effect of the spoiler and therefore reduce its effective height. These and similar details are discussed in the paper by Woods (1953b).

If the spoiler is at the trailing edge, $E = 1$, and (66) and (55) give $4a/Uc = 1$, $\lambda = 0$ and $x = c\sin^2\tfrac{1}{2}\gamma$. The load distribution in this case is $\Delta C_p = C_p(\gamma) - C_p(-\gamma)$. From (65) it follows that the load distribution due to the spoiler alone is

$$\Delta C_{p_s} = -\frac{4\xi\epsilon_0}{\beta_a\pi}\operatorname{cosec}\gamma.\tag{73}$$

The corresponding lift is

$$C_{L_s} = 2\xi F\left(\frac{h}{c}\right)^{\frac{1}{2}}.\tag{74}$$

Woods (1953b) has found (73) and (74) to be in close agreement with experiment. With trailing edge spoilers the wake pressures have an equal influence on each surface of the aerofoil, and so contribute nothing to Δp_s.

11.12   Curved obstacles

The examples so far given involve obstacles comprised of a number of straight surfaces. Curved obstacles are more difficult, as it is necessary to solve integral equations of a type similar to that described in §8.3 for aerofoils.

We shall follows Woods (1955c), and assume that Ω^+, Ω^- are given by (47). Substituting (47) into (8), and integrating the first term of (8) by parts we obtain

$$\Omega(\gamma) = \frac{\Omega_w}{1+k\cos\frac{1}{2}\gamma} - \frac{1}{\pi}\int_{-\pi}^{\pi} \ln\left|\frac{\sin\frac{1}{4}(\gamma^*-\gamma)}{\cos\frac{1}{4}(\gamma^*+\gamma)}\right| d\theta_s(\gamma^*). \tag{75}$$

Where θ_s is continuous (11) is applicable.

For simplicity we shall assume that the only discontinuity in θ_s is one of size θ_a at the front stagnation point, $\gamma = \lambda$. Then from (75) and (11)

$$\Omega(\gamma) = \frac{\Omega_w}{1+k\cos\frac{1}{2}\gamma} - \frac{\theta_a}{\pi}\ln\left|\frac{\sin\frac{1}{4}(\lambda-\gamma)}{\cos\frac{1}{4}(\lambda+\gamma)}\right|$$
$$+ \frac{1}{\pi}\int_0^{\pi}\left\{ C\ln\left|\frac{\sin\frac{1}{4}(\gamma^*-\gamma)\sin\frac{1}{4}(\gamma^*+\gamma)}{\cos\frac{1}{4}(\gamma^*-\gamma)\cos\frac{1}{4}(\gamma^*+\gamma)}\right|\right.$$
$$\left. + D\ln\left|\frac{\sin\frac{1}{4}(\gamma^*-\gamma)\cos\frac{1}{4}(\gamma^*-\gamma)}{\cos\frac{1}{4}(\gamma^*+\gamma)\sin\frac{1}{4}(\gamma^*+\gamma)}\right|\right\} d\gamma^*,$$

where

$$\left.\begin{aligned}
C &\equiv -\frac{2a}{Uc}\left[\left\{\frac{Uc}{Rq_s}(\gamma) - \frac{Uc}{Rq_s}(-\gamma)\right\}\tfrac{1}{2}\sin\gamma\right.\\
&\qquad\qquad \left. + \left\{\frac{Uc}{Rq_s}(\gamma) + \frac{Uc}{Rq_s}(-\gamma)\right\}\sin\tfrac{1}{2}\lambda\cos\tfrac{1}{2}\gamma\right],\\
D &\equiv -\frac{2a}{Uc}\left[\left\{\frac{Uc}{Rq_s}(\gamma) + \frac{Uc}{Rq_s}(-\gamma)\right\}\tfrac{1}{2}\sin\gamma\right.\\
&\qquad\qquad \left. + \left\{\frac{Uc}{Rq_s}(\gamma) - \frac{Uc}{Rq_s}(-\gamma)\right\}\sin\tfrac{1}{2}\lambda\cos\tfrac{1}{2}\gamma\right].
\end{aligned}\right\} \tag{76}$$

If matrices A, B, C and D are defined by

$$A_{ij} = -\frac{1}{\pi}\int_{\gamma_{i-\frac{1}{2}}}^{\gamma_{i+\frac{1}{2}}} \ln\left|\frac{\sin\frac{1}{4}(\gamma^*-\gamma_j)\cos\frac{1}{4}(\gamma^*-\gamma_j)}{\sin\frac{1}{4}(\gamma^*+\gamma_j)\cos\frac{1}{4}(\gamma^*+\gamma_j)}\right| d\gamma^*,$$

$$B_{ij} = -\frac{1}{\pi}\int_{\gamma_{i-\frac{1}{2}}}^{\gamma_{i+\frac{1}{2}}} \ln\left|\frac{\sin\frac{1}{4}(\gamma^*-\gamma_j)\sin\frac{1}{4}(\gamma^*+\gamma_j)}{\cos\frac{1}{4}(\gamma^*+\gamma_j)\cos\frac{1}{4}(\gamma^*-\gamma_j)}\right| d\gamma^*,$$

$$C_i = C(\gamma_i),$$

and $$D_i = D(\gamma_i),$$

the equation for Ω can be written approximately

$$\Omega(\pm\gamma_j) = -\sum_i C_i B_{ij} \mp \sum_i D_i A_{ij} - \frac{\theta_a}{\pi}\ln\left|\frac{\sin\frac{1}{4}(\lambda\mp\gamma_j)}{\cos\frac{1}{4}(\lambda\pm\gamma_j)}\right| + \frac{\Omega_w}{1+k\cos\frac{1}{2}\gamma_j}. \tag{77}$$

Let θ^+ and θ^- be the values of θ_s at the upper and lower separation points then it is self-evident that

$$\int_{-\pi}^{\pi} d\theta_s(\gamma) = \theta^+ - \theta^-. \tag{78}$$

this can be written in the approximate form

$$2 \sum_i C_i(\gamma_{i+\frac{1}{2}} - \gamma_{i-\frac{1}{2}}) = \theta_a - \theta^+ + \theta^-. \tag{79}$$

Similarly, after integration by parts, (17) can be expressed

$$-\sum_i D_i(\gamma^2_{i+\frac{1}{2}} - \gamma^2_{i-\frac{1}{2}}) = \pi(\theta^+ + \theta^-) - \lambda\theta_a. \tag{80}$$

Equations (79) and (80) can be regarded as fixing the values of θ_a and λ at each stage of the iteration described below. In the limit as the number of intervals into which the range of integration $(0, \pi)$ is subdivided tends to infinity, the value of θ_a given by (79) should tend to its geometrical value, which is π for a rounded-nose obstacle (cf. similar remarks in §7.6). Matrices A_{ij} and B_{ij}, which yield thirty-six points at which Ω is determined on the wetted surface, are given in the paper by Woods (1955c). These should be sufficient for most examples.

The steps in the solution of (77) are as follows:

(i) From the experimental value of the pressure behind the obstacle is calculated the corresponding value of Ω; this is Ω_w.

(ii) Likely points of flow separation are assigned.

(iii) A relation $q = q(\gamma)$ and a value of λ (e.g. $q = U$, $\lambda = 0$) are estimated, then s/c is calculated from (9). $R = R(\gamma)$ can now be determined since $R = R(s)$ is known, or can be deduced from the obstacle's coordinates.

(iv) Next C_i, D_i, θ_a and a new value of λ follow in turn from (76), (79) and (80). The constant k is calculated from

$$\frac{\Omega_w}{k} = -\frac{\theta_a}{\pi}\cos\tfrac{1}{2}\lambda + \frac{4}{\pi}\sum_i C_i(\sin\tfrac{1}{2}\gamma_{i+\frac{1}{2}} - \sin\tfrac{1}{2}\gamma_{i-\frac{1}{2}}),$$

which comes from (48).

(v) We can now calculate $\Omega(\gamma)$ from (77), and the first iteration is completed by using (6.3) to find the corresponding distribution $q(\gamma)$.

(vi) Steps (iii) to (v) are repeated until $q(\gamma)$ remains unchanged by an iteration.

In the case of the circular cylinder (§11.13) the convergence is found to be quite rapid—about five iterations being sufficient. With 'thinner' bodies the convergence should be even more rapid.

The separation points must be chosen so that

$$\frac{2}{\pi}\int_0^\pi (C\sec\tfrac{1}{2}\gamma + D\tan\tfrac{1}{2}\gamma)\,d\gamma - \frac{\theta_a}{\pi}\cot\tfrac{1}{2}(\pi - \lambda) + k\Omega_w \geqslant 0, \tag{81}$$

which follows from (47), (13), (76) and the physical reality condition given in §11.6. If smooth separation is required for given separation points, Ω_w must be determined from the equality in (81) at the end of each iteration. The final value obtained for Ω_w in this case will depend on the positions of the separation points initially assumed. If these positions are varied and the solution completed for each choice of separation points, a relation between Ω_w and these positions will be obtained. Hence by interpolation we can deduce the positions of the separation points giving smooth separation for an assigned (from experiment) value of Ω_w.

11.13   An example: the circular cylinder

The particular case of the symmetrical separating flow past a circular cylinder has the merit that good experimental data (Fage & Falkner, 1931) is available for comparison with theory. For this case $\lambda = 0$, and (76), (77), (79) and (9) become

$$C_i = -\frac{2a}{UR}\frac{U}{q}\sin\gamma_i,$$

$$\Omega(\gamma_j) = -\frac{\theta_a}{\pi}\ln\left|\tan\tfrac{1}{2}\gamma_j\right| - \sum_i C_i B_{ij} + \frac{\Omega_w}{1 + k\cos\tfrac{1}{2}\gamma_j},$$

$$\theta_a = 2(\tfrac{1}{2}\pi - \xi_s) + \tfrac{1}{9}\pi\sum_i C_i,$$

and
$$\frac{RU}{2a} = -\frac{1}{\xi_s}\int_0^\pi \frac{U}{q}\sin\gamma\,d\gamma,$$

respectively, where ξ_s is the value of ξ at the point of separation (see Fig. 11.13a), and the range of integration $(0, \pi)$, has been divided into eighteen equal parts. These equations have been solved by the method described in § 11.12, and with the aid of the tables published by Woods (1955c), for the following cases:

(i) $\xi_s = 56°,\ 60°,\ 80°,\ 100°$ and $120°,\quad M_a = 0,\quad \Omega_w = 0.$

(ii) $\xi_s = 120°,\quad M_a = 0\cdot40$ and $0\cdot45,\quad \Omega_w = 0.$

(iii) $\xi_s = 75°$ and $80°,\quad M_a = 0,\quad \Omega_w = 0\cdot41.$

The pressure coefficients are shown in Figs. 11.13b and 11.13c as functions of the angular coordinate ξ. The incompressible flow results are given in Fig. 11.13b, where, for comparison the non-separation flow, $C_p = 2\cos 2\xi - 1$ (see p. 149 of Milne-Thomson, 1949), is also shown. The two sets of experimental results plotted in the figure are taken from Fage & Falkner (1931). They are at different Reynolds numbers, and correspond to laminar and turbulent boundary-layer separation.

Consider first the results for which M_a and Ω_w vanish (Helmholtz flow). By extrapolation it was found that the equality sign in (81) holds at $\xi_s = 55\cdot1°$, which is thus the point of smooth separation in this case. This is in close agreement with the value of $55°$ obtained by Brodetsky (1923) for a nearly circular cylinder, by a different method based on the theory of Levi-Cività (1907). The left-hand side of (48) is found to be less than the right-hand side (cf. (25)) for values of ξ_s greater than 124°, so we can conclude that Helmholtz flows that separate downstream of $\xi_s = 124°$ are physically unreal.

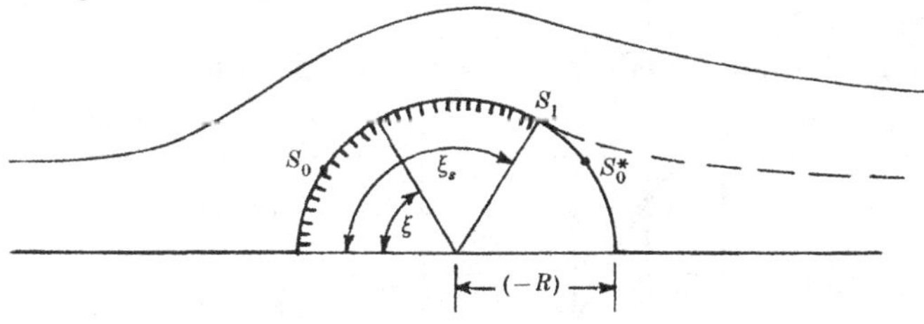

Fig. 11.13*a*

From the experimental value of $C_p = -1·26$ for laminar separation (see Fig. 11.13*b*) and (6.18) we find that $\Omega_w = -\tfrac{1}{2}\ln(1-C_p) \approx -0·41$. The curve $\xi_s = 80°$, $\Omega_s = -0·41$ shown in Fig. 11.13*b* is in good agreement with the experimental values for laminar separation, for which ξ_s is

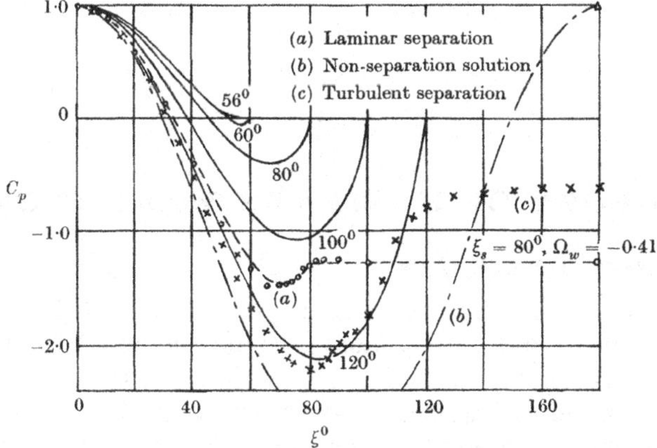

Fig. 11.13*b*

approximately 78°. It will be noticed that the corresponding Helmholtz flow ($\Omega_w = 0$, $\xi_s = 80°$) yields results very different from experiment. The fair agreement between the Helmholtz flow at $\xi_s = 120°$ and the experiment having separation at the same point is due, no doubt, to the much smaller value of $|\Omega_w|$ in this case. A final point of some significance is that for $\Omega_w = -0·41$ the left-hand side of (81) has the values $-0·016$ and $0·094$ for $\xi_s = 75$ and 80° respectively, hence by interpolation the point of smooth separation is at $\xi_s = 76°$. Thus the value of ξ_s for smooth separation is remarkably close to the experimental value of ξ_s for laminar separation. Agreement between these two values of ξ_s was predicted by Squire (1934), but at that time the only model of separation flow was the Helmholtz model for which, as remarked above, $\xi_s = 55 \cdot 1°$. The connexion between smooth separation and laminar separation has been explained at the end of § 11.2.

Fig. 11.13c shows the effect of compressibility on the separating flow.

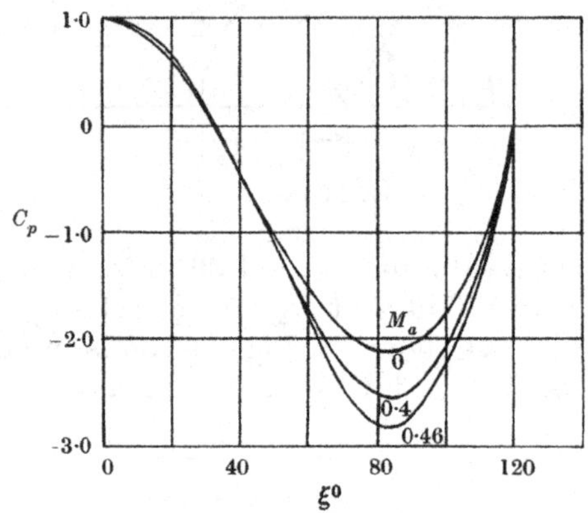

Fig. 11.13c

THEORY OF UNSTEADY WAKE FLOW

11.14 General description

Unsteady flows with free boundaries present several theoretical difficulties, the most serious of which is that the free boundary is not a streamline. Instead it is a *material* line, i.e. the locus of all particles which have passed through a given point; with free boundaries this generating point is the point of flow separation from the solid surface. In certain types of unsteady flow this difficulty can be overcome by adopting the approximation that the free boundary is a streamline. This is valid either if (i) the unsteady motion is a small perturbation of a basic steady flow (see discussion in § 2.21), or (ii) if the rate of change is relatively small. In each of these cases there is a corresponding steady flow with free streamlines lying very close to the material lines of the (unsteady) perturbed flow.

Type (i) flows have several important aerodynamic applications, and will receive detailed consideration in the following sections. A simple example is shown in Fig. 11.14a. The obstacle is a wedge, which is oscillating about a fixed point P. The flow separates from the wedge at S_1 and S_2, and because of the unsteady motion there will be vorticity of fluctuating strength on the free boundaries $S_1 B_\infty$ and $S_2 C_\infty$. These boundaries will vary in position, but if the amplitude of the oscillations is small, it is reasonable to assume that they do not deviate significantly from the positions occupied by the free boundaries of the mean steady flow, except perhaps well downstream of the obstacles. Therefore the boundary conditions of the unsteady flow can be applied at the boundaries of the mean steady flow. A similar situation occurs in unsteady aerofoil theory (see § 6.8). When the wedge angle ϵ tends to zero the two vortex sheets collapse into one sheet, and the flow becomes the same as that studied in § 9.6—

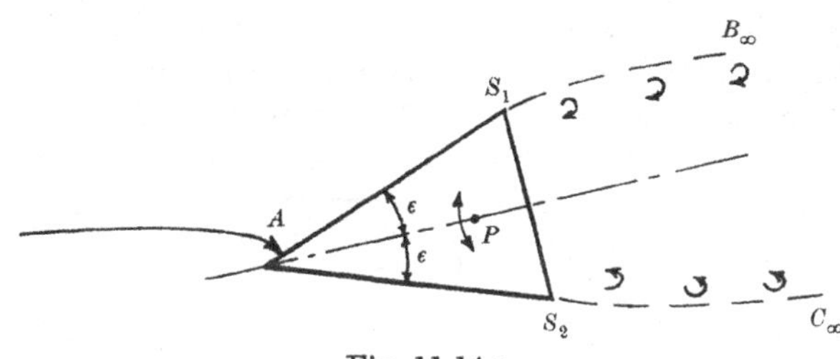

Fig. 11.14a

i.e. flow about an oscillating thin aerofoil. Indeed the theory of the following pages is distinguished from that of ch. 9 principally by the presence of *two* vortex sheets instead of one. This of course adds some complication to the algebra; for example, instead of having to solve one integral equation to determine the strength of the vortex sheet as in § 9.3, we must solve two simultaneous integral equations to obtain the strengths of the two vortex sheets.

The main interest in type (ii) flows is that they provide a model of cavity formation and growth behind obstacles moving through a liquid; they appear to have no aerodynamic application. Von Kármán (1949) and Gilbarg (1952) have considered a special class of cavity flows in which the cavity length is a function of time. Woods (1953a) has generalized their results, although it should be pointed out that all the models used by these authors are of doubtful validity as they incorporate some of the unsteady characteristics of the flow and neglect others (for example, as remarked by Lighthill, they neglect the sink at infinity which must arise with expanding wakes).

Fox & Morgan (1954) have studied the stability of various free boundary flows, by making small perturbations of the basic steady flows, and then linearizing in the physical plane. The boundary conditions are applied on the material lines, so that the approximation discussed above is not required. While the stability of the free boundaries is of considerable interest, it, or rather its absence, has little effect on the pressures acting on the 'wetted' surface of the obstacle, for this instability usually manifests itself some distance downstream of the obstacle. For an outline of the stability of surfaces of velocity discontinuity the reader can refer to pp. 251–7 of Birkhoff & Zarantonello (1957). Curle (1956 a, b) has also avoided the boundary approximation in a study of the initial development of flows involving free surfaces. However, neither of the methods used in these studies are relevant to the problem of finding the forces acting on an obstacle in unsteady (cavitating or aerodynamic) wake flow, and will not be discussed further.

We shall now complete our general description of type (i) flows. The theory of the following six sections is taken from Woods's (1955 f) paper. It is restricted to incompressible flow, but could be extended to slowly varying compressible flows by a method similar to that used in ch. 9.

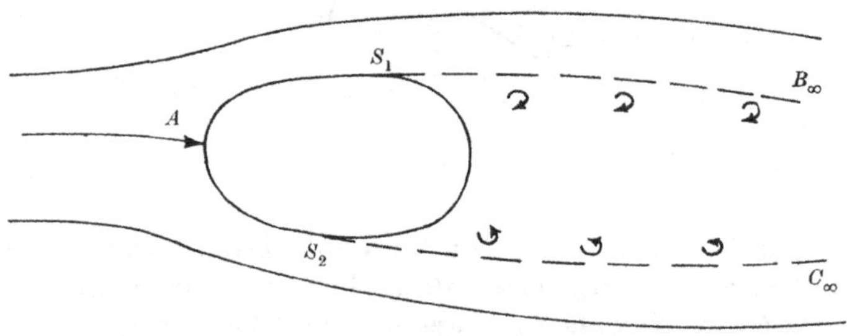

Fig. 11.14b

It is convenient to divide the flow into two parts, namely (a) the basic steady flow, and (b) the small unsteady perturbations superimposed on this basic flow. First the steady flow is calculated by the method described in the earlier part of this chapter, then the unsteady part is calculated by the method given below, and linearly superimposed on the steady flow. As mentioned above, this method involves the assumption that the perturbations are small, and that their second-order effects can be ignored.

Now the unsteady component of the flow can itself be divided into two distinct parts, viz. (i) flow in which the separation points S_1, S_2 (see Fig. 11.14b) are fixed in position, while the wetted surface $S_2 A S_1$ is perturbed about its mean position, and (ii) flow in which the surface is steady, but the separation points move about mean positions. An example of (i) alone occurs with the flow past a sharp-edged obstacle such as shown in Fig. 11.14b, as the separation points will remain fixed at the sharp corners. And an example of (ii) alone is the fluctuating flow at low

Reynolds numbers past a fixed circular cylinder. Of course with curved obstacles in unsteady motion both types of unsteady flow will occur, and it will be necessary to calculate each separately, and then to superimpose them. However, in order to superimpose these partial solutions realistically it is necessary to know how the motion of the separation points S_1, S_2 is related to the motion of the surface $S_2 A S_1$, e.g. with oscillatory motions the amplitude ratio and the phase difference will be required. To the author's knowledge no theoretical work has yet been carried out on this problem; it is extremely recondite, involving not only the theory of the following pages, but also the theory of unsteady boundary layers. And in any case a method of calculating the second type of unsteady flow has not yet been developed, although it should be possible to extend the method given below of dealing with type (i) flows to type (ii) flows. We must therefore be content to describe the method of calculating the first type of unsteady flow; this will be done in §§ 11.15–11.22.

11.15 The basic mathematical theory: fixed separation points

We shall distinguish quantities in the basic steady flow from corresponding quantities in the unsteady flow by a subscript s. Now define an auxiliary ζ-plane by

$$w_s = 4a_s(\sin \tfrac{1}{2}\zeta - \sin \tfrac{1}{2}\lambda_s)^2 \quad (\zeta = \eta + i\gamma), \tag{82}$$

where

$$4a_s = \tfrac{1}{4}(\sqrt{\phi_{2s}} + \sqrt{\phi_{1s}})^2, \quad \sin \tfrac{1}{2}\lambda_s = \frac{\sqrt{\phi_{2s}} - \sqrt{\phi_{1s}}}{\sqrt{\phi_{2s}} + \sqrt{\phi_{1s}}}. \tag{83}$$

This is of course exactly the same plane as that defined by (3)–(5)—we have just made its dependence on the steady flow explicit.

In accordance with the assumption, discussed in § 11.14, that the boundary conditions of the unsteady flow can be applied at the boundaries of the basic steady flow, we can now write down the solution for $\tau(\zeta, t)$ in terms of its boundary values in the ζ-plane. As the separation points are fixed in position, we have $\theta(\gamma, t)$ prescribed in $-\pi \leqslant \gamma \leqslant \pi$, and $\Omega^+(\eta, t)$, $\Omega^-(\eta, t)$ prescribed, or rather calculable, on $\gamma = \pm \pi$. Therefore the required solution is given by an equation identical in form with (8). Subtracting the steady flow solution from the unsteady flow solution we have

$$\tau(\zeta, t) - \tau_s(\zeta) = \frac{\cos \tfrac{1}{2}\zeta}{2\pi} \left\{ \int_{-\pi}^{\pi} \frac{\{\theta(\gamma, t) - \theta_s(\gamma)\}\, d\gamma}{\sin \tfrac{1}{2}\gamma - \sin \tfrac{1}{2}\zeta} \right.$$

$$\left. + \int_0^\infty \left(\frac{\Omega^+(\eta, t) - \Omega_s^+(\eta)}{\cosh \tfrac{1}{2}\eta - \sin \tfrac{1}{2}\zeta} + \frac{\Omega^-(\eta, t) - \Omega_s^-(\eta)}{\cosh \tfrac{1}{2}\eta + \sin \tfrac{1}{2}\zeta} \right) d\eta \right\}. \tag{84}$$

On the wetted surface, $\eta = 0$, (84) becomes

$$\Omega - \Omega_s = \frac{\cos\frac{1}{2}\gamma}{2\pi}\left\{\int_{-\pi}^{\pi}\frac{(\theta-\theta_s)\,d\gamma^*}{\sin\frac{1}{2}\gamma^*-\sin\frac{1}{2}\gamma}\right.$$
$$\left.+\int_0^{\infty}\left(\frac{\Omega^+-\Omega_s^+}{\cosh\frac{1}{2}\eta-\sin\frac{1}{2}\zeta}+\frac{\Omega^--\Omega_s^-}{\cosh\frac{1}{2}\eta+\sin\frac{1}{2}\zeta}\right)d\eta\right\}. \quad (85)$$

Note that in Helmholtz motions, $q_s = U$ on the free boundaries, and so Ω_s^+, Ω_s^- vanish; the pressure at any point in the wake satisfies $p_s = p_a$. We shall not yet restrict our theory to this case, but shall assume that the pressure varies as described in § 11.9.

To determine the strength of the vortex sheets a wake condition is needed. A reasonable *assumption* is that the pressure at any point in the wake in the unsteady flow, p_w say, remains constant and equal to its mean steady value, p_s, i.e.

$$p_w = p_s. \quad (86)$$

In the particular case of unsteady Helmholtz flow (86) becomes $p_w = p_a$, a condition which can be *deduced* from the obvious fact that in a wake extending to infinity the pressure, if constant, cannot differ from that at infinity outside the wake.

Equations (85) and (86) are still incomplete, that is they do not determine a unique unsteady flow. In the following we shall complete them by imposing suitable boundary conditions at infinity.

The first such boundary condition is clear enough; it is simply that the unsteady motion cannot alter the value of τ at infinity, i.e.

$$\lim_{\eta\to\infty}\{\tau(\gamma+i\eta,t)-\tau_s(\gamma+i\eta)\} = 0.$$

Applying this to (84) we get (cf. (17))

$$\int_{-\pi}^{\pi}(\theta-\theta_s)\,d\gamma+\int_0^{\infty}\{(\Omega^+-\Omega_s^+)-(\Omega^--\Omega_s^-)\}\,d\eta = 0. \quad (87)$$

This corresponds to (9·12) for an unsteady aerofoil; we also need equations corresponding to (9·15) and (9·16), which were derived from Kelvin's theorem on the circulation, and the closure condition respectively.

We can expand the right-hand side of (84) in a form similar to that given in (22), and then conclude that near infinity the free boundary slope is

$$\theta = \theta_s+(B-B_s)\left(\frac{a_s}{\phi_s}\right)^{\frac{1}{2}}-\{C-C_s+2(B-B_s)\sin\tfrac{1}{2}\lambda_s\}\frac{a_s}{\phi_s}+O\left(\frac{a_s}{\phi_s}\right)^{\frac{3}{2}},$$

using an obvious notation. If we use the approximations employed in the derivation of (24) at this point, it immediately follows that the unsteady free boundary deviates infinitely from the mean steady boundary unless $B = B_s$ and $C = C_s$. And as (84) is based on the assumption that these boundaries are always close, the theory breaks down unless these restrictions hold, i.e. unless

$$\int_{-\pi}^{\pi} (\theta - \theta_s)\sin \tfrac{1}{2}\gamma \, d\gamma + \int_0^{\infty} \{(\Omega^+ - \Omega_s^+) + (\Omega^- - \Omega_s^-)\}\cosh \tfrac{1}{2}\eta \, d\eta = 0, \quad (88)$$

and

$$\int_{-\pi}^{\pi} (\theta - \theta_s)\cos \gamma \, d\gamma - \int_0^{\infty} \{(\Omega^+ - \Omega_s^+) - (\Omega^- - \Omega_s^-)\}\cosh \eta \, d\eta = 0. \quad (89)$$

Physically these conditions mean that the unsteady perturbations cannot influence the shape of the wake near infinity. In § 11.18 (88) and (89) will be transformed into a pair of simultaneous integral equations for Ω^+ and Ω^-.

11.16   The wake conditions

The pressure equation in incompressible flow is

$$p + \tfrac{1}{2}\rho_a q^2 + \rho_a \frac{\partial \phi}{\partial t} = C(t), \quad (90)$$

from (1.19). In this and the following section it will be convenient to denote by C an arbitrary function of the time only (or a constant), not necessarily the same on each occurrence. Now from (6.17)

$$\Omega - \Omega_s = \ln\frac{U}{q} - \ln\frac{U}{q_s} = -\ln\left\{1 + \left(\frac{q}{q_s} - 1\right)\right\} \cong 1 - \frac{q}{q_s}, \quad (91)$$

to first order in the perturbation $(\Omega - \Omega_s)$. And since the steady form of Bernoulli's equation is $p_s + \tfrac{1}{2}\rho_a q_s^2 = C_s$, (90) can be written

$$p - p_s = \rho_a q_s^2(\Omega - \Omega_s) + \rho_a \frac{\partial}{\partial t}\left(\int q \, ds\right) + C.$$

As
$$q \, ds = \frac{q}{q_s} d\phi_s = \{1 - (\Omega - \Omega_s)\} d\phi_s,$$

then
$$p - p_s = \rho_a q_s^2(\Omega - \Omega_s) + \rho_a \int^{\phi} \frac{\partial}{\partial t}(\Omega - \Omega_s) \, d\phi_s + C. \quad (92)$$

Equation (86) shows that the left-hand side of (92) vanishes on each of the streamlines bounding the wake, and so by differentiation it follows that

$$\frac{\partial}{\partial \phi_s}\{q_s^2(\Omega - \Omega_s)\} + \frac{\partial}{\partial t}(\Omega - \Omega_s) = 0, \quad (93)$$

also holds on each of these streamlines. In the particular case of Helmholtz motion, for which $q_s = U$ and $\phi_s = Us$, (93) has the solution

$$U\Omega = f(s - Ut), \quad (94)$$

where f denotes an arbitrary function.

It follows from (82) that on the wake boundaries $\gamma = \pm\pi$,

$$\phi_s = 4a_s(\cosh\tfrac{1}{2}\eta \mp \sin\tfrac{1}{2}\lambda_s)^2, \tag{95}$$

hence (93) can be written

$$\left.\begin{aligned}
\frac{\partial}{\partial\eta}\{(q_s^+)^2\,(\Omega^+ - \Omega_s^+)\} + 2a_s(\sinh\eta - 2\sin\tfrac{1}{2}\lambda_s\sinh\tfrac{1}{2}\eta)\frac{\partial}{\partial t}(\Omega^+ - \Omega_s^+) = 0, \\[2mm]
\frac{\partial}{\partial\eta}\{(q_s^-)^2\,(\Omega^- - \Omega_s^-)\} + 2a_s(\sinh\eta + 2\sin\tfrac{1}{2}\lambda_s\sinh\tfrac{1}{2}\eta)\frac{\partial}{\partial t}(\Omega^- - \Omega_s^-) = 0.
\end{aligned}\right\} \tag{96}$$

These are the required wake conditions.

11.17 The pressure distribution and forces on the obstacle

First we shall obtain a convenient formula for the pressure increment, $p - p_s$, at a point $P(x, y)$ on the wetted surface, assumed to be moving relative to a frame fixed in space with small perturbation velocities n and m normal and tangential to the surface.

Now
$$\frac{\partial\phi}{\partial t} = \frac{\partial}{\partial t}\int q\,ds = \frac{\partial}{\partial t}\int\mathbf{q}\cdot d\mathbf{s} = \int\left(\frac{\partial\mathbf{q}}{\partial t}\right)\cdot d\mathbf{s} + \int\mathbf{q}\cdot\left(\frac{d}{dt}d\mathbf{s}\right),$$

where $\mathbf{q}$ and $d\mathbf{s}$ are parallel vectors. Thus if $\mathbf{v}$ represents the small perturbation vector, $\mathbf{v} = \mathbf{q} - \mathbf{q}_s$,

$$\frac{\partial}{\partial t}\int q\,ds = \int\frac{\partial q}{\partial t}\,ds + \int\mathbf{q}\cdot d\mathbf{v} \simeq \int\frac{\partial q}{\partial t}\,ds + \mathbf{v}\cdot\mathbf{q}_s,$$

to first order in $|\mathbf{v}|$. The last term is mq_s, $\mathbf{q}_s$ being parallel to the surface. Therefore on the wetted surface in place of (92) we have

$$p - p_s = C + \rho_a mq_s + \rho_a q_s^2(\Omega - \Omega_s) + \rho_a\int^{\phi_s}\frac{\partial}{\partial t}(\Omega - \Omega_s)\,d\phi_s.$$

(The additional term just introduced does not arise in (92) because the wake boundary is assumed to be stationary.) On this surface (82) gives

$$d\phi_s = 4a_s(\sin\tfrac{1}{2}\gamma - \sin\tfrac{1}{2}\lambda_s)\cos\tfrac{1}{2}\gamma\,d\gamma,$$

and so

$$\begin{aligned}
p - p_s = {}& C + \rho_a mq_s + \rho_a q_s^2(\Omega - \Omega_s) \\
& + 4a_s\rho_a\int_{\lambda_s}^{\gamma}(\sin\tfrac{1}{2}\gamma - \sin\tfrac{1}{2}\lambda_s)\cos\tfrac{1}{2}\gamma\frac{\partial}{\partial t}(\Omega - \Omega_s)\,d\gamma. \tag{97}
\end{aligned}$$

The following calculation is aimed at finding a more convenient form for this result.

Equation (85) can be expressed in the form

$$\Omega - \Omega_s = \frac{\cos\tfrac{1}{2}\gamma}{2\pi}\int_{-\pi}^{\pi}\frac{(\theta - \theta_s)\,d\gamma^*}{\sin\tfrac{1}{2}\gamma^* - \sin\tfrac{1}{2}\gamma} - \frac{1}{\cos\tfrac{1}{2}\gamma}(F + E\sin\tfrac{1}{2}\gamma)$$

$$+ \frac{E+G}{2\cos\tfrac{1}{2}\gamma\,(\sin\tfrac{1}{2}\gamma-\sin\tfrac{1}{2}\lambda_s)}$$

$$-\frac{1}{4\pi}\int_0^\infty \frac{(\sinh\eta-2\sin\tfrac{1}{2}\lambda_s\sinh\tfrac{1}{2}\eta)\sinh\tfrac{1}{2}\eta\,(\Omega^+-\Omega_s^+)}{\cos\tfrac{1}{2}\gamma(\sin\tfrac{1}{2}\gamma-\sin\tfrac{1}{2}\lambda_s)(\cosh\tfrac{1}{2}\eta-\sin\tfrac{1}{2}\gamma)}\,d\eta$$

$$+\frac{1}{4\pi}\int_0^\infty \frac{(\sinh\eta+2\sin\tfrac{1}{2}\lambda_s\sinh\tfrac{1}{2}\eta)\sinh\tfrac{1}{2}\eta\,(\Omega^--\Omega_s^-)}{\cos\tfrac{1}{2}\gamma(\sin\tfrac{1}{2}\gamma-\sin\tfrac{1}{2}\lambda_s)(\cosh\tfrac{1}{2}\eta+\sin\tfrac{1}{2}\gamma)}\,d\eta, \quad (98)$$

where in matrix notation

$$\begin{bmatrix} E \\ F \\ G \end{bmatrix} = -\frac{1}{2\pi}\int_0^\infty (\Omega^+-\Omega_s^+)\begin{bmatrix} 1 \\ \cosh\tfrac{1}{2}\eta \\ -\cosh\eta \end{bmatrix} d\eta - \frac{1}{2\pi}\int_0^\infty (\Omega^--\Omega_s^-)\begin{bmatrix} -1 \\ \cosh\tfrac{1}{2}\eta \\ \cosh\eta \end{bmatrix} d\eta.$$

$$(99)$$

The numbers E, G and F are functions of the time only and are related directly to the motion of the obstacle by (87)–(89). From (98)

$$\cos\tfrac{1}{2}\gamma\,(\sin\tfrac{1}{2}\gamma-\sin\tfrac{1}{2}\lambda_s)\frac{\partial}{\partial t}(\Omega-\Omega_s)$$

$$=\frac{\cos^2\tfrac{1}{2}\gamma}{2\pi}(\sin\tfrac{1}{2}\gamma-\sin\tfrac{1}{2}\lambda_s)\int_{-\pi}^{\pi}\frac{\dot\theta\,d\gamma^*}{\sin\tfrac{1}{2}\gamma^*-\sin\tfrac{1}{2}\gamma}$$

$$-(\sin\tfrac{1}{2}\gamma-\sin\tfrac{1}{2}\lambda_s)(\dot F+\dot E\sin\tfrac{1}{2}\gamma)$$

$$+\tfrac{1}{2}(\dot E+\dot G)-\frac{1}{4\pi}\int_0^\infty \frac{(\sinh\eta-2\sin\tfrac{1}{2}\lambda_s\sinh\tfrac{1}{2}\eta)\sinh\tfrac{1}{2}\eta}{\cosh\tfrac{1}{2}\eta-\sin\tfrac{1}{2}\gamma}\frac{\partial}{\partial t}(\Omega^+-\Omega_s^+)\,d\eta$$

$$+\frac{1}{4\pi}\int_0^\infty \frac{(\sinh\eta+2\sin\tfrac{1}{2}\lambda_s\sinh\tfrac{1}{2}\eta)\sinh\tfrac{1}{2}\eta}{\cosh\tfrac{1}{2}\eta+\sin\tfrac{1}{2}\gamma}\frac{\partial}{\partial t}(\Omega^--\Omega_s^-)\,d\eta,$$

where the dots indicate (real) time derivatives and not, as in ch. 9, *reduced time derivatives*. Making use of (96) in the last two integrals, and then integrating by parts we find

$$\int_{\lambda_s}^{\gamma}\cos\tfrac{1}{2}\gamma\,(\sin\tfrac{1}{2}\gamma-\sin\tfrac{1}{2}\lambda_s)\frac{\partial}{\partial t}(\Omega-\Omega_s)\,d\gamma$$

$$=C+\frac{1}{2\pi}\int_{-\pi}^{\pi}\dot\theta(\gamma^*)\,d\gamma^*\int_{\lambda_s}^{\gamma}\frac{\cos^2\tfrac{1}{2}\gamma(\sin\tfrac{1}{2}\gamma-\sin\tfrac{1}{2}\lambda_s)}{\sin\tfrac{1}{2}\gamma^*-\sin\tfrac{1}{2}\gamma}\,d\gamma+\tfrac{1}{2}(\gamma-\lambda_s)(\dot E+\dot G)$$

$$-\int_{\lambda_s}^{\gamma}(\dot F+\dot E\sin\tfrac{1}{2}\gamma)(\sin\tfrac{1}{2}\gamma-\sin\tfrac{1}{2}\lambda_s)\,d\gamma$$

$$-\frac{1}{8a_s\pi}\int_0^\infty (q_s^+)^2(\Omega^+-\Omega_s^+)\,d\eta\int_{\lambda_s}^{\gamma}\frac{\partial}{\partial\eta}\left(\frac{\sinh\tfrac{1}{2}\eta}{\cosh\tfrac{1}{2}\eta-\sin\tfrac{1}{2}\gamma}\right)d\gamma$$

$$+\frac{1}{8a_s\pi}\int_0^\infty (q_s^-)^2(\Omega^--\Omega_s^-)\,d\eta\int_{\lambda_s}^{\gamma}\frac{\partial}{\partial\eta}\left(\frac{\sinh\tfrac{1}{2}\eta}{\cosh\tfrac{1}{2}\eta+\sin\tfrac{1}{2}\gamma}\right)d\gamma.$$

With the aid of the alternative forms for E, F and G which can be obtained from (87) to (89) we can reduce the right-hand side of this equation to

$$C + \frac{U^2 \cos\frac{1}{2}\gamma}{8a_s\pi} \int_{-\pi}^{\pi} \frac{\dot{\Theta}\,d\gamma^*}{\sin\frac{1}{2}\gamma^* - \sin\frac{1}{2}\gamma}$$

$$- \frac{\cos\frac{1}{2}\gamma}{8a_s\pi} \int_{0}^{\infty} \left\{ \frac{(q_s^+)^2(\Omega^+ - \Omega_s^+)}{\cosh\frac{1}{2}\eta - \sin\frac{1}{2}\gamma} + \frac{(q_s^-)^2(\Omega^- - \Omega_s^-)}{\cosh\frac{1}{2}\eta + \sin\frac{1}{2}\gamma} \right\} d\eta,$$

where
$$\Theta = \frac{4a_s}{U} \int_{-\pi}^{\gamma} (\theta - \theta_s)\cos\frac{1}{2}\gamma(\sin\frac{1}{2}\gamma - \sin\frac{1}{2}\lambda_s)\,d\gamma. \tag{100}$$

This result and (85) enable us to write (97) as

$$p - p_s = C + \rho_a m q_s + \frac{\rho_a q_s^2 \cos\frac{1}{2}\gamma}{2\pi} \int_{-\pi}^{\pi} \frac{(\theta - \theta_s)\,d\gamma^*}{\sin\frac{1}{2}\gamma^* - \sin\frac{1}{2}\gamma}$$

$$+ \frac{\rho_a U^2 \cos\frac{1}{2}\gamma}{2\pi} \int_{-\pi}^{\pi} \frac{\dot{\Theta}\,d\gamma^*}{\sin\frac{1}{2}\gamma^* - \sin\frac{1}{2}\gamma}$$

$$+ \frac{\rho_a \cos\frac{1}{2}\gamma}{2\pi} \int_{0}^{\infty} \left\{ \frac{\{q_s^2 - (q_s^+)^2\}(\Omega^+ - \Omega_s^+)}{\cosh\frac{1}{2}\eta - \sin\frac{1}{2}\gamma} + \frac{\{q_s^2 - (q_s^-)^2\}(\Omega^- - \Omega_s^-)}{\cosh\frac{1}{2}\eta + \sin\frac{1}{2}\gamma} \right\} d\eta, \tag{101}$$

which gives the pressure increment at a point on the wetted surface of the obstacle. At points on the obstacle outside $-\pi < \gamma < \pi$ (in the wake) $p - p_s = 0$. The constant C in (101) obviously contributes nothing to the forces and moment acting on the body, and it can be ignored.

The velocity q_s is a function of γ and must be calculated as a preliminary step by the method described in § 11.12, but on the other hand the velocities q_s^+ and q_s^- on the separation streamlines cannot be calculated, but must be assigned in accordance with some plausible hypothesis, the simplest of which is that of Helmholtz, viz. that $q_s^+ = q_s^- = U$. Alternatively, we could adopt the distributions given in § 11.9; however, owing to the considerable labour involved in tabulating the integrals that would arise, we merely note here this possibility—in the following sections we shall consider only unsteady Helmholtz flows.

From (6.46), $\mathbf{n}.\mathbf{v} = n$ and $q = q_s + \mathbf{v}$, it follows to first order that

$$\theta = \theta_s + \frac{n}{q_s} - \alpha + \pi(\lambda_s - \lambda)\,\delta(\gamma - \lambda_s),$$

where $\gamma = \lambda$ is the position of the stagnation point in the unsteady flow, and δ is the delta function. We shall write

$$\theta - \theta_s = \theta_1 + \pi(\lambda_s - \lambda)\,\delta(\gamma - \lambda_s), \tag{102}$$

where
$$\theta_1 = \frac{n}{q_s} - \alpha \tag{103}$$

is that part of $\theta - \theta_s$ which is immediately defined by the motion of the obstacle (cf. (9·5)). The displacement $\lambda_s - \lambda$ of the front stagnation point from its mean (steady) position, which plays an important role in the theory, is not, in general, easy to calculate. It is considered in the next section.

Substituting (102) in (100) we find

$$\Theta = \frac{4a_s}{U^2} \int_{-\pi}^{\gamma} \theta_1 \cos \tfrac{1}{2}\gamma (\sin \tfrac{1}{2}\gamma - \sin \tfrac{1}{2}\lambda_s) \, d\gamma. \tag{104}$$

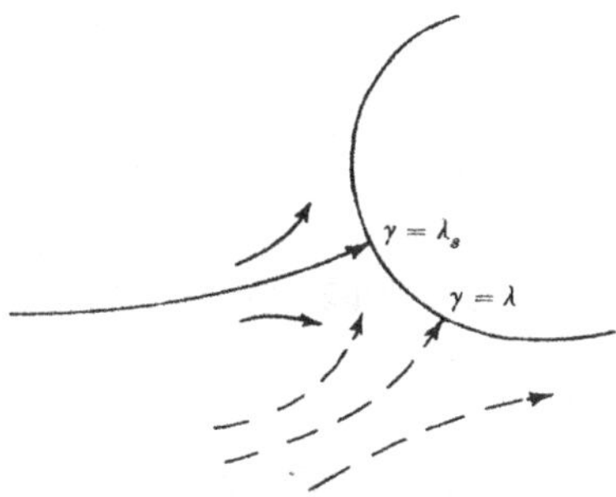

Fig. 11.17

Similarly, in the case of Helmholtz flow ($q_s^+ = q_s^- = U, \ \Omega^+ = \Omega_s^- = 0$) (101) yields

$$p - p_s = \rho_a m q_s + \tfrac{1}{2}\rho_a q_s^2 (\lambda_s - \lambda) \frac{\cos \tfrac{1}{2}\gamma}{\sin \tfrac{1}{2}\lambda_s - \sin \tfrac{1}{2}\gamma}$$

$$+ \frac{\rho_a \cos \tfrac{1}{2}\gamma}{2\pi} \int_{-\pi}^{\pi} \frac{\{q_s^2(\gamma)\,\theta_1(\gamma^*, t) + U^2 \dot\Theta(\gamma^*, t)\}\, d\gamma^*}{\sin \tfrac{1}{2}\gamma^* - \sin \tfrac{1}{2}\gamma}$$

$$+ \frac{\rho_a (q_s^2 - U^2)\cos \tfrac{1}{2}\gamma}{2\pi} \int_0^{\infty} \left(\frac{\Omega^+(\eta, t)}{\cosh \tfrac{1}{2}\eta - \sin \tfrac{1}{2}\gamma} + \frac{\Omega^-(\eta, t)}{\cosh \tfrac{1}{2}\eta + \sin \tfrac{1}{2}\gamma} \right) d\eta. \tag{105}$$

If the obstacle is thin, $q_s \approx U$, and the last term of (105) can be neglected.

A final point concerning the pressure distribution is that, since Ω^+ and Ω^- both satisfy (94), they vanish outside the ranges $\phi_{1s} \leqslant \phi_s \leqslant \phi_{1s} + U^2 t$ and $\phi_{2s} \leqslant \phi_s \leqslant \phi_{2s} + U^2 t$ on the free streamlines, where t is the time measured from the commencement of the unsteady motion. Thus the range $(0, \infty)$ occurring in the integrals in (87)–(89) and (105) can be reduced to $(0, k_1)$ for Ω^+ and $(0, k_2)$ for Ω^-, where from (95) k_1 and k_2 satisfy

$$\cosh \tfrac{1}{2}k_1 = \left(\frac{\phi_{1s} + U^2 t}{4a_s} \right)^{\frac{1}{2}} + \sin \tfrac{1}{2}\lambda_s,$$

$$\cosh \tfrac{1}{2}k_2 = \left(\frac{\phi_{2s} + U^2 t}{4a_s} \right)^{\frac{1}{2}} - \sin \tfrac{1}{2}\lambda_s. \tag{106}$$

When the pressure distribution is known on the obstacle surface, then the drag D, lift L, and nose-up moment M about the origin are given in matrix notation by

$$\begin{bmatrix} D \\ L \\ M \end{bmatrix} = \oint (p - p_a) \begin{bmatrix} \sin\theta_s \\ \cos\theta_s \\ -(x_s \cos\theta_s + y_s \sin\theta_s) \end{bmatrix} ds. \tag{107}$$

Thus if D_s, L_s and M_s denote the steady values, then by $ds = d\phi_s/q$ and (82)

$$\begin{bmatrix} D \\ L \\ M \end{bmatrix} = \begin{bmatrix} D_s \\ L_s \\ M_s \end{bmatrix}$$

$$+ \frac{4a_s}{U} \int_{-\pi}^{\pi} \frac{U}{q_s} (p - p_s) \cos\tfrac{1}{2}\gamma (\sin\tfrac{1}{2}\gamma - \sin\tfrac{1}{2}\lambda_s) \begin{bmatrix} -\sin\theta_s \\ -\cos\theta_s \\ x_s \cos\theta_s + y_s \sin\theta_s \end{bmatrix} d\gamma, \tag{108}$$

the change in sign arising because of the change in direction of the contour integration.

Before these equations can be used to determine values of D, L and M it is necessary to calculate the quantities $(\lambda_s - \lambda)$, Ω^+ and Ω^- occurring in (105).

11.18　Non-uniform motion (Helmholtz flow)

For a prescribed motion of the obstacle $(\lambda_s - \lambda)$, Ω^+ and Ω^- can be determined from (87) to (89) and (96). Substitution of (102) into (87) to (89) gives

$$\tfrac{1}{2}(\lambda_s - \lambda) \begin{bmatrix} 1 \\ \sin\tfrac{1}{2}\lambda_s \\ \cos\lambda_s \end{bmatrix} + \frac{1}{2\pi} \int_{-\pi}^{\pi} \theta_1 \begin{bmatrix} 1 \\ \sin\tfrac{1}{2}\gamma \\ \cos\gamma \end{bmatrix} d\gamma$$

$$= \frac{1}{2\pi} \int_0^{k_1} \Omega^+ \begin{bmatrix} -1 \\ -\cosh\tfrac{1}{2}\eta \\ \cosh\eta \end{bmatrix} d\eta + \frac{1}{2\pi} \int_0^{k_2} \Omega^- \begin{bmatrix} -1 \\ -\cosh\tfrac{1}{2}\eta \\ \cosh\eta \end{bmatrix} d\eta. \tag{109}$$

On eliminating $(\lambda_s - \lambda)$ from these equations we arrive at the simultaneous integral equations

$$\begin{bmatrix} b_1 \\ b_2 \end{bmatrix} = -\frac{1}{2\pi} \int_0^{k_1} \Omega^+ (\cosh\tfrac{1}{2}\eta - \sin\tfrac{1}{2}\lambda_s) \begin{bmatrix} 1 \\ \cosh\tfrac{1}{2}\eta \end{bmatrix} d\eta$$

$$- \frac{1}{2\pi} \int_0^{k_2} \Omega^- (\cosh\tfrac{1}{2}\eta + \sin\tfrac{1}{2}\lambda_s) \begin{bmatrix} 1 \\ -\cosh\tfrac{1}{2}\eta \end{bmatrix} d\eta, \tag{110}$$

where

$$\begin{bmatrix} b_1 \\ b_2 \end{bmatrix} \equiv \frac{1}{2\pi} \int_{-\pi}^{\pi} \theta_1 (\sin \tfrac{1}{2}\gamma - \sin \tfrac{1}{2}\lambda_s) \begin{bmatrix} 1 \\ \sin \tfrac{1}{2}\gamma \end{bmatrix} d\gamma. \tag{111}$$

From (94) Ω^+ and Ω^- also satisfy the functional equations:

$$\Omega^+(s,t) = \Omega^+\!\left(s_0, t - \frac{s - s_0}{U}\right), \quad \Omega^-(s,t) = \Omega^-\!\left(s_0, t - \frac{s - s_0}{U}\right),$$

or, since $\phi_s = Us$ on the free boundaries, we can write these relations

$$\begin{aligned} \Omega^+(\phi_s, U^2 t) &= \Omega^+(\phi_{1s}, U^2 t - \phi_s + \phi_{1s}), \\ \Omega^-(\phi_s, U^2 t) &= \Omega^-(\phi_{2s}, U^2 t - \phi_s + \phi_{2s}). \end{aligned} \tag{112}$$

These equations relate the general values of Ω^+ and Ω^- to their values at the separation points (cf. (9.10)).

If new independent variables are defined by

$$\xi = \frac{\phi_s}{4a_s}, \quad \xi_1 = \frac{\phi_{1s}}{4a_s}, \quad \xi_2 = \frac{\phi_{2s}}{4a_s}, \tag{113}$$

and

$$J = \frac{U^2 t}{4a_s} \tag{114}$$

(this differs from the reduced time defined in (9.9) by the factor $(4a/Uc)$), then from (95)

$$\begin{aligned} \cosh \tfrac{1}{2}\eta &= \xi^{\frac{1}{2}} - \xi_1^{\frac{1}{2}} + 1, \quad \xi_1^{\frac{1}{2}} = 1 - \sin \tfrac{1}{2}\lambda_s, \\ \cosh \tfrac{1}{2}\eta &= \xi^{\frac{1}{2}} - \xi_2^{\frac{1}{2}} + 1, \quad \xi_2^{\frac{1}{2}} = 1 + \sin \tfrac{1}{2}\lambda_s, \end{aligned} \tag{115}$$

and

hold on the upper and lower edges of the wake respectively. Further, it follows from (106) that $\xi = J + \xi_1$, when $\eta = k_1$ and that $\xi = J + \xi_2$, when $\eta = k_2$. Hence by (110) and (115)

$$\begin{bmatrix} b_1 \\ b_2 \end{bmatrix} = \frac{1}{2\pi} \int_{\xi_1 + J}^{\xi_1} \frac{\Omega^+(\xi, J)}{[(\xi^{\frac{1}{2}} - \xi_1^{\frac{1}{2}})(\xi^{\frac{1}{2}} - \xi_1^{\frac{1}{2}} + 2)]^{\frac{1}{2}}} \begin{bmatrix} 1 \\ (\xi^{\frac{1}{2}} - \xi_1^{\frac{1}{2}} + 1) \end{bmatrix} d\xi$$

$$- \frac{1}{2\pi} \int_{\xi_2 + J}^{\xi_2} \frac{\Omega^-(\xi, J)}{[(\xi^{\frac{1}{2}} - \xi_2^{\frac{1}{2}})(\xi^{\frac{1}{2}} - \xi_2^{\frac{1}{2}} + 2)]^{\frac{1}{2}}} \begin{bmatrix} -1 \\ (\xi^{\frac{1}{2}} - \xi_2^{\frac{1}{2}} + 1) \end{bmatrix} d\xi. \tag{116}$$

Equations (112) can be written

$$\Omega^+(\xi, J) = \Omega^+(\xi_1, J - \xi + \xi_1), \quad \Omega^-(\xi, J) = \Omega^-(\xi_2, J - \xi + \xi_2), \tag{117}$$

so that substituting $\sigma = \jmath - \xi + \xi_1$ in the first integral on the right-hand side of (116), and $\sigma = \jmath - \xi + \xi_2$ in the second, we find

$$\begin{bmatrix} b_1 \\ b_2 \end{bmatrix} = -\frac{1}{2\pi} \int_0^\jmath \frac{\Omega^+(\xi_1, \sigma)}{[\{(\jmath - \sigma + \xi_1)^{\frac{1}{2}} - \xi_1^{\frac{1}{2}}\}\{(\jmath - \sigma + \xi_1)^{\frac{1}{2}} - \xi_1^{\frac{1}{2}} + 2\}]^{\frac{1}{2}}}$$

$$\times \begin{bmatrix} 1 \\ (\jmath - \sigma + \xi_1)^{\frac{1}{2}} - \xi_1^{\frac{1}{2}} + 1 \end{bmatrix} d\sigma$$

$$+\frac{1}{2\pi} \int_0^\jmath \frac{\Omega^-(\xi_2, \sigma)}{[\{(\jmath - \sigma + \xi_2)^{\frac{1}{2}} - \xi_2^{\frac{1}{2}}\}\{(\jmath - \sigma + \xi_2)^{\frac{1}{2}} - \xi_2^{\frac{1}{2}} + 2\}]^{\frac{1}{2}}}$$

$$\times \begin{bmatrix} -1 \\ (\jmath - \sigma + \xi_2)^{\frac{1}{2}} - \xi_2^{\frac{1}{2}} + 1 \end{bmatrix} d\sigma. \qquad (118)$$

These integral equations can be solved by the Laplace transform method (see § 4.23). Taking the transforms of (118) with the aid of (4.199), and then solving the resulting equations for $\mathscr{L}(\Omega^+)$ and $\mathscr{L}(\Omega^-)$, we find

$$\mathscr{L}(\Omega^+) = -2\pi \frac{B_2 \mathscr{L}(b_1) + A_2 \mathscr{L}(b_2)}{A_1 B_2 + A_2 B_1}, \quad \mathscr{L}(\Omega^-) = -2\pi \frac{B_1 \mathscr{L}(b_1) - A_1 \mathscr{L}(b_2)}{A_1 B_2 + A_2 B_1}, \qquad (119)$$

where

$$\left. \begin{aligned} A_1 &\equiv \frac{1}{p} \mathscr{L}\left\{\frac{\mathbf{U}(\jmath)}{\mu_1^{\frac{1}{2}}(\mu_1 + 2)^{\frac{1}{2}}}; p\right\}, \quad A_2 \equiv \frac{1}{p} \mathscr{L}\left\{\frac{\mathbf{U}(\jmath)}{\mu_2^{\frac{1}{2}}(\mu_2 + 2)^{\frac{1}{2}}}; p\right\}, \\ B_1 &\equiv \frac{1}{p} \mathscr{L}\left\{\frac{(\mu_1 + 1)\,\mathbf{U}(\jmath)}{\mu_1^{\frac{1}{2}}(\mu_1 + 2)^{\frac{1}{2}}}; p\right\}, \quad B_2 \equiv \frac{1}{p} \mathscr{L}\left\{\frac{(\mu_2 + 1)\,\mathbf{U}(\jmath)}{\mu_2^{\frac{1}{2}}(\mu_2 + 2)^{\frac{1}{2}}}; p\right\}, \end{aligned} \right\} \qquad (120)$$

with

$$\mu_1 \equiv (\jmath + \xi_1)^{\frac{1}{2}} - \xi_1^{\frac{1}{2}}, \quad \mu_2 \equiv (\jmath + \xi_2)^{\frac{1}{2}} - \xi_2^{\frac{1}{2}}.$$

It only remains to take the inverse transform of (119) by a method similar to that used to derive (9.43) from (9.40) to obtain the formal solution for Ω^+ and Ω^-. It then follows from (109) and (115) that

$$\lambda(\jmath) - \lambda_s = \frac{1}{\pi} \int_0^\jmath \frac{(\jmath - \sigma + \xi_1)^{-\frac{1}{2}}\,\Omega^+(\xi_1, \sigma)\,d\sigma}{[\{(\jmath - \sigma + \xi_1)^{\frac{1}{2}} - \xi_1^{\frac{1}{2}}\}\{(\jmath - \sigma + \xi_1)^{\frac{1}{2}} - \xi_1^{\frac{1}{2}} + 2\}]^{\frac{1}{2}}} + \frac{1}{\pi} \int_{-\pi}^\pi \theta_1 \, d\gamma$$

$$-\frac{1}{\pi} \int_0^\jmath \frac{(\jmath - \sigma + \xi_2)^{-\frac{1}{2}}\,\Omega^-(\xi_2, \sigma)\,d\sigma}{[\{(\jmath - \sigma + \xi_2)^{\frac{1}{2}} - \xi_2^{\frac{1}{2}}\}\{(\jmath - \sigma + \xi_2)^{\frac{1}{2}} - \xi_2^{\frac{1}{2}} + 2\}]^{\frac{1}{2}}}. \qquad (121)$$

All the numbers appearing in (105) have now been evaluated, and the problem is formally solved, although of course for any particular case a great amount of computing labour would be required because of the somewhat recondite integrals which have emerged in the theory. We shall consider a number of important special cases of the theory in the following sections.

11.19　Separation points at the same steady potential

If the separation points are at the same steady potential in the mean steady flow, then by (83) $\lambda_s = 0$. Therefore (115) gives $\xi_1 = \xi_2 = 1$ and (118) reduce to

$$\begin{bmatrix} b_1 \\ b_2 \end{bmatrix} = -\frac{1}{2\pi} \int_0^J \frac{1}{(J-\sigma)^{\frac{1}{2}}} \begin{bmatrix} S(1,\sigma) \\ (J-\sigma+1)^{\frac{1}{2}} D(1,\sigma) \end{bmatrix} d\sigma, \tag{122}$$

where

$$S(1,\sigma) = \Omega^+(1,\sigma) + \Omega^-(1,\sigma), \quad D(1,\sigma) = \Omega^+(1,\sigma) - \Omega^-(1,\sigma), \tag{123}$$

and

$$2 \begin{bmatrix} b_1 \\ b_2 \end{bmatrix} = \frac{1}{2\pi} \int_{-\pi}^{\pi} \frac{n(\gamma,J)}{q_s(\gamma)} \begin{bmatrix} 2\sin\frac{1}{2}\gamma \\ 1-\cos\gamma \end{bmatrix} d\gamma - \begin{bmatrix} 0 \\ \alpha \end{bmatrix}, \tag{124}$$

by (103) and (111). Equation (121) reduces to

$$\lambda(J) = \frac{1}{\pi} \int_0^J \frac{D(1,\sigma)}{\{(J-\sigma)(J-\sigma+1)\}^{\frac{1}{2}}} + b_0, \tag{125}$$

where

$$b_0 \equiv \frac{1}{\pi} \int_{-\pi}^{\pi} \theta_1 \, d\gamma = \frac{1}{\pi} \int_{-\pi}^{\pi} \frac{n(\gamma,J)}{q_s(\gamma)} \, d\gamma - 2\alpha. \tag{126}$$

To take the transform of (122) and (125) we need (9.34) and the result (see van der Pol & Bremmer, 1950)

$$\mathcal{L}\{J^\nu \mathrm{U}(J); p\} = \frac{\Gamma(\nu+1)}{p^\nu}, \quad 0 < \mathcal{R}p < \infty \quad (\mathcal{R}\nu > -1). \tag{127}$$

Proceeding by the method used in §9.3 we find that (122) and (125) transform into

$$\mathcal{L}(b_1) = -\frac{1}{2\sqrt{\pi}} \mathcal{L}(S) \, p^{-\frac{1}{2}}, \tag{128}$$

$$\mathcal{L}(b_2) = -\frac{1}{4\pi} \mathcal{L}(D) \, e^{\frac{1}{2}p} \{K_0(\tfrac{1}{2}p) + K_1(\tfrac{1}{2}p)\}, \tag{129}$$

and

$$\mathcal{L}(\lambda) = \frac{1}{\pi} \mathcal{L}(D) \, e^{\frac{1}{2}p} K_0(\tfrac{1}{2}p) + \mathcal{L}(b_0). \tag{130}$$

Therefore by (127) and (4.198)

$$\mathcal{L}(S) = -2\mathcal{L}\{J^{-\frac{1}{2}}\mathrm{U}(J); p\} \mathcal{L}\{b_1(J); p\} = -\frac{2}{p} \mathcal{L}\{J^{-\frac{1}{2}}\mathrm{U}(J); p\} \mathcal{L}\left\{\frac{db_1}{dJ}(J); p\right\},$$

and so by an application of (4.199)

$$S(1,\sigma) = -2\int_{-\infty}^{\infty} \frac{db_1}{dJ} \frac{\mathrm{U}(\sigma-J)}{(\sigma-J)^{\frac{1}{2}}} dJ = -2\int_0^\sigma \frac{db_1(J)}{(\sigma-J)^{\frac{1}{2}}}, \tag{131}$$

as $db_1(J) = 0$ for $J < 0$ (the unsteady motion starts at $J = 0$). Incidentally with the second integral in (131) and similar integrals, the limits should always be interpreted as

$$\lim_{\epsilon \to 0} \int_{-\epsilon}^{\sigma+\epsilon} \dots \quad (\epsilon > 0),$$

see remarks following (9.43).

Eliminating $\mathscr{L}(D)$ from (129) and (130), and making use of (9.38) and (9.41), we find

$$\mathscr{L}(\lambda) = 2\mathscr{L}(b_2)\,\mathscr{L}\{k(t)\} + \mathscr{L}(b_0 - 4b_2).$$

Hence

$$\lambda(t) = 2\int_0^t k(t - t^*)\,db_2(t^*) + b_0(t) - 4b_2(t). \tag{132}$$

Denoting the last integral in (105) by J, and using (115), (117) and (123), the substitution $\sigma = t + 1 - \xi$, and the fact that in the present case $\xi_1 = \xi_2 = 1$, we obtain

$$J = 2\int_0^t \frac{S(1, \sigma)\,d\sigma}{(2t - 2\sigma + 1 + \cos\gamma)(t + \sigma)^{\frac{1}{2}}}$$
$$+ 2\sin\tfrac{1}{2}\gamma \int_0^t \frac{D(1, \sigma)\,d\sigma}{(2t - 2\sigma + 1 + \cos\gamma)\{(t - \sigma)(t - \sigma + 1)\}^{\frac{1}{2}}}. \tag{133}$$

Eliminating $\mathscr{L}(D)$ and $\mathscr{L}(S)$ from (128), (129) and the transform of (133) and then using the inversion formula, we arrive at an expression for J in terms of b_1 and b_2.

In the case of thin aerofoils we are spared this calculation of J because $q_0 \simeq U$, and the last term of (105) can be neglected. Also from (9.27), (9.32), (124) and (126) $2Ub_2 = a_0 - a_1 - \alpha U = -A$, $b_0 U = 2a_0 - 2\alpha U$, and apart from a change in the sign of λ (cf. (9.5) and (102)), (132) reduces to the special case $U = \text{constant}$ of (9.43).

Now consider the special case when $\alpha = 0$ and $n(\gamma, t)$ is an *odd* function of γ, e.g. as with a symmetrical body accelerated parallel to the undisturbed stream direction. In this case it follows from (122), (124) and (125) that b_2, D, and λ vanish. The transform of (133) is then

$$\mathscr{L}(J) = \frac{1}{p}\mathscr{L}(S)\,\mathscr{L}\left\{\frac{U(t)t^{-\frac{1}{2}}}{t + \cos^2\frac{1}{2}\gamma};\, p\right\}$$
$$= -\frac{2}{p}\mathscr{L}\left\{\frac{db_1}{dt}(t);\, p\right\}\frac{1}{p}\mathscr{L}\{t^{-\frac{1}{2}}U(t);\, p\}\,\mathscr{L}\left\{\frac{t^{-\frac{1}{2}}U(t)}{t + \cos^2\frac{1}{2}\gamma};\, p\right\},$$

by the equation following (130). Two applications of (4.199) give

$$\mathscr{L}(J) = -\frac{2}{p}\mathscr{L}\left\{\frac{db_1}{dt}(t);\, p\right\}\mathscr{L}\left\{\int_0^t \frac{\{\sigma(t - \sigma)\}^{-\frac{1}{2}}}{\sigma + \cos^2\frac{1}{2}\gamma}\,d\sigma;\, p\right\}$$
$$= -\frac{2\pi}{\cos\frac{1}{2}\gamma}\frac{1}{p}\mathscr{L}\left\{\frac{db_1}{dt}(t);\, p\right\}\mathscr{L}\{(t + \cos^2\tfrac{1}{2}\gamma)^{-\frac{1}{2}};\, p\}$$
$$= -\frac{2\pi}{\cos\frac{1}{2}\gamma}\mathscr{L}\left\{\int_0^t \frac{db_1(\sigma)}{(t - \sigma + \cos^2\frac{1}{2}\gamma)^{\frac{1}{2}}};\, p\right\},$$

and hence
$$J(\gamma,\jmath) = -\frac{2\pi}{\cos\frac{1}{2}\gamma}\int_0^{\jmath}\frac{db_1(\sigma)}{(\jmath-\sigma+\cos^2\frac{1}{2}\gamma)^{\frac{1}{2}}}.$$

Substitution in (105) now gives

$$p-p_s = \frac{2\rho_a\cos\frac{1}{2}\gamma}{\pi}\int_0^{\pi}\frac{\{q_s^2(\gamma)\,n(\gamma^*,t)+U^2q_s(\gamma^*)\,\dot{\Theta}(\gamma^*,t)\}}{q_s(\gamma^*)\,\{\cos\gamma-\cos\gamma^*\}}\sin\tfrac{1}{2}\gamma^*\,d\gamma^*$$
$$+\rho_a mq_s-\rho_a(q_s^2-U^2)\int_0^{\jmath}\frac{db_1(\sigma)}{(\jmath-\sigma+\cos^2\frac{1}{2}\gamma)^{\frac{1}{2}}}. \tag{134}$$

An example of this comparatively simple type of unsteady Helmholtz flow will be given in § 11.21.

11.20 Harmonic motions

If the unsteady perturbations are periodic, we can write

$$\theta_1(\gamma,t) = \mathbf{U}(t)\,e^{i\nu t}\tilde{\theta}_1(\gamma), \quad\text{or}\quad \theta_1(\gamma,\jmath) = \mathbf{U}(\jmath)\,e^{2i\omega\jmath}\tilde{\theta}_1(\gamma), \tag{135}$$

where from (114)
$$\omega = \frac{2a_s\nu}{U^2} = \left(\frac{4a_s}{Uc}\right)\left(\frac{c\nu}{2U}\right) \tag{136}$$

is the 'reduced' frequency of the oscillation. This reduced frequency differs from that used in aerofoil theory (see (9.59)) by the factor $(4a_s/Uc)$.

Let
$$\begin{bmatrix}\tilde{b}_1\\\tilde{b}_2\end{bmatrix} \equiv \frac{1}{2\pi}\int_{-\pi}^{\pi}\tilde{\theta}_1(\sin\tfrac{1}{2}\gamma-\sin\tfrac{1}{2}\lambda_s)\begin{bmatrix}1\\\dfrac{1}{\sin\frac{1}{2}\gamma}\end{bmatrix}d\gamma, \tag{137}$$

so that from (111) and (135) $\mathscr{L}\{b_1(\jmath);p\} = \tilde{b}_1\,\mathscr{L}\{\mathbf{U}(\jmath)\,e^{2i\omega\jmath};p\}$ and similarly for b_2. The first of (119) can now be expressed

$$\mathscr{L}(\Omega^+) = -\frac{2\pi}{p}\tilde{b}_1\,\mathscr{L}\{\mathbf{U}(\jmath)\,e^{2i\omega\jmath}\}\mathscr{L}(g_1)-\frac{2\pi}{p}\tilde{b}_2\,\mathscr{L}\{\mathbf{U}(\jmath)\,e^{2i\omega\jmath}\}\mathscr{L}(g_2),$$

where
$$\mathscr{L}(g_1) = \frac{pB_2}{A_1B_2+A_2B_1}, \quad \mathscr{L}(g_2) = \frac{pA_2}{A_1B_2+A_2B_1}. \tag{138}$$

Thus by (4.199)

$$e^{-2i\omega\jmath}\Omega^+ = -2\pi\tilde{b}_1\int_0^{\jmath}e^{2i\omega(\sigma-\jmath)}g_1(\jmath-\sigma)\,d\sigma - 2\pi\tilde{b}_2\int_0^{\jmath}e^{2i\omega(\sigma-\jmath)}g_2(\jmath-\sigma)\,d\sigma. \tag{139}$$

As the duration of the motion increases indefinitely it follows from (106) and (114) that the limits of the integrals in (139) tend to infinity, and in this case

$$\Omega^+ = -2\pi\tilde{b}_1\frac{e^{2i\omega\jmath}}{2i\omega}\left\{2i\omega\int_0^{\infty}e^{-2i\omega s}g_1(s)\,ds\right\} - 2\pi\tilde{b}_2\frac{e^{2i\omega\jmath}}{2i\omega}\left\{2i\omega\int_0^{\infty}e^{-2i\omega s}g_2(s)\,ds\right\}.$$

The bracketed terms are the values of the transforms of g_1 and g_2 at $p = 2i\omega$, and hence from (138)

$$\Omega^+(\xi_1, \mathfrak{t}) = -2\pi\tilde{b}_1 e^{2i\omega\mathfrak{t}}\left[\frac{B_2}{A_1 B_2 + A_2 B_1}\right]_{p=2i\omega}$$

$$-2\pi\tilde{b}_2 e^{2i\omega\mathfrak{t}}\left[\frac{A_2}{A_1 B_2 + A_2 B_1}\right]_{p=2i\omega}. \quad (140)$$

(This result could have been deduced directly from (119), the transform $\mathscr{L}\{U(\mathfrak{t})e^{2i\omega\mathfrak{t}}; p\} = p/(p - 2i\omega)$, and an application of the residue theorem— see p. 142 *et. seq.* of van der Pol & Bremmer, 1950.)

By (120) the value of A_1 at $p = 2i\omega$ is

$$A_1 = \int_0^\infty e^{-2i\omega\mathfrak{t}}\{(\mathfrak{t} + \xi_1)^{\frac{1}{2}} - \xi_1^{\frac{1}{2}}\}^{-\frac{1}{2}}\{(\mathfrak{t} + \xi_1)^{\frac{1}{2}} - \xi_1^{\frac{1}{2}} + 2\}^{-\frac{1}{2}}d\mathfrak{t}.$$

The transformation yields

$$(\mathfrak{t} + \xi_1)^{\frac{1}{2}} = \cosh\tfrac{1}{2}\epsilon + \xi_1^{\frac{1}{2}} - 1 = \cosh\tfrac{1}{2}\epsilon - \sin\tfrac{1}{2}\lambda_s,$$

$$A_1 = \exp\{i\omega(1 - 4\sin\tfrac{1}{2}\lambda_s)\}N_1(\omega, -\lambda_s),$$

where

$$N_1(\omega, \lambda_s) \equiv \int_0^\infty \exp\{-i\omega(\cosh\epsilon + 4\sin\tfrac{1}{2}\lambda_s\cosh\tfrac{1}{2}\epsilon)\}(\cosh\tfrac{1}{2}\epsilon + \sin\tfrac{1}{2}\lambda_s)\,d\epsilon.$$

$$(141)$$

Similarly, when $p = 2i\omega$,

$$A_2 = \exp\{i\omega(1 + 4\sin\tfrac{1}{2}\lambda_s)\}N_1(\omega, \lambda_s),$$

$$B_1 = \exp\{i\omega(1 - 4\sin\tfrac{1}{2}\lambda_s)\}N_2(\omega, -\lambda_s),$$

and

$$B_2 = \exp\{i\omega(1 + 4\sin\tfrac{1}{2}\lambda_s)\}N_2(\omega, \lambda_s),$$

where†

$$N_2(\omega, \lambda_s) \equiv \int_0^\infty \exp\{-i\omega(\cosh\epsilon + 4\sin\tfrac{1}{2}\lambda_s\cosh\tfrac{1}{2}\epsilon)\}$$

$$\times \cosh\tfrac{1}{2}\epsilon(\cosh\tfrac{1}{2}\epsilon + \sin\tfrac{1}{2}\lambda_s)\,d\epsilon. \quad (142)$$

We shall shortly also need the function

$$N_0(\omega, \lambda_s) = \int_0^\infty \exp\{-i\omega(\cosh\epsilon + 4\sin\tfrac{1}{2}\lambda_s\cosh\tfrac{1}{2}\epsilon)\}\,d\epsilon. \quad (143)$$

---

† This integral is not convergent in the usual sense, but if interpreted as the limit $\lim_{\mu\to+0}\int_0^\infty e^{-\mu\epsilon}\ldots d\epsilon$ it has a value. This value is termed by van der Pol & Bremmer (p. 103, 1950) the 'Cauchy' limit of the integral. It is clearly the appropriate value to use in (142) because it is really the *limiting* values of A_1, A_2, B_1 and B_2 as $p \to 2i\omega + 0$ that we require. An alternative approach is to replace ω in (135) by $\omega - i\mu$ ($\mu > 0$), i.e. to assume a slightly divergent motion. This renders (142) convergent, and the limit $\mu \to 0$ may be taken after the integral is evaluated. This method was used in § 9.7.

It now follows from (115), (117) and (140) that

$$\Omega^+(\eta,t) = -2\pi\, e^{2i\omega t} \exp\{-i\omega(\cosh\eta - 4\sin\tfrac{1}{2}\lambda_s \cosh\tfrac{1}{2}\eta)\}$$

$$\times \left\{ \frac{\tilde{b}_1 N_2(\omega,\lambda_s) + \tilde{b}_2 N_1(\omega,\lambda_s)}{N_1(\omega,-\lambda_s)\,N_2(\omega,\lambda_s) + N_1(\omega,\lambda_s)\,N_2(\omega,-\lambda_s)} \right\}. \quad (144)$$

A similar calculation applied to the second of (119) yields

$$\Omega^-(\eta,t) = -2\pi\, e^{2i\omega t} \exp\{-i\omega(\cosh\eta + 4\sin\tfrac{1}{2}\lambda_s \cosh\tfrac{1}{2}\eta)\}$$

$$\times \left\{ \frac{\tilde{b}_1 N_2(\omega,-\lambda_s) - \tilde{b}_2 N_1(\omega,-\lambda_s)}{N_1(\omega,-\lambda_s)\,N_2(\omega,\lambda_s) + N_1(\omega,\lambda_s)\,N_2(\omega,-\lambda_s)} \right\}. \quad (145)$$

The value of $(\lambda_s-\lambda)$ can now be obtained by substituting (144) and (145) into the first of (109), in which the limits k_1, k_2 are infinite in the present case. The result is

$$\lambda_s-\lambda = 2\,e^{2i\omega t}\left\{ \tilde{b}_1 F_2(\omega,\lambda_s) + \tilde{b}_2 F_1(\omega,\lambda_s) - \frac{1}{2\pi}\int_{-\pi}^{\pi}\tilde{\theta}_1\,d\gamma \right\}, \quad (146)$$

where $\quad F_1(\omega,\lambda_s) = \dfrac{N_1(\omega,\lambda_s)\,N_0(\omega,-\lambda_s) + N_1(\omega,-\lambda_s)\,N_0(\omega,\lambda_s)}{N_1(\omega,-\lambda_s)\,N_2(\omega,\lambda_s) + N_1(\omega,\lambda_s)\,N_2(\omega,-\lambda_s)},$

and $\quad F_2(\omega,\lambda_s) = \dfrac{N_2(\omega,\lambda_s)\,N_0(\omega,-\lambda_s) - N_2(\omega,-\lambda_s)\,N_0(\omega,\lambda_s)}{N_1(\omega,-\lambda_s)\,N_2(\omega,\lambda_s) + N_1(\omega,\lambda_s)\,N_2(\omega,-\lambda_s)}.$ $\qquad (147)$

The functions N_0, N_1 and N_2 have been tabulated by the Mathematics Division of the National Physical Laboratory, Teddington, England.

When the separation points are at the same steady potential, $\lambda_s = 0$, and with the help of the definition of $K_\nu(p)$ given in § 9.3 it is found that N_0, N_2 and N_1 reduce to

$$N_0(\omega,0) = K_0(i\omega), \quad N_2(\omega,0) = \tfrac{1}{2}\{K_1(i\omega) + K_0(i\omega)\},$$

and $\qquad N_1(\omega,0) = K_{\frac{1}{2}}(i\omega) = \left(\dfrac{\pi}{\omega}\right)^{\frac{1}{2}}\dfrac{e^{-i\omega}}{1+i}.$ $\qquad (148)$

If, in this case, the body is symmetrical, and accelerated parallel to the stream direction, then $n(\gamma,t)$ will be an odd function of γ and α will be zero. Hence $b_2 = 0$ from (124), and (144), (145) reduce to

$$\Omega^+(\eta,t) = \Omega^-(\eta,t) = -(1+i)\sqrt{\pi}\,e^{2i\omega t}\,\tilde{b}_1\sqrt{\omega}\,e^{-i\omega\cosh\eta}.$$

In this case (105) becomes

$$p-p_s = \frac{2\rho_a\cos\tfrac{1}{2}\gamma}{\pi}$$

$$\times \int_0^{\pi}\frac{\{q_s^2(\gamma)\,n(\gamma^*,t) + U^2 q_s(\gamma^*)\,\dot{\Theta}(\gamma^*,t)\}}{q_s(\gamma^*)(\cos\gamma - \cos\gamma^*)}\sin\tfrac{1}{2}\gamma^*\,d\gamma^* + \rho_a m q_s$$

$$+ \sqrt{(\pi\omega)}\,\rho_a(U^2 - q_s^2)(1+i)\tilde{b}_1\,e^{2i\omega t}e^{2i\omega\cos^2\frac{1}{2}\gamma}\{1 - \mathrm{erf}\sqrt{(2i\omega\cos^2\tfrac{1}{2}\gamma)}\},$$

$$(149)$$

where $\mathrm{erf}\,(x)$ is the error function $\dfrac{2}{\sqrt{\pi}}\displaystyle\int_0^x e^{-t^2}\,dt.$

11.21 Unsteady motion of a flat plate normal to the undisturbed stream

The simplest non-trivial example of the theory is that of a flat plate placed at right-angles to the undisturbed stream direction, with unsteady motions parallel to this direction.

The mean steady flow for the example (see Fig. 11.21a) can be calculated by putting $\epsilon = \tfrac{1}{2}\pi$ in (59) and (61) (see Fig. 11.10c). We get

$$\tau_s(\xi) = \tfrac{1}{2}i\pi - \ln(\tan\tfrac{1}{4}\zeta), \tag{150}$$

and

$$\frac{4a_s}{Uc} = \frac{1}{4+\pi}. \tag{151}$$

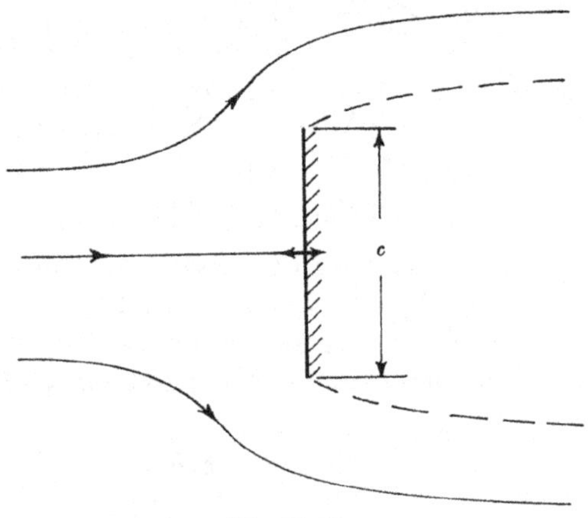

Fig. 11.21a

On the surface of the plate (150) yields

$$q_s = U|\tan\tfrac{1}{4}\gamma|. \tag{152}$$

From (114) and (151)

$$t = (4+\pi)\frac{Ut}{c}. \tag{153}$$

We shall consider two types of unsteady motion, namely (i) impulsive motion, and (ii) harmonic motion.

(i) *Impulsive motion*

Suppose that at $t = 0$ the plate's velocity relative to the fluid jumps from U to V, and that at $t = t_1$ it suddenly resumes its original value U. Of course both $V - U$ and t_1 must be small if the basic assumptions discussed in § 11.14 are not to be violated.

In the present case $\alpha = 0$ and

$$n = (V-U)\{\mathbf{U}(t) - \mathbf{U}(t-t_1)\},$$

where
$$t_1 = (4+\pi)\frac{Ut_1}{c}. \tag{154}$$

Hence by (103) and (152)

$$\theta_1 = h \cot\tfrac{1}{4}\gamma\{\mathbf{U}(t) - \mathbf{U}(t-t_1)\}, \tag{155}$$

where
$$h = \frac{V-U}{U}. \tag{156}$$

From (111), (155) and the fact that $\lambda_s = 0$ in the present example, we obtain $b_2 = 0$, and

$$b_1 = \left(1+\frac{2}{\pi}\right) h\{\mathbf{U}(t) - \mathbf{U}(t-t_1)\}, \tag{157}$$

while from (104) and (155), except for delta functions at $t = 0$, and $t = t_1$, which contribute nothing to the pressure or drag force at other values of t,

$$\dot{\Theta} = 0. \tag{158}$$

Substituting (152), (155), (157) and (158) into (134) we find that when $0 < t < t_1$

$$p - p_s = \frac{U^2\rho_a h}{\pi} \cot\tfrac{1}{2}\gamma \tan^2\tfrac{1}{4}\gamma \ln \frac{1-\sin\tfrac{1}{2}\gamma}{1+\sin\tfrac{1}{2}\gamma}$$

$$+ \rho_a U^2 h \frac{\pi+2}{\pi} \frac{\cos\tfrac{1}{2}\gamma}{\cos^2\tfrac{1}{4}\gamma}\{t + \cos^2\tfrac{1}{2}\gamma\}^{-\frac{1}{2}}, \tag{159}$$

and when $t_1 < t$

$$p - p_s = \rho_a U^2 h \frac{\pi+2}{\pi} \frac{\cos\tfrac{1}{2}\gamma}{\cos^2\tfrac{1}{4}\gamma}\{(t + \cos^2\tfrac{1}{2}\gamma)^{-\frac{1}{2}} - (t - t_1 + \cos^2\tfrac{1}{2}\gamma)^{-\frac{1}{2}}\}. \tag{160}$$

The drag increment follows from (108), (151), (152), (159) and (160). When $0 < t < t_1$

$$D - D_s = hcU^2\rho_a\left\{1 + \frac{8}{\pi}\left(\frac{\pi+2}{\pi+4}\right) F[(t+1)^{-\frac{1}{2}}]\right\},$$

and when $t_1 < t$

$$D - D_s = \frac{8hc}{\pi}\rho_a U^2 \left(\frac{\pi+2}{\pi+4}\right) \left\{\frac{F[(t+1)^{-\frac{1}{2}}]}{(t+1)^{\frac{1}{2}}} - \frac{F[(t-t_1+1)^{-\frac{1}{2}}]}{(t-t_1+1)^{\frac{1}{2}}}\right\},$$

where
$$F(k) \equiv \frac{1}{k^2} E(k) - \frac{1-k^2}{k^2} K(k),$$

$K(k)$ and $E(k)$ being the complete elliptic integrals of the first and second kind (see §§ 4.5 and 4.8). The value of D_s is shown in § 11.10 to be

$$D_s = \tfrac{1}{2}\rho_a c U^2 \frac{2\pi}{4+\pi}. \tag{161}$$

In terms of the drag coefficient $C_D \equiv D/\tfrac{1}{2}\rho_a c U^2$, our result can be written

$$C_D = \frac{2\pi}{4+\pi} + 2h \begin{cases} 1 + \dfrac{8}{\pi}\left(\dfrac{\pi+2}{\pi+4}\right)\eta_0 F(\eta_0) & (0 < \jmath < \jmath_1), \\[2ex] \dfrac{8}{\pi}\left(\dfrac{\pi+2}{\pi+4}\right)\{\eta_0 F(\eta_0) - \eta_1 F(\eta_1)\} & (\jmath_1 < \jmath), \end{cases}$$

where $\qquad \eta_0 \equiv (\jmath+1)^{-\frac{1}{2}} \quad$ and $\quad \eta_1 \equiv (\jmath-\jmath_1+1)^{-\frac{1}{2}}.$

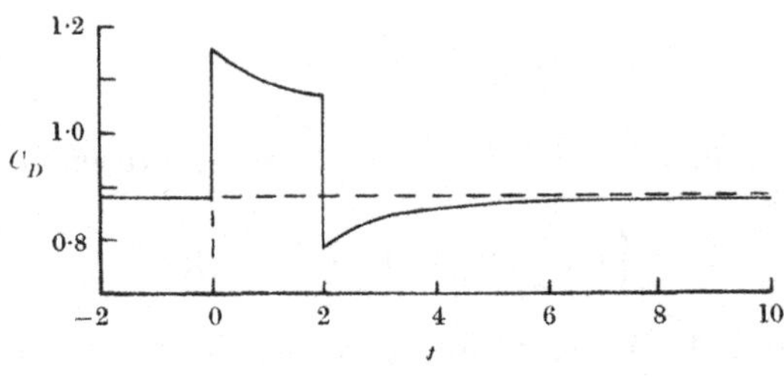

Fig. 11.21*b*

For example let $\jmath_1 = 2$ and $V - U = U/20$, so that $h = 1/20$. Then the resulting variation of C_D is as shown in Fig. 11.21*b*.

(ii) *Harmonic motion*

Suppose the small displacements of the plate about its mean position are described by $x = x^0 e^{i\nu t}$, so that the normal perturbation velocity is $\eta = i\nu x^0 e^{i\nu t}$, then from (103) and (152)

$$\theta_1 = \frac{i\nu x^0}{U} e^{i\nu t} \cot \tfrac{1}{2}\gamma.$$

From (104), (124) and (151)

$$b_1 = \frac{\pi+2}{2} \frac{i\nu x^0}{U} e^{i\nu t},$$

and $\qquad \Theta = -\dfrac{c}{4+\pi} \dfrac{\nu^2 x^0}{2U^2} e^{i\nu t} (4\sin\tfrac{1}{2}\gamma + \gamma + \sin\gamma + 4 + \pi).$

Substitution of these values in (149) yields

$$p - p_s = \frac{2\rho_a}{\pi} U i\nu x^0 e^{i\nu t} \cos\tfrac{1}{2}\gamma \tan^2\tfrac{1}{4}\gamma \int_0^\pi \frac{\cot\tfrac{1}{4}\gamma^* \sin\tfrac{1}{2}\gamma^* \, d\gamma^*}{\cos\gamma - \cos\gamma^*}$$

$$-\frac{\rho_a}{\pi}\frac{c}{4+\pi}\nu^2 x^0 e^{i\nu t} \cos\tfrac{1}{2}\gamma$$

$$\times \int_0^\pi \frac{\{4\sin\tfrac{1}{2}\gamma^* + \gamma^* + \sin\gamma^* + 4 + \pi\}\sin\tfrac{1}{2}\gamma^* \, d\gamma^*}{\cos\gamma - \cos\gamma^*}$$

$$+ \frac{\rho_a}{\sqrt{\pi}} U(\pi+2)(1+i) i\nu x^0 \sqrt{\omega}\, e^{i\nu t}$$

$$\times \frac{\cos\tfrac{1}{2}\gamma}{\cos^2\tfrac{1}{4}\gamma} e^{2i\omega \cos^2\frac{1}{2}\gamma}\{1 - \mathrm{erf}\sqrt{(2i\omega \cos^2\tfrac{1}{2}\gamma)}\}.$$

To calculate the drag increment for small values of the frequency parameter we make use of the expansion

$$e^x(1-\operatorname{erf}\sqrt{x}) = 1+x+\tfrac{1}{2}x^2+\ldots-2\sqrt{\left(\frac{\pi}{x}\right)}(1+\tfrac{1}{2}x+\tfrac{1}{10}x^2+\ldots).$$

From (108), (136) and (153) it is found that in this case the 'out of phase' or damping component of the drag increment is given by

$$D-D_0 = \frac{i}{2}\rho_a cU^2 e^{i\nu t}\frac{x^0}{l}\omega\epsilon,$$

where ϵ is a 'damping derivative' given by

$$\epsilon \simeq 2(4+\pi)\left\{1+\frac{2\pi+4}{\pi+4}[\sqrt{(\pi\omega)}(1-\tfrac{3}{2}\omega-\tfrac{5}{4}\omega^2+\tfrac{35}{48}\omega^3)+\tfrac{128}{15}\omega^2]\right\}.$$

11.22   An oscillating aerofoil fitted with a spoiler

In the classical unsteady aerofoil theory given in ch. 9 a single vortex sheet was assumed to be shed from the trailing edge, but if for some reason or other the flow does not remain attached to both surfaces right back to the trailing edge, and separates at some point on the upper surface say, then there will be two distinct vortex sheets as illustrated in Fig. 11.22. This is a special case of the flow studied above.

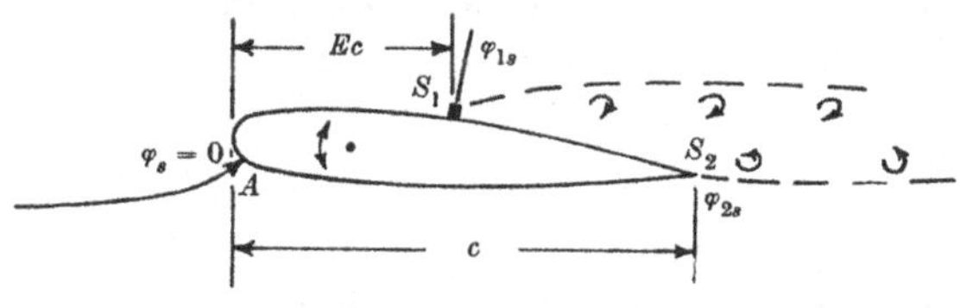

Fig. 11.22

We shall assume that the aerofoil is reasonably thin, so that we can write $q_s \simeq U$, $\phi_s \simeq Ux$, $\phi_{2s} \simeq Uc$, $\phi_{1s} \simeq EUs$, where Ec is the distance from the leading edge. The origin of x is at the leading edge. In this case (83) becomes

$$\frac{4a_s}{Uc} = \tfrac{1}{4}(1+E^{\frac{1}{2}})^2, \quad \sin\tfrac{1}{2}\lambda_s = \frac{1-E^{\frac{1}{2}}}{1+E^{\frac{1}{2}}}, \tag{162}$$

while on the aerofoil surface (82) gives

$$\frac{x}{c} = \frac{4a_s}{Uc}(\sin\tfrac{1}{2}\gamma-\sin\tfrac{1}{2}\lambda_s)^2. \tag{163}$$

From (136) and (162) we observe that the reduced frequency parameter ω is related to the classical parameter, ω_0 say, used in ch. 9, by

$$\omega = \tfrac{1}{4}(1+E^{\frac{1}{2}})^2\omega_0, \quad \omega_0 = \frac{c\nu}{2U}. \tag{164}$$

With non-separating flow, $E = 1$ and $\omega = \omega_0$.

From (108), (163) and the approximations $q_s \simeq U, \theta_s \simeq 0$ it follows that

$$L = L_s - c\left(\frac{4a_s}{Uc}\right) \int_{-\pi}^{\pi} (p-p_s) \cos \tfrac{1}{2}\gamma (\sin \tfrac{1}{2}\gamma - \sin \tfrac{1}{2}\lambda_s)\, d\gamma, \qquad (165)$$

$$M = M_s + c\left(\frac{4a_s}{Uc}\right)^2 \int_{-\pi}^{\pi} (p-p_s) \cos \tfrac{1}{2}\gamma (\sin \tfrac{1}{2}\gamma - \sin \tfrac{1}{2}\lambda_s)^3\, d\gamma, \qquad (166)$$

where from (105)

$$p - p_s = \tfrac{1}{2}\rho_a U^2 \frac{(\lambda_s - \lambda)\cos\tfrac{1}{2}\gamma}{\sin\tfrac{1}{2}\lambda_s - \sin\tfrac{1}{2}\gamma} + \frac{1}{2\pi}\rho_a U^2 \cos\tfrac{1}{2}\gamma$$

$$\times \int_{-\pi}^{\pi} \frac{\{\theta_1(\gamma^*,t) + \dot{\Theta}(\gamma^*,t)\}\, d\gamma^*}{\sin\tfrac{1}{2}\gamma^* - \sin\tfrac{1}{2}\gamma}. \qquad (167)$$

In this account we shall deal only with the case of a rigid aerofoil, but the theory of the flapped aerofoil, with the flow separating at the hinge, follows by a calculation similar to that given below—it is necessary only to make a slight modification to (168) for θ_1. With a rigid aerofoil the normal velocity on the aerofoil surface can be written $n = v - x\dot{\alpha}$, where v is the perturbation velocity of the leading edge ($x = 0$) in a direction at right angles to the undisturbed stream direction, and $\dot{\alpha}$ is the aerofoil's nose-up angular velocity. Hence from (103) and (163)

$$\theta_1 = \frac{v}{U} - \frac{\dot{\alpha}c}{U}\left(\frac{4a_s}{Uc}\right) (\sin\tfrac{1}{2}\gamma - \sin\tfrac{1}{2}\lambda_s)^2 - \alpha, \qquad (168)$$

where as the motion is assumed to be oscillatory we can write

$$\left.\begin{aligned}
y &= y^0 e^{i\nu t}, & \alpha &= \alpha^0 e^{i\nu t}, \\
\dot{y} &= v = i\nu y^0 e^{i\nu t}, & \dot{\alpha} &= i\nu\alpha^0 e^{i\nu t}, \\
\dot{v} &= -\nu^2 y^0 e^{i\nu t}, & \ddot{\alpha} &= -\nu^2 \alpha^0 e^{i\nu t}.
\end{aligned}\right\} \qquad (169)$$

Substitution of (168) into (135), (137) and (146) gives

$$\lambda - \lambda_s = 2\sin\tfrac{1}{2}\lambda_s \left\{\frac{\dot{\alpha}c}{2U}\left(\frac{4a_s}{Uc}\right)(3 + 2\sin^3\tfrac{1}{2}\lambda_s) - \left(\frac{v}{U} - \alpha\right)\right\} F_2(\omega, \lambda_s)$$

$$+ \left\{\frac{v}{U} - \alpha - \frac{3}{2}\frac{\dot{\alpha}c}{2U}\left(\frac{4a_s}{Uc}\right)(1 + 4\sin^2\tfrac{1}{2}\lambda_s)\right\} F_1(\omega, \lambda_s)$$

$$- 2\left\{\frac{v}{U} - \alpha - \frac{\dot{\alpha}c}{2U}\left(\frac{4a_s}{Uc}\right)(1 + 2\sin^2\tfrac{1}{2}\lambda_s)\right\}. \qquad (170)$$

Similarly, from (104), (167) and (168) it follows that the pressure increment is

$$p - p_s = \tfrac{1}{2}\rho_a U^2 \frac{(\lambda_s - \lambda)\cos\tfrac{1}{2}\gamma}{\sin\tfrac{1}{2}\lambda_s - \sin\tfrac{1}{2}\gamma} + \frac{\rho_a c^2}{16}\left(\frac{4a_s}{Uc}\right)^2 \ddot{\alpha}\sin 2\gamma$$

$$+ \rho_a c U\left(\frac{4a_s}{Uc}\right)\left(\frac{\dot{v}}{U} - 2\dot{\alpha} - \left(\frac{4a_s}{Uc}\right)\frac{\ddot{\alpha}c}{2U}\right)\cos\tfrac{1}{2}\gamma(\sin\tfrac{1}{2}\gamma - 2\sin\tfrac{1}{2}\lambda_s)$$

$$+ \rho_a c^2\left(\frac{4a_s}{Uc}\right)^2 \ddot{\alpha}\sin\tfrac{1}{2}\lambda_s\cos\tfrac{1}{2}\gamma(2\sin^2\tfrac{1}{2}\gamma$$

$$- 3\sin\tfrac{1}{2}\gamma\sin\tfrac{1}{2}\lambda_s + 2\sin^2\tfrac{1}{2}\lambda_s). \tag{171}$$

Equations (165), (166) and (171) give the lift and moment increments:

$$\frac{L - L_s}{\rho_a c U^2} = \frac{\pi}{2}\left(\frac{4a_s}{Uc}\right)(\lambda_s - \lambda) - \frac{\pi}{4}\left(\frac{4a_s}{Uc}\right)^2(1 + 8\sin^2\tfrac{1}{2}\lambda_s)\left\{\frac{\dot{v}c}{U^2} - \frac{2\dot{\alpha}c}{U} - \left(\frac{4a_s}{Uc}\right)\frac{\ddot{\alpha}c^2}{2U^2}\right\}$$

$$+ \frac{\pi}{4}\left(\frac{4a_s}{Uc}\right)^3\frac{\ddot{\alpha}c^2}{U^2}\sin^2\tfrac{1}{2}\lambda_s(5 + 8\sin^2\tfrac{1}{2}\lambda_s), \tag{172}$$

and

$$\frac{M - M_s}{\rho_a c^2 U^2} = -\frac{\pi}{8}\left(\frac{4a_s}{Uc}\right)^2(\lambda_s - \lambda)(1 + 4\sin^2\tfrac{1}{2}\lambda_s)$$

$$+ \frac{\pi}{8}\left(\frac{4a_s}{Uc}\right)^3\left\{\frac{\dot{v}c}{U^2} - \frac{2\dot{\alpha}c}{U} - \left(\frac{4a_s}{Uc}\right)\frac{\ddot{\alpha}c^2}{2U^2}\right\}(1 + 18\sin^2\tfrac{1}{2}\lambda_s + 16\sin^4\tfrac{1}{2}\lambda_s)$$

$$- \frac{\pi}{128}\left(\frac{4a_s}{Uc}\right)^3\frac{\ddot{\alpha}c^2}{U^2} - \frac{\pi}{8}\left(\frac{4a_s}{Uc}\right)^4\frac{\ddot{\alpha}c^2}{U^2}\sin^2\tfrac{1}{2}\lambda_s(9 + 34\sin^2\tfrac{1}{2}\lambda_s$$

$$+ 16\sin^4\tfrac{1}{2}\lambda_s). \tag{173}$$

The oscillatory derivatives (or air-load coefficients) l_{12}, l_{34}, m_{12} and m_{34} defined in (9.67) can now be deduced from (169), (172) and (173). Numerical values for the imaginary parts of these derivatives—the 'damping' derivatives— are to be found for a range of values of E and ω in the paper by Woods (1957b) from which the above theory is taken.

FINITE CAVITIES

11.23   Riabouchinsky flow

In § 8.32 we briefly indicated a generalization of Riabouchinsky's (1921) model of cavitating flow. The problem actually solved by Riabouchinsky is that shown in Fig. 11.23a, i.e. the flow past two equal, normal plates between which extends a constant pressure bubble. The reasons for adopting this model for finite cavities have been given in

§ 11.4. The very general case described in § 8.32 leads to very complicated algebra. We shall restrict attention to the symmetrical generalization of Riabouchinsky's problem shown in Fig. 11.23b. In this flow we have a constant pressure bubble between obstacles on the axis of a channel of width H.

The w-plane in Fig. 11.23b is mapped into the χ-plane by

$$\coth \chi = -\coth \frac{\pi \phi_0}{2h} \tanh \frac{\pi w}{2h}, \tag{174}$$

which follows by replacing ζ in (7.88) by $\tfrac{1}{2}\pi - \chi$ (cf. Fig. 7.13j and the w-plane in Fig. 11.23b).

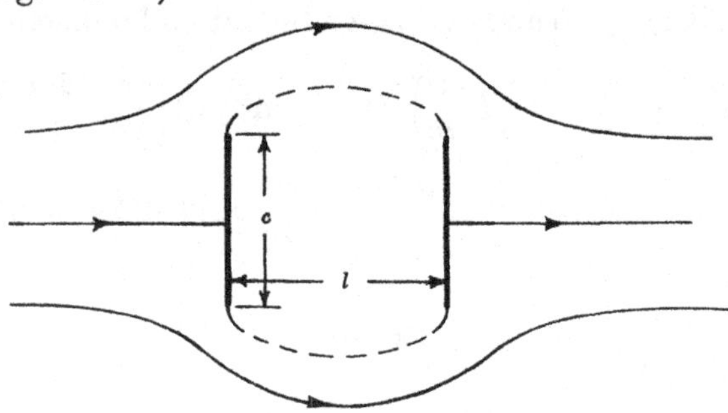

Fig. 11.23a

The general solution in the χ-plane is

$$\tau(\chi) = \tau(\infty) + \frac{2}{\pi} \int_{-\infty}^{\infty} \tanh^{-1} \exp(\xi - \chi)\, d\theta(\xi), \tag{175}$$

which is just the special case $\Omega_2 = \text{constant}$ of (7.87).

A more convenient plane is the semicircle shown in Fig. 11.23b. It is related to the χ-plane by

$$\coth \chi = \frac{1}{2}\left(\zeta + \frac{1}{\zeta}\right). \tag{176}$$

In this plane (175) assumes the form

$$\tau(\zeta) = \Omega_c + i\theta_c + \frac{1}{\pi} \int_{-1}^{1} \ln\left(\frac{1 - \zeta\gamma}{\zeta - \gamma}\right) d\theta(\gamma), \tag{177}$$

where Ω_c, θ_c are the values of Ω and θ at the separation point C.

The point upstream at infinity is at $\phi = -\infty$, which maps on to $\zeta = \gamma_0$, where from (174) and (176) $\gamma_0 + \gamma_0^{-1} = 2 \coth(\pi\phi_0/2h)$. Thus

$$\gamma_0 = \tanh \frac{\pi \phi_0}{4h}. \tag{178}$$

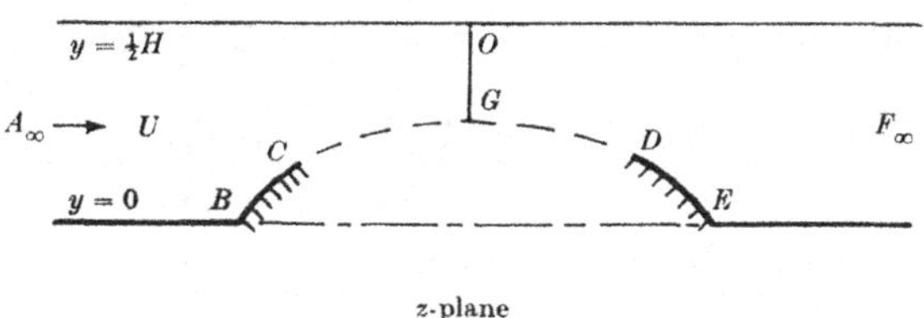

z-plane

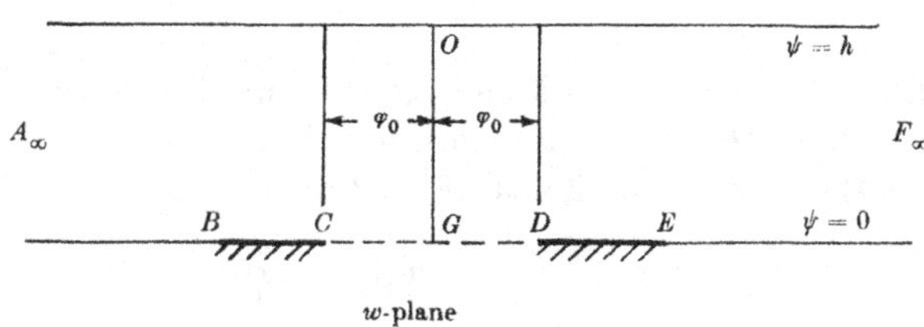

w-plane

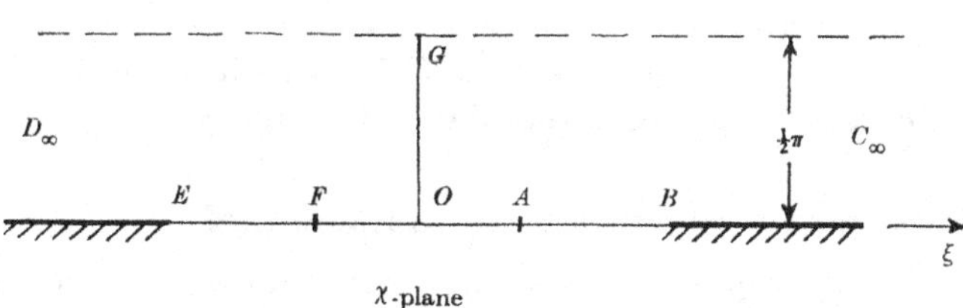

χ-plane

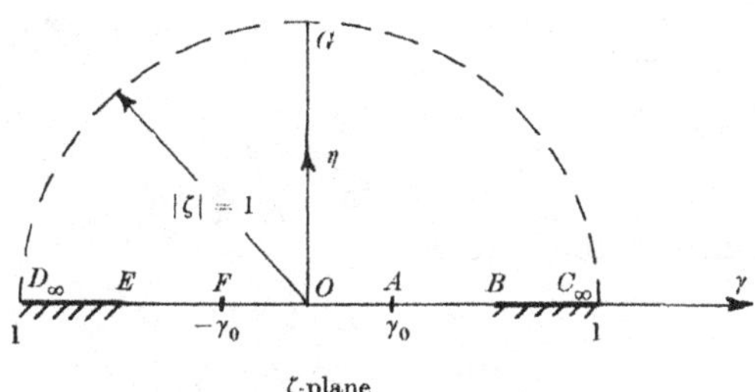

ζ-plane

Fig. 11.23b

Similarly, the point downstream at infinity maps on to $\zeta = -\gamma_0$. At these points we can put $q = U$ and $\theta = 0$; so $\tau(\gamma_0) = \tau(-\gamma_0) = 0$, and (177) yields

$$\Omega_c = \frac{1}{\pi} \int_{-1}^{1} \ln \left| \frac{\gamma - \gamma_0}{1 - \gamma_0 \gamma} \right| d\theta(\gamma) = \frac{1}{\pi} \int_{-1}^{1} \ln \left| \frac{\gamma + \gamma_0}{1 + \gamma_0 \gamma} \right| d\theta(\gamma), \qquad (179)$$

and

$$\int_{-1}^{1} \ln \left| \frac{\gamma - \gamma_0}{\gamma + \gamma_0} \frac{1 + \gamma_0 \gamma}{1 - \gamma_0 \gamma} \right| d\theta(\gamma) = 0. \qquad (180)$$

There is also the obvious restriction

$$\int_{-1}^{1} d\theta(\gamma) = 0. \qquad (181)$$

Let the flow be incompressible then from Bernoulli's equation $q_c^2/U^2 = Q + 1$, where Q is the cavitation number defined in (2), and q_c is the velocity at the surface of the cavity. Hence

$$\Omega_c = \ln \left(\frac{U}{q_c} \right) = -\tfrac{1}{2} \ln (1 + Q), \qquad (182)$$

and therefore by (179)

$$Q = \exp \left\{ -\frac{1}{\pi} \int_{-1}^{1} \ln \left| \frac{\gamma^2 - \gamma_0^2}{1 - \gamma^2 \gamma_0^2} \right| d\theta(\gamma) \right\} - 1. \qquad (183)$$

With curved obstacles the procedure is now similar to that of § 11.12— we substitute $d\theta = (Rq)^{-1} (d\phi/d\gamma)\, d\gamma$ in (177), which on the wetted surface then becomes a complicated integral equation for q.

11.24   Flow past a normal plate in a channel

The simplest example of the above theory is all we shall consider here. This is the flow past a normal plate in a channel; half the flow pattern is shown in Fig. 11.24.

In this case θ has discontinuities of $\tfrac{1}{2}\pi$ at each of $\gamma = \pm \gamma_1$ say, and (177), (179) become

$$\tau(\zeta) = \Omega_c + \tfrac{1}{2} i\pi + \tfrac{1}{2} \ln \frac{1 - \zeta \gamma_1}{\zeta - \gamma_1} + \tfrac{1}{2} \ln \frac{1 + \zeta \gamma_1}{\zeta + \gamma_1},$$

$$\Omega_c = \tfrac{1}{2} \ln \frac{\gamma_1^2 - \gamma_0^2}{1 - \gamma_0^2 \gamma_1^2}.$$

The flow is incompressible and so

$$e^{\tau} = \frac{U\, dz}{dw} = e^{\Omega_c} \left(\frac{1 - \zeta^2 \gamma_1^2}{\gamma_1^2 - \zeta^2} \right)^{\frac{1}{2}}.$$

In order to integrate this we introduce the substitution

$$\zeta = \frac{k \operatorname{sn}(u, k)}{1 + \operatorname{dn}(u, k)} = \frac{1 - \operatorname{dn}(u, k)}{k \operatorname{sn}(u, k)}, \qquad (184)$$

with
$$\gamma_1^2 = \frac{1-k'}{1+k'}, \quad \text{or} \quad k = \frac{2\gamma_1}{1+\gamma_1^2}. \tag{185}$$

Let u_0 be the value of u at $\zeta = \gamma_0$, then it follows from (174) and (176) that

$$\operatorname{ns} u = -\operatorname{ns} u_0 \tanh\frac{\pi w}{2h}, \tag{186}$$

where
$$k \operatorname{sn} u_0 = \tanh\frac{\pi\phi_0}{2h} = \frac{2\gamma_0}{1+\gamma_0^2}. \tag{187}$$

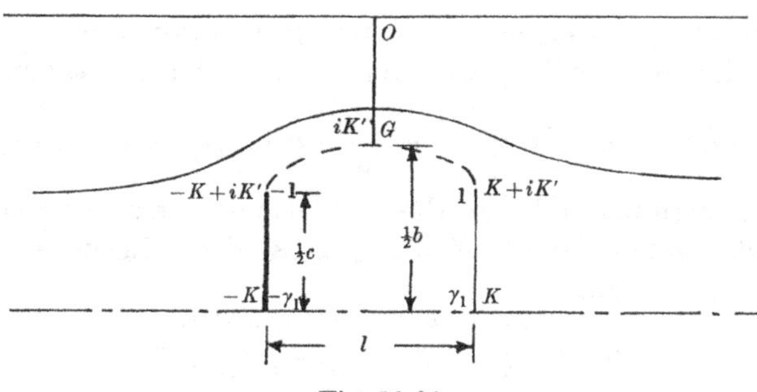

Fig. 11.24

Let q_c denote the velocity on the surface of the cavity, then

$$\frac{q_c}{U} = e^{-\Omega_c} = \frac{k \operatorname{cn} u_0}{\operatorname{dn} u_0 - k'}, \tag{188}$$

$$Q = \frac{2k'}{\operatorname{dn} u_0 - k'}, \tag{189}$$

and
$$e^\tau = \frac{U\,dz}{dw} = \frac{U}{q_c}\,\frac{k \operatorname{cn} u}{\operatorname{dn} u - k'}. \tag{190}$$

Equations (186) and (190) give

$$z = \frac{HU}{\pi q_c}\,\frac{\operatorname{sn} u_0}{k}\int^u \frac{\operatorname{dn}^2 u + k' \operatorname{dn} u}{\operatorname{sn}^2 u - \operatorname{sn}^2 u_0}\,du,$$

on replacing $2h$ by HU. This can be integrated with the aid of results given in §§ 4.6 and 4.8:

$$z = \frac{HU}{2\pi q_c k}\left\{-\frac{2u}{\operatorname{sn} u_0} + \frac{k'}{\operatorname{cn} u_0}\ln\frac{\operatorname{cs} u - \operatorname{cs} u_0}{\operatorname{cs} u + \operatorname{cs} u_0} + 2 \operatorname{dc} u_0 \,\Pi\,(u, u_0 + iK')\right\}, \tag{191}$$

the origins of z and u coinciding.

The points $\zeta = \pm\gamma_1$, $\zeta = \pm 1$ are at $u = \pm K$, $u = \pm K + iK'$.

Hence the length of the cavity is

$$l = z(-K+iK') - z(K+iK') = \frac{2KHU}{\pi q_c k}\{k^2 \operatorname{sn} u_0 - \operatorname{dc} u_0 Z_4(u_0)\}, \quad (192)$$

while the width of the plate is $c = -2i\{z(K+iK') - z(K)\}$, i.e.

$$c = \frac{2HU}{\pi q_c k}$$
$$\times \left\{\frac{\pi}{2K} u_0 \operatorname{dc} u_0 - K'k^2 \operatorname{sn} u_0 + K' \operatorname{dc} u_0 Z_4(u_0) + k' \operatorname{nc} u_0 \tan^{-1}(k' \operatorname{sc} u_0)\right\}.$$
$$(193)$$

The drag on the (front) plate is most readily obtained by an application of (2.102). Allowing for the pressure p_c over the back of the plate, we get

$$D - iL = -i\rho_a U \int_{-K}^{-K+iK'} e^{-\tau}\frac{dw}{du}du + c(p_a + \tfrac{1}{2}\rho_a U^2) - cp_c,$$

the change in sign being due to the fact that we are now on the opposite side of the streamline from that of Fig. 2.18. From Bernoulli's equation, (186) and (190) we find

$$D - iL = -i\rho_a q_c \frac{HU}{\pi k}\int_{-K}^{-K+iK'} \frac{\operatorname{dn}^2 u - k' \operatorname{dn} u}{\operatorname{sn}^2 u - \operatorname{sn}^2 u_0}du + \tfrac{1}{2}\rho_a q_c^2 c.$$

This gives

$$D = \rho_a q_c^2 \left(\frac{2HU}{\pi q_c k}\right)\left\{\frac{\pi}{2K} u_0 \operatorname{dc} u_0 - K'k^2 \operatorname{sn} u_0 + K' \operatorname{dc} u_0 Z_4(u_0)\right\}. \quad (194)$$

Given Q and c/H, we calculate k and u_0 from (189) and (193). The cavity shape, its length and the drag on the front plate then follow from (191), (192) and (194).

The special case $H = \infty$, the one studied by Riabouchinsky, is of some interest. In this case it follows from (187) that

$$\lim_{H\to\infty}\frac{u_0 HU}{\pi} = \frac{\phi_0}{k},$$

and with the help of this limit we find that (189) and (191)–(194) reduce to

$$Q = \frac{2k'}{1-k'}, \quad (195)$$

$$z = \frac{\phi_0}{q_c k^2}\{uk'^2 - E(u) - \operatorname{dn} u \operatorname{cs} u - k' \operatorname{cs} u\}, \quad (196)$$

$$l = \frac{2\phi_0}{q_c k^2}\{E - Kk'^2\}, \quad (197)$$

$$c = \frac{2\phi_0}{q_c k^2}\{E' - K'k^2 + k'^2\}, \quad (198)$$

$$C_D \equiv \frac{D}{\tfrac{1}{2}\rho_a c U^2} = 2\left(\frac{q_c}{U}\right)^2\left(\frac{E' - K'k^2}{E' - K'k^2 + k'^2}\right), \quad (199)$$

on taking advantage of Legendre's relation $EK' + E'K - KK' = \frac{1}{2}\pi$. The maximum width of the cavity, b, can be calculated from (see Fig. 11.24) $z(iK') - z(-K) = \frac{1}{2}(l+ib)$. It is found that

$$b = c + \frac{2\phi_0}{q_c k^2} k'(1-k'). \tag{200}$$

And finally from (188) the velocity on the cavity is

$$\frac{q_c}{U} = \frac{k}{1-k'}. \tag{201}$$

It should be noted that the ratio $2\phi_0/q_c$ equals the length of the curved surface of the cavity.

Suppose Q is small, then $Q \simeq 2k'$, and from the expansions of E, E', K and K' (see Byrd & Friedman, 1954, p. 297) we find that to first order in Q

$$\frac{l}{c} \simeq \frac{16}{4+\pi} Q^{-2}, \quad \frac{b}{c} \simeq \frac{8}{4+\pi} Q^{-1}, \tag{202}$$

and

$$C_D = \frac{2\pi}{4+\pi}(1+Q) \tag{203}$$

(cf. (63)).

11.25 Re-entrant jets

Another model of a finite cavity with a positive value of Q is shown in Fig. 11.25. This is the so-called re-entrant jet model due to Efros, Gilbarg & Rock (for full list of references see p. 56 of Birkhoff & Zarantonello, 1957). This model is physically unrealistic as the jet passes through the obstacle from the rear, producing a doubly-covered region of flow in front of it. Nevertheless, this flow more closely resembles real cavitating flow than the other models that have been proposed. Re-entrant jets have been observed in experiments, but with a horizontal cavity of reasonable length the action of gravity prevents them from reaching the back of the obstacle.

A simple application of the Schwarz–Christoffel theorem (§ 5.3) shows that the w-plane is related to the t-plane defined in the figure by $dw/dt = A(t^2 - t_0^2)t(t^2 - 1/k^2)^{-1}$. The t-plane is mapped into the rectangle in the ζ-plane by (see § 5.16) $t = \text{sn}\,(\zeta, k)$. Let $t_0 = \text{sn}\,(\gamma_0 + iK') = k^{-1}\,\text{ns}\,\gamma_0$, then we find that

$$\frac{dw}{d\zeta} = Ak^2(\text{ns}^2\,\gamma_0 - k^2\,\text{sn}^2\,\zeta)\,\text{sn}\,\zeta\,\text{cd}\,\zeta. \tag{204}$$

In the ζ-plane we can write down the general solution

$$\tau(\zeta) = \frac{1}{\pi}\,\text{cn}\,\zeta\,\text{dn}\,\zeta\left[\int_{-K}^{K}\left(\frac{\theta_0(\gamma)}{\text{sn}\,\gamma - \text{sn}\,\zeta} - \frac{k\theta_h(\gamma)}{\text{ns}\,\gamma - k\,\text{sn}\,\zeta}\right)d\gamma \right.$$
$$\left. + 2\Omega_c\int_0^{K'}\frac{\text{cd}\,(i\eta)\,d\eta}{\text{cd}^2\,(i\eta) - \text{sn}^2\,\zeta}\right],$$

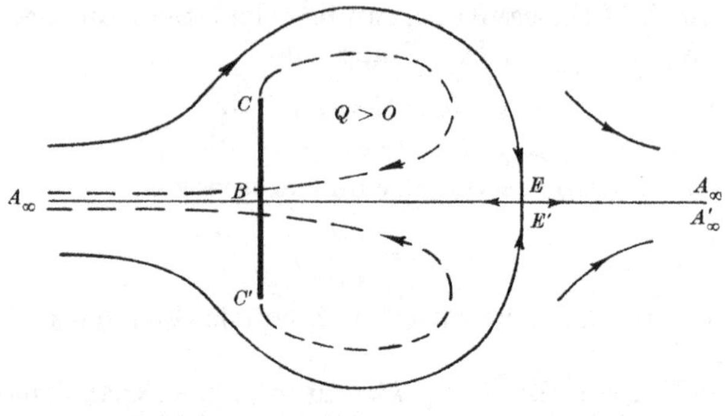

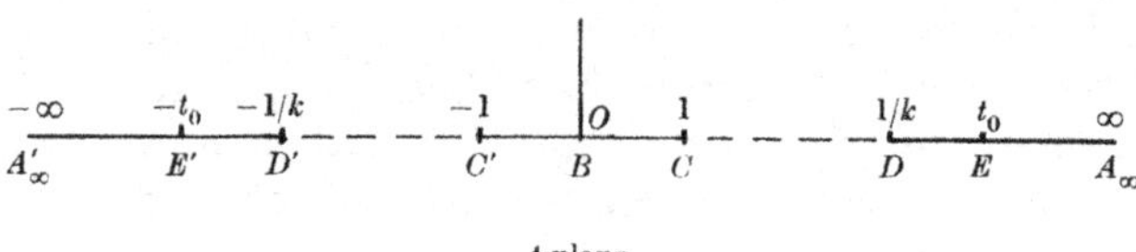

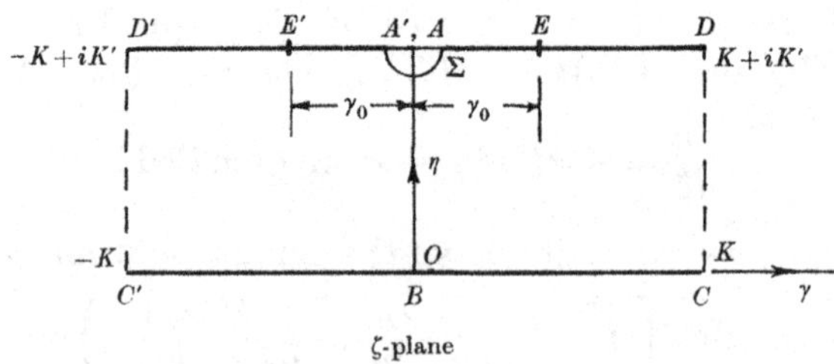

C
Q > 0
A_∞
B
E
E'
A_∞
A'_∞
C'
z-plane

ψ
E
A_∞
B
φ
C
A_∞
D_∞
C'
D'_∞
E'
A'_∞
w-plane

-∞
-t_0
-1/k
-1
O
1
1/k
t_0
∞
A'_∞
E'
D'
C'
B
C
D
E
A_∞
t-plane

D'
E'
A', A
E
D
-K + iK'
Σ
K + iK'
γ_0
γ_0
η
-K
O
K
C'
B
C
γ
ζ-plane

Fig. 11.25

which follows by applying (4.138) to the problem, and noting that Ω has the constant value Ω_c on $\gamma = \pm K$. The last integral is easily evaluated with the help of a result given in § 7.25. With symmetrical flows $\theta_0(\gamma)$ and $\theta_h(\gamma)$ are odd functions, and consequently the general solution can be written

$$\tau(\zeta) = \Omega_c + \frac{2}{\pi} \operatorname{cn} \zeta \operatorname{dn} \zeta \int_0^K \left(\frac{\theta_0(\gamma) \operatorname{sn} \gamma}{\operatorname{sn}^2 \gamma - \operatorname{sn}^2 \zeta} - \frac{k \theta_h(\gamma) \operatorname{ns} \gamma}{\operatorname{ns}^2 \gamma - k^2 \operatorname{sn}^2 \zeta} \right) d\gamma. \quad (205)$$

This result could be used to investigate re-entrant jet flows past curved obstacles, but we shall restrict attention to the flat plate shown in Fig. 11.25, for which $\theta_0 = \frac{1}{2}\pi \, (0 < \gamma < K)$ and $\theta_h = -\pi, (\gamma_0 < \gamma < K)$. In this case (use (4.81))

$$\tau(\zeta) = \Omega_c + \frac{1}{2} i\pi + \ln \left\{ \frac{k' \operatorname{sd} \zeta}{1 - \operatorname{cd} \zeta} \frac{\operatorname{dc} \gamma_0 + k \operatorname{cd} \zeta}{\operatorname{dc} \gamma_0 - k \operatorname{cd} \zeta} \right\}. \quad (206)$$

The point at infinity maps into $\zeta = iK'$. By writing $\zeta = \epsilon + iK'$, and expanding in powers of ϵ we find from (206) that

$$\frac{U dz}{dw} = e^\tau = e^{\Omega_c} \frac{k'}{1-k} \frac{\operatorname{dc} \gamma_0 + 1}{\operatorname{dc} \gamma_0 - 1} \{1 + (\operatorname{ds} \gamma_0 \operatorname{cs} \gamma_0 - \tfrac{1}{2}k) \epsilon^2 + \ldots\}. \quad (207)$$

At the point at infinity $q = U,\ \theta = 0$ so $\tau = 0$ and (207) gives

$$e^{-\Omega_c} = \frac{q_c}{U} = \frac{k'}{1-k} \frac{\operatorname{dc} \gamma_0 + 1}{\operatorname{dc} \gamma_0 - 1}. \quad (208)$$

Similarly, expanding (204) near infinity we get

$$dw = -\frac{A}{\epsilon^3}(1 - \operatorname{cs}^2 \gamma_0 \epsilon^2 + \ldots)\, d\epsilon, \quad (209)$$

and therefore from (207) to (209)

$$Uz = -A \int \frac{1}{\epsilon^3}\{1 + (\operatorname{ds} \gamma_0 \operatorname{cs} \gamma_0 - \operatorname{cs}^2 \gamma_0 - \tfrac{1}{2}k)\epsilon^2 + \ldots\}\, d\epsilon. \quad (210)$$

A glance at Fig. 11.25 shows that the small semicircular indentation Σ having $A,\ A'$ as centre, corresponds to a closed circuit in the z-plane. Thus $\int_\Sigma dz = 0$, which requires the coefficient of ϵ^{-1} in the integrand of (210) to be zero. Therefore $\operatorname{dc} \gamma_0 - 1 - \tfrac{1}{2}k \operatorname{sc}^2 \gamma_0 = 0$, which has the solution

$$\operatorname{dc} \gamma_0 = \frac{2k'^2}{k} - 1 = \frac{1}{k}(2 - k - 2k^2). \quad (211)$$

This leaves just one parameter (k or γ_0) in the theory. From (182), (208) and (211) we find that k is related to the cavitation number Q by

$$Q = \frac{(1-k)(1+k)^3}{(1-k-k^2)^2} - 1. \quad (212)$$

From (204) and (206)

$$Uz = iA \frac{U}{q_c} k^2 \int_0^\zeta \frac{k' \operatorname{sd} \zeta}{1 - \operatorname{cd} \zeta} \frac{\operatorname{dc} \gamma_0 + k \operatorname{cd} \zeta}{\operatorname{dc} \gamma_0 - k \operatorname{cd} \zeta} (\operatorname{ns}^2 \gamma_0 - k^2 \operatorname{sn}^2 \zeta) \operatorname{sn} \zeta \operatorname{cd} \zeta \, d\zeta$$

$$= iA \frac{U}{q_c} \frac{k^2}{k'^2} \operatorname{cs}^2 \gamma_0 \int_0^\zeta (1 + \operatorname{cd} \zeta)(\operatorname{dc} \gamma_0 + k \operatorname{cd} \zeta)^2 \operatorname{cn} \zeta \operatorname{dn}^2 \zeta \, d\zeta$$

$$= iA \frac{U}{q_c} \frac{k^2 \operatorname{cs}^2 \gamma_0}{2k'^2} \left\{ \frac{\operatorname{dc}^2 \gamma_0}{k} [\sin^{-1}(k \operatorname{sn} \zeta) + k \operatorname{sn} \zeta \operatorname{dn} \zeta] \right.$$

$$+ (2k + \operatorname{dc} \gamma_0) \operatorname{dc} \gamma_0 [\sin^{-1}(\operatorname{sn} \zeta) + \operatorname{sn} \zeta \operatorname{cn} \zeta] + \frac{1}{k^2}(3k^2 - 2)\sin^{-1}(\operatorname{sn} \zeta)$$

$$+ \frac{1}{k^2}(k + 2 \operatorname{dc} \gamma_0)[(2k^2 - 1)\sin^{-1}(k \operatorname{sn} \zeta) + k \operatorname{sn} \zeta \operatorname{dn} \zeta] + \operatorname{sn} \zeta \operatorname{cn} \zeta$$

$$\left. + \frac{2k'^3}{k^2} \tan^{-1}\left(\frac{k' \operatorname{sn} \zeta - \operatorname{cn} \zeta}{k' \operatorname{sn} \zeta + \operatorname{cn} \zeta}\right) + \frac{\pi}{2} \frac{k'^3}{k^2} \right\}. \tag{213}$$

The width of the plate is

$$c = -2iz(K) = \frac{A}{q_c} \frac{\operatorname{cs}^2 \gamma_0}{k'^2} \{ [k \operatorname{dc}^2 \gamma_0 + (k + 2 \operatorname{dc} \gamma_0)(2k^2 - 1)] \sin^{-1} k$$

$$+ [k \operatorname{dc}^2 \gamma_0 + k + 2 \operatorname{dc} \gamma_0] kk' + \tfrac{1}{2}\pi [k^2(2k + \operatorname{dc} \gamma_0) \operatorname{dc} \gamma_0 + 3k^2 - 2 + 2k'^3] \}, \tag{214}$$

from which A can be calculated.

The method used in §11.24 to compute the drag leads in the present case to

$$D = \tfrac{1}{2}\rho_a q_c^2 \left\{ c - \frac{2k^2 A \operatorname{cs}^2 \gamma_0}{k'^2 q_c} \int_0^K (1 - \operatorname{cd} \zeta)(\operatorname{dc} \gamma_0 - k \operatorname{cd} \zeta)^2 \operatorname{cn} \zeta \operatorname{dn}^2 \zeta \, d\zeta \right\}$$

$$= \tfrac{1}{2}\rho_a q_c^2 \frac{\pi A \operatorname{cs}^2 \gamma_0}{q_c k'^2} \{ k^2(2k + \operatorname{dc} \gamma_0) \operatorname{dc} \gamma_0 + 3k^2 - 2 + 2k'^3 \}. \tag{215}$$

When Q is small it follows from (211) and (212) that (215) yields the same drag coefficient as Riabouchinsky's model, i.e. the value given in (203). A numerical comparison of these two models of finite cavities is to be found in Birkhoff & Zarantonello (1957), to whose excellent text we refer the reader requiring a more detailed treatment of cavitating flow. The proceedings of a symposium (held at the National Physical Laboratory, Teddington, England) entitled *Cavitation in Hydrodynamics*, and published in 1956 by Her Majesty's Stationery Office, is also recommended.

CHAPTER 12

CASCADES OF AEROFOILS

12.1 Some general relations

By a cascade of aerofoils is meant an infinite set of similar aerofoils all at the same incidence and spaced at equal intervals along the y-axis as shown in Fig. 12.1. The flow through such an array of aerofoils is important because it is a valuable first approximation to the flow through the blading of an axial compressor, especially at the high-pressure end where the blade height is small compared with the radius. The flow also has other applications, especially if the flow is that past a *stalled* cascade (see

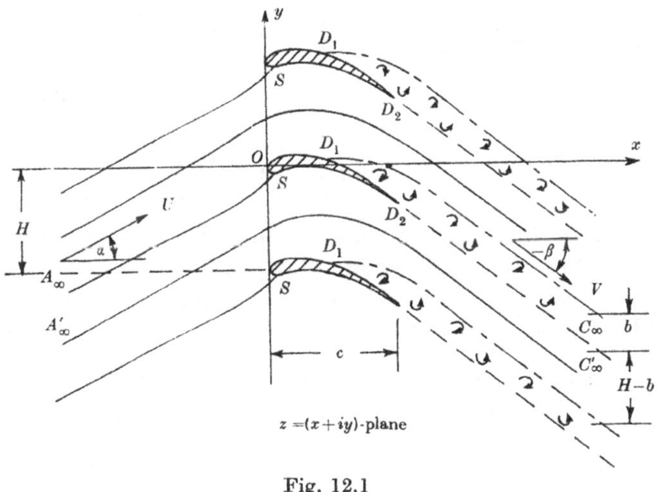

Fig. 12.1

Figs. 12.12b, c and d). In the limit as the gap : chord ratio, H/c, tends to infinity the theory of this chapter yields that of ch. 8 on single aerofoils as a special case. In this section we shall obtain some general results, based on the continuity and momentum equations, for the separating flow past an aerofoil cascade.

Consider the cascade shown in Fig. 12.1. The aerofoils are arranged along the Oy-axis at intervals of H, so that the flow conditions at points $(x, y+nH)$, $n = 0, \pm 1, \pm 2, \ldots$, are identical. Let the velocity vector (q, θ) be (U, α) upstream at infinity, and (V, β) downstream at infinity then α, β are termed the *inlet* and *outlet* angles respectively.

The flow is assumed to separate from the aerofoils at corresponding points D on their upper surfaces, with the result that a relatively large wake of slowly moving, turbulent fluid extends behind each aerofoil. We shall represent these wakes just as in ch. 11, i.e. by regions of stationary fluid, separated from the main stream by free streamlines along which, in general, the pressure can vary. Let the wake displacement thickness equal b downstream at infinity.

Consider the flow through the 'channel' bounded by the stagnation streamlines of two adjacent aerofoils. This channel has an upstream width of $H \cos \alpha$, and a downstream width of $(H - b) \cos \beta$, therefore the equation of continuity of mass yields

$$\rho_a U H \cos \alpha = \rho_\infty V (H - b) \cos \beta, \tag{1}$$

where ρ_a is the density at the reference velocity U, and ρ_∞ is the density at $\phi = \infty$. An expression for the ratio ρ_∞ / ρ_a can be obtained from (6.3), (6.4) and (6.38):

$$\left(\frac{\rho_\infty}{\rho_a}\right)^2 = \frac{1}{\beta_a^2} \tanh^2 (\Omega_\infty + \Omega^*) = \frac{1}{\beta_a^2} \{1 - \operatorname{sech}^2 (\Omega_\infty + \Omega^*)\}$$

$$= \frac{1}{\beta_a^2} \left\{1 - M_a^2 \left(\frac{\rho_\infty V}{\rho_a U}\right)^2\right\},$$

and hence by (1)

$$\frac{\rho_\infty}{\rho_a} = \frac{1}{\beta_a} \left\{1 - M_a^2 \left(\frac{H}{H - b}\right)^2 \left(\frac{\cos \alpha}{\cos \beta}\right)^2\right\}^{\frac{1}{2}}. \tag{2}$$

Let D, L be the Ox, Oy components of the force acting on an aerofoil of the cascade, then the Ox, Oy components of the momentum equation (see §1.13), when applied to the flow through a channel of width H bounded by similar streamlines, yield

$$D = \rho_a H U^2 \cos^2 \alpha + p_a H - \rho_\infty (H - b) V^2 \cos^2 \beta - p_\infty H,$$

and

$$L = \rho_a H U^2 \sin \alpha \cos \alpha - \rho_\infty (H - b) V^2 \sin \beta \cos \beta,$$

as the pressure contributions from the similar streamlines cancel out exactly.

From (6.3), (6.4), (6.8) and (2) we have

$$H(p_\infty - p_a) = \frac{\rho_a H U^2}{1 + \rho_a \coth \Omega_\infty} = \rho_a H U^2 \frac{\rho_a}{\rho_\infty} \left(\frac{\rho_\infty / \rho_a - 1}{M_a^2}\right)$$

$$= \rho_a H U^2 \frac{1}{\beta_a^2} \left(\frac{\rho_a / \rho_\infty}{\rho_\infty / \rho_a + 1}\right) \left\{1 - \left(\frac{H}{H - b}\right)^2 \left(\frac{\cos \alpha}{\cos \beta}\right)^2\right\},$$

and substituting this and (2) into the equations for D we get

$$D = \frac{\rho_a H U^2 \cos^2 \alpha}{\beta_a^2} \left(\frac{\rho_a / \rho_\infty}{\rho_\infty / \rho_a + 1}\right) \left[\left(\frac{H}{H - b}\right)^2 \left(\tan^2 \beta - M_a^2 \frac{\cos^2 \alpha}{\cos^2 \beta}\right) - \tan^2 \alpha \right.$$

$$\left. + \left(\frac{b}{H - b}\right)^2 + \left(\frac{b}{H - b}\right) \left(2 - \beta_a^2 \left\{1 + \frac{\rho_\infty}{\rho_a}\right\}\right) + M_a^2\right]. \tag{3}$$

Equation (1) enables us to write the equation for L in the form

$$L = \rho_a H U^2 \cos^2 \alpha \left\{ \tan \alpha - \frac{\rho_a}{\rho_\infty} \left(\frac{H}{H-b} \right) \tan \beta \right\}. \tag{4}$$

When $M_a = 0$, and $b = 0$ these reduce to

$$D = \tfrac{1}{2} \rho_a H U^2 \cos^2 \alpha \, \{ \tan^2 \beta - \tan^2 \alpha \}, \tag{5}$$

$$L = \rho_a H U^2 \cos^2 \alpha \, \{ \tan \alpha - \tan \beta \}, \tag{6}$$

which are the classical results for non-separating, incompressible flow (see Robinson & Laurmann, 1956, p. 149).

The circulation about an aerofoil of the cascade is easily calculated by taking a contour comprised of two similar streamlines distance H apart in the Oy-direction plus the two lines parallel to Oy joining these stream-lines at $\phi = -\infty$ and $\phi = \infty$. The contributions to the circulation integral from the similar streamlines cancel out, and there is left

$$\Gamma = HU \sin \alpha - (H-b)\, V \sin \beta - b \times 0 \times \sin \beta,$$

where here Γ is the *clockwise* circulation about an aerofoil. Combining this with (1) we have

$$\Gamma = HU \cos \alpha \left\{ \tan \alpha - \frac{\rho_a}{\rho_\infty} \tan \beta \right\}. \tag{7}$$

STEADY DIRICHLET FLOW

12.2   The conformal transformations

In Fig. 12.2a there are shown the z- and w-planes for Dirichlet flow past a cascade of 'stagger' angle α. Let h be the distance in the w-plane between corresponding points on adjacent aerofoils then integration of the second of (6.7) upstream at infinity gives

$$h = \beta_a U H, \tag{8}$$

after cancelling $\cos \alpha$ from each side of the equation. Notice that there is a discontinuity of Γ in ϕ about any closed circuit encircling an aerofoil (cf. Fig. 8.2), and that the angle of stagger α appears in the w-plane as the angle between $\psi = $ constant and lines joining corresponding points in the periodic flow pattern. Our first problem is to map the rather complicated w-plane shown in the figure on to a simple ζ-plane in which the solution of the boundary value problem can be written down.

The following method of deducing a suitable transformation is of some intrinsic interest. We first consider the particular case of the *incompressible* flow through the cascade for which both α and β are zero. (This is in effect a particular conformal mapping of the z-plane.) We shall distinguish the w-plane for this flow by the subscript 0. The z- and w_0-planes are shown in Fig. 12.2b. It follows from (7) that there is no circulation, and therefore the slits representing the aerofoils are closed. Equation (8) is still applicable, and h is now the jump in ψ_0 from one aerofoil to its neighbour. Let the slits be of length k, and take the origin of the w_0-plane to be at the leading edge of one of these slits. It is easily verified that the slits are mapped into a single slit of length $1 - \exp(-2\pi k/h)$ in the t-plane (not

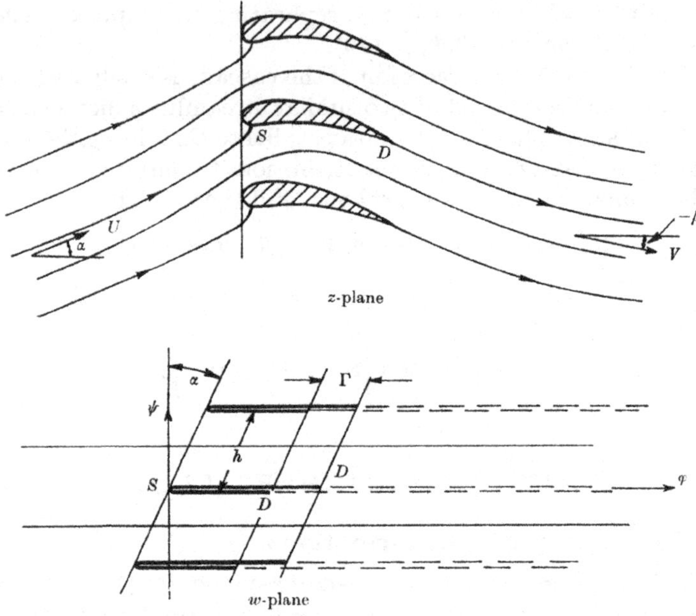

Fig. 12.2a

shown) by $t = 1 - \exp(-2\pi w_0/h)$, and in turn this is mapped into the ζ-plane shown in Fig. 12.2b by (cf. (5·91)) $t = \tfrac{1}{2}\{1 - \exp(-2\pi k/h)\}(1 - \cos\zeta)$. Hence

$$w_0 = \tfrac{1}{2}k - \frac{h}{2\pi}\ln\{\cosh r + \sinh r \cos \zeta\}, \tag{9}$$

where

$$r \equiv \frac{k\pi}{h}. \tag{10}$$

The point $\phi_0 = -\infty$ is mapped on to $\eta = \infty$, while the points $\phi_0 = \infty$, $\psi_0 > 0$ and $\phi_0 = \infty$, $\psi_0 < 0$ are mapped on to $\gamma = \pi$, $\eta = \mu$, and $\gamma = -\pi$, $\eta = \mu$, where

$$\cosh \mu = \coth r. \tag{11}$$

Note also that—focusing attention on the single aerofoil on $\psi_0 = 0$—the lines $\gamma = \pm\pi, 0 < \eta < \mu$ are opposite sides of the trailing edge streamline, $\psi_0 = 0, k \leqslant \phi_0 < \infty$, and that the lines $\gamma = \pm\pi, \mu < \eta < \infty$ are the corresponding streamlines $\psi_0 = \pm\tfrac{1}{2}h$.

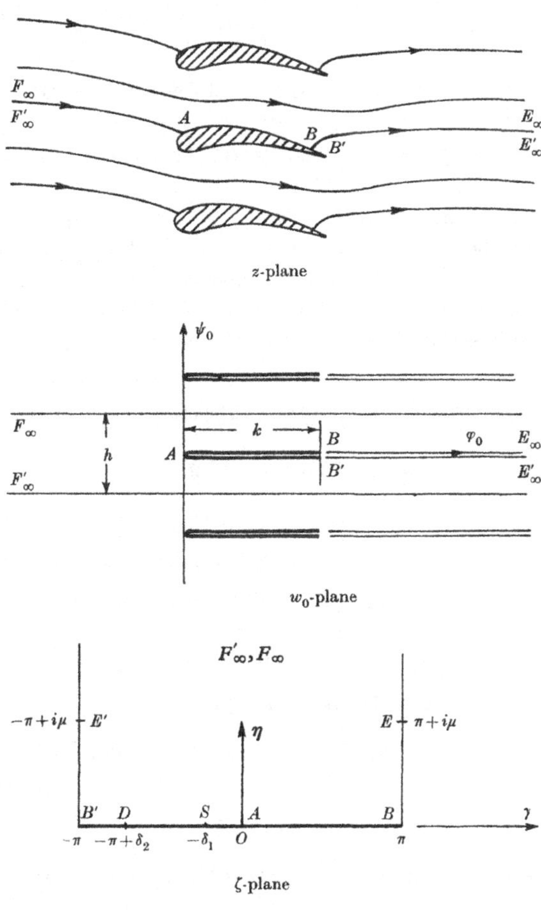

Fig. 12.2b

Now let the flow shown in Fig. 12.2a be an incompressible flow, then both w and w_0 will be analytic functions of z. Consequently w will be an analytic function of w_0 and by (9) of ζ. So in this case both

$$\tau \equiv \ln\frac{U\,dz}{dw}, \quad \tau_0 \equiv \ln\frac{U\,dz}{dw_0}, \tag{12}$$

are analytic functions of ζ. Furthermore, as the points (π, η), $(-\pi, \eta)$ represent either the same point of the z-plane (in $0 \leqslant \eta < \mu$) or corresponding points in the z-plane, i.e. points separated by a distance H in the Oy-direction, it is clear that the periodic relations $\tau(\pi, \eta) = \tau(-\pi, \eta)$ and $\tau_0(\pi, \eta) = \tau_0(-\pi, \eta)$ hold. Also, as upstream at infinity $q = U$,

$$\lim_{\eta \to \infty} \tau(\gamma, \eta) = i\alpha, \quad \lim_{\eta \to \infty} \tau_0(\gamma, \eta) = 0. \tag{13}$$

We are now able to apply the theory of §4.3 and write

$$\tau(\zeta) = \frac{1}{2\pi} \int_{-\pi}^{\pi} \theta_s(\gamma) \cot \tfrac{1}{2}(\gamma - \zeta) \, d\gamma, \tag{14}$$

with a similar equation for $\tau_0(\zeta)$, except that $\theta_s(\gamma)$ is replaced by $\theta_0(\gamma)$, the aerofoil slope for the non-circulating flow. A glance at Figs. 12.2a and 12.2b shows that θ_s and θ_0 differ only because of the different positions of the stagnation points in the two flows. Suppose the stagnation points S and D of the circulating flow map on to $\gamma = -\delta_1$ and $\gamma = -\pi + \delta_2$ respectively, then

$$\theta_s(\gamma) = \theta_0(\gamma) + \pi\{\mathrm{U}(\gamma - \delta_1) - \mathrm{U}(\gamma)\} - \pi\{\mathrm{U}(\gamma + \pi) - \mathrm{U}(\gamma + \pi - \delta_2)\}, \tag{15}$$

where $\mathrm{U}(\gamma)$ denotes the unit function (see §3.6 and (8.50)). Substituting (15) into (14) we find

$$\tau(\zeta) = \tau_0(\zeta) + \ln\left\{ \frac{\sin \tfrac{1}{2}\zeta \cos \tfrac{1}{2}\zeta}{\sin \tfrac{1}{2}(\delta_1 + \zeta) \cos \tfrac{1}{2}(\delta_2 - \zeta)} \right\}. \tag{16}$$

If the limits in (13) are employed in this expression there results

$$\alpha = \tfrac{1}{2}(\delta_1 - \delta_2). \tag{17}$$

It now follows from (12), (16) and (17) that

$$\frac{dw}{dw_0} = \frac{\sin \lambda + \cos \alpha \sin \zeta + \sin \alpha \cos \zeta}{\sin \zeta}, \tag{18}$$

where
$$\lambda = \tfrac{1}{2}(\delta_1 + \delta_2). \tag{19}$$

The next step is to differentiate (9) and combine the result with (18) to obtain

$$\frac{dw}{d\zeta} = \frac{h}{2\pi}\left\{ \frac{\sin \lambda + \cos \alpha \sin \zeta + \sin \alpha \cos \zeta}{\coth r + \cos \zeta} \right\}. \tag{20}$$

Integration gives

$$w = \frac{h}{2\pi}\left\{ \zeta \sin \alpha + 2(\sin \lambda \sinh r - \sin \alpha \cosh r) \tan^{-1}(e^{-r} \tan \tfrac{1}{2}\zeta) \right.$$
$$\left. - \cos \alpha \ln\left(\frac{1 + \tanh r \cos \zeta}{1 + \tanh r} \right) \right\}, \tag{21}$$

where the constant of integration has been chosen so that the origins of the w- and ζ-planes correspond.

The circulation can be calculated from (21) by keeping $\eta < \mu$, and determining the change in ϕ as γ varies from $-\pi$ to π. The result is

$$\Gamma = h\{\sin\lambda\sinh r - \sin\alpha\cosh r + \sin\alpha\}. \tag{22}$$

As a check on the theory note that when $\eta > \mu$ the change in w as γ varies from $-\pi$ to π is given by (21) to be $h(\sin\alpha + i\cos\alpha)$, and this is in agreement with Fig. 12.2a. Equation (22) enables us to eliminate the unknown quantity λ in favour of the circulation Γ and (21) can be replaced by

$$w = \frac{h}{2\pi}\zeta\sin\alpha + \frac{1}{\pi}(\Gamma - h\sin\alpha)\tan^{-1}(e^{-r}\tan\tfrac{1}{2}\zeta)$$
$$-\frac{h}{2\pi}\cos\alpha\ln\left(\frac{1+\tanh r\cos\zeta}{1+\tanh r}\right). \tag{23}$$

An alternative approach to this transformation would have been just to write (23) down (without any indication of its derivation) and to have asked the reader to verify that it does indeed map the w-plane of Fig.12.2a into the ζ-plane of Fig. 12.2b. This implies that (23) is really applicable to both compressible and incompressible flow; it can be regarded as being a conformal mapping of the w-plane into a convenient auxiliary ζ-plane (see §6.14), a mapping which has no connexion with the nature of the flow.

In compressible flow (12) will not hold, and w_0 will *not* be the complex stream function for the non-circulating flow past a cascade of zero stagger. Thus the number k will only be an approximation to the length of the slit in the non-circulating flow. However, this does not matter for the important number in cascade theory is c/H, the physical chord:gap ratio. The situation is similar to that described towards the end of §7.6. To apply the theory we first need an estimate of the ratio $r = k\pi/h$. Linear perturbation theory and (8) gives the approximation

$$r = \frac{\pi k}{h} \simeq \frac{\pi(cU)}{\beta_a H U} = \frac{\pi c}{\beta_a H}. \tag{24}$$

With this, or a better estimate, we calculate the surface velocities by the method described in the next section. Then integration of $ds = d\phi/q$ about the aerofoil surface enables us to determine the actual value of c/H achieved. If this is too small (big) r is proportionately increased (reduced) and the whole calculation repeated. In this way the relation between r and c/H can be established.

One further transformation proves to be very useful in the theory. This is

$$\sigma = -e^{-i\zeta}, \tag{25}$$

which maps the semi-infinite strip of the ζ-plane into the region $|\sigma| \geqslant 1$ shown in Fig. 12.2c. A similar transformation has already proved useful in aerofoil theory (see § 8.3), and it is introduced here for the same reasons. Its value lies in the fact that closed circuits in the z-plane about aerofoils of the cascade map into closed circuits about the unit circle in the ζ-plane. Aerofoils of the cascade are mapped on to $|\sigma| = 1$, and the point upstream at infinity is mapped on to $\sigma = \infty$. The point downstream at infinity $\zeta = \pm\pi + i\mu$ maps on to the point

$$\sigma = \sigma_1 = e^\mu = \coth \tfrac{1}{2}r, \tag{26}$$

by (11).

By using (22), (25) and (26) to eliminate λ, ζ and r from (20) we find

$$\frac{dw}{d\sigma} = \frac{h}{2\pi}\left\{\frac{e^{i\alpha}}{\sigma} - \frac{e^{i\alpha} - i\Gamma/h}{\sigma - 1/\sigma_1} - \frac{e^{-i\alpha} + i\Gamma/h}{\sigma - \sigma_1}\right\}, \tag{27}$$

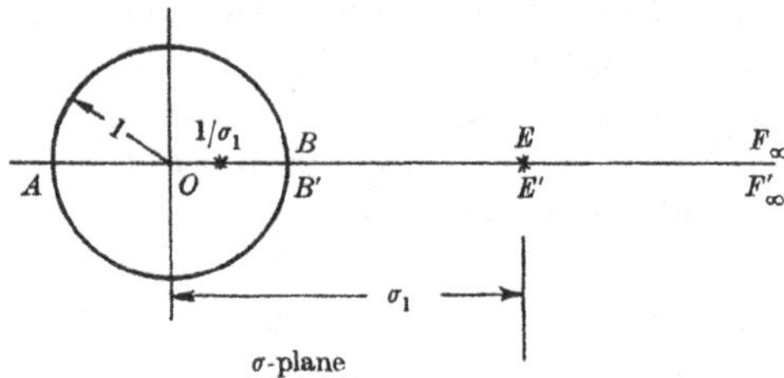

Fig. 12.2c

which shows that $dw/d\sigma$ has a singularity at the point $\sigma = \sigma_1$ outside the unit circle. Comparison of the form of (27) with the results given in § 5.10 reveals that if the σ-plane is regarded as being a real physical plane, then at $\sigma = \sigma_1$ there is superimposed a sink of strength $(h/2\pi)\cos\alpha$ and a vortex of strength $-(h/2\pi)\sin\alpha + \Gamma/2\pi$. While this physical interpretation of the transformation does not help us find the solution to the problem, in view of the closing remarks of ch. 6 it should warn us that closure difficulties can be expected due to the nature of the singularity at $\sigma = \sigma_1$. We find later (§ 12.5) that these difficulties do occur in tangent gas theory, and that as a consequence the theory of this chapter is valid only in the particular cases of (i) incompressible flow, and (ii) compressible flow provided either the stagger angle α is zero, or the angles α and β are small.

12.3 General solution

Now that the boundary value problem has been transformed from the w- to the ζ-plane it only remains to write down the appropriate general solution for the function $\tau(\zeta)$. As described in the paragraph preceding (14) the boundary conditions are periodic in the ζ-plane, and it remains only to prescribe the conditions holding on $\eta = 0$, $-\pi < \gamma < \pi$.

For example, for a cascade of given shape, (14), viz.

$$\tau(\zeta) = \frac{1}{2\pi} \int_{-\pi}^{\pi} \theta_s(\gamma) \cot \tfrac{1}{2}(\gamma - \zeta)\, d\gamma \tag{28}$$

is the appropriate solution. In cascade design we will need the conjugate of (28), i.e.

$$\tau(\zeta) = -\frac{i}{2\pi} \int_{-\pi}^{\pi} \Omega_s(\gamma) \cot \tfrac{1}{2}(\gamma - \zeta)\, d\gamma + i\alpha. \tag{29}$$

More generally if we have a cascade of porous aerofoils, or aerofoils having mixed boundary conditions, the equation to use for $\tau(\zeta)$ is (8.198). Obviously a great deal of the theory given in ch. 8 for single aerofoils can be extended to cover cascades of aerofoils. We shall not undertake this work here, but just outline a few of the salient points.

12.4　The inlet and outlet conditions

From (23) the point $\phi = \infty$ is mapped on to $\gamma = \pm\pi$, $\eta = \mu$, where μ is defined by (11). Thus the outlet conditions, $(q, \theta) = (V, \beta)$, can be expressed

$$\lim_{\eta \to \mu} \tau(\pm\pi + i\eta) = \Omega_\infty + i\beta, \tag{30}$$

where (see (8.15))

$$\Omega_\infty = \sinh^{-1}\left(\frac{U}{V} \sinh \Omega^*\right) - \Omega^*. \tag{31}$$

Applying this limit to (28) we find

$$\Omega_\infty = -\frac{1}{2\pi} \int_{-\pi}^{\pi} \frac{\theta_s(\gamma) \sin\gamma\, d\gamma}{\cos\gamma + \cosh\mu}, \tag{32}$$

and

$$\beta = \frac{\sinh\mu}{2\pi} \int_{-\pi}^{\pi} \frac{\theta_s(\gamma)\, d\gamma}{\cos\gamma + \cosh\mu}. \tag{33}$$

In addition to these relations involving $\theta_s(\gamma)$ there is also

$$\alpha = \frac{1}{2\pi} \int_{-\pi}^{\pi} \theta_s(\gamma)\, d\gamma, \tag{34}$$

which follows from (13) and (28).

Similarly, from (29) we can deduce the harmonic conjugates of (32)–(34), namely

$$\Omega_\infty = \frac{\sinh\mu}{2\pi} \int_{-\pi}^{\pi} \frac{\Omega_s(\gamma)\, d\gamma}{\cos\gamma + \cosh\mu}, \tag{35}$$

$$\beta - \alpha = \frac{1}{2\pi} \int_{-\pi}^{\pi} \frac{\Omega_s(\gamma) \sin\gamma\, d\gamma}{\cos\gamma + \cosh\mu}, \tag{36}$$

and
$$0 = \frac{1}{2\pi} \int_{-\pi}^{\pi} \Omega_s(\gamma)\, d\gamma. \tag{37}$$

These equations play the same role in cascade design as do (8.43) and (8.46) in aerofoil design.

12.5   The closure conditions

From (6.42) it follows that an aerofoil of the cascade is closed provided

$$(1 + \beta_a)\int_C e^{\tau(\sigma)}\frac{dw}{d\sigma}\,d\sigma = (1 - \beta_a)\overline{\int_C e^{-\tau(\sigma)}\frac{dw}{d\sigma}\,d\sigma}, \tag{38}$$

where C is unit circle in the σ-plane. As each of the integrands is analytic this contour can be deformed provided no singularities of the integrands are crossed during the deformation. The only singularity outside $|\sigma| = 1$

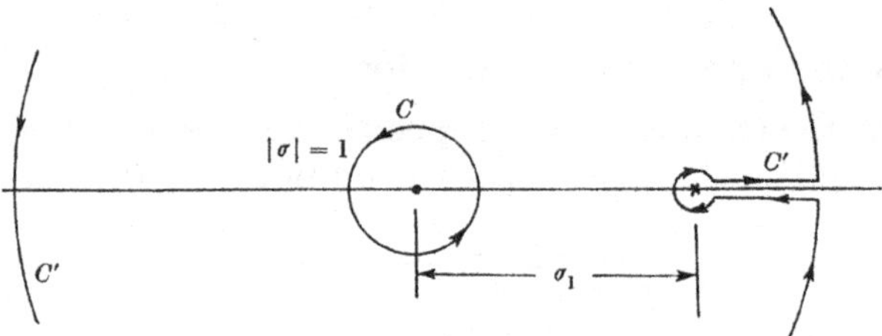

Fig. 12.5

is at $\sigma = \sigma_1$, and consequently we can deform C into the contour C' shown in Fig. 12.5. The integrands have the same value on the parallel lines of the contour, and so there is no contribution from this section; we are left with the contributions from the poles at $\sigma = \sigma_1$ and $\sigma = \infty$.

It follows from (13), (27) and (30) that the residues of $e^{\pm\tau}(dw/d\sigma)$ at $\sigma = \infty$ and $\sigma = \sigma_1$ are (see §§ 3.2, 3.3)

$$\frac{h}{2\pi}e^{\pm i\alpha - i\alpha} \quad \text{and} \quad -\frac{h}{2\pi}e^{\pm(\Omega_\infty + i\beta)}\left(e^{-i\alpha} + i\frac{\Gamma}{h}\right).$$

Hence (38) reduces to

$$(1 + \beta_a)\,ih\left\{e^{\Omega_\infty + i\beta}\left(e^{-i\alpha} + i\frac{\Gamma}{h}\right) - 1\right\}$$

$$= (1 - \beta_a)\,(-i)\,h\left\{e^{-\Omega_\infty + i\beta}\left(e^{i\alpha} - i\frac{\Gamma}{h}\right) - e^{2i\alpha}\right\}.$$

By using (6.4) to eliminate β_a, and rearranging this equation, we are able to write it as

$$\cosh(\Omega_\infty + \Omega^* - i\alpha) + i\frac{\Gamma}{h}\sinh(\Omega_\infty + \Omega^*) = e^{-i\beta}\cosh\Omega^* + i\,e^{-\Omega^*}\sin\alpha\,e^{i(\alpha - \beta)}. \tag{39}$$

We shall now show that this is not in agreement with the continuity equation (1) and (7) for Γ.

From (6.38)

$$\rho_\infty V = \rho_a U \frac{\cosh \Omega^*}{\cosh (\Omega_\infty + \Omega^*)},$$

and combining this with (1) (with $b = 0$) we obtain

$$\cosh (\Omega_\infty + \Omega^*) \cos \alpha = \cosh \Omega^* \cos \beta. \tag{40}$$

Again from (6.3), (6.4) and (8)

$$\Gamma = HU \sin \alpha - HV \sin \beta = \frac{h}{\beta_a} \left(\sin \alpha - \frac{V}{U} \sin \beta \right)$$

$$= h \coth \Omega^* \left(\sin \alpha - \frac{\sinh \Omega^*}{\sinh (\Omega_\infty + \Omega^*)} \sin \beta \right),$$

which can be combined with (40) to yield

$$\cosh (\Omega_\infty + \Omega^* - i\alpha) + i\frac{\Gamma}{h} \sinh (\Omega_\infty + \Omega^*) = e^{-i\beta} \cosh \Omega^*$$

$$+ i\, e^{-\Omega^*} \sin \alpha \frac{\sinh (\Omega_\infty + \Omega^*)}{\sinh \Omega^*}. \tag{41}$$

Comparison of (39) and (41) reveals that the closure conditions are satisfied exactly only if one of the following conditions applies: (i) $\Omega^* = \infty$, i.e. the flow is incompressible, (ii) $\alpha = 0$, i.e. no stagger, and (iii) if second-order terms in α and β can be ignored. (See comments at end of § 12.2.)

Note that in incompressible flow (39) and (41) both reduce to

$$e^{-(\Omega_\infty + i\beta)} = e^{-i\alpha} + i\frac{\Gamma}{h}. \tag{42}$$

12.6   The direct and indirect problems of cascade theory

First consider the direct problem: the theory of the flow through a cascade of given shape is quite similar to the aerofoil theory developed in §§ 8.3 and 8.23. We start with the form (28) takes on the cascade surface, viz.

$$\Omega(\gamma^*) = \frac{1}{2\pi} \int_{-\pi}^{\pi} \theta_s(\gamma) \cot \tfrac{1}{2}(\gamma - \gamma^*)\, d\gamma, \tag{43}$$

and then by an integration by parts transform it into an integral equation of a form similar to (8.16). This is solved by a method similar to that outlined in § 8.23.

Secondly the design (indirect) problem of cascade theory is based on

$$\theta(\gamma^*) = -\frac{1}{2\pi}\int_{-\pi}^{\pi} \Omega_s(\gamma)\cot\tfrac{1}{2}(\gamma-\gamma^*)\,d\gamma + \alpha, \tag{44}$$

from (29), plus the three restrictions contained in (35)–(37). For the details of this method, which is similar to that described in §8.6 for aerofoils, the reader can consult Rosenblat & Woods (1956).

It should be noted that the transformations introduced in §12.2 are by no means the only ones that are of value in cascade theory. An alternative method is first to reduce the set of aerofoils to a single closed region in a z_1-plane by a transformation of the form $z_1 = \exp(-2\pi z/H)$. The point downstream at infinity becomes a sink at $z_1 = 1$, and we now have the problem of finding the flow induced over a closed surface by a sink placed some distance from it. Lighthill (1945a) used this method in his work on cascade design. Another method, which was used by Garrick (1944) in the direct problem, is to replace w_0 in (9) by z. This maps the aerofoils into a single curve in the ζ-plane (see Fig. 8.22a), and we have the generalization of Theodorsen's method to a cascade of aerofoils. However, these methods are applicable only to incompressible flow or to linear perturbation flow, for only in these cases is the basic differential equation invariant to conformal transformations of the z-plane.

Finally we observe that most of the results of thin aerofoil theory given in §§8.14–8.21 can be immediately extended to cascade theory. The principal approximation involved is to replace (23) by

$$Z = -\frac{H\beta_a}{2\pi}\ln\left\{\frac{1+\tanh(\pi c/\beta_a H)\cos\zeta}{1+\tanh(\pi c/\beta_a H)}\right\} \quad (Z = x+i\beta_a y), \tag{45}$$

which follows from (8), (9), (24) and the approximation $w_0 = UZ$ (see §2.17). For another approach to thin 'cascade' theory, which avoids the usual leading edge singularity of linear perturbation theory, the reader can consult a recent paper by Fanti, Kemp & Nilson (1958).

12.7 The force and moment acting on a cascade blade

When the integral equation described above has been solved, and the function $\theta_s(\gamma)$ has consequently been determined, the lift and drag acting on a blade (i.e. aerofoil) of a cascade of given shape and inlet conditions can be calculated immediately from (3), (4), (32) and (33). (Equation (34) is used to fix the position of the front stagnation point during the solution of the integral equation.) In the next section this method is applied to the very simple case of a cascade of flat plates.

If α and β are small, then it follows from $b = 0$, (2), (3) and (4) that

$$C_L \equiv \frac{L}{\tfrac{1}{2}\rho_a c U^2} = \frac{2H}{c}(\alpha-\beta), \tag{46}$$

and

$$C_D \equiv \frac{D}{\frac{1}{2}\rho_a c U^2} = \frac{H}{\beta_a^2 c}(\beta^2 - \alpha^2). \tag{47}$$

Thus from (33) and (34)

$$C_L = \frac{H}{\pi c}\int_{-\pi}^{\pi} \theta_s(\gamma)\left\{1 - \frac{\sinh\mu}{\cos\gamma + \cosh\mu}\right\} d\gamma, \tag{48}$$

where μ is given by (11). If the blades are thin (24) is applicable, and

$$\sinh\mu = \operatorname{cosech}\left(\frac{\pi c}{\beta_a H}\right), \quad \cosh\mu = \coth\left(\frac{\pi c}{\beta_a H}\right). \tag{49}$$

An expression for the moment about the leading edge of a blade will now be obtained. For compressible flow the method given in § 8.11 for a single aerofoil could be generalized were it not for the fact established in § 12.5 that the cascade blade does not close exactly. This causes difficulty with the method of § 8.11, which can be observed from § 8.5 to depend on the closure conditions. We shall therefore restrict the following theory to a linear perturbation flow, in which case (45) and (49) give

$$x = -\frac{H\beta_a}{2\pi}\ln\left(\frac{\cosh\mu + \cos\gamma}{\cosh\mu + 1}\right), \tag{50}$$

for the distance measured along the blade from the leading edge.

With thin blades the nose-up moment about the leading edge is

$$M = \int_{-\pi}^{\pi}(p - p_a)\,x\,dx(\gamma) = \tfrac{1}{2}\rho_a U^2\int_{-\pi}^{\pi} C_p\,x\left(\frac{dx}{d\gamma}\right)d\gamma. \tag{51}$$

In the present case (see § 6.1) $C_p \simeq 2\Omega/\beta_a$, thus (43) and (50) enable (51) to be written

$$M = -\frac{1}{2\pi^2}\rho_a\beta_a H^2 U^2\int_{-\pi}^{\pi}\theta_s(\gamma)\,J(\mu,\gamma)\,d\gamma,$$

where

$$J(\mu,\gamma) \equiv \frac{1}{4\pi}\int_{-\pi}^{\pi}\cot\tfrac{1}{2}(\gamma - \gamma^*)\frac{\sin\gamma^*}{\cosh\mu + \cos\gamma^*}\ln\left(\frac{\cosh\mu + \cos\gamma^*}{\cosh\mu + 1}\right)d\gamma^*.$$

This integral can be evaluated by replacing μ in the logarithmic term by ν, differentiating with respect to ν, then integrating with respect to γ^* and ν in turn; finally let ν tend to μ. The result is

$$J(\mu,\gamma) = \ln(1 + e^{-\mu}) + \frac{\sinh\mu\ln(1 - e^{-\mu})}{\cosh\mu + \cos\gamma} + \frac{\sin\gamma}{\cosh\mu + \cos\gamma}\tan^{-1}\left\{\frac{\sin\gamma}{e^\mu + \cos\gamma}\right\}. \tag{52}$$

Substituting this in the integral for M, and taking advantage of (33) and (34) we arrive at

$$C_M \equiv \frac{M}{\frac{1}{2}\rho_a c^2 U^2} = -\frac{2\beta_a}{\pi}\left(\frac{H}{c}\right)^2 \left\{\alpha \ln\left(1+e^{-\mu}\right) + \beta \ln\left(1-e^{-\mu}\right)\right.$$

$$\left. + \frac{1}{2\pi}\int_{-\pi}^{\pi} \frac{\theta_s(\gamma)\sin\gamma}{\cosh\mu+\cos\gamma}\tan^{-1}\left(\frac{\sin\gamma}{e^{\mu}+\cos\gamma}\right)d\gamma\right\}. \quad (53)$$

To check this rather complicated result we can take the limit $H \to \infty$, which should agree with the result given for a single aerofoil in §8.11. From (24) and (49) $H \simeq \pi c/r\beta_a \simeq (\pi c/2\beta_a)\,e^{\mu}$, therefore

$$-\frac{2\beta_a}{\pi}\left(\frac{H}{c}\right)^2 \left\{\alpha \ln\left(1+e^{-\mu}\right) + \beta \ln\left(1-e^{-\mu}\right)\right\}$$

$$\simeq -\frac{2\beta_a}{\pi}\left(\frac{H}{c}\right)^2 (\alpha-\beta)\,e^{-\mu} \to -\tfrac{1}{2}C_L,$$

by (46). Let $\alpha \to 0$, then from the results just obtained and (34) we find the limit form of (53) to be

$$C_M = -\tfrac{1}{2}C_L + \frac{1}{4\beta_a}\int_{-\pi}^{\pi} \theta_s(\gamma)\cos 2\gamma\,d\gamma. \quad (54)$$

This agrees with the linear perturbation form ($T = 1$, $\alpha_0 = 0$) of (8.93) in which we must put $x'-x_c = -\tfrac{1}{2}c$, as (54) gives the moment about the leading edge.

12.8 An example: a cascade of flat plates in incompressible flow

The simplest non-trivial example of the theory is the incompressible flow through a cascade of flat plates set at an angle, α_0 say, to the x-axis. In Fig. 12.8a we show the non-circulating, zero stagger ($\alpha = 0$) flow through the cascade (cf. Fig. 12.2b). It is obvious from the anti-symmetry of this flow that if $\gamma = -\pi+\delta_2$ at the trailing edge then $\gamma = \delta_2$ at the sharp leading edge. Application of (34) to this flow gives $\delta_2 = \alpha_0$.

Now consider a flow with an inlet angle of α and an outlet angle, β, such that the flow is smooth past the trailing edge (see Fig. 12.8b). The front stagnation point is now at $\gamma = -\delta_1$ and the flow direction is given by

$$\theta_s(\gamma) = -\alpha_0 + \pi\{\mathrm{U}(\gamma+\delta_1) - \mathrm{U}(\gamma-\alpha_0)\}. \quad (55)$$

Substitution into (33) and (34) gives

$$\delta_1 = 2\alpha + \alpha_0, \quad (56)$$

and $\quad \beta = -\alpha_0 + \tan^{-1}(\tan\tfrac{1}{2}\alpha_0 \tanh\tfrac{1}{2}\mu) + \tan^{-1}(\tan\tfrac{1}{2}\delta_1 \tanh\tfrac{1}{2}\mu). \quad (57)$

Equations (43) and (55) give

$$e^{-\Omega} = \frac{q}{U} = \left| \frac{\sin \tfrac{1}{2}(\delta_1 + \gamma)}{\sin \tfrac{1}{2}(\alpha_0 - \gamma)} \right|, \tag{58}$$

so $\quad s = \dfrac{1}{U} \displaystyle\int_{\alpha_0}^{\gamma} \dfrac{U}{q} \dfrac{d\phi}{d\gamma} d\gamma = \dfrac{H}{2\pi} \displaystyle\int_{\alpha_0}^{\gamma} \dfrac{\sin \tfrac{1}{2}(\gamma - \alpha_0)}{\sin \tfrac{1}{2}(\delta_1 + \gamma)} \left(\dfrac{\sin \lambda + \sin(\gamma + \alpha)}{\cosh \mu + \cos \gamma} \right) d\gamma,$

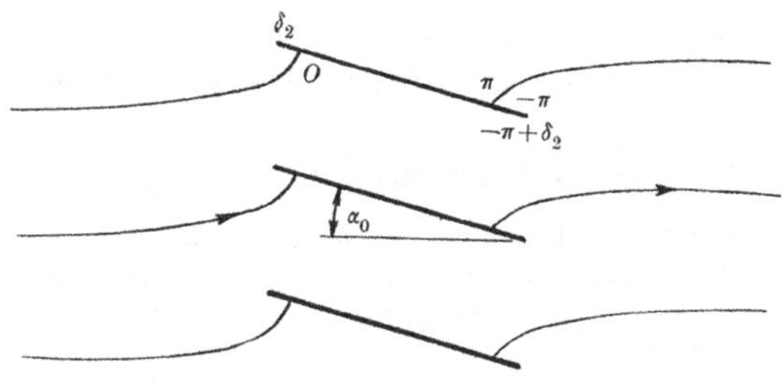

Fig. 12.8a

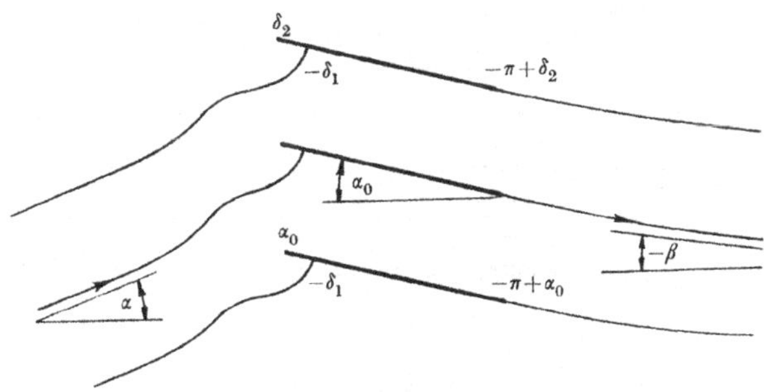

Fig. 12.8b

where s is the distance along the plate measured from the leading edge. Multiplying the integrand of this equation by $\cos \tfrac{1}{2}(\gamma - \alpha_0)/\cos \tfrac{1}{2}(\gamma - \alpha_0)$, and using (17), (19) and $\delta_2 = \alpha_0$, we get

$$s = \frac{H}{2\pi} \int_{\alpha_0}^{\gamma} \frac{\sin(\gamma - \alpha_0)}{\cosh \mu + \cos \gamma} d\gamma;$$

thus $\quad s = \dfrac{H}{2\pi} \Bigg[\cos \alpha_0 \ln \left(\dfrac{\cosh \mu + \cos \alpha_0}{\cosh \mu + \cos \gamma} \right) - (\gamma - \alpha_0) \sin \alpha_0$

$$+ 2 \sin \alpha_0 \coth \mu \, \{ \tan^{-1}(\tan \tfrac{1}{2}\gamma \tanh \tfrac{1}{2}\mu)$$

$$- \tan^{-1}(\tan \tfrac{1}{2}\alpha_0 \tanh \tfrac{1}{2}\mu) \} \Bigg]. \tag{59}$$

The chord length c is obtained by putting $\gamma = \pi + \alpha_0$ in this expression; the result is

$$\frac{c}{H} = \frac{1}{2\pi}\cos\alpha_0 \ln\left(\frac{\cosh\mu + \cos\alpha_0}{\cosh\mu - \cos\alpha_0}\right) + \tfrac{1}{2}\sin\alpha_0(\coth\mu - 1). \qquad (60)$$

Notice that this is independent of α. Given c/H and α_0, it serves to fix the value of μ. Equations (56) and (57) then determine the outlet angle β.

STEADY HELMHOLTZ FLOW

12.9   The conformal transformations

The w-plane for the separating flow of Fig. 12.1 is shown in Fig. 12.9, which is very similar to Fig. 12.2a. However the concept of circulation is not applicable with Helmholtz flow, and the upper and lower separation points D_1, D_2 do not correspond in the z-plane. Let $\phi = 0$ at the front

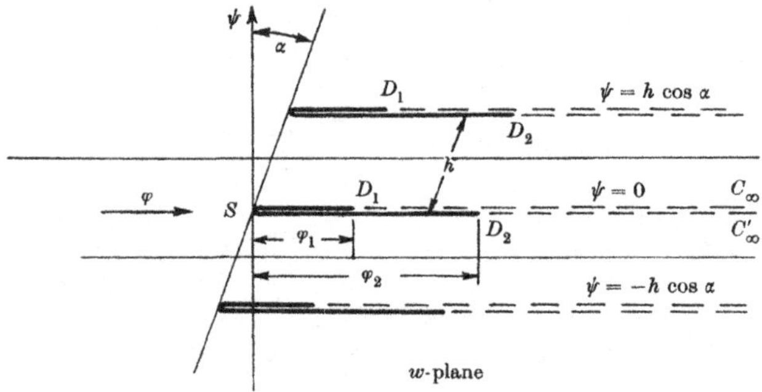

Fig. 12.9a

stagnation point S of the aerofoil lying on $\psi = 0$. (We shall refer only to this particular aerofoil in the sequel.) And let $\phi = \phi_1, \phi_2$ at the separation points D_1, D_2. The mixed boundary value problem we have to solve is θ given on the 'wetted' surface of the cascade of aerofoils, and Ω given on the free streamlines downstream of the separation points D_1 and D_2. Our first problem is to find a suitable ζ-plane. To do this we shall adopt a method similar to that employed in §12.2.

The w-plane of Fig. 12.9a can be transformed into the w_0-plane of Fig. 12.9b as follows. Suppose it is assumed that the slits of the w-plane are the solid walls of the cascade of channels shown in the z_0-plane of Fig. 12.9b. Let an incompressible flow be incident on these channels at such an angle that there is no stagger of the leading edges of the slits. This means that the flow upstream at infinity must be at right-angles to the line joining the leading edges of the slits, i.e. must make an angle $-\alpha$ with the real axis of the z_0-plane. If the velocity at infinity is unity then the slits in the w_0-plane will be a distance h apart.

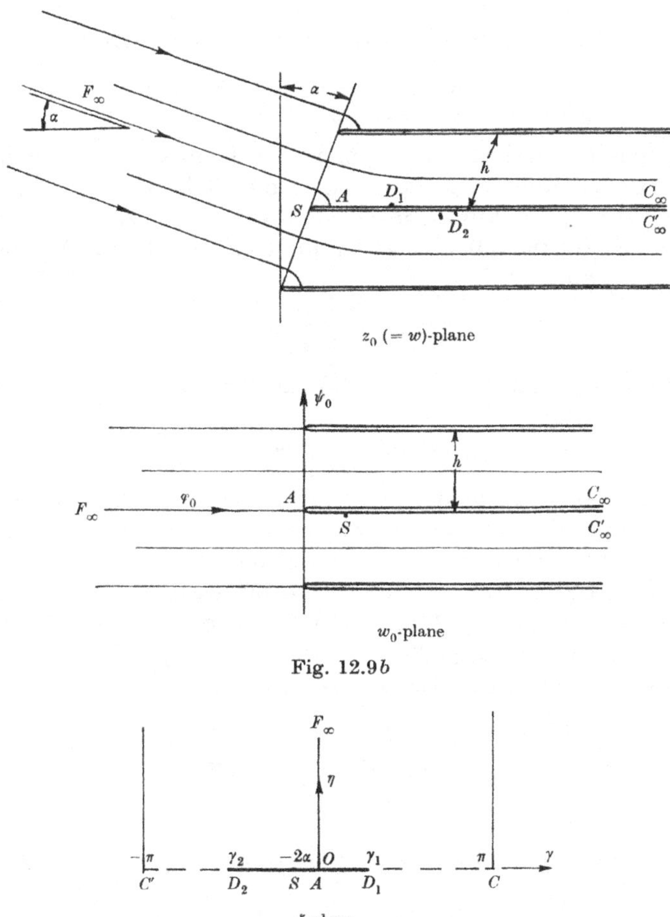

Fig. 12.9*b*

Fig. 12.9*c*

Next map the w_0-plane into the ζ-plane shown in Fig. 12.9*c*. In this the points C_∞, C'_∞ at $\phi_0 = \infty$ are mapped on to $\gamma = \pm\pi$, $\eta = 0$. A comparison with the corresponding transformation of Fig. 12.2*b* shows that the required mapping formula can be found by combining (9) and (10) and taking the limit $k \to \infty$. The result of this is

$$w_0 = -\frac{h}{2\pi}\ln\{\tfrac{1}{2}(1+\cos\zeta)\} = -\frac{h}{\pi}\ln(\cos\tfrac{1}{2}\zeta). \tag{61}$$

This maps the front stagnation point A of our incompressible, zero stagger flow on to $\gamma = 0$. The similar streamlines $\psi = \pm \frac{1}{2}h$ are mapped on to $\gamma = \pm \pi$, and consequently in the ζ-plane we have the usual periodic boundary conditions.

Hence the general solution for the incompressible flow is

$$\tau_0 \equiv \ln \frac{dz_0}{dw_0} = \frac{1}{2\pi} \int_{-\pi}^{\pi} \theta_0(\gamma) \cot \tfrac{1}{2}(\gamma - \zeta)\, d\gamma, \tag{62}$$

where $\theta_0(\gamma)$ refers to the boundary slope for the incompressible flow. Let the leading edge in the z_0-plane, i.e. the point S, be at $\gamma = -\delta_1$, then θ is zero everywhere except in the interval $-\delta_1 < \gamma < 0$, where it equals $-\pi$. This distribution in (62) plus the limit (see Fig. 12.9b) $\lim\limits_{\eta \to \infty} \tau_0 = -i\alpha$ yields $\delta_1 = 2\alpha$ and

$$\frac{dz_0}{dw_0} = \frac{\sin \tfrac{1}{2}(2\alpha - \zeta)}{\sin \tfrac{1}{2}\zeta}. \tag{63}$$

We can now eliminate w_0 from (61) and (63), and put $z_0 = w$ to find

$$w = \frac{h}{2\pi}(\zeta \sin \alpha - 2 \cos \alpha \ln \{\cos \tfrac{1}{2}\zeta\}), \tag{64}$$

which is the required transformation from the w- to the ζ-plane.

We can apply the same argument to (64) as we did to (23), viz. that it represents a transformation from the w-plane of Fig. 12.9a to the ζ-plane of Fig. 12.9c, and is applicable to both compressible and incompressible flow. To check the validity of this we need only to verify that the boundary $0 < \phi < \infty,\ \psi = 0$ is mapped on to $-\pi < \gamma < \pi,\ \eta = 0$.

It is apparent from Fig. 12.9a that

$$\tau(w) = \tau(w + nih\, e^{-i\alpha}), \quad n = \pm 1, \pm 2, \dots$$

but from (64)

$$w(\pi + i\eta) = w(-\pi + i\eta) + ih\, e^{-i\alpha},$$

and hence

$$\tau(\pi + i\eta) = \tau(-\pi + i\eta). \tag{65}$$

The separation points D_1, D_2 map on to $\eta = 0$, $\gamma = \gamma_1$, γ_2, where

$$\left.\begin{aligned}
\phi_2 &= \frac{h}{2\pi}\{\gamma_2 \sin \alpha - 2 \cos \alpha \ln (\cos \tfrac{1}{2}\gamma_2)\}, \\[2ex]
\phi_1 &= \frac{h}{2\pi}\{\gamma_1 \sin \alpha - 2 \cos \alpha \ln (\cos \tfrac{1}{2}\gamma_1)\},
\end{aligned}\right\} \tag{66}$$

and the upper and lower surfaces of the wake are mapped on to $\eta = 0,\ -\pi < \gamma < \gamma_2$, and $\eta = 0,\ \gamma_1 < \gamma < \pi$ respectively. The 'wetted' surface of the aerofoil—i.e. the surface over which θ_s is known—is mapped on to $\eta = 0,\ \gamma_2 < \gamma < \gamma_1$. The point downstream at infinity maps on to $\gamma = \pm \pi,\ \eta = 0$, while the point upstream at infinity ($\phi = -\infty$) maps on to the point $\eta = \infty$.

12.10 The boundary value problem and its solution

In the ζ-plane, in addition to (65), τ satisfies the following boundary conditions:

$$\left.\begin{aligned}\Omega = \Omega_s \quad (-\pi < \gamma < \gamma_2, \; \gamma_1 < \gamma < \pi; \quad \eta = 0),\\ \theta = \theta_s \quad (\gamma_2 < \gamma < \gamma_1; \quad \eta = 0),\end{aligned}\right\} \tag{67}$$

$$\lim_{\eta \to \infty} \tau(\gamma + i\eta) = i\alpha, \tag{68}$$

and

$$\lim_{\gamma \to \pm\pi} \tau(\gamma + i0) = \Omega_\infty + i\beta, \tag{69}$$

the last two of which follow from (13) and (30).

The solution to this mixed and periodic boundary value problem can be written down immediately as a special case of (8.198) for the flow past a porous aerofoil. It follows from (67) and the second paragraph of § 8.27 that in the present application $\epsilon = \frac{1}{2}$ in $-\pi < \gamma < \gamma_2$, $\epsilon = 0$ in $\gamma_2 < \gamma < \gamma_1$ and $\epsilon = -\frac{1}{2}$ in $\gamma_1 < \gamma < \pi$. With these values (8.198) gives

$$\begin{aligned}\tau(\zeta) = {} & \frac{\{\sin \frac{1}{2}(\zeta - \gamma_2)\sin\frac{1}{2}(\gamma_1 - \zeta)\}^{\frac{1}{2}}}{2\pi \cos\frac{1}{2}\zeta}\left[A + \int_{\gamma_2}^{\gamma_1} \theta_s(\gamma)\right.\\ & \times \frac{\cos\frac{1}{2}\gamma\cot\frac{1}{2}(\gamma - \zeta)\,d\gamma}{\{\sin\frac{1}{2}(\gamma - \gamma_2)\sin\frac{1}{2}(\gamma_1 - \gamma)\}^{\frac{1}{2}}} + \left(\int_{\gamma_1}^{\pi} - \int_{-\pi}^{\gamma_2}\right)\Omega_s(\gamma)\\ & \qquad\qquad \left. \times \frac{\cos\frac{1}{2}\gamma\cot\frac{1}{2}(\gamma - \zeta)\,d\gamma}{\{\sin\frac{1}{2}(\gamma - \gamma_2)\sin\frac{1}{2}(\gamma - \gamma_1)\}^{\frac{1}{2}}}\right], \end{aligned} \tag{70}$$

where A is a real constant.

The finite limit in (69) and the denominator term $\cos\frac{1}{2}\zeta$ in (70) require that the term in the square brackets in (70) vanishes at $\zeta = \pm\pi$. This enables us to calculate the constant A, then reduce (70) to

$$\begin{aligned}\tau(\zeta) = {} & \frac{1}{2\pi}\{\sin\frac{1}{2}(\zeta - \gamma_2)\sin\frac{1}{2}(\gamma_1 - \zeta)\}^{\frac{1}{2}}\left[\int_{\gamma_2}^{\gamma_1}\theta_s(\gamma)\frac{\operatorname{cosec}\frac{1}{2}(\gamma - \zeta)\,d\gamma}{\{\sin\frac{1}{2}(\gamma - \gamma_2)\sin\frac{1}{2}(\gamma_1 - \gamma)\}^{\frac{1}{2}}}\right.\\ & \left. + \left(\int_{\gamma_1}^{\pi} - \int_{-\pi}^{\gamma_2}\right)\Omega_s(\gamma)\frac{\operatorname{cosec}\frac{1}{2}(\gamma - \zeta)\,d\gamma}{\{\sin\frac{1}{2}(\gamma - \gamma_2)\sin\frac{1}{2}(\gamma - \gamma_1)\}^{\frac{1}{2}}}\right], \end{aligned} \tag{71}$$

and this is the required solution of our boundary value problem. Considerable care is required with the signs of the square roots in (71); in particular it should be noted that

$$\{\sin\tfrac{1}{2}(\gamma - \gamma_2)\sin\tfrac{1}{2}(\gamma - \gamma_1)\}^{\frac{1}{2}} = (-1)\{\sin\tfrac{1}{2}(\gamma_2 - \gamma)\sin\tfrac{1}{2}(\gamma_1 - \gamma)\}^{\frac{1}{2}}$$

and consequently this expression has a different sign in $(-\pi, \gamma_2)$ from that in (γ_1, π).

Equation (71) is a generalization of (11.8) for the separating flow past a single aerofoil. It can be converted into an integral equation and solved by a method similar to that of §11.12. Much of the theory of §11 can be extended to cascades, but the algebra is rather involved and we shall omit it.

If the Helmholtz–Kirchhoff hypothesis of constant wake pressure is adopted then $\Omega_s(\gamma) = \Omega_\infty$ and (71) can be simplified to

$$\tau(\zeta) = \Omega_\infty + \frac{1}{2\pi}\{\sin\tfrac{1}{2}(\zeta-\gamma_2)\sin\tfrac{1}{2}(\gamma_1-\zeta)\}^{\frac{1}{2}}\int_{\gamma_2}^{\gamma_1}\frac{\theta_s(\gamma)\operatorname{cosec}\tfrac{1}{2}(\gamma-\zeta)\,d\gamma}{\{\sin\tfrac{1}{2}(\gamma-\gamma_2)\sin\tfrac{1}{2}(\gamma_1-\gamma)\}^{\frac{1}{2}}}. \tag{72}$$

To evaluate the integral involved here we used the easily verified result:

$$-\frac{2}{\pi}\coth^{-1}\{F_0(\gamma,\zeta)\} = \frac{1}{2\pi}\{\sin\tfrac{1}{2}(\zeta-\gamma_2)\sin\tfrac{1}{2}(\gamma_1-\zeta)\}^{\frac{1}{2}}$$

$$\times\int\frac{\operatorname{cosec}\tfrac{1}{2}(\gamma-\zeta)\,d\gamma}{\{\sin\tfrac{1}{2}(\gamma-\gamma_2)\sin\tfrac{1}{2}(\gamma-\gamma_1)\}^{\frac{1}{2}}}, \tag{73}$$

where
$$F_0(\gamma,\zeta) \equiv \left\{\frac{\sin\tfrac{1}{2}(\gamma-\gamma_2)\sin\tfrac{1}{2}(\gamma_1-\zeta)}{\sin\tfrac{1}{2}(\gamma-\gamma_1)\sin\tfrac{1}{2}(\zeta-\gamma_2)}\right\}^{\frac{1}{2}}. \tag{74}$$

12.11 Inlet and outlet conditions

For algebraic simplicity we shall now restrict attention to Helmholtz flow, i.e. to the flow characterized by (72). The generalization of the following to (71) is trivial.

From (68), (69) and (72) it can be deduced that

$$\Omega_\infty = -\frac{1}{2\pi}\int_{\gamma_2}^{\gamma_1}\frac{\theta_s(\gamma)\sin\tfrac{1}{2}(\gamma-\lambda)\,d\gamma}{\{\sin\tfrac{1}{2}(\gamma-\gamma_2)\sin\tfrac{1}{2}(\gamma_1-\gamma)\}^{\frac{1}{2}}}, \tag{75}$$

$$\alpha = \frac{1}{2\pi}\int_{\gamma_2}^{\gamma_1}\frac{\theta_s(\gamma)\cos\tfrac{1}{2}(\gamma-\lambda)\,d\gamma}{\{\sin\tfrac{1}{2}(\gamma-\gamma_2)\sin\tfrac{1}{2}(\gamma_1-\gamma)\}^{\frac{1}{2}}}, \tag{76}$$

and
$$\beta = \frac{1}{2\pi}\{\cos\tfrac{1}{2}\gamma_2\cos\tfrac{1}{2}\gamma_1\}^{\frac{1}{2}}\int_{\gamma_2}^{\gamma_1}\frac{\theta_s(\gamma)\sec\tfrac{1}{2}\gamma\,d\gamma}{\{\sin\tfrac{1}{2}(\gamma-\gamma_2)\sin\tfrac{1}{2}(\gamma_1-\gamma)\}^{\frac{1}{2}}}, \tag{77}$$

where
$$\lambda \equiv \tfrac{1}{2}(\gamma_1+\gamma_2). \tag{78}$$

These results correspond to (32)–(34) for Dirichlet flow.

In the iterative solution of the integral equation for $\theta_s(\gamma)$ mentioned in §12.10 the numbers γ_2, γ_1 are calculated from (66), $s = \int^\phi d\phi/q$, and knowledge of the distances s_2, s_1 from the leading edge to the two separation points. Equation (76) is used to fix the position of the front stagnation point during this calculation. When this stage of the calculation is complete, the outlet conditions (V, β) can be calculated from (31), (75),

(77) and the known inlet conditions (U, α). Equations (1) and (2) fix the value of the wake width b, and finally the force acting on a blade is obtained from (3) and (4).

12.12   Example: a cascade of stalled flat plates

The simplest example of the theory of any interest occurs with a cascade of flat plates. Suppose the plates are at an angle of attack of $\alpha + \alpha_0$, so that α_0 is the angle between the plates and Ox (see Fig. 12.12a), then

$$\theta_s(\gamma) = -\alpha_0 + \pi\{\mathbf{U}(\gamma) - \mathbf{U}(\gamma - \delta)\},$$

where δ is the value assumed for γ at the sharp leading edge of the typical plate. Substitution of this value of $\theta_s(\gamma)$ into (72) gives

$$\tau(\zeta) = \Omega_\infty - i\alpha_0 - 2\{\coth^{-1} F_1(\delta, \zeta) - \coth^{-1} F_1(0, \zeta)\}, \tag{79}$$

where
$$F_1(\gamma, \zeta) \equiv \left\{ \frac{\sin\frac{1}{2}(\gamma - \gamma_2)\sin\frac{1}{2}(\gamma_1 - \zeta)}{\sin\frac{1}{2}(\gamma_1 - \gamma)\sin\frac{1}{2}(\zeta - \gamma_2)} \right\}^{\frac{1}{2}}. \tag{80}$$

The limits (68) and (69) give

$$\Omega_\infty - i\alpha = i\alpha_0 + 2i\cot^{-1}\left\{ e^{i(\gamma_1 - \gamma_2)/2} \frac{\sin\frac{1}{2}(\delta - \gamma_2)}{\sin\frac{1}{2}(\gamma_1 - \delta)} \right\}^{\frac{1}{2}}$$

$$- 2i\cot^{-1}\left\{ e^{i(\gamma_1 - \gamma_2)/2} \frac{\sin(-\frac{1}{2}\gamma_2)}{\sin\frac{1}{2}\gamma_1} \right\}^{\frac{1}{2}}, \tag{81}$$

and

$$\beta = -\alpha_0 - 2\cot^{-1}\left\{ \frac{\sin\frac{1}{2}(\delta - \gamma_2)\cos\frac{1}{2}\gamma_1}{\sin\frac{1}{2}(\gamma_1 - \delta)\cos\frac{1}{2}\gamma_2} \right\}^{\frac{1}{2}} + 2\cot^{-1}\left\{ \frac{\tan(-\frac{1}{2}\gamma_2)}{\tan\frac{1}{2}\gamma_1} \right\}^{\frac{1}{2}}. \tag{82}$$

When (U, α) and α_0 are given, these equations enable (V, β) and δ to be calculated.

With flat plates separation is very likely to occur at the sharp leading edge, $\gamma = \delta$. In this case (see Fig. 12.12a) $\gamma_1 = \delta$, and (79), (81) and (82) become

$$\tau(\zeta) = \Omega_\infty - i\alpha_0 + 2\coth^{-1}\left\{ \frac{\sin(-\frac{1}{2}\gamma_2)\sin\frac{1}{2}(\gamma_1 - \zeta)}{\sin\frac{1}{2}\gamma_1\sin\frac{1}{2}(\zeta - \gamma_2)} \right\}^{\frac{1}{2}}, \tag{83}$$

$$\Omega_\infty = -\coth^{-1}\left\{ \frac{\cos\frac{1}{4}(\gamma_1 + \gamma_2)}{\{\sin(-\frac{1}{2}\gamma_2)\sin\frac{1}{2}\gamma_1\}^{\frac{1}{2}}} \right\}, \tag{84}$$

$$\alpha = -\alpha_0 - \cot^{-1}\left\{ \frac{\sin\frac{1}{4}(\gamma_1 + \gamma_2)}{\{\sin(-\frac{1}{2}\gamma_2)\sin\frac{1}{2}\gamma_1\}^{\frac{1}{2}}} \right\}, \tag{85}$$

and
$$\beta = -\alpha_0 + 2\cot^{-1}\left\{ \frac{\tan(-\frac{1}{2}\gamma_2)}{\tan\frac{1}{2}\gamma_1} \right\}^{\frac{1}{2}}. \tag{86}$$

Several flow problems of practical interest can be calculated from these equations. The three examples shown in Figs. 12.12b, c and d are

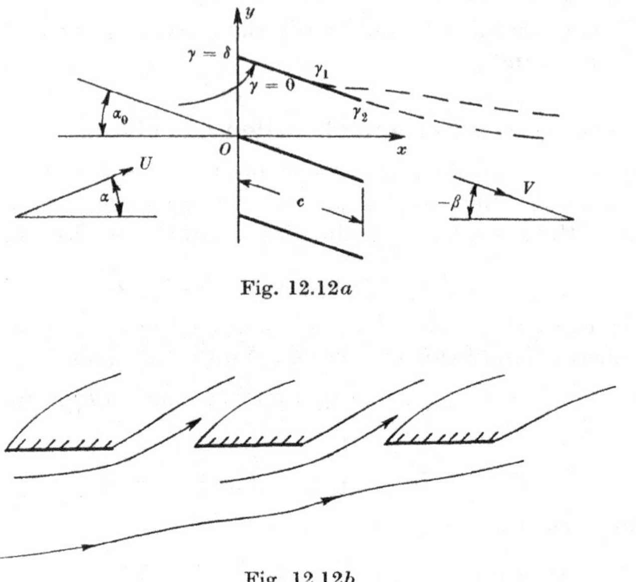

Fig. 12.12a

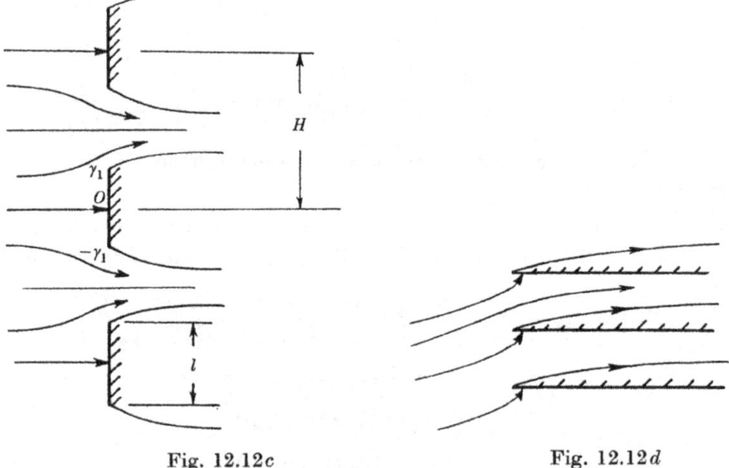

Fig. 12.12b

Fig. 12.12c Fig. 12.12d

the special cases $\alpha = -\frac{1}{2}\pi + \epsilon$, $\alpha_0 = \frac{1}{2}\pi$, ϵ small and positive (b), $\alpha = 0$, $\alpha_0 = \frac{1}{2}\pi$, $\gamma_1 = -\gamma_2$ (c), and $\alpha_0 = 0$, $\gamma_2 = -\pi$ (d). Thus the theory is applicable to (b) the flow past a slotted wall, (c) the flow through a grid of normal plates, and (d) the flow through a bank of tubes. We shall consider case (c) in more detail in the next section.

12.13　Incompressible flow through a grid of normal plates

For the flow shown in Fig. 12.12c $\alpha_0 = \tfrac{1}{2}\pi$, and by symmetry $\gamma_1 = -\gamma_2$. Equations (83)–(86) reduce to

$$\tau(\zeta) = \Omega_\infty - i(\tfrac{1}{2}\pi) + 2\coth^{-1}\left\{\frac{\sin\tfrac{1}{2}(\gamma_1-\zeta)}{\sin\tfrac{1}{2}(\gamma_1+\zeta)}\right\}^{\frac{1}{2}}, \tag{87}$$

$$\Omega_\infty = -\coth^{-1}(\operatorname{cosec}\tfrac{1}{2}\gamma_1) = \ln\left|\tan\tfrac{1}{4}(\pi-\gamma_1)\right|,$$

and $\alpha = \beta = 0$. As we are assuming the flow to be incompressible, $\Omega_\infty = \ln(U/V)$, whence

$$\frac{V}{U} = \cot\tfrac{1}{4}(\pi-\gamma_1). \tag{88}$$

At a point on the surface of the slotted wall ($\eta = 0$) (87) gives

$$\Omega(\gamma) = \ln\frac{U}{q} = \Omega_\infty + \ln\left|\frac{\{\sin\tfrac{1}{2}(\gamma_1+\gamma)\}^{\frac{1}{2}}+\{\sin\tfrac{1}{2}(\gamma_1-\gamma)\}^{\frac{1}{2}}}{\{\sin\tfrac{1}{2}(\gamma_1+\gamma)\}^{\frac{1}{2}}-\{\sin\tfrac{1}{2}(\gamma_1-\gamma)\}^{\frac{1}{2}}}\right|,$$

and so

$$q = V\left|\frac{\{\sin\tfrac{1}{2}(\gamma_1+\gamma)\}^{\frac{1}{2}}-\{\sin\tfrac{1}{2}(\gamma_1-\gamma)\}^{\frac{1}{2}}}{\{\sin\tfrac{1}{2}(\gamma_1+\gamma)\}^{\frac{1}{2}}+\{\sin\tfrac{1}{2}(\gamma_1-\gamma)\}^{\frac{1}{2}}}\right|. \tag{89}$$

The relation between the distance s measured from the centre line of one of the plates and the variable γ is found from

$$s = \frac{1}{V}\int_0^\gamma \frac{V}{q}\frac{d\phi}{d\gamma}\,d\gamma = \frac{H}{2\pi}\frac{U}{V}\int_0^\gamma \frac{1+y}{1-y}\tan\tfrac{1}{2}\gamma\,d\gamma,$$

where $d\phi/d\gamma$ is deduced from (8) ($\beta_a = 1$) and (64) ($\alpha = 0$), and where $y^2 \equiv \{\sin\tfrac{1}{2}(\gamma_1-\gamma)/\sin\tfrac{1}{2}(\gamma_1+\gamma)\}$. This substitution enables the integral to be evaluated without trouble.

Let l be the length of a plate of the grid (see figure), then

$$\tfrac{1}{2}l = \frac{H}{2\pi}\frac{U}{V}\int_0^{\gamma_1} \frac{1+y}{1-y}\tan\tfrac{1}{2}\gamma\,d\gamma.$$

So

$$\frac{l}{H} = \frac{1}{\pi}\frac{U}{V}\,8\sin^2\tfrac{1}{2}\gamma_1\int_0^1 \frac{(1+y)^2}{1+y^2}\frac{y\,dy}{1+2\cos\gamma_1 y^2+y^4}.$$

Evaluating this and eliminating U/V we find

$$\frac{l}{H} = \tan\tfrac{1}{4}(\pi-\gamma_1)\{\sec\tfrac{1}{2}\gamma_1 - 1 + \frac{\gamma_1}{\pi}\tan\tfrac{1}{2}\gamma_1\}. \tag{90}$$

Thus γ_1 can be computed once the geometry of the grid is known.

The drag force on a plate of the grid is

$$D = \frac{2}{V}\int_0^{\gamma_1} (p-p_\infty)\frac{V}{q}\frac{d\phi}{d\gamma}\,d\gamma,$$

where p_∞ is the pressure behind the plate, i.e. the pressure corresponding to the velocity V. Now from Bernoulli's theorem

$$\frac{V}{q}(p - p_\infty) = \tfrac{1}{2}\rho_a V^2\left(\frac{V}{q} - \frac{q}{V}\right) = \tfrac{1}{2}\rho_a V^2(e^\Omega - e^{-\Omega}),$$

and therefore $\qquad D = \dfrac{1}{\pi}\rho_a H U V \displaystyle\int_0^{\gamma_1} \sinh\Omega \tan\tfrac{1}{2}\gamma \, d\gamma.$

This can be evaluated by the same substitution as used above. The result is

$$D = \rho_a H U V(\sec\tfrac{1}{2}\gamma_1 - 1) = \rho_a H U^2 \cot\tfrac{1}{4}(\pi - \gamma_1)(\sec\tfrac{1}{2}\gamma_1 - 1). \tag{91}$$

We can verify that (91) correctly reduces to (11.63) in the limit $H \to \infty$ as follows. From (66)

$$4a = -\frac{HU}{\pi}\ln(\cos\tfrac{1}{2}\gamma_1), \tag{92}$$

where, in the notation of §11.5, $4a = \phi_1$. From this it follows that $\lim\limits_{H\to\infty} H\gamma_1^2 = 32\,\pi a/U$, which enables us to evaluate the limit forms of (90) and (91) immediately.

UNSTEADY FLOW

12.14   Description of problem

We shall now consider the unsteady flow of an incompressible fluid through a cascade of aerofoils or blades. An important application of the theory of such a flow is to the flutter problem in turbo-machinery. This is a serious problem to which a great deal of research is being devoted. An aerodynamic theory which can predict the air loads on an oscillating cascade member, together with a knowledge of the elastic properties of the member, will enable us to determine the flutter speeds of the cascade.

Another type of unsteady flow, which is not serious but always present under normal operating conditions, is due to the fact that each ring of rotor blades must move through the unsteady velocity field induced by all the stator rings, and vice versa. In most calculations this basic unsteadiness is ignored, being second order compared with the mean steady flow, however the induced fluctuating loads on each blade, even if small, will contribute to the fatigue and eventual failure of the blade. It is therefore worth while attempting to calculate these loads.

The stalling of cascade blades produces two distinct types of unsteady flow. First, and most serious, is the problem of 'stall flutter' which occurs when a row of blades reaches the critical stall angle. The problem is very difficult to investigate either theoretically or experimentally. Some experimental work has been reported by Pearson (1953), and an attempt at a theory has been made by Sisto (1953). It should be possible to extend the theory of § 12.12 to provide a theoretical treatment of stalled flutter. In this case the functions $\Omega_s(\gamma)$, $(\gamma_1 < \gamma < \pi)$ and $\Omega_s(\gamma)$, $(-\pi < \gamma < \gamma_2)$ appearing in (71) will be unknown functions to be determined in a manner similar to that used in § 11.18 for $\Omega^+(\eta)$ and $\Omega^-(\eta)$. A second type of unsteady flow due to blade stall is the phenomenon of the 'rotating stall' which occurs at an angle of attack smaller than the critical stall angle. (For a recent study of the rotating stall see Stenning & Kriebel (1957).) These stall problems are very difficult analytically, and so we shall confine our attention to the two (unstalled) unsteady problems described in the first two paragraphs.

The first of these is the classical flutter problem. At low incidences, and at airspeeds below the critical Mach number, the direct aerodynamic damping forces in flexure (vertical translation) and torsion (pitching) are in practice positive. However, it is possible to have unstable oscillations (i.e. energy extracted from the air stream) as a result of phase differences between the flexural and torsional motions (Duncan, 1949). Although flutter of this type apparently only rarely happens with turbine or compressor blades, it has been investigated in some detail, both experimentally (Belleriot & D'Epinay, 1950; Lilley, 1952) and theoretically (Timman, 1951; Legendre, 1954). The main experimental results are (i) that the motions of adjacent aerofoils are 180° out-of-phase, and the cascade does not flutter as a rigid group, and (ii) that the critical flutter speed decreases as the gap:chord ratio is reduced. A theoretical treatment of the problem is comparatively easy in the case when the cascade is unstaggered. Because the adjacent aerofoils move 180° out-of-phase, the mid-streamline between them is unaffected by the unsteady motion, which means that aerodynamically the problem is equivalent to that of an aerofoil oscillating between fixed tunnel walls. This latter problem was solved by Timman (1951) in order to obtain the wind-tunnel corrections to be applied to experimental results on single wing flutter, and as we shall study it in §§ 14.1–14.4, we can omit it here and pass on to the unsteady flow problem described in paragraph two of this section.

Theories enabling the calculation of the induced loading on cascades due to unsteady upwash distributions have been given by Billington (1949), Kemp & Sears (1953), Chang & Chu (1956) and Woods (1955d). Kemp & Sears studied the problem of a single stator cascade followed by a single-rotor cascade. They employed the method developed by Kármán & Sears (1938) for the case of a single aerofoil in unsteady motion. Bound vortices are distributed over each aerofoil chord, and the wake of each aerofoil is assumed to be a vortex sheet. Each aerofoil moves in the velocity field induced by (*a*) its own wake, (*b*) the bound vorticity of other members of the same cascade, (*c*) their wakes, (*d*) the bound vorticity of the aerofoils of the other cascade, and (*e*) their wakes. Kemp & Sears simplified this very complex problem by making two approximations. The first is that in calculating the unsteady component of the velocity field induced at each aerofoil it is permissible to ignore (*b*) and (*c*) provided the gap : chord ratio of the cascades is not too small. The second assumption is that the unsteady part of the circulation about any aerofoil is small compared with the steady part. These two assumptions, when valid, imply that the unsteady pressure distribution induced on an aerofoil can be calculated to first order by considering only the steady part of the bound vorticity over the aerofoils of the other cascade.

The other authors quoted above did not study the aerodynamic interference between moving blade rows, but concentrated on obtaining exact results for the induced loading on cascades due to unsteady upwash distributions. Billington's (1949) theory is developed for a harmonic upwash; it is rather complicated algebraically, with the main results expressed in infinite series form. The theories of Chang & Chu (1956) and Woods (1955*d*) are similar, but the latter is more concise and will be reproduced in the remaining sections of this chapter.

The exact theory just discussed, if applied to the interference problem studied by Kemp & Sears, would eliminate the need for their first approximation. If the second approximation of Kemp & Sears is accepted, the exact theory could be applied to their problem as follows. The unsteady upwash velocity induced at each member of the first cascade is calculated by steady cascade theory and the known relative velocity of the two cascade rings. The 'exact' theory (developed below) then permits the corresponding unsteady circulation and pressure distribution to be obtained for the first cascade. The roles of the cascades are now reversed and we find the unsteady circulation about each aerofoil of the second cascade, and the resulting vortex distribution in their wakes. The 'exact' theory then permits the unsteady upwash on the first cascade to be corrected for the increment due to the unsteady circulation about the aerofoils of the second cascade, and the resulting vortex distributions in their wakes. This leads to a modification of the pressure distribution and circulation about aerofoils of the first cascade, and so on. In this way we work from one cascade to the other, improving the solution each time.

Another obvious application of the theory of the following sections is to the determination of the transient loading on cascade blades due to a rapid change in inlet conditions.

12.15 General theory

We shall restrict attention to the problem of finding the induced loading due to an arbitrary unsteady upwash distribution on an unstaggered cascade of thin aerofoils in an incompressible flow. In this case (45) becomes

$$z = \tfrac{1}{2}c - \frac{c}{2r}\ln\left(\cosh r + \sinh r \cos \zeta\right), \tag{93}$$

where

$$r = \frac{\pi c}{H}. \tag{94}$$

The origin of the z-plane is at the leading edge, $\zeta = 0$, of a typical aerofoil of the cascade. The point downstream at infinity maps on to $\eta = \mu$, where from (11)

$$\coth r = \cosh \mu. \tag{95}$$

The upper and lower surfaces of the vortex sheet extending behind each aerofoil of the cascade map on to $\gamma = \pi, 0 < \eta < \mu$ and $\gamma = -\pi, 0 < \eta < \mu$ respectively (see Fig. 12.2b).

The boundary conditions in the ζ-plane are almost identical with those studied in §9.2 for the isolated aerofoil, and therefore the general solution is

$$\tau(\zeta,t) = \frac{1}{2\pi}\int_{-\pi}^{\pi} \theta(\gamma,t)\cot\tfrac{1}{2}(\gamma-\zeta)\,d\gamma + \frac{\sin\zeta}{2\pi}\int_0^{\mu}\frac{\varpi(\eta,t)\,d\eta}{\cosh\eta+\cos\zeta}, \tag{96}$$

which differs from (9.11) only in the upper limit of the second integral.

We shall set the inlet angle α equal to zero, then from (96) and the first of (13)

$$0 = \int_{-\pi}^{\pi} U\theta(\gamma,t)\,d\gamma + \int_0^{\mu}\chi(\eta,t)\,d\eta, \tag{97}$$

where

$$\chi \equiv \varpi U. \tag{98}$$

Equation (7) is still valid in unsteady flow, provided Γ is taken to be the circulation about the aerofoil plus its vortex sheet. In the present application $\alpha = 0$ and $\rho_a = \rho_\infty$, hence the circulation is zero only if the outlet angle β is zero. Assume that $\Gamma = 0$ at the start of the unsteady motion, then by Kelvin's theorem Γ will remain zero throughout the unsteady motion, i.e. β will vanish at all times. From (1), in which $b = 0$ in the present case, this means that $V = U$ at all times. These outlet conditions can be expressed $\lim_{\eta \to \mu} \tau(\pm\pi+i\eta) = 0$ (cf. (30)), and applying this limit to (96) we arrive at

$$\int_{-\pi}^{\pi}\frac{\theta\sin\gamma\,d\gamma}{\cos\gamma+\cosh\mu} = 0, \tag{99}$$

and

$$\int_{-\pi}^{\pi}\frac{U\theta\,d\gamma}{\cos\gamma+\cosh\mu} - \int_0^{\mu}\frac{\chi\,d\eta}{\cosh\eta-\cosh\mu} = 0, \tag{100}$$

which are the generalizations of (9.15) and (9.16).

12.16   The strength of the vortex sheet

The calculation in this section is very similar to that given in § 9.3 for an isolated aerofoil; we shall make use of this to shorten the explanation.

First observe that (9.6)–(9.10) also hold with a cascade of aerofoils. Thus

$$\frac{1}{c}\frac{\partial \chi}{\partial t} + \frac{\partial \chi}{\partial x} = 0, \tag{101}$$

where

$$t = \frac{1}{c}\int_0^t U(t)\,dt. \tag{102}$$

Equation (9.23), viz.

$$\chi(\xi,t) = \chi(1, t+1-\xi), \tag{103}$$

also holds provided (9.24) is replaced by

$$\xi = \frac{x}{c} = \frac{1}{2} - \frac{1}{2r}\ln(\cosh r - \sinh r \cosh \eta), \tag{104}$$

which follows by putting $\zeta = \pm\pi + i\eta$ in (93), and from the fact that the origin of the z-plane is now at the leading edge of a typical aerofoil. From (104)

$$d\eta = \frac{\sqrt{2r}\, e^{-r(\xi-\frac{1}{2})}\,d\xi}{\{\cosh r(2\xi-1) - \cosh r\}^{\frac{1}{2}}}. \tag{105}$$

The flow direction is given by (9.5), which in the present case reads

$$\theta = \theta_s + \frac{n}{U} - \alpha_0 + \pi\lambda\delta(\gamma), \tag{106}$$

where α_0 is the incidence of the cascade (not to be confused with the inlet angle α, see Fig. 12.8b). The steady component θ_s satisfies (97), (99) and (100) with $\chi = 0$, hence on substituting (106) into these equations, and after some algebra we get

$$\lambda U = 2\alpha_0 U - 2a_0 - \frac{1}{\pi}\int_0^\mu \chi(\eta,t)\,d\eta,$$

$$0 = \int_{-\pi}^\pi \frac{n(\gamma,t)\sin\gamma\,d\gamma}{\cos\gamma + \cosh\mu},$$

and

$$0 = (1 - e^{-\mu})\alpha_0 U - a_0 + a_1 + \frac{1}{2\pi}\int_0^\mu \chi(\eta,t)\frac{(\cosh\mu - 1)(1 + \cosh\eta)}{\cosh\eta - \cosh\mu}\,d\eta,$$

where

$$a_0 = \frac{1}{2\pi}\int_{-\pi}^\pi n(\gamma,t)\,d\gamma, \tag{107}$$

and

$$a_1 \equiv \frac{1}{2\pi}\int_{-\pi}^\pi n(\gamma,t)\left(\frac{1 + \cos\gamma\cosh\mu}{\cos\gamma + \cosh\mu}\right)d\gamma. \tag{108}$$

It is easily verified that these reduce to (9.20)–(9.22) in the limit $\mu \to \infty$, i.e. as the gap:chord ratio, H/c, tends to infinity.

From (103) to (105) and an argument similar to that used in the derivation of (9.29) and (9.30) we find

$$\lambda U = 2\alpha_0 U - 2a_0 - \frac{\sqrt{2}}{\pi} r\, e^{-\frac{1}{2}r} \int_0^f \frac{\chi(1,\sigma)\, e^{r(\sigma-f)}\, d\sigma}{[\cosh\{2r(f-\sigma)+r\} - \cosh r]^{\frac{1}{2}}}, \quad (109)$$

$$0 = A(f) - \frac{\sqrt{2}}{\pi}\, \frac{r\, e^{-\frac{1}{2}r}}{\sinh r} \int_0^f \frac{\chi(1,\sigma)\sinh\{r(f-\sigma)+r\}\, d\sigma}{[\cosh\{2r(f-\sigma)+r\} - \cosh r]^{\frac{1}{2}}}, \quad (110)$$

where
$$A(f) \equiv \{(1 - e^{-\mu})\alpha_0 U - a_0 + a_1\}\, U(f) \qquad (111)$$

(assuming, as usual, that the unsteady motion commences at $f = 0$).

The Laplace transform of (110) is (cf. derivation of (9.33))

$$\mathscr{L}\{A(f); p\} = \frac{\sqrt{2}}{\pi}\, \frac{r\, e^{-\frac{1}{2}r}}{\sinh r}\, \frac{1}{p}\, \mathscr{L}\{U(f)\,\chi(1,f); p\}\, \mathscr{L}\left\{\frac{U(f)\sinh(rf+r)}{\{\cosh(2rf+r) - \cosh r\}^{\frac{1}{2}}}; p\right\}. \quad (112)$$

Now van der Pol & Bremmer (1950) give the transform

$$\mathscr{L}\left\{\frac{(-\sinh r)^\nu\, e^{-\frac{1}{2}f}\, U(f-r)}{\Gamma(\frac{1}{2}-\nu)\,(\cosh f - \cosh r)^{\nu+\frac{1}{2}}}; p\right\}$$
$$= \left(\frac{2}{\pi}\right)^{\frac{1}{2}} p\, Q_p^\nu(\cosh r), \quad -1 - \mathscr{R}\nu < \mathscr{R}p < \infty \quad (\mathscr{R}\nu < \tfrac{1}{2}),$$

where $Q_p^\nu(\cosh r)$ is the associated Legendre function of order ν (see p. 325 of Whittaker & Watson, 1952). By applications of the rules given in (4.195)–(4.197) this can be modified to

$$\mathscr{L}[U(f)\{\cosh(2rf+r) - \cosh r\}^{-(\nu+\frac{1}{2})}; p]$$
$$= \frac{\Gamma(\frac{1}{2}-\nu)\,(-1)^\nu}{(\sinh r)^\nu}\, \frac{p\, e^{\frac{1}{2}p}}{r\sqrt{2\pi}}\, Q_N^\nu(\cosh r), \quad -r(1 + 2\mathscr{R}\nu) < \mathscr{R}p < \infty \quad (\mathscr{R}\nu < \tfrac{1}{2}), \quad (113)$$

where
$$N \equiv \frac{p-r}{2r}.$$

Hence by (4.197) and (113)

$$\mathscr{L}\left\{\frac{U(f)\sinh(rf+r)}{\{\cosh(2rf+r) - \cosh r\}^{\frac{1}{2}}}; p\right\}$$
$$= \tfrac{1}{2}e^r\, \frac{p}{r\sqrt{2}}\, e^{\frac{1}{2}(p-r)}\, Q_{(p/2r)-1}(\cosh r) - \tfrac{1}{2}e^{-r}\, \frac{p}{r\sqrt{2}}\, e^{\frac{1}{2}(p+r)}\, Q_{p/2r}(\cosh r)$$
$$(\mathscr{R}p > 0).$$

A more convenient form of these equations can be obtained by substituting

$$Q_{(p/2r)-1}(\cosh r) = \cosh r\, Q_{p/2r}(\cosh r) - \frac{2r}{p}\sinh r\, Q_{p/2r}^1(\cosh r), \quad (114)$$

a relation which is easily derived from equations given on pp. 318 and 325 of Whittaker & Watson (1952). These results enable us to write (112) in the form

$$\mathscr{L}(\chi) = \frac{2\pi\,\mathscr{L}(A)\,e^{-\frac{1}{2}p}}{Q_{p/2r} - \dfrac{2r}{p}\,Q^1_{p/2r}} \quad (\mathscr{R}p > 1), \tag{115}$$

where we have omitted the argument of the Legendre functions for brevity. This is the generalization of (9.36).

Similarly, the transform of (109) is

$$\mathscr{L}(\lambda U) = 2\mathscr{L}(A) - \frac{1}{\pi}\mathscr{L}(\chi)\,e^{p/2}\,Q_{p/2r} - 2\mathscr{L}(a_1 - e^{-\mu}\alpha_0 U),$$

and eliminating $\mathscr{L}(\chi)$ by (115),

$$\mathscr{L}(\lambda U) = \mathscr{L}(A)\,\mathscr{L}\{k(\mathcal{f},r); p\} - 2\mathscr{L}(a_1 - e^{-\mu}\alpha_0 U), \tag{116}$$

where

$$\mathscr{L}\{k(\mathcal{f},r); p\} = 2C(-\tfrac{1}{2}ip, r) = 2\frac{-\dfrac{2r}{p}\,Q^1_{p/2r}}{Q_{p/2r} - \dfrac{2r}{p}\,Q^1_{p/2r}}, \tag{117}$$

or from (4.194)

$$k(\mathcal{f},r) = \mathbf{U}(\mathcal{f})\frac{1}{\pi i}\int_{c_0-i\infty}^{c_0+i\infty} C(-\tfrac{1}{2}ip, r)\,e^{p\mathcal{f}}\frac{dp}{p}. \tag{118}$$

The inverse transform of (116) is (cf. (9.43))

$$\lambda(\mathcal{f})\,U(\mathcal{f}) = \int_0^{\mathcal{f}} \dot{A}(\mathcal{f}^*)\,k(\mathcal{f}-\mathcal{f}^*, r)\,d\mathcal{f}^* - 2a_1 + 2\,e^{-\mu}\alpha_0 U. \tag{119}$$

The function $k(\mathcal{f},r)$ is a generalization of the Wagner function, $k(\mathcal{f})$, defined in §9.3, and $C(-\tfrac{1}{2}ip, r)$ is the corresponding generalized Theodorsen function. In the limit $c/H \to 0$, or $r \to 0$ they tend to $k(\mathcal{f})$ and $C(-\tfrac{1}{2}ip)$. This can be verified with the aid of the formula (see p. 156 of Watson (1952))

$$\lim_{N\to\infty}\left[\frac{N^{-m}\sin N\pi}{\sin(N+m)\pi}\,Q^m_N\left(\cosh\frac{z}{N}\right)\right] = K_m(z). \tag{120}$$

The values $z = \tfrac{1}{2}p$, $N = p/2r$ in (120) give

$$\lim_{r\to 0} Q_{p/2r}(\cosh r) = K_0(\tfrac{1}{2}p), \quad \text{and} \quad \lim_{r\to 0} Q^1_{p/2r}(\cosh r) = -\tfrac{1}{2}pK_1(\tfrac{1}{2}p). \tag{121}$$

Substituting these limits in (117), and comparing the result with (9.38) we verify that

$$C(z, 0) = C(z) \quad \text{and} \quad k(\mathcal{f}, 0) = k(\mathcal{f}). \tag{122}$$

Similarly, from (115) we can write down the generalization of the function k_0 defined in (9.46).

12.17   The pressure distribution

As the flow is incompressible (9.17) can be written

$$p(x,t) = G(t) - \rho_a U\dot{U}\frac{x}{c} + \rho_a U^2\Omega + \frac{\rho_a U}{c}\int_0^x (U\dot{\Omega})\,dx,$$

where the dot denotes differentiation with respect to t. On the surface of an aerofoil of the cascade (93) and (95) give

$$\frac{x}{c} = \frac{1}{2}\left(1 - \frac{1}{r}\ln\sinh r\right) - \frac{1}{2r}\ln\left(\cosh\mu + \cos\gamma\right), \tag{123}$$

so that
$$dx = \frac{c}{2r}\frac{\sin\gamma\,d\gamma}{\cosh\mu + \cos\gamma}. \tag{124}$$

Hence $\quad p(\gamma,t) = G(t) - \rho_a U\dot{U}\dfrac{x}{c} + \rho_a U^2\Omega + \dfrac{\rho_a U}{2r}\displaystyle\int_0^\gamma \dfrac{(U\dot{\Omega})\sin\gamma\,d\gamma}{\cosh\mu + \cos\gamma}. \tag{125}$

On the vortex sheet, $\zeta = \pm\pi + i\eta$, and (124) is replaced by

$$dx = \frac{c}{2r}\frac{\sinh\eta\,d\eta}{\cosh\mu - \cosh\eta},$$

so (101) can be written

$$\frac{\partial\chi}{\partial\eta} = -\frac{1}{2r}\frac{\dot{\chi}\sinh\eta}{\cosh\mu - \cosh\eta}. \tag{126}$$

On an aerofoil surface it follows from (96) and (98) that

$$U\Omega = \frac{1}{2\pi}\int_{-\pi}^{\pi} U\theta\cot\tfrac{1}{2}(\gamma^* - \gamma)\,d\gamma^* + \frac{\sin\gamma}{2\pi}\int_0^\mu \frac{\chi\,d\eta}{\cosh\eta + \cos\gamma}. \tag{127}$$

We shall now simplify (125) with the help of (126) and (127) by a method that is a generalization of that given in § 9.4.

The last term in (127) can be written in the form

$$\frac{\cosh\mu + \cos\gamma}{\sin\gamma}A - \frac{\sinh^2\mu}{\sin\gamma}B + \frac{\cosh\mu + \cos\gamma}{2\pi\sin\gamma}$$

$$\times \int_0^\mu \frac{\chi\sinh^2\mu\,d\eta}{(\cosh\eta - \cosh\mu)(\cosh\eta + \cos\gamma)},$$

where $\quad A \equiv -\dfrac{1}{2\pi}\displaystyle\int_0^\mu \chi\,d\eta = \dfrac{1}{2\pi}\int_{-\pi}^{\pi} U\theta\,d\gamma,$

and $\quad B \equiv \dfrac{1}{2\pi}\displaystyle\int_0^\mu \dfrac{\chi\,d\eta}{\cosh\eta - \cosh\mu} = \dfrac{1}{2\pi}\int_{-\pi}^{\pi} \dfrac{U\theta\,d\gamma}{\cos\gamma + \cosh\mu},$

$$\tag{128}$$

using (97) and (100). Hence from (127)

$$\frac{(\dot U \Omega) \sin \gamma}{\cos \gamma + \cosh \mu} = \frac{1}{2\pi} \left(\frac{\sin \gamma}{\cos \gamma + \cosh \mu} \right) \int_{-\pi}^{\pi} (\dot U \theta) \cot \tfrac{1}{2}(\gamma^* - \gamma)\, d\gamma^* + \dot A$$

$$- \frac{\sinh^2 \mu\, \dot B}{\cos \gamma + \cosh \mu} + \frac{1}{2\pi} \int_0^\mu \left(\frac{\dot \chi \sinh \eta}{\cosh \eta - \cosh \mu} \right) \left(\frac{\sinh \eta}{\cosh \eta + \cos \gamma} \right) d\eta.$$

$$(129)$$

Equation (126) enables us to write the last term in the form

$$\frac{r}{\pi} \int_0^\mu \frac{\partial \chi}{\partial \eta} \frac{\sinh \eta\, d\eta}{\cosh \eta + \cos \gamma} = -\frac{r}{\pi} \int_0^\mu \chi \frac{\partial}{\partial \eta} \left(\frac{\sinh \eta}{\cosh \eta + \cos \gamma} \right) d\eta$$

$$= -\frac{r}{\pi} \int_0^\mu \chi \frac{\partial}{\partial \gamma} \left(\frac{\sin \gamma}{\cosh \mu + \cos \gamma} \right) d\eta,$$

on first integrating by parts, and then using an identity between the derivatives appearing in the integrands. Thus (129) yields

$$\int_0^\gamma \frac{(\dot U \Omega) \sin \gamma\, d\gamma}{\cos \gamma + \cosh \mu} = \frac{1}{2\pi} \int_{-\pi}^{\pi} (\dot U \theta) \int_0^\gamma \frac{\sin \gamma \cot \tfrac{1}{2}(\gamma^* - \gamma)}{\cosh \mu + \cos \gamma}\, d\gamma\, d\gamma^* + \int_0^\gamma \dot A\, d\gamma$$

$$- \sinh^2 \mu \int_0^\gamma \frac{\dot B\, d\gamma}{\cos \gamma + \cosh \mu} - \frac{r}{\pi} \int_0^\mu \frac{\chi \sin \gamma\, d\eta}{\cos \gamma + \cosh \eta}.$$

Substituting this result and (127) into the pressure equation (125) we arrive at

$$p = G(t) - \rho_a U \dot U \frac{x}{c} + \frac{\rho_a U^2}{2\pi} \int_{-\pi}^{\pi} \theta \cot \tfrac{1}{2}(\gamma^* - \gamma)\, d\gamma^*$$

$$+ \frac{\rho_a U}{4\pi r} \int_{-\pi}^{\pi} (\dot U \theta) \left[\int_0^\gamma \left\{ \frac{\sin \gamma \cot \tfrac{1}{2}(\gamma^* - \gamma)}{\cos \gamma + \cosh \mu} \right. \right.$$

$$\left. \left. + 1 - \frac{\sinh^2 \mu}{(\cos \gamma + \cosh \mu)(\cos \gamma^* + \cosh \mu)} \right\} d\gamma \right] d\gamma^*,$$

on eliminating A and B by using (128). The last term of this equation can be written as

$$I \equiv \frac{\rho_a U}{4\pi r} \int_0^\gamma \int_{-\pi}^{\pi} \frac{(\dot U \theta) \sin \gamma^*}{\cos \gamma^* + \cosh \mu} \left\{ \cot \tfrac{1}{2}(\gamma^* - \gamma) - \frac{\sin \gamma}{\cos \gamma + \cosh \mu} \right\} d\gamma^*\, d\gamma,$$

after some algebra. An integration by parts results in

$$I = \frac{\rho_a U}{2\pi} \int_0^\gamma \left(\left[\dot N \left\{ \cot \tfrac{1}{2}(\gamma^* - \gamma) - \frac{\sin \gamma}{\cos \gamma + \cosh \mu} \right\} \right]_{-\pi}^{\pi} \right.$$

$$\left. - \int_{-\pi}^{\pi} \dot N \frac{\partial}{\partial \gamma^*} \cot \tfrac{1}{2}(\gamma^* - \gamma)\, d\gamma^* \right) d\gamma,$$

where
$$N(\gamma^*,t) \equiv \frac{1}{2r}\int_0^{\gamma^*} \frac{U\theta\sin\gamma\,d\gamma}{\cos\gamma+\cosh\mu}. \tag{130}$$

From (99) $N(\pi)-N(-\pi)=0$, hence

$$I = \frac{\rho_a U}{2\pi}\int_0^\gamma\int_{-\pi}^\pi \dot{N}(\gamma^*,t)\frac{\partial}{\partial\gamma}\cot\tfrac{1}{2}(\gamma^*-\gamma)\,d\gamma^*\,d\gamma$$

$$= \frac{\rho_a U}{2\pi}\int_{-\pi}^\pi \dot{N}\cot\tfrac{1}{2}(\gamma^*-\gamma)\,d\gamma^*.$$

The pressure equation now assumes the simple form

$$p(\gamma,t) = G(t)-\rho_a U\dot{U}\frac{x}{c}+\frac{\rho_a U}{2\pi}\int_{-\pi}^\pi (U\theta+\dot{N})\cot\tfrac{1}{2}(\gamma^*-\gamma)\,d\gamma^*, \tag{131}$$

which is due to Woods (1955d).

The final step in this calculation is to eliminate θ by (106); the result is

$$p(\gamma,t) = p_s(\gamma)+\rho_a U\dot{U}\frac{x}{c}-\tfrac{1}{2}\rho_a\lambda U^2\cot\tfrac{1}{2}\gamma$$

$$+\frac{\rho_a U}{2\pi}\int_{-\pi}^\pi \{n(\gamma^*,t)+\dot{N}(\gamma^*,t)\}\cot\tfrac{1}{2}(\gamma^*-\gamma)\,d\gamma^*, \tag{132}$$

where
$$N(\gamma^*,t) = \frac{1}{2r}\int_0^{\gamma^*} \frac{\{n(\gamma,t)-\alpha_0 U\}\sin\gamma\,d\gamma}{\cos\gamma+\cosh\mu}, \tag{133}$$

and $p_s(\gamma)$ is the mean steady pressure.

12.18   The lift and moment acting on a blade of the cascade

The lift and moment acting about the mid-chord point are given by

$$L = L_s-\int_{-\pi}^\pi (p-p_s)\,dx(\gamma) = L_s-\frac{c}{2r}\int_{-\pi}^\pi \frac{(p-p_s)\sin\gamma\,d\gamma}{\cos\gamma+\cosh\mu},$$

and
$$M = M_s+\int_{-\pi}^\pi (p-p_s)(x-\tfrac{1}{2})\,dx(\gamma)$$

$$= M_s-\left(\frac{c}{2r}\right)^2\int_{-\pi}^\pi (p-p_s)\frac{\sin\gamma\ln(\cosh\mu+\cos\gamma)}{\cosh\mu+\cos\gamma}\,d\gamma$$

$$-\frac{c}{2r}(L-L_s)\ln\sinh\mu,$$

by (123) and (124). Substituting (132) into these equations, and making use of (52), we find after some algebra that the lift and moment coefficients are

$$C_L = C_{L_s}+\frac{H\lambda}{c}(1-\tanh\tfrac{1}{2}\mu)+\frac{H}{\pi cU}\int_{-\pi}^\pi (n+\dot{N})\left(1-\frac{\sinh\mu}{\cos\gamma+\cosh\mu}\right)d\gamma, \tag{134}$$

and
$$C_M = C_{M_s} - \left(\frac{H}{c}\right)^2 \frac{\lambda}{2\pi}(1 + \tanh \tfrac{1}{2}\mu)\ln(1 - e^{-2\mu})$$

$$- \left(\frac{H}{c}\right)^2 \frac{1}{2\pi^2 U}\int_{-\pi}^{\pi}(n + \dot{N})\left\{\ln(1 - e^{-2\mu})\left[1 + \frac{\sinh\mu}{\cos\gamma + \cosh\mu}\right]\right.$$

$$\left. + \frac{2\sin\gamma}{\cos\gamma + \cosh\mu}\tan^{-1}\left(\frac{\sin\gamma}{e^\mu + \cos\gamma}\right)\right\}d\gamma, \tag{135}$$

where $\cosh\mu = \coth(\pi c/H)$, and λ is given by (119). It is easily verified that in the limit $H \to \infty$ these equations reduce to (9.55) and (9.56).

12.19   The generalized Theodorsen function

The general theory of §§ 12.16–12.18 can be developed to deal with harmonic motions and sudden changes in inlet conditions ('gusts' and accelerations) by methods similar to those employed in §§ 9.6–9.12. We shall not pursue these details here, except to give a method of calculating the generalized Theodorsen function $C(\omega, r)$ for the case of harmonic oscillations of long duration.

In this case, by a method similar to that used in §§ 9.6 and 9.7, it can be shown that (119) tends to

$$\lambda U = 2AC(\omega, r) - 2(a_1 - e^{-\mu}\alpha_0 U),$$

where
$$C(\omega, r) = \frac{\dfrac{r}{i\omega}Q^1_{i\omega/r}(\cosh r)}{\dfrac{r}{i\omega}Q^1_{i\omega/r}(\cosh r) - Q_{i\omega/r}(\cosh r)}. \tag{136}$$

This function can be expressed in a suitable form for calculation as follows. On p. 317 of Whittaker & Watson (1952) is given the relation

$$Q_n(\rho) = \frac{\sqrt{\pi}}{2^{n+1}}\frac{\Gamma(n+1)}{\Gamma(n+\tfrac{3}{2})}z^{-(n+1)}F(\tfrac{1}{2}n + \tfrac{1}{2}, \tfrac{1}{2}n + 1; n + \tfrac{3}{2}; \rho^{-2}),$$

where $F(a, b; c; u)$ is the hypergeometric function,

$$F(a, b; c; u) \equiv 1 + \frac{a \cdot b}{1 \cdot c}u + \frac{a(a+1)b(b+1)}{1 \cdot 2\,c(c+1)}u^2$$

$$+ \frac{a(a+1)(a+2)b(b+1)(b+2)}{1 \cdot 2 \cdot 3\,c(c+1)(c+2)}u^3 + \dots.$$

An application of the transformation formula (see Erdelyi, 1953, p. 112)

$$F(a, a + \tfrac{1}{2}; c; u) = (1 + u^{\frac{1}{2}})^{-2a}F\left(2a, c - \tfrac{1}{2}; 2c - 1; \frac{2u^{\frac{1}{2}}}{1 + u^{\frac{1}{2}}}\right),$$

yields

$$Q_n(\rho) = \frac{\sqrt{\pi}}{2^{n+1}} \frac{\Gamma(n+1)}{\Gamma(n+\frac{3}{2})} (1+\rho)^{-(n+1)} F\left(n+1, n+1, 2n+2, \frac{2}{1+\rho}\right). \tag{137}$$

Finally, using the transformation (Erdelyi, 1953, p. 113)

$$F(a, a; 2a; u) = \{\tfrac{1}{2} + \tfrac{1}{2}(1-\mu)^{\frac{1}{2}}\}^{-2a} F\left[a, \tfrac{1}{2}; a+\tfrac{1}{2}; \left\{\frac{1-(1-u)^{\frac{1}{2}}}{1+(1-u)^{\frac{1}{2}}}\right\}^2\right],$$

on (137) and setting $\rho = \cosh r$ in the result, we arrive at

$$Q_n(\cosh r) = \sqrt{\pi} \frac{\Gamma(n+1)}{\Gamma(n+\frac{3}{2})} e^{-(n+1)r} F(n+1, \tfrac{1}{2}; n+\tfrac{3}{2}; e^{-2r}). \tag{138}$$

From this result and (114) it follows that

$$Q_n(\cosh r) - \frac{1}{n} Q_n^1(\cosh r) = \operatorname{cosech} r \{Q_{n-1}(\cosh r) - e^{-r} Q_n(\cosh r)\}$$

$$= \operatorname{cosech} r \sqrt{\pi} \frac{\Gamma(n)}{\Gamma(n+\frac{1}{2})} e^{-nr} \left\{ F(n, \tfrac{1}{2}; n+\tfrac{1}{2}; e^{-2r}) \right.$$

$$\left. - \frac{n}{n+\frac{1}{2}} e^{-2r} F(n+1, \tfrac{1}{2}; n+\tfrac{3}{2}; e^{-2r}) \right\}$$

$$= 2\sqrt{\pi} \frac{\Gamma(n)}{\Gamma(n+\frac{1}{2})} e^{-(n+1)r} F(n+1, \tfrac{1}{2}; n+\tfrac{1}{2}; e^{-2r}), \tag{139}$$

by an application of the Gauss relation

$$F(a-1, b; c; u) - \frac{c-b}{c} uF(a, b; c+1; u) = (1-u) F(a, b; c; u)$$

between contiguous hypergeometric functions. Equations (138) and (139) enable us to write (136) in the form

$$C(\omega, r) = 1 - \frac{\dfrac{i\omega}{r} F\left(\dfrac{i\omega}{r}+1, \tfrac{1}{2}; \dfrac{i\omega}{r}+\tfrac{3}{2}; e^{-2r}\right)}{\left(1+\dfrac{2i\omega}{r}\right) F\left(\dfrac{i\omega}{r}+1, \tfrac{1}{2}; \dfrac{i\omega}{r}+\tfrac{1}{2}; e^{-2r}\right)}, \tag{140}$$

which provides a convenient means of computing the function $C(\omega, r)$.

CHAPTER 13

LIFTING AEROFOILS IN CHANNELS AND JETS

THE CONFORMAL TRANSFORMATIONS

13.1 An aerofoil in a general position in a stream of finite width

In ch. 7 we studied flow through empty channels, and flow past symmetrical non-lifting obstacles and aerofoils placed on the channel axis. In this chapter these flows will be generalized to flow through channels occupied by aerofoils (or obstacles), which may be off-centre and experiencing a lifting force. The corresponding jet and aerofoil problem will also be considered. These flows occur through doubly-connected regions, and before the general solution can be written down, it is necessary to map these regions—or rather their w-planes—into singly-connected ζ-planes. These ζ-planes are usually rectangles with the aerofoil mapped on one side, the channel or jet boundaries on the opposite side, and each side of the surface of discontinuity of ϕ (assuming a circulation) on each of the remaining sides of the rectangle.

The case of an aerofoil in a general position in a stream of finite width is illustrated in Fig. 13.1 a. This figure is a copy of Fig. 5.15, and the reader will find a full description of the details of the transformation from the w- to the ζ-plane in §5.15. It is shown in that section that the transformation is defined by the equations

$$w = \frac{h_1+h_2}{\pi}[\{Z_4(\gamma_0-\gamma_1)-Z_4(\gamma_0+\gamma_1)+2Z_4(\gamma_1)\}\zeta - 2\Pi(\zeta,\gamma_1)], \qquad (1)$$

$$2b = \frac{h_1+h_2}{\pi}[\{Z_4(\gamma_0-\gamma_1)-Z_4(\gamma_0+\gamma_1)+2Z_4(\gamma_1)\}\gamma_0 - 2\Pi(\gamma_0,\gamma_1)], \quad (2)$$

$$\frac{\pi\Gamma}{2(h_1+h_2)} = \{Z_4(\gamma_0-\gamma_1)-Z_4(\gamma_0+\gamma_1)\}K, \qquad (3)$$

and

$$\frac{2\gamma_1}{K} = 1+\frac{h_2-h_1}{h_2+h_1}+\frac{\Gamma K'}{(h_1+h_2)K}, \qquad (4)$$

where all the symbols are defined in the figure. It also follows from equations given in § 5.15 that

$$\frac{dw}{d\zeta} = \frac{h_1+h_2}{\pi}\{Z_4(\gamma_0-\gamma_1)-Z_4(\gamma_0+\gamma_1)-Z_4(\zeta-\gamma_1)+Z_4(\zeta+\gamma_1)\}$$

$$= \frac{h_1+h_2}{\pi}\frac{2k^2\operatorname{sn}\gamma_1\operatorname{cn}\gamma_1\operatorname{dn}\gamma_1}{1-k^2\operatorname{sn}^2\gamma_0\operatorname{sn}^2\gamma_1}\left\{\frac{\operatorname{sn}^2\gamma_0-\operatorname{sn}^2\zeta}{1-k^2\operatorname{sn}^2\gamma_1\operatorname{sn}^2\zeta}\right\}. \qquad (5)$$

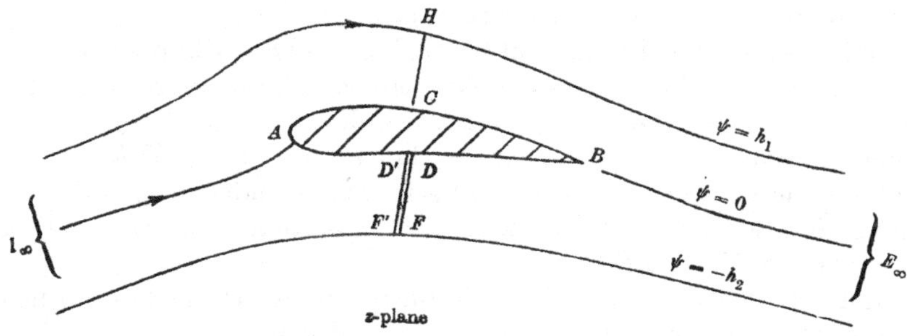

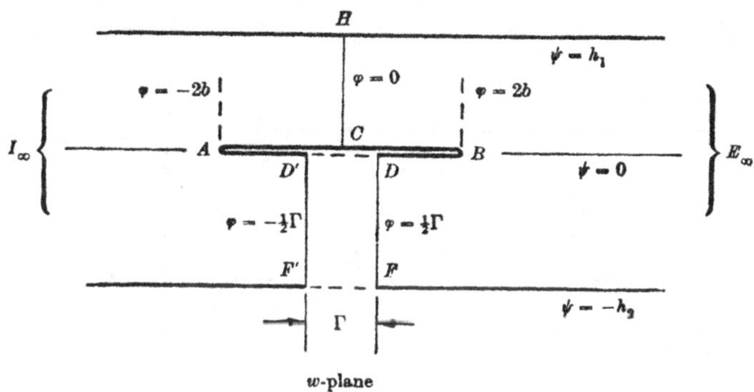

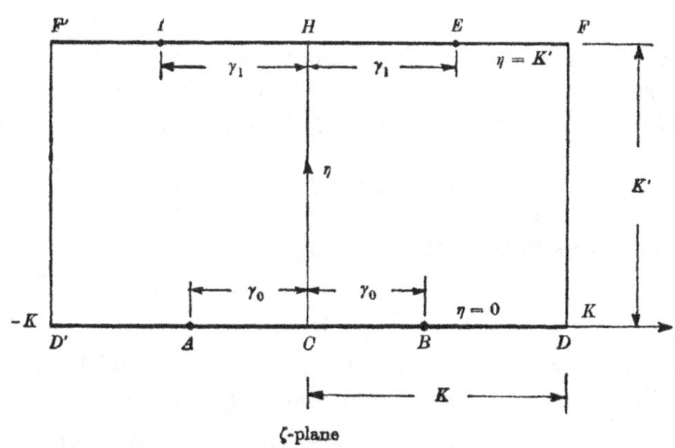

Fig. 13.1a

Note that the points $\gamma = \pm\gamma_0$ are at the stagnation points, A and B, while the points $\gamma = \pm\gamma_1 + iK'$ are at $\phi = \pm\infty$.

If it is assumed for the moment that b, Γ, h_1 and h_2 are known quantities, then the mapping into the ζ-plane introduces three unknown constants, viz. γ_0, γ_1 and k (K', K are determined uniquely by k, see §4.6), whose values are fixed by (2)–(4). It will be shown later that b, Γ, h_1 and h_2 are fixed by the values of the chord length c, the incidence α, and the distances H_1 and H_2 of a typical point on the aerofoil from the upper and lower walls of the channel.

Another auxiliary plane of some value in the theory can be obtained by applying Landen's transformation. This yields the ζ^1-plane ($\zeta^1 = \gamma^1 + i\eta^1$) shown in Fig. 13.1$b$; this is related to the ζ-plane by

$$\zeta^1 = \frac{2K_1}{K}\zeta, \quad \gamma_0^1 = \frac{2K_1}{K}\gamma_0, \quad \gamma_1^1 = \frac{2K_1}{K}\gamma_1, \quad K_1' = \frac{2K_1}{K}K'. \tag{6}$$

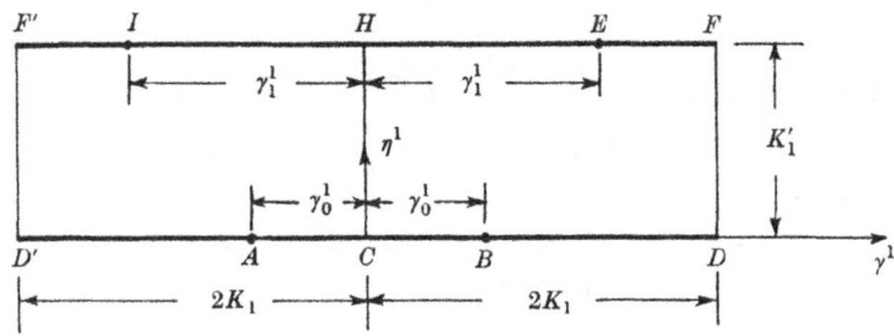

Fig. 13.1b

The form assumed by (5) in the ζ^1-plane can be simplified by applying Landen's transformation to the elliptic functions involved. First put $Kz = \zeta + iK'$ in (4.71), then use (4.35), (4.54), (4.48) and (6) to find

$$\frac{K}{K_1} Z_4(\zeta, k) = Z_4(\zeta^1, k_1) + k_1 \operatorname{sn}(\zeta^1, k_1), \tag{7}$$

where the modulus k_1 is related to k by

$$k_1 = \frac{1 - k'}{1 + k'} \tag{8}$$

(from the last equality in (4.54)). Transforming (5) with the help of (7), and then eliminating the zeta functions by (4.77) we get

$$\frac{dw}{d\zeta^1} = \frac{h_1 + h_2}{\pi} k_1 \operatorname{sn}\gamma_1^1 \left(\frac{k_1 \operatorname{sn}^2\gamma_0^1 \operatorname{cn}\gamma_1^1 \operatorname{dn}\gamma_1^1 - \operatorname{cn}\gamma_0^1 \operatorname{dn}\gamma_0^1}{1 - k_1^2 \operatorname{sn}^2\gamma_0^1 \operatorname{sn}^2\gamma_1^1} \right.$$
$$\left. - \frac{k_1 \operatorname{sn}^2\zeta^1 \operatorname{cn}\gamma_1^1 \operatorname{dn}\gamma_1^1 - \operatorname{cn}\zeta^1 \operatorname{dn}\zeta^1}{1 - k_1^2 \operatorname{sn}^2\gamma_1^1 \operatorname{sn}^2\zeta^1} \right), \tag{9}$$

where the elliptic functions are now all of modulus k_1. This is readily integrated with the help of (4.53) and (4.76) to give

$$\frac{\pi w}{h_1 + h_2} = k_1 \operatorname{sn} \gamma_1^1 \left(\frac{k_1 \operatorname{sn}^2 \gamma_0^1 \operatorname{cn} \gamma_1^1 \operatorname{dn} \gamma_1^1 - \operatorname{cn} \gamma_0^1 \operatorname{dn} \gamma_0^1}{1 - k_1^2 \operatorname{sn}^2 \gamma_0^1 \operatorname{sn}^2 \gamma_1^1} \right) \zeta^1$$

$$- \Pi(\zeta^1, \gamma_1^1) + \tfrac{1}{2} \ln \left\{ \frac{1 - k_1 \operatorname{sn} \gamma_1^1 \operatorname{sn} \zeta^1}{1 + k_1 \operatorname{sn} \gamma_1^1 \operatorname{sn} \zeta^1} \right\}. \tag{10}$$

The ζ^1-plane is particularly useful for the special case of a centrally placed aerofoil, which we consider in the following section.

13.2   An aerofoil centrally placed in a stream of finite width

A special case of considerable importance occurs when the aerofoil is centrally placed in the channel, and the circulation is small enough for its effect on the transformation to be neglected. In this case we can put

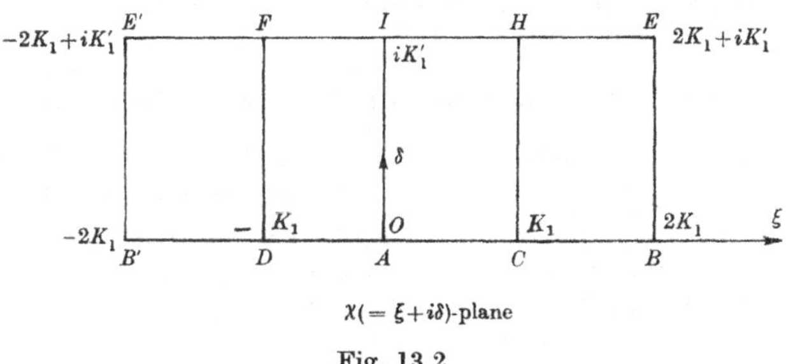

$$\chi(= \xi + i\delta)\text{-plane}$$

Fig. 13.2

$h_1 = h_2 = \tfrac{1}{2}h$ say, and $\Gamma = 0$, and it follows from (3) and (4) that $\gamma_0 = \gamma_1 = \tfrac{1}{2}K$. By (6) $\gamma_0^1 = \gamma_1^1 = K_1$ and (9) reduces to (see table of special values in §4.6)

$$\frac{dw}{d\zeta^1} = \frac{h k_1}{\pi} \operatorname{cd}(\zeta^1, k_1) = \frac{h k_1}{\pi} \operatorname{sn}(\chi, k_1), \tag{11}$$

where

$$\chi = \zeta^1 + K_1 = \xi + i\delta. \tag{12}$$

The χ-plane just defined is shown in Fig. 13.2.

From results given in §4.6

$$\int \operatorname{sn} x\, dx = \int \frac{k_1'^2 \operatorname{sc} x \operatorname{nc} x\, dx}{\operatorname{nc}^2 x} = \int \frac{d(\operatorname{dc} x)}{\operatorname{dc}^2 x - k_1^2} = \frac{1}{2k_1} \ln \left(\frac{\operatorname{dn} x - k_1 \operatorname{cn} x}{\operatorname{dn} x + k_1 \operatorname{cn} x} \right),$$

and therefore (11) can be integrated to give

$$\frac{2\pi w}{h} = \ln \left\{ \frac{\operatorname{dn}(\chi, k_1) - k_1 \operatorname{cn}(\chi, k_1)}{\operatorname{dn}(\chi, k_1) + k_1 \operatorname{cn}(\chi, k_1)} \right\}, \tag{13}$$

as w vanishes at $\chi = K_1$. With the help of the identity (see § 4.6) $\mathrm{dn}^2 x - k_1^2 \mathrm{cn}^2 x = k_1'^2$, we can transform (13) into

$$
\left.
\begin{aligned}
\mathrm{cn}\,(\chi, k_1) &= -\frac{k_1'}{k_1}\sinh\frac{\pi w}{h}, \\[2mm]
\mathrm{dn}\,(\chi, k_1) &= k_1'\cosh\frac{\pi w}{h}.
\end{aligned}
\right\}
\tag{14}
$$

or

Comparing Figs. 13.1a and 13.2 we notice that $w = 2b$ at $\chi = 2K_1$, whence (14) yield

$$
k_1 = \tanh\frac{2\pi b}{h}, \quad k_1' = \mathrm{sech}\frac{2\pi b}{h},
\tag{15}
$$

which correspond to (2).

An important feature of the χ-plane is that the stagnation streamline lies on $\xi = \pm 2K_1$, i.e. in unsteady flow about a centrally placed aerofoil the discontinuity in Ω is conveniently arranged on the boundaries of the χ-plane. We shall study this unsteady flow in ch. 14.

13.3   An aerofoil near a boundary in a stream of infinite width

Another special case of the general transformation of some importance occurs in the limit h_1 (or h_2) $\to \infty$. This is appropriate to the flow past an aerofoil placed near a single, solid boundary, a situation which occurs during the take-off of a monoplane aircraft.

Observing from (4) that $\lim\limits_{h_1 \to \infty} \gamma_1 = 0$, and $\lim\limits_{h_1 \to \infty} \gamma_1 h_1 = Kh_2 + \tfrac{1}{2}K'\Gamma$ we find that (5) has the limit form

$$
\frac{dw}{d\zeta} = \frac{k^2}{\pi}(2Kh_2 + K'\Gamma)(\mathrm{sn}^2\gamma_0 - \mathrm{sn}^2\zeta).
\tag{16}
$$

Equations (4.82) and (4.83) permit us to integrate this in the form

$$
w = \frac{1}{\pi}(2Kh_2 + K'\Gamma)\left\{Z_4(\zeta) + \frac{\zeta}{K}(E - K\,\mathrm{dn}^2\gamma_0)\right\}.
\tag{17}
$$

There is an increase of Γ in w as ζ increases by $2K$ (cf. Fig. 13.1a), and so on using the periodic property of the zeta function (4.34), we find from (17) that

$$
K\,\mathrm{dn}^2\gamma_0 = E - \frac{\pi\Gamma}{2(2Kh_2 + K'\Gamma)}.
\tag{18}
$$

Equation (17) can now be simplified to

$$
w = \frac{1}{\pi}(2Kh_2 + K'\Gamma)Z_4(\zeta) + \frac{\zeta\Gamma}{2K}.
\tag{19}
$$

At $w = 2b$, $\zeta = \gamma_0$ so

$$2b = \frac{1}{\pi}(2Kh_2 + K'\Gamma)\,Z_4(\gamma_0) + \frac{\gamma_0\Gamma}{2K}. \tag{20}$$

Equations (18) and (20) fix the values of γ_0 and k in terms of Γ and h_2.

AEROFOILS IN CHANNELS

13.4   The general solution

Notice that in the transformation given in §13.1 the aerofoil lies on $\eta = 0$ and the walls on $\eta = K'$; the surface of discontinuity in ϕ is on $\gamma = \pm K$. Furthermore, the symmetry of the transformation ensures that

$$z(K, \eta) = z(-K, \eta). \tag{21}$$

Let the subscripts 0, h refer to values on $\eta = 0$ and $\eta = K'$ respectively, i.e. the subscript 0 refers to the aerofoil, and the subscript h to the channel walls. And let the superscripts $+$, $-$ indicate values on $\gamma = K$ and $\gamma = -K$ respectively, then in the case when the shape of the aerofoil and channel are known, the following boundary conditions hold:

$$\left.\begin{array}{llll} (a) & \theta_0(\gamma)\ \text{prescribed}, & (b) & \theta_h(\gamma)\ \text{prescribed}, \\[4pt] (c) & \theta^+(\eta) = \theta^-(\eta), & (d) & \Omega^+(\eta) = \Omega^-(\eta), \end{array}\right\} \tag{22}$$

the last pair of which follow from (21).

The solution of this boundary value problem has been given in §4.9. It is the special case $\varpi = 0$ of (4.94). On transforming (4.94), (4.96) and (4.97) into the notation of this chapter we get

$$\tau(\zeta) = \Omega_a + \frac{1}{\pi}\int_{-K}^{K}\{\theta_0(\gamma)\,Z_1(\gamma - \zeta) - \theta_h(\gamma)\,Z_4(\gamma - \zeta)\}\,d\gamma, \tag{23}$$

and

$$0 = \int_{-K}^{K}\{\Omega_0(\gamma) - \Omega_h(\gamma)\}\,d\gamma = \int_{-K}^{K}\{\theta_0(\gamma) - \theta_h(\gamma)\}\,d\gamma, \tag{24}$$

where

$$\Omega_a \equiv \frac{1}{4K}\int_{-K}^{K}\{\Omega_0(\gamma) + \Omega_h(\gamma)\}\,d\gamma, \tag{25}$$

and the parameter of the zeta functions is $\mu = iK'/K$.

A more useful form for the constant Ω_a can be obtained by setting $\zeta = \pm\gamma_1 + iK'$ in (23). As these two points correspond to the points at infinity, $\phi = \pm\infty$, the real part of the left-hand side of (23) becomes $\Omega_{\pm\infty}$; the right-hand side of the resulting form of (23) can be simplified with the help of (4.35) and (24). There results

$$\Omega_\infty = \Omega_a + \frac{1}{\pi}\int_{-K}^{K}\{\theta_0(\gamma)\,Z_4(\gamma - \gamma_1) - \theta_h(\gamma)\,Z_1(\gamma - \gamma_1)\}\,d\gamma$$

and
$$\Omega_{-\infty} = \Omega_a + \frac{1}{\pi} \int_{-K}^{K} \{\theta_0(\gamma)\, Z_4(\gamma+\gamma_1) - \theta_h(\gamma)\, Z_1(\gamma+\gamma_1)\}\, d\gamma;$$

hence
$$\Omega_a = \tfrac{1}{2}(\Omega_{\infty}+\Omega_{-\infty}) - \frac{1}{2\pi}\int_{-K}^{K} \{\theta_0(\gamma)\,[Z_4(\gamma+\gamma_1)+Z_4(\gamma-\gamma_1)]$$
$$-\theta_h(\gamma)\,[Z_1(\gamma+\gamma_1)+Z_1(\gamma-\gamma_1)]\}\, d\gamma. \quad (26)$$

On the aerofoil, $\eta = 0$, (23) gives

$$\Omega_0(\gamma) = \Omega_a + \frac{1}{\pi}\int_{-K}^{K} \{\theta_0(\gamma^*)\, Z_1(\gamma^*-\gamma) - \theta_h(\gamma^*)\, Z_4(\gamma^*-\gamma)\}\, d\gamma^*, \quad (27)$$

while on the channel walls, $\eta = iK'$, and by (4.35) and (23)

$$\Omega_h(\gamma) = \Omega_a + \frac{1}{\pi}\int_{-K}^{K} \{\theta_0(\gamma^*)\, Z_4(\gamma^*-\gamma) - \theta_h(\gamma^*)\, Z_1(\gamma^*-\gamma)\}\, d\gamma^*. \quad (28)$$

These equations hold for an obstacle of given shape placed in a given channel. The next step is to convert (27) and (28) into integral equations for q_0 and q_h by an integration by parts (cf. (7.14) and (7.15)), but as the method is similar to that already described in §7.2 we shall omit the details.

Equations (4.98) and (6) yield an alternative formula to that given in (23), which is sometimes useful. This is

$$\tau(\zeta^1) = \Omega_b + \frac{1}{2\pi}\int_{-2K_1}^{2K_1} \left\{\theta_0(\gamma^1)\frac{\operatorname{cn}\zeta^1\operatorname{dn}\zeta^1 + \operatorname{cn}\gamma^1\operatorname{dn}\gamma^1}{\operatorname{sn}\gamma^1 - \operatorname{sn}\zeta^1}\right.$$
$$\left. -\theta_h(\gamma^1)\frac{k_1\operatorname{cn}\zeta^1\operatorname{dn}\zeta^1 - \operatorname{ds}\gamma^1\operatorname{cs}\gamma^1}{\operatorname{ns}\gamma^1 - k_1\operatorname{sn}\zeta^1}\right\} d\gamma^1, \quad (29)$$

where
$$\Omega_b \equiv \tfrac{1}{2}\Omega(2K_1, K_1') + \tfrac{1}{2}\Omega(0, K_1'). \quad (30)$$

13.5   Auxiliary conditions

There are several important auxiliary conditions on the distributions $\theta_0(\gamma)$ and $\theta_h(\gamma)$, which correspond to the 'closure' conditions given for isolated aerofoils in §8.4.

First there is (24). Secondly, there is

$$\Omega_{\infty} - \Omega_{-\infty} = \frac{1}{\pi}\int_{-K}^{K} [\theta_0(\gamma)\{Z_4(\gamma-\gamma_1)-Z_4(\gamma+\gamma_1)\}$$
$$-\theta_h(\gamma)\{Z_1(\gamma-\gamma_1)-Z_1(\gamma+\gamma_1)\}]\, d\gamma, \quad (31)$$

which follows from the pair of equations preceding (26). This equation is a generalization both of (7.9) for an empty channel, and of the imaginary part of (8.27) for an isolated aerofoil. With the aid of (5), (24) and (4.35) we can write it in the form

$$\Omega_{\infty} - \Omega_{-\infty} = \frac{1}{h_1+h_2}\left\{\int_{-K}^{K}\theta_h(\gamma)\, d\phi_h(\gamma) - \int_{-K}^{K}\theta_0(\gamma)\, d\phi_0(\gamma)\right\}. \quad (32)$$

In comparing this with (7.9) it should be remembered that the subscripts 0 and h have different meanings in the two equations—in (32) 'h' includes the ranges covered by both 0 and h in (7.9).

A further relation between Ω_∞ and $\Omega_{-\infty}$ is given in (7.6), viz.

$$\Omega_\infty = -\Omega^* + \cosh^{-1}\left\{\frac{H_\infty}{H_{-\infty}}\cosh\left(\Omega^* + \Omega_{-\infty}\right)\right\}, \tag{33}$$

where H is the physical width of the channel. If the reference speed U corresponds to $\Omega = \tfrac{1}{2}(\Omega_\infty + \Omega_{-\infty})$ then (see (7.25))

$$\Omega_\infty + \Omega_{-\infty} = 0. \tag{34}$$

It follows that the ratio $H_\infty/H_{-\infty}$ fixes the values of Ω_∞ and $\Omega_{-\infty}$, and in particular fixes the value of the left-hand side of (32).

Two further restrictions can be deduced from the aerofoil closure condition given in (6.13), which may be written as

$$(1+\beta_a)\int_{-K+i\eta}^{K+i\eta} e^{\tau(\zeta)}\frac{dw}{d\zeta}d\zeta = (1-\beta_a)\int_{-K+i\eta}^{K+i\eta} e^{-\overline{\tau(\zeta)}}\frac{d\overline{w}}{d\overline{\zeta}}d\overline{\zeta}, \tag{35}$$

where η has any value between 0 and K'. In §8.4 we were able to deduce a relatively simple restriction from (6.13) by evaluating the integrals from the residues at infinity. This is unfortunately not possible in the present application. However, (35) can be simplified when it is valid, as in linear perturbation theory, to ignore $|\tau|^2$ compared with $|\tau|$. In this case $e^{\tau(\zeta)} \simeq 1+\tau(\zeta)$, and (35) becomes

$$(1+\beta_a)\left\{\Gamma + \int_{-K+i\eta}^{K+i\eta}\tau(\zeta)\,dw(\zeta)\right\} = (1-\beta_a)\left\{\Gamma - \int_{-K+i\eta}^{K+i\eta}\overline{\tau(\zeta)\,dw(\zeta)}\right\},$$

as $\displaystyle\int_{-K}^{K}d\phi = \Gamma$. This can be written

$$\beta_a\Gamma + \mathscr{R}\int_{-K+i\eta}^{K+i\eta}\tau(\zeta)\,dw(\zeta) + i\beta_a\mathscr{I}\int_{-K+i\eta}^{K+i\eta}\tau(\zeta)\,dw(\zeta) = 0. \tag{36}$$

On $\eta = 0$, $dw(\zeta) = d\phi_0(\gamma)$, and (36) yields the pair of real equations

$$\beta_a\Gamma = -\int_{-K}^{K}\Omega_0(\gamma)\,d\phi_0(\gamma), \tag{37}$$

and

$$0 = \int_{-K}^{K}\theta_0(\gamma)\,d\phi_0(\gamma). \tag{38}$$

On $\eta = iK'$,

$$dw(\zeta) = d\phi_h(\gamma) + i(h_1+h_2)\{dU(\gamma+\gamma_1) - dU(\gamma-\gamma_1)\},$$

where the unit functions have been introduced to allow for the jump in ψ at the points at infinity, $\zeta = -\gamma_1 + iK'$ and $\zeta = \gamma_1 + iK'$ (cf. Fig. 13.1a).

　　　　　　　　　　　　　　　　　　　　　　　　　　　　　

At $\zeta = \pm\gamma_1 + iK'$, $\tau(\zeta) = \tau_{\pm\infty}$, and therefore evaluating the Stieltjes integrals in (36) in the usual way (see § 3.6) we get

$$\beta_a \Gamma + \int_{-K}^{K} \Omega_h(\gamma)\, d\phi_h(\gamma) - (h_1 + h_2)(\theta_{-\infty} - \theta_{\infty}) = 0, \tag{39}$$

and

$$\int_{-K}^{K} \theta_h(\gamma)\, d\phi_h(\gamma) + (h_1 + h_2)(\Omega_{-\infty} - \Omega_{\infty}) = 0. \tag{40}$$

It should be noted that as (38) and (40) are both derived from the imaginary part of (36), they will contain the same error term due to the use of the approximation $e^{\tau} \simeq 1 + \tau$. Therefore if we subtract one from the other, this term will be eliminated, and an exact result obtained. We have here an alternative derivation of (32). Similarly, from (37) and (39) we find the exact relation

$$\theta_{\infty} - \theta_{-\infty} = \frac{1}{h_1 + h_2}\left\{\int_{-K}^{K} \Omega_0(\gamma)\, d\phi_0(\gamma) - \int_{-K}^{K} \Omega_h(\gamma)\, d\phi_h(\gamma)\right\}, \tag{41}$$

which is the harmonic conjugate of (32).

Equation (38) enables us to reduce (32) to

$$\Omega_{\infty} - \Omega_{-\infty} = \frac{1}{h_1 + h_2}\int_{-K}^{K} \theta_h(\gamma)\, d\phi_h(\gamma), \tag{42}$$

which is in exact agreement with (7.9).

13.6 Fourier series expansions

From results given on page 489 of Whittaker & Watson (1952) and the definitions of the zeta functions given in § 4.5 it follows that $Z_4(x)$ and $Z_1(x)$ have the Fourier expansions:

$$Z_4(x, q) = \frac{2\pi}{K} \sum_{n=1}^{\infty} \frac{q^n}{1 - q^{2n}} \sin\frac{n\pi x}{K}, \tag{43}$$

$$Z_1(x, q) = \frac{\pi}{2K}\cot\frac{\pi x}{2K} + \frac{2\pi}{K}\sum_{n=1}^{\infty}\frac{q^{2n}}{1 - q^{2n}}\sin\frac{n\pi x}{K}, \tag{44}$$

where here q denotes the parameter

$$q = e^{\pi i \mu} = e^{-\pi K'/K}, \tag{45}$$

and should not be confused with the velocity q.

These expansions enable us to write (3), (5) and (23) (for $\Gamma = 0$) in the forms

$$0 = \sum_{n=1}^{\infty} \frac{q^n}{1 - q^{2n}} \sin\frac{n\pi\gamma_1}{K}\cos\frac{n\pi\gamma_0}{K}, \tag{46}$$

$$\frac{dw}{d\zeta} = \frac{4(h_1 + h_2)}{K}\sum_{n=1}^{\infty}\frac{q^n}{1 - q^{2n}}\sin\frac{n\pi\gamma_1}{K}\cos\frac{n\pi\zeta}{K}, \tag{47}$$

and

$$\tau(\zeta) = \Omega_a + \frac{1}{2K}\int_{-K}^{K}\theta_0(\gamma)\left\{\cot\frac{\pi}{2K}(\gamma-\zeta)+4\sum_{n=1}^{\infty}\frac{q^{2n}}{1-q^{2n}}\sin\frac{n\pi}{K}(\gamma-\zeta)\right\}d\gamma$$

$$-\frac{1}{2K}\int_{-K}^{K}\theta_h(\gamma)\,4\sum_{n=1}^{\infty}\frac{q^n}{1-q^{2n}}\sin\frac{n\pi}{K}(\gamma-\zeta)\,d\gamma. \quad (48)$$

Moses (1949) has used equations similar in form to (46), (47) and (48)—except that the displacement function defined in (6.20) was used in place of τ—to calculate the incompressible flow about aerofoils of arbitrary shape in wind tunnels. The numerical details and effort required to solve the integral equation implicit in (48) (cf. (8.16)) are rather extensive, and so we shall say only that an iterative method similar to that described in §8.23 can be used. The parallel work of Moses using the displacement function should be consulted by any reader attempting a problem of this type.

13.7　Design of a bend in a channel with a single turning vane

Consider for a moment the conjugate equations to (27) and (28):

$$\theta_0(\gamma) = \theta_a - \frac{1}{\pi}\int_{-K}^{K}\{\Omega_0(\gamma^*)\,Z_1(\gamma^*-\gamma)-\Omega_h(\gamma^*)\,Z_4(\gamma^*-\gamma)\}\,d\gamma^*, \quad (49)$$

$$\theta_h(\gamma) = \theta_a - \frac{1}{\pi}\int_{-K}^{K}\{\Omega_0(\gamma^*)\,Z_4(\gamma^*-\gamma)-\Omega_h(\gamma^*)\,Z_1(\gamma^*-\gamma)\}\,d\gamma^*, \quad (50)$$

where θ_a is defined by the conjugate equation to (26). These equations could be used to design a bend in a channel in which one turning vane is inserted (see Fig. 13.7). Given the velocity distributions over the channel walls and over the turning vane, (49) will yield the shape of the vane, and (50) will give the shape of the channel.

For example, to avoid boundary-layer separation on the walls we might set $q_h = U$, i.e. $\Omega_h = 0$; Ω_0 could then be chosen so that it gives the correct deflexion angle (see (41)), satisfies the first of (24) and minimizes the adverse pressure gradients inevitable near the trailing edge of the vane.

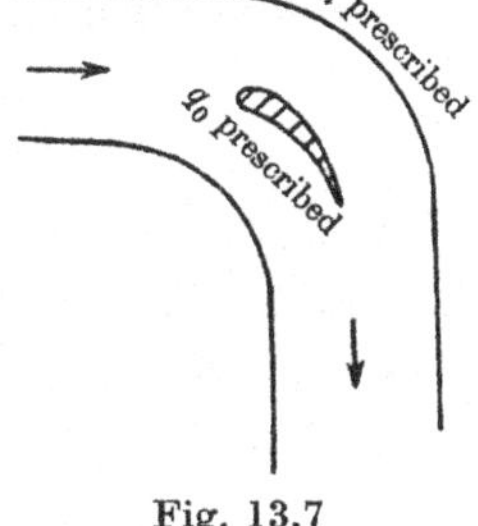

Fig. 13.7

This principle has been used by A. S. Thom (1952) to design channel bends. The design was carried out, not by the above equations, but by Thom's (1947) numerical method of calculating harmonic conjugates.

13.8 Exact theory for a flat plate in a straight channel

A classical problem, which was solved in the nineteen-thirties by a variety of methods (see Rosenhead, 1931; Tomotika, 1934–39; Havelock, 1938) is that of calculating incompressible flow past a flat plate at incidence in a straight, solid-walled channel. An exact theory for this problem is readily deduced from the equations given in §§ 13.4 and 13.5.

First consider the *non-circulating* flow about the plate shown in Fig. 13.8a (cf. Fig. 13.1a). The plate is at an incidence α, and the stagnation points are at $\gamma = \pm\gamma_0$. Let the leading and trailing edges of the plate be at $\gamma = -\gamma_0+\epsilon_n$ and $\gamma = \gamma_0+\epsilon_t$ respectively. The stagnation

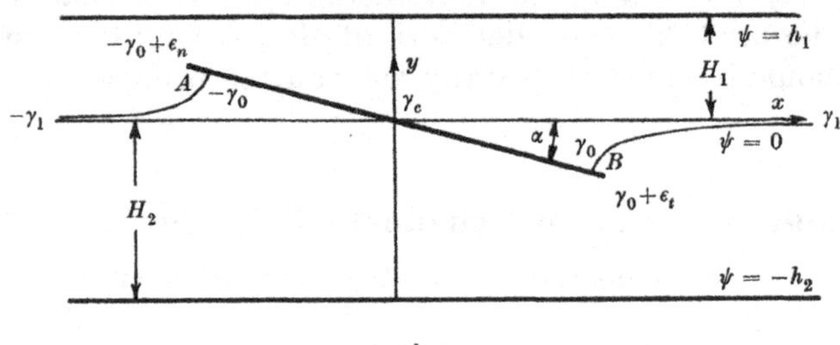

Fig. 13.8a

streamlines will be asymptotic to a line which we shall take to be the Ox-axis. Let this line intersect the upper surface of the plate at $\gamma = \gamma_c$. The speed at $x = \pm\infty$ is U and so $\Omega_\infty = \Omega_{-\infty} = 0$. As the flow is incompressible, $h_1 = H_1 U$, $h_2 = H_2 U$ where H_1, H_2 are the distances of the walls from the Ox-axis.

It follows from (3) and (4) that

$$\frac{\pi\gamma_1}{K} = \tfrac{1}{2}\pi + J, \quad \text{where} \quad J \equiv \tfrac{1}{2}\pi\frac{H_2-H_1}{H_2+H_1}, \tag{51}$$

and
$$Z_4(\gamma_0-\gamma_1) = Z_4(\gamma_0+\gamma_1). \tag{52}$$

Equations (4.45), (4.67) and (4.77) enable us to solve (52) for γ_0; we find that γ_0 is given by

$$\mathrm{ns}^2\gamma_0 = k^2\,\mathrm{sn}^2\gamma_1\frac{Z_1(\gamma_1)+Z_4(\gamma_1)}{2Z_4(\gamma_1)}. \tag{53}$$

It is apparent from Fig. 13.8a that the flow direction is

$$\theta_0 = -\alpha + \pi\{U(\gamma+\gamma_0) - U(\gamma+\gamma_0-\epsilon_n) + U(\gamma-\gamma_0) - U(\gamma-\gamma_0-\epsilon_t)\} \tag{54}$$

while on the walls, $\theta_h = 0$. Substitution of these values in (24) and (23) yields

$$\epsilon_n + \epsilon_t = \frac{2K\alpha}{\pi}, \tag{55}$$

and

$$\tau(\zeta) = \Omega_a - i\alpha + \ln \frac{\vartheta_1\!\left\{\frac{\pi}{2K}(\gamma_0 + \zeta - \epsilon_n)\right\} \vartheta_1\!\left\{\frac{\pi}{2K}(\gamma_0 - \zeta + \epsilon_l)\right\}}{\vartheta_1\!\left\{\frac{\pi}{2K}(\gamma_0 + \zeta)\right\} \vartheta_1\!\left\{\frac{\pi}{2K}(\gamma_0 - \zeta)\right\}}, \qquad (56)$$

where by (4.32) the parameter of the ϑ functions is $\mu = iK'/K$. As $\tau = \Omega + i\theta$ vanishes at $x = \pm\infty$, i.e. at $\zeta = \pm\gamma_1 + iK'$, it follows from (56) and (4.26) that

$$\Omega_a = \ln \frac{\vartheta_4\!\left\{\frac{\pi}{2K}(\gamma_0 + \gamma_1)\right\} \vartheta_4\!\left\{\frac{\pi}{2K}(\gamma_0 - \gamma_1)\right\}}{\vartheta_4\!\left\{\frac{\pi}{2K}(\gamma_0 + \gamma_1 - \epsilon_n)\right\} \vartheta_4\!\left\{\frac{\pi}{2K}(\gamma_0 - \gamma_1 + \epsilon_l)\right\}}$$

$$= \ln \frac{\vartheta_4\!\left\{\frac{\pi}{2K}(\gamma_0 + \gamma_1)\right\} \vartheta_4\!\left\{\frac{\pi}{2K}(\gamma_0 - \gamma_1)\right\}}{\vartheta_4\!\left\{\frac{\pi}{2K}(\gamma_0 - \gamma_1 - \epsilon_n)\right\} \vartheta_4\!\left\{\frac{\pi}{2K}(\gamma_0 + \gamma_1 + \epsilon_l)\right\}} \qquad (57)$$

whence

$$\vartheta_4\!\left\{\frac{\pi}{2K}(\gamma_0 + \gamma_1 - \epsilon_n)\right\} \vartheta_4\!\left\{\frac{\pi}{2K}(\gamma_0 - \gamma_1 + \epsilon_l)\right\}$$

$$= \vartheta_4\!\left\{\frac{\pi}{2K}(\gamma_0 - \gamma_1 - \epsilon_n)\right\} \vartheta_4\!\left\{\frac{\pi}{2K}(\gamma_0 + \gamma_1 + \epsilon_l)\right\}, \quad (58)$$

which can also be deduced from (1), (38), (52) and (54). With the help of the addition formulae given for the theta functions on p. 487 of Whittaker & Watson (1952), and (55) we can solve (58) to find

$$k\,\mathrm{sn}^2\!\left\{\frac{\pi\gamma_0}{2K} + \frac{\pi}{4K}(\epsilon_l - \epsilon_n)\right\} = \frac{\vartheta_4^2\!\left(\frac{\pi\gamma_1}{2K} + \frac{\alpha}{2}\right) - \vartheta_4^2\!\left(\frac{\pi\gamma_1}{2K} - \frac{\alpha}{2}\right)}{\vartheta_1^2\!\left(\frac{\pi\gamma_1}{2K} + \frac{\alpha}{2}\right) - \vartheta_1^2\!\left(\frac{\pi\gamma_1}{2K} - \frac{\alpha}{2}\right)}. \qquad (59)$$

Equations (51), (53), (55) and (59) enable us to calculate the values of γ_0, γ_1, ϵ_n and ϵ_l provided we know the value of k, the modulus of the elliptic functions. A relation between k and c, the length of the plate, can be found from $c = \{z(\gamma_0 + \epsilon_l) - z(-\gamma_0 + \epsilon_n)\} e^{i\alpha}$, and

$$Uz = \int_{\gamma_c}^{\zeta} \frac{U\,dz}{dw}\frac{dw}{d\zeta}\,d\zeta = \int_{\gamma_c}^{\zeta} e^{\tau(\zeta)}\frac{dw}{d\zeta}\,d\zeta, \qquad (60)$$

where $dw/d\zeta$ and $\tau(\zeta)$ are given by (5) and (56). Finally a relation for γ_0 can be found from $H_1 = \mathscr{I}z(\gamma + iK')$, or

$$UH_1 = \mathscr{I}\int_{\gamma_c}^{\gamma + iK'} e^{\tau(\zeta)}\frac{dw}{d\zeta}\,d\zeta. \qquad (61)$$

The distance x_c from the leading edge of the plate to γ_c is then $x_c = \{z(\gamma_c) - z(\gamma_0 + \epsilon_t)\}\, e^{i\alpha}$. It does not appear to be possible to evaluate the integrals in (60) and (61) in terms of tabulated functions, and so approximate methods may have to be adopted here.

Having established the relation between the z-plane and the ζ-plane for zero circulation we can use this ζ-plane to study the circulating flow about the plate. This is so because (i) the ζ-plane defined by the inverse of (60) can be regarded as being a (convenient) conformal mapping of the z-plane, (ii) the z-plane is not affected by the circulation, Γ, and (iii) in incompressible flow τ (with $\Gamma \neq 0$) is an analytic function of z. In compressible flow (iii) does not hold and we must use the ζ-plane corresponding to the actual circulating flow. The disadvantage of this is that the presence of the last term of (4) makes it impossible to disentangle the theory into an explicit form unless approximations are introduced.

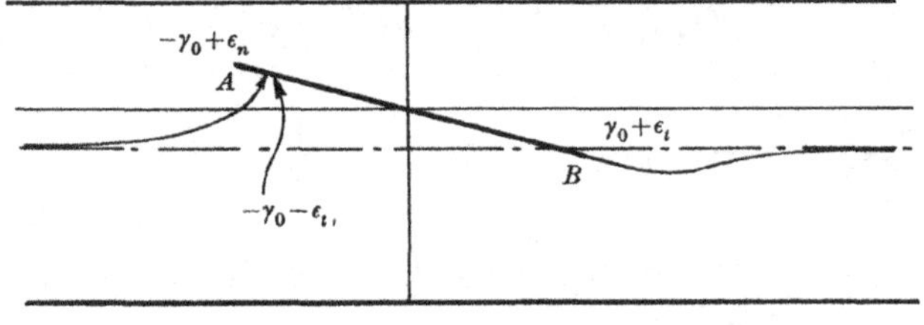

Fig. 13.8b

The effect of circulation on the flow about the plate is shown in Fig. 13.8b. The rear stagnation point moves to the trailing edge by Joukowski's condition, while the front stagnation point moves to $\gamma = -\gamma_0 - \epsilon_t$ in order that the new distribution of θ_0, viz. (an asterisk will be used to denote values in the circulating flow)

$$\theta_0^* = -\alpha + \pi\{\mathbf{U}(\gamma + \gamma_0 + \epsilon_t) - \mathbf{U}(\gamma + \gamma_0 - \epsilon_n)\} \tag{62}$$

should satisfy (24). From (23) and (62)

$$\tau^* = \Omega_a^* - i\alpha + \ln \frac{\vartheta_1\left\{\dfrac{\pi}{2K}(\gamma_0 + \zeta - \epsilon_n)\right\}}{\vartheta_1\left\{\dfrac{\pi}{2K}(\gamma_0 + \zeta + \epsilon_t)\right\}}, \tag{63}$$

where in place of (57),

$$\Omega_a^* = \ln \frac{\vartheta_4\left\{\dfrac{\pi}{2K}(\gamma_0 - \gamma_1 + \epsilon_t)\right\}}{\vartheta_4\left\{\dfrac{\pi}{2K}(\gamma_0 - \gamma_1 - \epsilon_n)\right\}} = \ln \frac{\vartheta_4\left\{\dfrac{\pi}{2K}(\gamma_0 + \gamma_1 + \epsilon_t)\right\}}{\vartheta_4\left\{\dfrac{\pi}{2K}(\gamma_0 + \gamma_1 - \epsilon_n)\right\}} \tag{64}$$

(cf. (58)).

The lift and moment acting on the plate can now be calculated from

$$L + iD = \tfrac{1}{2}\rho_a U \int_{-K}^{K} e^{-\tau^*} \frac{dw^*}{dz}\frac{dz}{dw}\frac{dw}{d\gamma} d\gamma = \tfrac{1}{2}\rho_a U \int_{-K}^{K} e^{-2\tau^*+\tau}\left(\frac{d\phi_0}{d\gamma}\right) d\gamma, \quad (65)$$

and

$$M = -\tfrac{1}{2}\rho_a U \mathscr{R} \int_{-K}^{K} z\, e^{-\tau^*}\frac{dw^*}{dz}\frac{dz}{dw}\frac{dw}{d\gamma} d\gamma = -\tfrac{1}{2}\rho_a U \mathscr{R} \int_{-K}^{K} z\, e^{-2\tau^*+\tau}\left(\frac{d\phi_0}{d\gamma}\right) d\gamma \tag{66}$$

(cf. similar equations given in §§ 8.10 and 8.11).

This completes the exact theory for the flat plate. Probably the best method of using it is to estimate the modulus k from the approximate relation between k and c/H ($H \equiv H_1 + H_2$) given below in § 13.10, then calculate $\gamma_c, \gamma_1, \epsilon_n$ and ϵ_t from (51), (53), (55) and (59). Then the values of $\gamma_c, x_c/H$ and c/H are deduced from (60) and (61). This final value of c/H will probably differ slightly from that initially assumed, but it is necessary only to repeat the process with a modified value of k to obtain the required c/H ratio—or something very close to it.

13.9   Approximate theory for the flat plate

A useful approximate theory for the flat plate can be derived when it is valid to ignore terms $O(q^3)$ and higher in the Fourier Series expansions given above. As we show below this is equivalent to neglecting terms of order $(c/H)^3$ and higher.

In place of (52) we have

$$0 = \sin\frac{\pi\gamma_1}{K}\cos\frac{\pi\gamma_0}{K} + q\sin\frac{2\pi\gamma_1}{K}\cos\frac{2\pi\gamma_0}{K}$$
$$+ q^2\left\{\sin\frac{3\pi\gamma_1}{K}\cos\frac{3\pi\gamma_0}{K} + \sin\frac{\pi\gamma_1}{K}\cos\frac{\pi\gamma_0}{K}\right\} + O(q^3),$$

the solution of which is readily found to be

$$\frac{\pi\gamma_0}{K} = \tfrac{1}{2}\pi + 2q\sin J + O(q^3), \tag{67}$$

where γ_1 has been eliminated by (51). Similarly, from (38), (51) and (47)

$$(1+q^2)\int_{-K}^{K}\theta_0(\gamma)\cos\frac{\pi\gamma}{K}d\gamma - 2q\sin J\int_{-K}^{K}\theta_0(\gamma)\cos\frac{2\pi\gamma}{K}d\gamma$$
$$+ q^2(3 - 4\cos^2 J)\int_{-K}^{K}\theta_0(\gamma)\cos\frac{3\pi\gamma}{K}d\gamma + O(q^3) = 0.$$

On substituting (54) into this equation, and using (55) we find that

$$\sin\alpha\cos Y - q\sin J\sin 2\alpha\cos 2Y + q^2\{(1 - \tfrac{4}{3}\cos^2 J)\sin 3\alpha\sin 3Y$$
$$+ \sin\alpha\cos Y\} + O(q^3) = 0,$$

where
$$Y \equiv \frac{\pi}{2K}(\epsilon_n - \epsilon_t) - \frac{\pi\gamma_0}{K}.$$

Solving this equation and eliminating γ_0 by (67) we obtain

$$\frac{\pi}{2K}(\epsilon_n - \epsilon_t) = 2q \sin J(1 - \cos \alpha) + O(q^3). \tag{68}$$

From (48) and the fact that $\theta_h = 0$,

$$\tau(\zeta) = \Omega_a + \frac{1}{2K}\int_{-K}^{K} \theta_0(\gamma)\left\{\cot\frac{\pi}{2K}(\gamma - \zeta) + 4q^2 \sin\frac{\pi}{K}(\gamma - \zeta)\right\} d\gamma + O(q^4). \tag{69}$$

Substituting (54) into this equation we get

$$\tau(\zeta) = \Omega_a - i\alpha + \ln\left\{\frac{\sin\dfrac{\pi}{2K}(\gamma_0 + \zeta - \epsilon_n)\sin\dfrac{\pi}{2K}(\gamma_0 - \zeta + \epsilon_t)}{\sin\dfrac{\pi}{2K}(\gamma_0 + \zeta)\sin\dfrac{\pi}{2K}(\gamma_0 - \zeta)}\right\} + O(q^3), \tag{70}$$

where the remainder term is $O(q^3)$ because it is found from (55), (67) and (68) that the *coefficient* of the $O(q^2)$ term arising from the corresponding term in (69) is itself $O(q)$.

An equation for Ω_a is most conveniently deduced from (26), (34), (43) and (51). The result is

$$\Omega_a = \frac{2q}{K}\int_{-K}^{K} \theta_0(\gamma)\left(\sin J \sin\frac{\pi\gamma}{K} + q \cos 2J \sin\frac{2\pi\gamma}{K}\right) d\gamma + O(q^3). \tag{71}$$

From this and (54) we can obtain an expression for Ω_a, which can be simplified with the help of (55), (67) and (68) to

$$\Omega_a = -4q^2 \sin^2 \alpha + O(q^3). \tag{72}$$

By subtracting (46) from (47), and using elementary trigonometry we find

$$\frac{dw}{d\zeta} = \frac{8HU}{K} q \cos J \sin\frac{\pi}{2K}(\gamma_0 - \zeta)\sin\frac{\pi}{2K}(\gamma_0 + \zeta)\left[1 - 8q \sin J \cos\frac{\pi}{2K}\right.$$

$$\times (\gamma_0 - \zeta)\cos\frac{\pi}{2K}(\gamma_0 + \zeta) + q^2\left\{(3 - 4\cos^2 J)\left(1 + 2\cos\frac{\pi}{K}(\gamma_0 - \zeta)\right.\right.$$

$$\left.\left.\left. + 2\cos\frac{\pi}{K}(\gamma_0 + \zeta) + 2\cos\frac{2\pi\zeta}{K} + 2\cos\frac{2\pi\gamma_0}{K}\right) + 1\right\}\right],$$

where
$$h_1 + h_2 = U(H_1 + H_2) = UH, \tag{73}$$

and γ_1 has been eliminated by (51). Elimination of γ_0 from the last factor of $dw/d\zeta$ by (67) gives

$$\frac{dw}{d\zeta} = \frac{8HU}{K} q \cos J \sin \frac{\pi}{2K} (\gamma_0 - \zeta) \sin \frac{\pi}{2K} (\gamma_0 + \zeta) \left\{ 1 - 4q \sin J \cos \frac{\pi \zeta}{K} \right.$$

$$\left. + 2q^2 \left(1 + 2\sin^2 J - \cos \frac{2\pi \zeta}{K} + 4\sin^2 J \cos \frac{2\pi \zeta}{K} \right) \right\}. \quad (74)$$

Substituting (70) and (74) into (60) and making use of (55), (67), (68) and (72) we get

$$z = \frac{4H}{\pi} q \cos J \, e^{-i\alpha} \left[\sin \left(\frac{\pi \zeta}{K} - \alpha \right) - q \sin J \sin \left(\frac{2\pi \zeta}{K} - \alpha \right) \right.$$

$$+ q^2 \left\{ (1 - 4\sin^2 \alpha) \sin \frac{\pi \zeta}{K} \cos \alpha - (3 - 4\sin^2 \alpha) \cos \frac{\pi \zeta}{K} \sin \alpha \right.$$

$$\left. \left. - \tfrac{1}{3}(1 - 4\sin^2 J) \sin \left(\frac{3\pi \zeta}{K} - \alpha \right) \right\} \right]. \quad (75)$$

The chord length is $c = \{z(\gamma_0 + \epsilon_t) - z(-\gamma_0 + \epsilon_n)\} e^{i\alpha}$ (see Fig. 13.8a), so

$$c = \frac{8H}{\pi} q \cos J \{ 1 + \tfrac{2}{3} q^2 (2 - 4\sin^2 \alpha + \sin^2 J + \sin^2 J \sin^2 \alpha) \},$$

by (55), (67), (68) and (75). Hence

$$q = \frac{\pi c}{8H} \sec J \left\{ 1 - \frac{1}{96} \left(\frac{\pi c}{H} \right)^2 \sec^2 J (2 - 4\sin^2 \alpha + \sin^2 J + \sin^2 J \sin^2 \alpha) \right\}. \quad (76)$$

With the help of (4.23), (55), (67) and (68) we can expand (63) and (64) and find

$$e^{\tau^*} = e^{\Omega_a^* - i\alpha} \frac{\sin \dfrac{\pi}{2K} (\gamma_0 + \zeta - \epsilon_n)}{\sin \dfrac{\pi}{2K} (\gamma_0 + \zeta + \epsilon_t)} \left(1 - 4q^2 \sin \alpha \cos \frac{\pi \zeta}{K} \right), \left. \right\} \quad (77)$$

and $\qquad e^{\Omega_a^*} = 1 - 4q \sin \alpha \sin J + 8q^2 \sin^2 \alpha \sin^2 J.$

Substitution of (70), (74) and (77) into (65), and reduction of the result by (55), (67) and (68) now yields

$$C_L \equiv \frac{L}{\tfrac{1}{2} c \rho_a U^2} = 16 \left(\frac{H}{c} \right) q \cos J \sin \alpha \{ 1 + 4q \sin J \sin \alpha$$

$$+ q^2 (4 + 2\sin^2 J + 10 \sin^2 J \sin^2 \alpha) \}.$$

Finally elimination of q by (76) yields the result

$$C_L = 2\pi \sin \alpha \left[1 + \frac{\pi c}{2H} \tan J \sin \alpha \right.$$
$$\left. + \left(\frac{\pi c}{4H}\right)^2 \{\sec^2 J - \tfrac{1}{3} + (3 \tan^2 J + \tfrac{2}{3}) \sin^2 \alpha\} \right], \quad (78)$$

the error term being $O(c/H)^3$.

Similarly, from (66) it can be shown that the moment coefficient about the origin is given by

$$C_M = \tfrac{1}{4}\pi \sin 2\alpha \left[1 + \frac{\pi c}{2H} \tan J \sin \alpha \right.$$
$$\left. + \frac{1}{2}\left(\frac{\pi c}{4H}\right)^2 \{\sec^2 J - \tfrac{1}{3} + (4 \sec^2 J + 6 \tan^2 J) \sin^2 \alpha\} \right]. \quad (79)$$

Equations (78) and (79) are due to Havelock (1938), who has carried the expansion up to terms of order $(c/H)^4$ by a somewhat different approach to that given above.

The above theory can be readily extended to deal with thin aerofoils by adding a term $(dy/dx)_s$ (cf. (8.118) and (8.121)) to the right-hand side of (54). This extension will not be given here, but the reader can consult a paper by Goldstein (1942), which gives results up to order $(c/H)^2$ for aerofoils of small camber and thickness in any position in a wind tunnel, and without restriction on the incidence.

13.10   Linear perturbation theory

The most important example of the theory given in §§ 13.4 and 13.5 is the case of a thin aerofoil at a small incidence between straight, solid channel walls (see Fig. 13.10). The following theory is also valid for curved walls provided these walls remain fairly close to the straight lines $y = H_1$, $y = -H_2$, assumed to pass through their mean positions.

The linear perturbation approximations (see § 6.1) viz. $w = UZ$, $\Omega = \tfrac{1}{2}\beta_a C_p$ and $\theta = dy/dx$ will now be introduced. Because of the circulation Γ, there is a discontinuity in w about any contour enclosing the aerofoil, but of course this discontinuity is not shared by Z ($Z = x + i\beta_a y$). Consequently, when we replace w in (1) by UZ, we must also put $\Gamma = 0$ in (3). In other words, the linearization $w = UZ$ must be applied to the w-plane for zero circulation. Let the channel have the same width H at $\phi = \infty$ and $\phi = -\infty$, and let $q = U$ at these points, then

$$\Omega_\infty = \Omega_{-\infty} = 0. \quad (80)$$

By integrating (6.11) across the tunnel at infinity we obtain

$$\int_0^{h_1} d\psi = \beta_a U \int_0^{H_1} dn,$$

i.e. $h_1 = \beta_a U H_1$ and similarly $h_2 = \beta_a U H_2$. Also the chord length c is approximately $4b/U$. Equations (1)–(4) can now be replaced by the following linear theory equivalents:

$$Z = x + i\beta_a y = \frac{2\beta_a H}{\pi}\{\zeta Z_4(\gamma_1) - \Pi(\zeta, \gamma_1)\}, \tag{81}$$

$$c = \frac{4\beta_a H}{\pi}\{\gamma_0 Z_4(\gamma_1) - \Pi(\gamma_0, \gamma_1)\}, \tag{82}$$

$$0 = Z_4(\gamma_0 - \gamma_1) - Z_4(\gamma_0 + \gamma_1), \tag{83}$$

and

$$\gamma_1 = \tfrac{1}{2}K\left(1 + \frac{m}{H}\right), \tag{84}$$

where

$$H \equiv H_1 + H_2, \quad m \equiv H_2 - H_1. \tag{85}$$

The points $-\gamma_0$, γ_0 are now the leading and trailing edges of the aerofoil, rather than the front and rear stagnation points.

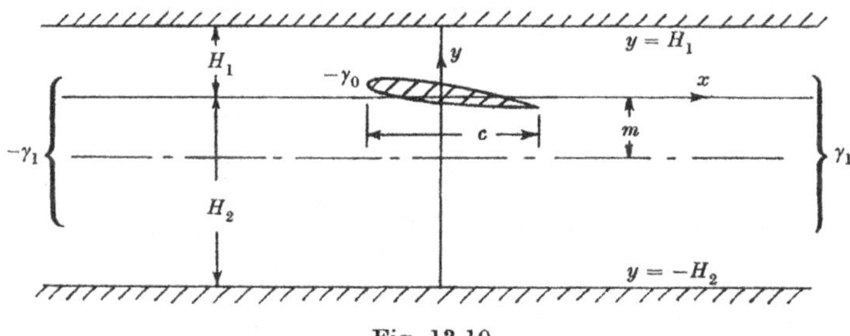

Fig. 13.10

A general expression for the lift force acting on a thin aerofoil can be calculated as follows. The lift and drag forces acting on the aerofoil are given by (see (8.77))

$$L + iD = -\frac{\rho_a U}{1 + \beta_a}\int_C e^{-\tau(\zeta)}dw(\zeta),$$

where τ and w refer to circulating, compressible flow, and C is a contour enclosing the aerofoil. (The positive direction for C is from $\gamma = K$ to $\gamma = -K$.) On $\eta = 0$ this becomes

$$L + iD = \frac{\rho_a U}{1 + \beta_a}\int_{-K}^{K} e^{-\tau_0(\gamma)}d\phi_0(\gamma) \simeq \frac{\rho_a U}{1 + \beta_a}\left\{\Gamma - \int_{-K}^{K}(\Omega_0 + i\theta_0)\,d\phi_0(\gamma)\right\},$$

adopting the linearization employed in the derivation of (37) and (38). It follows from these last-named equations that $D = 0$, and $L = \rho_a U\Gamma$. Hence

$$C_L = \frac{2\Gamma}{cU} = -\frac{2}{cU\beta_a}\int_{-K}^{K}\Omega_0(\gamma)\,d\phi_0(\gamma). \tag{86}$$

By substitution of (27) into (86) one finds

$$C_L = -\frac{2}{cU\beta_a}\left\{\Omega_a\Gamma + \frac{1}{\pi}\int_{-K}^{K}\theta_0(\gamma^*)\,I_0(\gamma^*)\,d\gamma^* - \frac{1}{\pi}\int_{-K}^{K}\theta_h(\gamma^*)\,I_h(\gamma^*)\,d\gamma^*\right\},$$

(87)

where

$$I_0(\gamma^*) \equiv \int_{-K}^{K}Z_1(\gamma^*-\gamma)\,d\phi_0(\gamma), \quad I_h(\gamma^*) \equiv \int_{-K}^{K}Z_4(\gamma^*-\gamma)\,d\phi_h(\gamma).$$

Now

$$I_0(\gamma^*) = \Gamma Z_4(\gamma^*) + \int_0^K\{Z_1(\gamma^*-\gamma)+Z_1(\gamma^*+\gamma)-2Z_4(\gamma^*)\}\left(\frac{d\phi_0}{d\gamma}\right)d\gamma,$$

as $d\phi_0/d\gamma$ is an even function of γ (put $\zeta = \gamma$ in (5)). Put $z = \gamma + iK'$, $y = \gamma^*$ in (4.77) and use (4.35) and (4.48) to obtain

$$Z_1(\gamma^*-\gamma)+Z_1(\gamma^*+\gamma)-2Z_4(\gamma^*) = -\frac{2\,\mathrm{sn}\,\gamma^*\,\mathrm{cn}\,\gamma^*\,\mathrm{dn}\,\gamma^*}{\mathrm{sn}^2\gamma-\mathrm{sn}^2\gamma^*}.$$

(88)

Thus

$$I_0(\gamma^*) = \Gamma Z_4(\gamma^*) - \frac{h_1+h_2}{\pi}\frac{4k^2\,\mathrm{sn}\,\gamma_1\,\mathrm{cn}\,\gamma_1\,\mathrm{dn}\,\gamma_1\,\mathrm{sn}\,\gamma^*\,\mathrm{cn}\,\gamma^*\,\mathrm{dn}\,\gamma^*}{1-k^2\,\mathrm{sn}^2\gamma_0\,\mathrm{sn}^2\gamma_1}$$

$$\times\int_0^K\frac{(\mathrm{sn}^2\gamma_0-\mathrm{sn}^2\gamma)\,d\gamma}{(1-k^2\,\mathrm{sn}^2\gamma_1\,\mathrm{sn}^2\gamma)(\mathrm{sn}^2\gamma-\mathrm{sn}^2\gamma^*)}.$$

To evaluate this integral we note first that

$$\int_0^K\frac{d\gamma}{\mathrm{sn}^2\gamma-\mathrm{sn}^2\gamma^*} = -\mathrm{ns}^2\gamma^*\int_0^K\left(1+\frac{\mathrm{sn}^2\gamma}{\mathrm{sn}^2\gamma^*-\mathrm{sn}^2\gamma}\right)d\gamma$$

$$= -\mathrm{ns}^2\gamma^*\left\{K-\frac{\mathrm{sn}\,\gamma^*}{\mathrm{cn}\,\gamma^*\,\mathrm{dn}\,\gamma^*}\Pi(K,\gamma^*+iK')\right\}$$

$$= -K\,\mathrm{ns}^2\gamma^*\left\{1-\frac{\mathrm{sn}\,\gamma^*}{\mathrm{cn}\,\gamma^*\,\mathrm{dn}\,\gamma^*}Z_4(\gamma^*+iK')\right\}$$

$$= -K\,\mathrm{ns}^2\gamma^*\left\{1-\frac{\mathrm{sn}\,\gamma^*}{\mathrm{cn}\,\gamma^*\,\mathrm{dn}\,\gamma^*}Z_1(\gamma^*)\right\}$$

$$= \frac{KZ_4(\gamma^*)}{\mathrm{sn}\,\gamma^*\,\mathrm{cn}\,\gamma^*\,\mathrm{dn}\,\gamma^*},$$

(89)

on making use of (4.35), (4.48), (4.67) and relations given in § 4.8. (In this calculation we have dropped an imaginary term not relevant in the final result.) And secondly that

$$\int_0^K\frac{d\gamma}{1-k^2\,\mathrm{sn}^2\gamma_1\,\mathrm{sn}^2\gamma} = K\left\{1+\frac{\mathrm{sn}\,\gamma_1}{\mathrm{cn}\,\gamma_1\,\mathrm{dn}\,\gamma_1}Z_4(\gamma_1)\right\} = \frac{K\,\mathrm{sn}\,\gamma_1\,Z_1(\gamma_1)}{\mathrm{cn}\,\gamma_1\,\mathrm{dn}\,\gamma_1}.$$

(90)

The integral in I_0 can be evaluated with the help of these two results.

$$I_0(\gamma^*) = \Gamma Z_4(\gamma^*) + \frac{h_1+h_2}{\pi}\,\frac{4Kk^2\operatorname{sn}^2\gamma_1}{1-k^2\operatorname{sn}^2\gamma_1\operatorname{sn}^2\gamma^*}\left\{\operatorname{sn}\gamma^*\operatorname{cn}\gamma^*\operatorname{dn}\gamma^* Z_1(\gamma_1)\right.$$
$$\left. + \frac{\operatorname{cn}\gamma_1\operatorname{dn}\gamma_1}{\operatorname{sn}\gamma_1}\,\frac{\operatorname{sn}^2\gamma^*-\operatorname{sn}^2\gamma_0}{1-k^2\operatorname{sn}^2\gamma_0\operatorname{sn}^2\gamma_1}Z_4(\gamma^*)\right\},$$

or by (3), (5) and (4.77)

$$I_0(\gamma^*) = 2K\frac{h_1+h_2}{\pi}\{[Z_4(\gamma^*-\gamma_1)-Z_4(\gamma^*+\gamma_1)]Z_4(\gamma^*)$$
$$-[Z_4(\gamma^*-\gamma_1)+Z_4(\gamma^*+\gamma_1)-2Z_4(\gamma^*)]Z_1(\gamma_1)\}. \quad (91)$$

Similarly, it can be shown that

$$I_h(\gamma^*) = 2K\frac{h_1+h_2}{\pi}\{[Z_4(\gamma^*-\gamma_1)-Z_4(\gamma^*+\gamma_1)]Z_1(\gamma^*)$$
$$-[Z_4(\gamma^*-\gamma_1)+Z_4(\gamma^*+\gamma_1)-2Z_1(\gamma^*)]Z_4(\gamma_1)\}. \quad (92)$$

The product $\Omega_a\Gamma$ in (87) is a second-order term, and so it can be neglected in a linear theory. Therefore from (87), (91), (92) and $h_1+h_2 = \beta_a U(H_1+H_2) = \beta_a UH$,

$$C_L = \frac{4K}{\pi^2}\left(\frac{H}{c}\right)\left[\int_{-K}^{K}\theta_0\{Z_1(\gamma_1)[Z_4(\gamma-\gamma_1)+Z_4(\gamma+\gamma_1)-2Z_4(\gamma)]\right.$$
$$-Z_4(\gamma)[Z_4(\gamma-\gamma_1)-Z_4(\gamma+\gamma_1)]\}\,d\gamma$$
$$-\int_{-K}^{K}\theta_h\{Z_4(\gamma_1)[Z_4(\gamma-\gamma_1)+Z_4(\gamma+\gamma_1)-2Z_1(\gamma)]$$
$$\left.-Z_1(\gamma)[Z_4(\gamma-\gamma_1)-Z_4(\gamma+\gamma_1)]\}\,d\gamma\right]. \quad (93)$$

Consider the case of a straight channel, i.e. the case $\theta_h = 0$. By (24) and (38) θ_0 must satisfy

$$\int_{-K}^{K}\theta_0(\gamma)\,d\gamma = 0,$$

and

$$\int_{-K}^{K}\theta_0(\gamma)\,d\phi_0(\gamma) = 0. \quad (94)$$

As $\theta_0 d\phi_0 \simeq \dfrac{dy}{dx}d(Ux) = Udy$, the second of these is automatically satisfied in linear perturbation theory and needs no further consideration.

Let α^* denote the incidence measured from the aerofoil chord line, and let $(dy/dx)_s$ be the aerofoil slope relative to this line, then as the leading edge is at $\gamma = -\gamma_0$ we can set (cf. (8.159))

$$\theta_0(\gamma) = \left(\frac{dy}{dx}\right)_s -\alpha^* + 2K(\alpha^*-\alpha_i)\,\delta(\gamma+\gamma_0), \quad (95)$$

where α_i is the ideal incidence of the aerofoil, and δ is the delta function. The number γ_0 is related to γ_1 by (53). To verify (95) we notice that it satisfies (94) at all values of α^* provided

$$\alpha_i = \frac{1}{2K} \int_{-K}^{K} \left(\frac{dy}{dx}\right)_s d\gamma, \tag{96}$$

which should be compared with (8.122).

A formula for the no-lift angle α' can be deduced from (93), (95) and (96)—in fact a thin aerofoil (in tunnel) theory can be developed in exact parallel to the theory for an isolated aerofoil given in §§ 8.14–8.20. We shall leave this to the reader.

13.11   Aerofoil in the centre of the channel

When the aerofoil is in the centre of the tunnel, and the circulation is zero, or relatively small, it is convenient to use the χ-plane shown in Fig. 13.2 as the plane of independent variables. This plane is related to w by (see § 13.2)

$$\frac{dw}{d\chi} = \frac{hk_1}{\pi} \operatorname{sn}\chi, \quad w = \frac{h}{2\pi}\ln\left(\frac{1 - k_1 \operatorname{cd}\chi}{1 + k_1 \operatorname{cd}\chi}\right), \tag{97}$$

where
$$k_1 = \tanh\frac{2\pi b}{h}, \quad k_1' = \operatorname{sech}\frac{2\pi b}{h} \tag{98}$$

is the modulus of the elliptic functions. The points at infinity are at $\chi = iK_1'$, and $\chi = 2K_1 + iK_1'$.

Suppose that the channel is straight ($\theta_h = 0$) then the solution in this plane is

$$\tau(\chi) = \frac{1}{2\pi}\int_{-2K_1}^{2K_1} \theta_0(\xi)\frac{\operatorname{cn}\chi\operatorname{dn}\chi + \operatorname{cn}\xi\operatorname{dn}\xi}{\operatorname{sn}\xi - \operatorname{sn}\chi}\,d\xi, \tag{99}$$

which follows from (4.98) and (34). Equations (94) transform into

$$\int_{-2K_1}^{2K_1} \theta_0(\xi)\,d\xi = 0, \quad \int_{-2K_1}^{2K_1} \theta_0(\xi)\,d\phi_0(\xi) = 0. \tag{100}$$

An exact theory for the incompressible flow past a flat plate can be based on (97)–(100), a theory which is very similar, but rather easier algebraically, than the theory given in § 13.8. This can be left to the reader to develop if he requires it; we shall just outline the corresponding linear perturbation theory.

In linear theory (97) and (98) are replaced by

$$\frac{dZ}{d\chi} = \frac{\beta_a H k_1}{\pi}\operatorname{sn}\chi, \quad Z = \frac{\beta_a H}{2\pi}\ln\left(\frac{1 - k_1\operatorname{cd}\chi}{1 + k_1\operatorname{cd}\chi}\right), \tag{101}$$

and
$$k_1 = \tanh\left(\frac{\pi c}{2\beta_a H}\right), \quad k_1' = \operatorname{sech}\left(\frac{\pi c}{2\beta_a H}\right). \tag{102}$$

The lift coefficient is given by

$$C_L = -\frac{1}{c}\int_{-2K_1}^{2K_1} C_{p_0}(\xi)\, dx(\xi) = -\frac{2k_1 H}{\pi c}\int_{-2K_1}^{2K_1} \Omega_0(\xi)\,\mathrm{sn}\,\xi\, d\xi,$$

by the linearizations $C_p = 2\Omega/\beta_a$, and (101). On the aerofoil surface the real part of (99) gives

$$\Omega_0(\xi) = \frac{1}{2\pi}\int_{-2K_1}^{2K_1} \theta_0(\xi^*)\frac{\mathrm{cn}\,\xi\,\mathrm{dn}\,\xi + \mathrm{cn}\,\xi^*\,\mathrm{dn}\,\xi^*}{\mathrm{sn}\,\xi^* - \mathrm{sn}\,\xi}\, d\xi^*, \tag{103}$$

consequently

$$C_L = \frac{k_1 H}{\pi^2 c}\int_{-2K_1}^{2K_1} \theta_0(\xi)\, I(\xi)\, d\xi,$$

where

$$I(\xi) \equiv \int_{-2K_1}^{2K_1} \mathrm{sn}\,\xi^*\frac{\mathrm{cn}\,\xi^*\,\mathrm{dn}\,\xi^* + \mathrm{cn}\,\xi\,\mathrm{dn}\,\xi}{\mathrm{sn}\,\xi^* - \mathrm{sn}\,\xi}\, d\xi^*. \tag{104}$$

With the aid of (4.53), (4.67), (4.80) and (4.82) it is found that $I(\xi) = 4K_1\,\mathrm{sn}\,\xi\, Z_1(\xi)$, and hence

$$C_L = \frac{4H}{\pi^2 c}k_1 K_1\int_{-2K_1}^{2K_1} \theta_0(\xi)\,\mathrm{sn}\,\xi\, Z_1(\xi)\, d\xi. \tag{105}$$

The origin of the Z-plane is at $\xi = \pm K_1$, and in linear theory this is approximately at the mid-chord point of the aerofoil (cf. Figs. 13.1a and 13.2). The nose-up moment coefficient about this point is

$$C_M = \frac{1}{c^2}\int_{-2K_1}^{2K_1} C_p(\xi)\, x(\xi)\, dx(\xi)$$

$$= \frac{k_1}{\beta_a}\left(\frac{H\beta_a}{\pi c}\right)^2\int_{-2K_1}^{2K_1} \Omega_0(\xi)\,\mathrm{sn}\,\xi\,\ln\left(\frac{1 - k_1\,\mathrm{cd}\,\xi}{1 + k_1\,\mathrm{cd}\,\xi}\right) d\xi.$$

Hence by (103)

$$C_M = \frac{-k_1}{2\pi\beta_a}\left(\frac{H\beta_a}{\pi c}\right)^2\int_{-2K_1}^{2K_1} \theta_0(\xi)\, J(\xi)\, d\xi, \tag{106}$$

where

$$J(\xi) \equiv \int_{-2K_1}^{2K_1} \frac{\mathrm{cn}\,\xi^*\,\mathrm{dn}\,\xi^* + \mathrm{cn}\,\xi\,\mathrm{dn}\,\xi}{\mathrm{sn}\,\xi^* - \mathrm{sn}\,\xi}\,\mathrm{sn}\,\xi^*\,\ln\left(\frac{1 - k_1\,\mathrm{cd}\,\xi^*}{1 + k_1\,\mathrm{cd}\,\xi^*}\right) d\xi^*.$$

To evaluate $J(\xi)$ we need the following results:

$$\int_{-2K_1}^{2K_1} \ln\left(\frac{1 - k_1\,\mathrm{cd}\,\xi^*}{1 + k_1\,\mathrm{cd}\,\xi^*}\right) d\xi^* = 0,$$

$$\int_{-2K_1}^{2K_1} \mathrm{cn}\,\xi^*\,\mathrm{dn}\,\xi^*\,\ln\left(\frac{1 - k_1\,\mathrm{cd}\,\xi^*}{1 + k_1\,\mathrm{cd}\,\xi^*}\right) d\xi^* = \frac{8}{k_1}(E_1 - K_1),$$

and

$$\int_{-2K_1}^{2K_1} \frac{\mathrm{cn}\,\xi^*\,\mathrm{dn}\,\xi^* + \mathrm{cn}\,\xi\,\mathrm{dn}\,\xi}{\mathrm{sn}\,\xi^* - \mathrm{sn}\,\xi}\ln\left(\frac{1 - k_1\,\mathrm{cd}\,\xi^*}{1 + k_1\,\mathrm{cd}\,\xi^*}\right) d\xi^* = 8K_1 k_1\int_0^{\xi} \mathrm{sn}\,\xi\, Z_1(\xi)\, d\xi.$$

The first of these can be verified by replacing $\int_{-2K_1}^{2K_1} \ldots$ by $2\int_0^{2K_1} \ldots$, and putting $\xi = u + K_1$. As $\mathrm{cd}\,\xi = \mathrm{sn}\,u$, this transformation yields an odd integrand over the range $-K_1 < u < K_1$, and so the integral vanishes. The second is established by an integration by parts. The derivative of the logarithmic term is $2k_1 \,\mathrm{sn}\,\xi^*$ (cf. (101)), so the integral in question equals

$$-2k_1 \int_{-2K_1}^{2K_1} \mathrm{sn}^2\,\xi^* \, d\xi^* = \frac{4}{k_1} \int_0^{2K_1} (\mathrm{dn}^2\,\xi^2 - 1)\, d\xi^* = \frac{8}{k_1}(E_1 - K_1),$$

by (4.82) and (4.83). Let $L(\xi)$ denote the third of our integrals, then by the same method as employed to evaluate the first we can show that the term involving $\mathrm{cn}\,\xi \,\mathrm{dn}\,\xi$ vanishes. Integration by parts then gives

$$L(\xi) = -2k_1 \int_{-2K_1}^{2K_1} \mathrm{sn}\,\xi^* \ln |\mathrm{sn}\,\xi^* - \mathrm{sn}\,\xi|\, d\xi^*.$$

Hence
$$L'(\xi) = 2k_1 \,\mathrm{cn}\,\xi \,\mathrm{dn}\,\xi \int_{-2K_1}^{2K_1} \frac{\mathrm{sn}\,\xi^* \, d\xi^*}{\mathrm{sn}\,\xi^* - \mathrm{sn}\,\xi}$$

$$= 2k_1 \,\mathrm{cn}\,\xi \,\mathrm{dn}\,\xi \left\{ 4K_1 + 2\,\mathrm{sn}^2\xi \int_0^{2K_1} \frac{d\xi^*}{\mathrm{sn}^2\,\xi^* - \mathrm{sn}^2\xi} \right\}$$

$$= 8k_1 K_1 \,\mathrm{sn}\,\xi \left\{ \frac{\mathrm{cn}\,\xi \,\mathrm{dn}\,\xi}{\mathrm{sn}\,\xi} + Z_4(\xi) \right\}$$

(cf. (89)). Thus by (4.67)
$$L'(\xi) = 8k_1 K_1 \,\mathrm{sn}\,\xi \, Z_1(\xi).$$

It is apparent that $L(0) = 0$, and so integration yields the value of $L(\xi)$ given above.

With the aid of the integrals just evaluated we obtain

$$J(\xi) = \frac{8}{k_1}(E_1 - K_1) + 8K_1 k_1 \,\mathrm{sn}\,\xi \int_0^\xi \mathrm{sn}\,\xi^* \, Z_1(\xi^*)\, d\xi^*.$$

Substituting this into (106) and taking advantage of (100), we arrive at

$$C_M = -\frac{4k_1^2 K_1}{\pi \beta_a}\left(\frac{H\beta_a}{\pi c}\right)^2 \int_{-2K_1}^{2K_1} \theta_0(\xi) \,\mathrm{sn}\,\xi \int_0^\xi \mathrm{sn}\,\xi^* \, Z_1(\xi^*)\, d\xi^* \, d\xi. \quad (107)$$

As
$$\theta_0(\xi) \,\mathrm{sn}\,\xi\, d\xi = \frac{\pi}{H\beta_a k_1}\left(\frac{dy}{dx}\right)_0 dx(\xi) = \frac{\pi}{H\beta_a k_1}\, dy\,(\xi),$$

integration by parts yields the alternative formula

$$C_M = \frac{4k_1 K_1}{\pi^2}\left(\frac{H}{c}\right) \int_{-2K_1}^{2K_1} \frac{y}{c}(\xi) \,\mathrm{sn}\,\xi \, Z_1(\xi)\, d\xi. \quad (108)$$

The leading edge of the aerofoil is at $\xi = 0$ so in place of (95) there is

$$\theta_0(\xi) = \left(\frac{dy}{dx}\right)_s - \alpha^* + 4K_1(\alpha^* - \alpha_i)\,\delta(\xi), \qquad (109)$$

where
$$\alpha_i = \frac{1}{4K_1}\int_{-2K_1}^{2K_1}\left(\frac{dy}{dx}\right)_s d\xi. \qquad (110)$$

By substituting (109) into (105), and putting $\alpha^* = -\alpha'$, the no-lift angle (see §8.7), we find that

$$\alpha' = \frac{1}{4K_1}\int_{-2K_1}^{2K}\left(\frac{dy}{dx}\right)_s \{\mathrm{sn}\,\xi\,Z_1(\xi) - 1\}\,d\xi. \qquad (111)$$

Now
$$\int_{-2K_1}^{2K_1}\mathrm{sn}\,\xi\,Z_1(\xi)\,d\xi = 2\int_0^{2K_1}\mathrm{sn}\,\xi\,Z_1(\xi)\,d\xi$$

$$= 2\int_{-K_1}^{K_1}\mathrm{cd}\,\xi\,Z_2(\xi)\,d\xi = 0, \qquad (112)$$

as the integrand is an odd function. Also as $\lim\limits_{\xi\to 0}\mathrm{sn}\,\xi\,Z_1(\xi) = \lim\limits_{\xi\to 0}\xi\times\dfrac{1}{\xi} = 1$, then

$$\int_{-2K_1}^{2K_1}\delta(\xi)\,\mathrm{sn}\,\xi\,Z_1(\xi)\,d\xi = 1. \qquad (113)$$

Consequently if (109) is substituted into (105) and use is made of (110)–(113) there results

$$C_L = 4k_1\left(\frac{H}{c}\right)\left(\frac{2K_1}{\pi}\right)^2(\alpha^* + \alpha'). \qquad (114)$$

From the expansions given in (4.65)

$$\frac{2K_1}{\pi} = 1 + \tfrac{1}{4}k_1^2 + \tfrac{9}{64}k_1^4 + O(k_1^6),$$

where k_1 is related to (c/H) by (102). On eliminating k_1 from (114), we obtain the expansion

$$C_L = \frac{2\pi}{\beta_a}(\alpha^* + \alpha')\left\{1 + \frac{1}{24}\left(\frac{\pi c}{H\beta_a}\right)^2 - \frac{11}{7680}\left(\frac{\pi c}{H\beta_a}\right)^4 + O\left(\frac{c}{H}\right)^6\right\},$$

which is due to Tomotika (1934).

Similarly, from (107) and (109) we can derive the expansion

$$C_M = C_M(\alpha^* = 0) + \frac{\pi\alpha^*}{2\beta_a}\left\{1 + \frac{1}{48}\left(\frac{\pi c}{H\beta_a}\right)^2 - \frac{11}{23040}\left(\frac{\pi c}{H\beta_a}\right)^4 + O\left(\frac{c}{H}\right)^6\right\},$$

where the first term is the coefficient at zero incidence.

From these equations we can derive expressions for the corrections that must be applied to wind-tunnel measurements of C_L and C_M. These corrections must, of course, not only include the terms in $(c/H)^2$ given above, but also the effect of the increase in the average velocity at the aerofoil caused by tunnel blockage, and this introduces an additional factor $\{1-(2-M_a^2)\bar{\epsilon}\}$ into the correction (cf. (7.69)). It must also be remembered that the no-lift angle $-\alpha'$, and the ideal incidence α_i will be altered by the presence of the walls.

13.12   Ground effect on aerofoils

During the take-off and landing of an aircraft, the wing surface moves closely across a solid surface. The effect of this boundary on the forces acting on the wing is of some interest, and if the aircraft has a conventional high aspect ratio wing, this effect can be deduced from equations given in §§ 13.3 and 13.4.

We shall omit the exact theory, and immediately linearize (16), (18), (19) and (20) to find

$$\frac{dZ}{d\zeta} = \frac{2KH_2\beta_a}{\pi}\left(\mathrm{dn}^2\,\zeta - \frac{E}{K}\right), \quad Z = \frac{2KH_2\beta_a}{\pi}Z_4(\zeta),$$

and

$$\mathrm{dn}\,\gamma_0 = \sqrt{\left(\frac{E}{K}\right)}, \quad \frac{\pi c}{H_2\beta_a} = 4KZ_4(\gamma_0).$$

In the ζ-plane just defined the solution is given by (23) and (26). The velocities up and downstream at infinity must now be equal, so $\Omega_\infty = \Omega_{-\infty}$.

All the results of §§ 13.4–13.9 can now be applied to the present case by taking the limit $h_1, H_1 \to \infty$. Thus, for example, from (51), (73) and (78) we find that

$$C_L = 2\pi \sin\alpha \left\{1 - \frac{c}{2H_2}\sin\alpha + \frac{1}{4}\left(\frac{c}{2H_2}\right)^2(1 + 3\sin^2\alpha) + \ldots\right\}.$$

The subject of ground effect on thick, cambered aerofoils has been studied by several authors, among whom are Green (1940, 1947) and Tomotika, Tamada & Umemoto (1950); further references can be found in these papers.

The above theory could be developed to obtain results more general than those to which we have just referred. It could also be used to calculate the effect of a variation in the ground shape on the forces acting on the aerofoil, e.g. as occurs in the approach of an aircraft to the deck of a carrier, or in the flight of a sailplane in the neighbourhood of a cliff.

AEROFOILS IN JETS

13.13   The general solution

Free jets of air and water are frequently used to test the performance of aero- and hydrofoils, and so an 'exact' theory for dealing with this type of flow is of some interest. Another application of such a theory would be to flexible-walled wind tunnels (see §7.22). The existing theory for a lifting aerofoil is based on the theory of a vortex in a free jet (Lock & Beavan, 1944), and this could be considerably improved with the aid of the theory given below.

The z-plane for the flow of a jet past an aerofoil is shown in Fig. 13.13. The aerofoil is at an incidence α and the jet is assumed to be deflected through an angle β. The velocity at the free surface of the jet must be constant, and equal to U say. The w-plane of the flow is as shown in Fig. 13.1a, and this is transformed into the ζ-plane by (1)–(5). In this plane the boundary conditions are (cf. (22))

$$(a)\ \theta_0(\gamma)\ \text{prescribed},\quad (b)\ \Omega_h = 0\ (q = U\ \text{on}\ \eta = K'),$$
$$(c)\ \theta^+(\eta) = \theta^-(\eta),\quad (d)\ \Omega^+(\eta) = \Omega^-(\eta). \tag{115}$$

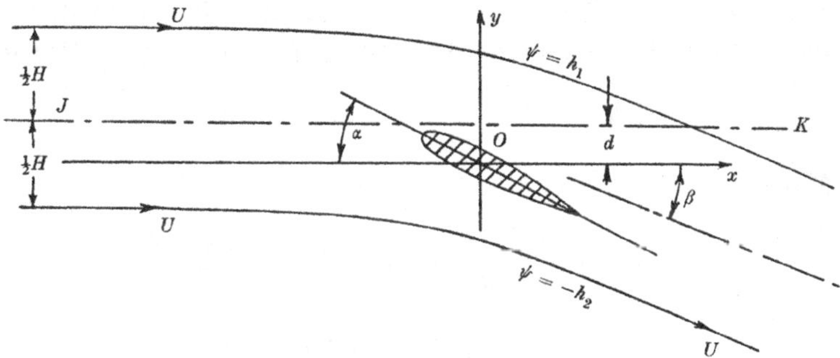

Fig. 13.13

The solution to this mixed boundary problem is

$$\tau(\zeta) = \frac{K_1}{\pi K}\int_{-K}^{K}\theta_0(\gamma)\frac{\mathrm{sn}'}{\mathrm{sn}}\left\{\frac{K_1}{K}(\gamma^* - \zeta), k_1\right\}d\gamma^*, \tag{116}$$

which follows from (4.114) (with appropriate changes in the independent variable) and (115). A more convenient form for this result can be deduced as follows.

From (4.54) and (4.62)

$$\left\{\mathrm{sn}\left(\frac{K_1 u}{K}, k_1\right)\right\}^2 = \left(\frac{2K_1}{K}\right)^2\left\{\mathrm{sn}\left(\tfrac{1}{2}u, k\right)\,\mathrm{cd}\left(\tfrac{1}{2}u, k\right)\right\}^2 = \left(\frac{2K_1}{Kk}\right)^2\frac{1 - \mathrm{dn}\,(u, k)}{1 + \mathrm{dn}\,(u, k)},$$

and hence by taking logarithms and differentiating

$$\frac{2K_1}{K}\frac{\mathrm{sn}'}{\mathrm{sn}}\left(\frac{K_1 u}{K}, k_1\right) = k^2\,\mathrm{sn}\,(u, k)\,\mathrm{cn}\,(u, k)\left\{\frac{1}{1 + \mathrm{dn}\,(u, k)} + \frac{1}{1 - \mathrm{dn}\,(u, k)}\right\}$$
$$= 2\,\mathrm{cs}\,(u, k).$$

Therefore (116) can be written in the concise form

$$\tau(\zeta) = \frac{1}{\pi} \int_{-K}^{K} \theta_0(\gamma^*)\, \mathrm{cs}\,(\gamma^* - \zeta)\, d\gamma^*, \tag{117}$$

where the modulus is understood to be k. This result is due to Woods (1958b).

13.14   The jet deflexion angle and the closure conditions

Select the direction of the x-axis so that upstream at infinity the flow is parallel to it, then at $\phi = -\infty$, $\theta = 0$, and as β is the (downwards) deflexion angle, $\theta = -\beta$ at $\phi = \infty$. In the ζ-plane these results can be expressed

$$\theta_{-\infty} = \theta(\zeta = -\gamma_1 + iK') = 0, \quad \theta_\infty = \theta(\zeta = \gamma_1 + iK') = -\beta. \tag{118}$$

On the free surface $\zeta = \gamma + iK'$, and so substituting this value in (117) and making use of (4.48) we find that the slope of the free surface is given by

$$\theta_h(\gamma) = \frac{1}{\pi} \int_{-K}^{K} \theta_0(\gamma^*)\, \mathrm{dn}\,(\gamma^* - \gamma)\, d\gamma^*. \tag{119}$$

It follows from (118) that

$$\beta = -\frac{1}{\pi} \int_{-K}^{K} \theta_0(\gamma^*)\, \mathrm{dn}\,(\gamma^* - \gamma_1)\, d\gamma^*, \tag{120}$$

and

$$O = \frac{1}{\pi} \int_{-K}^{K} \theta_0(\gamma^*)\, \mathrm{dn}\,(\gamma^* + \gamma_1)\, d\gamma^*. \tag{121}$$

The first of these relates the deflexion angle to the shape and incidence of the aerofoil, while the second plays the same role as the second of (24), i.e. that of defining the position of the front stagnation point on the aerofoil surface.

The closure conditions can be obtained from results given in §13.5, results which are quite general and not dependent on the character of the boundary conditions. First note that as $q = U$ in the jet at $\phi = \pm\infty$, then

$$\Omega_\infty = \Omega_{-\infty} = 0. \tag{122}$$

The closure conditions are

$$\int_{-K}^{K} \theta_0(\gamma)\, d\phi_0(\gamma) = \int_{-K}^{K} \theta_h(\gamma)\, d\phi_h(\gamma), \tag{123}$$

and

$$-\beta(h_1 + h_2) = \int_{-K}^{K} \Omega_0(\gamma)\, d\phi_0(\gamma), \tag{124}$$

which follow from (32), (41), (118) and (122). These are exact relations. In addition we have the two approximate relations given in (37) and (38).

From (119), (123) can be written

$$\int_{-K}^{K} \theta_0(\gamma)\, d\phi_0(\gamma) = \frac{1}{\pi} \int_{-K}^{K} \theta_0(\gamma^*)\, I(\gamma^*)\, d\gamma^*,$$

where

$$I(\gamma^*) \equiv \int_{-K}^{K} \mathrm{dn}\,(\gamma^* - \gamma)\, d\phi_h(\gamma).$$

By (4.61) and (5) (with $\zeta = \gamma + iK'$)

$$I(\gamma^*) = A\,\mathrm{dn}\,\gamma^* \int_0^{K} \frac{\mathrm{dn}\,\gamma(1 - k^2\,\mathrm{sn}^2\gamma_0\,\mathrm{sn}^2\gamma)\,d\gamma}{(\mathrm{sn}^2\gamma - \mathrm{sn}^2\gamma_1)\,(1 - k^2\,\mathrm{sn}^2\gamma^*\,\mathrm{sn}^2\gamma)},$$

where A is a constant. The integrand of I can be transformed into an elementary form by the substitution $x = \mathrm{cs}\,\gamma$; its value is then found to be zero. Hence

$$\int_{-K}^{K} \theta_0(\gamma)\, d\phi_0(\gamma) = 0, \tag{125}$$

so that in the present case (38) is exactly true. Unfortunately this does not extend to (37).

When the linearization $e^\tau \simeq 1 + \tau$ is valid it follows from (37) and (124) that

$$\Gamma = \beta(h_1 + h_2)/\beta_a = \beta H U. \tag{126}$$

It is a simple calculation from the equation of linear momentum (§ 1.13) to show that the force L on the aerofoil makes an angle of $\tfrac{1}{2}(\pi - \beta)$ with the Ox-axis, and that its magnitude is

$$L = 2H\rho_a U^2 \sin\tfrac{1}{2}\beta. \tag{127}$$

Thus to first order in β, $L = \rho_a U\Gamma$ in agreement with the classical result for an infinite stream.

13.15　Linear perturbation theory

In all the linear perturbation theories of fluid flow given hitherto in this book, we were able to replace w in $\tau(w)$ by UZ (see § 2.17). However this approximation—or rather the usual method of applying the boundary conditions (see (2.50) *et seq.*)—requires that the boundaries of the flow should not deviate appreciably from straight lines parallel to the x-axis. But with the deflected flow of a jet past an aerofoil, even if the aerofoil is at a small incidence, the jet itself will deviate further and further from $x = 0$ downstream of the aerofoil, until near $\phi = \infty$ the deviation will be nearly infinite. This means that we cannot put $w = UZ$ in (1) and still accept (117) as the solution in the ζ-plane. The correct linearization is to replace e^τ by $1 + \tau$ in the exact relation between z and w. This yields the relation given in (6.24), viz.

$$\tfrac{1}{2}(1 + \beta_a^{-2})\,(UZ - w) + \tfrac{1}{2}(1 - \beta_a^{-2})\,(U\bar{Z} - \bar{w}) = \frac{1}{\beta_a}\int_0^{\zeta} \tau(\zeta)\, dw(\zeta), \tag{128}$$

where we have taken the origin of the Z-plane to coincide with those of w and ζ.

From (5) and (117)

$$\int_0^\zeta \tau(\zeta)\,dw(\zeta) = \frac{1}{\pi}\int_{-K}^{K} \theta_0(\gamma)\,J(\gamma,\zeta)\,d\gamma, \tag{129}$$

where

$$J(\gamma,\zeta) = A\int_0^\zeta \operatorname{cs}(\gamma-\zeta)\left(\frac{\operatorname{sn}^2\gamma_0 - \operatorname{sn}^2\zeta}{1 - k^2\operatorname{sn}^2\gamma_1\operatorname{sn}^2\zeta}\right)d\zeta,$$

and

$$A = \frac{h_1+h_2}{\pi}\,\frac{2k^2\operatorname{sn}\gamma_1\operatorname{cn}\gamma_1\operatorname{dn}\gamma_1}{1 - k^2\operatorname{sn}^2\gamma_0\operatorname{sn}^2\gamma_1}.$$

Now (4.48) and (4.61) give

$$\operatorname{cs}(\gamma-\zeta) = i\operatorname{dn}(\gamma-\zeta+iK') = \frac{\operatorname{sn}\zeta\operatorname{cn}\zeta\operatorname{dn}\gamma^* + \operatorname{sn}\gamma^*\operatorname{cn}\gamma^*\operatorname{dn}\zeta}{\operatorname{sn}^2\gamma^* - \operatorname{sn}^2\zeta},$$

which enables us to separate the integral $J(\gamma,\zeta)$ into two integrals. The integrand containing $\operatorname{sn}\zeta\operatorname{cn}\zeta$ is reduced to an elementary form by the substitution $x = \operatorname{dn}\zeta$, while that involving $\operatorname{dn}\zeta$ is similarly reduced by putting $x = \operatorname{cs}\zeta$. The final result is

$$\begin{aligned}
J(\gamma,\zeta) = A\Bigg[&\frac{\operatorname{sn}^2\gamma - \operatorname{sn}^2\gamma_0}{1 - k^2\operatorname{sn}^2\gamma_1\operatorname{sn}^2\gamma}\ln\left\{\frac{1 - \operatorname{dn}(\gamma-\zeta)}{1 + \operatorname{dn}(\gamma+\zeta)}\frac{1 + \operatorname{dn}\gamma}{1 - \operatorname{dn}\gamma}\right\} \\
&+ \frac{2(1 - k^2\operatorname{sn}^2\gamma_1\operatorname{sn}^2\gamma_0)}{1 - k^2\operatorname{sn}^2\gamma_1\operatorname{sn}^2\gamma}\left\{\frac{\operatorname{sn}\gamma\operatorname{cn}\gamma}{\operatorname{dn}\gamma_1}\tan^{-1}(\operatorname{sc}\zeta\operatorname{dn}\gamma_1)\right. \\
&\left. - \frac{\operatorname{dn}\gamma}{k^2\operatorname{sn}\gamma_1\operatorname{cn}\gamma_1}[\tan^{-1}(\operatorname{dn}\zeta\operatorname{sc}\gamma_1) - \tan^{-1}(\operatorname{sc}\gamma_1)]\right\}\Bigg]. \tag{130}
\end{aligned}$$

The required relation between Z and ζ follows from (128) to (130).

The actual position of the aerofoil relative to the jet will obviously depend on both the aerofoil incidence, and the position of the aerofoil centre relative to the jet position upstream at infinity. Let JK in Fig. 13.13 be the centre-line of the jet in its undisturbed, upstream position, and let the vertical distance from O (origin of the Z-plane) to JK be d. Then if d is fixed h_1/h_2 will clearly decrease with increasing α. The relation between these variables can be deduced from the imaginary part of (128), i.e. from

$$Uy = \frac{\psi}{\beta_a} + \frac{1}{\pi}\mathscr{I}\int_{-K}^{K}\theta_0(\gamma)\,J(\gamma,\zeta)\,d\gamma.$$

Let ϵ be a small *positive* number, then $\lim_{\epsilon\to 0} y(\zeta = \epsilon - \gamma_1 + iK')$ is the value of y on the *upper* free surface upstream at infinity. The centre-line of the jet is a distance $\frac{1}{2}H = (h_1 + h_2)/2U\beta_a$ below this free surface, hence

$$
d = \lim_{\epsilon \to 0} y(\zeta = \epsilon - \gamma_1 + iK') = \frac{h_1 + h_2}{2U\beta_a}
$$

$$
= \frac{h_1 - h_2}{2\beta_a U} + \frac{1}{\pi U} \int_{-K}^{K} \theta_0(\gamma) J(\gamma, -\gamma_1 + iK')\, d\gamma. \tag{131}
$$

Another important geometric number is the chord to jet-width ratio, c/H. A simple expression for this can be derived from (128) if it is assumed that $c \simeq$ the distance between the front and rear stagnation points. In this case the real part of (128) yields

$$
\frac{c}{H} = \frac{4b}{HU} + \frac{1}{\pi HU\beta_a} \int_{-K}^{K} \theta_0(\gamma) \{J(\gamma, \gamma_0) - J(\gamma, -\gamma_0)\}\, d\gamma, \tag{132}
$$

where b is the number defined in (2).

In § 13.1 it was stated that it would be shown later that b, Γ, h_1 and h_2 are fixed by the values of c, α, d and H. Sufficient relations for this are contained in (121), (126), (131) and (132).

13.16   A flat plate in a jet of incompressible fluid

The simplest example of the theory, which will indicate the kind of results obtained without introducing complicated algebra, is provided by a flat plate at a small incidence in a jet of incompressible fluid.

In this case (cf. (95))
$$
\theta_0(\gamma) = -\alpha + A\delta(\gamma + \gamma_0), \tag{133}
$$

where A is a constant to be determined. Substitution in (120) and (121) gives

$$
\beta = \frac{\alpha}{\pi} \left[\sin^{-1}\{\operatorname{sn}(\gamma^* - \gamma_1)\} \right]_{-K}^{K} - \frac{A}{\pi} \operatorname{dn}(\gamma_1 + \gamma_0) = \alpha - \frac{A}{\pi} \operatorname{dn}(\gamma_1 + \gamma_0),
$$

and
$$
0 = -\alpha + \frac{A}{\pi} \operatorname{dn}(\gamma_1 - \gamma_0). \tag{134}
$$

Therefore
$$
\beta = \alpha\{1 - \operatorname{dn}(\gamma_1 + \gamma_0)\operatorname{nd}(\gamma_1 - \gamma_0)\}. \tag{135}
$$

Similarly, (117), (133) and (134) give

$$
U e^{-\tau} = \frac{dw}{dz} = U\{1 + i\alpha + \alpha \operatorname{cs}(\gamma_0 + \zeta)\operatorname{nd}(\gamma_0 - \gamma_1)\}, \tag{136}
$$

and in particular the velocity on the surface of the plate is given by

$$
\frac{q}{U} = 1 + \alpha \operatorname{cs}(\gamma_0 + \gamma)\operatorname{nd}(\gamma_0 - \gamma_1).
$$

Integrating (136) by the method suggested above for $J(\gamma, \zeta)$ we find

$$Uz = w(1 - i\alpha) + \frac{(h_1 + h_2)(\beta - 2\alpha)}{\pi}\{\tan^{-1}(\mathrm{dn}\,\zeta\,\mathrm{sc}\,\gamma_1) - \tan^{-1}(\mathrm{sc}\,\gamma_1)\}$$

$$- \frac{(h_1 + h_2)\beta}{\pi}\tan^{-1}(\mathrm{sc}\,\zeta\,\mathrm{dn}\,\gamma_1). \tag{137}$$

From (131), (133) and some algebra it is found that

$$\frac{2d}{H} = \frac{h_1 - h_2}{h_1 + h_2} + \frac{\beta}{\pi}\ln\left[\frac{(1 - \mathrm{dn}\,2\gamma_1)\,\vartheta_4(\pi\gamma_1/K)}{2\vartheta_4(0)}\right]$$

$$+ \frac{2\alpha - \beta}{\pi}\ln\left[\frac{(1 + \mathrm{dn}\,2\gamma_1)\,\vartheta_4(\pi\gamma_1/K)}{2\vartheta_4(0)}\right]. \tag{138}$$

Similarly, (132) yields

$$\frac{c}{H} = \frac{4b}{HU} - \frac{2\beta}{\pi}\tan^{-1}(\mathrm{sc}\,\gamma_0\,\mathrm{dn}\,\gamma_1), \tag{139}$$

where from (2), (126) and (4.78)

$$\frac{4b}{HU} = \frac{\beta\gamma_0}{K} + \frac{2}{\pi}\ln\left|\frac{\vartheta_4\left\{\frac{\pi}{2K}(\gamma_1 + \gamma_0)\right\}}{\vartheta_4\left\{\frac{\pi}{2K}(\gamma_1 - \gamma_0)\right\}}\right|.$$

Thus to first order $$\frac{c\pi}{2H} = \ln\left|\frac{\vartheta_4\left\{\frac{\pi}{2K}(\gamma_1 + \gamma_0)\right\}}{\vartheta_4\left\{\frac{\pi}{2K}(\gamma_1 - \gamma_0)\right\}}\right|. \tag{140}$$

Finally from (3), (4) and (126)

$$\beta = \frac{2K}{\pi}\{Z_4(\gamma_0 - \gamma_1) - Z_4(\gamma_0 + \gamma_1)\}, \tag{141}$$

and $$\frac{h_1 - h_2}{h_1 + h_2} = 1 - \frac{2\gamma_1}{K} + \frac{\beta K'}{K}. \tag{142}$$

With a given geometrical arrangement of jet and plate, c/H, d/H and α will be known, and (135), (138), (140)–(142) fix the values of the five unknowns β, K, γ_0, γ_1 and h_1/h_2.

Suppose the plate is in the centre of the jet, then $h_1 = h_2$, and the first-order solution of (141) and (142) is $\gamma_0 = \gamma_1 = \frac{1}{2}K$. Then (140) gives

$$\frac{c\pi}{2H} = \ln\left|\frac{\vartheta_3(0)}{\vartheta_4(0)}\right| = \frac{1}{2}\ln\frac{1}{k'}.$$

If the plate is now set at a small incidence, γ_0 and γ_1 will differ from $\frac{1}{2}K$ by terms of order α. Thus to first order in α, (135) gives

$$\beta = \alpha(1 - \mathrm{dn}\,K\,\mathrm{nd}\,0) = \alpha(1 - k') = \alpha(1 - e^{-\pi c/H}).$$

By (127) $C_L = 2H\beta/c$ to first order, so

$$C_L = 2\pi\alpha \left(\frac{H}{\pi c}\right)(1 - e^{-\pi c/H}), \tag{143}$$

a result due to Woods (1958b). If terms $O(c/H)^3$ are neglected, then

$$C_L = 2\pi\alpha \left\{1 - \frac{\pi c}{2H} + \frac{1}{6}\left(\frac{\pi c}{H}\right)^2\right\}, \tag{144}$$

which was first derived by Glauert (1933). Tomotika (1939) has developed an accurate theory for the flat plate, similar (although somewhat more complicated) to that given above. He finds the result

$$C_L = 2\pi \sin\alpha \left\{1 - \frac{\pi c}{2H}\cos\alpha + \tfrac{1}{24}(4 - 11\sin^2\alpha)\left(\frac{\pi c}{H}\right)^2 + O\left(\frac{c}{H}\right)^3\right\},$$

in which there is no restriction on the magnitude of α. This could be derived from (117) by developing an accurate theory similar to that given in § 13.8.

13.17   Aerofoil in the centre of a jet

If the aerofoil is centrally placed in the jet, and if the circulation is small, it is more convenient to work in the χ-plane defined in § 13.2. In this plane it follows from (4.114), (4.115) and (115) that

$$\tau(\chi) = \frac{1}{2\pi}\int_{-2K_1}^{2K_1} \theta_0(\xi)\frac{\mathrm{sn}'}{\mathrm{sn}}\{\tfrac{1}{2}(\xi - \chi)\}\,d\xi$$

$$= \frac{1}{2\pi}\int_{-2K_1}^{2K_1} \theta_0(\xi)\frac{\mathrm{cn}\,(\xi - \chi) + \mathrm{dn}\,(\xi - \chi)}{\mathrm{sn}\,(\xi - \chi)}\,d\xi, \tag{145}$$

where the modulus of the elliptic functions is now k_1.

From the addition formulae for the elliptic functions (see § 4.6),

$$(\mathrm{sn}\,\xi - \mathrm{sn}\,\chi)\frac{\mathrm{cn}\,(\xi - \chi) + \mathrm{dn}\,(\xi - \chi)}{\mathrm{sn}\,(\xi - \chi)} = \mathrm{dn}\,\xi\,\mathrm{cn}\,\chi + \mathrm{dn}\,\chi\,\mathrm{cn}\,\xi, \tag{146}$$

so that yet another version of the general solution is

$$\tau(\chi) = \frac{1}{2\pi}\int_{-2K_1}^{2K_1} \theta_0(\xi)\frac{\mathrm{dn}\,\xi\,\mathrm{cn}\,\chi + \mathrm{dn}\,\chi\,\mathrm{cn}\,\xi}{\mathrm{sn}\,\xi - \mathrm{sn}\,\chi}\,d\xi. \tag{147}$$

which is remarkably similar in form to (99) for the solid-walled channel.

The points $\phi = -\infty$, $\phi = \infty$ are at $\chi = iK_1'$ and $\chi = 2K_1 + iK_1'$ (cf. Figs. 13.1a and 13.2), hence substitution of these values in (147) gives

$$0 = \int_{-2K_1}^{2K_1} \theta_0(\xi)\,(\mathrm{dn}\,\xi + k\,\mathrm{cn}\,\xi)\,d\xi, \tag{148}$$

and
$$-\beta = \frac{1}{2\pi} \int_{-2K_1}^{2K_1} \theta_0(\xi)\,(\mathrm{dn}\,\xi - k\,\mathrm{cn}\,\xi)\,d\xi, \tag{149}$$

by (118) and (122).

The further development of this theory is quite straightforward, and so we shall leave it at this point. We shall be considering more general equations than those above when we deal with the unsteady motion of an aerofoil in a jet (see §§ 14.6–14.9).

We must also omit any account of the theory of the flow about an aerofoil (or more appropriately, 'hydrofoil') near the free surface of an infinitely deep fluid. The reader should be able to develop this theory for himself from (117) and the equations given in § 13.2.

AEROFOILS BETWEEN POROUS WALLS

13.18 The mixed boundary value problem

The work of the preceding sections of this chapter will now be generalized to apply to the case when sections of the channel or tunnel walls are porous or closely perforated, and have the linear characteristic described in §§ 6.3 and 6.4. The type of tunnel we shall study is shown in Fig. 13.18; it is the same as that already considered in § 7.28 with symmetrical, non-lifting aerofoils, and the results we shall obtain below for thin lifting aerofoils complement those given in § 7.28 for thickness effects. The theory is due to Woods (1957 c).

Fig. 13.18 shows an aerofoil at incidence midway between two porous intervals JN, LM on the otherwise solid tunnel walls. The porous sections are jacketed, and it is assumed that air can be removed from or pumped into the jackets—so as to control the jacket pressures p_{s1}, p_{s2} along the outsides of JN and LM. The porosities of JN and LM are assumed to be equal and constant.

We shall adopt linear perturbation theory, so that the χ-plane shown in Fig. 13.18 will be related to the Z- $(= x + i\beta_a y)$ plane by (101) and (102). The symmetry of the mapping about AI permits us to assume that the upper and lower porous surfaces of the tunnel fall into intervals $\xi_0 < \xi < \xi_1$ and $-\xi_1 < \xi < -\xi_0$ (on $\delta = K'$) respectively.

The second of (101) is equivalent to (cf. (14))

$$\mathrm{cn}\,(\chi, k_1) = -\frac{k_1'}{k_1}\sinh\frac{\pi Z}{H\beta_a}. \tag{150}$$

On $Z = x \pm \tfrac{1}{2}i\beta_a H$ this equation gives $\mathrm{ds}\,(\xi, k_1) = \pm k_1'\cosh\,(\pi x/H\beta_a)$, and as $x = \tfrac{1}{2}R$ when $\xi = \pm\xi_1$, and $x = -\tfrac{1}{2}R$ when $\xi = \pm\xi_0$ it follows that

$\mathrm{ds}\,\xi_0 = k_1'\cosh\,(\pi R/2H\beta_0) = \mathrm{ds}\,\xi_1$. Because of the periodic properties of $\mathrm{ds}\,\xi$ we can write

$$\xi_1 = K_1 + \nu, \quad \xi_0 = K_1 - \nu, \tag{151}$$

where

$$\mathrm{nc}\,\nu = \cosh\left(\frac{\pi R}{2H\beta_a}\right). \tag{152}$$

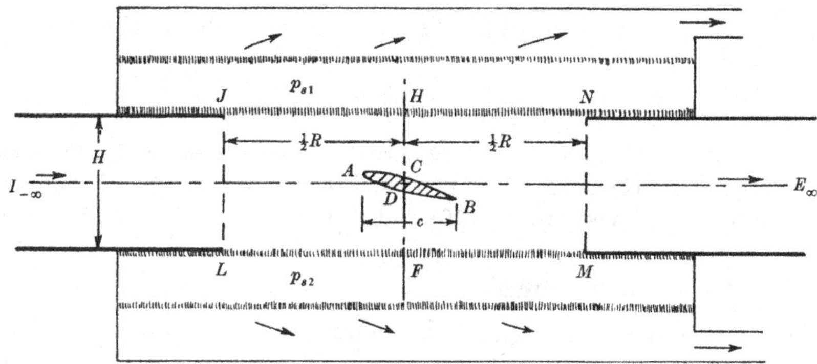

z-plane

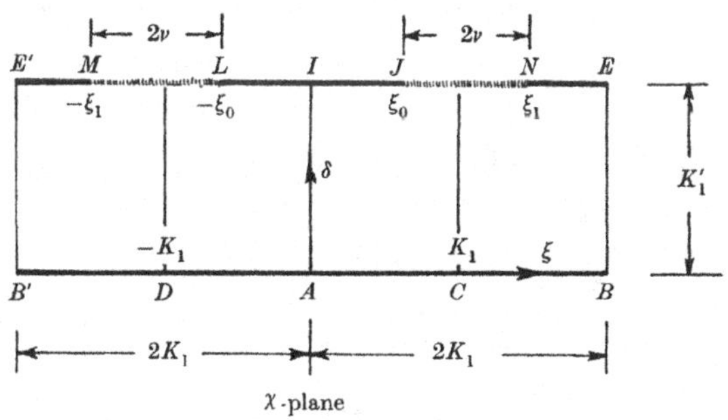

χ-plane

Fig. 13.18

It now follows from Fig. 13.18, (6.33)–(6.35) and the remark about the sign of ϵ following (6.35) that the mixed boundary conditions in the χ-plane are

$$\theta\cos\pi\epsilon - \Omega\sin\pi\epsilon = \theta_s\cos\pi\epsilon - \Omega_s\sin\pi\epsilon,$$

with

$$\left.\begin{array}{lll}
\epsilon = 0,\ \theta_s = \theta_0, & \text{on} & \delta = 0,\ -2K_1 < \xi < 2K_1, \\[4pt]
\epsilon = 0,\ \theta_s = 0, & \text{on} & \delta = K_1',\ -2K_1 < \xi < -K_1 - \nu, \\[2pt]
& & \qquad -K_1 + \nu < \xi < K_1 - \nu, \\[2pt]
& & \qquad K_1 + \nu < \xi < 2K_1, \\[4pt]
\epsilon = \epsilon_1,\ \theta_s = 0,\ \Omega_s = \Omega_{s1}, & \text{on} & \delta = K_1',\ K_1 - \nu < \xi < K_1 + \nu, \\[4pt]
\epsilon = -\epsilon_1,\ \theta_s = 0,\ \Omega_s = \Omega_{s2}, & \text{on} & \delta = K_1',\ -K_1 - \nu < \xi < -K_1 + \nu,
\end{array}\right\} \tag{153}$$

where (cf. (7.178))

$$\Omega_{s1} = \beta_a \frac{p_{s1} - p_a}{\rho_a U^2}, \quad \Omega_{s2} = \beta_a \frac{p_{s2} - p_a}{\rho_a U^2}, \tag{154}$$

$$\epsilon_1 = \frac{1}{\pi} \tan^{-1}(\lambda_s/\beta_a), \tag{155}$$

and λ_s is the porosity factor defined in §1.20. In addition there are the conditions at infinity:

$$\tau(iK_1') = \tau_{-\infty} = \Omega_{-\infty}, \quad \tau(2K_1 + iK_1') = \tau_{\infty} = \Omega_{\infty}, \tag{156}$$

as $\theta_{-\infty} = \theta_{\infty} = 0$. One of $\Omega_{-\infty}$ and Ω_{∞} can be chosen arbitrarily, then the other depends on the mass of air removed from the tunnel and the wake displacement thickness (see (7.176) and (7.177)).

13.19   The general solution

The general solution, $\tau(\chi)$, of the boundary value problem defined above can be written down with the help of the very general theory given in §4.19. But instead of the form for Q given in (4.169) we shall use the equivalent form appearing in (4.98). With $l = 2K_1$, and in the χ-plane, this form is

$$Q(\chi, t) = \frac{1}{2\pi} \left(\frac{\operatorname{cn} \chi \operatorname{dn} \chi + \operatorname{cn} t \operatorname{dn} t}{\operatorname{sn} t - \operatorname{sn} \chi} \right),$$

where $t = \xi$ on $\delta = 0$, and $t = \xi + iK_1'$ on $\delta = iK_1'$. The modulus of the elliptic functions is k_1.

To evaluate the first factor in (4.176) we need

$$I \equiv \frac{1}{2} \int_{C_1} \epsilon(t) Q(\chi, t) \, dt,$$

where C_1 are the sides $\delta = 0, \delta = iK_1'$ of the basic rectangle (cf. Fig. 4.19). From (153)

$$I = \tfrac{1}{2}\epsilon \left(\int_{-K_1 - \nu}^{-K_1 + \nu} + \int_{K_1 - \nu}^{K_1 + \nu} \right) \frac{k_1 \operatorname{cn} \chi \operatorname{dn} \chi - \operatorname{ds} \xi \operatorname{cs} \xi}{\operatorname{ns} \xi - k_1 \operatorname{sn} \chi} \, d\xi,$$

where the subscript '1' on ϵ is omitted for later typographical convenience.

The two integrals in I are made to cover the same range by putting $\xi = -K_1 + y$ in the first, and $\xi = K_1 + y$ in the second. The result is

$$I = \epsilon k_1 \operatorname{cn} \chi \operatorname{dn} \chi \int_{-\nu}^{\nu} \frac{\operatorname{dc} y \, dy}{k_1^2 \operatorname{sn}^2 \chi - \operatorname{dc}^2 y},$$

which can be integrated by the substitution $x = \operatorname{ns} y$; we get

$$I = \epsilon \ln \left(\frac{1 - k_1 \sigma \operatorname{cd} \chi}{1 + k_1 \sigma \operatorname{cd} \chi} \right),$$

where
$$\sigma \equiv \operatorname{sn} \nu = \tanh \frac{\pi R}{2\beta_a H}, \tag{157}$$

by (152). Equation (157) defines the same variable σ as used in § 7.28.

The first factor in (4.176) is therefore

$$\left(\frac{1 - k_1 \sigma \operatorname{cd} \chi}{1 + k_1 \sigma \operatorname{cd} \chi}\right)^\epsilon,$$

and with the aid of this value and (153) we can complete our reduction of (4.176) to find

$$\tau(\chi) = \frac{1}{2\pi}\left(\frac{1 - k_1 \sigma \operatorname{cd} \chi}{1 + k_1 \sigma \operatorname{cd} \chi}\right)^\epsilon \left[A + \tfrac{1}{2}\sin \pi\epsilon \left\{(\Omega_{s1} - \Omega_{s2})J_2 + (\Omega_{s1} + \Omega_{s2})J_1\right\}\right.$$
$$\left. + \int_{-2K_1}^{2K_1} \theta_0(\xi)\left(\frac{1 + k_1 \sigma \operatorname{cd} \xi}{1 - k_1 \sigma \operatorname{cd} \xi}\right)^\epsilon \frac{\operatorname{cn}\chi \operatorname{dn}\chi + \operatorname{cn}\xi \operatorname{dn}\xi}{\operatorname{sn}\xi - \operatorname{sn}\chi}\,d\xi\right], \tag{158}$$

where
$$\{J_2, J_1\} \equiv 2\int_{K_1-\nu}^{K_1+\nu}\left[\frac{\sigma \operatorname{dc}\xi + 1}{\sigma \operatorname{dc}\xi - 1}\right]^\epsilon \frac{k_1 \operatorname{cn}\chi \operatorname{dn}\chi - \operatorname{ds}\xi \operatorname{cs}\xi}{\operatorname{ns}^2\xi - k_1^2 \operatorname{sn}^2\chi}\{k_1 \operatorname{sn}\chi, \operatorname{ns}\xi\}\,d\xi, \tag{159}$$

and A is a constant.

The integral J_1 (but not J_2) can be evaluated by means of the substitution

$$x = (\sigma \operatorname{dc}\xi + 1)/(\sigma \operatorname{dc}\xi - 1),$$

and the result
$$\int_0^\infty \frac{x^{\epsilon-1}}{1 + x}\,dx = \frac{\pi}{\sin \pi\epsilon} \quad (0 < \epsilon < 1).$$

It is found that
$$J_1 = \frac{2\pi}{\sin \pi\epsilon}\left\{\left(\frac{1 + k_1 \sigma \operatorname{cd}\chi}{1 - k_1 \sigma \operatorname{cd}\chi}\right)^\epsilon - \frac{1}{2}\left(\frac{1 + \sigma}{1 - \sigma}\right)^\epsilon - \frac{1}{2}\left(\frac{1 - \sigma}{1 + \sigma}\right)^\epsilon\right\}. \tag{160}$$

To evaluate A put $\chi = \mu + iK_1'$ in (158), and take the limit $\mu \to \infty$ of the resulting expression. By (156) the left-hand side tends to the finite number $\Omega_{-\infty}$. However, as

$$\operatorname{cn}(\mu + iK_1') = -\frac{i}{k_1\mu}\{1 + O(\mu^2)\}, \quad \operatorname{dn}(\mu + iK_1')$$
$$= -\frac{i}{\mu}\{1 + O(\mu^2)\}, \quad \operatorname{sn}(\mu + iK_1') = \frac{1}{k_1\mu}\{1 + O(\mu^2)\},$$

the right-hand side of (158) becomes infinite unless

$$J_0(\Omega_{s2} - \Omega_{s1})\sin \pi\epsilon = \int_{-2K_1}^{2K_1} \theta_0(\xi)\left(\frac{1 + k_1 \sigma \operatorname{cd}\xi}{1 - k_1 \sigma \operatorname{cd}\xi}\right)^\epsilon d\xi, \tag{161}$$

where
$$J_0 \equiv \int_{K_1-\nu}^{K_1+\nu}\left[\frac{\sigma \operatorname{dc}\xi + 1}{\sigma \operatorname{dc}\xi - 1}\right]^\epsilon d\xi. \tag{162}$$

With (161) satisfied the limit of (158) becomes

$$\Omega_{-\infty} = \frac{1}{2\pi}\left(\frac{1-\sigma}{1+\sigma}\right)^{\epsilon}\left[A + \tfrac{1}{2}\pi(\Omega_{s1}+\Omega_{s2})\left\{\left(\frac{1+\sigma}{1-\sigma}\right)^{\epsilon} - \left(\frac{1-\sigma}{1+\sigma}\right)^{\epsilon}\right\}\right.$$
$$\left. + \int_{-2K_1}^{2K_1} \theta_0(\xi)\left(\frac{1+k_1\sigma\,\mathrm{cd}\,\xi}{1-k_1\sigma\,\mathrm{cd}\,\xi}\right)^{\epsilon} k_1\,\mathrm{sn}\,\xi\,d\xi\right].$$

Similarly, by putting $\chi = \mu + 2K_1 + iK_1'$ in (158) and taking the limit $\mu \to 0$ we obtain

$$\Omega_{\infty} = \frac{1}{2\pi}\left(\frac{1+\sigma}{1-\sigma}\right)^{\epsilon}\left[A - \frac{\pi}{2}(\Omega_{s1}+\Omega_{s2})\left\{\left(\frac{1+\sigma}{1-\sigma}\right)^{\epsilon} - \left(\frac{1-\sigma}{1+\sigma}\right)^{\epsilon}\right\}\right.$$
$$\left. - \int_{-2K_1}^{2K_1} \theta_0(\xi)\left(\frac{1+k_1\sigma\,\mathrm{cd}\,\xi}{1-k_1\sigma\,\mathrm{cd}\,\xi}\right)^{\epsilon} k_1\,\mathrm{sn}\,\xi\,d\xi\right].$$

Hence
$$A = 2\pi\left\{\Omega_{\infty}\left(\frac{1-\sigma}{1+\sigma}\right)^{\epsilon} + \Omega_{-\infty}\left(\frac{1+\sigma}{1-\sigma}\right)^{\epsilon}\right\}, \tag{163}$$

and
$$\{\Omega_{-\infty} - \tfrac{1}{2}(\Omega_{s1}+\Omega_{s2})\}\left(\frac{1+\sigma}{1-\sigma}\right)^{\epsilon} - \{\Omega_{\infty} - \tfrac{1}{2}(\Omega_{s1}+\Omega_{s2})\}\left(\frac{1-\sigma}{1+\sigma}\right)^{\epsilon}$$
$$= \frac{1}{\pi}\int_{-2K_1}^{2K_1} \theta_0(\xi)\left(\frac{1+k_1\sigma\,\mathrm{cd}\,\xi}{1-k_1\sigma\,\mathrm{cd}\,\xi}\right)^{\epsilon} k_1\,\mathrm{sn}\,\xi\,d\xi. \tag{164}$$

Equations (159), (160) and (163) enable (158) to be written

$$\tau(\chi) = \tfrac{1}{2}(\Omega_{s1}+\Omega_{s2}) + \left(\frac{1-k_1\sigma\,\mathrm{cd}\,\chi}{1+k_1\sigma\,\mathrm{cd}\,\chi}\right)^{\epsilon}\left[\tfrac{1}{2}\{\Omega_{-\infty} - \tfrac{1}{2}(\Omega_{s1}+\Omega_{s2})\}\left(\frac{1+\sigma}{1-\sigma}\right)^{\epsilon}\right.$$
$$+ \tfrac{1}{2}\{\Omega_{\infty} - \tfrac{1}{2}(\Omega_{s1}+\Omega_{s2})\}\left(\frac{1-\sigma}{1+\sigma}\right)^{\epsilon} + \frac{\sin\pi\epsilon}{4\pi}(\Omega_{s1}-\Omega_{s2})J_2$$
$$+ \frac{1}{2\pi}\int_{-2K_1}^{2K_1} \theta_0(\xi)\left(\frac{1+k_1\sigma\,\mathrm{cd}\,\xi}{1-k_1\sigma\,\mathrm{cd}\,\xi}\right)^{\epsilon}\frac{\mathrm{cn}\,\chi\,\mathrm{dn}\,\chi + \mathrm{cn}\,\xi\,\mathrm{dn}\,\xi}{\mathrm{sn}\,\xi - \mathrm{sn}\,\chi}\,d\xi\right]. \tag{165}$$

The general solution of our problem is now contained in (161)–(165), plus one further relation between Ω_{∞} and $\Omega_{-\infty}$ (see last sentence of §13.18).

13.20   Some special cases

The equations given above are very general, and contain as special cases the basic theoretical results for the flow about an aerofoil, (i) between solid tunnel walls, (ii) in a porous wind tunnel having infinitely long porous sections, (iii) in the centre of an open section of finite length, (iv) in a free jet, and (v) in an infinite stream. Case (i) is obtained by putting $\epsilon = 0$ (zero porosity, see (155)), case (ii) by letting $R \to \infty$, case (iii) by putting $\epsilon = \frac{1}{2}$ (see §6.4), case (iv) by letting $R \to \infty$ in case (iii), and case (v) by letting $H \to \infty$.

Case (i). *The solid wind tunnel*

If $\epsilon = 0$, (161), (164) and (165) reduce correctly to (100) and (99) given in §13.11 for the solid wind tunnel.

Case (ii). *Uniformly porous tunnel*

If $R \to \infty$ it is clear from (157) that $\sigma \to 1$. In this limit (164) gives $\frac{1}{2}(\Omega_{s1} + \Omega_{s2}) = \Omega_{-\infty}$, and so puts a restriction on the achievable average jacket pressure. If the jacket pressures are equal, $\Omega_{s1} = \Omega_{s2}$ and (161), (164) and (165) give

$$0 = \int_{-2K_1}^{2K_1} \theta_0(\xi) \left(\frac{1 + k_1 \,\mathrm{cd}\, \xi}{1 - k_1 \,\mathrm{cd}\, \xi}\right)^{\epsilon} d\xi, \tag{166}$$

and

$$\tau(\chi) = \Omega_{-\infty} + \frac{1}{2\pi} \left(\frac{1 - k_1 \,\mathrm{cd}\, \chi}{1 + k_1 \,\mathrm{cd}\, \chi}\right)^{\epsilon}$$

$$\times \int_{-2K_1}^{2K_1} \theta_0(\xi) \left(\frac{1 + k_1 \,\mathrm{cd}\, \xi}{1 - k_1 \,\mathrm{cd}\, \xi}\right)^{\epsilon} \left(\frac{\mathrm{cn}\, \chi \,\mathrm{dn}\, \chi + \mathrm{cn}\, \xi \,\mathrm{dn}\, \xi}{\mathrm{sn}\, \xi - \mathrm{sn}\, \chi} + k_1 \,\mathrm{sn}\, \xi\right) d\xi. \tag{167}$$

Case (iii). *Open-section wind tunnel*

The limit $\epsilon \to \frac{1}{2}$, which yields an open (constant pressure) section instead of a porous wall, makes no great simplification to the equations so we will not write them down.

Case (iv). *The free jet*

It is of some interest to verify that the equations for the flow of a free jet about a centrally placed aerofoil (i.e. the equations given in §13.17) are obtained by putting $\epsilon = \frac{1}{2}$ in (166) and (167). This verification is easily made with the aid of the identity

$$\left(\frac{1 - k_1 \,\mathrm{cd}\, \chi}{1 + k_1 \,\mathrm{cd}\, \chi}\right)^{\frac{1}{2}} \left(\frac{1 + k_1 \,\mathrm{cd}\, \xi}{1 - k_1 \,\mathrm{cd}\, \xi}\right)^{\frac{1}{2}} \left(\frac{\mathrm{cn}\, \chi \,\mathrm{dn}\, \chi + \mathrm{cn}\, \xi \,\mathrm{dn}\, \xi}{\mathrm{sn}\, \xi - \mathrm{sn}\, \chi} + k_1 \,\mathrm{sn}\, \xi\right)$$

$$= \frac{\mathrm{cn}\, \xi \,\mathrm{dn}\, \chi + \mathrm{cn}\, \chi \,\mathrm{dn}\, \xi}{\mathrm{sn}\, \xi - \mathrm{sn}\, \chi} - \frac{k_1 \,\mathrm{sn}\, \xi}{\mathrm{dn}\, \xi + k_1 \,\mathrm{cn}\, \xi}(\mathrm{dn}\, \chi + k_1 \,\mathrm{cn}\, \chi),$$

which can be established from relations given in §4.6.

Case (v). *The infinite stream*

If $H \to \infty$ then $k_1 \to 0$ by (102) and the elliptic functions degenerate to the circular functions as shown in (4.57). Equations (161), (164) and (165) reduce to

$$0 = \int_{-\pi}^{\pi} \theta_0(\xi)\, d\xi = \int_{-\pi}^{\pi} \theta_0(\xi) \sin \xi\, d\xi,$$

and

$$\tau(\chi) = \tfrac{1}{2}(\Omega_\infty + \Omega_{-\infty}) + \frac{1}{2\pi}\int_{-\pi}^{\pi}\theta_0(\xi)\frac{\cos\xi + \cos\chi}{\sin\xi - \sin\chi}d\xi,$$

which agree with the linear perturbation forms of (8.11), (8.28) and (8.10).

13.21   The lift and moment coefficients

In § 13.11 it was shown that the lift coefficient is

$$C_L = -\frac{2k_1}{\pi}\left(\frac{H}{c}\right)\int_{-2K_1}^{2K_1}\Omega_0(\xi)\,\mathrm{sn}\,\xi\,d\xi, \tag{168}$$

and the nose-up moment coefficient about the mid-chord point is

$$C_M = \frac{k_1}{\beta_a}\left(\frac{H\beta_a}{\pi c}\right)^2\int_{-2K_1}^{2K_1}\Omega_0(\xi)\,\mathrm{sn}\,\xi\ln\left(\frac{1-k_1\,\mathrm{cd}\,\xi}{1+k_1\,\mathrm{cd}\,\xi}\right)d\xi, \tag{169}$$

where

$$k_1 = \tanh\left(\frac{\pi c}{2\beta_a H}\right). \tag{170}$$

The value of $\Omega_0(\xi)$ to be inserted in these expressions is the real part of (165) at $\chi = \xi$, but the resulting integrals cannot be evaluated exactly. To obtain results convenient for practical use we shall expand the integrands in powers of c/H or in powers of k_1 for from (170)

$$k_1 = \frac{\pi c}{2\beta_a H}\left\{1 - \frac{1}{12}\left(\frac{\pi c}{\beta_a H}\right)^2 + O\left(\frac{c}{H}\right)^4\right\}. \tag{171}$$

When k_1 is small it follows from (4.64) and (4.65) that

$$\mathrm{sn}\,\xi = \sin\frac{\pi\xi}{2K_1}\left(1 + \tfrac{1}{4}k_1^2\cos^2\frac{\pi\xi}{2K_1}\right) + O(k_1^4),$$

$$\mathrm{cn}\,\xi = \cos\frac{\pi\xi}{2K_1}\left(1 - \tfrac{1}{4}k_1^2\sin^2\frac{\pi\xi}{2K_1}\right) + O(k_1^4),$$

$$\mathrm{dn}\,\xi = 1 - \tfrac{1}{2}k_1^2\sin^2\frac{\pi\xi}{2K_1} + O(k_1^4),$$

$$2K_1 = \pi(1 + \tfrac{1}{4}k_1^2) + O(k_1^4),$$

and with the aid of these expansions it is found that

$$\left(\frac{1+k_1\sigma\,\mathrm{cd}\,\xi^*}{1-k_1\sigma\,\mathrm{cd}\,\xi^*}\right)^\epsilon = 1 + 2k_1\epsilon\sigma\cos\frac{\pi\xi^*}{2K_1} + 2\left(k_1\epsilon\sigma\cos\frac{\pi\xi^*}{2K_1}\right)^2 + O(k_1^3),$$

and

$$\frac{\mathrm{cn}\,\xi\,\mathrm{dn}\,\xi + \mathrm{cn}\,\xi^*\,\mathrm{dn}\,\xi^*}{\mathrm{sn}\,\xi^* - \mathrm{sn}\,\xi} = \cot\frac{\pi}{4K_1}(\xi^* - \xi)\left[1 - \tfrac{1}{4}k_1^2\right.$$

$$\left.\times\left\{2 - 2\left(\cos\frac{\pi\xi^*}{2K_1} - \cos\frac{\pi\xi}{2K_1}\right)^2 - \cos\frac{\pi}{2K_1}(\xi - \xi^*)\right\} + O(k_1^4)\right].$$

As far as the lift and moment are concerned the aerofoil thickness can be ignored to first order, and we can therefore write (see (109))

$$\theta_0(\xi) = -\alpha + A\boldsymbol{\delta}(\xi), \tag{172}$$

where α is the aerofoil's incidence, and A is a constant to be determined by (161). From (172) and the expansions given above it is found that

$$A = 2\pi\alpha\{1 - 2k_1\epsilon\sigma + 3(k_1\epsilon\sigma)^2\}$$
$$+ J_0(\Omega_{s2} - \Omega_{s1})\sin\pi\epsilon\{1 - 2k_1\epsilon\sigma + 2(k_1\epsilon\sigma)^2 - \tfrac{1}{4}k_1^2 + O(k_1^3)\}. \tag{173}$$

The expansion for $\Omega_0(\xi)$ can now be calculated with the aid of (171)–(173), and then substituted into (168) and (169) to give

$$C_L = \frac{2\pi\alpha}{\beta_a}\left\{1 - 2\epsilon\sigma\left(\frac{\pi c}{2\beta_a H}\right) + \tfrac{1}{6}(1 + 12\epsilon^2\sigma^2)\left(\frac{\pi c}{2\beta_a H}\right)^2\right\}$$
$$+ \frac{1}{\beta_a}(\Omega_{s2} - \Omega_{s1})\left[J_0\sin\pi\epsilon\left\{1 - \epsilon\sigma\left(\frac{\pi c}{2\beta_a H}\right) - \tfrac{1}{12}(1 - 12\epsilon^2\sigma^2)\right.\right.$$
$$\left.\left.\times\left(\frac{\pi c}{2\beta_a H}\right)^2\right\} - \left(\frac{\pi c}{2\beta_a H}\right)\pi I_0\right] + O\left(\frac{c}{H}\right)^3, \tag{174}$$

and

$$C_M = \frac{\pi\alpha}{2\beta_a}\left\{1 - 2\epsilon\sigma\left(\frac{\pi c}{2\beta_a H}\right) + \tfrac{1}{12}(1 + 36\epsilon^2\sigma^2)\left(\frac{\pi c}{2\beta_a H}\right)^2\right\}$$
$$+ \frac{1}{4\beta_a}(\Omega_{s2} - \Omega_{s1})\left[J_0\sin\pi\epsilon\left\{1 - 2\epsilon\sigma\left(\frac{\pi c}{2\beta_a H}\right) - \tfrac{1}{24}(7 - 36\epsilon^2\sigma^2)\right.\right.$$
$$\left.\left.\times\left(\frac{\pi c}{2\beta_a H}\right)^2\right\} - \frac{\pi}{2}\left(\frac{\pi c}{2\beta_a H}\right)^2(I_1 + 2\epsilon\sigma I_0)\right] + O\left(\frac{c}{H}\right)^3, \tag{175}$$

where

$$I_0 \equiv \frac{\sigma^2}{2\pi}\sin\pi\epsilon\int_{-1}^{1}\left(\frac{1+x}{1-x}\right)^{\epsilon}\frac{x\,dx}{(1-\sigma^2 x^2)^{\frac{1}{2}}}, \tag{176}$$

and

$$I_1 \equiv \frac{\sigma}{2\pi}\sin\pi\epsilon\int_{-1}^{1}\left(\frac{1+x}{1-x}\right)^{\epsilon}(1-\sigma^2 x^2)^{\frac{1}{2}}\,dx. \tag{177}$$

The integrals J_0, I_0 and I_1 must be evaluated numerically, but this is unnecessary and the theory is considerably simplified if the jacket pressures p_{s1} and p_{s2} are equal. In this case $\Omega_{s1} = \Omega_{s2}$ and the equations reduce to

$$C_L = \frac{2\pi\alpha}{\beta_a}\left\{1 - \epsilon\sigma\left(\frac{\pi c}{\beta_a H}\right) + \tfrac{1}{24}(1 + 12\epsilon^2\sigma^2)\left(\frac{\pi c}{\beta_a H}\right)^2\right\}, \tag{178}$$

and

$$C_M = \frac{\pi\alpha}{2\beta_a}\left\{1 - \epsilon\sigma\left(\frac{\pi c}{\beta_a H}\right) + \tfrac{1}{48}(1 + 36\epsilon^2\sigma^2)\left(\frac{\pi c}{\beta_a H}\right)^2\right\}, \tag{179}$$

correct to $O(c/H)^2$.

The special values $\epsilon = 0$ and $\epsilon = \frac{1}{2}$ at $\sigma = 1$ yield results already obtained (see §§ 13.11 and 13.17) for the solid wind tunnel and the free jet respectively. A uniformly porous tunnel is obtained by putting $\sigma = 1$. If we put $\epsilon = \frac{1}{2}$ in (178) and (179) we obtained the coefficients for an open-section wind tunnel (see Fig. 7.28b). The interference on the lift vanishes in this case provided

$$\sigma \equiv \tanh\frac{\pi R}{2\beta_a H} = 6\left(\frac{H\beta_a}{\pi c}\right)\left[1 - \left\{1 - \frac{1}{24}\left(\frac{\pi c}{\beta_a H}\right)^2\right\}^{\frac{1}{2}}\right] \simeq \frac{\pi c}{4\beta_a H},$$

i.e. if $R \simeq \frac{1}{2}c$, which is a surprisingly narrow section.

CHAPTER 14

UNSTEADY MOTION OF AEROFOILS IN CHANNELS AND JETS

UNSTEADY MOTION OF AN AEROFOIL IN A CLOSED WIND TUNNEL

14.1 The basic equations

The theory of the unsteady motion of an aerofoil between solid tunnel walls has two main applications, namely (i) to the evaluation of the wind-tunnel corrections which need to be applied to experimental results on the flutter and stability of aerofoils, and (ii) to the study of the flutter of a cascade of aerofoils, when the mode is such that adjacent aerofoils are moving $180°$ out of phase (see discussion in §12.14).

Several authors have studied the incompressible flow problem, notably Reissner (1947), who gave a theory valid only for small values of c/H, Timman (1951), whose exact theory is based on the Green's function for a rectangle, Lilley (1952) who used the method of summing images to obtain results similar to Timman's, and Rosenblat (1957), whose solution is based on Woods's (1955e) equation for $\tau(\chi)$ which has been given in §4.9. We shall follow Rosenblat's treatment here, generalizing it slightly by allowing for accelerating motion.

Compressible unsteady flow past an aerofoil in a wind tunnel is very difficult to treat analytically, and approximate methods such as used by Jones (1953) seem to be inevitable. Incompressible flow only will be studied below. We shall now collect the basic equations of the theory.

First notice that (9.6) and (9.10), viz.

$$\frac{1}{c}\frac{\partial \chi}{\partial t}+\frac{\partial \chi}{\partial x}=0, \tag{1}$$

and

$$\chi(x,t) = \chi\left(\tfrac{1}{2}c, t+\tfrac{1}{2}-\frac{x}{c}\right), \tag{2}$$

where

$$\chi \equiv \varpi U, \tag{3}$$

and

$$t \equiv \frac{1}{c}\int_0^t U\,dt \tag{4}$$

are also applicable in the present case.

To avoid confusion between the symbol χ defined above and that defined in (13.150), and also to keep a closer parallel to the symbols used in § 9 for the unsteady motion of an isolated aerofoil, we shall rename the χ-plane to be the 'ζ-plane'. We shall also take the opportunity to simplify the notation by dropping the subscript '1' from the quarter-periods and modulus of the elliptic functions. This will cause no confusion with the notation of § 13.1, which will not be used again. The z-plane and (new) ζ-plane are shown in Fig. 14.1.

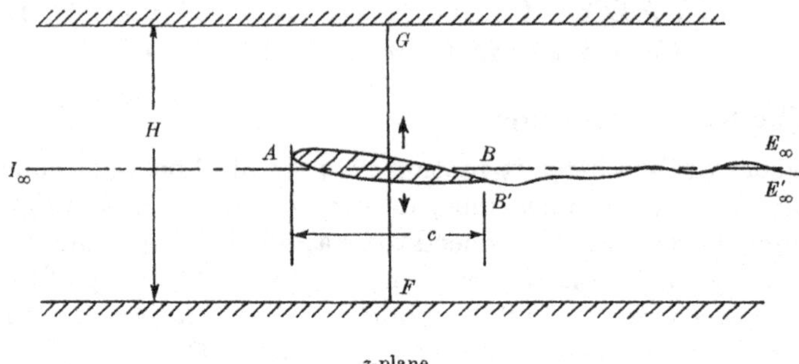

z-plane

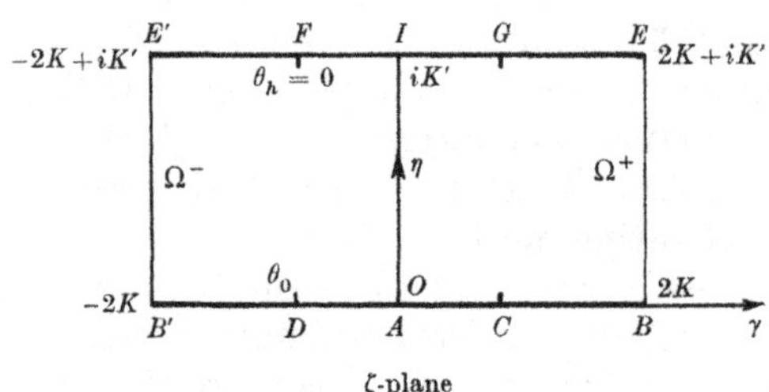

ζ-plane

Fig. 14.1

As the flow will be assumed incompressible, $\beta_a = 1$ and $Z \equiv x + i\beta_a y$ will equal $z = x + iy$. Hence (13.101), (13.102) and (13.150) will now be replaced by

$$\frac{dz}{d\zeta} = \frac{Hk}{\pi}\,\mathrm{sn}\,\zeta, \quad z = \frac{H}{2\pi}\ln\left(\frac{1 - k\,\mathrm{cd}\,\zeta}{1 + k\,\mathrm{cd}\,\zeta}\right), \tag{5}$$

$$k = \tanh\left(\frac{\pi c}{2H}\right), \tag{6}$$

and

$$\mathrm{cn}\,\zeta = -\frac{k'}{k}\sinh\left(\frac{\pi z}{H}\right). \tag{7}$$

This transformation maps one side of the vortex sheet extending behind the aerofoil on to $\gamma = 2K$ and the other side on to $\gamma = -2K$. Across this sheet we have a discontinuity in velocity to which corresponds a jump ϖ in Ω.

The boundary conditions in the ζ-plane are

(a) $\theta_0(\gamma, t)$ prescribed,   (b) $\theta_h(\gamma, t) = 0$,

(c) $\theta^+(\eta, t) = \theta^-(\eta, t)$,   (d) $\Omega^+(\eta, t) - \Omega^-(\eta, t) = \varpi(\eta, t)$,

where $\varpi(\eta, t) \equiv \chi(\eta, t)/U(t)$ is determined by (2) and Kelvin's circulation theorem. There is also the boundary condition at infinity which we can take to be

$$\tau(iK') = \Omega_{-\infty} + i\theta_{-\infty} = 0, \quad \tau(2K + iK') = \Omega_{\infty} + i\theta_{\infty} = 0. \tag{8}$$

These boundary conditions are the same as those studied in §4.9, and therefore on changing the notation used in (4.98) to that of the present section, we arrive at the general solution

$$\tau(\zeta, t) = \frac{1}{2\pi} \int_{-2K}^{2K} \theta_0(\gamma, t) \frac{\operatorname{cn}\zeta \operatorname{dn}\zeta + \operatorname{cn}\gamma \operatorname{dn}\gamma}{\operatorname{sn}\gamma - \operatorname{sn}\zeta} d\gamma$$

$$+ \frac{\operatorname{sn}\zeta}{2\pi U} \int_0^{K'} \chi(\eta, t) \frac{\operatorname{cn}\zeta \operatorname{dn}\zeta - \operatorname{cn} i\eta \operatorname{dn} i\eta}{\operatorname{sn}^2 i\eta - \operatorname{sn}^2 \zeta} d\eta. \tag{9}$$

By taking the limits $\zeta \to iK'$, $\zeta \to 2K + iK'$ in (9) (use the expansions following (13.160)) we find that

$$O = \int_{-2K}^{2K} U\theta_0 d\gamma + \int_0^{K'} \chi d\eta \tag{10}$$

and

$$O = \int_{-2K}^{2K} \theta_0 \operatorname{sn}\gamma \, d\gamma, \tag{11}$$

which should be compared with (9.12) and (9.16).

On the aerofoil surface (9) becomes

$$\Omega(\gamma, t) = \frac{1}{2\pi} \int_{-2K}^{2K} \theta_0 \frac{\operatorname{cn}\gamma \operatorname{dn}\gamma + \operatorname{cn}\gamma^* \operatorname{dn}\gamma^*}{\operatorname{sn}\gamma^* - \operatorname{sn}\gamma} d\gamma^*$$

$$+ \frac{\operatorname{sn}\gamma}{2\pi U} \int_0^{K'} \chi \frac{\operatorname{cn}\gamma \operatorname{dn}\gamma - \operatorname{cn} i\eta \operatorname{dn} i\eta}{\operatorname{sn}^2 i\eta - \operatorname{sn}^2 \gamma} d\eta. \tag{12}$$

The last basic equation of the theory is Kelvin's circulation theorem, $d\Gamma/dt = 0$. It will be assumed that prior to the start of the unsteady motion $\Gamma = 0$ then we will have $\Gamma = 0$ throughout the unsteady motion. Thus, with the usual approximations of linear perturbation theory,

$$\Gamma = \int_{-2K}^{2K} q \, dx(\gamma) = U \int_{-2K}^{2K} (1 - \Omega) \, dx(\gamma) = -\frac{HkU}{\pi} \int_{-2K}^{2K} \Omega \operatorname{sn}\gamma \, d\gamma = 0.$$

Substitution of (12) into this gives

$$\int_{-2K}^{2K} U\theta_0 I(\gamma) \, d\gamma + \int_0^{K'} \tfrac{1}{2}\chi\{I(2K + i\eta) + I(-2K + i\eta)\} \, d\eta = 0,$$

where, apart from the changed notation, I is the integral defined in (13.104). Thus by (4.34), (4.39) and $I(\gamma) = 4K \operatorname{sn} \gamma Z_1(\gamma)$, we arrive at

$$\int_{-2K}^{2K} U\theta_0 \sin \gamma \, Z_1(\gamma) \, d\gamma - \int_0^{K'} \chi \operatorname{sn} i\eta \, Z_1(i\eta) \, d\eta = 0, \tag{13}$$

which is the required generalization of (9.15).

14.2   The strength of the vortex sheet

The calculation of the strength of the vortex sheet runs parallel to the simpler calculation given in § 9.3 for an isolated aerofoil.

First note from (9.5) that we can set

$$\theta_0 = \theta_s + \frac{n}{U} - \alpha + 2K\lambda\boldsymbol{\delta}(\gamma), \tag{14}$$

where θ_s is the steady component of the flow direction, n is the unsteady perturbation velocity of the aerofoil surface normal to itself, λ is an unknown number related to the position of the front stagnation point, and $\boldsymbol{\delta}(\gamma)$ is the delta function. The steady component θ_s satisfies (10), (11) and (13) with $\chi = 0$. Consequently corresponding to (9.20)–(9.22) there is

$$\lambda U = 2\alpha U - 2a_0 - \frac{1}{2K} \int_0^{K'} \chi(\eta, t) \, d\eta, \tag{15}$$

$$0 = 2\alpha U - 2a_0 + 2a_1 - \frac{1}{2K} \int_0^{K'} \chi(\eta, t) \{1 + \operatorname{sn} i\eta \, Z_1(i\eta)\} \, d\eta, \tag{16}$$

$$0 = \int_{-2K}^{2K} n(\gamma, t) \operatorname{sn} \gamma \, d\gamma, \tag{17}$$

where

$$a_0 \equiv \frac{1}{4K} \int_{-2K}^{2K} n(\gamma, t) \, d\gamma, \tag{18}$$

$$a_1 \equiv \frac{1}{4K} \int_{-2K}^{2K} n(\gamma, t) \operatorname{sn} \gamma \, Z_1(\gamma) \, d\gamma, \tag{19}$$

and use has been made of (13.112) and (13.113).

On the wake $\zeta = \pm 2K + i\eta$ and $z = x$ so (5) and (7) become

$$\frac{dx}{d\eta} = \frac{iHk}{\pi} \operatorname{sn} i\eta, \quad x = \frac{H}{2\pi} \ln\left(\frac{1 + k \operatorname{cd} i\eta}{1 - k \operatorname{cd} i\eta}\right), \tag{20}$$

and

$$\operatorname{cn} i\eta = \frac{k'}{k} \sinh \frac{\pi x}{H}.$$

Hence

$$\operatorname{sn} i\eta = \left\{1 - \frac{1 - k^2}{k^2} \sinh^2 \frac{\pi x}{H}\right\}^{\frac{1}{2}} = -\frac{ik'}{\sqrt{2}\,k}\left(\cosh \frac{2\pi x}{H} - \cosh \frac{\pi c}{H}\right)^{\frac{1}{2}},$$

by (6), and so

$$dη = \frac{π\sqrt{2}}{Hk'} \frac{dx}{\left(\cosh\dfrac{2πx}{H} - \cosh\dfrac{πc}{H}\right)^{\frac{1}{2}}}. \tag{21}$$

Let

$$r \equiv \frac{πc}{H}, \tag{22}$$

and

$$\frac{x}{c} = ξ - \tfrac{1}{2} = \mathit{f} + \tfrac{1}{2} - σ, \tag{23}$$

then by an argument parallel to that used to deduce (9.29) and (9.30) from (9.20) and (9.21) we can transform (15) and (16) into

$$λU = 2αU - 2a_0 - \frac{r}{\sqrt{2\,k'K}} \int_0^{\mathit{f}} \frac{χ(1,σ)\,dσ}{[\cosh\{2r(\mathit{f}-σ)+r\} - \cosh r]^{\frac{1}{2}}}, \tag{24}$$

and

$$O = 2A - \frac{r}{\sqrt{2\,k'K}} \int_0^{\mathit{f}} \frac{χ(1,σ)\,dσ}{[\cosh\{2r(\mathit{f}-σ)+r\} - \cosh r]^{\frac{1}{2}}}$$

$$- \frac{r}{2Kk} \int_0^{\mathit{f}} χ(1,σ)\,iZ_1\{m(\mathit{f}-σ)\}\,dσ, \tag{25}$$

where

$$A = \begin{cases} α(\mathit{f})\,U(\mathit{f}) - a_0(\mathit{f}) + a_1(\mathit{f}) & (\mathit{f} > 0), \\ 0 & (\mathit{f} < 0) \end{cases} \tag{26}$$

(we are assuming as in §9.1 that the unsteady motion commences at $\mathit{f} = t = 0$),

and

$$m(\mathit{f}-σ) \equiv \operatorname{sn}^{-1}\left(\frac{ik'}{\sqrt{2\,k}}[\cosh\{2r(\mathit{f}-σ)+r\} - \cosh r]^{\frac{1}{2}}\right).$$

The Laplace transform of (25) is (cf. derivation of (9.33))

$$\mathscr{L}\{A(\mathit{f}); p\} = \frac{r}{2\sqrt{2\,k'Kp}} \mathscr{L}\{U(\mathit{f})\,χ(1,\mathit{f}); p\}$$

$$\times \mathscr{L}[U(\mathit{f})\{\cosh(2r\mathit{f}+r) - \cosh r\}^{-\frac{1}{2}}]$$

$$- \frac{r}{4kKp} \mathscr{L}\{U(\mathit{f})\,χ(1,\mathit{f}); p\}\,\mathscr{L}[U(\mathit{f})\,iZ_1\{m(\mathit{f})\}; p]. \tag{27}$$

The last transform in this equation can be changed into a standard type as follows. By (4.198)

$$\mathscr{L}[U(\mathit{f})\,iZ_1\{m(\mathit{f})\}; p] = \frac{1}{p}\mathscr{L}\left[U(\mathit{f})\,i\frac{d}{d\mathit{f}}Z_1\{m(\mathit{f})\}; p\right]. \tag{28}$$

Now from (4.53), (4.67), (4.82), (4.83) and (20) we find that

$$i\frac{d}{dx}Z_1(i\eta) = i\left\{dn^2\,i\eta - \frac{E}{K} + \frac{d}{d(i\eta)}\left(\frac{\operatorname{cn} i\eta\,\operatorname{dn} i\eta}{\operatorname{sn} i\eta}\right)\right\}\frac{d(i\eta)}{dx}$$

$$= -\left\{\left(1-\frac{E}{K}\right) - \operatorname{ns}^2 i\eta\right\}\frac{d\eta}{dx}$$

$$= -\frac{\sqrt{2}}{k'}\frac{r}{c}\left[\left(1-\frac{E}{K}\right)\left\{\cosh 2r\frac{x}{c} - \cosh r\right\}^{-\frac{1}{2}}\right.$$

$$\left. + \frac{2k^2}{k'^2}\left\{\cosh 2r\frac{x}{c} - \cosh r\right\}^{-\frac{3}{2}}\right],$$

and therefore, as $m(t) = -i\eta \left(\dfrac{x}{c} = t + \tfrac{1}{2}\right)$, it follows that

$$i\frac{d}{dt}Z_1[m(t)] = \frac{\sqrt{2}\,r}{k'}\left[\left(1-\frac{E}{K}\right)\{\cosh(2rt + r) - \cosh r\}^{-\frac{1}{2}}\right.$$

$$\left. + \frac{2k^2}{k'^2}\{\cosh(2rt+r) - \cosh r\}^{-\frac{3}{2}}\right]. \quad (29)$$

The transform of this relation could be obtained with the aid of (12.113) but for the restriction $\mathscr{R}v < \tfrac{1}{2}$ given in that equation. This limitation is unnecessarily restrictive, for the right-hand side of (12.113) exists for $v = 0$ and all real integer values of v, and although the integral on the left-hand side is improper if $\mathscr{R}v > \tfrac{1}{2}$ it can be interpreted as a principal value (see §§ 3.5 and 3.9), i.e. as the derivative of a convergent integral. It will be noted from (29) that the value $v = 1$ arises in just this way. We are therefore justified in using (12.113) to evaluate the transform of (29).

From (27) to (30) we can now deduce that (cf. (9.36))

$$\mathscr{L}(\chi) = \frac{4k'K\,\mathscr{L}(A)\,e^{-p/2}}{\left\{1 + \dfrac{r(E-K)}{pkK}\right\}Q_N(\rho) - \dfrac{2r}{p}Q_N^1(\rho)} \quad (-r < \mathscr{R}p < \infty), \quad (30)$$

where

$$N \equiv \frac{p-r}{2r}, \quad (31)$$

and

$$\rho \equiv \cosh r = \frac{1+k^2}{1-k^2}, \quad (32)$$

from (6) and (22). Similarly, by taking the transform of (24) and using (30) to eliminate $\mathscr{L}(\chi)$ we obtain

$$\mathscr{L}\{\lambda(t)\,U(t); p\} = \mathscr{L}\{A(t); p\}\,\mathscr{L}\{k(t,r); p\} - 2\mathscr{L}\{a_1(t); p\}, \quad (33)$$

where

$$\mathscr{L}\{k(\mathit{t},r); p\} = 2C\left(-\frac{i}{2}p,r\right) = \frac{\dfrac{2r}{p}\dfrac{E-K}{kK}Q_N(\rho) - \dfrac{4r}{p}Q_N^1(\rho)}{\left[1+\dfrac{r(E-K)}{pkK}\right]Q_N(\rho) - \dfrac{2r}{p}Q_N^1(\rho)}. \tag{34}$$

This function is, of course, not the same as that defined in (12.117), although the same notation has been employed. That the generalized growth of lift function $k(\mathit{t},r)$ and the generalized Theodorsen function $C\left(-\dfrac{i}{2}p,r\right)$ defined in (34), correctly reduce to $k(\mathit{t})$ and $C\left(-\dfrac{i}{2}p\right)$ in the limit $r \to 0$ easily follows from (12.121) and $\lim\limits_{r\to 0} r(E-K)/kK = 0$, the latter of which holds as both E and K tend to $\tfrac{1}{2}\pi$ and r/k tends to 2 by (6).

The solution of (33) has the same form as (9.43), except that $k(\mathit{t}-\mathit{t}^*)$ is replaced by $k(\mathit{t}-\mathit{t}^*,r)$. Similarly, from (30) we can write down the generalization of the function k_0 defined in (9.46).

14.3 The pressure distribution

The calculation of the pressure distribution on the aerofoil surface is similar to the calculations given in §§ 9.4 and 12.17.

In place of (12.123)–(12.126) there is

$$\frac{x}{c} = \frac{1}{2r}\ln\left(\frac{1-k\,\mathrm{cd}\,\gamma}{1+k\,\mathrm{cd}\,\gamma}\right), \tag{35}$$

$$dx = \frac{ck}{r}\mathrm{sn}\,\gamma\,d\gamma, \tag{36}$$

$$p(\gamma,\mathit{t}) = G(\mathit{t}) - \rho_a U\dot{U}\frac{x}{c} + \rho_a U^2\Omega + \frac{\rho_a Uk}{r}\int_0^\gamma (U\dot{\Omega})\,\mathrm{sn}\,\gamma\,d\gamma, \tag{37}$$

and

$$\frac{\partial\chi}{\partial\eta} = \frac{ik}{r}\mathrm{sn}\,i\eta\,\dot{\chi}, \tag{38}$$

by (1), (5) and (22).

Equation (12) can be written

$$U\Omega = \frac{1}{2\pi}\int_{-2K}^{2K} U\theta_0(\gamma^*)\left\{\frac{(\mathrm{cn}\,\gamma\,\mathrm{dn}\,\gamma + \mathrm{cn}\,\gamma^*\,\mathrm{dn}\,\gamma^*)\,\mathrm{sn}\,\gamma^*}{(\mathrm{sn}\,\gamma^* - \mathrm{sn}\,\gamma)\,\mathrm{sn}\,\gamma}\right.$$

$$\left.- \frac{\mathrm{cn}\,\gamma\,\mathrm{dn}\,\gamma + \mathrm{cn}\,\gamma^*\,\mathrm{dn}\,\gamma^*}{\mathrm{sn}\,\gamma}\right\}d\gamma^*$$

$$+ \frac{1}{2\pi}\int_0^{K'} \chi(\eta)\left\{\frac{(\mathrm{cn}\,\gamma\,\mathrm{dn}\,\gamma - \mathrm{cn}\,i\eta\,\mathrm{dn}\,i\eta)\,\mathrm{sn}^2 i\eta}{(\mathrm{sn}^2 i\eta - \mathrm{sn}^2\gamma)\,\mathrm{sn}\,\gamma}\right.$$

$$\left.- \frac{\mathrm{cn}\,\gamma\,\mathrm{dn}\,\gamma - \mathrm{cn}\,i\eta\,\mathrm{dn}\,i\eta}{\mathrm{sn}\,\gamma}\right\}d\eta.$$

By (10) the coefficient of $\operatorname{cn}\gamma\,\operatorname{dn}\gamma/\operatorname{sn}\gamma$ vanishes; then adding $(2\pi\operatorname{sn}\gamma)^{-1}$ times (13) and taking advantage of the relation (see §4.7)

$$Z_4(\gamma) = Z_1(\gamma) - \frac{\operatorname{cn}\gamma\,\operatorname{dn}\gamma}{\operatorname{sn}\gamma}, \tag{39}$$

we get

$$U\Omega = \frac{1}{2\pi}\int_{-2K}^{2K} U\theta_0(\gamma^*)\left\{\frac{(\operatorname{cn}\gamma\,\operatorname{dn}\gamma + \operatorname{cn}\gamma^*\operatorname{dn}\gamma^*)\operatorname{sn}\gamma^*}{(\operatorname{sn}\gamma^* - \operatorname{sn}\gamma)\operatorname{sn}\gamma} + \frac{\operatorname{sn}\gamma^*\, Z_4(\gamma^*)}{\operatorname{sn}\gamma}\right\}d\gamma^*$$

$$+ \frac{1}{2\pi}\int_0^{K'}\chi(\eta)\left\{\frac{(\operatorname{cn}\gamma\,\operatorname{dn}\gamma - \operatorname{cn}i\eta\,\operatorname{dn}i\eta)\operatorname{sn}^2 i\eta}{(\operatorname{sn}^2 i\eta - \operatorname{sn}^2\gamma)\operatorname{sn}\gamma} - \frac{\operatorname{sn}i\eta\, Z_4(i\eta)}{\operatorname{sn}\gamma}\right\}d\eta.$$

Hence by (38)

$$\frac{k}{r}\int_0^\gamma \operatorname{sn}\gamma\,(U\dot\Omega)\,d\gamma$$

$$= \frac{1}{2\pi}\int_{-2K}^{2K}\frac{\partial\dot N}{\partial\gamma^*}d\gamma^*\int_0^\gamma\left\{\frac{\operatorname{cn}\gamma\,\operatorname{dn}\gamma + \operatorname{cn}\gamma^*\operatorname{dn}\gamma^*}{\operatorname{sn}\gamma^* - \operatorname{sn}\gamma} + Z_4(\gamma^*)\right\}d\gamma$$

$$- \frac{i}{2\pi}\int_0^{K'}\frac{\partial\chi}{\partial\eta}d\eta\int_0^\gamma\left\{\frac{\operatorname{sn}i\eta(\operatorname{cn}\gamma\,\operatorname{dn}\gamma - \operatorname{cn}i\eta\,\operatorname{dn}i\eta)}{\operatorname{sn}^2 i\eta - \operatorname{sn}^2\gamma} - Z_4(i\eta)\right\}d\gamma, \tag{40}$$

where

$$N(\gamma,\mathnormal{f}) \equiv \frac{k}{r}\int_0^\gamma U\theta_0(\gamma)\operatorname{sn}\gamma\,d\gamma. \tag{41}$$

By direct differentiation (use (4.53), (4.82) and (4.83))

$$i\frac{\partial}{\partial\eta}\left\{\frac{\operatorname{sn}i\eta(\operatorname{cn}\gamma\,\operatorname{dn}\gamma - \operatorname{cn}i\eta\,\operatorname{dn}i\eta)}{\operatorname{sn}^2 i\eta - \operatorname{sn}^2\gamma} - Z_4(i\eta)\right\}$$

$$= -\frac{\partial}{\partial\gamma}\left\{\frac{\operatorname{sn}\gamma(\operatorname{cn}\gamma\,\operatorname{dn}\gamma - \operatorname{cn}i\eta\,\operatorname{dn}i\eta)}{\operatorname{sn}^2 i\eta - \operatorname{sn}^2\gamma} - Z_4(\gamma)\right\},$$

and

$$\frac{\partial}{\partial\gamma^*}\left\{\frac{\operatorname{cn}\gamma\,\operatorname{dn}\gamma + \operatorname{cn}\gamma^*\operatorname{dn}\gamma^*}{\operatorname{sn}\gamma^* - \operatorname{sn}\gamma} + Z_4(\gamma^*)\right\}$$

$$= -\frac{\partial}{\partial\gamma}\left\{\frac{\operatorname{cn}\gamma\,\operatorname{dn}\gamma + \operatorname{cn}\gamma^*\operatorname{dn}\gamma^*}{\operatorname{sn}\gamma^* - \operatorname{sn}\gamma} - Z_4(\gamma)\right\}.$$

It follows from (11) that $N(2K,\mathnormal{f}) = N(-2K,\mathnormal{f})$, consequently if the right-hand side of (40) is integrated by parts the integrated term from the first integral vanishes. The integrated term from the second integral also vanishes as $\chi(K') = 0$. Thus

$$\frac{k}{r}\int_0^\gamma (\dot U\Omega)\operatorname{sn}\gamma\,d\gamma = \frac{1}{2\pi}\int_{-2K}^{2K}\dot N(\gamma^*)\left\{\frac{\operatorname{cn}\gamma\,\operatorname{dn}\gamma + \operatorname{cn}\gamma^*\operatorname{dn}\gamma^*}{\operatorname{sn}\gamma^* - \operatorname{sn}\gamma} - Z_4(\gamma)\right\}d\gamma^*$$

$$- \frac{1}{2\pi}\int_0^{K'}\chi(\eta)\left\{\frac{\operatorname{sn}\gamma(\operatorname{cn}\gamma\,\operatorname{dn}\gamma - \operatorname{cn}i\eta\,\operatorname{dn}i\eta)}{\operatorname{sn}^2 i\eta - \operatorname{sn}^2\gamma} - Z_4(\gamma)\right\}d\eta. \tag{42}$$

Our final step is to substitute (12) and (42) into (37) and to eliminate χ by (10) from the result to obtain

$$p(\gamma,\prime) = G(\prime) - \rho_a U\dot{U}\frac{x}{c}$$
$$+ \frac{\rho_a U}{2\pi}\int_{-2K}^{2K}(U\theta_0 + \dot{N})\left\{\frac{\mathrm{cn}\,\gamma\,\mathrm{dn}\,\gamma + \mathrm{cn}\,\gamma^*\,\mathrm{dn}\,\gamma^*}{\mathrm{sn}\,\gamma^* - \mathrm{sn}\,\gamma} - Z_4(\gamma)\right\}d\gamma^*,$$
$$(43)$$

which is due to Rosenblat (1957). From (14) and (39) an alternative formula is

$$p(\gamma,\prime) = p_s(\gamma) - \rho_a U\dot{U}\frac{x}{c} - \frac{\rho_a U K\lambda}{\pi}\{\mathrm{ns}\,\gamma + Z_1(\gamma)\}$$
$$+ \frac{\rho_a U}{2\pi}\int_{-2K}^{2K}\{n(\gamma^*,\prime) + \dot{N}(\gamma^*,\prime)\}$$
$$\times\left\{\frac{\mathrm{cn}\,\gamma\,\mathrm{dn}\,\gamma + \mathrm{cn}\,\gamma^*\,\mathrm{dn}\,\gamma^*}{\mathrm{sn}\,\gamma^* - \mathrm{sn}\,\gamma} - Z_4(\gamma)\right\}d\gamma^*, \quad (44)$$

where
$$N(\gamma,\prime) = \frac{k}{r}\int_0^\gamma (n - \alpha U)\,\mathrm{sn}\,\gamma\,d\gamma, \tag{45}$$

and the subscript s denotes values in the mean steady flow.

14.4   The lift and moment

The lift coefficient is given by

$$C_L = -\frac{1}{c}\int_{-2K}^{2K}C_p(\gamma)\,dx(\gamma) = -\frac{2}{\rho_a U^2}\frac{k}{r}\int_{-2K}^{2K}p(\gamma,\prime)\,\mathrm{sn}\,\gamma\,d\gamma,$$

by (36), and substituting (44) into this one gets

$$C_L = C_{L_s} + \frac{4}{\pi^2}\frac{H}{c}\frac{kK}{U}\left\{2K\lambda + \int_{-2K}^{2K}\{n(\gamma,\prime) + \dot{N}(\gamma,\prime)\}\,\mathrm{sn}\,\gamma\,Z_1(\gamma)\,d\gamma\right\}, \tag{46}$$

on making use of (13.104), (13.112) and (39).

From (35) and (36) the nose-up moment coefficient about the mid-chord point is

$$C_M = \frac{1}{c^2}\int_{-2K}^{2K}C_p(\gamma)\,x(\gamma)\,dx(\gamma)$$
$$= \left(\frac{2}{\rho_a U^2}\right)\frac{k}{2r^2}\int_{-2K}^{2K}p(\gamma,\prime)\,\mathrm{sn}\,\gamma\,\ln\left(\frac{1-k\,\mathrm{cd}\,\gamma}{1+k\,\mathrm{cd}\,\gamma}\right)d\gamma,$$

and it follows from (44) and some integrals we have evaluated in § 13.11 that

$$C_M = C_{M_s} + \frac{k}{2\pi}\left(\frac{H}{\pi c}\right)^2\left\{2K\lambda I_a + \int_{-2K}^{2K}\{\eta(\gamma,\prime) + \dot{N}(\gamma,\prime)\}\{I(\gamma) - I_a\}\,d\gamma\right\}, \tag{47}$$

where
$$I(\gamma) \equiv 8Kk\,\mathrm{sn}\,\gamma \int_0^\gamma \mathrm{sn}\,\gamma\, Z_1(\gamma)\,d\gamma,$$

and
$$I_a \equiv \frac{1}{4K}\int_{-2K}^{2K} I(\gamma)\,d\gamma = \int_{-2K}^{2K}\mathrm{sn}\,\gamma\, Z_1(\gamma)\ln\left(\frac{1+k\,\mathrm{cd}\,\gamma}{1-k\,\mathrm{cd}\,\gamma}\right)d\gamma.$$

14.5 Harmonic motion of a rigid aerofoil

A method similar to that used in §§ 9.6 and 9.7 gives that in the case of harmonic oscillations of very long duration the inverse of (33) tends to

$$\lambda U = 2AC(\omega,r) - 2a_1, \tag{48}$$

where $C(\omega,r)$ is obtained by putting $p = 2i\omega$ in (34). An equation due to Timman (1951), which expresses $C(\omega,r)$ in terms of hypergeometric functions, is easily found with the aid of results given in § 12.19.

In the notation of § 9.6 the perturbation velocity n for a rigid aerofoil can be written

$$n = i\frac{2U}{c}\omega\left\{y_0 - \frac{c\alpha_0}{2r}\ln\left(\frac{1-k\,\mathrm{cd}\,\gamma}{1+k\,\mathrm{cd}\,\gamma}\right)\right\}e^{2i\omega t}. \tag{49}$$

Also $\alpha = \alpha_0 e^{2i\omega t}$. Substitution of these values into (18), (19) and (45) gives

$$a_0 = \frac{2i\omega}{c}U\,e^{2i\omega t}y_0, \quad a_1 = \frac{2i\omega}{c}\frac{cUI_a}{8Kr}e^{2i\omega t}\alpha_0,$$

and
$$N(\gamma,t) = \frac{U}{2r}e^{2i\omega t}\ln\left(\frac{1-k\,\mathrm{cd}\,\gamma}{1+k\,\mathrm{cd}\,\gamma}\right)\left\{\frac{2i\omega}{c}y_0 - \alpha_0 - \frac{i\omega\alpha_0}{2r}\ln\left(\frac{1-k\,\mathrm{cd}\,\gamma}{1+k\,\mathrm{cd}\,\gamma}\right)\right\},$$

on taking advantage of some integrals evaluated in § 13.11. Thus from (26) and (48)

$$\lambda = 2C(\omega,r)\left\{\alpha_0\left(1+\frac{i\omega I_a}{4Kr}\right) - 2i\omega\frac{y_0}{c}\right\}e^{2i\omega t} - \frac{i\omega I_a}{2Kr}\alpha_0\,e^{2i\omega t}.$$

The pressure, lift and moment now follow immediately on substituting these values in (44), (46) and (47). New untabulated integrals arise, and so if numerical values are required considerable computing effort is necessary. Some numerical results have been obtained by de Jager (1953), whose calculations are based on Timman's (1951) theory. Rosenblat (Ph.D. Thesis, Sydney University, 1957) has also calculated values of the lift and moment for various values of ω and c/H and finds his results to be at variance with those of de Jager for $\omega > \frac{1}{2}$. The probable reason for this was stated at the end of Rosenblat's (1957) paper.

UNSTEADY MOTION OF AN AEROFOIL
IN A FREE JET

14.6 The basic equations

Now let the aerofoil be placed in the centre of a free jet, and assume this jet to be undeflected in the steady motion that exists prior to $t = 0$. When the unsteady perturbations commence, the deflexion remains zero, a result which can be deduced from (13.39) as follows. First note that the derivation of (13.39) is not affected by the unsteadiness of the flow provided ϕ_h is interpreted as applying to the mean steady flow. Secondly, as the pressure and velocity at the jet surface is constant, we can set $\Omega_h = 0$ and this leaves the circulation, Γ, proportional to the deflexion of the jet. As this deflexion is zero before $t = 0$ it follows from Kelvin's theorem that it remains zero throughout the unsteady motion.

As the jet remains undeflected, we are free to make the usual linear perturbation approximations and in particular to apply the free surface boundary condition, $\Omega_h = 0$ on the straight lines $y = \pm \tfrac{1}{2}H$. Thus Fig. 14.1 is applicable to the jet problem, provided the solid walls of the channel are replaced by the free surfaces of the jet. Equations (1)–(8) are also applicable, but in place of (9) there is

$$\tau(\zeta, t) = \frac{1}{2\pi} \int_{-2K}^{2K} \theta_0(\gamma, t) \frac{\operatorname{cn}(\gamma - \zeta) + \operatorname{dn}(\gamma - \zeta)}{\operatorname{sn}(\gamma - \zeta)} d\gamma$$

$$+ \frac{k'^2}{2\pi U} \int_0^{K'} \frac{\operatorname{sn}\zeta \, \chi(\gamma, t) \, d\gamma}{\operatorname{cn} i\gamma \operatorname{dn}\zeta + \operatorname{dn} i\gamma \operatorname{cn}\zeta}, \qquad (50)$$

which is obtained by transforming (4.115) into the notation of this chapter.

On taking the limits $\zeta \to iK'$, $\zeta \to 2K + iK'$ of (50) we find

$$0 = \int_{-2K}^{2K} U\theta_0(\operatorname{dn}\gamma \pm k \operatorname{cn}\gamma) \, d\gamma + \int_0^{K'} \chi(\operatorname{dn} i\eta \mp k \operatorname{cn} i\eta) \, d\eta, \qquad (51)$$

by (8). In addition to these two equations there is also the restriction contained in (11), which applies to the present problem because it is equivalent to the aerofoil closure condition $\int_{-2K}^{2K} dy(\gamma) = 0$.

In the next section we shall deduce the strength of the vortex sheet from (51) by an analysis similar to that used in § 14.2.

14.7 The strength of the vortex sheet

From (51)

$$0 = \int_{-2K}^{2K} U\theta_0 \,\mathrm{dn}\,\gamma \, d\gamma + \int_0^{K'} \chi \,\mathrm{dn}\, i\eta \, d\eta,$$
$$\text{and} \qquad 0 = \int_{-2K}^{2K} U\theta_0 \,\mathrm{cn}\,\gamma \, d\gamma - \int_0^{K'} \chi \,\mathrm{cn}\, i\eta \, d\eta. \tag{52}$$

On substituting (14) into these equations, and noting that θ_s will satisfy them if $\chi = 0$ we find after a little rearranging that

$$\lambda U = 2\frac{1+k}{1-k}A - 2a_1 - \frac{1}{2K}\int_0^{K'} \chi \,\mathrm{dn}\, i\eta \, d\eta, \tag{53}$$

$$0 = \frac{1+k}{1-k}A - \frac{1}{2K}\int_0^{K'} \chi(\mathrm{dn}\, i\eta + \mathrm{cn}\, i\eta) \, d\eta, \tag{54}$$

where

$$A \equiv \frac{1-k}{1+k}\left\{\frac{\pi}{2K}\alpha U - a_0 + a_1\right\} \mathbf{U}(\mathit{t}), \tag{55}$$

$$a_0 \equiv \frac{1}{4K}\int_{-2K}^{2K} n(\gamma,\mathit{t}) \,\mathrm{dn}\,\gamma \, d\gamma, \qquad a_1 \equiv \frac{1}{4K}\int_{-2K}^{2K} n(\gamma,\mathit{t}) \,\mathrm{cn}\,\gamma \, d\gamma. \tag{56}$$

A calculation almost identical to that used in the derivations of (24) and (25) now gives

$$\lambda U = 2\frac{1+k}{1-k}A - 2a_1 - \frac{r}{\sqrt{2}\,K}\int_0^{\mathit{t}} \frac{\chi(1,\sigma)\cosh r(\mathit{t}-\sigma+\frac{1}{2})\,d\sigma}{[\cosh\{2r(\mathit{t}-\sigma)+r\} - \cosh r]^{\frac{1}{2}}},$$

and $\quad 0 = \frac{1+k}{1-k}A$

$$-\frac{r}{2\sqrt{2}\,K}\int_0^{\mathit{t}} \frac{\chi(1,\sigma)\{\cosh \tau(\mathit{t}-\sigma+\frac{1}{2}) + k^{-1}\sinh r(\mathit{t}-\sigma+\frac{1}{2})\}\,d\sigma}{[\cosh\{2r(\mathit{t}-\sigma)+r\} - \cosh r]^{\frac{1}{2}}}$$

(cf. (12.109) and (12.110)). Using results given in § 12.16 we find that these equations transform into

$$\mathscr{L}(\lambda U) = 2\frac{1+k}{1-k}\mathscr{L}(A) - 2\mathscr{L}(a_1) - \frac{1}{4K}\mathscr{L}(\chi)\,e^{p/2}\{Q_{(p-2r)/2r}(\rho) + Q_{p/2r}(\rho)\},$$

$$0 = \frac{1+k}{1-k}\mathscr{L}(A) - \frac{1}{8Kk}\mathscr{L}(\chi)\,e^{p/2}\{(1+k)\,Q_{(p-2r)/2r}(\rho) - (1-k)\,Q_{p/2r}(\rho)\},$$

where ρ is the number defined in (32). Eliminating $Q_{(p-2r)/2r}(\rho)$ we have

$$\mathscr{L}(\lambda U) = 2\mathscr{L}(A) - 2\mathscr{L}\left(a_1 - \frac{k}{1-k}A\right) - \frac{1}{2K}\mathscr{L}(\chi)\frac{e^{p/2}Q_{p/2r}(\rho)}{1+k}. \tag{57}$$

From (12.114) and (6) it follows that

$$(1+k)\,Q_{(p-2r)/2r}(\rho) - (1-k)\,Q_{p/2r}(\rho) = \frac{2k}{1-k}\left\{Q_{p/2r}(\rho) - \frac{2r}{p}\,Q^1_{p/2r}(\rho)\right\},$$

and therefore on eliminating $\mathscr{L}(\lambda U)$ we find

$$\mathscr{L}(\chi) = \frac{4K(1+k)\,\mathscr{L}(A)\,e^{-p/2}}{Q_{p/2r}(\rho) - \dfrac{2r}{p}\,Q^1_{p/2r}(\rho)}, \tag{58}$$

the inverse transform of which is the required vortex sheet strength. Eliminating $\mathscr{L}(\chi)$ from (57) and (58),

$$\mathscr{L}(\lambda U) = \mathscr{L}(A)\,\mathscr{L}\{k(t,r);\,p\} - 2\mathscr{L}\left(a_1 - \frac{k}{1-k}A\right), \tag{59}$$

where $\quad \mathscr{L}\{k(t,r);\,p\} = 2C\left(-\frac{i}{2}p,\,r\right) = 2\,\dfrac{-\dfrac{2r}{p}\,Q^1_{p/2r}(\rho)}{Q_{p/2r}(\rho) - \dfrac{2r}{p}\,Q^1_{p/2r}(\rho)}. \tag{60}$

The solution of (59) has a similar form to (12.119).

It is remarkable that the generalized Theodorsen function defined in (60) is identical with that found in § 12.16 for a cascade of aerofoils. The theory given in § 12.19 also holds in the present case.

14.8   The pressure distribution

On the aerofoil surface, $\eta = 0$, it follows from (50) and (52) that

$$U\Omega = \frac{1}{2\pi}\int_{-2K}^{2K} U\theta_0(\gamma^*)\,\frac{\operatorname{cn}(\gamma^*-\gamma) + \operatorname{dn}(*-\gamma)}{\operatorname{sn}(\gamma^*-\gamma)}\,d\gamma^* + F\operatorname{ds}\gamma + E\operatorname{cs}\gamma$$

$$+ \frac{k'^2}{2\pi}\int_0^{K'} \frac{\chi(\eta)\operatorname{sn}^2 i\eta\,d\eta}{\operatorname{sn}\gamma\{\operatorname{cn}i\eta\operatorname{dn}\gamma + \operatorname{dn}i\eta\operatorname{cn}\gamma\}}, \tag{61}$$

where $\quad E = \dfrac{1}{2\pi}\displaystyle\int_{-2K}^{2K} U\theta_0(\gamma)\operatorname{dn}\gamma\,d\gamma, \quad F = \dfrac{1}{2\pi}\int_{-2K}^{2K} U\theta_0(\gamma)\operatorname{cn}\gamma\,d\gamma.$

We shall just summarize the subsequent steps, which are similar to those used in § 14.3. These are: (i) differentiate (61) with respect to t, (ii) eliminate $\dot{\chi}$ by (38), (iii) integrate both integrals by parts, (iv) use

$$\frac{\partial}{\partial\eta}\left(\frac{i\operatorname{sn}i\eta}{\operatorname{cn}i\eta\operatorname{dn}\gamma + \operatorname{dn}i\eta\operatorname{cn}\gamma}\right) = -\frac{\partial}{\partial\gamma}\left(\frac{\operatorname{sn}\gamma}{\operatorname{cn}i\eta\operatorname{dn}\gamma + \operatorname{dn}i\eta\operatorname{cn}\gamma}\right),$$

and a similar result involving $\partial/\partial\gamma^*$ to eliminate $\partial/\partial\eta$ and $\partial/\partial\gamma^*$, (v) multiply by $\operatorname{sn}\gamma$ and integrate over $(0,\gamma)$, and (vi) substitute the result of this calculation and the real part of (50) on $\eta = 0$ into (37). The final result is

$$p(\gamma,t) = G(t) - \rho_a U\dot{U}\frac{x}{c} + \frac{\rho_a U}{2\pi}\int_{-2K}^{2K}(U\theta_0 + \dot{N})\frac{\mathrm{cn}\,(\gamma^* - \gamma) + \mathrm{dn}\,(\gamma^* - \gamma)}{\mathrm{sn}\,(\gamma^* - \gamma)}\,d\gamma^*,$$

where N is the function defined in (41).

From (14) we find

$$p(\gamma,t) = p_s(\gamma) - \rho_a U\dot{U}\frac{x}{c} - \frac{\rho_a U K\lambda}{\pi}(\mathrm{cs}\,\gamma + \mathrm{ds}\,\gamma)$$

$$+ \frac{\rho_a U}{2\pi}\int_{-2K}^{2K}\{n(\gamma^*,t) + \dot{N}(\gamma^*,t)\}\{\mathrm{cs}\,(\gamma^* - \gamma) + \mathrm{ds}(\gamma^* - \gamma)\}\,d\gamma^*,$$

$$\tag{62}$$

where N is now given by (45). An alternative form of the integrand in (62) follows from the relation given in (13.146).

These results and the theory of §§ 14.6 and 14.7 are due to Woods (1955g).

14.9   The lift and moment

Substitution of (62) into the first equation of § 14.4 yields

$$C_L = C_{L_s} + \frac{4kK\lambda}{rU} - \frac{k}{\pi rU}\int_{-2K}^{2K}(n + \dot{N})\,I(\gamma)\,d\gamma,$$

where
$$I(\gamma) \equiv \int_{-2K}^{2K}\mathrm{sn}\,\gamma^*\{\mathrm{cs}\,(\gamma - \gamma^*) + \mathrm{ds}\,(\gamma - \gamma^*)\}\,d\gamma^*,$$

and use has been made of

$$\int_{-2K}^{2K}\mathrm{cn}\,\gamma\,d\gamma = \left[\frac{1}{k}\cos^{-1}\mathrm{dn}\,\gamma\right]_{-2K}^{2K} = 0,$$

$$\int_{-2K}^{2K}\mathrm{dn}\,\gamma\,d\gamma = \left[\sin^{-1}\mathrm{sn}\,\gamma\right]_{-2K}^{2K} = 2\pi.$$

By (13.146) and (4.116)

$$I(\gamma) = \mathrm{sn}\,\gamma\int_{-2K}^{2K}\{\mathrm{cs}\,(\gamma - \gamma^*) + \mathrm{ds}\,(\gamma - \gamma^*)\}\,d\gamma^*$$

$$- \int_{-2K}^{2K}(\mathrm{dn}\,\gamma^*\,\mathrm{cn}\,\gamma + \mathrm{dn}\,\gamma\,\mathrm{cn}\,\gamma^*)\,d\gamma^*$$

$$= 2\,\mathrm{sn}\,\gamma\left[\ln|\mathrm{sn}\,\tfrac{1}{2}(\gamma^* - \gamma)|\right]_{-2K}^{2K} - 2\pi\,\mathrm{cn}\,\gamma = -2\pi\,\mathrm{cn}\,\gamma.$$

Therefore
$$C_L = C_{L_s} + \frac{4kK\lambda}{rU} + \frac{2k}{rU}\int_{-2K}^{2K}(n + \dot{N})\,\mathrm{cn}\,\gamma\,d\gamma. \tag{63}$$

Substitution of (62) in the equation for C_M given in § 14.4 gives

$$C_M = C_{M_s} - \frac{4K\lambda}{r^2 U}\ln k' + \frac{k}{2\pi r^2 U}\int_{-2K}^{2K}(n+\dot{N})\,J(\gamma)\,d\gamma,$$

where

$$J(\gamma) \equiv \int_{-2K}^{2K} \operatorname{sn}\gamma^*\{\operatorname{cs}(\gamma-\gamma^*)+\operatorname{ds}(\gamma-\gamma^*)\}\ln\left(\frac{1-k\operatorname{cd}\gamma^*}{1+k\operatorname{cd}\gamma^*}\right)d\gamma^*,$$

and the readily established results

$$\left.\begin{aligned}
\int_{-2K}^{2K}\operatorname{cn}\gamma\ln\left(\frac{1-k\operatorname{cd}\gamma}{1+k\operatorname{cd}\gamma}\right)d\gamma &= \frac{2\pi}{k}\ln(1-k^2),\\[2mm]
\int_{-2K}^{2K}\operatorname{dn}\gamma\ln\left(\frac{1-k\operatorname{cd}\gamma}{1+k\operatorname{cd}\gamma}\right)d\gamma &= 0,
\end{aligned}\right\} \tag{64}$$

have been employed. To evaluate $J(\gamma)$ we first use (13.146) and (64) to reduce it to

$$J(\gamma) = \operatorname{sn}\gamma\int_{-2K}^{2K}\{\operatorname{cs}(\gamma-\gamma^*)+\operatorname{ds}(\gamma-\gamma^*)\}\ln\left(\frac{1-k\operatorname{cd}\gamma^*}{1+k\operatorname{cd}\gamma^*}\right)d\gamma^*$$

$$-\frac{2\pi}{k}\operatorname{cn}\gamma\ln(1-k^2).$$

The remaining integral can be evaluated by integration by parts, followed by a differentiation under the integral sign with respect to γ and two integrations. The final result is

$$J(\gamma) = -4\pi\operatorname{sn}\gamma\sin^{-1}(k\operatorname{sn}\gamma)-\frac{2\pi}{k}\operatorname{cn}\gamma\ln(1-k^2),$$

so that

$$C_M = C_{M_s} - \frac{4K\lambda}{r^2 U}\ln k'$$

$$-\frac{2k}{r^2 U}\int_{-2K}^{2K}(n+\dot{N})\,[k^{-1}\ln k'\operatorname{cn}\gamma+\operatorname{sn}\gamma\sin^{-1}(k\operatorname{sn}\gamma)]\,d\gamma. \tag{65}$$

The harmonic oscillations of a rigid aerofoil are treated just as in § 14.5, the perturbation velocity being given by (49).

CONCLUDING REMARKS

Lack of time and space has prevented my giving several other important applications of the theory of parts I and II of this book. Some of the more important omissions are as follows:

(1) *The theory of biplanes.* This can be developed by inserting a barrier connecting the two aerofoils, and mapping the resulting domain on a rectangle, with the aerofoils on opposite sides. The barrier maps on to each of the remaining sides of the rectangle, thus providing periodic boundary conditions of the type studied in § 4.9. The reader is referred to the account by Robinson & Laurmann (1956).

(2) *Linear shear flow.* As shown in § 2.4 the theory depends on Laplace's equation, so that a complex variable theory similar to that for irrotational flow could be readily obtained. A contribution by James (1951) on this subject is worth studying.

(3) *Small vorticity flow.* The theory of § 2.22 lends itself to a complex variable treatment. Newman (1957) and the author have used the method of § 5.17 to solve (2.128) for the rotational flow past an aerofoil.

(4) *Surface waves.* With the boundary conditions of § 1.22 and the analysis of § 4.21 it is a simple matter to develop a useful approach to the theory of two-dimensional surface waves.

(5) *Slender body theory.* It was intended to give a treatment of the cross-flow (which occurs in the flow past a long slender body) based on the approach of § 2.23. For an account of the use of complex variable methods in this connexion the reader can consult the volume edited by Sears (1954).

REFERENCES

Numerals in square brackets refer to pages in this book.

Abbot, H. J. & von Doenhoff, A. E. (1949). *Theory of Wing Sections*. McGraw-Hill. [302, 335].

Allen, H. J. (1945). General theory of airfoil sections having arbitrary shape or pressure distribution. *Rep. Nat. Adv. Comm. Aero., Wash.*, no. 833. [302, 345.]

Allen, H. J. & Vincenti, W. G. (1944). Wall interference in a two-dimensional flow wind tunnel, with consideration of the effect of compressibility. *Rep. Nat. Adv. Comm. Aero., Wash.*, no. 782. [242.]

Armstrong, A. H. (1953). Abrupt and smooth separation in plane and axisymmetric cavity flow. *Memor. Arm. Res. Est., Gt. Br.* no. 22/53. [429.]

Baldwin, B. S., Turner, J. B. & Knechtel, E. D. (1954). Wall interference in wind tunnels with slotted and porous boundaries at subsonic speeds. *Tech. Notes Nat. Adv. Comm. Aero., Wash.*, no. 3176. [206, 287, 296.]

Beckenback, E. F. (ed.) (1952). *Construction and Applications of Conformal Maps*. *Nat. Bureau Stand., Appl. Math. Series*, no. 18, U.S. Government Printing Office. [33.]

Belleriot, C. & D'Epinay, J. L. (1950). Self-induced vibrations of turbo-machine blades. *Brown Boveri Rev.* [511.]

Bergman, S. & Schiffer, M. (1953). *Kernel Functions and Elliptic Differential Equations in Mathematical Physics*. Academic Press, N.Y. [33, 105.]

Billington, A. E. (1949). Aerodynamic lift and moment for oscillating aerofoils in cascade. *Rep. A.R.L., Aust.*, E 63. [511, 512.]

Birkhoff, G. (1950). *Hydrodynamics*. Princeton University Press. [430.]

Birkhoff, G. & Zarantonello, E. H. (1957). *Jets, Wakes and Cavities*. Academic Press, N.Y. [33, 428, 429, 431, 456, 483, 486.]

Blasius, H. (1910). Funktionentheoretische Methoden in der Hydrodynamik. *Z. angew. Math. Mech.* **58**, 90. [58.]

Brillouin, M. (1911). Les surfaces de glissement de Helmholtz et la resistance des fluides. *Ann. Chimie (Phys.)*, **23**, 145–230. [432.]

Brodetsky, S. (1923). Discontinuous fluid motion past circular and elliptic cylinders. *Proc. Roy. Soc. A*, **102**, 542–53. [443, 452.]

Busemann, A. (1937). Hodographenmethode der Gasdynamik. *Z. angew. Math. Mech.* **17**, 73–9. [46.]

Byrd, F. P. & Friedman, M. D. (1954). *Handbook of Elliptic Integrals for Engineers and Physicists*. Lange, Maxwell and Springer. [121, 483.]

Chang, C. C. & Chu, W. H. (1956). Aerodynamic interference of cascade blades in synchronized oscillation. *J. Appl. Mech.* **77**, 503–8. [511, 512.]

Chaplygin, S. A. (1902). On gas jets. *Tech. Memor. Nat. Adv. Comm. Aero., Wash.*, no. 1063. Translated from *Sci. Memor. Math. Phys. Soc.* no. 21, 1–121. Moscow University. [5, 41, 46.]

Cherry, T. M. (1947). Flow of an incompressible fluid about a cylinder. *Proc. Roy. Soc. A*, **192**, 45–79. See also *Proc. Roy. Soc. A*, **196** (1949), 1–36; **202** (1950), 507–22; and *Phil. Trans. A*, **245** (1953), 583–624. [41.]

Comrie, L. J. (1949). *Chamber's Six-Figure Mathematical Tables*, Vol. 2. Chambers, London. [236.]

Copson, E. T. (1935). *An Introduction to the Theory of Functions of a Complex Variable*. Oxford University Press. [73, 75, 95.]

Curle, N. (1956a). Unsteady two-dimensional flows with free boundaries. I. General Theory. *Proc. Roy. Soc.* A, **235**, 375–81. [456.]

Curle, N. (1956b). Unsteady two-dimensional flows with free boundaries. II. The incompressible, inviscid jet. *Proc. Roy Soc.* A, **235**, 381–95. [456.]

Davidson, I. M. (1956). The jet flap. *J. Roy. Aero. Soc.* **60**, 25–41. [389, 408.]

Davidson, I. M. & Stratford, B. S. (1954). An introduction to the jet flap. Unpublished Report of N.G.T.E., Great Britain. [420.]

Demtchenko, B. (1932). Sur les movements lents des fluides compressibles. *C.R. Acad. Sci. Paris*, **194**, 1215. [46.]

Dimmock, N. A. (1957). Some early jet flap experiments.. *Aero. Quart.* **8**, 331–45. [418.]

Duncan, W. J. (1949). Flutter and stability. *J. Roy. Aero. Soc.* **53**, 529–57. [511.]

Durand, W. F. (ed.) (1935). *Aerodynamic Theory*, Vol. 3. Springer, Berlin. [21.]

Ehlers, F. (1946). On the influence of sinks on the lift and pressure distribution of airfoils with suction slots. *Aero. Res. Coun., Lond.*, A.R.C. no. 10466. [356.]

Erdelyi, A. (1953). *Higher Transcendental Functions*, Vol. 1. McGraw-Hill. [520, 521.]

Fage, A. & Falkner, V. M. (1931). Further experiments on the flow around a circular cylinder. *Rep. Memor. Aero. Res. Coun., Lond.*, no. 1369. [452.]

Fage, A. & Johansen, F. C. (1927). On the flow behind an inclined plate of infinite span. *Proc. Roy. Soc.* A, **116**, 170. [430, 431.]

Fage, A. & Johansen, F. C. (1928). The structure of vortex sheets. *Phil. Mag.* **5**, 417. [431.]

Fanti, R. A., Kemp, N. H. & Nilson, E. N. (1958). A theory of thin airfoils, isolated and in cascade, yielding finite pressures at smooth leading edges. *J. Aero./Space Sci.* **25**, 409–24. [498.]

Fox, J. L. & Morgan, G. W. (1954). On the stability of some flows of an ideal fluid with free surfaces. *Quart. Appl. Math.* **11**, 439. [455.]

Garrick, J. E. (1944). On the plane potential flow past a lattice of arbitrary airfoils. *Rep. Nat. Adv. Comm. Aero., Wash.*, no. 778. [498.]

Garrick, J. E. (1952). Conformal mapping in aerodynamics, with emphasis on the method of successive conjugates. Beckenback (ed.). *Construction and Applications of Conformal Maps*, pp. 137–47. [178.]

Gilbarg, D. (1952). Unsteady flow with free boundaries. *Z. angew. Math. Phys.* **3**, 34–42. [455.]

Gilbarg, D. & Rock, D. H. (1945). On two theories of plane potential flows with finite cavities. *Memor. U.S. Naval Ordnance Lab.* no. 8718. [432.]

Gilbarg, D. & Serrin, J. (1950). Free boundaries and jets in the theory of cavitation. *J. Math. Phys.* **29**, 1–12. [432.]

Glauert, H. (1924). A theory of thin aerofoils. *Rep. Memor. Aero. Res. Coun., Lond.*, no. 910. [302.]

Glauert, H. (1933). Wind tunnel interference on wings, bodies and airscrews. *Rep. Memor. Aero. Res. Coun., Lond.*, no. 1566. [553.]

Glauert, H. (1948). *The Elements of Aerofoil and Airscrew Theory* (2nd. ed.). Cambridge University Press. [157, 302.]

Glauert, M. B. (1947). The application of the exact method of aerofoil design. *Rep. Memor. Aero. Res. Coun., Lond.*, no. 2683. [345.]

Goldstein, S. (ed.) (1938). *Modern Developments in Fluid Dynamics*, 2 vols. Oxford University Press. [18, 20, 23, 25, 28, 427.]

Goldstein, S. (1942). Steady two-dimensional flow past a solid cylinder in a non-uniform stream and two-dimensional wind tunnel interference. *Rep. Memor. Aero. Res. Coun., Lond.*, no. 1902. [538.]

Goldstein, S. (1952). Approximate two-dimensional aerofoil theory, Parts 1–6. *Curr. Papers, Aero. Res. Coun., Lond.*, nos. 68–73. [302, 327, 338, 341, 345.]

Goldstein, S. (1948). Low drag and suction airfoils. *J. Aero. Sci.* **15**, 189–220. [25, 26.]

Goldstein, S. & Lighthill, M. J. (1943). Two-dimensional compressible flow past a solid body in an unlimited fluid or symmetrically placed in a channel. *Aero. Res. Coun., Lond.*, A.R.C. no. 7184. [230.]

Green, A. E. (1940). The forces acting on a circular-arc aerofoil in a stream bounded by a plane wall. *Proc. Lond. Math. Soc.* **46**, 19–54. [546.]

Green, A. E. (1947). The two-dimensional aerofoil in a bounded stream. *Quart. J. Math.* **18**, 167–77. [546.]

Greidamus, G. H. & Van Heemert, A. (1948). Theory of the oscillating aerofoil in two-dimensional incompressible flow. *Rep. Nat. Luchtvaartlaboratorium, Amst.*, no. F 41. [366.]

Hagedorn, H. & Ruden, P. (1938). Windkanaluntersuchungen an einem Junkers-Doppelflügel mit ausblaseschlitz am Heck des Hauptflügels. *Lilienthal-Gesellschaft für Luftfahrtforschung Bericht*, A 64. [408.]

Havelock, T. H. (1938). The lift and moment on a flat plate in a stream of finite width. *Proc. Roy. Soc. A*, **925**, 178–96. [532, 538.]

Helmholtz, H. von. (1868.) Über discontinuierliche Flüssigkeitsbewegungen. *Mber. Ber. Akad. Wiss.* **23**, 215–28. [426.]

Hilbert, D. (1924). *Grundzüge einer allgemeinen Theorie der linearen Integralgleichungen* (2nd ed.). Leipzig, Berlin. [85.]

Howarth, L. (ed.). (1953). *Modern Developments in Fluid Dynamics: High Speed Flow*, Vol. 2. Oxford University Press. [19, 20, 267].

Hurley, D. G. (1957). An application of free streamline theory to aerofoil design. Unpublished M.A. dissertation, Melbourne University Library. [362.]

de Jager, E. M. (1953). The aerodynamic forces and moments on an oscillating aerofoil with control-surface between two parallel walls. *Rep. Nat. Luchtvaartlaboratorium, Amst.*, no. F 140. [572.]

James, D. J. (1951). Two-dimensional airfoils in shear flow. Part 1. *Quart. J. Mech. Appl. Math.* **4**, 407. [578.]

Jones, R. T. (1946). Properties of low-aspect-ratio pointed wings at speeds below and above the speed of sound. *Tech. Notes Nat. Adv. Comm. Aero., Wash.*, no. 1032. [64.]

Jones, W. P. (1953). Wind-tunnel wall interference effects on oscillating aerofoils in subsonic flow. *Rep. Memor. Aero. Res. Coun., Lond.*, no. 2943. [563.]

Jones, W. P. (1956). The oscillating aerofoil in subsonic flow. *Rep. Memor. Aero. Res. Coun., Lond.*, no. 2921. [383.]

Kármán, Th. von. (1941). Compressibility effects in aerodynamics. *J. Aero. Sci.* **8**, 337–56. [5, 45, 302.]

Kármán, Th. von. (1949). Les sillages non stationnaires. *Ann. mat.* **29**, 247–9. [455.]

Kármán, Th. von & Sears, W. R. (1938). Airfoil theory for non-uniform motion. *J. Aero. Sci.* **5**, 379–90. [511.]

Kemp, N. H. & Sears, W. R. (1953). Aerodynamic interference between moving blade rows. *J. Aero. Sci.* **20**, 585. [511.]

Küssner, H. G. (1940). Das zweidimensionale Problem der beliebig bewegten Tragfläche unter Berücksichtigung von Partialbewegung der Flüssigkeit. *Luftfahrtforsch.* **17**, 355–61. [371–2.]

Lamb, H. (1932). *Hydrodynamics* (6th ed.). Cambridge University Press. [446.]

Legendre, R. (1954). Premiers elements d'un calcul de l'amortissement aéro-dynamic des vibrations d'aubes de compressors. *Rech. aéro.* **27**, 3. [511.]

Leray, J. (1935). Les problèmes de représéntation conforme de Helmholz, théorie des sillages et des proues. *Commentarii Math. Helv.* **8**, 149–80, 250–63. [437.]

Levi-Cività, T. (1907). Scie e leggi de resistenzia. *R.C. Circ. mat. Palermo*, **23**, 1–37. [442, 453.]

Lighthill, M. J. (1945a). A mathematical method of cascade design. *Rep. Memor. Aero. Res. Coun., Lond.*, no. 2104. [302, 498.]

Lighthill, M. J. (1945b). Notes on the deflexion of jets by insertion of curved surfaces, and on the design of bends in wind tunnels. *Rep. Memor. Aero. Res. Coun., Lond.*, no. 2105. [26, 247, 302.]

Lighthill, M. J. (1945c). A new method of aerodynamic design. *Rep. Memor. Aero. Res. Coun., Lond.*, no. 2112. [246, 302, 345.]

Lighthill, M. J. (1945d). A theoretical discussion of wings with leading edge suction. *Rep. Memor. Aero. Res. Coun., Lond.*, no. 2162. [302.]

Lighthill, M. J. (1947). The hodograph transformation in transonic flow. *Proc. Roy. Soc.* A, **191**, I, 323–41; II, 341–51; III, 352–69. [41.]

Lighthill, M. J. (1949a). A note on cusped cavities. *Rep. Memor. Aero. Res. Coun., Lond.*, no. 2328. [433.]

Lighthill, M. J. (1949b). A technique for rendering approximate solutions to physical problems uniformly valid. *Phil. Mag.* **40**, 1179. See also *Aero. Quart.* **3** (1951), 193. [60, 302.]

Lighthill, M. J. (1950). Reflection at a laminar boundary layer of a weak steady disturbance to a supersonic stream, neglecting viscosity and heat conduction. *Quart. J. Mech. Appl. Math.* **3**, 303–25. [63.]

Lighthill, M. J. (1951). A new approach to thin aerofoil theory. *Aero. Quart.* **3**, 193. [302, 327].

Lighthill, M. J. (1953). On boundary layers and upstream influence: I. A comparison between subsonic and supersonic flows. *Proc. Roy. Soc.* A, **217**, 344–57. [256, 428.]

Lilley, G. M. (1952). An investigation of the flexure-torsion flutter characteristics of aerofoils in cascade. *Rep. Coll. Aero. Cranfield*, no. 60. [511, 563.]

Lock, C. N. H. & Bevan, J. A. (1944). Tunnel interference at compressibility speeds using the flexible walls of the rectangular high speed wind tunnel. *Rep. Memor. Aero. Res. Coun., Lond.*, no. 2005. [547.]

Maeder, P. F. & Wood, A. D. (1956). Transonic wind tunnel test sections. *Z. angew. Math. Phys.* **7**, 177–212. [284, 287.]

McCullough, G. B. & Gault, D. E. (1951). Examples of three representative types of airfoil-section stall at low speed. *Tech. Notes Nat. Adv. Comm. Aero., Wash.*, no. 2502. [357.]

Milne-Thomson, L. M. (1948). *Theoretical Aerodynamics*. Macmillan, London. [302, 320, 331, 344.]

Milne-Thomson, L. M. (1949). *Theoretical Hydrodynamics* (2nd ed.). Macmillan, London. [7, 33, 157, 238, 429, 431, 452.]

Milne-Thomson, L. M. (1950). *Jacobian Elliptic Function Tables*. Dover, N.Y. [121.]

Mimura, Y. (1958). Phenomenological free streamline theory. Part I. Flow past a normal plate. *J. Phys. Soc. Japan*, **13**, 1361–74. [444.]

Minhinnick, I. T. (1950). Tables of functions for evaluation of wing and control surface flutter derivatives for incompressible flow. *Aero. Res. Coun., Lond.*, A.R.C. no. 13730. [378.]

Moses, H. E. (1949). Velocity distributions on arbitrary airfoils by conformal mapping. *Tech. Notes Nat. Adv. Comm. Aero.*, *Wash.*, no. 1899. [531.]

Munk, M. M. (1924). Elements of the wing section theory and of the wing theory. *Rep. Nat. Adv. Comm. Aero.*, *Wash.*, no. 191. [64.]

Muskhelishvili, N. I. (1953). *Singular Integral Equations.* Trans. and ed. J. R. M. Radok. T. Noordhoff, N. V., Holland. [4, 92, 95, 111.]

Nehari, Z. (1952). *Conformal Mapping.* McGraw-Hill. [4, 52, 73, 105, 106, 157, 159, 160, 178.]

Newman, M. (1957). The force and moment on an aerofoil in shear flow. Unpublished M.Sc. dissertation. Sydney University Library. [578.]

Newmark, S. (1955). Velocities on two-dimensional closed and semi-infinite aerofoils at zero incidence. *Roy. Aircraft Est. Rep. Aero.* no. 2362. [390, 392.]

Ostrowski, A. M. (1952). On the convergence of Theodorsen's and Garrick's method of conformal mapping. Beckenback (ed.). *Construction and Applications of Conformal Maps*, pp. 149–63. [178.]

Pankhurst, R. C. & Squire, H. B. (1950). Calculated pressure distributions for the RAE 100–104 aerofoil sections. *Curr. Paper, Aero. Res. Coun., Lond.*, no. 80. [322.]

Pearcey, H. H., Pankhurst, R. C. & Lee, G. F. (1952). Further results from NPL high-speed wind-tunnel test of spoilers on an aerofoil with 0·25c. flaps. *Aero. Res. Coun., Lond.*, A.R.C. no. 15415. [358.]

Pearson, H. (1953). The aerodynamics of compressor blade vibration. *Proc. 4th Anglo-American Aero. Conference.* [511.]

Piercy, N. A. V. (1947). *Aerodynamics* (2nd ed.). English University Press, London. [25.]

Pinkerton, R. M. (1936). Calculated and measured pressure distributions over the midspan section of a NACA 4412 airfoil. *Rep. Nat. Adv. Comm. Aero.*, *Wash.*, no. 563. [338, 341.]

Plemelj, J. (1908). Ein Ergänzungssatz zur Cauchyshen Integraldarstellung analytischer Functionen, Randwerte betreffend. *Mh. Math. Phys.* XIX Jahrgang, 205–10. [82.]

Poincaré, H. (1910). *Leçons de Mécanique Céleste*, Vol. 3, ch. x. Paris. [112.]

Pope, A. (1951). *Basic Wing and Airfoil Theory.* McGraw-Hill. [338.]

Preston, J. H. (1953). The calculation of lift taking account of the boundary layer. *Rep. Memor. Aero. Res. Coun., Lond.*, no. 2725. [302].

Preston, J. H. & Rawcliffe, A. G. (1950). Note on sintered metal with a view to its use as a porous surface in distributed suction experiments. *Curr. Paper, Aero. Res. Coun., Lond.*, no. 9. [26, 205.]

Riabouchinsky, D. (1921). On unsteady fluid motions with free surfaces. *Proc. Lond. Math. Soc.* 19, 206–15. [362, 432, 477.]

Riegels, F. (1948). Das Umströmungsproblem bei inkompressiblen Potential-strömungen. *Ing.-Archiv.* 16, 373–6. [327.]

Riegels, F. (1949). Das Umströmungsproblem bei inkompressiblen Potential-strömungen. *Ing.-Archiv.* 17, 94–106. [327.]

Riessner, E. (1947). Wind tunnel corrections for the two-dimensional theory of oscillating aerofoils. *Rep. Cornell Aero. Lab.* no. SB–318–S–3. [563.]

Ringleb, F. (1940). Exakte Lösungen der Differentialgleichungen einer adiabatischen Gasströmung. *Z. angew. Math. Mech.* 20, 185–9. [41.]

Robinson, A. & Laurmann, J. A. (1956). *Wing Theory.* Cambridge University Press. [18, 21, 302, 323, 329, 341, 489, 578.]

Rosenblat, S. (1957). Some problems in aerofoil theory. Unpublished Ph.D. dissertation, Sydney University Library. [563, 571, 572.]

Rosenblat, S. (1957). The aerodynamic forces on an aerofoil in non-uniform unsteady motion in a closed tunnel. *Phil. Trans.* A, **250**, 247–78. [129, 572.]

Rosenblat, S. & Woods, L. C. (1956). A method of cascade design for two-dimensional incompressible flow. *Rep. Aust. Aero. Res. Comm.* ACA–58. [498.]

Rosenhead, L. (1931). The lift on a flat plate between parallel walls. *Proc. Roy. Soc.* A, **132**, 127–52. [532.]

Roshko, A. (1954). A new hodograph for free streamline theory. *Tech. Notes Nat. Adv. Comm. Aero., Wash.*, no. 3168. [443.]

Sears, W. R. (ed.) (1954). *General Theory of High Speed Aerodynamics.* Vol. VI. *High Speed Aerodynamics and Jet Propulsion.* Princeton University Press. [41, 49, 84, 238, 287, 578.]

Sisto F. (1953). Stall-flutter in cascades. *J. Aero. Sci.* **20**, 598–604. [511.]

Söhngen, H. (1940). Bestimmung der Auftriebsverteilung für beliebige instationäre Bewegungen (Ebenes Problem). *Luftfahrtforsch.* **17**, 379. [372.]

Southwell, R. (1948). *Relaxation Methods in Mathematical Physics.* Oxford University Press. [42.]

Southwell, R. & Vaisey, G. (1946). Fluid motions characterized by 'free' streamlines. *Phil. Trans.* A, **240**, 117–61. [433.]

van Speigel, E. & van de Vooren, A. J. (1953). On the theory of the oscillating wing in two-dimensional subsonic flow. *Rep. Nat. Luchtvaartlaboratorium, Amst.*, no. F 142. [3, 383.]

Spence, D. A. (1954). Prediction of the characteristics of two-dimensional aerofoils. *J. Aero. Sci.* **21**, 577–87. [23.]

Spence, D. A. (1956). The lift coefficient of a thin, jet-flapped wing. *Proc. Roy. Soc.* A, **238**, 46–68. [413, 418.]

Spence, D. A. & Routledge, N. A. (1955). Velocity calculations by conformal mapping for two-dimensional aerofoils. *Roy. Aircraft Est. Rep. Aero.* no. 2539. [344.]

Squire, H. B. (1934). On the laminar flow of a viscous fluid with vanishing viscosity. *Phil. Mag.* **7**, 1150–60. [429, 454.]

Squire, H. B. & Young, A. D. (1937). The calculation of the profile drag of aerofoils. *Rep. Memor. Aero. Res. Coun., Lond.*, no. 1838. [447.]

Squire, H. B. *et. al.* (1945). Wind tunnel tests of oblique jet units. *Roy. Aircraft Est. Rep. Aero.* no. 2007. [253.]

Stenning, A. H. & Kriebel, A. R. (1958). Stall propagation in a cascade of aerofoils. *A.S.M.E.* Paper no. 57–SA–29, pp. 1–13. [511.]

Stewartson, K. (1953). On the flow downstream of separation in an incompressible fluid. *Proc. Camb. Phil. Soc.* **49**, 561–9. [428, 431].

Stratford, B. S. (1956). Early thoughts on the jet flap. *Aero. Quart.* **7**, 45–59. [408, 420.]

Theodorsen, Th. (1931). On the theory of wing sections with particular reference to the lift distribution. *Rep. Nat. Adv. Comm. Aero., Wash.*, no. 383. [325.]

Theodorsen, Th. (1932). Theory of wing sections of arbitrary shape. *Rep. Nat. Adv. Comm. Aero., Wash.* no. 411. [302, 338.]

Theodorsen, Th. (1935). General theory of aerodynamic instability and the mechanism of flutter. *Rep. Nat. Adv. Comm. Aero., Wash.*, no. 496. [371.]

Theodorsen, Th. (1944). Airfoil contour modifications based on ε-curve method of calculating pressure distribution. *Rep. Nat. Adv. Comm. Aero., Wash.*, no. W.R.L. 135. [345.]

Theodorsen, Th. & Garrick, J. E. (1933). General potential theory of arbitrary wing sections. *Rep. Nat. Adv. Comm. Aero., Wash.*, no. 452. [176, 178, 338.]

Thom, A. (1933). Flow past circular cylinders at low speeds. *Proc. Roy. Soc.* A, **141**, 651–69. [42.]

Thom, A. (1943). Blockage corrections in a closed high speed tunnel. *Rep. Memor. Aero. Res. Coun., Lond.*, no. 2033. [21.]

Thom, A. (1948). A cambered aerofoil treated by squares method. *Aero. Res. Coun., Lond.*, no. 11652. [343.]

Thom, A. & Klanfer, L. (1947). The method of influence factors in arithmetical solutions of certain field problems. *Aero. Res. Coun., Lond.*, no. 2440. [229, 344, 531.]

Thom, A. S. (1952). Design of a right-angled bend with constant velocity at the walls. *Curr. Paper, Aero. Res. Coun., Lond.*, no. 135. [531.]

Thwaites, B. (1945a). A method of aerofoil design. Part I. Symmetrical aerofoils. *Rep. Memor. Aero. Res. Coun., Lond.*, no. 2166. [302, 345.]

Thwaites, B. (1945b). A method of aerofoil design. Part II. Cambered aerofoils. *Rep. Memor. Aero. Res. Coun., Lond.*, no. 2167. [302, 345.]

Thwaites, B. (1954). A note on Kücheman's aerofoil. *Aero. Res. Coun., Lond.*, A.R.C. no. 17158. [390, 392.]

Thwaites, B. (ed.) (1960). *Incompressible Aerodynamics*. Oxford University Press. [302, 428.]

Timman, R. (1946). Beschouwingen over de luchtkrachten op trillends vliegtuigsilengels. Unpublished Ph.D. dissertation, Delft University Library. [383.]

Timman, R. (1948). The direct and the inverse problem of aerofoil theory. A method with complete auxiliary tables to obtain numerical solutions. *Rep. Nat. Luchtvaartlaboratorium, Amst.*, no. F 16. [302, 343.]

Timman, R. (1951). The aerodynamic forces on an oscillating aerofoil between two parallel walls. *Appl. Sci. Res.* A, **3**. [511, 563, 572.]

Tomlinson, R. C. (1954). The theoretical interference velocity on the axis of a two-dimensional wind tunnel with slotted walls. *Curr. Paper, Aero. Res. Coun., Lond.*, no. 181. [287.]

Tomotika, S. *et al. Rep. Aero. Res. Inst., Tokyo Imp. Univ.* nos. 97 (1933), 101 (1934), 120 (1935), 153 (1937), 170 (1938), 182 (1939). [532, 545.]

Tomotika, S. & Tamada, K. (1950). Studies on two-dimensional transonic flows of compressible fluid. I. *Quart. Appl. Math.* **7**, 381–97. [42.]

Tomotika, S., Tamada, K. & Umemoto, H. (1950). The lift and moment acting on a circular-arc aerofoil in a stream bounded by a plane wall. *Quart. J. Mech. Appl. Math.* **4**, 1–22. [546.]

Tomotika, S. & Umemoto, H. (1939). The forces on a plane aerofoil in a wind tunnel of the Göttingen type, with special reference to approximate formulae for the lift. *Rep. Aero. Res. Inst., Tokyo Imp. Univ.* no. 185. [553.]

Tsien, H. S. (1939). Two-dimensional subsonic flow of compressible fluids. *J. Aero. Sci.* **6**, 399–407. [5, 45, 302.]

Van Dyke, M. D. (1954). Subsonic edges in thin-wing and slender-body theory. *Tech. Notes Nat. Adv. Comm. Aero., Wash.*, no. 3343. [329.]

van der Pol, B. & Bremmer, H. (1950). *Operational Calculus Based on the Two-sided Laplace Integral*. Cambridge University Press. [81, 155, 156, 295, 370, 383, 467, 470, 515.]

Wagner, H. (1925). Über die Entstehung des dynamischen Auftriebes von Tragflügeln. *Z. angew. Math. Mech.* **5**, 17–35. [382.]

Watson, G. N. (1952). *A Treatise on the Theory of Bessel Functions* (2nd ed.). Cambridge University Press. [516.]

Weber, J. (1953). The calculation of the pressure distribution over the surface of two-dimensional and swept wings with symmetrical aerofoil sections. *Roy. Aircraft Est. Rep. Aero.* no. 2497; see also *Ibid.* no. 2548. [344.]

Weinstein, A. (1949). Non-linear problems in the theory of fluid motion with free boundaries. *Proc. Sym. Appl. Math. (Amer. Math. Soc.)*, **1**, 1–18. [431.]

Whitham, G. B. (1950). The behaviour of supersonic flow past a body of revolution far from the axis. *Proc. Roy. Soc.* A, **201**, 89. [60.]

Whittaker, E. T. & Watson, G. N. (1952). *A Course of Modern Analysis* (4th ed.). Cambridge University Press. [100, 112, 116, 117, 118, 119, 120, 124, 125, 126, 191, 286, 352, 417, 515, 516, 520, 530, 533.]

Williams, J. (1950). Some improvements on the design of thick suction aerofoils. *Curr. Paper, Aero. Res. Coun., Lond.*, no. 31. [345.]

Woods, L. C. (1950). The two-dimensional subsonic flow of an inviscid fluid about an aerofoil of arbitrary shape. *Rep. Memor. Aero. Res. Coun., Lond.*, no. 2811. [342, 343, 344.]

Woods, L. C. (1952 a). The application of the polygon method to the calculation of the compressible subsonic flow round two-dimensional profiles. *Curr. Paper, Aero. Res. Coun., Lond.*, no. 115. [342.]

Woods, L. C. (1952 b). The theory of aerofoils with hinged flaps in two-dimensional compressible flow. *Curr. Paper, Aero. Res. Coun., Lond.*, no. 138. [343.]

Woods, L. C. (1953 a). Unsteady cavitating flow past curved obstacles. *Curr. Paper, Aero. Res. Coun., Lond.*, no. 149. [455.]

Woods, L. C. (1953 b). Theory of aerofoil spoilers. *Rep. Memor. Aero. Res. Coun., Lond.*, no. 2969. [449.]

Woods, L. C. (1954 a). Compressible subsonic flow in two-dimensional channels with mixed boundary conditions. *Quart. J. Mech. Appl. Math.* **7**, 263–82. [252.]

Woods, L. C. (1954 b). The lift and moment acting on a thick aerofoil in unsteady motion. *Phil. Trans.* A, **247**, 131–62. [365, 366.]

Woods, L. C. (1954 c). Some contributions to jet-flap theory, and to the theory of source-flow from aerofoils. *Curr. Paper, Aero. Res. Coun., Lond.*, no. 388. [417.]

Woods, L. C. (1955 a). The design of two-dimensional aerofoils with mixed boundary conditions. *Quart. Appl. Math.* **13**, 139–46. [362.]

Woods, L. C. (1955 b). Compressible subsonic flow in two-dimensional channels. Part. I. Basic mathematical theory. *Aero. Quart.* **6**, 205–20. Part II. The application of the theory to problems of channel flow, 254–76. [238, 265, 268.]

Woods, L. C. (1955 c). Two-dimensional flow of a compressible fluid past given curved obstacles with infinite wakes. *Proc. Roy. Soc.* A, **227**, 367–86. [426, 443, 449, 451, 452.]

Woods, L. C. (1955 d). On unsteady flow through a cascade of aerofoils. *Proc. Roy. Soc.* A, **228**, 50–65. [511, 512, 519.]

Woods, L. C. (1955 e). Subsonic plane flow in an annulus or a channel with spacewise periodic boundary conditions. *Proc. Roy. Soc.* A, **229**, 63–85. [563.]

Woods, L. C. (1955 f). Unsteady plane flow past curved obstacles with infinite wakes. *Proc. Roy. Soc.* A, **229**, 152–80. [456.]

Woods, L. C. (1955 g). The aerodynamic forces on an oscillating aerofoil in a free jet. *Proc. Roy. Soc.* A, **229**, 235–50. [576.]

Woods, L. C. (1955 h). On the theory of two-dimensional wind tunnels with porous walls. *Proc. Roy. Soc.* A, **233**, 74–90. [279.]

Woods, L. C. (1956 a). Generalized aerofoil theory. *Proc. Roy. Soc.* A, **238**, 358–88. [346, 359, 362.]

Woods, L. C. (1956 b). On the theory of source-flow from aerofoils. *Quart. J. Mech. Appl. Math.* **9**, 441–56. [357.]

Woods, L. C. (1956c). On the thrust due to an air jet flowing from a wing placed in a wind tunnel. *J. Fluid Mech.* **1**, 54–60. [408.]

Woods, L. C. (1956d). On the flow past semi-infinite bodies, with an application to the theory of jet-flaps. Unpublished paper read at the inaugural meeting of the *Aust. Math. Soc.* [417.]

Woods, L. C. (1957a). On harmonic functions satisfying a mixed boundary condition with an application to the flow past a porous wall. *Appl. Sci. Res.* A, **6**, 351–64. [286.]

Woods, L. C. (1957b). Aerodynamic forces on an oscillating aerofoil fitted with a spoiler. *Proc. Roy. Soc.* A, **239**, 328–37. [477.]

Woods, L. C. (1957c). On the lifting aerofoil in a wind tunnel with porous walls. *Proc. Roy. Soc.* A, **242**, 351–54. [554.]

Woods, L. C. (1958a). Some generalizations of the Schwarz–Christoffel mapping formula. *Appl. Sci. Res.* B, **7**, 89–101. [178.]

Woods, L. C. (1958b). On the deflexion of jets by aerofoils. *Quart. J. Mech. Appl. Math.* **11**, 24–38. [548, 553.]

INDEX OF SUBJECTS

Made in the USA
Monee, IL
07 July 2026

56552151R00360